DATE DUE

JUL 3 0 2007

GAYLORD   PRINTED IN U.S.A.

Finite Mathematics

An Applied Approach

Eighth Edition

Abe Mizrahi
Indiana University Northwest

Michael Sullivan
Chicago State University

John Wiley & Sons, Inc.
New York / Chichester / Weinheim / Brisbane / Singapore / Toronto

EXECUTIVE EDITOR: Ruth Baruth
ACQUISITIONS EDITOR: Sharon Prendergast
SENIOR MARKETING MANAGER: Peter Ryttel
FULL SERVICE MANAGER: Jeanine Furino
SENIOR DESIGNER: Kevin Murphy
SENIOR ILLUSTRATION EDITOR: Sigmund Malinowski
COVER AND INTERIOR ART: Marjory Dressler
FREELANCE PRODUCTION MANAGEMENT: Ingrao Associates

This book was set in Times Roman by Progressive Information Technologies and printed and bound by Von Hoffmann Press. The cover was printed by Lehigh Press.

This book is printed on acid-free paper.

Library of Congress Cataloging in Publication Data:
Mizrahi, Abe.
    Finite mathematics : an applied approach / Abe Mizrahi, Michael
Sullivan.—8th ed.
       p.   cm.
    Includes index.
    ISBN 0-471-32202-4 (cloth : alk. paper)
    1. Business mathematics.   2. Social science—Mathematics.
3. Mathematics.   I. Sullivan, Michael, 1942–
II. Title.
HF5695.M66   2000
519—dc21   99-29009
                                                                 CIP

Printed in the United States of America

10 9 8 7 6 5

*To The Memory of Our Parents
With Gratitude*

Abe Mizrahi received his doctorate in mathematics from the Illinois Institute of Technology in 1965. He is currently Professor of Mathematics at Indiana University Northwest. Dr. Mizrahi is a member of the Mathematics Association of America. Articles he has written explore topics in math education and the applications of mathematics to economics. Professor Mizrahi has served on many CUPM committees and has been a panel member on the CUPM Committee on Applied Mathematics in the Undergraduate Curriculum. Professor Mizrahi is a recipient of many NSF grants and has served as a consultant to a number of businesses and federal agencies.

Michael Sullivan received his doctorate in mathematics from the Illinois Institute of Technology in 1967. Since 1965, he has been Professor of Mathematics and Computer Science at Chicago State University. Dr. Sullivan is a member of the American Mathematical Society, the Mathematics Association of America, and the Text and Academic Authors Association (Treasurer). He has served on CUPM curriculum committees and is a member of the Illinois Section MAA High School Lecture Committee. Professor Sullivan has written a variety of scholarly articles and has served as a curriculum consultant to high schools, colleges, professional organizations, and government agencies.

Drs. Mizrahi and Sullivan are also the coauthors of *Mathematics: An Applied Approach,* Seventh Edition, John Wiley & Sons, Inc., 2000.

Preface

The first edition of this text was published in 1973. At that time our purpose was to present an accessible approach to the mathematics required of business and social science majors, giving emphasis to real-world applications from these fields. In subsequent editions this has remained our purpose, as it does in this Eighth Edition. Also, in this edition, extra effort has been made to further enhance the pedagogy. Explanations have been rewritten to increase clarity and eliminate unnecessary verbage. Wherever prudent, reasons are given for steps used and an increased use of boxed formulas and boxed steps now make it easier for students to identify important procedures. Further, every example is now labeled, making it easier for both student and instructor to identify what the example is about.

ORGANIZATION

This text is divided into three independent parts: Linear Algebra, Probability, and Discrete Mathematics. Depending on the curriculum, an instructor can start a course at the beginning of any of the three parts. The chart on page ix illustrates the interaction of the chapters.

Part One, Linear Algebra, contains five chapters: Chapter 1 lays the foundation for subsequent chapters and is considered review. Chapter 2 presents matrices, beginning with systems of linear equations solved by substitution and elimination. Chapters 3 and 4 present Linear Programming and Chapter 5 provides an introduction to finance.

Part Two, Probability, also contains five chapters: Chapter 6 introduces sets and counting techniques. Chapter 7 deals with probability. Chapter 8 continues the study of probability. Chapter 9 introduces statistics, including bar charts and pie charts and Chapter 10 covers the topics of Markov chains and games.

Part Three, Discrete Mathematics, contains three chapters: Chapter 11 covers logic and logic circuits. Chapter 12 deals with relations, functions, and induction and Chapter 13 covers graphs and trees. This chapter contains new material dealing with models in Sociology and Business.

FEATURES FOR THE STUDENT

A variety of features work together to help students use this book.

Style and Pedagogy

The writing style is clear and concise. Students can easily follow the straightforward explanations. At the same time, the text is mathematically accurate and pedagogically sound. Topics are introduced in simple algebraic settings so that students can more easily concentrate on the underlying concepts.

Examples

An abundance of examples is used to thoroughly illustrate concepts. Some exhibit a basic idea, whereas others gradually raise the complexity of the idea. Where appropriate, examples are used to illustrate applications of the material. In all cases, sufficient detail is provided so that the mathematics is easy to follow.

Interaction

To encourage students to interact with the text, the instruction "Now Work Problem XX" appears in the margin in selected places in each section marked with an ✍. These references direct students to problems in the exercise set at the end of the section that relate to the concepts and examples just introduced. By working a related problem in the exercises, the student is able to discover whether the idea just presented is understood, or whether it needs further study before he or she continues to the next concept.

Exercises

At the end of each section, problems are given that progress from easy (to instill confidence) to ones that are more difficult (to challenge better students). Motivation is sustained through application-type exercises that deal with business and the life and social sciences.

Technology

At the end of most sections, technology-based exercises have been included, marked with an (▶), that demonstrated the usefulness of a graphing utility for solving certain types of exercises. These are used as an instructional tool in order to facilitate the crucial combination of graphical, numerical, and algebraic viewpoints. These exercises, which are appropriate for use with a programmable and/or graphing calculator or graphing software, have been placed in exercise sets where applicable and may be omitted by instructors who prefer that their students not use calculators or computers.

Appendix B contains examples of solving linear programming problems using the LINDO software package.

Chapter Review

Each chapter concludes with a summary of important terms and formulas, followed by their page reference. A variety of problems affords another check as to whether the material of the chapter has been mastered.

Professional Exams

When appropriate, mathematical questions from CPA, CMA, and Actuary exams are given. This feature adds to the realism of the text by providing students with a preview of those questions that might appear on professional exams.

Functional Appendix

Prerequisite material from algebra, including exercise sets, is presented in Appendix A. This review focuses precisely on topics from earlier courses that are used and needed in this course. Timely references to Appendix A appear at appropriate places in the text.

Purposeful Design

Important terms are given in boldface type when first introduced. Those requiring more emphasis are set off with their definitions in boldface type. Formulas and properties are enclosed in boxes and highlighted with color. Procedures are also highlighted in boxes. Numerous illustrations have been used, each utilizing a second color for clarity and emphasis.

Answers

A section of Answers to Odd-Numbered Problems appears at the end of the book.

CHANGES AND IMPROVEMENTS TO THIS EDITION

Content

In the chapter on linear programming, the simplex method, the section on mixed constraints has been redone, utilizing a slightly different algorithm that students will find easier to follow. For those interested, references on the validity of this algorithm have been included.

The explanation of effective rate of interest has been expanded.

The new introduction to the second chapter on probability should better motivate students. The explanation of expected value was rewritten for clarity.

In statistics, the explanations involving grouped data, median for grouped data, and standard deviation were rewritten to improve clarity.

Organization

In Chapter 2, the use of parameters now appears in the first section. Also, least squares now appears in the last section as another application, making the earlier section easier to cover in a single period.

In Chapter 3, section 3.2 now appears as two sections. By breaking out the applications, the material will be easier to teach and to learn.

In Chapter 6, section 6.1 now appears as two sections to make the material easier to cover in one period each.

Section 7.3 was reorganized to improve the flow of material.

Sections 9.1 and 9.2 were interchanged to keep like material clustered.

Exercises

Many of the exercises have been revised and, where appropriate, new exercises have been added. For example, new technology exercises now appear in Section 1.1. In finance, new problems involving inflation and purchasing power appear. In statistics, the exercises now revisit real data, each time utilizing a new statistical concept.

Examples

Where appropriate, new examples have been included. For example, in Chapter 1, two examples to further explain vertical lines were added. In Chapter 2, new examples were inserted to explain the relationship between augmented matrices and systems, to highlight the row-echelon and reduced row-echelon form of a matrix, and to solve systems using the inverse matrix. In statistics, new examples utilizing pie charts and bar graphs are given. In Chapter 6 on counting, a new example to explain combinations was added. In Chapter 7 a new example has been included to ease students into probability using counting.

OUTLINE

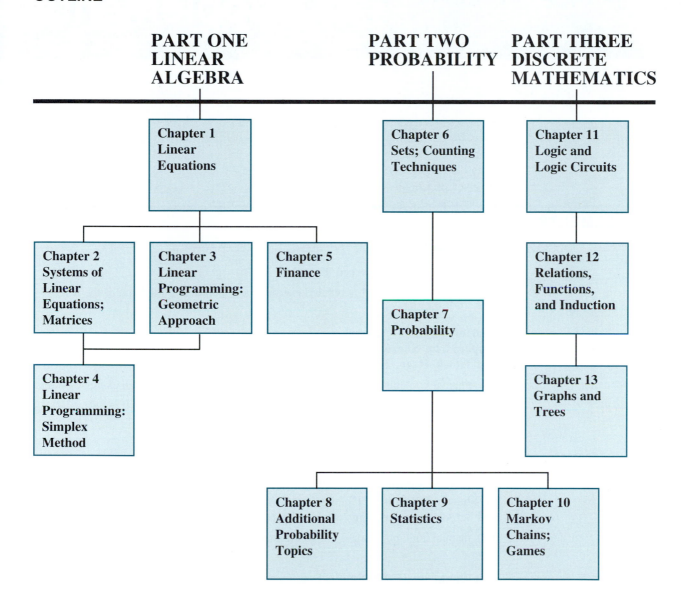

PART ONE
LINEAR
ALGEBRA

PART TWO
PROBABILITY

PART THREE
DISCRETE
MATHEMATICS

Chapter 1
Linear
Equations

Chapter 6
Sets; Counting
Techniques

Chapter 11
Logic and
Logic Circuits

Chapter 2
Systems of
Linear
Equations;
Matrices

Chapter 3
Linear
Programming:
Geometric
Approach

Chapter 5
Finance

Chapter 7
Probability

Chapter 12
Relations,
Functions,
and Induction

Chapter 4
Linear
Programming:
Simplex
Method

Chapter 13
Graphs and
Trees

Chapter 8
Additional
Probability
Topics

Chapter 9
Statistics

Chapter 10
Markov
Chains;
Games

Supplements

SUPPLEMENTS

Student Aids

Answers to Odd-Numbered Exercises

Answers to all of the odd-numbered exercises are included at the back of the text.

**Student Solutions Manual,
Prepared by Sharon O'Donnell, Chicago State University
and Iztock Hozo, Indiana University-Northwest**

This manual contains worked-out solutions to all of the odd-numbered exercises and is available through your bookstore.
ISBN 0-471-35508-9

Instructor Aids

**Instructor's Solutions Manual,
by Sharon O'Donnell, Chicago State University
and Iztock Hozo, Indiana University-Northwest**

This manual contains worked-out solutions to all of the problems in the text.

Test Bank, by Henry Smith, Southeastern Louisiana University

A wide range of problems and their solutions are keyed to the text and exercise sets.

Computerized Test Bank

Available in both IBM and Macintosh formats, the Computerized Test Bank allows instructors to create, customize, and print a test containing any combinations of questions from the test bank. Instructors can also edit the questions or add their own.

Instructor's Resource CD-ROM

This CD-ROM, accessible from both IBM and Mactintosh platforms, includes the Instructor's Solutions Manual and the Test Bank.

Acknowledgments

W e thank the following individuals who made valuable contributions to this edition.

REVIEWERS

Dona Boccio, *Queensborough Community College*

Sam Councilman, *California State University-Long Beach*

Morteza Ebneshahrashoob, *California State University-Long Beach*

Benjamin Eichorn, *Rider University*

Allen Emerson, *St. John Fisher College*

Michael J. Failla, *Harford Community College*

Theodore Faticoni, *Fordham University*

Klara Grodzinsky, *Georgia Institute of Technology*

Irvin Roy Hentzel, *Iowa State University*

John O. Herzog, *Pacific Lutheran University*

Eleanor Kendrick, *San Jose City College*

Nigel Ray, *University of Oregon*

Dan Rinne, *California State University-San Bernardino*

Lora Titus, *St. John Fisher College*

David Vella, *Skidmore College*

Peter Williams, *California State University-San Bernardino*

Joseph Zund, *New Mexico State University*

ACCURACY READERS AND ACCURACY CHECKERS

Sharon O'Donnell, *Chicago State University*

Stella Pudar-Hozo, *Indiana University Northwest*

Iztock Hozo, *Indiana University Northwest*

Contents

Part One
Linear Algebra   1

Chapter 1   Linear Equations   3
1.1   Rectangular Coordinates; Lines   3
1.2   Parallel and Intersecting Lines   19
1.3   Applications   25
Chapter Review   33

Chapter 2   Systems of Linear Equations; Matrices   37
2.1   Systems of Linear Equations: Substitution; Elimination   38
2.2   Systems of Linear Equations: Matrix Method   51
2.3   Systems of m Linear Equations Containing n Variables   66
2.4   Matrix Algebra   78
2.5   Multiplication of Matrices   90
2.6   Inverse of a Matrix   98
2.7   Applications: Leontief Model; Cryptography; Accounting; The Method of Least Squares   108
Chapter Review   126

Chapter 3   Linear Programming: Geometric Approach   130
3.1   Linear Inequalities   130
3.2   A Geometric Approach to Linear Programming Problems   142

3.3   Applications   151
Chapter Review   159

Chapter 4   Linear Programming: Simplex Method   166
4.1   The Simplex Tableau; Pivoting   167
4.2   The Simplex Method: Solving Maximum Problems in Standard Form   179
4.3   Solving Minimum Problems in Standard Form; The Duality Principle   196
4.4   The Simplex Method with Mixed Constraints   205
Chapter Review   220

Chapter 5   Finance   224
5.1   Interest   224
5.2   Compound Interest   231
5.3   Annuities; Sinking Funds   240
5.4   Present Value of an Annuity; Amortization   249
5.5   Applications: Leasing; Capital Expenditure; Bonds   257
Chapter Review   260

Part Two
Probability   265

Chapter 6   Sets; Counting Techniques   267
6.1   Sets   267
6.2   The Number of Elements in a Set   277

6.3   The Multiplication Principle   282

6.4   Permutations   287

6.5   Combinations   292

6.6   More Counting Problems   299

6.7   The Binomial Theorem   305

Chapter Review   312

Chapter 7   Probability   316

7.1   Sample Spaces and Assignment of Probabilities   317

7.2   Properties of the Probability of an Event   328

7.3   Probability Problems Using Counting Techniques   339

7.4   Conditional Probability   345

7.5   Independent Events   355

Chapter Review   363

Chapter 8   Additional Probability Topics   367

8.1   Bayes' Formula   367

8.2   The Binomial Probability Model   379

8.3   Expected Value   387

8.4   Applications   396

8.5   Random Variables   402

Chapter Review   406

Chapter 9   Statistics   411

9.1   Bar Graphs; Pie Charts   412

9.2   Organization of Data   418

9.3   Measures of Central Tendency   426

9.4   Measures of Dispersion   433

9.5   The Normal Distribution   441

Chapter Review   451

Chapter 10   Markov Chains; Games   454

10.1   Markov Chains and Transition Matrices   454

10.2   Regular Markov Chains   463

10.3   Absorbing Markov Chains   472

10.4   Two-Person Games   481

10.5   Mixed Strategies   485

10.6   Optimal Strategy in Two-Person Zero-Sum Games with 2×2 Matrices   488

Chapter Review   494

Part Three
Discrete Mathematics   499

Chapter 11   Logic and Logic Circuits   501

11.1   Propositions   502

11.2   Truth Tables   508

11.3   Implications; The Biconditional Connective; Tautologies   517

11.4   Arguments   524

11.5   Logic Circuits   530

Chapter Review   535

Chapter 12   Relations, Functions, and Induction   538

12.1   Relations   538

12.2   Functions   543

12.3   Sequences   550

12.4   Mathematical Induction   553

12.5   Recurrence Relations   558

Chapter Review   564

Chapter 13   Graphs and Trees   566

13.1   Graphs   566

13.2   Paths and Connectedness   572

13.3   Eulerian and Hamiltonian Circuits   576

13.4   Trees   584

13.5   Directed Graphs   592

13.6   Matrix Applications to Graphs and Directed Graphs   600

Chapter Review   610

Appendix A   Review Topics from Algebra and Geometry   A-1

A.1   Basic Algebra   A-1

A.2   Exponents and Logarithms   A-9

A.3   Geometric Sequences   A-14

Appendix B   Using LINDO to Solve Linear Programming Problems   A-20

Tables   A-28

Answers to Odd-Numbered Problems   AN-1

Index   I-1

Finite Mathematics

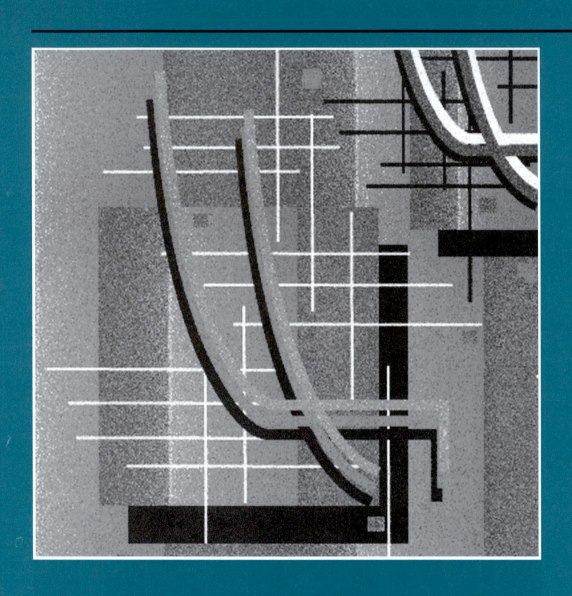

PART ONE

Linear Algebra

Chapter 1   Linear Equations
Chapter 2   Systems of Linear Equations; Matrices
Chapter 3   Linear Programming: Geometric Approach
Chapter 4   Linear Programming: Simplex Method
Chapter 5   Finance

Chapter 1

Linear Equations

1.1 **Rectangular Coordinates; Lines**

1.2 **Parallel and Intersecting Lines**

1.3 **Applications**

Chapter Review

Much of the material presented here will already be familiar to you. We begin with a review of rectangular coordinates, followed by a discussion of lines, including parallel and intersecting lines. We close the chapter with some applications involving lines.

1.1 RECTANGULAR COORDINATES; LINES

Rectangular Coordinates

We begin with two lines located in the same plane: one horizontal and the other vertical. We call the horizontal line the **x-axis,** the vertical line the **y-axis,** and the point of intersection the **origin O.** We assign coordinates to every point on these lines as shown in Figure 1, using a convenient scale. (The scales are usually, but not necessarily, the same on both axes.) Also, the origin O has a value of 0 on both the x-axis and the y-axis. We follow the usual convention that points on the x-axis to the right of O are associated with positive real numbers, and those to the left of O are associated with negative real numbers. Those on the y-axis above O are associated with positive real numbers, and those below O are associated with negative real numbers. In Figure 1, the x-axis and y-axis are labeled as x and y, respectively, and we have used an arrow at the end of each axis to denote the positive direction.

Figure 1

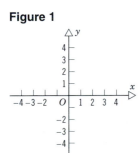

The coordinate system described here is called a **rectangular,** or **Cartesian, coordinate system.** The plane containing the x-axis and y-axis is sometimes called the **xy-plane,** and the x-axis and y-axis are referred to as the **coordinate axes.**

Any point P in the xy-plane can then be located by using an **ordered pair (x, y)** of real numbers, also called the **coordinates** of P. The number x denotes the signed dis-

3

Figure 2

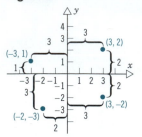

Figure 3

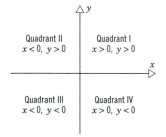

tance of P from the y-axis (*signed* in the sense that if P is to the right of the y-axis, then $x > 0$, and if P is to the left of the y-axis, then $x < 0$); y denotes the signed distance of P from the x-axis. If (x, y) are the coordinates of a point P, then x is called the **x-coordinate**, or **abscissa**, of P and y is the **y-coordinate**, or **ordinate**, of P. We identify the point P by its coordinates (x, y) by writing $P = (x, y)$. Usually, we will simply say "the point (x, y)" rather than "the point whose coordinates are (x, y)."

To locate the point $(-3, 1)$, we go 3 units along the x-axis to the left of O and then go straight up 1 unit. We **plot** this point by placing a dot at this location. See Figure 2, in which the points with coordinates $(-3, 1)$, $(-2, -3)$, $(3, -2)$, and $(3, 2)$ are plotted.

The origin has coordinates $(0, 0)$. Any point on the x-axis has coordinates of the form $(x, 0)$, and any point on the y-axis has coordinates of the form $(0, y)$.

The coordinate axes divide the xy-plane into four sections, called **quadrants,** as shown in Figure 3. In quadrant I both the x-coordinate and the y-coordinate are positive; in quadrant II x is negative and y is positive; in quadrant III both x and y are negative; and in quadrant IV x is positive and y is negative. Points on the coordinate axes belong to no quadrant.

Graphs of Linear Equations in Two Variables

A **linear equation in two variables** is of the form

$$Ax + By = C \qquad (1)$$

where A and B are not both zero. Examples of linear equations are

$3x - 5y - 6 = 0$ This equation can be written as
$3x - 5y = 6 \qquad A = 3, B = -5, C = 6$

$-3x = 2y - 1$ This equation can be written as
$-3x - 2y = -1 \qquad A = -3, B = -2, C = -1$

$y = \frac{3}{4}x - 5$ Here we can write
$-\frac{3}{4}x + y = -5 \qquad$ or $\qquad 3x - 4y = 20$
$A = 3, B = -4, C = 20$

$y = -5$ Here we can write
$0 \cdot x + y = -5 \qquad A = 0, B = 1, C = -5$

$x = 4$ Here we can write
$x + 0 \cdot y = 4 \qquad A = 1, B = 0, C = 4$

The **graph** of an equation is the set of all points (x, y) whose coordinates satisfy the equation. For example, $(0, 4)$ is a point on the graph of the equation $3x + 4y = 16$, because when we substitute 0 for x and 4 for y in the equation, we get

$$3 \cdot 0 + 4 \cdot 4 = 16$$

which is a true statement.

It can be shown that if A, B, C are real numbers, with A and B not both zero, then the graph of the equation

$$Ax + By = C$$

is a *line*. This is the reason we call it a *linear* equation.

Conversely, any line is the graph of an equation of the form (1).

✏ **Now Work Problem 1**

Figure 4

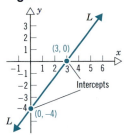

Since any line can be written as an equation in the form $Ax + By = C$, we call this form the **general equation** of a line.

Given a linear equation, we can obtain its graph by plotting two points that satisfy its equation and connecting them with a line. The easiest two points to plot are the *intercepts*. For example, the line shown in Figure 4 has the intercepts $(0, -4)$ and $(3, 0)$.

Intercepts

The points at which the graph of a line crosses the axes are called **intercepts.** The **x-intercept** is the point at which the graph crosses the x-axis; the **y-intercept** is the point at which the graph crosses the y-axis.

Steps for Finding the Intercepts of a Line

To find the intercepts of a line $Ax + By = C$, with $A \neq 0$ or $B \neq 0$, follow these steps:

Step 1 Set $y = 0$ and solve for x. This determines the x-intercept of the line.

Step 2 Set $x = 0$ and solve for y. This determines the y-intercept of the line.

EXAMPLE 1   Finding the Intercepts of an Equation

Find the intercepts of the equation $2x + 3y = 6$. Graph the equation.

SOLUTION

Step 1 To find the x-intercept, we need to find the number x for which $y = 0$. Thus we set $y = 0$ to get

$$2x + 3(0) = 6$$
$$2x = 6$$
$$x = 3$$

The x-intercept is $(3, 0)$.

Step 2 To find the y-intercept, we set $x = 0$ and solve for y:

$$2(0) + 3y = 6$$
$$3y = 6$$
$$y = 2$$

The y-intercept is $(0, 2)$.

Figure 5

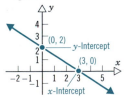

Since the equation is a linear equation, its graph is a line. We use the two intercepts $(3, 0)$ and $(0, 2)$ to graph it. See Figure 5.

EXAMPLE 2   Graphing an Equation

Graph the equation: $y = 2x + 5$

SOLUTION   This equation can be written as

$$-2x + y = 5$$

This is a linear equation, so its graph is a line. The intercepts are $(0, 5)$ and $(-\frac{5}{2}, 0)$, which you should verify. For reassurance we'll find a third point. Arbitrarily, we let $x = 10$. Then $y = 2(10) + 5 = 25$, so $(10, 25)$ is a point on the graph. See Figure 6.

Figure 6

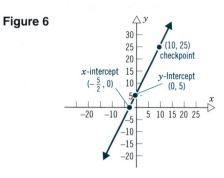

x	y
0	5
$-\frac{5}{2}$	0
10	25

When a line passes through the origin, it has only one intercept. To graph such lines, we need to locate an additional point on the graph.

EXAMPLE 3   Graphing an Equation

Graph the equation: $-x + y = 0$

SOLUTION   This is a linear equation, so its graph is a line. The only intercept is $(0, 0)$. To locate another point on the graph, let $x = 3$. Then $y = 3$ and $(3, 3)$ is a point on the graph. See Figure 7.

Figure 7

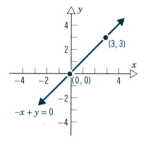

 **Now Work Problem 5**

EXAMPLE 4   Graphing an Equation (a Vertical Line)

Graph the equation: $x = 3$

Figure 8 $x = 3$

SOLUTION We are looking for all points (x, y) in the plane for which $x = 3$. Thus, no matter what y-coordinate is used, the corresponding x-coordinate always equals 3. Consequently, the graph of the equation $x = 3$ is a vertical line with x-intercept $(3, 0)$ as shown in Figure 8. ∎

As suggested by Example 4, we have the following result:

Equation of a Vertical Line

A vertical line is given by an equation of the form

$$x = a$$

where $(a, 0)$ is the x-intercept.

EXAMPLE 5 **Finding the Equation of a Vertical Line**

Find an equation for the vertical line that passes through the point $(3, 4)$.

SOLUTION The x-coordinate of any point on a vertical line is always the same. Since $(3, 4)$ is a point on the vertical line, its equation is $x = 3$. ∎

Slope of a Line

An important characteristic of a line, called its *slope,* is best defined by using rectangular coordinates.

Slope of a Line

Let $P = (x_1, y_1)$ and $Q = (x_2, y_2)$ be two distinct points with $x_1 \neq x_2$. The **slope** m of the nonvertical line L containing P and Q is defined by the formula

$$m = \frac{y_2 - y_1}{x_2 - x_1} \qquad x_1 \neq x_2 \tag{2}$$

If $x_1 = x_2$, L is a vertical line and the slope m of L *is undefined* (since this results in division by 0).

Figure 9(a) provides an illustration of the slope of a nonvertical line; Figure 9(b) illustrates a vertical line.

Figure 9

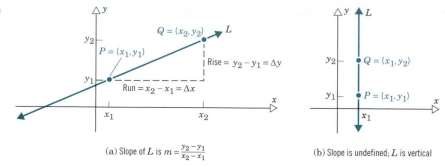

(a) Slope of L is $m = \frac{y_2 - y_1}{x_2 - x_1}$

(b) Slope is undefined; L is vertical

As Figure 9(a) illustrates, the slope m of a nonvertical line may be viewed as

$$m = \frac{y_2 - y_1}{x_2 - x_1} = \frac{\text{Rise}}{\text{Run}} = \frac{\text{Change in } y}{\text{Change in } x}$$

The change in y may be denoted by Δy, read "delta y," and the change in x by Δx.

The slope m of a nonvertical line L measures the amount y changes, Δy, as x changes from x_1 to x_2, Δx. This is called the **average rate of change of y with respect to x.** Then, the slope m is

$$m = \frac{\Delta y}{\Delta x} = \text{average rate of change of } y \text{ with respect to } x$$

EXAMPLE 6   Finding and Interpreting the Slope of a Line

The slope m of the line containing the points $(3, -2)$ and $(1, 5)$ is

$$m = \frac{5 - (-2)}{1 - 3} = \frac{7}{-2} = \frac{-7}{2}$$

We interpret the slope to mean that for every 2-unit change in x, y will change by -7 units. That is, if x increases by 2 units, then y decreases by 7 units. The average rate of change of y with respect to x is $\frac{-7}{2}$. ◼

Two comments about computing the slope of a nonvertical line may prove helpful:

1. Any two distinct points on the line can be used to compute the slope of the line. (See Figure 10 for justification.)
2. The slope of a line may be computed from $P = (x_1, y_1)$ to $Q = (x_2, y_2)$ or from Q to P because

$$\frac{y_2 - y_1}{x_2 - x_1} = \frac{y_1 - y_2}{x_1 - x_2}$$

Figure 10 Triangles *ABC* and *PQR* are similar (equal angles). Hence, ratios of corresponding sides are proportional. Thus:

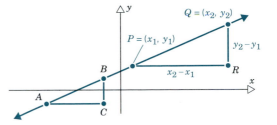

Slope using *P* and $Q = \dfrac{y_2 - y_1}{x_2 - x_1}$ = Slope using *A* and $B = \dfrac{d(B, C)}{d(A, C)}$

where $d(B, C)$ denotes the distance from *B* to *C* and $d(A, C)$ denotes the distance from *A* to *C*.

🖎 **Now Work Problem 9**

To get a better idea of the meaning of the slope *m* of a line *L*, consider the following example.

EXAMPLE 7   **Finding the Slopes of Various Lines Containing the Same Point (2, 3)**

Compute the slopes of the lines L_1, L_2, L_3, and L_4 containing the following pairs of points. Graph all four lines on the same set of coordinate axes.

$$L_1: \quad P = (2, 3) \qquad Q_1 = (-1, -2)$$
$$L_2: \quad P = (2, 3) \qquad Q_2 = (3, -1)$$
$$L_3: \quad P = (2, 3) \qquad Q_3 = (5, 3)$$
$$L_4: \quad P = (2, 3) \qquad Q_4 = (2, 5)$$

SOLUTION   Let m_1, m_2, m_3, and m_4 denote the slopes of the lines L_1, L_2, L_3, and L_4, respectively. Then

$$m_1 = \frac{-2 - 3}{-1 - 2} = \frac{-5}{-3} = \frac{5}{3} \qquad \text{A rise of 5 divided by a run of 3}$$

$$m_2 = \frac{-1 - 3}{3 - 2} = \frac{-4}{1} = -4 \qquad \text{A rise (drop) of } -4 \text{ divided by a run of 1}$$

$$m_3 = \frac{3 - 3}{5 - 2} = \frac{0}{3} = 0 \qquad \text{A rise of 0 divided by a run of 3}$$

m_4 is undefined

The graphs of these lines are given in Figure 11 on page 10.   ■

As Figure 11 illustrates,

1. When the slope *m* of a line is positive, the line slants upward from left to right (L_1).
2. When the slope *m* is negative, the line slants downward from left to right (L_2).
3. When the slope *m* is 0, the line is horizontal (L_3).
4. When the slope *m* is undefined, the line is vertical (L_4).

Figure 11

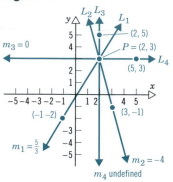

Figure 12

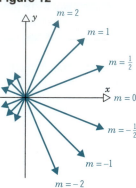

Figure 12 illustrates the slopes of several lines. Note the pattern. Do you see that the slope of a line is an indicator of its steepness?

Now Work Problem 13

EXAMPLE 8   **Graphing a Line When its Slope and a Point Are Given**

Draw a graph of the line that contains the point (3, 2) and has a slope of

(a) $\frac{3}{4}$ (b) $-\frac{4}{5}$

SOLUTION

(a) Slope = rise/run. The fact that the slope is $\frac{3}{4}$ means that for every horizontal movement (run) of 4 units to the right, there will be a vertical movement (rise) of 3 units. If we start at the given point (3, 2) and move 4 units to the right and 3 units up, we reach the point (7, 5). By drawing the line through this point and the point (3, 2), we have the graph (see Figure 13).

(b) The fact that the slope is $-\frac{4}{5} = \frac{-4}{5}$ means that for every horizontal movement of 5 units to the right, there will be a corresponding vertical movement of -4 units (a downward movement). If we start at the given point (3, 2) and move 5 units to the right and then 4 units down, we arrive at the point (8, -2). By drawing the line through these points, we have the graph (see Figure 14).

Figure 13   Slope $= \frac{3}{4}$

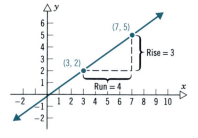

Figure 14   Slope $= -\frac{4}{5}$

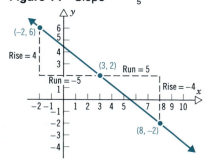

Alternatively, we can set $-\frac{4}{5} = \frac{4}{-5}$ so that for every horizontal movement of -5 units (a movement to the left), there will be a corresponding vertical movement of 4 units (upward). This approach brings us to the point $(-2, 6)$, which is also on the graph shown in Figure 14.

 **Now Work Problem 23**

Other Forms of the Equation of a Line

Figure 15

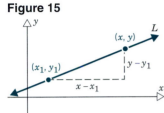

Let L be a nonvertical line with slope m and containing the point (x_1, y_1). See Figure 15. Any two distinct points on L can be used to compute slope. Thus, for any other point (x, y) on L, we have

$$m = \frac{y - y_1}{x - x_1} \quad \text{or} \quad y - y_1 = m(x - x_1)$$

Point–Slope Form of an Equation of a Line

An equation of a nonvertical line of slope m that contains the point (x_1, y_1) is

$$y - y_1 = m(x - x_1) \tag{3}$$

EXAMPLE 9   Using the Point–Slope Form of a Line

An equation of the line with slope 4 and containing the point $(1, 2)$ can be found by using the point–slope form with $m = 4$, $x_1 = 1$, and $y_1 = 2$:

$$y - y_1 = m(x - x_1)$$
$$y - 2 = 4(x - 1)$$
$$y = 4x - 2$$

See Figure 16.

Figure 16 $y = 4x - 2$

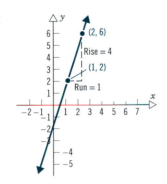

EXAMPLE 10   Finding the Equation of a Horizontal Line

Find an equation of the horizontal line containing the point (3, 2).

SOLUTION The slope of a horizontal line is 0. To get an equation, we use the point–slope form with $m = 0$, $x_1 = 3$, and $y_1 = 2$:

$$y - y_1 = m(x - x_1)$$
$$y - 2 = 0 \cdot (x - 3)$$
$$y - 2 = 0$$
$$y = 2$$

Figure 17 $y = 2$

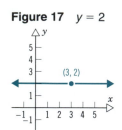

See Figure 17 for the graph.

As suggested by Example 10, we have the following result:

> **Equation of a Horizontal Line**
>
> A horizontal line is given by an equation of the form
>
> $$y = b$$
>
> where $(0, b)$ is the y-intercept.

EXAMPLE 11   Finding the Equation of a Line Given Two Points

Find the equation of the line L containing the points (2, 3) and $(-4, 5)$. Graph the line L.

SOLUTION Since two points are given, we first compute the slope of the line:

$$m = \frac{5 - 3}{-4 - 2} = \frac{2}{-6} = \frac{-1}{3}$$

We use the point (2, 3) and the fact that the slope $m = -\frac{1}{3}$ to get the point–slope form of the equation of the line:

$$y - 3 = -\tfrac{1}{3}(x - 2)$$

See Figure 18 for the graph.

Figure 18
$y - 3 = -\tfrac{1}{3}(x - 2)$

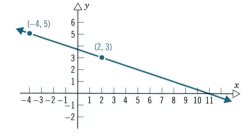

In the solution to Example 11 we could have used the point $(-4, 5)$ instead of the point $(2, 3)$. The equation that results, although it looks different, is equivalent to the equation we obtained in the example. (Try it for yourself.)

The general form of the equation of the line in Example 11 can be obtained by multiplying both sides of the point–slope equation by 3 and collecting terms:

$$y - 3 = -\tfrac{1}{3}(x - 2)$$
$$3(y - 3) = 3(-\tfrac{1}{3})(x - 2) \qquad \text{Multiply by 3.}$$
$$3y - 9 = -1(x - 2)$$
$$3y - 9 = -x + 2$$
$$x + 3y = 11$$

This is the general form of the equation.

 **Now Work Problem 35**

Another useful equation of a line is obtained when the slope m and y-intercept $(0, b)$ are known. In this case we know both the slope m of the line and a point $(0, b)$ on the line; thus, we may use the point–slope form, Equation (3), to obtain the following equation:

$$y - b = m(x - 0) \qquad \text{or} \qquad y = mx + b$$

Slope–Intercept Form of an Equation of a Line

An equation of a line L with slope m and y-intercept $(0, b)$ is

$$y = mx + b \qquad\qquad (4)$$

When an equation of a line is written in slope–intercept form, it is easy to find the slope m and y-intercept $(0, b)$ of the line. For example, suppose the equation of the line is

$$y = -2x + 3$$

Compare it to $y = mx + b$:

$$y = -2x + 3$$
$$\uparrow \qquad \uparrow$$
$$y = \;\; mx + b$$

The slope of this line is -2 and its y-intercept is $(0, 3)$.

Let's look at another example.

EXAMPLE 12 **Finding the Slope and y-Intercept of a Line**

Find the slope m and y-intercept $(0, b)$ of the line $2x + 4y - 8 = 0$. Graph the line.

SOLUTION To obtain the slope and y-intercept, we transform the equation into its slope–intercept form. To do this, we need to solve for y:

$$2x + 4y - 8 = 0$$
$$4y = -2x + 8$$
$$y = -\tfrac{1}{2}x + 2$$

The coefficient of x, $-\frac{1}{2}$, is the slope, and the y-intercept is $(0, 2)$.

We can graph the line in either of two ways:

Figure 19

$2x + 4y - 8 = 0$

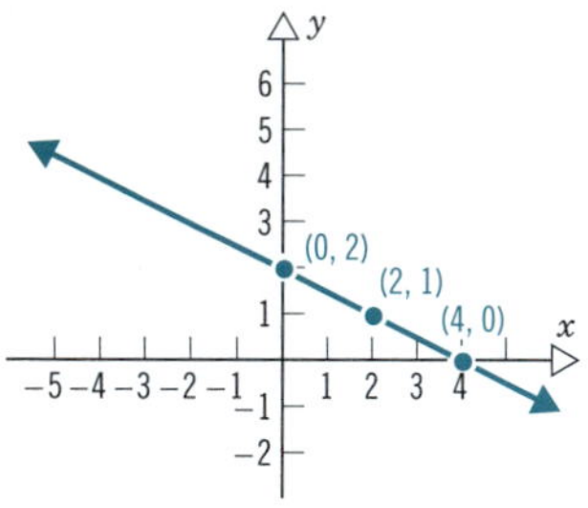

1. Use the fact that the y-intercept is $(0, 2)$ and the slope is $-\frac{1}{2}$. Then, starting at the point $(0, 2)$, go to the right 2 units and then down 1 unit to the point $(2, 1)$.
2. Locate the intercepts. The y-intercept is $(0, 2)$. To obtain the x-intercept, we let $y = 0$ and solve for x. When $y = 0$, we have

$$2x + 4 \cdot 0 - 8 = 0$$
$$2x - 8 = 0$$
$$x = 4$$

Thus, the intercepts are $(4, 0)$ and $(0, 2)$.

See Figure 19.

[*Note:*   The second method, locating the intercepts, only produces one point when the line passes through the origin. In this case some other point on the line must be found in order to graph the line. Refer back to Example 3.]

✍ **Now Work Problem 53**

EXAMPLE 13   **Daily Cost of Production**

A certain factory has daily fixed overhead expenses of $2000, while each item produced costs $100. Find an equation that relates the daily cost C to the number x of items produced each day.

SOLUTION   The fixed overhead expense of $2000 represents the fixed cost, the cost incurred no matter how many items are produced. Since each item produced costs $100, the variable cost of producing x items is $100x$. Thus the total daily cost C of production is

Figure 20

$$C = 100x + 2000$$

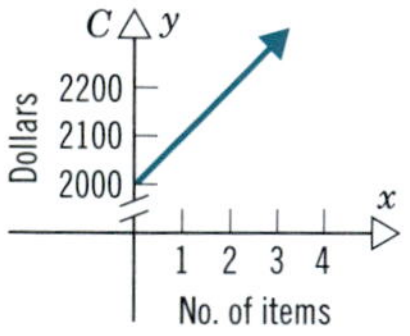

The graph of this equation is given by the line in Figure 20. Notice that the fixed cost $2000 is represented by the y-intercept, while the $100 cost of producing each item is the slope. Also notice that a different scale is used on each axis.

SUMMARY

The graph of a linear equation, $Ax + By = C$, where A and B are not both zero, is a line. In this form it is referred to as the general equation of a line.

1. Given the general equation of a line, information can be found about the line:
   (a) Place the equation in slope–intercept form $y = mx + b$ to find the slope m and y-intercept $(0, b)$.
   (b) Let $x = 0$ and solve for y to find the y-intercept.
   (c) Let $y = 0$ and solve for x to find the x-intercept.

2. Given information about a line, an equation of the line can be found. The form of the equation to use depends on the given information.

Given	Use	Equation
Point (x_1, y_1), slope m	Point–slope form	$y - y_1 = m(x - x_1)$

Two points $(x_1, y_1), (x_2, y_2)$	If $x_1 = x_2$:
	  The line is vertical
	If $x_1 \neq x_2$:
	  Find the slope m,
	  $m = \dfrac{y_2 - y_1}{x_2 - x_1}$
	  Then use the point–slope form
		$x = x_1$
		
		$y - y_1 = m(x - x_1)$

Slope m, y-intercept $(0, b)$	Slope–intercept form	$y = mx + b$

EXERCISE 1.1   Answers to odd-numbered problems begin on page AN-1.

1. Give the coordinates of each point in the following figure. Assume each coordinate is an integer.

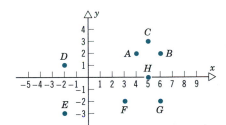

2. Plot each point in the xy-plane. Tell in which quadrant or on what coordinate axis each point lies.

  (a) $A = (-3, 2)$ (b) $B = (6, 0)$
  (c) $C = (-2, -2)$ (d) $D = (6, 5)$
  (e) $E = (0, -3)$ (f) $F = (6, -3)$

3. Plot the points $(2, 0)$, $(2, -3)$, $(2, 4)$, $(2, 1)$, and $(2, -1)$. Describe the collection of all points of the form $(2, y)$, where y is a real number.

4. Plot the points $(0, 3)$, $(1, 3)$, $(-2, 3)$, $(5, 3)$, and $(-4, 3)$. Describe the collection of all points of the form $(x, 3)$, where x is a real number.

In Problems 5–8 copy the tables given and fill in the missing values using the given equations. Use these points to graph each equation.

5. $y = 2x + 4$

x	0		2	-2	4	-4
y		0				

6. $y = -3x + 6$

x	0		2	-2	4	-4
y		0				

7. $2x - y = 6$

x	0		2	-2	4	-4
y		0				

8. $x + 3y = 9$

x	0		2	-2	4	-4
y		0				

In Problems 9–12 find the slope of the line. Give an interpretation of the slope.

9.

10.

11.
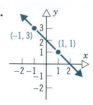

12.

In Problems 13–22 plot each pair of points and determine the slope of the line containing them. Graph the line.

13. $(2, 3)$; $(1, 0)$

14. $(1, 2)$; $(3, 4)$

15. $(-2, 3)$; $(2, 1)$

16. $(-1, 1)$; $(2, 3)$

17. $(-3, -1)$; $(2, -1)$

18. $(4, 2)$; $(-5, 2)$

19. $(-1, 2)$; $(-1, -2)$

20. $(2, 0)$; $(2, 2)$

21. $(\sqrt{2}, 3)$; $(1, \sqrt{3})$

22. $(-2\sqrt{2}, 0)$; $(4, \sqrt{5})$

In Problems 23–30 graph the line containing the point P and having slope m.

23. $P = (1, 2)$; $m = 2$

24. $P = (2, 1)$; $m = 3$

25. $P = (2, 4)$; $m = -\frac{3}{4}$

26. $P = (1, 3)$; $m = -\frac{2}{5}$

27. $P = (-1, 3)$; $m = 0$

28. $P = (2, -4)$; $m = 0$

29. $P = (0, 3)$;   slope undefined

30. $P = (-2, 0)$;   slope undefined

In Problems 31–34 find a general equation of each line.
Write the equation in the form $Ax + By = C$.

31.

32.

33.

34.

In Problems 35–52 find a general equation for the line with the given properties.
Write the equation in the form $Ax + By = C$.

35. Slope $= 2$;   containing the point $(-4, 1)$

36. Slope $= 3$;   containing the point $(-3, 4)$

37. Slope $= -\frac{2}{3}$;   containing the point $(1, -1)$

38. Slope $= \frac{1}{2}$;   containing the point $(3, 1)$

39. Containing the points $(1, 3)$ and $(-1, 2)$

40. Containing the points $(-3, 4)$ and $(2, 5)$

41. Slope $= -2$; y-intercept $= (0, 3)$

42. Slope $= -3$; y-intercept $= (0, -2)$

43. Slope $= 3$; x-intercept $= (-4, 0)$

44. Slope $= -4$; x-intercept $= (2, 0)$

45. Slope $= \frac{4}{5}$;   containing the point $(0, 0)$

46. Slope $= \frac{7}{3}$;   containing the point $(0, 0)$

47. x-intercept $= (2, 0)$; y-intercept $= (0, -1)$

48. x-intercept $= (-4, 0)$; y-intercept $= (0, 4)$

49. Slope undefined;   containing the point $(1, 4)$

50. Slope undefined;   containing the point $(2, 1)$

51. Slope $= 0$;   containing the point $(1, 4)$

52. Slope $= 0$;   containing the point $(2, 1)$

In Problems 53–68 find the slope and y-intercept of each line. Graph the line.

53. $y = 2x + 3$ **54.** $y = -3x + 4$ **55.** $\frac{1}{2}y = x - 1$ **56.** $\frac{1}{3}x + y = 2$

57. $2x - 3y = 6$ **58.** $3x + 2y = 6$ **59.** $x + y = 1$ **60.** $x - y = 2$

61. $x = -4$ **62.** $y = -1$ **63.** $y = 5$ **64.** $x = 2$

65. $y - x = 0$ **66.** $x + y = 0$ **67.** $2y - 3x = 0$ **68.** $3x + 2y = 0$

69. Find a general equation of the *x*-axis.

70. Find a general equation of the *y*-axis.

71. Find the equation of the horizontal line containing the point $(-1, -3)$.

72. Find the equation of the vertical line containing the point $(-2, 5)$.

73. Temperature Conversion The relationship between Celsius (°C) and Fahrenheit (°F) degrees for measuring temperature is linear. Find an equation relating °C and °F if 0°C corresponds to 32°F and 100°C corresponds to 212°F. Use the equation to find the Celsius measure of 70°F.

74. Temperature Conversion The Kelvin (K) scale for measuring temperature is obtained by adding 273 to the Celsius temperature.

   (a) Write an equation relating K and °C.
   (b) Write an equation relating K and °F (see Problem 73).

75. Profit from Selling Newspapers Each Sunday a newspaper agency sells *x* copies of a certain newspaper for $1.00 per copy. The cost to the agency of each newspaper is $0.50. The agency pays a fixed cost for storage, delivery, and so on, of $100 per Sunday. Write an equation that relates the profit *P*, in dollars, to the number *x* of copies sold. Graph this equation.

76. Repeat Problem 75 if the cost to the agency is $0.45 per copy and the fixed cost is $125 per Sunday.

77. Electricity Rates Commonwealth Edison Company supplies electricity to residential customers for a monthly customer charge of $9.36 plus 10.494 cents per kilowatt-hour for up to 400 kilowatt-hours.* Write an equation that relates the monthly charge *C*, in dollars, to the number *x* of kilowatt-hours used in a month, $0 \leq x \leq 400$. Graph this equation. What is the monthly charge for using 100 kilowatt-hours? For using 300 kilowatt-hours? Interpret the slope of the line.

78. Cost of Operating a Car The average cost of operating a standard-size car is $0.45 per mile. Write an equation that relates the average cost *C* of operating a standard-size car and the number *x* of miles it is driven.

79. Cost of Renting a Truck The cost of renting a truck is $280 per week plus a charge of $0.20 per mile driven. Write an equation that relates the cost *C* for a weekly rental in which the truck is driven *x* miles.

80. Wages of a Car Salesperson Dan receives $300 per week for selling used cars. As part of his weekly salary he also receives 10% of the sales he generates. Write an equation that relates Dan's weekly salary *S* when he generates sales of *x* dollars.

81. Weight–Height Relation Suppose the weights (*w*) of male college students are linearly related to their heights (*h*). If a student 62 inches tall weighs 120 pounds and a student 72 inches tall weighs 170 pounds, write an equation to express weight in terms of height.

82. Disease Propagation Research indicates that in a controlled environment, the number of diseased mice will increase linearly each day after one of the mice in the cage is infected with a particular type of disease-causing germ. There were 8 diseased mice 4 days after the first exposure and 14 diseased mice after 6 days. Write an equation that will give the number of diseased mice after any given number of days. If there were 40 mice in the cage, how long will it take until they are all infected?

83. Water Preservation Since the beginning of the month a local reservoir has been losing water at a constant rate. On the 10th of the month, the reservoir held 300 million gallons of water and on the 18th it held only 262 million gallons.

   (a) Write an equation that will give the amount of water in the reservoir at any time.
   (b) How much water was in the reservoir on the 14th of the month?
   (c) Interpret the slope.

* *Source:* Commonwealth Edison Rates for Residential Service, 1998.

84. Product Promotion   A cereal company finds that the number of people who will buy one of its products the first month it is introduced is linearly related to the amount of money it spends on advertising. If it spends $400,000 on advertising, 100,000 boxes of cereal will be sold, and if it spends $600,000, 140,000 boxes will be sold.

(a) Write an equation describing the relation between the amount spent on advertising and the number of boxes sold.
(b) How much advertising is needed to sell 200,000 boxes of cereal?
(c) Interpret the slope.

Technology Exercises

In Problems 1–8 graph each linear equation. Be sure to use a viewing rectangle that shows the intercepts. Then locate each intercept rounded to two decimal places.

1. $1.2x + 0.8y = 20$
2. $-1.3x + 2.7y = 81.1$
3. $215x - 0.1y = 53$
4. $0.5x - 313y = 82$
5. $\frac{4}{17}x + \frac{6}{23}y = \frac{2}{3}$
6. $\frac{9}{14}x - \frac{43}{8}y = \frac{22}{7}$
7. $\pi x - \sqrt{3}\,y = \sqrt{6}$
8. $x + \pi y = \sqrt{15}$

9. Seeing the Concept   On the same screen graph each of the following lines:

(a) $y = 0$ (slope 0)
(b) $y = \frac{1}{2}x$ (slope $\frac{1}{2}$)
(c) $y = x$ (slope 1)
(d) $y = 2x$ (slope 2)

10. Seeing the Concept   On the same screen graph each of the following lines:

(a) $y = 0$ (slope 0)
(b) $y = -\frac{1}{2}x$ (slope $-\frac{1}{2}$)
(c) $y = -x$ (slope -1)
(d) $y = -2x$ (slope -2)

11. Exploration   On the same screen graph each of the following lines:

(a) $y = 2x - 3$
(b) $y = 2x - 1$
(c) $y = 2x$
(d) $y = 2x + 2$

What do you conclude about lines of the form $y = 2x + b$?

12. Exploration   On the same screen, graph each of the following lines:

(a) $y = 2x + 3$
(b) $y = -3x + 3$
(c) $y = 3$
(d) $y = 8x + 3$

What do you conclude about lines of the form $y = mx + 3$?

In Problems 13–16, match each graph with the correct equation:
(a) $y = x$;   (b) $y = 2x$;   (c) $y = x/2$;   (d) $y = 4x$.

13.

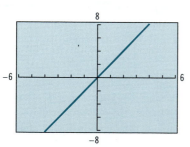

14.

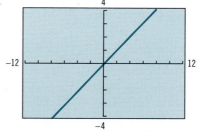

15.

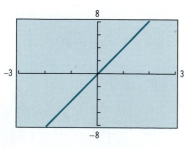

16.

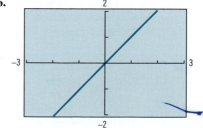

In Problems 17–22, write an equation of each line. Express your answer using either the general form or the slope–intercept form of the equation of a line, whichever you prefer.

17.

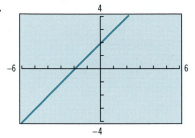

18.

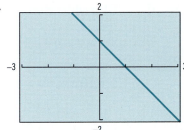

19.

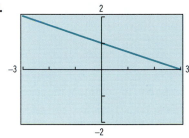

20.

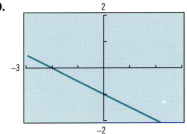

21.

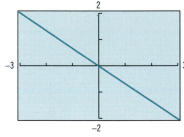

22.

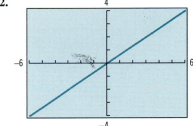

1.2 PARALLEL AND INTERSECTING LINES

Let L and M be two lines. Exactly one of the following three relationships must hold for these two lines:

1. All the points on L are the same as the points on M.
2. L and M have no points in common.
3. L and M have exactly one point in common.

If the first relationship holds, the lines L and M are identical. Such lines are called **coincident lines.**

> **Coincident Lines**
>
> Coincident lines that are vertical have undefined slope and the same x-intercept. Coincident lines that are nonvertical have the same slope and the same intercepts.

To show that two lines are coincident only requires that you show they have the same slope and the same y-intercept. Do you see why?

EXAMPLE 1   Showing that Two Lines Are Coincident

Show that the lines given by the equations below are coincident.

$$L: \quad 2x - y = 5 \qquad M: \quad -4x + 2y = -10$$

Figure 21

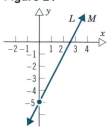

SOLUTION   We put each equation into slope–intercept form:

$$L: \quad 2x - y = 5 \qquad\qquad M: \quad -4x + 2y = -10$$
$$-y = -2x + 5 \qquad\qquad\qquad 2y = 4x - 10$$
$$y = 2x - 5 \qquad\qquad\qquad\qquad y = 2x - 5$$

The lines L and M have the same slope 2 and the same y-intercept $(0, -5)$. Hence, they are coincident. See Figure 21.

Figure 22

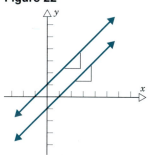

If two lines L and M (in the same plane) have no points in common, they are called **parallel lines.**

Look at Figure 22. For the two nonvertical parallel lines, equal runs result in equal rises. As a result, nonvertical parallel lines have equal slopes. Since they also have no points in common, they will also have different x- and y-intercepts.

> **Parallel Lines**
>
> Parallel lines that are vertical have undefined slope and different x-intercepts. Parallel lines that are nonvertical have the same slope and different intercepts.

To show that two lines are parallel only requires that you show they have the same slope and different y-intercepts. Do you see why?

EXAMPLE 2   Showing that Two Lines Are Parallel

Show that the lines given by the equations below are parallel.

$$L: \quad 2x + 3y = 6 \qquad M: \quad 4x + 6y = 0$$

SOLUTION   To see if these lines have equal slopes, we put each equation into slope–intercept form:

$$L: \quad 2x + 3y = 6 \qquad\qquad M: \quad 4x + 6y = 0$$
$$3y = -2x + 6 \qquad\qquad\qquad 6y = -4x$$
$$y = \frac{-2}{3}x + 2 \qquad\qquad\qquad y = \frac{-2}{3}x$$
$$\text{Slope} = \frac{-2}{3} \qquad\qquad\qquad \text{Slope} = \frac{-2}{3}$$
$$y\text{-intercept} = (0, 2) \qquad\qquad y\text{-intercept} = (0, 0)$$

Figure 23

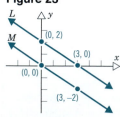

Since each has slope $-\frac{2}{3}$ and different y-intercepts, the lines are parallel. See Figure 23.

If two lines L and M have exactly one point in common, they are said to **intersect,** and the common point is called the **point of intersection.** Necessarily, the slopes of intersecting lines are unequal.

> **Intersecting Lines**
>
> Intersecting lines have different slopes.

EXAMPLE 3   Showing that Two Lines Intersect

Show that the lines given by the equations below intersect.

$$L:\quad 2x - y = 5 \qquad M:\quad x + y = 4$$

SOLUTION   We put each equation into slope–intercept form.

$$
\begin{aligned}
L:\quad 2x - y &= 5 & M:\quad x + y &= 4\\
-y &= -2x + 5 & y &= -x + 4\\
y &= 2x - 5
\end{aligned}
$$

The slope of L is 2 and the slope of M is -1. Hence, the lines intersect.

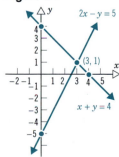 **Now Work Problem 1**

EXAMPLE 4   Finding the Point of Intersection

Find the point of intersection of the two lines

$$L:\quad 2x - y = 5 \qquad M:\quad x + y = 4$$

SOLUTION   The slope–intercept form of each line is

$$L:\quad y = 2x - 5 \qquad M:\quad y = -x + 4$$

If (x_0, y_0) denotes the point of intersection, then (x_0, y_0) is a point on both L and M. As a result, we must have

$$y_0 = 2x_0 - 5 \qquad \text{and} \qquad y_0 = -x_0 + 4$$

We set these equal and proceed to solve for x_0.

$$
\begin{aligned}
2x_0 - 5 &= -x_0 + 4\\
3x_0 &= 9\\
x_0 &= 3
\end{aligned}
$$

Substituting $x_0 = 3$ in $y_0 = 2x_0 - 5$ (or in $y_0 = -x_0 + 4$), we find

$$y_0 = 2x_0 - 5 = 2(3) - 5 = 1$$

Thus the point of intersection of L and M is $(3, 1)$. See Figure 24.

Check: We verify that the point $(3, 1)$ is on both L and M.

$$L:\quad 2x - y = 2(3) - 1 = 6 - 1 = 5 \qquad M:\quad x + y = 3 + 1 = 4$$

Figure 24

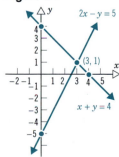

 Now Work Problem 13

Perpendicular Lines

When two lines intersect and form a right angle, they are said to be **perpendicular.** For example, a vertical line and a horizontal line are perpendicular.

Perpendicular Lines

Two distinct nonvertical lines L and M with slopes m_1 and m_2, respectively, are **perpendicular** if and only if $m_1 m_2 = -1$. That is, the product of their slopes is -1.

EXAMPLE 5   Showing that Two Lines Are Perpendicular

Show that the lines given by the equations below are perpendicular.

$$L: \quad x - 2y = 6 \qquad M: \quad 2x + y = 1$$

SOLUTION To see if these lines are perpendicular, find the slope of each:

$$
\begin{aligned}
x - 2y &= 6 & 2x + y &= 1 \\
-2y &= -x + 6 & y &= -2x + 1 \\
y &= \tfrac{1}{2}x - 3 & \text{Slope} = m_2 &= -2 \\
\text{Slope} = m_1 &= \tfrac{1}{2}
\end{aligned}
$$

Since $m_1 m_2 = \tfrac{1}{2} \cdot (-2) = -1$, the lines are perpendicular.   ■

**EXAMPLE 6   Finding the Equation of Two Lines:
One Parallel to and the Other Perpendicular to a Given Line**

Given the line $x - 4y = 8$, find an equation for the line that contains the point $(2, 1)$ and is

(a) Parallel to the given line   (b) Perpendicular to the given line

SOLUTION First find the slope of the line $x - 4y = 8$ by putting it in slope–intercept form $y = mx + b$:

$$
\begin{aligned}
x - 4y &= 8 \\
-4y &= -x + 8 \\
y &= \tfrac{1}{4}x - 2
\end{aligned}
$$

The slope of the line is $\tfrac{1}{4}$.

(a) We seek a line parallel to the given line that contains the point $(2, 1)$. The slope of this line must be $\tfrac{1}{4}$. (Do you see why?) Using the point–slope form of the equation of a line, we have

$$
\begin{aligned}
y - y_1 &= m(x - x_1) \qquad m = \tfrac{1}{4},\ x_1 = 2,\ y_1 = 1 \\
y - 1 &= \tfrac{1}{4}(x - 2) \\
y &= \tfrac{1}{4}x + \tfrac{1}{2}
\end{aligned}
$$

(b) We seek a line perpendicular to the given line, whose slope is $\frac{1}{4}$. The slope m of this line obeys

$$m \cdot \tfrac{1}{4} = -1$$
$$m = -4$$

Since the line we seek contains the point $(2, 1)$, we have

$$y - y_1 = m(x - x_1) \quad m = -4;\ x_1 = 2, y_1 = 1$$
$$y - 1 = -4(x - 2)$$
$$y = -4x + 9$$

Figure 25 illustrates these solutions.

Figure 25

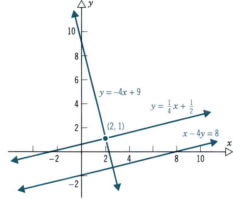

✎ **Now Work Problems 35 and 41**

SUMMARY

Pair of Lines	Conclusion: The lines are
Both vertical	(a) Coincident, if they have the same x-intercept
	(b) Parallel, if they have different x-intercepts
One vertical, one nonvertical	(a) Intersecting
	(b) Perpendicular, if the nonvertical line is horizontal
Neither vertical	Write the equation of each line in slope–intercept form: $y = m_1x + b_1$, $y = m_2x + b_2$
	(a) Coincident, if $m_1 = m_2$, $b_1 = b_2$
	(b) Parallel, if $m_1 = m_2$, $b_1 \neq b_2$
	(c) Intersecting, if $m_1 \neq m_2$
	(d) Perpendicular, if $m_1 m_2 = -1$

EXERCISE 1.2   Answers to odd-numbered problems begin on page AN-4.

In Problems 1–12 determine whether the given pairs of lines are parallel, coincident, or intersecting.

1. L: $x + \ y = 10$
     M: $3x + 3y = \ 6$

2. L: $x - \ y = 5$
     M: $-2x + 2y = 8$

3. L: $2x + y = 4$
     M: $2x - y = 8$

4. L: $2x + y = 8$
   M: $2x - y = -4$

5. L: $-x + y = 2$
   M: $2x - 2y = -4$

6. L: $x + y = -4$
   M: $3x + 3y = -12$

7. L: $2x - 3y = -8$
   M: $6x - 9y = -2$

8. L: $4x - 2y = -7$
   M: $-2x + y = -2$

9. L: $3x - 4y = 1$
   M: $x - 2y = -4$

10. L: $4x + 3y = 2$
   M: $2x - y = -1$

11. L: $x = 3$
   M: $y = -2$

12. L: $x = 4$
   M: $x = -2$

In Problems 13–24 the given pairs of lines intersect. Find the point of intersection. Graph each pair of lines.

13. L: $x + y = 5$
   M: $3x - y = 7$

14. L: $2x + y = 7$
   M: $x - y = -4$

15. L: $x - y = 2$
   M: $2x + y = 7$

16. L: $2x - y = -1$
   M: $x + y = 4$

17. L: $4x + 2y = 4$
   M: $4x - 2y = 4$

18. L: $4x - 2y = 8$
   M: $6x + 3y = 0$

19. L: $3x - 4y = 2$
   M: $x + 2y = 4$

20. L: $4x + 3y = 2$
   M: $2x - y = -1$

21. L: $3x - 2y = -5$
   M: $3x + y = -2$

22. L: $4x + y = 6$
   M: $4x - 2y = 0$

23. L: $x = 4$
   M: $y = -2$

24. L: $x = 0$
   M: $y = 0$

In Problems 25–30 show that the lines are perpendicular.

25. $-x + 3y = 2$
   $6x + 2y = 5$

26. $2x + 3y = 4$
   $9x - 6y = 1$

27. $x + 2y = 7$
   $-2x + y = 15$

28. $4x + 12y = 3$
   $15x - 5y = -1$

29. $3x + 12y = 2$
   $4x - y = -2$

30. $20x - 2y = -7$
   $x + 10y = 8$

In Problems 31–34, find an equation for the line L. Express the answer using the general form or the slope–intercept form, whichever you prefer.

31.

$y = 2x$ L
L is parallel to $y = 2x$

32.

$y = -x$ L
L is parallel to $y = -x$

33.

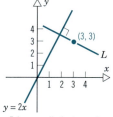

$y = 2x$
L is perpendicular to $y = 2x$

34.

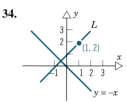

$y = -x$
L is perpendicular to $y = -x$

In Problems 35–44 find an equation for the line with the given properties. Express the answer using the general form or the slope–intercept form, whichever you prefer.

35. Parallel to the line $y = 4x$; containing the point $(-1, 2)$

36. Parallel to the line $y = -3x$; containing the point $(-1, 2)$

37. Parallel to the line $2x - y + 2 = 0$; containing the point $(0, 0)$

38. Parallel to the line $x - 2y + 5 = 0$; containing the point $(0, 0)$

39. Parallel to the line $x = 3$; containing the point $(4, 2)$

40. Parallel to the line $y = 3$; containing the point $(4, 2)$

41. Perpendicular to the line $y = 2x - 5$; containing the point $(-1, -2)$

42. Perpendicular to the line $6x - 2y + 5 = 0$; containing the point $(-1, -2)$

43. Perpendicular to the line $y = 2x - 5$; containing the point $(\frac{-1}{3}, \frac{4}{5})$

44. Perpendicular to the line $y = 3x - 15$; containing the point $(\frac{-2}{3}, \frac{3}{5})$

45. Find t so that $tx - 4y + 3 = 0$ is perpendicular to the line $2x + 2y - 5 = 0$.

46. Find t so that $(1, 2)$ is a point on the line $tx - 3y + 4 = 0$.

47. Find the equation of the line containing the point $(-2, -5)$ and perpendicular to the line containing the points $(-2, 9)$ and $(3, -10)$.

48. Find the equation of the line containing the point $(-2, -5)$ and perpendicular to the line containing the points $(-4, 5)$ and $(2, -1)$.

1.3 APPLICATIONS

Prediction

Linear equations are sometimes used as predictors of future results. Let's look at an example.

EXAMPLE 1   Predicting the Cost of a Home

In 1994 the cost of an average home in the Midwest was \$85,000.* In 1998 the cost was \$97,000. Assuming that the relationship between time and cost is linear, develop a formula for predicting the cost of an average home in 2003.

SOLUTION   We agree to let x represent the year and y represent the cost. We seek a relationship between x and y. Two points on the graph of the equation relating x and y are

$$(1994, 85{,}000) \qquad \text{and} \qquad (1998, 97{,}000)$$

The assumption is that the equation relating x and y is linear. The slope of this line is

$$\frac{97{,}000 - 85{,}000}{1998 - 1994} = 3000$$

Using this fact and the point $(1994, 85{,}000)$, the point–slope form of the equation of the line is

$$y - 85{,}000 = 3000(x - 1994)$$
$$y = 85{,}000 + 3000(x - 1994)$$

For $x = 2003$ we predict the cost of an average home to be

$$\begin{aligned}
y &= 85{,}000 + 3000(x - 1994) \\
&= 85{,}000 + 3000(2003 - 1994) \\
&= 85{,}000 + 3000(9) \\
&= \$112{,}000
\end{aligned}$$

Figure 26 on page 26 illustrates the situation.

Source: National Association of Realtors.

Figure 26

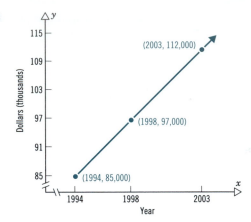

This prediction of future cost is based on the assumption that annual increases remain the same. In this example, the assumption is that each year the cost of a house will go up $3000 (the slope of the line). Of course, if this assumption is not correct, the predicted cost will also not be correct.

Now Work Problem 1

Break-Even Point

In many businesses the cost C of production and the number x of items produced can be expressed as a linear equation. Similarly, sometimes the revenue R obtained from sales and the number x of items produced can be expressed as a linear equation. When the cost C of production exceeds the revenue R from the sales, the business is operating at a loss; when the revenue R exceeds the cost C, there is a profit; and when the revenue R and the cost C are equal, there is no profit or loss. The point at which $R = C$, that is, the point of intersection of the two lines, is usually referred to as the **break-even point** in business.

EXAMPLE 2   Finding the Break-Even Point

Sweet Delight Candies, Inc., has daily fixed costs from salaries and building operations of $300. Each pound of candy produced costs $1 and is sold for $2.

(a) Find the cost C of production for x pounds of candy.
(b) Find the revenue R from selling x pounds of candy.
(c) What is the break-even point? That is, how many pounds of candy must be sold daily to guarantee no loss and no profit?

SOLUTION

(a) The cost C of production is the fixed cost of $300 plus the variable cost of producing x pounds of candy at $1 per pound. Thus

$$C = \$1 \cdot x + \$300 = x + 300$$

(b) The revenue R realized from the sale of x pounds of candy at $2 per pound is

$$R = \$2 \cdot x = 2x$$

(c) The break-even point is the point where $R = C$. Setting $R = C$, we find

$$2x = x + 300$$

$$x = 300$$

That is, 300 pounds of candy must be sold to break even. ∎

In Figure 27 we see a graphical interpretation of the break-even point for Example 2. Note that for $x > 300$, the revenue R always exceeds the cost C so that a profit results. Similarly, for $x < 300$, the cost exceeds the revenue, resulting in a loss.

Figure 27

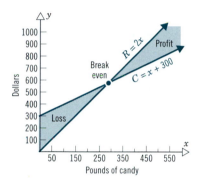

EXAMPLE 3   Analyzing Break-Even Points

After negotiations with employees of Sweet Delight Candies and an increase in the price of sugar, the daily cost C of production for x pounds of candy is

$$C = \$1.05x + \$330$$

(a) If each pound of candy is sold for $2.00, how many pounds must be sold daily to make a profit?
(b) If the selling price is increased to $2.25 per pound, what is the break-even point?
(c) If it is known that 325 pounds of candy can be sold daily, what price should be charged per pound to guarantee no loss?

SOLUTION

(a) If each pound is sold for $2.00, the revenue R from sales is

$$R = \$2x$$

where x represents the number of pounds sold. When we set $R = C$, we find that the break-even point is the solution of

$$2x = 1.05x + 330$$

$$0.95x = 330$$

$$x = \frac{330}{0.95} = 347.37$$

Thus, if 347 pounds or less of candy are sold, a loss is incurred; if 348 pounds or more are sold, a profit results.

(b) If the selling price is increased to $2.25 per pound, the revenue R from sales is

$$R = \$2.25x$$

The break-even point is the solution of

$$2.25x = 1.05x + 330$$
$$1.2x = 330$$
$$x = \frac{330}{1.2} = 275$$

With the new selling price, the break-even point is 275 pounds.

(c) If we know that 325 pounds of candy will be sold daily, the price per pound p needed to guarantee no loss (that is, to guarantee at worst a break-even point) is the solution of

$$R = C$$
$$325p = (1.05)(325) + 330$$
$$325p = 671.25$$
$$p = \$2.07$$

We should charge $2.07 per pound to guarantee no loss, provided 325 pounds will be sold.

∎

EXAMPLE 4   Analyzing Break-Even Points

A producer sells items for $0.30 each.

(a) If the cost for production is

$$C = \$0.15x + \$105$$

where x is the number of items sold, find the break-even point.

(b) If the cost can be changed to

$$C = \$0.12x + \$110$$

would it be advantageous?

SOLUTION   The revenue R received is

$$R = \$0.3x$$

(a) If the cost for production is $C = \$0.15x + \$105$, then the break-even point is the solution of the equation

$$0.3x = 0.15x + 105$$
$$0.15x = 105$$
$$x = 700$$

Thus the break-even point is 700 items.

(b) If the revenue received remains at $R = \$0.3x$, but the cost for production changes to $C = \$0.12x + \$110$, then the break-even point is the solution of the equation

$$0.3x = 0.12x + 110$$
$$0.18x = 110$$
$$x = 611.11$$

The break-even point for the cost in (a) was 700 items. Since the cost in (b) will require fewer items to be sold in order to break even, management should probably change over to the new cost. See Figure 28.

Figure 28

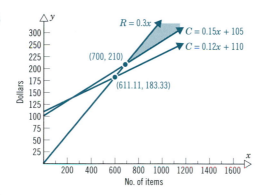

Now Work Problem 13

Mixture Problems

The next example is often referred to as a **mixture problem.**

EXAMPLE 5   Mixing Peanuts

A store that specializes in selling nuts sells cashews for $5 per pound and peanuts for $2 per pound. At the end of the month the manager finds that the peanuts are not selling well. In order to sell 30 pounds of peanuts more quickly, the manager decides to mix the 30 pounds of peanuts with some cashews and sell the mixture of peanuts and cashews for $3 a pound. How many pounds of cashews should be mixed with the peanuts so that the revenue remains the same?

SOLUTION   There are two unknowns: the number of pounds of cashews (call this x) and the number of pounds of the mixture (call this y). Since we know that the number of pounds of cashews plus 30 pounds of peanuts equals the number of pounds of the mixture, we can write

$$y = x + 30$$

Also, in order to keep revenue the same, we must have

$$\begin{pmatrix} \text{Price} \\ \text{per} \\ \text{pound} \end{pmatrix} \cdot \begin{pmatrix} \text{Pounds} \\ \text{of} \\ \text{cashews} \end{pmatrix} + \begin{pmatrix} \text{Price} \\ \text{per} \\ \text{pound} \end{pmatrix} \cdot \begin{pmatrix} \text{Pounds} \\ \text{of} \\ \text{peanuts} \end{pmatrix} = \begin{pmatrix} \text{Price} \\ \text{per} \\ \text{pound} \end{pmatrix} \cdot \begin{pmatrix} \text{Pounds} \\ \text{of} \\ \text{mixture} \end{pmatrix}$$

That is,

$$5x + 2(30) = 3y \quad \text{Divide both sides by 3}$$
$$\tfrac{5}{3}x + 20 = y$$

Thus we have the two equations

$$y = x + 30 \quad \text{and} \quad y = \tfrac{5}{3}x + 20$$

Since the number of pounds of the mixture is the same in each case, we have

$$\tfrac{5}{3}x + 20 = x + 30$$
$$\tfrac{2}{3}x = 10$$
$$x = 15$$

Figure 29

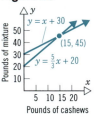

The manager should mix 15 pounds of cashews with 30 pounds of peanuts. See Figure 29. Notice that the point of intersection (15, 45) represents the pounds of cashews (15) in the mixture (45 pounds).

✎ **Now Work Problem 15**

Economics

The **supply equation** in economics is used to specify the amount of a particular commodity that sellers are willing to offer in the market at various prices. The **demand equation** specifies the amount of a particular commodity that buyers are willing to purchase at various prices.

An increase in price p usually causes an increase in the supply S and a decrease in demand D. On the other hand, a decrease in price brings about a decrease in supply and an increase in demand. The **market price** is defined as the price at which supply and demand are equal (the point of intersection).

Figure 30 illustrates a typical supply/demand situation.

Figure 30

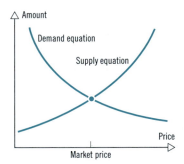

EXAMPLE 6   Supply and Demand

The supply and demand for flour have been estimated as being given by the equations

$$S = 0.8p + 0.5 \qquad D = -0.4p + 1.5$$

where p is measured in dollars and S and D are measured in pound units of flour. Find the market price and graph the supply and demand equations.

SOLUTION The market price p is the solution of the equation

$$S = D$$
$$0.8p + 0.5 = -0.4p + 1.5$$
$$1.2p = 1$$
$$p = 0.83$$

Thus, at a price of $0.83 per pound, supply and demand for flour are equal. The graphs are shown in Figure 31.

Figure 31

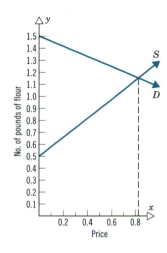

Now Work Problem 27

EXERCISE 1.3   Answers to odd-numbered problems begin on page AN-4.

Problems 1–6 involve prediction.

1. Suppose the sales of a company are given by

   $$S = \$5000x + \$80,000$$

   where x is measured in years and $x = 0$ corresponds to the year 1998.

   (a) Find S when $x = 0$.
   (b) Find S when $x = 3$.
   (c) Find the predicted sales in 2003, assuming this trend continues.
   (d) Find the predicted sales in 2006, assuming this trend continues.

2. Rework Problem 1 if the sales of the company are given by

   $$S = \$3000x + \$60,000$$

3. **Predicting the Cost of a Compact Car** In 1995, the cost of a compact car averaged $8000. In 1998, the cost of a compact car averaged $9500. Assuming that the relationship between time and cost is linear, develop a formula for predicting the average cost of a compact car in the future. What do you predict the average cost of a compact car will be in 2000? Interpret the slope.

4. **Oil Depletion** Oil is pumped from an oil field at a constant rate each year, so that its oil reserves have been decreasing linearly with time. Geologists estimate that the field reserves were 400,000 barrels in 1980 and 320,000 barrels in 1990.

   (a) Write an equation describing the amount of oil left in the field at any time.
   (b) If the trend continues, when will the oil well dry out?
   (c) Interpret the slope.

5. **SAT Scores** The average SAT scores of incoming freshmen at a midwestern college have been declining at a constant rate in recent years. In 1994 the average SAT score was 592, while in 1998 the average SAT score was only 564.

   (a) Write an equation that will give the average SAT score at any time.
   (b) If the trend continues, what will the average SAT score of incoming freshmen be in 2000?

6. **College Degrees** The percent of people over 55 who have college degrees is summarized in the following table:

Year (t)	1970	1975	1980	1985	1990	1995
Percent with college degree (y)	30	34	38	42	46	50

(a) Find an equation of the line containing the points (1970, 30) (1990, 46).
(b) If the trend continues, estimate the percentage of people over 55 who will have a college degree by the year 2000.
(c) Interpret the slope.

Problems 7–14 involve break-even points. In Problems 7–10 find the break-even point for the cost C of production and the revenue R. Graph each result.

7. $C = \$10x + \$600$ $R = \$30x$
8. $C = \$5x + \$200$ $R = \$8x$
9. $C = \$0.2x + \$50$ $R = \$0.3x$
10. $C = \$1800x + \$3000$ $R = \$2500x$

11. **Break-Even Point** A manufacturer produces items at a daily cost of \$0.75 per item and sells them for \$1 per item. The daily operational overhead is \$300. What is the break-even point? Graph your result.

12. **Break-Even Point** If the manufacturer of Problem 11 is able to reduce the cost per item to \$0.65, but with a resultant increase to \$350 in operational overhead, is it advantageous to do so? Graph your result.

13. **Profit from Selling Newspapers** Each Sunday, a newspaper agency sells x copies of a certain newspaper for \$2.00 per copy. The cost to the agency for each newspaper is \$1.00. The agency pays a fixed cost for storage, delivery, and so on, of \$200 per Sunday. How many newspapers need to be sold for the agency to break even?

14. **Profit from Selling Newspapers** Repeat Problem 13 if the cost to the agency is \$1.25 per copy and the fixed cost is \$225 per Sunday.

Problems 15–22 involve mixture problems.

15. **Mixture Problem** Sweet Delight Candies sells boxes of candy consisting of creams and caramels. Each box sells for \$8.00 and holds 50 pieces of candy (all pieces are the same size). If the caramels cost \$0.10 to produce and the creams cost \$0.20 to produce, how many caramels and creams should be in each box for no profit or loss? Would you increase or decrease the number of caramels in order to obtain a profit?

16. **Mixture Problem** The manager of Nutt's Nuts regularly sells cashews for \$6.50 per pound, pecans for \$7.50 per pound, and peanuts for \$2.00 per pound. How many pounds of cashews and pecans should be mixed with 40 pounds of peanuts to obtain a mixture of 100 pounds that will sell for \$4.89 a pound so that the revenue is unchanged?

17. **Investment Problem** Mr. Nicholson has just retired and needs \$10,000 per year in supplementary income. He has \$150,000 to invest and can invest in AA bonds at 10% annual interest or in Savings and Loan Certificates at 5% interest per year. How much money should be invested in each so that he realizes exactly \$10,000 in extra income per year?

18. **Investment Problem** Mr. Nicholson finds after 2 years that because of inflation he now needs \$12,000 per year in supplementary income. How should he transfer his funds to achieve this amount? (Use the data from Problem 17.)

19. **Theater Attendance Problem** The Star Theater wants to know whether the majority of its patrons are adults or children. During a week in July 2600 tickets were sold and the receipts totaled \$16,440. The adult admission is \$8 and the children's admission is \$4. How many adult patrons were there?

20. **Mixture Problem** A coffee manufacturer wants to market a new blend of coffee that will cost \$6.00 per pound by mixing \$5.00 per pound coffee and \$7.50 per pound coffee. What amounts of the \$5.00 per pound coffee and \$7.50 per pound coffee should be blended to obtain the desired mixture? [*Hint:* Assume the total weight of the desired blend is 100 pounds.]

21. Mixture Problem   One solution is 15% acid and another is 5% acid. How many cubic centimeters of each should be mixed to obtain 100 cubic centimeters of a solution that is 8% acid?

22. Investment Problem   A bank loaned $10,000, some at an annual rate of 8% and some at an annual rate of 12%. If the income from these loans was $1000, how much was loaned at 8%? How much at 12%?

Problems 23–30 involve economics. In Problems 23–26 find the market price for each pair of supply and demand equations.

23. $S = p + 1$ $\qquad$ $D = 3 - p$

24. $S = 2p + 3$ $\qquad$ $D = 6 - p$

25. $S = 20p + 500$ $\qquad$ $D = 1000 - 30p$

26. $S = 40p + 300$ $\qquad$ $D = 1000 - 30p$

27. Market Price of Sugar   The supply and demand equations for sugar have been estimated to be given by the equations

$$S = 0.7p + 0.4 \qquad D = -0.5p + 1.6$$

Find the market price. What quantity of supply is demanded at this market price? Graph both the supply and demand equations. Interpret the point of intersection of the two lines.

28. Supply and Demand Problem   The market price for a certain product is $5.00 per unit and occurs when 14,000 units are produced. At a price of $1, no units are manu-

factured and, at a price of $19.00, no units will be purchased. Find the supply and demand equations, assuming they are linear.

29. Supply and Demand Problem   For a certain commodity the supply equation is given by

$$S = 2p + 5$$

At a price of $1, 19 units of the commodity are demanded. If the demand equation is linear and the market price is $3, find the demand equation.

30. Supply and Demand Problem   For a certain commodity the demand equation is given by

$$D = -3p + 20$$

At a price of $1, four units of the commodity are supplied. If the supply equation is linear and the market price is $4, find the supply equation.

CHAPTER REVIEW

IMPORTANT TERMS AND CONCEPTS

rectangular coordinates 3
linear equation 4
finding intercepts 5
slope of a line 7
average rate of change 8

point–slope form 11
slope–intercept form 13
parallel lines 20
intersecting lines 21

perpendicular lines 22
break-even point 26
supply and demand equations 30

IMPORTANT FORMULAS

Linear Equation, General Form

$$Ax + By = C \quad A, B \text{ not both zero}$$

Slope of a Line

$$m = \frac{y_2 - y_1}{x_2 - x_1} \quad \text{if } x_1 \neq x_2$$

Point–Slope Form of the Equation of a Line

$$y - y_1 = m(x - x_1)$$

Slope–Intercept Form of the Equation of a Line

$$y = mx + b$$

Pair of Lines	*Conclusion: The lines are*
Both vertical	(a) Coincident, if they have the same x-intercept
	(b) Parallel, if they have different x-intercepts
One vertical, one nonvertical	(a) Intersecting
	(b) Perpendicular, if the nonvertical line is horizontal
Neither vertical	Write the equation of each line in slope–intercept form:
	$y = m_1 x + b_1$, $y = m_2 x + b_2$
	(a) Coincident, if $m_1 = m_2$, $b_1 = b_2$
	(b) Parallel, if $m_1 = m_2$, $b_1 \neq b_2$
	(c) Intersecting, if $m_1 \neq m_2$
	(d) Perpendicular, if $m_1 m_2 = -1$

TRUE–FALSE ITEMS   Answers are on page AN-5.

T_____ F_____ **1.** In the slope–intercept equation of a line, $y = mx + b$, m is the slope and $(0, b)$ is the x-intercept.

T_____ F_____ **2.** The graph of the equation $Ax + By = C$, where A, B, C are real numbers and A, B are not both zero, is a line.

T_____ F_____ **3.** The y-intercept of the line $2x - 3y + 6 = 0$ is $(0, 2)$.

T_____ F_____ **4.** The slope of the line $2x - 4y + 7 = 0$ is $-\frac{1}{2}$.

T_____ F_____ **5.** Parallel lines always have the same intercepts.

T_____ F_____ **6.** Intersecting lines have different slopes.

T_____ F_____ **7.** Perpendicular lines have slopes that are reciprocals of each other.

T_____ F_____ **8.** A linear relation between two variables can always be graphed as a line.

T_____ F_____ **9.** All lines with equal slopes are distinct.

T_____ F_____ **10.** All vertical lines have positive slope.

FILL IN THE BLANKS   Answers are on page AN-5.

1. If (x, y) are rectangular coordinates of a point, the number x is called the _____ and y is called the _____ .

2. The slope of a vertical line is _____ ; the slope of a horizontal line is _____ .

3. If a line slants downward as it moves from left to right, its slope will be a _____ number.

4. If two lines have the same slope but different y-intercepts, they are _____ .

5. If two lines have the same slope and the same y-intercept, they are said to be _____ .

6. Lines that intersect at right angles are said to be _____ to each other.

7. Distinct lines that have different slopes are _____ lines.

REVIEW EXERCISES   Answers to odd-numbered problems begin on page AN-5.

In Problems 1–4 graph each equation.

1. $y = -2x + 3$

2. $y = 6x - 2$

3. $2y = 3x + 6$

4. $3y = 2x + 6$

In Problems 5–8 find a general equation for the line containing each pair of points.

5. $P = (1, 2)$ $Q = (-3, 4)$

6. $P = (-1, 3)$ $Q = (1, 1)$

7. $P = (0, 0)$ $Q = (-2, 3)$

8. $P = (-2, 3)$ $Q = (0, 0)$

In Problems 9–20 find an equation of the line having the given characteristics. Express the answer using the general form or the slope–intercept form, whichever you prefer.

9. Slope $= -3$;   containing the point $(2, -1)$

10. Slope $= 4$;   containing the point $(-1, -3)$

11. Slope $= 0$;   containing the point $(-3, 4)$

12. Slope undefined;   containing the point $(-3, 4)$

13. x-intercept $= (2, 0)$;   containing the point $(4, -5)$

14. y-intercept $= (0, -2)$;   containing the point $(5, -3)$

15. x-intercept $= (-3, 0)$; y-intercept $= (0, -4)$

16. Containing the points $(3, -4)$ and $(2, 1)$

17. Parallel to the line $2x + 3y + 4 = 0$;
containing the point $(-5, 3)$

18. Parallel to the line $x + y - 2 = 0$;
containing the point $(1, -3)$

19. Perpendicular to the line $2x + 3y + 4 = 0$;
containing the point $(-5, 3)$

20. Perpendicular to the line $x + y - 2 = 0$;
containing the point $(1, -3)$

In Problems 21–24 find the slope and y-intercept of each line. Graph each line.

21. $-9x - 2y + 18 = 0$

22. $-4x - 5y + 20 = 0$

23. $4x + 2y - 9 = 0$

24. $3x + 2y - 8 = 0$

In Problems 25–30 determine whether the two lines are parallel, coincident, or intersecting.

25. $3x - 4y + 12 = 0$
$6x - 8y + 9 = 0$

26. $2x + 3y + 5 = 0$
$4x + 6y + 10 = 0$

27. $x - y + 2 = 0$
$3x - 4y + 12 = 0$

28. $2x + 3y - 5 = 0$
$x + y - 2 = 0$

29. $4x + 6y + 12 = 0$
$2x + 3y + 6 = 0$

30. $-3x + y = 0$
$6x - 2y + 5 = 0$

In Problems 31–36 the given pair of lines intersect. Find the point of intersection. Graph the lines.

31. L: $x - y = 4$
M: $x + 2y = 7$

32. L: $x + y = 4$
M: $x - 2y = 1$

33. L: $x - y = -2$
M: $x + 2y = 7$

34. L: $2x + 4y = 4$
M: $2x - 4y = 8$

35. L: $2x - 4y = -8$
M: $3x + 6y = 0$

36. L: $3x + 4y = 2$
M: $x - 2y = 1$

37. Investment Problem   Mr. and Mrs. Byrd have just retired and find that they need $10,000 per year to live on. Fortunately, they have a nest egg of $90,000, which they can invest in somewhat risky B-rated bonds at 12% interest per year or in a well-known bank at 5% per year. How much money should they invest in each so that they realize exactly $10,000 in interest income each year?

38. Mixture Problem   One solution is 20% acid and another is 12% acid. How many cubic centimeters of each solution should be mixed to obtain 100 cubic centimeters of a solution that is 15% acid?

39. Car Sales   The annual sales of Motors, Inc., over a period of 5 years are listed in the table on page 36.

Year	Units Sold (in thousands)
1994	3400
1995	3200
1996	3100
1997	2800
1998	2200

(a) Graph this information using the *x*-axis for years and the *y*-axis for units sold. (For convenience, use different scales on the axes.)

(b) Draw a line *L* that contains two of the points and comes close to passing through the remaining points.

(c) Find the equation of this line *L*.

(d) Using this equation of the line, what is your estimate for units sold in 1999?

(e) Interpret the slope.

40. Attendance at a Dance   A church group is planning a dance in the school auditorium to raise money for its school. The band they will hire charges $500; the advertising costs are estimated at $100; and food will be supplied at the rate of $5 per person. The church group would like to clear at least $900 after expenses.

(a) Determine how many people need to attend the dance for the group to break even if tickets are sold at $10 each.

(b) Determine how many people need to attend in order to achieve the desired profit if tickets are sold for $10 each.

(c) Answer the above two questions if the tickets are sold for $12 each.

MATHEMATICAL QUESTIONS FROM PROFESSIONAL EXAMS

1. CPA Exam   The Oliver Company plans to market a new product. Based on its market studies, Oliver estimates that it can sell 5500 units in 1992. The selling price will be $2 per unit. Variable costs are estimated to be 40% of the selling price. Fixed costs are estimated to be $6000. What is the break-even point?

(a) 3750 units   (b) 5000 units
(c) 5500 units   (d) 7500 units

2. CPA Exam   The Breiden Company sells rodaks for $6 per unit. Variable costs are $2 per unit. Fixed costs are $37,500. How many rodaks must be sold to realize a profit before income taxes of 15% of sales?

(a) 9375 units   (b) 9740 units
(c) 11,029 units   (d) 12,097 units

3. CPA Exam   Given the following notations, what is the break-even sales level in units?

$$SP = \text{Selling price per unit}$$
$$FC = \text{Total fixed cost}$$
$$VC = \text{Variable cost per unit}$$

(a) $\dfrac{SP}{FC \div VC}$ (b) $\dfrac{FC}{VC \div SP}$

(c) $\dfrac{VC}{SP - FC}$ (d) $\dfrac{FC}{SP - VC}$

4. CPA Exam   At a break-even point of 400 units sold, the variable costs were $400 and the fixed costs were $200. What will the 401st unit sold contribute to profit before income taxes?

(a) $0   (b) $0.50
(c) $1.00   (d) $1.50

5. CPA Exam   A graph is set up with "depreciation expense" on the vertical axis and "time" on the horizontal axis. Assuming linear relationships, how would the graphs for straight-line and sum-of-the-year's-digits depreciation, respectively, be drawn?

(a) Vertically and sloping down to the right
(b) Vertically and sloping up to the right
(c) Horizontally and sloping down to the right
(d) Horizontally and sloping up to the right

The following statement applies to Questions 6–8:

In analyzing the relationship of total factory overhead with changes in direct labor hours, the following relationship was found to exist: $Y = \$1000 + \$2X.$

6. CMA Exam   The relationship as shown above is

(a) Parabolic   (b) Curvilinear   (c) Linear
(d) Probabilistic   (e) None of the above

7. CMA Exam   *Y* in the above equation is an estimate of

(a) Total variable costs   (b) Total factory overhead
(c) Total fixed costs   (d) Total direct labor hours
(e) None of the above

8. CMA Exam   The $2 in the equation is an estimate of

(a) Total fixed costs
(b) Variable costs per direct labor hour
(c) Total variable costs
(d) Fixed costs per direct labor hour
(e) None of the above

Chapter 2

Systems of Linear Equations; Matrices

2.1 **Systems of Linear Equations: Substitution; Elimination**

2.2 **Systems of Linear Equations: Matrix Method**

2.3 **Systems of m Linear Equations Containing n Variables**

2.4 **Matrix Algebra**

2.5 **Multiplication of Matrices**

2.6 **Inverse of a Matrix**

2.7 **Applications: Leontief Model; Cryptography; Accounting; The Method of Least Squares**

Chapter Review

In this chapter we take up the problem of solving systems of linear equations containing two or more variables. As the section titles suggest, there are various ways to solve such problems. The *method of substitution* for solving equations in several unknowns goes back to ancient times. The *method of elimination,* though it had existed for centuries, was put into systematic order by Karl Friedrich Gauss (1777–1855) and by Camille Jordan (1838–1922). This method led to the *matrix method* that is now used for solving large systems by computer.

The theory of *matrices* was developed in 1857 by Arthur Cayley (1821–1895), though only later were matrices used as we use them in this chapter. Matrices have become a very flexible instrument, invaluable in almost all areas of mathematics.

2.1* SYSTEMS OF LINEAR EQUATIONS: SUBSTITUTION; ELIMINATION

We begin with an example.

EXAMPLE 1   Movie Theater Ticket Sales

A movie theater sells adult tickets for $8 each and child tickets for $5 each. One Saturday the theater took in $3255 in revenue. If x represents the number of adult tickets sold and y the number of child tickets sold, write an equation that relates these variables.

SOLUTION   Each adult ticket brings in $8, so x adult tickets will bring in $8x$ dollars. Similarly, y child tickets bring in $5y$ dollars. If the total brought in is $3255, then we have the equation

$$8x + 5y = 3255$$

■

The equation found in Example 1 is an example of a **linear equation in two variables.** Some other examples of linear equations are

$$2x + 3y = 2 \qquad 5x - 2y + 3z = 10 \qquad 8x_1 + 8x_2 - 2x_3 + 5x_4 = 0$$
$$\text{2 variables} \qquad\qquad \text{3 variables} \qquad\qquad\qquad \text{4 variables†}$$

In general, an equation in n variables is said to be **linear** if it can be written in the form

$$a_1x_1 + a_2x_2 + \cdots + a_nx_n = b$$

where $x_1, x_2, \ldots, x_n$ are n distinct variables, $a_1, a_2, \ldots, a_n, b$ are constants, and at least one of the a_i's is not 0.

In Example 1 suppose we also know that 525 tickets were sold on that Saturday. Then we have another equation relating the variables x and y, namely,

$$x + y = 525$$

The two linear equations

$$x + y = 525$$
$$8x + 5y = 3255$$

form a *system* of linear equations. In general, a **system of linear equations** is a collection of two or more linear equations, each containing one or more variables. Example 2 gives some samples of systems of linear equations.

---

* Based on material from *Precalculus,* 5th ed., by Michael Sullivan. Used here with the permission of the author and Prentice-Hall, Inc.

† The notation x_n is read as "x sub n." The number n is called a **subscript** and should not be confused with an exponent. We use subscripts to distinguish one variable from another when a large or undetermined number of variables is required.

EXAMPLE 2   **Examples of Systems of Linear Equations**

(a) $\begin{cases} 2x + y = 5 \\ -4x + 6y = -2 \end{cases}$ (1)   Two equations containing
(2)   two variables, x and y

(b) $\begin{cases} x + y + z = 6 \\ 3x - 2y + 4z = 9 \\ x - y - z = 0 \end{cases}$ (1)   Three equations containing
(2)   three variables, x, y, and z
(3)

(c) $\begin{cases} x + y + z = 5 \\ x - y = 2 \end{cases}$ (1)   Two equations containing
(2)   three variables, x, y, and z

(d) $\begin{cases} x + y + z = 6 \\ 2x + 2z = 4 \\ y + z = 2 \\ x = 4 \end{cases}$ (1)   Four equations containing
(2)   three variables, x, y, and z
(3)
(4)

(e) $\begin{cases} x_1 - 2x_2 + x_3 - x_4 = 5 \\ 3x_1 + x_2 - x_3 - 5x_4 = 2 \end{cases}$ (1)   Two equations containing four
(2)   variables x_1, x_2, x_3, and x_4

We use a brace, as shown above, to remind us that we are dealing with a system of equations. We also will find it convenient to number each equation in the system.

A **solution** of a system of equations consists of values for the variables that are solutions of each equation of the system. To **solve** a system of equations means to find all solutions of the system.

For example, $x = 2$, $y = 1$ is a solution of the system in Example 2(a) because

$$2(2) + 1 = 5 \quad \text{and} \quad -4(2) + 6(1) = -2$$

A solution of the system in Example 2(b) is $x = 3$, $y = 2$, $z = 1$ because

$$\begin{cases} 3 & + 2 & + 1 & = 6 & (1) \\ 3(3) & - 2(2) & + 4(1) & = 9 & (2) \\ 3 & - 2 & - 1 & = 0 & (3) \end{cases}$$

Note that $x = 3$, $y = 3$, $z = 0$ is not a solution of the system in Example 2(b):

$$\begin{cases} 3 & + 3 & + 0 & = 6 & (1) \\ 3(3) & - 2(3) & + 4(0) & = 3 \neq 9 & (2) \\ 3 & - 3 & - 0 & = 0 & (3) \end{cases}$$

Although the given values satisfy Equations (1) and (3), they do not satisfy Equation (2). Any solution of the system must satisfy *each* equation of the system.

When a system of equations has at least one solution, it is said to be **consistent;** otherwise, it is called **inconsistent.**

Now Work Problem 1

Two Linear Equations Containing Two Variables

We can view the problem of solving a system of two linear equations containing two variables as a geometry problem. The graph of each equation in such a system is a line.

Thus, a system of two equations containing two variables represents a pair of lines. The lines either (1) are parallel or (2) are intersecting or (3) are coincident (that is, identical).

1. If the lines are parallel, then the system of equations has no solution, because the lines never intersect. The system is **inconsistent.**
2. If the lines intersect, then the system of equations has one solution, given by the point of intersection. The system is **consistent** and the equations are **independent.**
3. If the lines are coincident, then the system of equations has infinitely many solutions, represented by the totality of points on the line. The system is **consistent** and the equations are **dependent.**

Thus a system of equations is either

  (I)   Inconsistent; has no solution

or

  (II) Consistent; with
      (a)   One solution (equations are independent)
          or
      (b)   Infinitely many solutions (equations are dependent)

Figure 1 illustrates these conclusions.

Figure 1

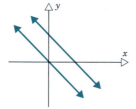 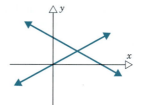 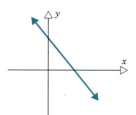

(a) Parallel lines; system has no solution and is inconsistent

(b) Intersecting lines; system has one solution and is consistent; the equations are independent

(c) Coincident lines; system has infinitely many solutions and is consistent; the equations are dependent

---

EXAMPLE 3   Graphing a System of Linear Equations

Figure 2

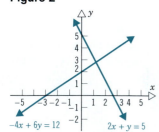

Graph the system: $\begin{cases} 2x + y = 5 & (1) \\ -4x + 6y = 12 & (2) \end{cases}$

SOLUTION   Equation (1) is a line with x-intercept $(\frac{5}{2}, 0)$ and y-intercept $(0, 5)$. Equation (2) is a line with x-intercept $(-3, 0)$ and y-intercept $(0, 2)$.
Figure 2 shows their graphs.

■

From the graph in Figure 2 we see that the lines intersect, so the system is consistent and the equations are independent. We can also use the graph as a means of approximating the solution. For this system the solution would appear to be close to the point $(1, 3)$. The actual solution, which you should verify, is $(\frac{9}{8}, \frac{11}{4})$.

To obtain the exact solution, we use algebraic methods. The first algebraic method we take up is the *method of substitution*.

Method of Substitution

We illustrate the method of substitution by solving the system of Example 3.

EXAMPLE 4   **Solving a System of Equations Using Substitution**

Solve: $\begin{cases} 2x + y = 5 & (1) \\ -4x + 6y = 12 & (2) \end{cases}$

SOLUTION   We solve the first equation for y, obtaining

$$y = -2x + 5$$

We substitute this result for y in the second equation. This results in an equation containing one variable, which we can solve.

$$-4x + 6y = 12$$
$$-4x + 6(-2x + 5) = 12$$
$$-4x - 12x + 30 = 12$$
$$-16x = -18$$
$$x = \frac{-18}{-16} = \frac{9}{8}$$

Once we know that $x = \frac{9}{8}$, we can easily find the value of y by **back-substitution,** that is, by substituting $\frac{9}{8}$ for x in one of the previous equations. We will substitute $x = \frac{9}{8}$ into the equation $y = -2x + 5$. The result is

$$y = -2x + 5 = -2(\tfrac{9}{8}) + 5 = -\tfrac{9}{4} + \tfrac{20}{4} = \tfrac{11}{4}$$

The solution of the system is $x = \frac{9}{8}$, $y = \frac{11}{4}$.

Check:
$$2x + y = 2(\tfrac{9}{8}) + \tfrac{11}{4} = \tfrac{9}{4} + \tfrac{11}{4} = \tfrac{20}{4} = 5$$
$$-4x + 6y = -4(\tfrac{9}{8}) + 6(\tfrac{11}{4}) = -\tfrac{18}{4} + \tfrac{66}{4} = \tfrac{48}{4} = 12$$

■

The method used to solve the system in Example 4 is called **substitution.** The steps used are outlined in the box below.

Steps for Solving by Substitution

Step 1 Pick one of the equations and solve for one of the variables in terms of the remaining variables.

Step 2 Substitute the result in the remaining equations.

Step 3 If one equation in one variable results, solve this equation. Otherwise, repeat Steps 1 and 2 until a single equation with one variable remains.

Step 4 Find the values of the remaining variables by back-substitution.

Step 5 Check the solution found.

EXAMPLE 5 **Solving a System of Equations by Substitution**

Solve: $\begin{cases} 3x - 2y = 5 & (1) \\ 5x - y = 6 & (2) \end{cases}$

SOLUTION

Step 1 After looking at the two equations, we conclude that it is easiest to solve for the variable y in Equation (2):

$$5x - y = 6$$
$$y = 5x - 6$$

Step 2 We substitute this result into Equation (1) and simplify:

$$3x - 2y = 5$$
$$3x - 2(5x - 6) = 5$$

Step 3
$$-7x + 12 = 5$$
$$-7x = -7$$
$$x = 1$$

Step 4 Knowing $x = 1$, we can find y from the equation

$$y = 5x - 6 = 5(1) - 6 = -1$$

Step 5 *Check:* $\begin{cases} 3(1) - 2(-1) = 3 + 2 = 5 \\ 5(1) - (-1) = 5 + 1 = 6 \end{cases}$

The solution of the system is $x = 1$, $y = -1$.

 **Now Work Problem 7**

Using the Method of

Substitution

Method of Elimination

A second method for solving a system of linear equations is the *method of elimination.* This method is usually preferred over substitution if substitution leads to fractions or if the system contains more than two variables. Elimination also provides the necessary motivation for solving systems using matrices (the subject of the next section).

The idea behind the method of elimination is to keep replacing the original equations in the system with equivalent equations until a system of equations with an obvious solution is reached. When we proceed in this way, we obtain **equivalent systems of equations,** namely, systems of equations that have the same solutions. The rules for obtaining equivalent equations are given below:

Rules for Obtaining an Equivalent System of Equations

1. Interchange any two equations of the system.

2. Multiply (or divide) each side of an equation by the same nonzero constant.

3. Replace any equation in the system by the sum (or difference) of that equation and any other equation in the system.

An example will give you the idea. As you work through the example, pay particular attention to the pattern being followed.

EXAMPLE 6   Solving a System of Linear Equations by Elimination

Solve: $\begin{cases} 2x + 3y = 1 & (1) \\ -x + y = -3 & (2) \end{cases}$

SOLUTION   The idea behind the method of elimination is to rewrite the system so that adding two of the equations eliminates a variable. For this system, we can accomplish this by multiplying equation (2) by 2. The result is the equivalent system

$$\begin{cases} 2x + 3y = 1 & (1) \\ -2x + 2y = -6 & (2) \end{cases}$$

Now we add the two equations and solve the resulting equation for y.

$$\begin{cases} 2x + 3y = 1 & (1) \\ -2x + 2y = -6 & (2) \end{cases}$$

$$5y = -5 \qquad \text{Add equations (1) and (2)}$$

$$y = -1 \qquad \text{Solve for } y$$

We **back-substitute** $y = -1$ in Equation (1) and simplify, to get

$$2x + 3(-1) = 1 \qquad 2x + 3y = 1; \, y = -1$$

$$2x = 4$$

$$x = 2$$

Thus the solution of the original system is $x = 2$, $y = -1$. We leave it to you to check the solution. ∎

The procedure used in Example 6 is called the **method of elimination.** Notice the pattern of the solution. First, we eliminated the variable x from the two equations. Then we solved for y and back-substituted; that is, we substituted the value found for y back into the first equation to find x.

Steps for Solving by Elimination

Step 1 Select two equations from the system and eliminate a variable from them.

Step 2 If there are additional equations in the system, pair off equations and eliminate the same variable from them.

Step 3 Continue Steps 1 and 2 on successive systems until one equation containing one variable remains.

Step 4 Solve for this variable and back-substitute in previous equations until all the variables have been found.

Let's do another example.

EXAMPLE 7 Solving a System of Linear Equations by Elimination

Use the method of elimination to solve the system of equations:

$$\begin{cases} \frac{1}{3}x + 5y = -4 & (1) \\ 2x + 3y = 3 & (2) \end{cases}$$

SOLUTION We begin by multiplying each side of Equation (1) by 3 in order to remove the fraction $\frac{1}{3}$:

$$\begin{cases} \frac{1}{3}x + 5y = -4 & (1) \\ 2x + 3y = 3 & (2) \end{cases}$$

$$\begin{cases} x + 15y = -12 & (1) \quad \text{Multiply Equation (1) by 3} \\ 2x + 3y = 3 & (2) \end{cases}$$

Thinking ahead, we decide to multiply each side of Equation (1) above by -2, because then the sum of the two equations will result in an equation with the variable x eliminated. [Note that we also could multiply each side of Equation (1) by 2 and then replace Equation (1) by the difference of the two equations. However, because subtracting is more likely to result in a calculation error, we follow the safer practice of adding equations.]

$$\begin{cases} -2x - 30y = 24 & \text{Multiply Equation (1) by } -2. \\ 2x + 3y = 3 \end{cases}$$

$$-27y = 27 \quad \text{Add the two equations.}$$

$$y = -1 \quad \text{Solve for } y.$$

Now back-substitute $y = -1$ in Equation (2). Then,

$$2x + 3(-1) = 3 \quad 2x + 3y = 3; y = -1$$

$$2x = 6$$

$$x = 3$$

The solution of the original system is $x = 3$, $y = -1$, which you should check.

 Now Work Problem 7

Using the Method of

Elimination   We have already seen several examples in which the system of equations has exactly one solution. The next example illustrates what happens when the method of substitution is used to solve a system of equations that has no solution.

EXAMPLE 8 An Inconsistent System of Linear Equations

Solve: $\begin{cases} 2x + y = 5 & (1) \\ 4x + 2y = 8 & (2) \end{cases}$

SOLUTION We choose to use the method of substitution and solve Equation (1) for y:

$$2x + y = 5$$

$$y = 5 - 2x$$

Substituting in Equation (2), we get

$$4x + 2y = 8$$
$$4x + 2(5 - 2x) = 8 \qquad y = 5 - 2x$$
$$4x + 10 - 4x = 8$$
$$0 \cdot x = -2$$

This equation has no solution. Thus we conclude that the system itself has no solution and is therefore inconsistent.

Figure 3 illustrates the pair of lines whose equations form the system in Example 8. Notice that the graphs of the two equations are lines, each with slope -2; one line has y-intercept $(0, 5)$, the other has y-intercept $(0, 4)$. Thus, the lines are parallel and have no point of intersection. This geometric statement is equivalent to the algebraic statement that the system has no solution.

Figure 3

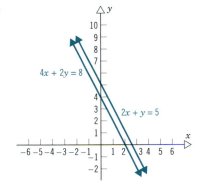

The next example is an illustration of a system with infinitely many solutions.

EXAMPLE 9   Solving a System of Linear Equations with Infinitely Many Solutions

Solve: $\begin{cases} 2x + y = 4 & (1) \\ -6x - 3y = -12 & (2) \end{cases}$

SOLUTION   We choose to use the method of elimination:

$$\begin{cases} 2x + y = 4 & (1) \\ -6x - 3y = -12 & (2) \end{cases}$$

$$\begin{cases} 6x + 3y = 12 & (1) \quad \text{Multiply each side of Equation (1) by 3.} \\ -6x - 3y = -12 & (2) \end{cases}$$

$$\begin{cases} 6x + 3y = 12 & (1) \quad \text{Replace Equation (2) by the sum of} \\ 0 = 0 & (2) \quad \text{Equations (1) and (2).} \end{cases}$$

The original system is thus equivalent to a system containing one equation, so the equations are dependent. This means that any values of x and y for which $6x + 3y = 12$ (or, equivalently, $2x + y = 4$) are solutions. For example, $x = 2$, $y = 0$; $x = 0$, $y = 4$; $x = -2$, $y = 8$; $x = 4$, $y = -4$; and so on, are solutions. There are, in fact, infinitely many values of x and y for which $2x + y = 4$, so the original system has infinitely many solutions. We will write the solutions of the original system either as

$$y = 4 - 2x$$

where x can be any real number, or as

$$x = 2 - \tfrac{1}{2}y$$

where y can be any real number.

Figure 4 illustrates the situation presented in Example 9. Notice that the graphs of the two equations are lines, each with slope -2 and each with y-intercept $(0, 4)$. Thus the lines are coincident. Notice also that Equation (2) in the original system is just -3 times Equation (1). This indicates that the two equations are dependent.

Figure 4

$$\begin{cases} 2x + y = 4 \\ -6x - 3y = -12 \end{cases}$$

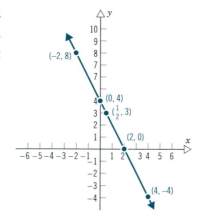

For the system in Example 9 we can find some of the infinite number of solutions by assigning values to x and then finding $y = 4 - 2x$.

When we express the solution in this way, we call x a **parameter.** Thus:

If $x = 4$, then $y = -4$.   This is the point $(4, -4)$ on the graph.

If $x = 0$, then $y = 4$.   This is the point $(0, 4)$ on the graph.

If $x = \tfrac{1}{2}$, then $y = 3$.   This is the point $(\tfrac{1}{2}, 3)$ on the graph.

Alternatively, if we express the solution in the form $x = 2 - \tfrac{1}{2}y$, then y is the parameter and we can assign values to y in order to find x.

$$\text{If } y = -4, \text{ then } x = 2 - \tfrac{1}{2}(-4) = 4$$

$$\text{If } y = 0, \text{ then } x = 2 - \tfrac{1}{2}(0) = 2$$

$$\text{If } y = 4, \text{ then } x = 2 - \tfrac{1}{2}(4) = 0$$

✍ **Now Work Problems**

15 and 19

EXAMPLE 10   Movie Theater Ticket Sales

The movie theater of Example 1 charges $8 for each adult admission and $5 for each child. One Saturday 525 tickets were sold, bringing in a total of $3255. How many of each type of ticket were sold?

SOLUTION   If x represents the number of adult tickets and y the number of child tickets, then, as mentioned at the beginning of this chapter, the given information leads to the system of equations

$$\begin{cases} x + y = 525 \\ 8x + 5y = 3255 \end{cases}$$

We use elimination and multiply the first equation by -5 and then add the two equations:

$$\begin{cases} -5x - 5y = -2625 \\ 8x + 5y = 3255 \end{cases}$$
$$3x = 630$$
$$x = 210$$

Since $x + y = 525$, then $y = 525 - x = 525 - 210 = 315$. Thus 210 adult tickets and 315 child tickets were sold that Saturday.

 **Now Work Problem 55**

Three Linear Equations Containing Three Variables

Just as with a system of two linear equations containing two variables, a system of three linear equations containing three variables also has either no solution, or one solution, or infinitely many solutions.

Let's see how elimination works on a system of three equations containing three variables.

EXAMPLE 11   Solving a System of Three Linear Equations with Three Variables

Use the method of elimination to solve the system of equations:

$$\begin{cases} x + y - z = -1 & (1) \\ 4x - 3y + 2z = 16 & (2) \\ 2x - 2y - 3z = 5 & (3) \end{cases}$$

SOLUTION   For a system of three equations we attempt to eliminate one variable at a time, using pairs of equations. Our plan of attack on this system will be to first eliminate the variable x from Equations (2) and (3) and then from Equations (1) and (2). Next we will eliminate the variable y from Equation (3), leaving only the variable z. Back-substitution can then be used to obtain the values of y and then x.

We begin by multiplying each side of Equation (1) by -2, in anticipation of eliminating the variable x from Equation (3) by adding Equations (1) and (3):

$$\begin{cases} -2x - 2y + 2z = \ \ 2 & (1) \\ \ \ 4x - 3y + 2z = 16 & (2) \\ \ \ 2x - 2y - 3z = \ \ 5 & (3) \end{cases}$$ Multiply each side of Equation (1) by -2.

$$\begin{cases} -2x - 2y + 2z = \ \ 2 & (1) \\ \ \ 4x - 3y + 2z = 16 & (2) \\ \ \ \ \ \ \ \ \ -4y - \ \ z = \ \ 7 & (3) \end{cases}$$ Replace Equation (3) by the sum of Equations (1) and (3).

We now eliminate the same variable x from Equation (2):

$$\begin{cases} -4x - 4y + 4z = \ \ 4 & (1) \\ \ \ 4x - 3y + 2z = 16 & (2) \\ \ \ \ \ \ \ \ \ -4y - \ \ z = \ \ 7 & (3) \end{cases}$$ Multiply each side of Equation (1) by 2.

$$\begin{cases} -4x - 4y + 4z = \ \ 4 & (1) \\ \ \ \ \ \ \ \ \ -7y + 6z = 20 & (2) \\ \ \ \ \ \ \ \ \ -4y - \ \ z = \ \ 7 & (3) \end{cases}$$ Replace Equation (2) by the sum of Equations (1) and (2).

We now eliminate y from Equation (3):

$$\begin{cases} -4x - \ \ 4y + \ \ 4z = \ \ \ \ 4 & (1) \\ \ \ \ \ \ \ \ \ \ \ -28y + 24z = \ \ \ 80 & (2) \\ \ \ \ \ \ \ \ \ \ \ \ \ 28y + \ \ 7z = -49 & (3) \end{cases}$$ Multiply each side by 4.
Multiply each side by -7.

$$\begin{cases} -4x - \ \ 4y + \ \ 4z = \ \ \ \ 4 & (1) \\ \ \ \ \ \ \ \ \ \ \ -28y + 24z = \ \ \ 80 & (2) \\ 31z = \ \ \ 31 & (3) \end{cases}$$ Replace Equation (3) by the sum of Equations (2) and (3).

$$\begin{cases} -4x - \ \ 4y + \ \ 4z = \ \ \ \ 4 & (1) \\ \ \ \ \ \ \ \ \ \ \ -28y + 24z = \ \ \ 80 & (2) \\ z = \ \ \ \ 1 & (3) \end{cases}$$ Multiply each side by $\frac{1}{31}$.

$$\begin{cases} -4x - \ \ 4y + \ \ 4 \ = \ \ \ \ 4 & (1) \\ \ \ \ \ \ \ \ \ \ \ -28y + 24 \ = \ \ \ 80 & (2) \\ z = \ \ \ \ 1 & (3) \end{cases}$$ Back-substitute; replace z by 1 in Equations (1) and (2).

$$\begin{cases} -4x - \ \ 4y \ \ \ \ \ \ \ \ \ = \ \ \ \ 0 & (1) \\ \ \ \ \ \ \ \ \ \ \ \ \ \ y \ \ \ \ \ \ \ \ \ = \ -2 & (2) \\ z = \ \ \ \ 1 & (3) \end{cases}$$ Solve Equation (2) for y.

$$\begin{cases} -4x + \ \ 8 \ \ \ \ \ \ \ \ \ = \ \ \ \ 0 & (1) \\ \ \ \ \ \ \ \ \ \ \ \ \ \ y \ \ \ \ \ \ \ \ \ = \ -2 & (2) \\ z = \ \ \ \ 1 & (3) \end{cases}$$ Back-substitute; replace y by -2.

$$\begin{cases} x = \ \ \ \ 2 & (1) \\ y = \ -2 & (2) \\ z = \ \ \ \ 1 & (3) \end{cases}$$

The solution of the original system is $x = 2$, $y = -2$, $z = 1$. (You should check this.)

■

Look back over the solution given in Example 11. Note the pattern of making Equation (3) contain only the variable z, followed by making Equation (2) contain only the variable y and Equation (1) contain only the variable x. Although which variables to isolate is your choice, the methodology remains the same for all systems.

✍ **Now Work Problem 37**

EXERCISE 2.1   Answers to odd-numbered problems begin on page AN-6.

In Problems 1–6 verify that the values of the variables listed are solutions of the system of equations.

1. $\begin{cases} 2x - y = 5 \\ 5x + 2y = 8 \end{cases}$

$x = 2, y = -1$

2. $\begin{cases} 3x + 2y = 2 \\ x - 7y = -30 \end{cases}$

$x = -2, y = 4$

3. $\begin{cases} 3x - 4y = 4 \\ x - 3y = \frac{1}{2} \end{cases}$

$x = 2, y = \frac{1}{2}$

4. $\begin{cases} 2x + y = 0 \\ 5x - 4y = -\frac{13}{2} \end{cases}$

$x = -\frac{1}{2}, y = 1$

5. $\begin{cases} 3x + 3y + 2z = 4 \\ x - y - z = 0 \\ 2y - 3z = -8 \end{cases}$

$x = 1, y = -1, z = 2$

6. $\begin{cases} 4x - z = 7 \\ 8x + 5y - z = 0 \\ -x - y + 5z = 6 \end{cases}$

$x = 2, y = -3, z = 1$

In Problems 7–42 solve each system of equations. If the system has no solution, say it is inconsistent. Use either substitution or elimination.

7. $\begin{cases} x + y = 9 \\ x - y = 3 \end{cases}$

8. $\begin{cases} x + 2y = 4 \\ x + y = 3 \end{cases}$

9. $\begin{cases} 5x - y = 13 \\ 2x + 3y = 12 \end{cases}$

10. $\begin{cases} x + 3y = 5 \\ 2x - 3y = -8 \end{cases}$

11. $\begin{cases} 3x = 24 \\ x + 2y = 0 \end{cases}$

12. $\begin{cases} 4x + 5y = -3 \\ -2y = -4 \end{cases}$

13. $\begin{cases} 3x - 6y = 24 \\ 5x + 4y = 12 \end{cases}$

14. $\begin{cases} 2x + 4y = 16 \\ 3x - 5y = -9 \end{cases}$

15. $\begin{cases} 2x + y = 1 \\ 4x + 2y = 6 \end{cases}$

16. $\begin{cases} x - y = 5 \\ -3x + 3y = 2 \end{cases}$

17. $\begin{cases} 2x - 4y = -2 \\ 3x + 2y = 3 \end{cases}$

18. $\begin{cases} 3x + 3y = 3 \\ 4x + 2y = \frac{8}{3} \end{cases}$

19. $\begin{cases} x + 2y = 4 \\ 2x + 4y = 8 \end{cases}$

20. $\begin{cases} 3x - y = 7 \\ 9x - 3y = 21 \end{cases}$

21. $\begin{cases} 2x - 3y = -1 \\ 10x + 10y = 5 \end{cases}$

22. $\begin{cases} 3x - 2y = 0 \\ 5x + 10y = 4 \end{cases}$

23. $\begin{cases} 2x + 3y = 6 \\ x - y = \frac{1}{2} \end{cases}$

24. $\begin{cases} \frac{1}{2}x + y = -2 \\ x - 2y = 8 \end{cases}$

25. $\begin{cases} 2x + 3y = 5 \\ 4x + 6y = 10 \end{cases}$

26. $\begin{cases} 2x + 3y = 5 \\ 4x + 6y = 6 \end{cases}$

27. $\begin{cases} 3x - 5y = 3 \\ 15x + 5y = 21 \end{cases}$

28. $\begin{cases} 2x - y = -1 \\ x + \frac{1}{2}y = \frac{3}{2} \end{cases}$

29. $\begin{cases} x - y = 6 \\ 2x - 3z = 16 \\ 2y + z = 4 \end{cases}$

30. $\begin{cases} 2x + y = -4 \\ -2y + 4z = 0 \\ 3x - 2z = -11 \end{cases}$

31. $\begin{cases} x - 2y + 3z = 7 \\ 2x + y + z = 4 \\ -3x + 2y - 2z = -10 \end{cases}$

32. $\begin{cases} 2x + y - 3z = 0 \\ -2x + 2y + z = -7 \\ 3x - 4y - 3z = 7 \end{cases}$

33. $\begin{cases} x - y - z = 1 \\ 2x + 3y + z = 2 \\ 3x + 2y = 0 \end{cases}$

34. $\begin{cases} 2x - 3y - z = 0 \\ -x + 2y + z = 5 \\ 3x - 4y - z = 1 \end{cases}$

35. $\begin{cases} x - y - z = 1 \\ -x + 2y - 3z = -4 \\ 3x - 2y - 7z = 0 \end{cases}$

36. $\begin{cases} 2x - 3y - z = 0 \\ 3x + 2y + 2z = 2 \\ x + 5y + 3z = 2 \end{cases}$

37. $\begin{cases} 2x - 2y + 3z = 6 \\ 4x - 3y + 2z = 0 \\ -2x + 3y - 7z = 1 \end{cases}$

38. $\begin{cases} 3x - 2y + 2z = 6 \\ 7x - 3y + 2z = -1 \\ 2x - 3y + 4z = 0 \end{cases}$

39. $\begin{cases} x + y - z = 6 \\ 3x - 2y + z = -5 \\ x + 3y - 2z = 14 \end{cases}$

40. $\begin{cases} x - y + z = -4 \\ 2x - 3y + 4z = -15 \\ 5x + y - 2z = 12 \end{cases}$

41. $\begin{cases} x + 2y - z = -3 \\ 2x - 4y + z = -7 \\ -2x + 2y - 3z = 4 \end{cases}$

42. $\begin{cases} x + 4y - 3z = -8 \\ 3x - y + 3z = 12 \\ x + y + 6z = 1 \end{cases}$

43. The sum of two numbers is 81. The difference of twice one number and three times the other is 62. Find the two numbers.

44. The difference of two numbers is 40. Six times the smaller number less the larger number is 5. Find the two numbers.

45. Dimensions of a Floor   The perimeter of a rectangular floor is 90 feet. Find the dimensions of the floor if the length is twice the width.

46. Dimensions of a Field   The length of fence required to enclose a rectangular field is 3000 meters. What are the dimensions of the field if it is known that the difference between its length and width is 50 meters?

47. Cost of Fast Food   Four large cheeseburgers and two chocolate shakes cost a total of $7.90. Two shakes cost 15 cents more than one cheeseburger. What is the cost of a cheeseburger? A shake?

48. Movie Theater Tickets   A movie theater charges $7 for adults and $4 for children under 12. On a day when 325 people paid for admission, the total receipts were $1900. How many who paid were adults? How many were under 12?

49. Investment   Mr. Nicholson has just retired and needs $6000 per year in income to live on. He has $50,000 to invest and can invest in AA bonds at 15% annual interest or in Savings and Loan Certificates at 7% interest per year. How much money should be invested in each so that he realizes exactly $6000 in income per year?

50. Investment   Mr. Nicholson finds after 2 years that because of inflation he now needs $7000 per year to live on. How should he transfer his funds to achieve this amount? (Use the data from Problem 49.)

51. Mixture   Joan has $1.65 in her piggy bank. She knows she placed only nickels and quarters in the bank and she knows that, in all, she put 13 coins in the bank. Can she find out how many nickels she has without breaking her bank?

52. Mixture   A coffee manufacturer wants to market a new blend of coffee that will cost $5 per pound by mixing $3.75 per pound coffee and $8 per pound coffee. What amounts of the $3.75 per pound coffee and $8 per pound coffee should be blended to obtain the desired mixture? [*Hint:* Assume the total weight of the desired blend is 100 pounds.]

53. Mixture   One solution is 30% acid and another is 10% acid. How many cubic centimeters of each should be mixed to obtain 100 cc of a solution that is 18% acid?

54. Investment   A bank loaned $10,000, some at an annual rate of 8% and some at an annual rate of 18%. If the income from these loans was $1000, how much was loaned at 8%? How much at 18%?

55. Receipts   The Star Theater wants to know whether the majority of its patrons are adults or children. During a week in July 5200 tickets were sold and the receipts totaled $33,840. The adult admission is $8 and the children's admission is $4. How many adult patrons were there?

56. Mixture   A candy manufacturer sells a candy mix for $1.10 a pound. It is composed of two candies, one worth $0.90 a pound, the other worth $1.50 a pound. How many pounds of each candy is in 60 pounds of mixture?

57. Finance   Two investments are made totaling $50,000. In 1 year the first investment yields a profit of 10%, whereas the second yields a profit of 12%. Total profit for this year is $5250. Find the amount initially put into each investment.

58. Finance   Laura invested part of her money at 8% and the rest at 10%. The income from both investments totaled $3200. If she interchanged her investments, her income would have totaled $2800. How much did she have in each investment?

59. Agriculture   The Smith farm has 1000 acres of land to be used to raise corn and soybeans. The cost of cultivating the corn and the soybeans is $62 and $44 per acre, respectively. Mr. Smith has a budget of $45,800 to use for cultivating these crops. If Mr. Smith wishes to use all the land and all the money budgeted, how many acres of each crop should he plant?

60. Diet Preparation   A farmer prepares feed for livestock by combining two grains. Each unit of the first grain contains 2 units of protein and 5 units of iron, while each unit of the second grain contains 4 units of protein and 1 unit of iron. Determine the number of units of each kind of grain the farmer needs to feed each animal daily if each animal must have 10 units of protein and 16 units of iron each day.

61. Purchasing   David owns a restaurant and wants to order 200 dinner sets. One design costs $20 per set and the other costs $15 per set. He has $3200 with which to buy the sets. How many of each type of set should he buy if he is to use all the money and acquire 200 sets?

Technology Exercises

In Problems 1–4 graph each system of linear equations. Be sure to use a viewing rectangle that shows the point of intersection. Then approximate the point of intersection rounded to two decimal places.

1. $\begin{cases} 1.2x + 0.8y = 20 \\ -1.3x + 2.7y = 81.1 \end{cases}$
2. $\begin{cases} 215x - 0.1y = 53 \\ 0.5x - 313y = 82 \end{cases}$
3. $\begin{cases} \frac{4}{17}x + \frac{6}{23}y = \frac{2}{3} \\ \frac{9}{14}x - \frac{43}{8}y = \frac{22}{7} \end{cases}$
4. $\begin{cases} \pi x - \sqrt{3}\,y = \sqrt{6} \\ x + \pi y = \sqrt{15} \end{cases}$

2.2 SYSTEMS OF LINEAR EQUATIONS: MATRIX METHOD

The systematic approach of the method of elimination for solving a system of linear equations provides another method of solution that involves a simplified notation using a *matrix*.

A **matrix** is defined as a rectangular array of numbers, enclosed by brackets. The numbers are referred to as the **entries** of the matrix. A matrix is further identified by naming its *rows* and *columns*. Some examples of matrices are

$$
\begin{array}{c}
\text{Column 1}\ \text{Column 2} \\
\begin{array}{c} \text{Row 1} \\ \text{Row 2} \\ \text{Row 3} \end{array}
\begin{bmatrix} 8 & 0 \\ 1 & 3 \\ -2 & 4 \end{bmatrix}
\end{array}
\qquad
\begin{array}{c}
\text{Column 1}\ \text{Column 2}\ \text{Column 3} \\
\begin{array}{c} \text{Row 1} \\ \text{Row 2} \end{array}
\begin{bmatrix} 4 & 1 & -3 \\ 2 & 1 & 2 \end{bmatrix}
\end{array}
\qquad
\begin{array}{c}
\text{Column 1}\ \text{Column 2} \\
\text{Row 1}\ \begin{bmatrix} 4 & 3 \end{bmatrix}
\end{array}
$$

$$\text{(a)} \qquad\qquad\qquad \text{(b)} \qquad\qquad\qquad \text{(c)}$$

Matrix Representation of a System of Linear Equations

Consider the following two systems of two linear equations containing two variables

$$\begin{cases} x + 4y = 14 \\ 3x - 2y = 0 \end{cases} \quad \text{and} \quad \begin{cases} u + 4v = 14 \\ 3u - 2v = 0 \end{cases}$$

We observe that, except for the symbols used to represent the variables, these two systems are identical. As a result, we can dispense altogether with the letters used to symbolize the variables, provided we have some means of keeping track of them. A matrix serves us well in this regard.

When a matrix is used to represent a system of linear equations, it is called the **augmented matrix** of the system. For example, the system

$$\begin{cases} x + 4y = 14 \\ 3x - 2y = 0 \end{cases}$$

can be represented by the augmented matrix

	Column 1 (x)	Column 2 (y)	Column 3 (right-hand side)
Row 1	1	4	14
Row 2	3	−2	0

Here it is understood that column 1 contains the coefficients of the variable x, column 2 contains the coefficients of the variable y, and column 3 contains the numbers to the right of the equal sign. Each row of the matrix represents an equation of the system. Although not required, it has become customary to place a vertical bar in the matrix as a reminder of the equal sign.

In this book we shall follow the practice of using x and y to denote the variables for systems containing two variables. We will use x, y, and z for systems containing three variables; we will use subscripted variables (x_1, x_2, x_3, x_4, etc.) for systems containing four or more variables.

EXAMPLE 1   Writing the Augmented Matrix of a System of Linear Equations

System of Linear Equations	*Augmented Matrix*
(a) $\begin{cases} 5x - 2y = 5 \\ 2x - y = -4 \end{cases}$	$\begin{bmatrix} 5 & -2 & 5 \\ 2 & -1 & -4 \end{bmatrix}$
(b) $\begin{cases} 3x + 4y + 3z = 10 \\ x + y - z = 1 \\ x + 6y = 4 \end{cases}$	$\begin{bmatrix} 3 & 4 & 3 & 10 \\ 1 & 1 & -1 & 1 \\ 1 & 6 & 0 & 4 \end{bmatrix}$
(c) $\begin{cases} 3x_1 - x_2 + x_3 + x_4 = 5 \\ 2x_1 + 6x_3 = 2 \end{cases}$	$\begin{bmatrix} 3 & -1 & 1 & 1 & 5 \\ 2 & 0 & 6 & 0 & 2 \end{bmatrix}$

■

Now Work Problem 1

Given an augmented matrix, we can write the corresponding system of equations.

EXAMPLE 2   Writing the System of Linear Equations from the Augmented Matrix

Write a system of linear equations for each augmented matrix.

(a) $\begin{bmatrix} 4 & -1 & 6 \\ 3 & 1 & 2 \end{bmatrix}$ (b) $\begin{bmatrix} 3 & 2 & 1 & 1 & 5 \\ -1 & 0 & 1 & 4 & 0 \end{bmatrix}$

SOLUTION

(a) The matrix has two rows and so represents a system of two equations. The two columns to the left of the vertical bar indicate that the system has two variables. If x and y are used to denote these variables, the system of equations is

$$\begin{cases} 4x - y = 6 \\ 3x + y = 2 \end{cases}$$

(b) This matrix represents a system of two equations containing four variables. If we use $x_1, x_2, x_3,$ and x_4 as variables, the system of equations is

$$\begin{cases} 3x_1 + 2x_2 + x_3 + x_4 = 5 \\ \qquad\quad -x_1 + x_3 + 4x_4 = 0 \end{cases}$$

■

The use of matrices to solve a system of linear equations requires the use of *row operations* on the augmented matrix of the system.

Row Operations

We begin by going through the solution of a system of two linear equations in two variables. Although this system is more easily solved using the methods of substitution or elimination (Section 2.1), the *pattern* of the solution shown below will lead to the formulation of a third method for solving systems of linear equations.

EXAMPLE 3   **Solving a System of Equations**

Solve: $\begin{cases} x + 4y = 14 \\ 3x - 2y = 0 \end{cases}$ (1)

SOLUTION   We multiply the first equation by -3, obtaining $-3x - 12y = -42$, and then add it to the second equation.

$$\begin{cases} -3x - 12y = -42 \\ \underline{\ \ 3x - \ 2y = \quad 0} \\ \qquad\quad -14y = -42 \end{cases}$$

The original system of Equations (1) may now be written as the equivalent system

$$\begin{cases} x + \ 4y = \quad 14 \\ \quad\ -14y = -42 \end{cases}$$

Dividing the second equation by -14 or, equivalently, multiplying by $-\frac{1}{14}$, we get

$$\begin{cases} x + 4y = 14 \\ \qquad\ y = \ 3 \end{cases}$$

To find x, we multiply the second equation by -4 and add it to the first. The result is

$$\begin{cases} x = 2 \\ y = 3 \end{cases}$$ (2)

This system of equations has the obvious solution $x = 2$, $y = 3$ and is equivalent to the original system (1), so that the solution of the original system (1) is also $x = 2$, $y = 3$.

We obtained the final system (2) from the original system (1) by a series of operations. We chose the above operations because the *pattern* of the solution has certain advantages:

1. It is algorithmic in character. This means it consists of repetitive steps that can be programmed on a computer.

2. It works on any system of linear equations.

Let's repeat the steps we took to solve Example 3, except now we will manipulate the augmented matrix instead of the equations. For convenience, the equations are listed next to the augmented matrix.

$$\begin{bmatrix} 1 & 4 & \bigm| & 14 \\ 3 & -2 & \bigm| & 0 \end{bmatrix} \qquad \begin{cases} x + 4y = 14 \\ 3x - 2y = 0 \end{cases}$$

Multiplying the first *equation* by -3 and adding it to the second *equation* corresponds to multiplying each entry in the first *row* of the matrix by -3 and adding the result to each corresponding entry in the second *row*. The result is the matrix

$$\begin{bmatrix} 1 & 4 & \bigm| & 14 \\ 0 & -14 & \bigm| & -42 \end{bmatrix} \qquad \begin{cases} x + 4y = 14 \\ -14y = -42 \end{cases}$$

Next, divide the second row of this last matrix by -14 or, equivalently, multiply it by $-\frac{1}{14}$, to obtain

$$\begin{bmatrix} 1 & 4 & \bigm| & 14 \\ 0 & 1 & \bigm| & 3 \end{bmatrix} \qquad \begin{cases} x + 4y = 14 \\ y = 3 \end{cases}$$

Finally, multiply the second row of this matrix by -4 and add it to the first row to obtain

$$\begin{bmatrix} 1 & 0 & \bigm| & 2 \\ 0 & 1 & \bigm| & 3 \end{bmatrix} \qquad \begin{cases} x = 2 \\ y = 3 \end{cases}$$

When manipulations such as the above are performed on a matrix, they are called *row operations*. We now list the three basic types of row operations:

Row Operations

1. Interchange any two rows.

2. Replace any row by a nonzero constant multiple of that row.

3. Replace any row by the sum of that row and a constant multiple of some other row.

An example of each type of row operation is given next.

EXAMPLE 4   Examples of Row Operations

$$A = \begin{bmatrix} 3 & 4 & -3 \\ 7 & -\frac{1}{2} & 0 \end{bmatrix}$$

1. The matrix obtained by interchanging the first and second rows of A is

$$\begin{bmatrix} 7 & -\frac{1}{2} & 0 \\ 3 & 4 & -3 \end{bmatrix}$$

2. The matrix obtained by multiplying row 2 of A by 5 is

$$\begin{bmatrix} 3 & 4 & -3 \\ 35 & -\frac{5}{2} & 0 \end{bmatrix}$$

We denote this operation by writing $R_2 = 5r_2$, where r_2 denotes the "old" row 2 and R_2 denotes the "new" row 2.

3. The matrix obtained from A by replacing row 2 of A by row 2 plus 3 times row 1 is

$$\begin{bmatrix} 3 & 4 & -3 \\ 7 + 3 \cdot 3 & (-\frac{1}{2}) + 3 \cdot 4 & 0 + 3 \cdot (-3) \end{bmatrix} = \begin{bmatrix} 3 & 4 & -3 \\ 16 & \frac{23}{2} & -9 \end{bmatrix}$$

We denote this operation by writing $R_2 = r_2 + 3r_1$.

 **Now Work Problem 13**

A word about the notation we have introduced. A row operation such as $R_1 = r_1 + 2r_2$ changes the entries in row 1. Note also that for this type of row operation we change the entries in a given row by adding the original entries of the row to be changed and a constant multiple of the entries of some other row.

Compare the three types of row operations to the Rules for Obtaining an Equivalent System of Equations (see page 42). The three types of row operations, when applied to the augmented matrix of a system of equations, result in a new augmented matrix that represents a system of equations equivalent to the original. With this in mind, let's see how row operations are used to solve a system of equations.

EXAMPLE 5   Solving a System of Linear Equations Using Matrices

Solve: $\begin{cases} 4x + 3y = 11 \\ x - 3y = -1 \end{cases}$

SOLUTION   First, we write the augmented matrix that represents this system:

$$\begin{bmatrix} 4 & 3 & | & 11 \\ 1 & -3 & | & -1 \end{bmatrix} \qquad \begin{cases} 4x + 3y = 11 \\ x - 3y = -1 \end{cases}$$

The first step is to place a 1 in row 1, column 1. (This will make the back-substitution that comes later a little easier.) An interchange of rows 1 and 2 is the easiest way to do this.

$$\begin{bmatrix} 1 & -3 & | & -1 \\ 4 & 3 & | & 11 \end{bmatrix} \qquad \begin{cases} x - 3y = -1 \\ 4x + 3y = 11 \end{cases}$$

Next we want a 0 under the entry 1 in column 1. (This eliminates the variable x from the second equation.) We use the row operation $R_2 = r_2 + (-4)r_1$.

$$\begin{bmatrix} 1 & -3 & | & -1 \\ 0 & 15 & | & 15 \end{bmatrix} \qquad \begin{cases} x - 3y = -1 \\ 15y = 15 \end{cases}$$

Now we want the entry 1 in row 2, column 2. (This makes it easy to solve for y.) We use $R_2 = \frac{1}{15}r_2$.

$$\begin{bmatrix} 1 & -3 & | & -1 \\ 0 & 1 & | & 1 \end{bmatrix} \qquad \begin{cases} x - 3y = -1 \\ y = 1 \end{cases}$$

The second row of the matrix on the right represents the equation $y = 1$. Thus, using $y = 1$, we back-substitute into the equation $x - 3y = -1$ (from the first row) to get

$$x - 3(1) = -1 \quad y = 1$$
$$x = 2$$

The solution of the system is $x = 2$, $y = 1$.

 **Now Work Problem 33**

The steps we used to solve the system of linear equations in Example 5 can be summarized as follows:

Steps for Solving a System of Linear Equations Using Matrices

Step 1 Write the augmented matrix that represents the system.

Step 2 Perform row operations that place the entry 1 in row 1, column 1.

Step 3 Perform row operations that leave the entry 1 in row 1, column 1 unchanged, while causing 0's to appear below it in column 1.

Step 4 Perform row operations that place the entry 1 in row 2, column 2 and leave the entries in columns to the left unchanged. If it is impossible to place a 1 in row 2, column 2, then proceed to place a 1 in row 2, column 3. Once a 1 is in place, perform row operations to place 0's under it.

Step 5 Repeat step 4, placing a 1 in the next row, but one column to the right. Continue until the bottom row or the vertical bar is reached.

Step 6 If any rows are obtained that contain only 0's on the left side of the vertical bar, then place such rows at the bottom of the matrix.

After steps 1 to 6 have been completed, the matrix is said to be in **row-echelon form.** A little thought should convince you that a matrix is in row-echelon form when the following conditions are met.

Conditions for the Row-Echelon Form of a Matrix

1. The entry in row 1, column 1 is a 1, and 0's appear below it.

2. The first nonzero entry in each row after the first row is a 1, 0's appear below it, and it appears to the right of the first nonzero entry in any row above.

3. Any rows that contain all 0's to the left of the vertical bar appear at the bottom.

Examples of matrices that are in row-echelon form are:

$$\begin{bmatrix} 1 & 2 & | & 3 \\ 0 & 1 & | & 2 \end{bmatrix} \quad \begin{bmatrix} 1 & 6 & 2 & | & 1 \\ 0 & 1 & 0 & | & 2 \\ 0 & 0 & 1 & | & 3 \end{bmatrix} \quad \begin{bmatrix} 1 & 6 & 0 & | & 2 \\ 0 & 0 & 1 & | & 3 \\ 0 & 0 & 0 & | & 0 \end{bmatrix} \quad \begin{bmatrix} 1 & 0 & 3 & 0 & | & 3 \\ 0 & 1 & 2 & 0 & | & 4 \\ 0 & 0 & 0 & 1 & | & 6 \\ 0 & 0 & 0 & 0 & | & 1 \end{bmatrix}$$

Here are some examples of matrices that are *not* in row-echelon form:

$$\begin{bmatrix} 2 & 4 & 1 & | & 1 \\ 0 & 1 & 2 & | & 2 \end{bmatrix} \qquad \begin{bmatrix} 1 & 4 & 3 & | & 0 \\ 0 & 0 & 1 & | & 1 \\ 0 & 1 & 0 & | & 2 \end{bmatrix} \qquad \begin{bmatrix} 1 & 3 & 2 & | & 1 \\ 0 & 0 & 0 & | & 1 \\ 0 & 1 & 2 & | & 3 \end{bmatrix}$$

The entry in row 1, column 1 is not 1.   The first nonzero entry in row 3 is not to the right of the first nonzero entry in row 2 above.   Any row with 0's to the left of the vertical bar should appear at the bottom.

The next example shows how to write a matrix in row-echelon form.

EXAMPLE 6   Solving a System of Linear Equations Using Matrices

Solve: $\begin{cases} x + y + z = 6 \\ 3x + 2y - z = 4 \\ 3x + y + 2z = 11 \end{cases}$

SOLUTION

Step 1 The augmented matrix corresponding to the system is

$$\begin{bmatrix} 1 & 1 & 1 & | & 6 \\ 3 & 2 & -1 & | & 4 \\ 3 & 1 & 2 & | & 11 \end{bmatrix}$$

Step 2 Since a 1 appears in row 1, column 1, we can skip to Step 3.

Step 3 Perform the row operations*

$$R_2 = r_2 + (-3)r_1$$
$$R_3 = r_3 + (-3)r_1$$

Notice that these operations leave row 1 unchanged, while getting 0s in the rest of column 1:

$$\begin{bmatrix} 1 & 1 & 1 & | & 6 \\ 0 & -1 & -4 & | & -14 \\ 0 & -2 & -1 & | & -7 \end{bmatrix}$$

Step 4 To place a 1 in row 2, column 2, we use

$$R_2 = (-1)r_2$$

Notice that column 1 remains unchanged:

$$\begin{bmatrix} 1 & 1 & 1 & | & 6 \\ 0 & 1 & 4 & | & 14 \\ 0 & -2 & -1 & | & -7 \end{bmatrix}$$

* You should convince yourself that doing both of these simultaneously is the same as doing the first followed by the second.

To place 0s below row 2 in column 2, we use

$$R_3 = r_3 + 2r_2$$

$$\begin{bmatrix} 1 & 1 & 1 & | & 6 \\ 0 & 1 & 4 & | & 14 \\ 0 & 0 & 7 & | & 21 \end{bmatrix}$$

Step 5 Continuing, we seek a 1 in row 3, column 3, so we use

$$R_3 = \tfrac{1}{7} r_3$$

Notice that this is the only choice available to us, since any other choice would change the 0s and 1s already obtained:

$$\begin{bmatrix} 1 & 1 & 1 & | & 6 \\ 0 & 1 & 4 & | & 14 \\ 0 & 0 & 1 & | & 3 \end{bmatrix}$$

Because we have reached the bottom row, the matrix is in row-echelon form and we can stop.

The system of equations represented by the matrix in row-echelon form is

$$\begin{cases} x + y + z = 6 \\ \quad\;\; y + 4z = 14 \\ \quad\qquad z = 3 \end{cases}$$

Using $z = 3$, we back-substitute to obtain

$$\begin{cases} x + y + 3 = 6 \\ \quad\; y + 4(3) = 14 \end{cases} \quad \text{or equivalently} \quad \begin{cases} x + y = 3 \\ \quad\;\; y = 2 \end{cases}$$

Thus, $y = 2$, and back-substituting into $x + y = 3$, we find $x = 1$. The solution of the system is $x = 1$, $y = 2$, $z = 3$. ∎

Sometimes, it is advantageous to write a matrix in **reduced row-echelon form.** In this form, row operations are used to obtain entries that are 0 above (as well as below) the leading 1 in a row. For example, the row-echelon form obtained in the solution to Example 6 is

$$\begin{bmatrix} 1 & 1 & 1 & | & 6 \\ 0 & 1 & 4 & | & 14 \\ 0 & 0 & 1 & | & 3 \end{bmatrix}$$

To write this matrix in reduced row-echelon form, we proceed as follows:

$$\begin{bmatrix} 1 & 1 & 1 & | & 6 \\ 0 & 1 & 4 & | & 14 \\ 0 & 0 & 1 & | & 3 \end{bmatrix} \rightarrow \begin{bmatrix} 1 & 0 & -3 & | & -8 \\ 0 & 1 & 4 & | & 14 \\ 0 & 0 & 1 & | & 3 \end{bmatrix} \rightarrow \begin{bmatrix} 1 & 0 & 0 & | & 1 \\ 0 & 1 & 0 & | & 2 \\ 0 & 0 & 1 & | & 3 \end{bmatrix}$$

$$\uparrow \qquad\qquad\qquad\qquad \uparrow$$
$$R_1 = r_1 + (-1)r_2 \qquad\quad R_1 = r_1 + 3r_3$$
$$R_2 = r_2 + (-4)r_3$$

The matrix is now written in reduced row-echelon form. The advantage of writing the matrix in this form is that the solution to the system, $x = 1$, $y = 2$, $z = 3$, is readily found without the need to back-substitute. Another advantage will be seen in Section 2.6, where the inverse of a matrix is discussed.

✍ **Now Work Problem 49**

EXAMPLE 7 **Calculating Production Output**

FoodPerfect Corporation manufactures three models of the Perfect Foodprocessor. Each Model X processor requires 30 minutes of electrical assembly, 40 minutes of mechanical assembly, and 30 minutes of testing; each Model Y requires 20 minutes of electrical assembly, 50 minutes of mechanical assembly, and 30 minutes of testing; and each Model Z requires 30 minutes of electrical assembly, 30 minutes of mechanical assembly, and 20 minutes of testing. If 2500 minutes of electrical assembly, 3500 minutes of mechanical assembly, and 2400 minutes of testing are used in one day, how many of each model will be produced?

SOLUTION The table below summarizes the given information:

	Model			Time Used
	X	**Y**	**Z**	
Electrical Assembly	30	20	30	2500
Mechanical Assembly	40	50	30	3500
Testing	30	30	20	2400

We assign variables to represent the unknowns:

$$x = \text{Number of Model X produced}$$
$$y = \text{Number of Model Y produced}$$
$$z = \text{Number of Model Z produced}$$

Based on the table, we obtain the following system of equations:

$$\begin{cases} 30x + 20y + 30z = 2500 \\ 40x + 50y + 30z = 3500 \\ 30x + 30y + 20z = 2400 \end{cases} \quad \text{or} \quad \begin{cases} 3x + 2y + 3z = 250 & (1) \\ 4x + 5y + 3z = 350 & (2) \\ 3x + 3y + 2z = 240 & (3) \end{cases}$$

The augmented matrix of this system is

$$\begin{bmatrix} 3 & 2 & 3 & | & 250 \\ 4 & 5 & 3 & | & 350 \\ 3 & 3 & 2 & | & 240 \end{bmatrix}$$

We could obtain a 1 in row 1, column 1 by using the row operation $R_1 = \frac{1}{3}r_1$, but the introduction of fractions is best avoided. Instead, we use

$$R_2 = r_2 + (-1)r_1$$

to place a 1 in row 2, column 1. The result is

$$\begin{bmatrix} 3 & 2 & 3 & | & 250 \\ 1 & 3 & 0 & | & 100 \\ 3 & 3 & 2 & | & 240 \end{bmatrix}$$

Next, interchange row 1 and row 2:

$$\begin{bmatrix} 1 & 3 & 0 & | & 100 \\ 3 & 2 & 3 & | & 250 \\ 3 & 3 & 2 & | & 240 \end{bmatrix}$$

Use $R_2 = r_2 + (-3)r_1$, $R_3 = r_3 + (-3)r_1$ to obtain

$$\left[\begin{array}{ccc|c} 1 & 3 & 0 & 100 \\ 0 & -7 & 3 & -50 \\ 0 & -6 & 2 & -60 \end{array}\right]$$

We use $R_2 = (-1)r_2$ followed by $R_2 = r_2 + r_3$:

$$\left[\begin{array}{ccc|c} 1 & 3 & 0 & 100 \\ 0 & 7 & -3 & 50 \\ 0 & -6 & 2 & -60 \end{array}\right] \qquad \left[\begin{array}{ccc|c} 1 & 3 & 0 & 100 \\ 0 & 1 & -1 & -10 \\ 0 & -6 & 2 & -60 \end{array}\right]$$

Next, use $R_3 = r_3 + 6r_2$ to obtain

$$\left[\begin{array}{ccc|c} 1 & 3 & 0 & 100 \\ 0 & 1 & -1 & -10 \\ 0 & 0 & -4 & -120 \end{array}\right]$$

Next, use $R_3 = (-\frac{1}{4})r_3$. The result is

$$\left[\begin{array}{ccc|c} 1 & 3 & 0 & 100 \\ 0 & 1 & -1 & -10 \\ 0 & 0 & 1 & 30 \end{array}\right]$$

The matrix is in row-echelon form. We find $z = 30$. From row 2, we have $y - z = -10$ so that $y = z - 10 = 30 - 10 = 20$. Finally, from row 1, we have $x + 3y = 100$ so $x = -3y + 100 = -60 + 100 = 40$. The solution of the system is $x = 40$, $y = 20$, $z = 30$. Thus in one day 40 Model X, 20 Model Y, and 30 Model Z processors were produced. ∎

The matrix method also reveals systems that have no solution or an infinite number of solutions.

EXAMPLE 8 **An Inconsistent System**

Solve: $\begin{cases} 3x - 6y = 4 \\ 6x - 12y = 5 \end{cases}$

SOLUTION The augmented matrix representing this system is

$$\left[\begin{array}{cc|c} 3 & -6 & 4 \\ 6 & -12 & 5 \end{array}\right]$$

To place a 1 in row 1, column 1, we use $R_1 = \frac{1}{3}r_1$. (Note that no other choice is possible, since 6 is a multiple of 3.)

$$\left[\begin{array}{cc|c} 1 & -2 & \frac{4}{3} \\ 6 & -12 & 5 \end{array}\right]$$

To place a 0 in row 2, column 1, we use $R_2 = r_2 + (-6)r_1$:

$$\left[\begin{array}{cc|c} 1 & -2 & \frac{4}{3} \\ 0 & 0 & -3 \end{array}\right]$$

Let's stop here to look at the actual system of equations:

$$\begin{cases} x - 2y = \frac{4}{3} \\ 0x + 0y = -3 \end{cases}$$

The second equation can never be true—no matter what the choice of x and y. Hence, there are no numbers x and y that can obey both equations. That is, the system has no solution.

■

EXAMPLE 9   **A System with Infinitely Many Solutions**

Solve: $\begin{cases} 2x - 3y = 5 \\ 4x - 6y = 10 \end{cases}$

SOLUTION   The augmented matrix representing this system is

$$\begin{bmatrix} 2 & -3 & \bigm| & 5 \\ 4 & -6 & \bigm| & 10 \end{bmatrix}$$

To place a 1 in row 1, column 1, we use $R_1 = \frac{1}{2}r_1$:

$$\begin{bmatrix} 1 & -\frac{3}{2} & \bigm| & \frac{5}{2} \\ 4 & -6 & \bigm| & 10 \end{bmatrix}$$

To place a 0 in column 1, row 2, we use $R_2 = r_2 + (-4)r_1$:

$$\begin{bmatrix} 1 & -\frac{3}{2} & \bigm| & \frac{5}{2} \\ 0 & 0 & \bigm| & 0 \end{bmatrix}$$

The system of equations looks like

$$\begin{cases} x - \frac{3}{2}y = \frac{5}{2} \\ 0x + 0y = 0 \end{cases}$$

The second equation is true for any choice of x and y. Hence, all numbers x and y that obey the first equation are solutions of the system. Since any point on the line $x - \frac{3}{2}y = \frac{5}{2}$ is a solution, there are an infinite number of solutions.

Using y as parameter, we can list some of these solutions by assigning values to y and then calculating x from the equation $x = \frac{3}{2}y + \frac{5}{2}$.

If $y = 0$, then $x = \frac{5}{2}$. Thus $x = \frac{5}{2}$, $y = 0$ is a solution.

If $y = 1$, then $x = 4$. Thus $x = 4$, $y = 1$ is a solution.

If $y = 5$, then $x = 10$. Thus $x = 10$, $y = 5$ is a solution.

If $y = -3$, then $x = -2$. Thus $x = -2$, $y = -3$ is a solution.

And so on.

✐ Now Work Problem 59

■

We close with a capsule summary of what to expect from any system of two equations containing two variables.

SUMMARY

After finding the row-echelon form of the augmented matrix of a system of two linear equations containing two variables, one of the following matrices will result.

$$\begin{bmatrix} 1 & a & | & c \\ 0 & 1 & | & d \end{bmatrix}$$

Unique solution: $y = d$
Back-substitute to find x.

$$\begin{bmatrix} a & b & | & c \\ 0 & 0 & | & 0 \end{bmatrix}$$

Infinite number of solutions: $ax + by = c$
Either x or y can be used as parameter.

$$\begin{bmatrix} a & b & | & c \\ 0 & 0 & | & \text{nonzero number} \end{bmatrix}$$

No solution

EXERCISE 2.2 Answers to odd-numbered problems begin on page AN-6.

In Problems 1–12 write the augmented matrix of each system.

1. $\begin{cases} 2x - 3y = 5 \\ x - y = 3 \end{cases}$

2. $\begin{cases} 4x + y = 5 \\ 2x + y = 5 \end{cases}$

3. $\begin{cases} 2x + y + 6 = 0 \\ 3x + y = -1 \end{cases}$

4. $\begin{cases} -3x - y = -3 \\ 4x - y + 2 = 0 \end{cases}$

5. $\begin{cases} 2x - y - z = 0 \\ x - y + z = 1 \\ 3x - y = 2 \end{cases}$

6. $\begin{cases} x + y + z = 3 \\ 2x + z = 0 \\ 3x - y - z = 1 \end{cases}$

7. $\begin{cases} 2x - 3y + z - 7 = 0 \\ x + y - z = 1 \\ 2x + 2y - 3z + 4 = 0 \end{cases}$

8. $\begin{cases} 5x - 3y + 6z + 1 = 0 \\ -x - y + z = 1 \\ 2x + 3y + 5 = 0 \end{cases}$

9. $\begin{cases} 4x_1 - x_2 + 2x_3 - x_4 = 4 \\ x_1 + x_2 + 6 = 0 \\ 2x_2 - x_3 + x_4 = 5 \end{cases}$

10. $\begin{cases} 3x_1 - 5x_2 + x_3 = 2 \\ x - x_2 + x_3 = 6 \\ 2x_1 + x_3 + 4 = 0 \end{cases}$

11. $\begin{cases} x_1 - x_2 + x_3 - x_4 = 0 \\ 2x_1 + 3x_2 - x_3 + 4x_4 = 5 \end{cases}$

12. $\begin{cases} x_1 + x_2 + x_3 + x_4 = 4 \\ x_1 - 2x_2 + 3x_3 - 4x_4 = 5 \end{cases}$

In Problems 13–20 perform, in order, (a), followed by (b), followed by (c), on the given augmented matrix.

13. $\begin{bmatrix} 1 & -3 & -5 & | & -2 \\ 2 & -5 & -4 & | & 5 \\ -3 & 5 & 4 & | & 6 \end{bmatrix}$
(a) $R_2 = r_2 + (-2)r_1$
(b) $R_3 = r_3 + 3r_1$
(c) $R_3 = r_3 + 4r_2$

14. $\begin{bmatrix} 1 & -3 & -3 & | & -3 \\ 2 & -5 & 2 & | & -4 \\ -3 & 2 & 4 & | & 6 \end{bmatrix}$
(a) $R_2 = r_2 + (-2)r_1$
(b) $R_3 = r_3 + 3r_1$
(c) $R_3 = r_3 + 7r_2$

15. $\begin{bmatrix} 1 & -3 & 4 & | & 3 \\ 2 & -5 & 6 & | & 6 \\ -3 & 3 & 4 & | & 6 \end{bmatrix}$
(a) $R_2 = r_2 + (-2)r_1$
(b) $R_3 = r_3 + 3r_1$
(c) $R_3 = r_3 + 6r_2$

16. $\begin{bmatrix} 1 & -3 & 3 & | & -5 \\ 2 & -5 & -3 & | & -5 \\ -3 & -2 & 4 & | & 6 \end{bmatrix}$
(a) $R_2 = r_2 + (-2)r_1$
(b) $R_3 = r_3 + 3r_1$
(c) $R_3 = r_3 + 11r_2$

17. $\begin{bmatrix} 1 & -3 & 2 & | & -6 \\ 2 & -5 & 3 & | & -4 \\ -3 & -6 & 4 & | & 6 \end{bmatrix}$
(a) $R_2 = r_2 + (-2)r_1$
(b) $R_3 = r_3 + 3r_1$
(c) $R_3 = r_3 + 15r_2$

18. $\begin{bmatrix} 1 & -3 & -4 & | & -6 \\ 2 & -5 & 6 & | & -6 \\ -3 & 1 & 4 & | & 6 \end{bmatrix}$
(a) $R_2 = r_2 + (-2)r_1$
(b) $R_3 = r_3 + 3r_1$
(c) $R_3 = r_3 + 8r_2$

19. $\begin{bmatrix} 1 & -3 & 1 & | & -2 \\ 2 & -5 & 6 & | & -2 \\ -3 & 1 & 4 & | & 6 \end{bmatrix}$
(a) $R_2 = r_2 + (-2)r_1$
(b) $R_3 = r_3 + 3r_1$
(c) $R_3 = r_3 + 8r_2$

20. $\begin{bmatrix} 1 & -3 & -1 & | & 2 \\ 2 & -5 & 2 & | & 6 \\ -3 & -6 & 4 & | & 6 \end{bmatrix}$
(a) $R_2 = r_2 + (-2)r_1$
(b) $R_3 = r_3 + 3r_1$
(c) $R_3 = r_3 + 15r_2$

In Problems 21–32, the row-echelon form of a system of linear equations is given. Write the system of equations corresponding to the given matrix. Use x, y; or x, y, z; or x_1, x_2, x_3, x_4 as variables. Determine whether the system is consistent or inconsistent. If it is consistent, give the solution.

21. $\begin{bmatrix} 1 & 2 & | & 5 \\ 0 & 1 & | & -1 \end{bmatrix}$

22. $\begin{bmatrix} 1 & -3 & | & -4 \\ 0 & 1 & | & 0 \end{bmatrix}$

23. $\begin{bmatrix} 1 & 2 & 3 & | & 1 \\ 0 & 1 & 4 & | & 2 \\ 0 & 0 & 0 & | & 3 \end{bmatrix}$

24. $\begin{bmatrix} 1 & 2 & -1 & | & 0 \\ 0 & 1 & -1 & | & 1 \\ 0 & 0 & 0 & | & 2 \end{bmatrix}$

25. $\begin{bmatrix} 1 & 0 & 2 & | & -1 \\ 0 & 1 & -4 & | & -2 \\ 0 & 0 & 0 & | & 0 \end{bmatrix}$

26. $\begin{bmatrix} 1 & 0 & 4 & | & 4 \\ 0 & 1 & 3 & | & 2 \\ 0 & 0 & 0 & | & 0 \end{bmatrix}$

27. $\begin{bmatrix} 1 & 2 & -1 & 1 & | & 1 \\ 0 & 1 & 4 & 1 & | & 2 \\ 0 & 0 & 1 & 2 & | & 3 \end{bmatrix}$

28. $\begin{bmatrix} 1 & 2 & 4 & 0 & | & 1 \\ 0 & 1 & -1 & 2 & | & 2 \\ 0 & 0 & 1 & 3 & | & 0 \end{bmatrix}$

29. $\begin{bmatrix} 1 & 2 & 0 & 4 & | & 2 \\ 0 & 1 & 1 & 3 & | & 3 \\ 0 & 0 & 0 & 0 & | & 0 \end{bmatrix}$

30. $\begin{bmatrix} 1 & 0 & 3 & 0 & | & 1 \\ 0 & 1 & 4 & 3 & | & 2 \\ 0 & 0 & 1 & 2 & | & 3 \end{bmatrix}$

31. $\begin{bmatrix} 1 & -2 & 0 & 1 & | & -2 \\ 0 & 1 & -3 & 2 & | & 2 \\ 0 & 0 & 1 & -1 & | & 0 \\ 0 & 0 & 0 & 0 & | & 0 \end{bmatrix}$

32. $\begin{bmatrix} 1 & 3 & 0 & 4 & | & 1 \\ 0 & 1 & 2 & -1 & | & 2 \\ 0 & 0 & 1 & 2 & | & 3 \\ 0 & 0 & 0 & 1 & | & 0 \end{bmatrix}$

In Problems 33–56 solve each system of equations using matrices.

33. $\begin{cases} x + y = 6 \\ 2x - y = 0 \end{cases}$

34. $\begin{cases} x - y = 2 \\ 2x + y = 1 \end{cases}$

35. $\begin{cases} 2x + y = 5 \\ x - y = 1 \end{cases}$

36. $\begin{cases} 3x + 2y = 7 \\ x + y = 3 \end{cases}$

37. $\begin{cases} 2x + 3y = 7 \\ 3x - y = 5 \end{cases}$

38. $\begin{cases} 2x - 3y = 5 \\ 3x + y = 2 \end{cases}$

39. $\begin{cases} 5x - 7y = 31 \\ 3x + 2y = 0 \end{cases}$

40. $\begin{cases} 2x + 8y = 17 \\ 3x - y = 1 \end{cases}$

41. $\begin{cases} 2x - 3y = 0 \\ 4x + 9y = 5 \end{cases}$

42. $\begin{cases} 3x - 4y = 3 \\ 6x + 2y = 1 \end{cases}$

43. $\begin{cases} 4x - 3y = 4 \\ 2x + 6y = 7 \end{cases}$

44. $\begin{cases} 3x - 5y = 3 \\ 6x + 10y = 10 \end{cases}$

45. $\begin{cases} \frac{1}{2}x + \frac{1}{3}y = 2 \\ x + y = 5 \end{cases}$

46. $\begin{cases} x - \frac{1}{4}y = 0 \\ \frac{1}{2}x + \frac{1}{2}y = \frac{5}{2} \end{cases}$

47. $\begin{cases} x + y = 1 \\ 3x - 2y = \frac{4}{3} \end{cases}$

48. $\begin{cases} 4x - y = \frac{11}{4} \\ 3x + y = \frac{5}{2} \end{cases}$

49. $\begin{cases} 2x + y + z = 6 \\ x - y - z = -3 \\ 3x + y + 2z = 7 \end{cases}$

50. $\begin{cases} x + y + z = 5 \\ 2x - y + z = 2 \\ x + 2y - z = 3 \end{cases}$

51. $\begin{cases} x + y - z = -2 \\ 3x + y + z = 0 \\ 2x - y + 2z = 1 \end{cases}$

52. $\begin{cases} 2x - y - z = -5 \\ x + y + z = 2 \\ x + 2y + 2z = 5 \end{cases}$

53. $\begin{cases} 2x + y - z = 2 \\ x + 3y + 2z = 1 \\ x + y + z = 2 \end{cases}$

54. $\begin{cases} 2x + 2y + z = 6 \\ x - y - z = -2 \\ x - 2y - 2z = -5 \end{cases}$

55. $\begin{cases} x + y - z = 0 \\ 2x + 4y - 4z = -1 \\ 2x + y + z = 2 \end{cases}$

56. $\begin{cases} x + y - z = 0 \\ 4x + 2y - 4z = 0 \\ x + 2y + z = 0 \end{cases}$

57. $\begin{cases} 3x + y - z = \frac{2}{3} \\ 2x - y + z = 1 \\ 4x + 2y = \frac{8}{3} \end{cases}$

58. $\begin{cases} x + y = 1 \\ 2x - y + z = 1 \\ x + 2y + z = \frac{8}{3} \end{cases}$

In Problems 59–70 discuss each system of equations. Determine whether the system has a unique solution, no solution, or infinitely many solutions. Use matrix techniques.

59. $\begin{cases} x - y = 5 \\ 2x - 2y = 6 \end{cases}$

60. $\begin{cases} 4x + y = 5 \\ 8x + 2y = 10 \end{cases}$

61. $\begin{cases} 2x - 3y = 6 \\ 4x - 6y = 12 \end{cases}$

62. $\begin{cases} 2x - 3y = 6 \\ 4x - 6y = 8 \end{cases}$

63. $\begin{cases} 5x - 6y = 1 \\ -10x + 12y = 0 \end{cases}$

64. $\begin{cases} 3x + 4y = 7 \\ x - y = 2 \end{cases}$

65. $\begin{cases} 2x + 3y = 5 \\ 4x + 4y = 8 \end{cases}$

66. $\begin{cases} 2x - y = 0 \\ 4x - 2y = 0 \end{cases}$

67. $\begin{cases} x + 2y = 6 \\ 2x + y = 6 \end{cases}$

68. $\begin{cases} 2x - y = 3 \\ x - 2y = -1 \end{cases}$

69. $\begin{cases} x = y + 3 \\ -2x + 2y = 4 \end{cases}$

70. $\begin{cases} y = 2x + 3 \\ 2x - y + 3 = 0 \end{cases}$

71. Mixture   A store sells cashews for $5 per pound and peanuts for $1.50 per pound. The manager decides to mix 30 pounds of peanuts with some cashews and sell the mixture for $3 per pound. How many pounds of cashews should be mixed with the peanuts so that the mixture will produce the same revenue as would selling the nuts separately?

72. Mixture   A store sells almonds for $6 per pound, cashews for $5 per pound, and peanuts for $2 per pound. One week the manager decides to prepare 100 16-ounce packages of nuts by mixing 40 pounds of peanuts with some almonds and cashews. Each package will be sold for $4. How many pounds of almonds and cashews should be mixed with the peanuts so that the mixture will produce the same revenue as selling the nuts separately?

73. Laboratory Work Stations   A chemistry laboratory can be used by 38 students at one time. The laboratory has 16 work stations, some set up for 2 students each and the others set up for 3 students each. How many are there of each kind of work station?

74. Cost of Fast Food   One group of people purchased 10 hot dogs and 5 soft drinks at a cost of $12.50. A second group bought 7 hot dogs and 4 soft drinks at a cost of $9. What is the cost of a single hot dog? A single soft drink?

75. Refunding a Purchase   The grocery store we use does not mark prices on its goods. My wife went to this store, bought three 1-pound packages of bacon and two cartons of eggs, and paid a total of $7.45. Not knowing she went to the store, I also went to the same store, purchased two 1-pound packages of bacon and three cartons of eggs, and paid a total of $6.45. Now we want to return two 1-pound packages of bacon and two cartons of eggs. How much will be refunded?

76. Coin Collections   A coin collection consists of 37 coins—nickels, dimes, and quarters. If the collection has a face value of $3.25 and there are 5 more dimes than there are nickels, how many of each coin are in the collection?

77. Theater Seating   A Broadway theater has 500 seats, divided into orchestra, main, and balcony seating. Orchestra seats sell for $75, main seats for $50, and balcony seats for $35. If all the seats are sold, the gross revenue to the theater is $23,000. If all the main and balcony seats are sold, but only half the orchestra seats are sold, the gross revenue is $21,500. How many are there of each kind of seat?

78. Investment   An amount of $5000 is put into three investments at rates of 6%, 7%, and 8% per annum, respectively. The total annual income is $358. The income from the first two investments is $70 more than the income from the third investment. Find the amount of each investment.

79. Investment   An amount of $6500 is placed in three investments at rates of 6%, 8%, and 9% per annum, respectively. The total annual income is $480. If the income from the third investment is $60 more than the income from the second investment, find the amount of each investment.

80. Mixture   Sally's Girl Scout troop is selling cookies for the Christmas season. There are three different kinds of cookies in three different containers: *bags* that hold 1 dozen chocolate chip and 1 dozen oatmeal; *gift boxes* that hold 2 dozen chocolate chip, 1 dozen mints, and 1 dozen oatmeal; and *cookie tins* that hold 3 dozen mints and 2 dozen chocolate chip. Sally's mother is having a Christmas party and wants 6 dozen oatmeal cookies, 10 dozen mints, and 14 dozen chocolate chip cookies. How can Sally fill her mother's order?

81. Production   A citrus company completes the preparation of its products by cleaning, filling, and labeling bottles. Each case of orange juice requires 10 minutes in the cleaning machine, 4 minutes in the filling machine, and 2 minutes in the labeling machine. For each case of tomato juice, the times are 12 minutes of cleaning, 4 minutes for filling, and 1 minute to label. Pineapple juice requires 9 minutes of cleaning, 6 minutes of filling, and 1 minute of labeling per case. If the company runs the cleaning machine for 398 minutes, the filling machine for 164 minutes, and the labeling machine for 58 minutes, how many cases of each type of juice are prepared?

82. Production   The manufacture of an automobile requires painting, drying, and polishing. The Rome Motor Company produces three types of cars: the Centurion, the Tribune, and the Senator. Each Centurion requires 8 hours for painting, 2 hours for drying, and 1 hour for polishing. A Tribune needs 10 hours for painting, 3 hours for drying, and 2 hours for polishing. It takes 16 hours of painting, 5 hours of drying, and 3 hours of polishing to prepare a Senator. If the company uses 240 hours for painting, 69 hours for drying, and 41 hours for polishing in a given month, how many of each type of car are produced?

83. **Inventory Control**   An art teacher finds that colored paper can be bought in three different packages. The first package has 20 sheets of white paper, 15 sheets of blue paper, and 1 sheet of red paper. The second package has 3 sheets of blue paper and 1 sheet of red paper. The last package has 40 sheets of white paper and 30 sheets of blue paper. Suppose he needs 200 sheets of white paper, 180 sheets of blue paper, and 12 sheets of red paper. How many of each type of package should he order?

84. **Inventory Control**   An interior decorator has ordered 12 cans of sunset paint, 35 cans of brown, and 18 cans of fuchsia. The paint store has special pair packs, containing 1 can each of sunset and fuchsia; darkening packs containing 2 cans of sunset, 5 cans of brown, and 2 cans of fuchsia; and economy packs, containing 3 cans of sunset, 15 cans of brown, and 6 cans of fuchsia. How many of each type of pack should the paint store send to the interior decorator?

85. **Diet Preparation**   A hospital dietician is planning a meal consisting of three foods whose ingredients are summarized as follows:

	Units of		
	Food I	**Food II**	**Food III**
Units of Protein	10	5	15
Units of Carbohydrates	3	6	3
Units of Iron	4	4	6

Determine the number of units of each food needed to create a meal containing 100 units of protein, 50 units of carbohydrates, and 50 units of iron.

86. **Production**   A luggage manufacturer produces three types of luggage: economy, standard, and deluxe. The company produces 1000 pieces of luggage at a cost of $20, $25, and $30 for the economy, standard, and deluxe luggage, respectively. The manufacturer has a budget of $20,700. Each economy luggage requires 6 hours of labor, each standard luggage requires 10 hours of labor, and each deluxe model requires 20 hours of labor. The manufacturer has a maximum of 6800 hours of labor available. If the manufacturer sells all the luggage, consumes the entire budget, and uses all the available labor, how many of each type of luggage should be produced?

87. **Packaging**   A recreation center wants to purchase compact discs (CD'S) to be used in the center. There is no requirement as to the artists. The only requirement is that they purchase 40 rock CD's, 32 western CD's, and 14 blues CD's. There are three different shipping packages offered by the company. They are an *assorted* carton, containing 2 rock CD's, 4 western CD's, and 1 blues CD; a *mixed* carton containing 4 rock and 2 western CD's; and a *single* carton containing 2 blues CD's. What combination of these packages is needed to fill the center's order?

88. **Mixture**   Suppose that a store has three sizes of cans of nuts. The *large* size contains 2 pounds of peanuts and 1 pound of cashews. The *mammoth* size contains 1 pound of walnuts, 6 pounds of peanuts, and 2 pounds of cashews. The *giant* size contains 1 pound of walnuts, 4 pounds of peanuts, and 2 pounds of cashews. Suppose that the store receives an order for 5 pounds of walnuts, 26 pounds of peanuts, and 12 pounds of cashews. How can it fill this order with the given sizes of cans?

89. **Mixture**   Suppose that the store in Problem 88 receives a new order for 6 pounds of walnuts, 34 pounds of peanuts, and 15 pounds of cashews. How can this order be filled with the given cans?

Technology Exercises

Some graphing calculators are able to perform elementary row operations on matrices stored in the calculator memory. Check your user's manual to see how to perform these functions with your graphing calculator.

In Problems 1–4 use the elementary row operations available on your graphing calculator to perform the given operations on the matrix

$$A = \begin{bmatrix} 1 & 0 & 4 & | & 0 \\ -2 & 1 & -4 & | & 1 \\ 3 & -1 & 0 & | & 1 \end{bmatrix}$$

1. $R_1 = -2r_1$.

2. $R_1 = r_1 + (-2)r_2$

3. $R_2 = r_2 + 3r_1$

4. $R_3 = r_3 + (-2)r_1$

Some graphing calculators are capable of finding the row-echelon form (ref) and the reduced row-echelon form (rref) of a matrix. Find the ref and rref of the augmented matrix of each of the following systems of equations. Express your answer both in decimal form and in fraction form.

5. $\begin{cases} 2x - 2y + z = 2 \\ x - \frac{1}{2}y + 2z = 1 \\ 2x + \frac{1}{3}y - z = 0 \end{cases}$

6. $\begin{cases} x + y = -1 \\ x - z = 0 \\ y - z = 1 \end{cases}$

7. $\begin{cases} x + y + z = 4 \\ x - y - z = 0 \\ y - z = -4 \end{cases}$

8. $\begin{cases} 2x + y + z = 6 \\ x - y - z = -3 \\ 3x + y + 2z = 7 \end{cases}$

9. $\begin{cases} x_1 + x_2 + x_3 + x_4 = 20 \\ x_2 + x_3 + x_4 = 0 \\ x_3 + x_4 = 13 \\ x_2 - 2x_4 = -5 \end{cases}$

10. $\begin{cases} x_1 - 2x_2 + 3x_3 - 4x_4 = 40 \\ 4x_2 + 6x_4 = -10 \\ x_3 - x_4 = 12 \\ x_2 + 2x_4 = -10 \end{cases}$

2.3 SYSTEMS OF m LINEAR EQUATIONS CONTAINING n VARIABLES

We learned in the previous two sections that a system of linear equations will have either one solution, no solution, or infinitely many solutions. As it turns out, no matter how many equations are in a system of linear equations and no matter how many variables a system has, only these three possibilities can arise.

Thus, for example, the system of three linear equations containing four variables

$$x_1 + 3x_2 + 5x_3 + x_4 = 2$$
$$2x_1 + 3x_2 + 4x_3 + 2x_4 = 1$$
$$x_1 + 2x_2 + 3x_3 + x_4 = 1$$

will have either no solution, one solution, or infinitely many solutions.

System of m Equations Containing n Variables

A *system of m linear equations containing n variables* $x_1, x_2, \ldots, x_n$ is of the form

$$\begin{cases} a_{11}x_1 + a_{12}x_2 + \cdots + a_{1n}x_n = b_1 \\ a_{21}x_1 + a_{22}x_2 + \cdots + a_{2n}x_n = b_2 \\ a_{31}x_1 + a_{32}x_2 + \cdots + a_{3n}x_n = b_3 \\ \quad \cdot \qquad\quad \cdot \qquad\qquad\quad \cdot \qquad\quad \cdot \\ \quad \cdot \qquad\quad \cdot \qquad\qquad\quad \cdot \qquad\quad \cdot \\ \quad \cdot \qquad\quad \cdot \qquad\qquad\quad \cdot \qquad\quad \cdot \\ a_{i1}x_1 + a_{i2}x_2 + \cdots + a_{in}x_n = b_i \\ \quad \cdot \qquad\quad \cdot \qquad\qquad\quad \cdot \qquad\quad \cdot \\ \quad \cdot \qquad\quad \cdot \qquad\qquad\quad \cdot \qquad\quad \cdot \\ \quad \cdot \qquad\quad \cdot \qquad\qquad\quad \cdot \qquad\quad \cdot \\ a_{m1}x_1 + a_{m2}x_2 + \cdots + a_{mn}x_n = b_m \end{cases}$$

where a_{ij} and b_i are real numbers, $i = 1, 2, \ldots, m, j = 1, 2, \ldots, n$.

A **solution** of a system of m equations containing n variables $x_1, x_2, \ldots, x_n$ is any ordered set $(x_1, x_2, \ldots, x_n)$ of real numbers for which *each* of the m equations of the system is satisfied.

Reduced Row-Echelon Form

A system of m linear equations containing n variables will have either no solution, one solution, or infinitely many solutions. We can determine which of these possibilities occurs and, if solutions exist, find them by performing row operations on the augmented matrix of the system until we arrive at the reduced row-echelon form of the augmented matrix.

Let's review the conditions required for the reduced row-echelon form:

Conditions for the Reduced Row-Echelon Form of a Matrix

1. The first nonzero entry in each row is 1 and it has 0s above it and below it.

2. The leftmost 1 in any row is to the right of the leftmost 1 in the row above.

3. Any rows that contain all 0's to the left of the vertical bar appear at the bottom.

The next two examples will help you recognize when an augmented matrix is in reduced row-echelon form.

EXAMPLE 1   Examples of Matrices that Are in Reduced Row-Echelon Form

(a) $\left[\begin{array}{cc|c} 1 & 0 & 2 \\ 0 & 1 & 3 \end{array}\right]$ (b) $\left[\begin{array}{ccc|c} 1 & 0 & -3 & 4 \\ 0 & 1 & -2 & 2 \\ 0 & 0 & 0 & 0 \end{array}\right]$

(c) $\left[\begin{array}{ccc|c} 1 & -2 & 0 & 1 \\ 0 & 0 & 1 & 3 \\ 0 & 0 & 0 & 0 \end{array}\right]$

■

EXAMPLE 2   Examples of Matrices that Are Not in Reduced Row-Echelon Form

(a) $\left[\begin{array}{cc|c} 1 & 0 & 0 \\ 0 & 0 & 0 \\ 0 & 1 & 0 \end{array}\right]$ The second row contains all 0s and the third does not—this violates the rule that states that any rows with all 0s are at the bottom.

(b) $\left[\begin{array}{ccc|c} 1 & 0 & 2 & 4 \\ 0 & 2 & 4 & 3 \\ 0 & 0 & 0 & 1 \end{array}\right]$ The first nonzero entry in row 2 is not a 1.

$$(c) \begin{bmatrix} 1 & 0 & 0 & | & 0 \\ 0 & 0 & 1 & | & 1 \\ 0 & 1 & 0 & | & 0 \end{bmatrix}$$

The leftmost 1 in the third row is not to the right of the leftmost 1 in the row above it.

 **Now Work Problem 1**

Now let's analyze the reduced row-echelon form of an augmented matrix.

EXAMPLE 3   Analyzing the Reduced Row-Echelon Form of an Augmented Matrix

The matrix

$$\begin{bmatrix} 1 & 0 & 0 & | & 3 \\ 0 & 1 & 0 & | & 8 \\ 0 & 0 & 1 & | & -4 \end{bmatrix}$$

is the reduced row-echelon form of the augmented matrix of a system of three linear equations containing three variables. If the variables are x, y, z, this matrix represents the system of equations

$$\begin{cases} x = 3 \\ y = 8 \\ z = -4 \end{cases}$$

Thus the system has the one solution: $x = 3$, $y = 8$, $z = -4$.

EXAMPLE 4   Example of a Reduced Row-Echelon Form (No Solution)

The matrix

$$\begin{bmatrix} 1 & 0 & 3 & | & 0 \\ 0 & 1 & 2 & | & 0 \\ 0 & 0 & 0 & | & 1 \\ 0 & 0 & 0 & | & 0 \end{bmatrix}$$

is the reduced row-echelon form of the augmented matrix of a system of four equations containing three variables. If the variables are x, y, z, the equation represented by the third row is

$$0 \cdot x + 0 \cdot y + 0 \cdot z = 1 \qquad \text{or} \qquad 0 = 1$$

Since $0 = 1$ is a contradiction, we conclude the system has no solution.

EXAMPLE 5   Example of a Reduced Row-Echelon Form (Infinitely Many Solutions)

The matrix

$$\begin{bmatrix} 1 & 0 & 0 & 2 & | & 5 \\ 0 & 1 & 0 & 1 & | & 2 \\ 0 & 0 & 1 & 3 & | & 4 \end{bmatrix}$$

is the reduced row-echelon form of a system of three equations containing four varia-bles. If x_1, x_2, x_3, x_4 are the variables, the system of equations is

$$\begin{cases} x_1 + 2x_4 = 5 \\ x_2 + x_4 = 2 \\ x_3 + 3x_4 = 4 \end{cases} \quad \text{or} \quad \begin{cases} x_1 = -2x_4 + 5 \\ x_2 = -x_4 + 2 \\ x_3 = -3x_4 + 4 \end{cases}$$

Thus the system has infinitely many solutions. In this form, the variable x_4 is the parameter. We assign values to the parameter x_4 from which the variables x_1, x_2, x_3 can be calculated. Some of the possibilities are

If $x_4 = 0$, then $x_1 = 5$, $x_2 = 2$, $x_3 = 4$.
If $x_4 = 1$, then $x_1 = 3$, $x_2 = 1$, $x_3 = 1$.
If $x_4 = 2$, then $x_1 = 1$, $x_2 = 0$, $x_3 = -2$.

And so on.

 **Now Work Problem 17**

Let's review the procedure for obtaining the reduced row-echelon form of a matrix. Then we will use this procedure to solve systems of linear equations.

EXAMPLE 6   Finding the Reduced Row-Echelon Form of a Matrix

Find the reduced row-echelon form of

$$A = \left[\begin{array}{cc|c} 1 & -1 & 2 \\ 2 & -3 & 2 \\ 3 & -5 & 2 \end{array} \right]$$

SOLUTION   The entry in row 1, column 1 is 1. Thus we proceed to obtain a matrix in which all the remaining entries in column 1 are 0s. We can obtain such a matrix by performing the row operations

$$R_2 = r_2 + (-2)r_1$$
$$R_3 = r_3 + (-3)r_1$$

The new matrix is

$$\left[\begin{array}{cc|c} 1 & -1 & 2 \\ 0 & -1 & -2 \\ 0 & -2 & -4 \end{array} \right]$$

We want the entry in row 2, column 2 (now -1), to be 1. By multiplying row 2 by -1, we obtain

$$\begin{bmatrix} 1 & -1 & | & 2 \\ 0 & 1 & | & 2 \\ 0 & -2 & | & -4 \end{bmatrix}$$

Now we want the entry in row 1, column 2 and in row 3, column 2 to be 0. This can be accomplished by applying the row operations

$$R_1 = r_1 + r_2$$
$$R_3 = r_3 + 2r_2$$

The new matrix is

$$\begin{bmatrix} 1 & 0 & | & 4 \\ 0 & 1 & | & 2 \\ 0 & 0 & | & 0 \end{bmatrix}$$

This is the reduced row-echelon form of A.

EXAMPLE 7   Solving a System of Three Linear Equations Containing Two Variables

Solve: $\begin{cases} x - y = 2 \\ 2x - 3y = 2 \\ 3x - 5y = 2 \end{cases}$

SOLUTION   The augmented matrix of this system is

$$\begin{bmatrix} 1 & -1 & | & 2 \\ 2 & -3 & | & 2 \\ 3 & -5 & | & 2 \end{bmatrix}$$

Using the solution to Example 6, the reduced row-echelon form of this augmented matrix is

$$\begin{bmatrix} 1 & 0 & | & 4 \\ 0 & 1 & | & 2 \\ 0 & 0 & | & 0 \end{bmatrix}$$

We conclude that the system has the solution $x = 4$, $y = 2$.

EXAMPLE 8   Solving a System of Four Linear Equations Containing Three Variables

Solve: $\begin{cases} x - y + 2z = 2 \\ 2x - 3y + 2z = 1 \\ 3x - 5y + 2z = -3 \\ -4x + 12y + 8z = 10 \end{cases}$

SOLUTION   We need to find the reduced row-echelon form of the augmented matrix of this system, namely,

$$\begin{bmatrix} 1 & -1 & 2 & | & 2 \\ 2 & -3 & 2 & | & 1 \\ 3 & -5 & 2 & | & -3 \\ -4 & 12 & 8 & | & 10 \end{bmatrix}$$

The entry 1 is already present in row 1, column 1. To obtain 0s elsewhere in column 1, we use the row operations

$$R_2 = r_2 + (-2)r_1 \qquad R_3 = r_3 + (-3)r_1 \qquad R_4 = r_4 + 4r_1$$

The new matrix is

$$\begin{bmatrix} 1 & -1 & 2 & | & 2 \\ 0 & -1 & -2 & | & -3 \\ 0 & -2 & -4 & | & -9 \\ 0 & 8 & 16 & | & 18 \end{bmatrix}$$

To obtain the entry 1 in row 2, column 2, we use $R_2 = -r_2$, obtaining

$$\begin{bmatrix} 1 & -1 & 2 & | & 2 \\ 0 & 1 & 2 & | & 3 \\ 0 & -2 & -4 & | & -9 \\ 0 & 8 & 16 & | & 18 \end{bmatrix}$$

To obtain 0s elsewhere in column 2, we use

$$R_1 = r_1 + r_2 \qquad R_3 = r_3 + 2r_2 \qquad R_4 = r_4 + (-8)r_2$$

The new matrix is

$$\begin{bmatrix} 1 & 0 & 4 & | & 5 \\ 0 & 1 & 2 & | & 3 \\ 0 & 0 & 0 & | & -3 \\ 0 & 0 & 0 & | & -6 \end{bmatrix}$$

We can stop here even though the matrix is not in reduced row-echelon form. Because the third row yields the equation

$$0 \cdot x + 0 \cdot y + 0 \cdot z = -3$$

we conclude the system has no solution.

Now Work Problem 31

Infinite Number of Solutions

We have seen several examples of systems of linear equations that have an infinite number of solutions. Let's look at a few more examples.

EXAMPLE 9   **Solving a System of Two Linear Equations Containing Three Variables**

Solve: $\begin{cases} x + y + z = 7 \\ x - y - 3z = 1 \end{cases}$

SOLUTION The augmented matrix of the system is

$$\begin{bmatrix} 1 & 1 & 1 & | & 7 \\ 1 & -1 & -3 & | & 1 \end{bmatrix}$$

The reduced row-echelon form (as you should verify) is

$$\begin{bmatrix} 1 & 0 & -1 & | & 4 \\ 0 & 1 & 2 & | & 3 \end{bmatrix}$$

The system of equations represented by this matrix is

$$\begin{cases} x - z = 4 \\ y + 2z = 3 \end{cases} \quad \text{or} \quad \begin{cases} x = z + 4 \\ y = -2z + 3 \end{cases} \tag{1}$$

Thus the system has infinitely many solutions. In the form (1), the variable z is the parameter. We can assign any value to z and use it to compute values of x and y.

∎

We check the solution to Example 9 as follows:

$$x + y + z = (z + 4) + (-2z + 3) + z = 7 + z - 2z + z = 7$$
$$x - y - 3z = (z + 4) - (-2z + 3) - 3z = 1 + z + 2z - 3z = 1$$

The solution is verified.

The next example illustrates a system having an infinite number of solutions with two parameters.

EXAMPLE 10 Solving a System of Three Linear Equations Containing Four Variables

Solve: $\begin{cases} x_1 + x_2 + 2x_3 + 2x_4 = 2 \\ x_1 + x_3 + x_4 = 0 \\ x_2 + x_3 + x_4 = 2 \end{cases}$

SOLUTION The augmented matrix of the system is

$$\begin{bmatrix} 1 & 1 & 2 & 2 & | & 2 \\ 1 & 0 & 1 & 1 & | & 0 \\ 0 & 1 & 1 & 1 & | & 2 \end{bmatrix}$$

The reduced row-echelon form (as you should verify) is

$$\begin{bmatrix} 1 & 0 & 1 & 1 & | & 0 \\ 0 & 1 & 1 & 1 & | & 2 \\ 0 & 0 & 0 & 0 & | & 0 \end{bmatrix}$$

The equations represented by this system are

$$\begin{cases} x_1 + x_3 + x_4 = 0 \\ x_2 + x_3 + x_4 = 2 \end{cases}$$

We can rewrite this system in the form

$$\begin{cases} x_1 = -x_3 - x_4 \\ x_2 = -x_3 - x_4 + 2 \end{cases} \tag{2}$$

The system has infinitely many solutions. In this form, (2), the system has two parameters x_3 and x_4. Solutions are obtained by assigning the two parameters x_3 and x_4 arbitrary values. Some choices are shown in the table below.

x_3	x_4	x_1	x_2	Solution (x_1, x_2, x_3, x_4)
0	0	0	2	$(0, 2, 0, 0)$
1	0	-1	1	$(-1, 1, 1, 0)$
0	2	-2	0	$(-2, 0, 0, 2)$

The variables used as parameters are not unique. We could have chosen x_1 and x_4 as parameters by rewriting Equations (2) in the following manner.

From the first equation

$$x_3 = -x_1 - x_4$$

We can replace the parameter x_3 in the second equation by the above to produce

$$x_2 = 2 - x_3 - x_4 = 2 + x_1 + x_4 - x_4 = 2 + x_1$$

We then obtain the system

$$\begin{cases} x_2 = x_1 + 2 \\ x_3 = -x_1 - x_4 \end{cases}$$

showing the solution with x_1 and x_4 as parameters.

Now Work Problem 35

SUMMARY

To solve a system of *m* linear equations containing *n* variables:

> **Steps for Solving a System of *m* Linear Equations Containing *n* Variables**
>
> **Step 1** Write the augmented matrix.
>
> **Step 2** Find the reduced row-echelon form of the augmented matrix.
>
> **Step 3** Analyze this matrix to determine if the system has no solution, one solution, or infinitely many solutions.

Application

EXAMPLE 11   Mixing Chemicals

In a chemistry laboratory one solution contains 10% hydrochloric acid (HCl), a second solution contains 20% HCl, and a third contains 40% HCl. How many liters of each should be mixed to obtain 100 liters of 25% HCl?

SOLUTION Let x, y, and z represent the number of liters of 10%, 20%, and 40% solutions of HCl, respectively. Since we want 100 liters in all and the amount of HCl obtained from each solution must sum to 25% of 100, or 25 liters, we must have

$$\begin{cases} x + y + z = 100 \\ 0.1x + 0.2y + 0.4z = 25 \end{cases}$$

Thus our problem is to solve a system of two equations containing three variables. By matrix techniques we obtain the solution

$$\begin{cases} x = 2z - 50 \\ y = -3z + 150 \end{cases} \tag{3}$$

where z is the parameter. Now the practical considerations of this problem lead us to the conditions that $x \geq 0$, $y \geq 0$, $z \geq 0$. From (3) we see that we must have $z \geq 25$ and $z \leq 50$, since otherwise $x < 0$ or $y < 0$. Some possible solutions are listed in the table. The final determination by the chemistry laboratory will more than likely be based on the amount and availability of one acid solution versus the others.

No. of Liters 10% Solution	0	10	12	16	20	25	26	30	36	38	46	50
No. of Liters 20% Solution	75	60	57	51	45	37.5	36	30	21	18	6	0
No. of Liters 40% Solution	25	30	31	33	35	37.5	38	40	43	44	48	50

 **Now Work Problem 53**

Matrices in Practice

Systems of linear equations arise in business, economics, sociology, chemistry—in fact, in any field that has a quantitative side to it. The systems encountered in practice are often quite large, with 100 equations containing 100 variables not unusual, making hand calculations with such systems out of the question. Thus computer routines are used to implement work such as row-reducing the augmented matrix. A very popular collection of such routines is the LINPACK package. In more advanced treatments it is shown that solving n equations containing n variables requires roughly n^3 multiplications and additions. Thus a 100 by 100 system will require $(100)^3 = 10^6$ arithmetic operations. But if we have access to a machine that can perform 100,000 operations per second, the task seems less formidable since our 100 by 100 system would be solved in 10 seconds. The availability of high-speed computing has greatly enhanced the applicability of matrices since they can now be used in large-scale problems. See the article by Kolata in *Science*.*

*Gina Kolata, "Solving Linear Systems Faster." *Science* (June 14, 1985).

There is an aspect of computer-performed matrix calculations that can at times be potentially troublesome in applications. Computers by their nature can perform arithmetic only on decimals that have finitely many nonzero terms. Decimals that are infinite in length are rounded. For example, $\frac{2}{3}$ has the nonterminating decimal expansion 0.6666 . . . . A machine would round this and store it as, say, 0.6666667 (the actual number of significant places would vary with the machine). This can have consequences for matrix calculations, as we show in the following example.

EXAMPLE 12   Examining Difficulties With Rounding

Suppose we wish to find the reduced row-echelon form of the matrix

$$A = \begin{bmatrix} 1 & \frac{1}{3} \\ 2 & \frac{2}{3} \end{bmatrix}$$

A direct hand calculation shows that the reduced row-echelon form is

$$\begin{bmatrix} 1 & \frac{1}{3} \\ 0 & 0 \end{bmatrix}$$

Now assume we did this on a computer that rounded and stored, say, two significant digits.

Our matrix A would then be represented in the machine as

$$A = \begin{bmatrix} 1 & 0.33 \\ 2 & 0.67 \end{bmatrix}$$

If a program were now called upon to row-reduce A, the following steps would result:

$$R_2 = r_2 + (-2)r_1 \qquad \begin{bmatrix} 1 & 0.33 \\ 0 & 0.01 \end{bmatrix}$$

$$R_2 = 100 r_2 \qquad \begin{bmatrix} 1 & 0.33 \\ 0 & 1 \end{bmatrix}$$

$$R_1 = r_1 + (-0.33)r_2 \qquad \begin{bmatrix} 1 & 0 \\ 0 & 1 \end{bmatrix}$$

Note that the end result of the computer calculation differs drastically from the actual reduced row-echelon form. The problem clearly lies in the rounded representation of the numbers $\frac{1}{3}$ and $\frac{2}{3}$. ■

Though the above example is a bit simplistic, errors in matrix calculations introduced by rounding or truncation can and do occur and can have disastrous consequences. Were the matrix in the example above the coefficient matrix of a system, then the computer program would conclude that the system has a unique solution, while in reality the system has either infinitely many solutions or no solution. The subject of error propagation when doing matrix arithmetic on a computer is of great importance to people who use matrices in practice, since they want to be assured that their results are meaningful. It is also a subject where much current work is being done by computer scientists and mathematicians.

EXERCISE 2.3   Answers to odd-numbered problems begin on page AN-8.

In Problems 1–12 tell whether the given matrix is in reduced row-echelon form.

1. $\begin{bmatrix} 1 & 2 & | & 3 \\ 0 & 0 & | & 0 \\ 0 & 0 & | & 1 \end{bmatrix}$ 2. $\begin{bmatrix} 1 & 2 & | & 3 \\ 0 & 0 & | & 0 \\ 0 & 0 & | & 0 \end{bmatrix}$ 3. $\begin{bmatrix} 1 & 1 & | & 0 \\ 0 & 1 & | & 0 \end{bmatrix}$

4. $\begin{bmatrix} 1 & 0 & | & 3 \\ 0 & 1 & | & 0 \end{bmatrix}$ 5. $\begin{bmatrix} 0 & | & 1 \\ 1 & | & 0 \end{bmatrix}$ 6. $\begin{bmatrix} 0 & 1 & | & 0 \\ 0 & 0 & | & 1 \\ 0 & 0 & | & 0 \end{bmatrix}$

7. $\begin{bmatrix} 1 & 2 & | & 1 \\ 0 & 0 & | & 0 \end{bmatrix}$ 8. $\begin{bmatrix} 1 & 0 & | & 8 \\ 0 & 2 & | & 9 \end{bmatrix}$ 9. $\begin{bmatrix} 1 & 0 & 0 & 0 & 0 \\ 0 & 0 & 1 & 2 & 0 \\ 0 & 0 & 0 & 0 & 1 \\ 0 & 0 & 0 & 0 & 0 \end{bmatrix}$

10. $\begin{bmatrix} 1 & 1 & | & 0 \\ 0 & 0 & | & 2 \end{bmatrix}$ 11. $\begin{bmatrix} 1 & 0 & | & 1 \\ 0 & 1 & | & 2 \\ 0 & 0 & | & 0 \end{bmatrix}$ 12. $\begin{bmatrix} 1 & 0 & 2 \\ 0 & 1 & 2 \end{bmatrix}$

In Problems 13–28 the reduced row-echelon form of the augmented matrix of a system of linear equations is given. Tell whether the system has one solution, no solution, or infinitely many solutions.

13. $\begin{bmatrix} 1 & 1 & | & 1 \\ 0 & 0 & | & 0 \end{bmatrix}$ 14. $\begin{bmatrix} 1 & 0 & | & 0 \\ 0 & 0 & | & 1 \end{bmatrix}$ 15. $\begin{bmatrix} 1 & 0 & | & 4 \\ 0 & 1 & | & 5 \end{bmatrix}$

16. $\begin{bmatrix} 1 & 0 & 0 & | & 0 \\ 0 & 1 & 0 & | & 0 \\ 0 & 0 & 1 & | & 6 \end{bmatrix}$ 17. $\begin{bmatrix} 1 & 0 & -2 & | & 6 \\ 0 & 1 & 3 & | & 1 \end{bmatrix}$ 18. $\begin{bmatrix} 1 & 0 & 0 & | & 0 \\ 0 & 1 & 0 & | & 5 \\ 0 & 0 & 0 & | & 0 \end{bmatrix}$

19. $\begin{bmatrix} 1 & 2 & 0 & | & 1 \\ 0 & 0 & 1 & | & 2 \\ 0 & 0 & 0 & | & 0 \end{bmatrix}$ 20. $\begin{bmatrix} 1 & 2 & 0 & | & 0 \\ 0 & 0 & 1 & | & 0 \\ 0 & 0 & 0 & | & 1 \end{bmatrix}$ 21. $\begin{bmatrix} 1 & 0 & 0 & | & 1 \\ 0 & 1 & 0 & | & 2 \\ 0 & 0 & 0 & | & 0 \end{bmatrix}$

22. $\begin{bmatrix} 1 & 0 & 1 & -1 & | & 0 \\ 0 & 1 & 2 & 1 & | & 1 \\ 0 & 0 & 0 & 0 & | & 0 \end{bmatrix}$ 23. $\begin{bmatrix} 1 & 0 & 0 & | & -1 \\ 0 & 1 & 0 & | & 3 \\ 0 & 0 & 1 & | & 4 \\ 0 & 0 & 0 & | & 0 \end{bmatrix}$ 24. $\begin{bmatrix} 1 & 2 & 0 & 0 & | & -4 \\ 0 & 0 & 1 & 0 & | & -3 \\ 0 & 0 & 0 & 1 & | & 2 \\ 0 & 0 & 0 & 0 & | & 0 \end{bmatrix}$

25. $\begin{bmatrix} 1 & 0 & -1 & | & 1 \\ 0 & 1 & 2 & | & 1 \end{bmatrix}$ 26. $\begin{bmatrix} 1 & 0 & | & 1 \\ 0 & 1 & | & 1 \\ 0 & 0 & | & 0 \end{bmatrix}$ 27. $\begin{bmatrix} 1 & 0 & 0 & -1 & | & 0 \\ 0 & 1 & 2 & 3 & | & 0 \end{bmatrix}$

28. $\begin{bmatrix} 1 & 0 & 2 & 4 & | & -1 \\ 0 & 1 & 3 & 5 & | & -2 \end{bmatrix}$

In Problems 29–52 solve each system of equations by finding the reduced row-echelon form of the augmented matrix. If there is no solution, say the system is inconsistent.

29. $\begin{cases} x + y = 3 \\ 2x - y = 3 \end{cases}$ 30. $\begin{cases} x - y = 5 \\ 2x + 3y = 15 \end{cases}$ 31. $\begin{cases} 3x - 3y = 12 \\ 3x + 2y = -3 \\ 2x + y = 4 \end{cases}$

32. $\begin{cases} 6x + y = 8 \\ x - 3y = -5 \\ 2x + y = 2 \end{cases}$ 33. $\begin{cases} 2x - 4y = 8 \\ x - 2y = 4 \\ -x + 2y = -4 \end{cases}$ 34. $\begin{cases} 3x + y = 8 \\ 6x + 2y = 16 \\ -9x - 3y = -24 \end{cases}$

35. $\begin{cases} 2x + y + 3z = -1 \\ -x + y + 3z = 8 \\ 2x - 2y - 6z = -16 \end{cases}$

36. $\begin{cases} x + 2y + 3z = 5 \\ -2x + 6y + 4z = 0 \\ 2x + 4y + 6z = 10 \end{cases}$

37. $\begin{cases} x - y = 1 \\ y - z = 6 \\ x + z = -1 \end{cases}$

38. $\begin{cases} 2x - y + 3z = 0 \\ x + 2y - z = 5 \\ 2y + z = 1 \end{cases}$

39. $\begin{cases} x_1 + x_2 = 7 \\ x_2 - x_3 + x_4 = 5 \\ x_1 - x_2 + x_3 + x_4 = 6 \\ x_2 - x_4 = 10 \end{cases}$

40. $\begin{cases} x_1 + x_2 + x_3 + x_4 = 0 \\ 2x_1 - x_2 - x_3 + x_4 = 0 \\ x_1 - x_2 - x_3 + x_4 = 0 \\ x_1 + x_2 - x_3 - x_4 = 0 \end{cases}$

41. $\begin{cases} x_1 + 2x_2 + 3x_3 - x_4 = 0 \\ 3x_1 - x_4 = 4 \\ x_2 - x_3 - x_4 = 2 \end{cases}$

42. $\begin{cases} 2x - 3y + 4z = 7 \\ x - 2y + 3z = 2 \end{cases}$

43. $\begin{cases} x - y + z = 5 \\ 2x - 2y + 2z = 8 \end{cases}$

44. $\begin{cases} x + y + z = 3 \\ x - y + z = 7 \\ x - y - z = 1 \end{cases}$

45. $\begin{cases} 3x - y + 2z = 3 \\ 3x + 3y + z = 3 \\ 3x - 5y + 3z = 12 \end{cases}$

46. $\begin{cases} x + y - z = 12 \\ 3x - y = 1 \\ 2x - 3y + 4z = 3 \end{cases}$

47. $\begin{cases} x_1 + x_2 + x_3 + x_4 = 4 \\ 2x_1 - x_2 + x_3 = 0 \\ 3x_1 + 2x_2 + x_3 - x_4 = 6 \\ x_1 - 2x_2 - 2x_3 + 2x_4 = -1 \end{cases}$

48. $\begin{cases} x_1 + x_2 + x_3 + x_4 = 4 \\ -x_1 + 2x_2 + x_3 = 0 \\ 2x_1 + 3x_2 + x_3 - x_4 = 6 \\ -2x_1 + x_2 - 2x_3 + 2x_4 = -1 \end{cases}$

49. $\begin{cases} 2x - y - z = 0 \\ x - y - z = 1 \\ 3x - y - z = 2 \end{cases}$

50. $\begin{cases} x + y + z = 3 \\ 2x + y + z = 0 \\ 3x + y + z = 1 \end{cases}$

51. $\begin{cases} 2x - y + z = 6 \\ 3x - y + z = 6 \\ 4x - 2y + 2z = 12 \end{cases}$

52. $\begin{cases} x - y + z = 2 \\ 2x - 3y + z = 0 \\ 3x - 3y + 3z = 6 \end{cases}$

53. Mixing Chemicals A chemistry laboratory has available three kinds of hydrochloric acid (HCl): 10%, 30%, and 50% solutions. How many liters of each should be mixed to obtain 100 liters of 25% HCl? Provide a table showing at least six of the possible solutions.

54. Repeat Problem 53 if the mixture is to be 100 liters of 40% HCl.

55. Cost of Fast Food One group of customers bought 8 deluxe hamburgers, 6 orders of large fries, and 6 large colas for $26.10. A second group ordered 10 deluxe hamburgers, 6 large fries, and 8 large colas and paid $31.60. Is there sufficient information to determine the price of each food item? If not, construct a table showing the various possibilities. Assume the hamburgers cost between $1.75 and $2.25, the fries between $0.75 and $1.00, and the colas between $0.60 and $0.90.

56. Use the information given in Problem 55 and add a third group that purchased 3 deluxe hamburgers, 2 large fries, and 4 colas for $10.95. Is there now sufficient information to determine the price of each food item?

57. Investment Goals A retired couple has $25,000 available to invest. They require a return on their investment of $2000 per year. As their financial consultant, you recommend they invest some money in Treasury bills that yield 7%, some money in corporate bonds that yield 9%, and some in junk bonds that yield 11%. Prepare a table showing the various ways this couple can achieve their goal.

58. (a) Rework Problem 57 if the couple has only $20,000 to invest.
(b) Rework Problem 57 if the couple has $30,000 to invest.
(c) What general conclusions can you make regarding the amount to invest and the choices available to the couple?

59. (a) Rework Problem 57 if the couple requires a return of only $1500 per year.
(b) Rework Problem 57 if the couple requires a return of $2500 per year.
(c) What general conclusions can you make regarding the required return per year and the choices available to the couple?

60. Inventory Control Three species of bacteria will be kept in one test tube and will feed on three resources. Each member of the first species consumes three units of the first resource and one unit of the third. Each bacterium of the second type consumes one unit of the first resource and two units each of the second and third. Each bacterium of the third type consumes two units of the first resource and four each of the second and third. If the test tube is supplied daily with 12,000 units of the first resource, 12,000 units of the second resource, and 14,000 units of the third, how many of each species can coexist in equilibrium in the test tube so that all of the supplied resources are consumed?

Technology Exercises

Some graphing calculators are capable of finding the row-echelon form (ref) and the reduced row-echelon form (rref) of a matrix. Find the ref and rref of the augmented matrix of each of the following systems of equations. Express your answer both in decimal form and in fraction form.

1. $\begin{cases} 2x_1 + x_2 + x_3 + x_4 = 6 \\ x_1 - x_2 - 2x_4 = 2 \\ -x_1 + x_4 = 0 \\ x_2 - x_3 + 2x_4 = -1 \end{cases}$

2. $\begin{cases} x_1 + x_2 + x_3 + x_4 = 20 \\ -x_1 + 2x_2 + x_4 = 0 \\ 2x_1 + x_4 = 0 \\ x_2 + 2x_4 = -5 \end{cases}$

3. $\begin{cases} x_1 + x_2 - x_3 - x_4 = 6 \\ x_1 + x_2 = -4 \\ -2x_1 - x_2 - x_3 + x_4 = 0 \\ 2x_3 + x_4 = -1 \end{cases}$

4. $\begin{cases} x_1 + x_3 = 6 \\ x_2 + x_4 = 4 \\ -x_3 + x_4 = 0 \\ x_2 - x_3 = -2 \end{cases}$

5. $\begin{cases} x_1 - 2x_2 - 5x_3 = -25 \\ x_2 + x_3 = 15 \\ 2x_2 + 3x_3 + 4x_4 = 15 \\ 3x_2 - 2x_4 = -10 \end{cases}$

6. $\begin{cases} x_1 + x_2 - 5x_3 + x_5 = -25 \\ x_2 + x_3 - x_5 = 15 \\ 2x_2 + 3x_3 + 4x_4 = 15 \\ 3x_2 - 2x_4 - x_5 = -10 \\ x_1 + x_4 - 4x_5 = 18 \end{cases}$

2.4 MATRIX ALGEBRA

Matrices can be added, subtracted, and multiplied. They also possess many of the algebraic properties of real numbers. Matrix algebra is the study of these properties. Its importance lies in the fact that many situations in both pure and applied mathematics involve rectangular arrays of numbers. In fact, in many branches of business and the biological and social sciences, it is necessary to express and use data in a rectangular array. Let's look at an example.

EXAMPLE 1   Writing Data as a Matrix

Motors, Inc., produces three models of cars: a sedan, a hardtop, and a station wagon. If the company wishes to compare the units of raw material and the units of labor involved in 1 month's production of each of these models, the rectangular array displayed below may be used to present the data:

	Sedan Model	Hardtop Model	Station Wagon Model
Units of Material	23	16	10
Units of Labor	7	9	11

The same information may be written concisely as the matrix

$$\begin{bmatrix} 23 & 16 & 10 \\ 7 & 9 & 11 \end{bmatrix}$$

This matrix has 2 rows (the units) and 3 columns (the models). The first row represents units of material and the second row represents units of labor. The first, second, and third columns represent the sedan, hardtop, and station wagon models, respectively.

A **matrix** is defined as a rectangular array of the form:

$$
\begin{array}{c}
\\
\text{Row 1} \\
\text{Row 2} \\
\vdots \\
\text{Row } i \\
\vdots \\
\text{Row } m
\end{array}
\begin{array}{ccccccc}
\text{Column 1} & \text{Column 2} & & \text{Column } j & & \text{Column } n \\
\begin{bmatrix}
a_{11} & a_{12} & \cdots & a_{1j} & \cdots & a_{1n} \\
a_{21} & a_{22} & \cdots & a_{2j} & \cdots & a_{2n} \\
\vdots & \vdots & & \vdots & & \vdots \\
a_{i1} & a_{i2} & \cdots & a_{ij} & \cdots & a_{in} \\
\vdots & \vdots & & \vdots & & \vdots \\
a_{m1} & a_{m2} & \cdots & a_{mj} & \cdots & a_{mn}
\end{bmatrix}
\end{array}
$$

The symbols $a_{11}, a_{12}, \ldots$ of a matrix are referred to as the **entries** (or **elements**) of the matrix.

Each entry a_{ij} of the matrix A has two indices: the **row index,** i, and the **column index,** j. The symbols $a_{i1}, a_{i2}, \ldots, a_{in}$ represent the entries in the ith row, and the symbols $a_{1j}, a_{2j}, \ldots, a_{mj}$ represent the entries in the jth column. The matrix A, above, which has m rows and n columns, can be abbreviated by

$$A = [a_{ij}] \qquad i = 1, 2, \ldots, m; \qquad j = 1, 2, \ldots, n$$

Dimension

The **dimension of a matrix** A is determined by the number of rows and the number of columns of the matrix. If a matrix A has m rows and n columns, we denote the dimension of A by $m \times n$, read as "m by n."

For a 2×3 matrix, remember that the first number 2 denotes the number of rows and the second number 3 is the number of columns. A matrix with 3 rows and 2 columns is of dimension 3×2.

Square Matrix

If a matrix A has the same number of rows as it has columns, it is called a **square matrix.**

In a square matrix $A = [a_{ij}]$ the entries for which $i = j$, namely $a_{11}, a_{22}, a_{33}, a_{44}$, and so on, are the **diagonal entries** of A.

EXAMPLE 2   Writing Data as a Matrix

In the recent U.S. census the following figures were obtained with regard to the city of Glenwood. Each year 7% of city residents move to the suburbs and 1% of the people in the suburbs move to the city. This situation can be represented by the matrix

$$
P = \begin{array}{c} \\ \text{City} \\ \text{Suburbs} \end{array}
\begin{array}{c}
\begin{array}{cc} \text{City} & \text{Suburbs} \end{array} \\
\begin{bmatrix} 0.93 & 0.07 \\ 0.01 & 0.99 \end{bmatrix}
\end{array}
$$

Here, the entry in row 1, column 2—0.07—indicates that 7% of city residents move to the suburbs. The matrix P is a square matrix and its dimension is 2×2. The diagonal entries are 0.93 and 0.99.

◼

> ### Row and Column Matrix
>
> A **row matrix** is a matrix with 1 row of entries. A **column matrix** is a matrix with 1 column of entries. Row matrices and column matrices are also referred to as **row vectors** and **column vectors,** respectively.

EXAMPLE 3 **Examples of Row Matrices and Column Matrices**

The matrices

$$A = [23 \quad 16 \quad 10] \qquad B = [7 \quad 9] \qquad C = \begin{bmatrix} 23 \\ -1 \\ 7 \end{bmatrix}$$

$$D = \begin{bmatrix} 16 \\ 9 \end{bmatrix} \qquad E = [9]$$

have the following dimensions: A, 1×3; B, 1×2; C, 3×1; D, 2×1; E, 1×1. Here A and B are row matrices and C and D are column matrices. E is both a row matrix and a column matrix.

◼

The matrix $E = [9]$ in Example 3 is a 1×1 matrix and, as such, can be treated simply as a number.

Now Work Problem 3

Equality of Matrices

As with most mathematical quantities, we now want to define various relationships. We might ask, "When, if at all, are two matrices equal?"

Let's try to arrive at a sound definition for equality of matrices by requiring equal matrices to have certain desirable properties. First, it would seem necessary that two equal matrices have the same dimension—that is, that they both be $m \times n$ matrices. Next, it would seem necessary that their entries be identical numbers. With these two restrictions, we define equality of matrices.

> ### Equality of Matrices
>
> Two matrices A and B are **equal** if they are of the same dimension and if corresponding entries are equal. In this case we write $A = B$, read as "matrix A is equal to matrix B."

EXAMPLE 4   Determining Equality of Matrices

In order for the two matrices

$$\begin{bmatrix} p & q \\ 1 & 0 \end{bmatrix} \quad \text{and} \quad \begin{bmatrix} 2 & 4 \\ n & 0 \end{bmatrix}$$

to be equal, we must have $p = 2$, $q = 4$, and $n = 1$.

■

EXAMPLE 5   Determining Equality of Matrices

Let A and B be two matrices given by

$$A = \begin{bmatrix} x + y & 6 \\ 2x - 3 & 2 - y \end{bmatrix} \qquad B = \begin{bmatrix} 5 & 5x + 2 \\ y & x - y \end{bmatrix}$$

Determine if there are values of x and y so that A and B are equal.

SOLUTION   Both A and B are 2×2 matrices. Thus $A = B$ if

(a) $x + y = 5$ (b) $6 = 5x + 2$
(c) $2x - 3 = y$ (d) $2 - y = x - y$

Here we have four equations containing the two variables x and y. From Equation (d) we see that $x = 2$. Using this value in Equation (a), we obtain $y = 3$. But $x = 2$, $y = 3$ do not satisfy either (b) or (c). Hence, there are *no* values for x and y satisfying all four equations. This means A and B can never be equal.

 Now Work Problem 13

■

Addition of Matrices

Can two matrices be added? And, if so, what is the rule or law for addition of matrices?

Let's return to Example 1. In that example we recorded 1 month's production of Motors, Inc., by the matrix

$$A = \begin{bmatrix} 23 & 16 & 10 \\ 7 & 9 & 11 \end{bmatrix}$$

Suppose the next month's production is

$$B = \begin{bmatrix} 18 & 12 & 9 \\ 14 & 6 & 8 \end{bmatrix}$$

in which the pattern of recording units and models remains the same.

The total production for the 2 months can be displayed by the matrix

$$C = \begin{bmatrix} 41 & 28 & 19 \\ 21 & 15 & 19 \end{bmatrix}$$

since the number of units of material for sedan models is $41 = 23 + 18$, the number of units of material for hardtop models is $28 = 16 + 12$, and so on.

This leads us to define the sum $A + B$ of two matrices A and B as the matrix consisting of the sum of corresponding entries from A and B.

Addition of Matrices

Let $A = [a_{ij}]$ and $B = [b_{ij}]$ be two $m \times n$ matrices. The **sum** $A + B$ is defined as the $m \times n$ matrix $[a_{ij} + b_{ij}]$.

EXAMPLE 6   Adding Two Matrices

(a) $\begin{bmatrix} 23 & 16 & 10 \\ 7 & 9 & 11 \end{bmatrix} + \begin{bmatrix} 18 & 12 & 9 \\ 14 & 6 & 8 \end{bmatrix} = \begin{bmatrix} 23 + 18 & 16 + 12 & 10 + 9 \\ 7 + 14 & 9 + 6 & 11 + 8 \end{bmatrix}$

$\qquad\qquad\qquad\qquad\qquad = \begin{bmatrix} 41 & 28 & 19 \\ 21 & 15 & 19 \end{bmatrix}$

(b) $\begin{bmatrix} 0.6 & 0.4 \\ 0.1 & 0.9 \end{bmatrix} + \begin{bmatrix} 2.3 & 0.6 \\ 1.8 & 5.2 \end{bmatrix} = \begin{bmatrix} 0.6 + 2.3 & 0.4 + 0.6 \\ 0.1 + 1.8 & 0.9 + 5.2 \end{bmatrix}$

$\qquad\qquad\qquad\qquad\qquad = \begin{bmatrix} 2.9 & 1.0 \\ 1.9 & 6.1 \end{bmatrix}$

■

Notice that it is possible to add two matrices only if their dimensions are the same. Also, the dimension of the sum of two matrices is the same as that of the two original matrices.

The following pairs of matrices cannot be added since they are of different dimensions:

$$A = \begin{bmatrix} 1 & 2 \\ 7 & 2 \end{bmatrix} \qquad \text{and} \qquad B = \begin{bmatrix} 1 \\ -3 \end{bmatrix}$$

$$A = [2 \quad 3] \qquad \text{and} \qquad B = [1 \quad 1 \quad 1]$$

$$A = \begin{bmatrix} -1 & 7 & 0 \\ 2 & \frac{1}{2} & 0 \end{bmatrix} \qquad \text{and} \qquad B = \begin{bmatrix} -1 & 2 \\ 3 & 0 \\ 1 & 5 \end{bmatrix}$$

✍ **Now Work Problem 25**

It turns out that the usual rules for the addition of real numbers (such as the commutative property and associative property) are also valid for matrix addition.

EXAMPLE 7   Demonstrating the Commutative Property

Let $\qquad\qquad A = \begin{bmatrix} 1 & 5 \\ 7 & -3 \end{bmatrix} \qquad \text{and} \qquad B = \begin{bmatrix} 3 & -2 \\ 4 & 1 \end{bmatrix}$

Then

$$A + B = \begin{bmatrix} 1 & 5 \\ 7 & -3 \end{bmatrix} + \begin{bmatrix} 3 & -2 \\ 4 & 1 \end{bmatrix} = \begin{bmatrix} 1 + 3 & 5 + (-2) \\ 7 + 4 & -3 + 1 \end{bmatrix}$$

$$= \begin{bmatrix} 4 & 3 \\ 11 & -2 \end{bmatrix}$$

$$B + A = \begin{bmatrix} 3 & -2 \\ 4 & 1 \end{bmatrix} + \begin{bmatrix} 1 & 5 \\ 7 & -3 \end{bmatrix} = \begin{bmatrix} 4 & 3 \\ 11 & -2 \end{bmatrix}$$

This leads us to formulate the following property for addition.

> **Commutative Property for Addition**
>
> If A and B are two matrices of the same dimension, then
>
> $$A + B = B + A$$

The associative property for addition of matrices is also true.

> **Associative Property for Addition**
>
> If A, B, and C are three matrices of the same dimension, then
>
> $$A + (B + C) = (A + B) + C$$

The fact that addition of matrices is associative means that the notation $A + B + C$ is *not* ambiguous, since $(A + B) + C = A + (B + C)$.

 **Now Work Problems**

45 and 46

> **Zero Matrix**
>
> A matrix in which all entries are 0 is called a **zero matrix.** We use the symbol **0** to represent a zero matrix of any dimension.

For real numbers, 0 has the property that $x + 0 = x$ for any x. A property of a zero matrix is that $A + \mathbf{0} = A$, provided the dimension of **0** is the same as that of A.

EXAMPLE 8   Demonstrating a Property of the Zero Matrix

Let

$$A = \begin{bmatrix} 3 & 4 & -\frac{1}{2} \\ \sqrt{2} & 0 & 3 \end{bmatrix}$$

Then

$$A + \mathbf{0} = \begin{bmatrix} 3 & 4 & -\frac{1}{2} \\ \sqrt{2} & 0 & 3 \end{bmatrix} + \begin{bmatrix} 0 & 0 & 0 \\ 0 & 0 & 0 \end{bmatrix}$$

$$= \begin{bmatrix} 3+0 & 4+0 & -\frac{1}{2}+0 \\ \sqrt{2}+0 & 0+0 & 3+0 \end{bmatrix} = \begin{bmatrix} 3 & 4 & -\frac{1}{2} \\ \sqrt{2} & 0 & 3 \end{bmatrix} = A$$

If A is any matrix, the **additive inverse** of A, denoted by $-A$, is the matrix obtained by replacing each entry in A by its negative.

EXAMPLE 9 Finding the Additive Inverse of a Matrix

If $\qquad A = \begin{bmatrix} -3 & 0 \\ 5 & -2 \\ 1 & 3 \end{bmatrix} \qquad$ then $\qquad -A = \begin{bmatrix} 3 & 0 \\ -5 & 2 \\ -1 & -3 \end{bmatrix}$

Additive Inverse Property

For any matrix A, we have the property that

$$A + (-A) = \mathbf{0}$$

Now Work Problem 47

Subtraction of Matrices

Now that we have defined the sum of two matrices and the additive inverse of a matrix, it is natural to ask about the *difference* of two matrices. As you will see, subtracting matrices and subtracting real numbers are much the same kind of process.

Difference of Two Matrices

Let $A = [a_{ij}]$ and $B = [b_{ij}]$ be two $m \times n$ matrices. The **difference** $A - B$ is defined as the $m \times n$ matrix $[a_{ij} - b_{ij}]$.

EXAMPLE 10 Subtracting Two Matrices

Let $\qquad A = \begin{bmatrix} 2 & 3 & 4 \\ 1 & 0 & 2 \end{bmatrix} \qquad$ and $\qquad B = \begin{bmatrix} -2 & 1 & -1 \\ 3 & 0 & 3 \end{bmatrix}$

Then

$$A - B = \begin{bmatrix} 2 & 3 & 4 \\ 1 & 0 & 2 \end{bmatrix} - \begin{bmatrix} -2 & 1 & -1 \\ 3 & 0 & 3 \end{bmatrix}$$

$$= \begin{bmatrix} 2 - (-2) & 3 - 1 & 4 - (-1) \\ 1 - 3 & 0 - 0 & 2 - 3 \end{bmatrix} = \begin{bmatrix} 4 & 2 & 5 \\ -2 & 0 & -1 \end{bmatrix}$$

■

Notice that the difference $A - B$ is nothing more than the matrix formed by subtracting the entries in B from the corresponding entries in A.

Using the matrices A and B from Example 10, we find that

$$B - A = \begin{bmatrix} -2 & 1 & -1 \\ 3 & 0 & 3 \end{bmatrix} - \begin{bmatrix} 2 & 3 & 4 \\ 1 & 0 & 2 \end{bmatrix}$$

$$= \begin{bmatrix} -2 - 2 & 1 - 3 & -1 - 4 \\ 3 - 1 & 0 - 0 & 3 - 2 \end{bmatrix} = \begin{bmatrix} -4 & -2 & -5 \\ 2 & 0 & 1 \end{bmatrix}$$

Observe that $A - B \neq B - A$, illustrating that matrix subtraction, like subtraction of real numbers, is not commutative.

Multiplying a Matrix by a Number

Let's return to the production of Motors Inc., during the month specified in Example 1. The matrix A describing this production is

$$A = \begin{bmatrix} 23 & 16 & 10 \\ 7 & 9 & 11 \end{bmatrix}$$

Let's assume that for 3 consecutive months, the monthly production remained the same. Then the total production for the 3 months is simply the sum of the matrix A taken 3 times. If we represent the total production by the matrix T, then

$$T = \begin{bmatrix} 23 & 16 & 10 \\ 7 & 9 & 11 \end{bmatrix} + \begin{bmatrix} 23 & 16 & 10 \\ 7 & 9 & 11 \end{bmatrix} + \begin{bmatrix} 23 & 16 & 10 \\ 7 & 9 & 11 \end{bmatrix}$$

$$= \begin{bmatrix} 23 + 23 + 23 & 16 + 16 + 16 & 10 + 10 + 10 \\ 7 + 7 + 7 & 9 + 9 + 9 & 11 + 11 + 11 \end{bmatrix}$$

$$= \begin{bmatrix} 3 \cdot 23 & 3 \cdot 16 & 3 \cdot 10 \\ 3 \cdot 7 & 3 \cdot 9 & 3 \cdot 11 \end{bmatrix} = \begin{bmatrix} 69 & 48 & 30 \\ 21 & 27 & 33 \end{bmatrix}$$

In other words, when we add the matrix A 3 times, we multiply each entry of A by 3. This leads to the following definition.

Scalar Multiplication

Let $A = [a_{ij}]$ be an $m \times n$ matrix and let c be a real number, called a **scalar.** The product of the matrix A by the scalar c, called **scalar multiplication,** is the $m \times n$ matrix $cA = [ca_{ij}]$.

When multiplying a matrix by a real number, each entry of the matrix is multiplied by the number. Notice that the dimension of A and the dimension of the product cA are the same.

EXAMPLE 11   Computing Scalar Multiples

For
$$A = \begin{bmatrix} 2 \\ 5 \\ -7 \end{bmatrix} \quad \text{and} \quad B = \begin{bmatrix} 20 & 0 \\ 18 & 8 \end{bmatrix}$$
compute   (a) $3A$ (b) $\frac{1}{2}B$

SOLUTION

(a) $3A = 3\begin{bmatrix} 2 \\ 5 \\ -7 \end{bmatrix} = \begin{bmatrix} 3 \cdot 2 \\ 3 \cdot 5 \\ 3 \cdot (-7) \end{bmatrix} = \begin{bmatrix} 6 \\ 15 \\ -21 \end{bmatrix}$

(b) $\frac{1}{2}B = \frac{1}{2}\begin{bmatrix} 20 & 0 \\ 18 & 8 \end{bmatrix} = \begin{bmatrix} \frac{1}{2} \cdot 20 & \frac{1}{2} \cdot 0 \\ \frac{1}{2} \cdot 18 & \frac{1}{2} \cdot 8 \end{bmatrix} = \begin{bmatrix} 10 & 0 \\ 9 & 4 \end{bmatrix}$

■

EXAMPLE 12   Illustrating a Property of Matrices

For
$$A = \begin{bmatrix} 3 & 1 \\ 4 & 0 \\ 2 & -3 \end{bmatrix} \quad \text{and} \quad B = \begin{bmatrix} 2 & -3 \\ -1 & 1 \\ 1 & 0 \end{bmatrix}$$
compute   (a) $A - B$ (b) $A + (-1) \cdot B$

SOLUTION

(a) $A - B = \begin{bmatrix} 1 & 4 \\ 5 & -1 \\ 1 & -3 \end{bmatrix}$

(b) $A + (-1) \cdot B = \begin{bmatrix} 3 & 1 \\ 4 & 0 \\ 2 & -3 \end{bmatrix} + \begin{bmatrix} -2 & 3 \\ 1 & -1 \\ -1 & 0 \end{bmatrix} = \begin{bmatrix} 1 & 4 \\ 5 & -1 \\ 1 & -3 \end{bmatrix}$

■

The above example illustrates the result that

$$A - B = A + (-1) \cdot B = A + (-B)$$

✎ **Now Work Problem 35**

We continue our study of scalar multiplication by listing some of its properties in the box that follows.

> **Properties of Scalar Multiplication**
>
> Let k and h be two real numbers and let A and B be two matrices of dimension $m \times n$. Then
>
> $$\textbf{(I)} \qquad k(hA) = (kh)A$$
> $$\textbf{(II)} \qquad (k + h)A = kA + hA$$
> $$\textbf{(III)} \qquad k(A + B) = kA + kB$$

Properties I–III are illustrated in the following example.

EXAMPLE 13   Using Properties of Scalar Multiplication

For
$$A = \begin{bmatrix} 2 & -3 & -1 \\ 5 & 6 & 4 \end{bmatrix} \quad \text{and} \quad B = \begin{bmatrix} -3 & 0 & 4 \\ 2 & -1 & 5 \end{bmatrix}$$
show that

(a) $5[2A] = 10A$ (b) $(4 + 3)A = 4A + 3A$ (c) $3[A + B] = 3A + 3B$

SOLUTION

(a) $5[2A] = 5 \begin{bmatrix} 4 & -6 & -2 \\ 10 & 12 & 8 \end{bmatrix} = \begin{bmatrix} 20 & -30 & -10 \\ 50 & 60 & 40 \end{bmatrix}$

$\quad 10A = \begin{bmatrix} 20 & -30 & -10 \\ 50 & 60 & 40 \end{bmatrix}$

(b) $(4 + 3)A = 7A = \begin{bmatrix} 14 & -21 & -7 \\ 35 & 42 & 28 \end{bmatrix}$

$\quad 4A + 3A = \begin{bmatrix} 8 & -12 & -4 \\ 20 & 24 & 16 \end{bmatrix} + \begin{bmatrix} 6 & -9 & -3 \\ 15 & 18 & 12 \end{bmatrix}$

$\qquad\qquad = \begin{bmatrix} 14 & -21 & -7 \\ 35 & 42 & 28 \end{bmatrix}$

(c) $3[A + B] = 3 \begin{bmatrix} -1 & -3 & 3 \\ 7 & 5 & 9 \end{bmatrix} = \begin{bmatrix} -3 & -9 & 9 \\ 21 & 15 & 27 \end{bmatrix}$

$\quad 3A + 3B = \begin{bmatrix} 6 & -9 & -3 \\ 15 & 18 & 12 \end{bmatrix} + \begin{bmatrix} -9 & 0 & 12 \\ 6 & -3 & 15 \end{bmatrix}$

$\qquad\qquad = \begin{bmatrix} -3 & -9 & 9 \\ 21 & 15 & 27 \end{bmatrix}$

EXERCISE 2.4   Answers to odd-numbered problems begin on page AN-9.

In Problems 1–12 write the dimension of each matrix.

1. $\begin{bmatrix} 3 & 2 \\ -1 & 3 \end{bmatrix}$ **2.** $\begin{bmatrix} -1 & 0 \\ 0 & 5 \end{bmatrix}$ **3.** $\begin{bmatrix} 2 & 1 & -3 \\ 1 & 0 & -1 \end{bmatrix}$ **4.** $\begin{bmatrix} 1 & 2 \\ 2 & 1 \\ 0 & -3 \end{bmatrix}$

5. $\begin{bmatrix} 4 & 0 \\ -1 & 2 \\ 5 & 8 \end{bmatrix}$ 6. $\begin{bmatrix} 0 & -3 & 6 \\ 0 & 5 & 2 \\ 1 & 8 & 7 \end{bmatrix}$ 7. $\begin{bmatrix} 1 & 4 \\ -2 & 8 \\ 0 & 0 \end{bmatrix}$ 8. $\begin{bmatrix} 8 & 1 & 0 & 0 \\ 3 & -4 & 0 & 0 \end{bmatrix}$

9. $\begin{bmatrix} 4 \\ 1 \end{bmatrix}$ 10. $[2 \quad 1 \quad -3]$ 11. $[2]$ 12. $[0]$

In Problems 13–24 determine whether the given statements are true or false. If false, tell why.

13. $\begin{bmatrix} 0 \\ 1 \end{bmatrix} = [0 \quad 1]$ 14. $\begin{bmatrix} 3 & 2 \\ -1 & 0 \end{bmatrix} = \begin{bmatrix} 3 & 2 \\ -1 & 4 \end{bmatrix}$

15. $\begin{bmatrix} 5 & 0 \\ 0 & 1 \end{bmatrix}$ is square   16. $\begin{bmatrix} 3 & 2 & 1 \\ 4 & -1 & 0 \end{bmatrix}$ is 3×2

17. $\begin{bmatrix} x & 2 \\ 4 & 0 \end{bmatrix} = \begin{bmatrix} 3 & 2 \\ 4 & 0 \end{bmatrix}$ if $x = 3$ 18. $\begin{bmatrix} x & y \\ 0 & 0 \end{bmatrix} = [x \quad y]$

19. $\begin{bmatrix} 5 & 0 \\ 1 & 1 \end{bmatrix} = \begin{bmatrix} 2+3 & 0 \\ 1 & 1 \end{bmatrix}$ 20. $\begin{bmatrix} 1 & 0 \\ 0 & 1 \end{bmatrix} = \begin{bmatrix} 3-2 & 3-3 \\ 3-3 & 3-2 \end{bmatrix}$

21. $2\begin{bmatrix} 1 & 0 \\ 0 & 2 \end{bmatrix} = \begin{bmatrix} 2 & 0 \\ 0 & 4 \end{bmatrix}$ 22. $-\begin{bmatrix} 8 & 0 \\ 5 & -1 \end{bmatrix} = \begin{bmatrix} -8 & 0 \\ 5 & 1 \end{bmatrix}$

23. $\begin{bmatrix} 8 \\ 1 \end{bmatrix} + \begin{bmatrix} 2 \\ 9 \end{bmatrix} = [10]$ 24. $[6 \quad 0] + [1] = [7 \quad 1]$

In Problems 25–32 perform the indicated operations. Express your answer as a single matrix.

25. $\begin{bmatrix} 3 & -1 \\ 4 & 2 \end{bmatrix} + \begin{bmatrix} -2 & 2 \\ 2 & 5 \end{bmatrix}$ 26. $\begin{bmatrix} 2 & 4 \\ 3 & -4 \end{bmatrix} - \begin{bmatrix} 8 & 9 \\ 0 & 7 \end{bmatrix}$

27. $3\begin{bmatrix} 2 & 6 & 0 \\ 4 & -2 & 1 \end{bmatrix}$ 28. $-3\begin{bmatrix} 2 & 1 \\ -2 & 1 \\ 0 & 3 \end{bmatrix}$

29. $2\begin{bmatrix} 1 & -1 & 8 \\ 2 & 4 & 1 \end{bmatrix} - 3\begin{bmatrix} 0 & -2 & 8 \\ 1 & 4 & 1 \end{bmatrix}$ 30. $6\begin{bmatrix} 2 & 1 \\ 3 & 1 \\ -1 & 0 \end{bmatrix} + 4\begin{bmatrix} 6 & -4 \\ -2 & -3 \\ 0 & 1 \end{bmatrix}$

31. $3\begin{bmatrix} a & 8 \\ b & 1 \\ c & -2 \end{bmatrix} + 5\begin{bmatrix} 2a & 6 \\ -b & -2 \\ -c & 0 \end{bmatrix}$ 32. $2\begin{bmatrix} 2x & y & z \\ 2 & -4 & 8 \end{bmatrix} - 3\begin{bmatrix} -3x & 4y & 2z \\ 6 & -1 & 4 \end{bmatrix}$

In Problems 33–50 use the matrices below. For Problems 33–44 perform the indicated operation(s); for Problems 45–50 verify the indicated property.

$$A = \begin{bmatrix} 2 & -3 & 4 \\ 0 & 2 & 1 \end{bmatrix} \quad B = \begin{bmatrix} 1 & -2 & 0 \\ 5 & 1 & 2 \end{bmatrix} \quad C = \begin{bmatrix} -3 & 0 & 5 \\ 2 & 1 & 3 \end{bmatrix}$$

33. $A + B$ 34. $B + C$ 35. $2A - 3C$ 36. $3C - 4B$

37. $(A + B) - 2C$ 38. $4C + (A - B)$ 39. $3A + 4(B + C)$ 40. $(A + B) + 3C$

41. $2(A - B) - C$ 42. $2A - 5(B + C)$ 43. $3A - B - 6C$ 44. $3A + 2B - 4C$

45. Verify the commutative property for addition by finding $A + B$ and $B + A$.

46. Verify the associative property for addition by finding $(A + B) + C$ and $A + (B + C)$.

47. Verify the additive inverse property by showing that $A + (-A) = 0$.

48. Verify Property I of scalar multiplication by finding $2(3A)$ and $6A$.

49. Verify Property II of scalar multiplication by finding $2B + 3B$ and $5B$.

50. Verify Property III of scalar multiplication by finding $2(A + C)$ and $2A + 2C$.

51. Find x and z so that

$$\begin{bmatrix} x \\ 4 \end{bmatrix} = \begin{bmatrix} -4 \\ z \end{bmatrix}$$

52. Find x, y, and z so that

$$\begin{bmatrix} x + y & -2 \\ 4 & 10 \end{bmatrix} = \begin{bmatrix} 6 & x - y \\ 4 & z \end{bmatrix}$$

53. Find x and y so that

$$\begin{bmatrix} x - 2y & 0 \\ -2 & 6 \end{bmatrix} = \begin{bmatrix} 3 & 0 \\ -2 & x + y \end{bmatrix}$$

54. Find x, y, and z so that

$$\begin{bmatrix} x - 2 & 3 & 2z \\ 6y & x & 2y \end{bmatrix} = \begin{bmatrix} y & z & 6 \\ 18z & y + 2 & 6z \end{bmatrix}$$

55. Find x, y, and z so that

$$[2 \quad 3 \quad -4] + [x \quad 2y \quad z] = [6 \quad -9 \quad 2]$$

56. Find x and y so that

$$\begin{bmatrix} 3 & -2 & 2 \\ 1 & 0 & -1 \end{bmatrix} + \begin{bmatrix} x - y & 2 & -2 \\ 4 & x & 6 \end{bmatrix}$$
$$= \begin{bmatrix} 6 & 0 & 0 \\ 5 & 2x + y & 5 \end{bmatrix}$$

57. Nail Production XYZ Company produces steel and aluminum nails. One week 25 gross of $\frac{1}{2}$-inch steel nails and 45 gross of 1-inch steel nails were produced. Suppose 13 gross of $\frac{1}{2}$-inch aluminum nails, 20 gross of 1-inch aluminum nails, 35 gross of 2-inch steel nails, and 23 gross of 2-inch aluminum nails were also made. Write a 2 × 3 matrix depicting this. Could you also write a 3 × 2 matrix for this situation?

58. Katy, Mike, and Danny go to the candy store. Katy buys 5 sticks of gum, 2 ice cream cones, and 20 jelly beans. Mike buys 2 sticks of gum, 15 jelly beans, and 3 candy bars. Danny buys 1 stick of gum, 1 ice cream cone, and 4 candy bars. Write a matrix depicting this situation.

59. Use a matrix to display the information given below, which was obtained in a survey of voters. Label the rows and columns.

351	Democrats earning under $25,000
271	Republicans earning under $25,000
73	Independents earning under $25,000
203	Democrats earning $25,000 or more
215	Republicans earning $25,000 or more
55	Independents earning $25,000 or more

60. Listing Stocks One day on the New York Stock Exchange, 800 issues went up and 600 went down. Of the 800 up issues, 200 went up more than $1 per share. Of the 600 down issues, 50 went down more than $1 per share. Express this information in a 2 × 2 matrix. Label the rows up and down and the columns more than $1 and less than $1.

61. Surveys In a survey of 1000 college students, the following information was obtained: 500 were Liberal Arts and Sciences (LAS) majors, of which 50% were female; 300 were Engineering (ENG) majors, of which 75% were male; and the remaining were Education (EDUC) majors, of which 60% were female. Express this information using a 2 × 3 matrix. Label the rows male and female and the columns LAS, ENG, and EDUC.

62. Sales of Cars The sales figures for two car dealers during June showed that dealer A sold 100 compacts, 50 intermediates, and 40 full-size cars, while dealer B sold 120 compacts, 40 intermediates, and 35 full-size cars. During July, dealer A sold 80 compacts, 30 intermediates, and 10 full-size cars, while dealer B sold 70 compacts, 40 intermediates, and 20 full-size cars. Total sales over the 3-month period of June–August revealed that dealer A sold 300 compacts, 120 intermediates, and 65 full-size cars. In the same 3-month period, dealer B sold 250 compacts, 100 intermediates, and 80 full-size cars.

(a) Write 2 × 3 matrices summarizing sales data for June, July, and the 3-month period for each dealer.
(b) Use matrix addition to find the sales over the 2-month period for June and July for each dealer.
(c) Use matrix subtraction to find the sales in August for each dealer.

Technology Exercises

Graphing calculators can make the sometimes tedious process of matrix algebra easy. In fact, most graphing calculators can handle matrices as large as 9 by 9, some even larger ones.

In Problems 1–6 use your graphing calculator to perform the indicated operations on the matrices below.

$$A = \begin{bmatrix} -1 & -1 & 3 & 0 \\ 2 & 6 & 2 & 2 \\ -4 & 2 & 3 & 2 \\ 7 & 0 & 5 & -1 \end{bmatrix} \qquad B = \begin{bmatrix} -1 & 2 & 4 & 5 \\ 2 & 0 & 5 & 3 \\ 0.5 & 6 & -7 & 11 \\ 5 & -1 & 2 & 7 \end{bmatrix} \qquad C = \begin{bmatrix} 13 & -8 & 7 & 0 \\ 0 & 5 & 0 & -2 \\ 5 & 0 & 7 & 0 \\ 7 & 7 & 7 & 7 \end{bmatrix}$$

1. $A + B$

2. $3C - 2B$

3. $C - 3(A + B)$

4. $2(A - B) + \frac{1}{2}C$

5. $3(B + C) - A$

6. $\frac{1}{3}(A + 2C) - B$

2.5 MULTIPLICATION OF MATRICES

While addition of matrices and scalar multiplication are fairly straightforward, defining the *product $A \cdot B$* of the two matrices A and B requires a bit more detail.

We explain first what we mean by the product of a row vector (a matrix with one row) with a column vector (a matrix with one column).

Let's look at a simple example. In a given month suppose 23 units of material and 7 units of labor were required to manufacture 4-door sedans. We can represent this by the column matrix

$$\begin{bmatrix} 23 \\ 7 \end{bmatrix}$$

Also, suppose the cost per unit of material is $450 and the cost per unit of labor is $600. We represent these costs by the row matrix

$$[450 \quad 600]$$

The total cost of producing the sedans is calculated as follows:

$$\text{Total cost} = (\text{Cost per unit of material}) \times (\text{Units of material})$$
$$+ (\text{Cost per unit of labor}) \times (\text{Units of labor})$$
$$= 450 \times 23 + 600 \times 7 = 14{,}550$$

In terms of the matrix representations,

$$\text{Total cost} = [450 \quad 600] \begin{bmatrix} 23 \\ 7 \end{bmatrix} = 450 \times 23 + 600 \times 7 = 14{,}550$$

This leads us to formulate the following definition for multiplying a row matrix times a column matrix:

If $R = [r_1 \; r_2 \; \ldots \; r_n]$ is a row matrix of dimension $1 \times n$ and $C = \begin{bmatrix} c_1 \\ c_2 \\ \cdot \\ \cdot \\ \cdot \\ c_n \end{bmatrix}$ is a column matrix of dimension $n \times 1$, then by the **product of R and C** we mean the number

$$r_1 c_1 + r_2 c_2 + r_3 c_3 + \; \ldots \; + r_n c_n$$

EXAMPLE 1   **Finding the Product of a 1 × 3 Row Matrix and a 3 × 1 Column Matrix**

If $\qquad R = [1 \quad 5 \quad 3] \qquad$ and $\qquad C = \begin{bmatrix} 2 \\ -1 \\ 4 \end{bmatrix}$

then the product of R and C is

$$R \cdot C = 1 \cdot 2 + 5 \cdot (-1) + 3 \cdot 4 = 9$$

■

Notice that for the product of a row matrix R and a column matrix C to make sense, if R is a $1 \times n$ row matrix, then C must have dimension $n \times 1$.

EXAMPLE 2   **Finding the Product of a 1 × 4 Row Matrix and a 4 × 1 Column Matrix**

Let $\qquad R = [1 \quad 0 \quad 1 \quad 5] \qquad$ and $\qquad C = \begin{bmatrix} 0 \\ -11 \\ 0 \\ 8 \end{bmatrix}$

Then the product of R and C is

$$R \cdot C = 1 \cdot 0 + 0 \cdot (-11) + 1 \cdot 0 + 5 \cdot 8 = 40$$

✎ **Now Work Problem 1**

■

Given two matrices A and B, the rows of A can be thought of as row matrices, while the columns of B can be thought of as column matrices. This observation will be used in the following main definition.

Matrix Multiplication

Let $A = [a_{ij}]$ be a matrix of dimension $m \times r$ and let $B = [b_{ij}]$ be a matrix of dimension $r \times n$. The **product** $A \cdot B$ is the matrix of dimension $m \times n$, whose ijth entry is the sum of the products of corresponding elements of the ith row of A and the jth column of B. That is, the ijth entry of $A \cdot B$ is

$$a_{i1}b_{1j} + a_{i2}b_{2j} + a_{i3}b_{3j} + \cdots + a_{ir}b_{rj}$$

The rule for multiplication of matrices is best illustrated by an example.

EXAMPLE 3   Finding the Product of Two Matrices

Find the product $A \cdot B$ if

$$A = \begin{bmatrix} 1 & 3 & -2 \\ 4 & -1 & 5 \end{bmatrix} \quad \text{and} \quad B = \begin{bmatrix} 2 & -3 & 4 & 1 \\ -1 & 2 & 2 & 0 \\ 4 & 5 & 1 & 1 \end{bmatrix}$$

SOLUTION   Since A is 2×3 and B is 3×4, the product $A \cdot B$ will be 2×4. To obtain, for example, the entry in row 2, column 3 of $A \cdot B$, we form the product of the second row of A with the third column of B. That is,

$$\begin{bmatrix} 1 & 3 & -2 \\ \boxed{4} & \boxed{-1} & \boxed{5} \end{bmatrix} \qquad \begin{bmatrix} 2 & -3 & \boxed{4} & 1 \\ -1 & 2 & \boxed{2} & 0 \\ 4 & 5 & \boxed{1} & 1 \end{bmatrix}$$

We compute

$$4 \cdot 4 + (-1) \cdot 2 + 5 \cdot 1 = 19$$

Thus far, we have

$$A \cdot B = \begin{bmatrix} - & - & - & - \\ - & - & 19 & - \end{bmatrix}$$

To obtain the entry in row 1, column 2, we compute

$$[1 \quad 3 \quad -2] \begin{bmatrix} -3 \\ 2 \\ 5 \end{bmatrix} = 1 \cdot (-3) + 3 \cdot 2 + (-2) \cdot 5 = -7$$

The other entries of $A \cdot B$ are obtained in a similar fashion. The final result is shown here. You should verify it.

$$A \cdot B = \begin{bmatrix} -9 & -7 & 8 & -1 \\ 29 & 11 & 19 & 9 \end{bmatrix}$$

EXAMPLE 4   Manufacturing Cost

One month's production at Motors, Inc., may be given in matrix form as

	Sedan	Hardtop	Station Wagon	
$A = $	23	16	10	Units of material
	7	9	11	Units of labor

Suppose that in this month's production, the cost for each unit of material is \$450 and the cost for each unit of labor is \$600. What is the total cost to manufacture the sedans, the hardtops, and the station wagons?

SOLUTION For sedans, the cost is 23 units of material at $450 each, plus 7 units of labor at $600 each, for a total cost of

$$23 \cdot \$450 + 7 \cdot \$600 = 10{,}350 + 4200 = \$14{,}550$$

Similarly, for hardtops, the total cost is

$$16 \cdot \$450 + 9 \cdot \$600 = 7200 + 5400 = \$12{,}600$$

Finally, for station wagons, the total cost is

$$10 \cdot \$450 + 11 \cdot \$600 = 4500 + 6600 = \$11{,}100$$

If we represent the cost of units of material and units of labor by the 1×2 matrix

$$U = \begin{matrix} \text{Unit cost of} & \text{Unit cost of} \\ \text{material} & \text{labor} \\ [450 & 600] \end{matrix}$$

then the total cost is the product $U \cdot A$.

$$U \cdot A = [450 \quad 600] \begin{bmatrix} 23 & 16 & 10 \\ 7 & 9 & 11 \end{bmatrix}$$
$$= [450 \cdot 23 + 600 \cdot 7 \quad 450 \cdot 16 + 600 \cdot 9 \quad 450 \cdot 10 + 600 \cdot 11]$$
$$= [14{,}550 \quad 12{,}600 \quad 11{,}100]$$

Properties of Matrix Multiplication

If A is a matrix of dimension $m \times r$ (which has r columns) and B is a matrix of dimension $p \times n$ (which has p rows) and if $r \neq p$, the product $A \cdot B$ is not defined.

> Multiplication of matrices is possible only if the number of columns of the matrix on the left equals the number of rows of the matrix on the right.
>   If A is of dimension $m \times r$ and B is of dimension $r \times n$, then the product $A \cdot B$ is of dimension $m \times n$.

Figure 5 illustrates the result stated above.

Figure 5

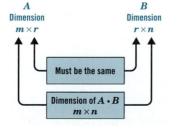

In Example 3, A is of dimension 2×3, B is of dimension 3×4, and we found the product $A \cdot B$ to be of dimension 2×4. Observe that the product $B \cdot A$ is not defined.

 **Now Work Problem 17**

EXAMPLE 5   Finding the Product of Two Square Matrices

For

$$A = \begin{bmatrix} 2 & 0 \\ 1 & 5 \end{bmatrix} \quad \text{and} \quad B = \begin{bmatrix} 3 & 2 \\ 1 & 4 \end{bmatrix}$$

compute $A \cdot B$ and $B \cdot A$.

SOLUTION   We observe that the products $A \cdot B$ and $B \cdot A$ are both defined, and so

$$A \cdot B = \begin{bmatrix} 6 & 4 \\ 8 & 22 \end{bmatrix} \qquad B \cdot A = \begin{bmatrix} 8 & 10 \\ 6 & 20 \end{bmatrix}$$

■

We observe in the above example that even if both $A \cdot B$ and $B \cdot A$ are defined, they may not be equal. We conclude that

---

Matrix multiplication is not commutative. That is, in general,

$$A \cdot B \qquad \text{is not always equal to} \qquad B \cdot A$$

In listing some of the properties of matrix multiplication in this book, we will follow the usual convention and write $A \cdot B$ as AB.

Associative Property of Multiplication

Let A be a matrix of dimension $m \times r$, let B be a matrix of dimension $r \times p$, and let C be a matrix of dimension $p \times n$. Then matrix multiplication is **associative.** That is,

$$A(BC) = (AB)C$$

The resulting matrix ABC is of dimension $m \times n$.

Notice the limitations that are placed on the dimensions of the matrices in order for multiplication to be associative.

Distributive Property

Let A be a matrix of dimension $m \times r$. Let B and C be matrices of dimension $r \times n$. Then the **distributive property** states that

$$A(B + C) = AB + AC$$

The resulting matrix $AB + AC$ is of dimension $m \times n$.

Now Work Problems

41 and 42

Identity Matrix

A special type of square matrix is the **identity matrix,** which is denoted by I_n. It is the matrix whose diagonal entries are 1s and all other entries are 0s. Thus,

$$I_n = \begin{bmatrix} 1 & 0 & \cdots & 0 & 0 \\ 0 & 1 & \cdots & 0 & 0 \\ \cdot & \cdot & & \cdot & \cdot \\ \cdot & \cdot & & \cdot & \cdot \\ \cdot & \cdot & & \cdot & \cdot \\ 0 & 0 & \cdots & 1 & 0 \\ 0 & 0 & \cdots & 0 & 1 \end{bmatrix}$$

where the subscript n implies that I_n is of dimension $n \times n$.

EXAMPLE 6 **Finding the Product of a Matrix and the Identity Matrix I_2**

For
$$A = \begin{bmatrix} 3 & 2 \\ -4 & 5 \end{bmatrix}$$

compute   (a) AI_2 (b) I_2A

SOLUTION

(a) $AI_2 = \begin{bmatrix} 3 & 2 \\ -4 & 5 \end{bmatrix}\begin{bmatrix} 1 & 0 \\ 0 & 1 \end{bmatrix} = \begin{bmatrix} 3 & 2 \\ -4 & 5 \end{bmatrix} = A$

(b) $I_2A = \begin{bmatrix} 1 & 0 \\ 0 & 1 \end{bmatrix}\begin{bmatrix} 3 & 2 \\ -4 & 5 \end{bmatrix} = \begin{bmatrix} 3 & 2 \\ -4 & 5 \end{bmatrix} = A$

■

Example 6 demonstrates a more general result.

If A is a square matrix of dimension $n \times n$, then $AI_n = I_nA = A$.

Thus for square matrices the identity matrix plays the role that the number 1 plays in the set of real numbers.

When the matrix A is not square, care must be taken when forming the products AI and IA. For example, if

$$A = \begin{bmatrix} 1 & 2 \\ 3 & 2 \\ 1 & 1 \end{bmatrix}$$

then A is of dimension 3×2 and

$$AI_2 = A\begin{bmatrix} 1 & 0 \\ 0 & 1 \end{bmatrix} = \begin{bmatrix} 1 & 2 \\ 3 & 2 \\ 1 & 1 \end{bmatrix}\begin{bmatrix} 1 & 0 \\ 0 & 1 \end{bmatrix} = \begin{bmatrix} 1 & 2 \\ 3 & 2 \\ 1 & 1 \end{bmatrix} = A$$

Although the product I_2A is not defined, we can calculate the product I_3A as follows:

$$I_3A = \begin{bmatrix} 1 & 0 & 0 \\ 0 & 1 & 0 \\ 0 & 0 & 1 \end{bmatrix} \begin{bmatrix} 1 & 2 \\ 3 & 2 \\ 1 & 1 \end{bmatrix} = \begin{bmatrix} 1 & 2 \\ 3 & 2 \\ 1 & 1 \end{bmatrix} = A$$

The above observations can be generalized as follows:

Identity Property

If A is a matrix of dimension $m \times n$ and if I_n denotes the identity matrix of dimension $n \times n$, and I_m denotes the identity matrix of dimension $m \times m$, then

$$I_mA = A \qquad \text{and} \qquad AI_n = A$$

EXERCISE 2.5   Answers to odd-numbered problems begin on page AN-10.

In Problems 1–16 find the product.

1. $[1 \quad 3] \begin{bmatrix} 2 \\ 4 \end{bmatrix}$

2. $[-1 \quad 4] \begin{bmatrix} 5 \\ 2 \end{bmatrix}$

3. $[1 \quad -2 \quad 3] \begin{bmatrix} 0 \\ 1 \\ 2 \end{bmatrix}$

4. $[-1 \quad 1 \quad 0] \begin{bmatrix} 1 \\ -1 \\ 1 \end{bmatrix}$

5. $[1 \quad 4] \begin{bmatrix} 2 & 0 \\ 4 & -2 \end{bmatrix}$

6. $\begin{bmatrix} 2 & 0 \\ 4 & -2 \end{bmatrix} \begin{bmatrix} 2 \\ 4 \end{bmatrix}$

7. $\begin{bmatrix} 2 & 0 \\ 4 & -2 \end{bmatrix} \begin{bmatrix} 2 & 1 \\ 3 & -2 \end{bmatrix}$

8. $\begin{bmatrix} 1 & 4 \\ -1 & 2 \end{bmatrix} \begin{bmatrix} 3 & 0 \\ -2 & 2 \end{bmatrix}$

9. $[1 \quad -2 \quad 3] \begin{bmatrix} 0 & 1 \\ 1 & 2 \\ 2 & 3 \end{bmatrix}$

10. $\begin{bmatrix} 1 & -2 & 3 \\ 4 & 0 & 6 \end{bmatrix} \begin{bmatrix} 0 \\ 1 \\ 2 \end{bmatrix}$

11. $\begin{bmatrix} 1 & -2 & 3 \\ 4 & 0 & 6 \end{bmatrix} \begin{bmatrix} 0 & -2 \\ 1 & 0 \\ 2 & -4 \end{bmatrix}$

12. $\begin{bmatrix} 1 & -2 & 3 \\ 4 & 0 & 6 \end{bmatrix} \begin{bmatrix} -1 & 2 & 1 \\ 1 & 3 & 0 \\ 0 & 4 & -2 \end{bmatrix}$

13. $\begin{bmatrix} 2 & 0 \\ 4 & -2 \\ 6 & -1 \end{bmatrix} \begin{bmatrix} 2 & 1 \\ 3 & -2 \end{bmatrix}$

14. $\begin{bmatrix} 1 & 4 \\ -1 & 2 \end{bmatrix} \begin{bmatrix} 2 & 0 & 6 \\ -1 & 4 & 1 \end{bmatrix}$

15. $\begin{bmatrix} 1 & -1 & 6 \\ 2 & 0 & -1 \\ 3 & 1 & 2 \end{bmatrix} \begin{bmatrix} 3 & 2 \\ 0 & 1 \\ 1 & 0 \end{bmatrix}$

16. $\begin{bmatrix} 2 & 0 & 1 \\ -1 & 1 & 1 \end{bmatrix} \begin{bmatrix} 1 & 0 & 0 \\ 2 & 1 & 2 \\ 3 & 2 & 4 \end{bmatrix}$

In Problems 17–26 let A be of dimension 3 × 4, B be of dimension 3 × 3, C be of dimension 2 × 3, and D be of dimension 3 × 2. Determine which of the following expressions are defined and, for those that are, give the dimension.

17. BA **18.** CD **19.** AB **20.** DC **21.** $(BA)C$

22. $A(CD)$ **23.** $BA + A$ **24.** $CD + BA$ **25.** $DC + B$ **26.** $CB - A$

In Problems 27–42 use the matrices given at the top of page 97. For Problems 27– 40 perform the indicated operation(s); for Problems 41 and 42 verify the indicated property.

$$A = \begin{bmatrix} 1 & 2 \\ 0 & 4 \end{bmatrix} \qquad B = \begin{bmatrix} 1 & 2 & 3 \\ -1 & 4 & -2 \end{bmatrix} \qquad C = \begin{bmatrix} 3 & 1 \\ 4 & -1 \\ 0 & 2 \end{bmatrix} \qquad D = \begin{bmatrix} 1 & 0 & 4 \\ 0 & 1 & 2 \\ 0 & -1 & 1 \end{bmatrix} \qquad E = \begin{bmatrix} 3 & -1 \\ 4 & 2 \end{bmatrix}$$

27. AB

28. DC

29. BC

30. AA

31. $(D + I_3)C$

32. $DC + C$

33. EI_2

34. I_3D

35. $(2E)B$

36. $E(2B)$

37. $-5E + A$

38. $3A + 2E$

39. $3CB + 4D$

40. $2EA - 3BC$

41. Verify the associative property of matrix multiplication by finding $D(CB)$ and $(DC)B$.

42. Verify the distributive property by finding $(A + E)B$ and $AB + EB$.

43. For

$$A = \begin{bmatrix} 1 & -1 \\ 2 & 0 \end{bmatrix} \quad \text{and} \quad B = \begin{bmatrix} 3 & 2 \\ -2 & 4 \end{bmatrix}$$

find AB and BA. Notice that $AB \neq BA$.

44. Show that, for all values a, b, c, and d, the matrices

$$A = \begin{bmatrix} a & b \\ -b & a \end{bmatrix} \quad \text{and} \quad B = \begin{bmatrix} c & d \\ -d & c \end{bmatrix}$$

are commutative; that is, $AB = BA$.

45. If possible, find a matrix A such that

$$A \begin{bmatrix} 0 & 1 \\ 2 & -1 \end{bmatrix} = \begin{bmatrix} 2 & 1 \\ -1 & 0 \end{bmatrix}$$

Hint: Let $A = \begin{bmatrix} a & b \\ c & d \end{bmatrix}$

46. For what numbers x will the following be true?

$$[x \quad 4 \quad 1] \begin{bmatrix} 2 & 1 & 0 \\ 1 & 0 & 2 \\ 0 & 2 & 4 \end{bmatrix} \begin{bmatrix} x \\ -7 \\ \frac{5}{4} \end{bmatrix} = 0$$

47. Let

$$A = \begin{bmatrix} 1 & 2 & 5 \\ 2 & 4 & 10 \\ -1 & -2 & -5 \end{bmatrix}$$

Show that $A \cdot A = A^2 = 0$. Thus the rule in the real number system that if $a^2 = 0$, then $a = 0$ does not hold for matrices.

48. What must be true about a, b, c, and d, if we demand that $AB = BA$ for the following matrices?

$$A = \begin{bmatrix} a & b \\ c & d \end{bmatrix} \qquad B = \begin{bmatrix} 1 & 1 \\ -1 & 1 \end{bmatrix}$$

Assume that

$$\begin{bmatrix} a & b \\ c & d \end{bmatrix} \neq \begin{bmatrix} 1 & 0 \\ 0 & 1 \end{bmatrix}$$

49. Let

$$A = \begin{bmatrix} a & b \\ b & a \end{bmatrix}$$

Find a and b such that $A^2 + A = 0$, where $A^2 = AA$.

50. For the matrix

$$A = \begin{bmatrix} a & 1 - a \\ 1 + a & -a \end{bmatrix}$$

show that $A^2 = AA = I_2$.

51. Find the row vector $[x_1 \quad x_2]$ for which

$$[x_1 \quad x_2] \begin{bmatrix} \frac{1}{2} & \frac{1}{2} \\ \frac{1}{4} & \frac{3}{4} \end{bmatrix} = [x_1 \quad x_2]$$

under the condition that $x_1 + x_2 = 1$. Here, the row vector $[x_1 \quad x_2]$ is called a **fixed vector** of the matrix

$$\begin{bmatrix} \frac{1}{2} & \frac{1}{2} \\ \frac{1}{4} & \frac{3}{4} \end{bmatrix}$$

52. Department Store Purchases Lee went to a department store and purchased 6 pairs of pants, 8 shirts, and 2 jackets. Chan purchased 2 pairs of pants, 5 shirts, and 3 jackets. If pants are \$25 each, shirts are \$18 each, and jackets are \$39 each, use matrix multiplication to find the amounts spent by Lee and Chan.

53. Factory Production Suppose a factory is asked to produce three types of products, which we will call P_1, P_2, P_3. Suppose the following purchase order was received: $P_1 = 7$, $P_2 = 12$, $P_3 = 5$. Represent this order by a row vector and call it P:

$$P = [7 \quad 12 \quad 5]$$

To produce each of the products, raw material of four kinds is needed. Call the raw material M_1, M_2, M_3, and M_4. The matrix below gives the amount of material needed for each product:

$$\begin{array}{c} \\ \\ P_1 \\ Q = P_2 \\ P_3 \end{array} \begin{array}{cccc} M_1 & M_2 & M_3 & M_4 \\ \begin{bmatrix} 2 & 3 & 1 & 12 \\ 7 & 9 & 5 & 20 \\ 8 & 12 & 6 & 15 \end{bmatrix} \end{array}$$

Suppose the cost for each of the materials M_1, M_2, M_3, and M_4 is \$10, \$12, \$15, and \$20, respectively. The cost vector is

$$C = \begin{bmatrix} 10 \\ 12 \\ 15 \\ 20 \end{bmatrix}$$

Compute each of the following and interpret each one:

(a) PQ (b) QC (c) PQC

54. For a square matrix A it is possible to find $A \cdot A = A^2$. It is also clear that we can compute

$$A^n = \underbrace{A \cdot A \cdot \ldots \cdot A}_{n \text{ factors}}$$

Find A^2, A^3, and A^4 for each of the following square matrices:

(a) $A = \begin{bmatrix} 1 & 0 \\ 3 & 2 \end{bmatrix}$ (b) $A = \begin{bmatrix} 3 & 1 \\ -2 & -1 \end{bmatrix}$

(c) $A = \begin{bmatrix} 1 & 0 \\ 0 & 1 \end{bmatrix}$ (d) $A = \begin{bmatrix} \frac{1}{2} & \frac{1}{2} \\ \frac{1}{4} & \frac{3}{4} \end{bmatrix}$

Can you guess what A^n looks like for Part (c)? For Part (d)?

Technology Exercises

In Problems 1–8 use your graphing calculator and the matrices given below to perform the indicated operations.

$$A = \begin{bmatrix} -1 & -1 & 3 & 0 \\ 2 & 6 & 2 & 2 \\ -4 & 2 & 3 & 2 \\ 7 & 0 & 5 & -1 \end{bmatrix} \qquad B = \begin{bmatrix} -1 & 2 & 4 & 5 \\ 2 & 0 & 5 & 3 \\ 0.5 & 6 & -7 & 11 \\ 5 & -1 & 2 & 7 \end{bmatrix} \qquad C = \begin{bmatrix} 13 & -8 & 7 & 0 \\ 0 & 5 & 0 & -2 \\ 5 & 0 & 7 & 0 \\ 7 & 7 & 7 & 7 \end{bmatrix}$$

1. AB **2.** BA **3.** $(AB)\,C$ **4.** $A(BC)$

5. $B(A + C)$ **6.** $(A + C)\,B$ **7.** $A(2B - 3C)$ **8.** $(A + B)(A - B)$

9. For a square matrix A it is possible to find $A \cdot A = A^2$. It is also clear that we can compute

$$A^n = \underbrace{A \cdot A \cdot \ldots \cdot A}_{n \text{ factors}}$$

Use your graphing calculator to compute A^2, A^{10}, and A^{15} for each of the following matrices:

(a) $A = \begin{bmatrix} -0.5 & -1 & 0.3 & 0 & 0.3 \\ 2 & 1.6 & 1 & -1 & 0.4 \\ -4 & 2 & 0.7 & 2 & 0.2 \\ 1 & 0 & 0 & -1 & 0 \\ 0 & 0 & -0.9 & 0 & 0 \end{bmatrix}$

(b) $A = \begin{bmatrix} -1 & 0.02 & 0.24 & 0 \\ 2 & 0 & 0 & 1.3 \\ 0.5 & 6 & -0.7 & 1.1 \\ 2.5 & -1 & 0.02 & 0.7 \end{bmatrix}$

(c) $A = \begin{bmatrix} 0 & 1 & 0 \\ 1 & 0 & 1 \\ 1 & 1 & 1 \end{bmatrix}$

2.6 INVERSE OF A MATRIX

The *inverse* of a matrix, if it exists, plays the role in matrix algebra that the reciprocal of a number plays in the set of real numbers. For example, the product of 2 and its reciprocal, $\frac{1}{2}$, equals 1, the multiplicative identity. The product of a matrix and its inverse equals the identity matrix.

> **Inverse**
>
> Let A be a matrix of dimension $n \times n$. A matrix B of dimension $n \times n$ is called the **inverse** of A if $AB = BA = I_n$. We denote the inverse of a matrix A, if it exists, by A^{-1}.

EXAMPLE 1   **Verifying that One Matrix Is the Inverse of Another Matrix**

Show that $\begin{bmatrix} \frac{1}{2} & -\frac{1}{2} \\ 0 & 1 \end{bmatrix}$ is the inverse of $\begin{bmatrix} 2 & 1 \\ 0 & 1 \end{bmatrix}$.

SOLUTION   Since

$$\begin{bmatrix} 2 & 1 \\ 0 & 1 \end{bmatrix} \begin{bmatrix} \frac{1}{2} & -\frac{1}{2} \\ 0 & 1 \end{bmatrix} = \begin{bmatrix} 1 & 0 \\ 0 & 1 \end{bmatrix}$$

and

$$\begin{bmatrix} \frac{1}{2} & -\frac{1}{2} \\ 0 & 1 \end{bmatrix} \begin{bmatrix} 2 & 1 \\ 0 & 1 \end{bmatrix} = \begin{bmatrix} 1 & 0 \\ 0 & 1 \end{bmatrix}$$

the required condition is met.

 **Now Work Problem 1**

The next example provides a technique for finding the inverse of a matrix. Although this technique is not the one we shall ultimately use, it is illustrative.

EXAMPLE 2   **Finding the Inverse of a Matrix**

Find the inverse of the matrix $A = \begin{bmatrix} 2 & 1 \\ 0 & 1 \end{bmatrix}$.

SOLUTION   We begin by assuming that this matrix has an inverse of the form

$$A^{-1} = \begin{bmatrix} a & b \\ c & d \end{bmatrix}$$

Then the product of A and A^{-1} must be the identity matrix:

$$\begin{bmatrix} 2 & 1 \\ 0 & 1 \end{bmatrix} \begin{bmatrix} a & b \\ c & d \end{bmatrix} = \begin{bmatrix} 1 & 0 \\ 0 & 1 \end{bmatrix}$$

Multiplying the matrices on the left side, we get

$$\begin{bmatrix} 2a + c & 2b + d \\ c & d \end{bmatrix} = \begin{bmatrix} 1 & 0 \\ 0 & 1 \end{bmatrix}$$

The condition for equality requires that

$$2a + c = 1 \qquad 2b + d = 0 \qquad c = 0 \qquad d = 1$$

Using $c = 0$ and $d = 1$ in the first two equations, we find

$$a = \tfrac{1}{2} \qquad b = -\tfrac{1}{2} \qquad c = 0 \qquad d = 1$$

Hence, the inverse of

$$A = \begin{bmatrix} 2 & 1 \\ 0 & 1 \end{bmatrix} \quad \text{is} \quad A^{-1} = \begin{bmatrix} a & b \\ c & d \end{bmatrix} = \begin{bmatrix} \tfrac{1}{2} & -\tfrac{1}{2} \\ 0 & 1 \end{bmatrix}$$

■

Sometimes a square matrix does not have an inverse.

EXAMPLE 3   Showing a Matrix Has No Inverse

Show that the matrix below does not have an inverse.

$$A = \begin{bmatrix} 0 & 1 \\ 0 & 0 \end{bmatrix}$$

SOLUTION   We proceed as in Example 2 by assuming that A does have an inverse. It will be of the form

$$A^{-1} = \begin{bmatrix} a & b \\ c & d \end{bmatrix}$$

The product of A and A^{-1} must be the identity matrix. Thus

$$\begin{bmatrix} 0 & 1 \\ 0 & 0 \end{bmatrix}\begin{bmatrix} a & b \\ c & d \end{bmatrix} = \begin{bmatrix} 1 & 0 \\ 0 & 1 \end{bmatrix}$$

Performing the multiplication on the left side, we have

$$\begin{bmatrix} c & d \\ 0 & 0 \end{bmatrix} = \begin{bmatrix} 1 & 0 \\ 0 & 1 \end{bmatrix}$$

But these two matrices can never be equal (look at row 2, column 2: $0 \neq 1$). We conclude that our assumption that A has an inverse is false. That is, A does not have an inverse.

✍ **Now Work Problem 21**

■

So far, we have shown that a square matrix may or may not have an inverse. What about nonsquare matrices? Can they have inverses? The answer is "No." By definition, whenever a matrix has an inverse, it will commute with its inverse under multiplication. So if the nonsquare matrix A had the alleged inverse B, then AB would have to be equal to BA. But the fact that A is not square causes AB and BA to have different dimensions and prevents them from being equal. So such a B could not exist.

> A nonsquare matrix has no inverse.

The procedure used in Example 2 to find the inverse, if it exists, of a square matrix becomes quite involved as the dimension of the matrix gets larger. A more efficient method that uses the reduced row-echelon form of a matrix is provided next.

Reduced Row-Echelon Technique for Finding Inverses

We will introduce this technique by looking at an example.

EXAMPLE 4   Finding the Inverse of a Matrix

Find the inverse of the matrix

$$A = \begin{bmatrix} 4 & 2 \\ 3 & 1 \end{bmatrix}$$

SOLUTION   Assuming A has an inverse, we will denote it by

$$X = \begin{bmatrix} x_1 & x_2 \\ x_3 & x_4 \end{bmatrix}$$

Then the product of A and X is the identity matrix of dimension 2×2. That is,

$$AX = I_2$$

$$\begin{bmatrix} 4 & 2 \\ 3 & 1 \end{bmatrix}\begin{bmatrix} x_1 & x_2 \\ x_3 & x_4 \end{bmatrix} = \begin{bmatrix} 1 & 0 \\ 0 & 1 \end{bmatrix}$$

Performing the multiplication on the left yields

$$\begin{bmatrix} 4x_1 + 2x_3 & 4x_2 + 2x_4 \\ 3x_1 + x_3 & 3x_2 + x_4 \end{bmatrix} = \begin{bmatrix} 1 & 0 \\ 0 & 1 \end{bmatrix}$$

This matrix equation can be written as the following system of four equations containing four variables:

$$\begin{cases} 4x_1 + 2x_3 = 1 & \quad 4x_2 + 2x_4 = 0 \\ 3x_1 + x_3 = 0 & \quad 3x_2 + x_4 = 1 \end{cases}$$

We find the solution to be

$$x_1 = -\tfrac{1}{2} \qquad x_2 = 1 \qquad x_3 = \tfrac{3}{2} \qquad x_4 = -2$$

Thus the inverse of A is

$$A^{-1} = \begin{bmatrix} -\tfrac{1}{2} & 1 \\ \tfrac{3}{2} & -2 \end{bmatrix}$$

∎

Let's look at this example more closely. The system of four equations containing four variables can be written as two systems:

(a) $\begin{cases} 4x_1 + 2x_3 = 1 \\ 3x_1 + x_3 = 0 \end{cases}$ (b) $\begin{cases} 4x_2 + 2x_4 = 0 \\ 3x_2 + x_4 = 1 \end{cases}$

Their augmented matrices are

$$
\text{(a)} \begin{bmatrix} 4 & 2 & | & 1 \\ 3 & 1 & | & 0 \end{bmatrix} \qquad \text{(b)} \begin{bmatrix} 4 & 2 & | & 0 \\ 3 & 1 & | & 1 \end{bmatrix}
$$

Since the matrix A appears in both (a) and (b), any row operation we perform on (a) and (b) can be performed more easily on the single augmented matrix that combines the two right-hand columns. We denote this matrix by $A|I_2$ and write

$$
[A|I_2] = \begin{bmatrix} 4 & 2 & | & 1 & 0 \\ 3 & 1 & | & 0 & 1 \end{bmatrix}
$$

If we perform the row operations on $[A|I_2]$ needed to obtain the reduced row-echelon form of A, we get

$$
\begin{bmatrix} 1 & 0 & | & -\frac{1}{2} & 1 \\ 0 & 1 & | & \frac{3}{2} & -2 \end{bmatrix}
$$

The 2×2 matrix on the right-hand side of the vertical bar is A^{-1}.

This example illustrates the general procedure:

To find the inverse, if it exists, of a square matrix of dimension $n \times n$, follow these steps:

Steps for Finding the Inverse of a Matrix

Step 1 Write the augmented matrix $[A|I_n]$.

Step 2 Using row operations, write $[A|I_n]$ in reduced row-echelon form.

Step 3 If the resulting matrix is of the form $[I_n|B]$, that is, if the identity matrix appears on the left side of the bar, then B is the inverse of A. Otherwise, A has no inverse.

Let's work another example.

EXAMPLE 5   Finding the Inverse of a Matrix

Find the inverse of

$$
A = \begin{bmatrix} 1 & 1 & 2 \\ 2 & 1 & 0 \\ 1 & 2 & 2 \end{bmatrix}
$$

SOLUTION   Since A is of dimension 3×3, we use the identity matrix I_3. The matrix $[A|I_3]$ is

$$
\begin{bmatrix} 1 & 1 & 2 & | & 1 & 0 & 0 \\ 2 & 1 & 0 & | & 0 & 1 & 0 \\ 1 & 2 & 2 & | & 0 & 0 & 1 \end{bmatrix}
$$

We proceed to obtain the reduced row-echelon form of this matrix:

Use $\begin{array}{l} R_2 = r_2 + (-2)r_1 \\ R_3 = r_3 + (-1)r_1 \end{array}$ to obtain $\left[\begin{array}{ccc|ccc} 1 & 1 & 2 & 1 & 0 & 0 \\ 0 & -1 & -4 & -2 & 1 & 0 \\ 0 & 1 & 0 & -1 & 0 & 1 \end{array}\right]$

Use $R_2 = (-1)r_2$ to obtain $\left[\begin{array}{ccc|ccc} 1 & 1 & 2 & 1 & 0 & 0 \\ 0 & 1 & 4 & 2 & -1 & 0 \\ 0 & 1 & 0 & -1 & 0 & 1 \end{array}\right]$

Use $\begin{array}{l} R_1 = r_1 + (-1)r_2 \\ R_3 = r_3 + (-1)r_2 \end{array}$ to obtain $\left[\begin{array}{ccc|ccc} 1 & 0 & -2 & -1 & 1 & 0 \\ 0 & 1 & 4 & 2 & -1 & 0 \\ 0 & 0 & -4 & -3 & 1 & 1 \end{array}\right]$

Use $R_3 = (-\frac{1}{4})r_3$ to obtain $\left[\begin{array}{ccc|ccc} 1 & 0 & -2 & -1 & 1 & 0 \\ 0 & 1 & 4 & 2 & -1 & 0 \\ 0 & 0 & 1 & \frac{3}{4} & -\frac{1}{4} & -\frac{1}{4} \end{array}\right]$

Use $\begin{array}{l} R_1 = r_1 + 2r_3 \\ R_2 = r_2 + (-4)r_3 \end{array}$ to obtain $\left[\begin{array}{ccc|ccc} 1 & 0 & 0 & \frac{1}{2} & \frac{1}{2} & -\frac{1}{2} \\ 0 & 1 & 0 & -1 & 0 & 1 \\ 0 & 0 & 1 & \frac{3}{4} & -\frac{1}{4} & -\frac{1}{4} \end{array}\right]$

Since the identity matrix I_3 appears on the left side, the matrix appearing on the right is the inverse. That is,

$$A^{-1} = \left[\begin{array}{ccc} \frac{1}{2} & \frac{1}{2} & -\frac{1}{2} \\ -1 & 0 & 1 \\ \frac{3}{4} & -\frac{1}{4} & -\frac{1}{4} \end{array}\right]$$

■

Now Work Problem 11   You should verify that, in fact, $AA^{-1} = I_3$.

EXAMPLE 6   Showing that a Matrix Has No Inverse

Show that the matrix given below has no inverse.

$$\left[\begin{array}{cc} 3 & 2 \\ 6 & 4 \end{array}\right]$$

SOLUTION   We set up the matrix

$$\left[\begin{array}{cc|cc} 3 & 2 & 1 & 0 \\ 6 & 4 & 0 & 1 \end{array}\right]$$

Use $R_1 = \frac{1}{3}r_1$ to obtain $\left[\begin{array}{cc|cc} 1 & \frac{2}{3} & \frac{1}{3} & 0 \\ 6 & 4 & 0 & 1 \end{array}\right]$

Use $R_2 = r_2 + (-6)r_1$ to obtain $\left[\begin{array}{cc|cc} 1 & \frac{2}{3} & \frac{1}{3} & 0 \\ 0 & 0 & -2 & 1 \end{array}\right]$

The 0s in row 2 tell us we cannot get the identity matrix. This, in turn, tells us the original matrix has no inverse.

■

Solving a System of n Linear Equations Containing n Variables Using Inverses

The inverse of a matrix can also be used to solve a system of n linear equations containing n variables. We begin with a system of 3 linear equations containing 3 variables:

$$\begin{cases} x + y + 2z = 1 \\ 2x + y = 2 \\ x + 2y + 2z = 3 \end{cases}$$

If we let

$$A = \begin{bmatrix} 1 & 1 & 2 \\ 2 & 1 & 0 \\ 1 & 2 & 2 \end{bmatrix} \qquad X = \begin{bmatrix} x \\ y \\ z \end{bmatrix} \qquad B = \begin{bmatrix} 1 \\ 2 \\ 3 \end{bmatrix}$$

the above system can be written as

$$AX = B$$

In general, any system of n linear equations containing n variables $x_1, x_2, \ldots , x_n$, can be written in the form

$$AX = B$$

where A is the $n \times n$ matrix of the coefficients of the unknowns, B is an $n \times 1$ column matrix whose entries are the numbers appearing to the right of each equal sign in the system, and X is an $n \times 1$ column matrix containing the n variables.

We start with the matrix equation $AX = B$ and use properties of matrices. Our assumption is that the $n \times n$ matrix A has an inverse A^{-1}.

$$AX = B \qquad \text{\small\color{blue} A has an inverse } A^{-1}$$
$$A^{-1}(AX) = A^{-1}B \qquad \text{\small\color{blue} Multiply both sides by } A^{-1}$$
$$(A^{-1}A)X = A^{-1}B \qquad \text{\small\color{blue} Apply the Associative Property on the left side}$$
$$I_nX = A^{-1}B \qquad \text{\small\color{blue} Apply the Inverse Property: } A^{-1}A = I_n$$
$$X = A^{-1}B \qquad \text{\small\color{blue} Apply the Identity Property: } I_nX = X$$

This leads us to formulate the following result:

A system of n linear equations containing n variables

$$AX = B$$

for which A is a square matrix and A^{-1} exists, always has a unique solution that is given by

$$X = A^{-1}B$$

We use the above result in the next example.

EXAMPLE 7   Solving a System of Equations Using Inverses

Solve the system of equations: $\begin{cases} x + y + 2z = 1 \\ 2x + y = 2 \\ x + 2y + 2z = 3 \end{cases}$

SOLUTION   Here

$$A = \begin{bmatrix} 1 & 1 & 2 \\ 2 & 1 & 0 \\ 1 & 2 & 2 \end{bmatrix} \qquad X = \begin{bmatrix} x \\ y \\ z \end{bmatrix} \qquad B = \begin{bmatrix} 1 \\ 2 \\ 3 \end{bmatrix}$$

From Example 5 we know A has the inverse.

$$A^{-1} = \begin{bmatrix} \frac{1}{2} & \frac{1}{2} & -\frac{1}{2} \\ -1 & 0 & 1 \\ \frac{3}{4} & -\frac{1}{4} & -\frac{1}{4} \end{bmatrix}$$

Based on the result just stated,

$$X = A^{-1}B$$

$$\begin{bmatrix} x \\ y \\ z \end{bmatrix} = \begin{bmatrix} \frac{1}{2} & \frac{1}{2} & -\frac{1}{2} \\ -1 & 0 & 1 \\ \frac{3}{4} & -\frac{1}{4} & -\frac{1}{4} \end{bmatrix} \begin{bmatrix} 1 \\ 2 \\ 3 \end{bmatrix} = \begin{bmatrix} 0 \\ 2 \\ -\frac{1}{2} \end{bmatrix}$$

Thus the solution is $x = 0$, $y = 2$, $z = -\frac{1}{2}$.

The method used to solve the system in Example 7 requires that A have an inverse. If, for a system of equations $AX = B$, the matrix A has no inverse, then the system must be analyzed using the methods discussed in Section 2.3.

The method used in Example 7 for solving a system of equations is particularly useful for applications in which the constants appearing to the right of the equal sign change while the coefficients of the variables on the left side do not. See Problems 39–50 for an illustration. See also the discussion of Leontief models in Section 2.7.

EXERCISE 2.6   Answers to odd-numbered problems begin on page AN-12.

In Problems 1–6 show that the given matrices are inverses of each other.

1. $\begin{bmatrix} 1 & 2 \\ 2 & 3 \end{bmatrix} \begin{bmatrix} -3 & 2 \\ 2 & -1 \end{bmatrix}$

2. $\begin{bmatrix} 1 & 5 \\ 2 & 0 \end{bmatrix} \begin{bmatrix} 0 & \frac{1}{2} \\ \frac{1}{5} & -\frac{1}{10} \end{bmatrix}$

3. $\begin{bmatrix} -1 & -2 \\ 3 & 4 \end{bmatrix} \begin{bmatrix} 2 & 1 \\ -\frac{3}{2} & -\frac{1}{2} \end{bmatrix}$

4. $\begin{bmatrix} 1 & 3 \\ 2 & -1 \end{bmatrix} \begin{bmatrix} \frac{1}{7} & \frac{3}{7} \\ \frac{2}{7} & -\frac{1}{7} \end{bmatrix}$

5. $\begin{bmatrix} 1 & 2 & 3 \\ 2 & 3 & 4 \\ 1 & 2 & 1 \end{bmatrix} \begin{bmatrix} -\frac{5}{2} & 2 & -\frac{1}{2} \\ 1 & -1 & 1 \\ \frac{1}{2} & 0 & -\frac{1}{2} \end{bmatrix}$

6. $\begin{bmatrix} 1 & 3 & 3 \\ 1 & 4 & 3 \\ 1 & 3 & 4 \end{bmatrix} \begin{bmatrix} 7 & -3 & -3 \\ -1 & 1 & 0 \\ -1 & 0 & 1 \end{bmatrix}$

In Problems 7–20 find the inverse of each matrix using the reduced row-echelon technique.

7. $\begin{bmatrix} 3 & 7 \\ 2 & 5 \end{bmatrix}$

8. $\begin{bmatrix} 4 & 1 \\ 3 & 1 \end{bmatrix}$

9. $\begin{bmatrix} 1 & -1 \\ 3 & -4 \end{bmatrix}$

10. $\begin{bmatrix} 5 & 3 \\ 3 & 2 \end{bmatrix}$

11. $\begin{bmatrix} 2 & 1 \\ 4 & 3 \end{bmatrix}$

12. $\begin{bmatrix} 2 & 3 \\ 2 & -2 \end{bmatrix}$

13. $\begin{bmatrix} 0 & 0 & 1 \\ 0 & 1 & 0 \\ 1 & 0 & 0 \end{bmatrix}$

14. $\begin{bmatrix} -1 & 1 & 0 \\ 1 & 0 & 2 \\ 3 & 1 & 0 \end{bmatrix}$

15. $\begin{bmatrix} 1 & 1 & -1 \\ 3 & -1 & 0 \\ 2 & -3 & 4 \end{bmatrix}$

16. $\begin{bmatrix} 1 & 1 & 1 \\ 2 & 1 & 1 \\ 1 & 1 & 2 \end{bmatrix}$

17. $\begin{bmatrix} 1 & 1 & -1 \\ 2 & 1 & 1 \\ 1 & 0 & 1 \end{bmatrix}$

18. $\begin{bmatrix} 2 & 3 & -1 \\ 1 & 1 & 1 \\ 0 & 2 & -1 \end{bmatrix}$

19. $\begin{bmatrix} 1 & 1 & 0 & 0 \\ 0 & 1 & -1 & 1 \\ 1 & -1 & 1 & 1 \\ 0 & 1 & 0 & -1 \end{bmatrix}$

20. $\begin{bmatrix} 1 & 2 & -3 & -2 \\ 0 & 1 & 4 & -2 \\ 3 & -1 & 4 & 0 \\ 2 & 1 & 0 & 3 \end{bmatrix}$

In Problems 21–26 show that each matrix has no inverse.

21. $\begin{bmatrix} 4 & 6 \\ 2 & 3 \end{bmatrix}$

22. $\begin{bmatrix} -1 & 2 \\ 3 & -6 \end{bmatrix}$

23. $\begin{bmatrix} -8 & 4 \\ -4 & 2 \end{bmatrix}$

24. $\begin{bmatrix} 2 & 10 \\ 1 & 5 \end{bmatrix}$

25. $\begin{bmatrix} 1 & 1 & 1 \\ 3 & -4 & 2 \\ 0 & 0 & 0 \end{bmatrix}$

26. $\begin{bmatrix} -1 & 2 & 3 \\ 5 & 2 & 0 \\ 2 & -4 & -6 \end{bmatrix}$

In Problems 27–34 find the inverse, if it exists, of each matrix.

27. $\begin{bmatrix} 1 & 1 \\ 1 & 2 \end{bmatrix}$

28. $\begin{bmatrix} 2 & 1 \\ 1 & 1 \end{bmatrix}$

29. $\begin{bmatrix} 3 & -2 \\ 0 & 4 \end{bmatrix}$

30. $\begin{bmatrix} 4 & -1 \\ -2 & 0 \end{bmatrix}$

31. $\begin{bmatrix} 3 & 2 \\ 6 & 4 \end{bmatrix}$

32. $\begin{bmatrix} 4 & 2 \\ 2 & 1 \end{bmatrix}$

33. $\begin{bmatrix} 1 & -2 & -1 \\ -2 & 5 & 4 \\ 3 & -8 & -5 \end{bmatrix}$

34. $\begin{bmatrix} 1 & 1 & -1 \\ -2 & -1 & 4 \\ 3 & 2 & -8 \end{bmatrix}$

35. Find the inverse of both

$$A = \begin{bmatrix} 1 & 2 \\ 2 & -1 \end{bmatrix} \quad \text{and} \quad B = \begin{bmatrix} 1 & 3 \\ 2 & 1 \end{bmatrix}$$

to determine $A^{-1} - B^{-1}$.

36. Find the inverse of

$$A = \begin{bmatrix} 1 & -4 \\ 2 & -3 \end{bmatrix} \quad \text{and} \quad B = \begin{bmatrix} 2 & -2 \\ 3 & 2 \end{bmatrix}$$

Determine $A^{-1} - B^{-1}$.

37. Write the matrix product $A^{-1}B$ used to find the solution to the system $AX = B$.

$$\begin{cases} x + 3y + 2z = 2 \\ 2x + 7y + 3z = 1 \\ x \quad\quad + 6z = 3 \end{cases}$$

38. Write the matrix product $A^{-1}B$ used to find the solution to the system $AX = B$.

$$\begin{cases} x + 2y + 2z = 3 \\ 2x + 5y + 7z = 2 \\ 2x + y - 4z = 4 \end{cases}$$

In Problems 39–50 solve each system of equations by the method of Example 7. For Problems 39–44, use the inverse found in Problem 7. For Problems 45–50, use the inverse found in Problem 15.

39. $\begin{cases} 3x + 7y = 10 \\ 2x + 5y = 2 \end{cases}$

40. $\begin{cases} 3x + 7y = -4 \\ 2x + 5y = -3 \end{cases}$

41. $\begin{cases} 3x + 7y = 13 \\ 2x + 5y = 9 \end{cases}$

42. $\begin{cases} 3x + 7y = 0 \\ 2x + 5y = 14 \end{cases}$

43. $\begin{cases} 3x + 7y = 12 \\ 2x + 5y = -4 \end{cases}$

44. $\begin{cases} 3x + 7y = -2 \\ 2x + 5y = 10 \end{cases}$

45. $\begin{cases} x + y - z = 3 \\ 3x - y = -4 \\ 2x - 3y + 4z = 6 \end{cases}$

46. $\begin{cases} x + y - z = 6 \\ 3x - y = 8 \\ 2x - 3y + 4z = -3 \end{cases}$

47. $\begin{cases} x + y - z = 12 \\ 3x - y = -4 \\ 2x - 3y + 4z = 16 \end{cases}$

48. $\begin{cases} x + y - z = -8 \\ 3x - y = 12 \\ 2x - 3y + 4z = -2 \end{cases}$

49. $\begin{cases} x + y - z = 0 \\ 3x - y = -8 \\ 2x - 3y + 4z = -6 \end{cases}$

50. $\begin{cases} x + y - z = 21 \\ 3x - y = 12 \\ 2x - 3y + 4z = 14 \end{cases}$

51. Show that the inverse of

$$A = \begin{bmatrix} a & b \\ c & d \end{bmatrix}$$

is given by the formula

$$A^{-1} = \begin{bmatrix} \dfrac{d}{\Delta} & \dfrac{-b}{\Delta} \\ \dfrac{-c}{\Delta} & \dfrac{a}{\Delta} \end{bmatrix}$$

where $\Delta = ad - bc \neq 0$. The number Δ is called the **determinant** of A.

In Problems 52–55 use the result of Problem 51 to find the inverse of each matrix.

52. $\begin{bmatrix} 1 & 2 \\ 2 & 3 \end{bmatrix}$

53. $\begin{bmatrix} 1 & 5 \\ 2 & 0 \end{bmatrix}$

54. $\begin{bmatrix} -1 & -2 \\ 3 & 4 \end{bmatrix}$

55. $\begin{bmatrix} 1 & 2 \\ 8 & 15 \end{bmatrix}$

Technology Exercises

A graphing calculator can be used to find the inverse of a matrix. If no inverse exists, an error message is given.

In Problems 1–4 use a graphing calculator to find the inverse, if it exists, of each matrix.

1. $\begin{bmatrix} 25 & 61 & -12 \\ 18 & -2 & 4 \\ 8 & 35 & 21 \end{bmatrix}$

2. $\begin{bmatrix} 18 & -3 & 4 \\ 6 & -20 & 14 \\ 10 & 25 & -15 \end{bmatrix}$

3. $\begin{bmatrix} 44 & 21 & 18 & 6 \\ -2 & 10 & 15 & 5 \\ 21 & 12 & -12 & 4 \\ -8 & -16 & 4 & 9 \end{bmatrix}$

4. $\begin{bmatrix} 16 & 22 & -3 & 5 \\ 21 & -17 & 4 & 8 \\ 2 & 8 & 27 & 20 \\ 5 & 15 & -3 & -10 \end{bmatrix}$

Use the idea behind Example 7 and a graphing calculator to solve the following systems of equations.

5. $\begin{cases} 25x + 61y - 12z = 10 \\ 18x - 12y + 7z = -9 \\ 3x + 4y - z = 12 \end{cases}$

6. $\begin{cases} 25x + 61y - 12z = 5 \\ 18x - 12y + 7z = -3 \\ 3x + 4y - z = 12 \end{cases}$

7. $\begin{cases} 25x + 61y - 12z = 21 \\ 18x - 12y + 7z = 7 \\ 3x + 4y - z = -2 \end{cases}$

8. $\begin{cases} 25x + 61y - 12z = 25 \\ 18x - 12y + 7z = 10 \\ 3x + 4y - z = -4 \end{cases}$

9. Use your graphing calculator to find the inverse of the matrix

$$A = \begin{bmatrix} 3 & 0 & 2 & -1 & 3 \\ -2 & 1 & 2 & 3 & 0 \\ 2 & 2 & 1 & 1 & -1 \\ 1 & 2 & 0 & 2 & -3 \\ 4 & 0 & -1 & 1 & -1 \end{bmatrix}$$

Check your result by multiplying your answer by the matrix A. What should the result be?

10. Try to find the inverse of the matrix

$$A = \begin{bmatrix} 0 & 0 & 2 & -1 & 3 \\ -2 & 0 & 2 & 3 & 0 \\ 2 & 2 & 0 & 0 & -1 \\ 1 & 2 & 0 & 2 & -3 \\ 4 & 4 & 0 & 0 & -2 \end{bmatrix}$$

How does your calculator (or computer) respond?

2.7 APPLICATIONS: LEONTIEF MODEL; CRYPTOGRAPHY; ACCOUNTING; THE METHOD OF LEAST SQUARES*

Application 1: Leontief Models

The Leontief models in economics are named after Wassily Leontief, who received the Nobel prize in economics in 1973. These models can be characterized as a description of an economy in which input equals output or, in other words, consumption equals production. That is, the models assume that whatever is produced is always consumed.

Leontief models are of two types: closed, in which the entire production is consumed by those participating in the production; and open, in which some of the production is consumed by those who produce it and the rest of the production is consumed by external bodies.

In the *closed model* we seek the relative income of each participant in the system. In the *open model* we seek the amount of production needed to achieve a forecast demand, when the amount of production needed to achieve current demand is known.

THE CLOSED MODEL

We begin with an example to illustrate the idea.

EXAMPLE 1   Analyzing a Closed Leontief Model

Three homeowners, Juan, Luis, and Carlos, each with certain skills, agreed to pool their talents to make repairs on their houses. As it turned out, Juan spent 20% of his time on his own house, 40% of his time on Luis' house, and 40% on Carlos' house. Luis spent 10% of his time on Juan's house, 50% of his time on his own house, and 40% on Carlos' house. Of Carlos' time, 60% was spent on Juan's house, 10% on Luis', and 30% on his own. Now that the projects are finished, they need to figure out how much money each should get for his work, including the work performed on his own house, so that the amount paid by each person equals the amount received by each one. They agreed in advance that the payment to each one should be approximately $3000.00.

*Each application is optional, and the applications given are independent of one another.

SOLUTION We place the information given in the problem in a 3×3 matrix, as follows:

Work done by

	Juan	Luis	Carlos

Proportion of work done on Juan's house		$\begin{bmatrix} 0.2 & 0.1 & 0.6 \\ 0.4 & 0.5 & 0.1 \\ 0.4 & 0.4 & 0.3 \end{bmatrix}$
Proportion of work done on Luis' house		
Proportion of work done on Carlos' house		

Next, we define the variables:

$$x = \text{Juan's wages}$$
$$y = \text{Luis' wages}$$
$$z = \text{Carlos' wages}$$

We require that the amount paid out by each one equals the amount received by each one. Let's analyze this requirement, by looking just at the work done on Juan's house. Juan's wages are x. Juan's expenditures for work done on his house are $0.2x + 0.1y + 0.6z$. Juan's wages and expenditures are required to be equal, so

$$x = 0.2x + 0.1y + 0.6z$$

Similarly,

$$y = 0.4x + 0.5y + 0.1z$$
$$z = 0.4x + 0.4y + 0.3z$$

These three equations can be written compactly as

$$\begin{bmatrix} x \\ y \\ z \end{bmatrix} = \begin{bmatrix} 0.2 & 0.1 & 0.6 \\ 0.4 & 0.5 & 0.1 \\ 0.4 & 0.4 & 0.3 \end{bmatrix} \begin{bmatrix} x \\ y \\ z \end{bmatrix}$$

Some computation reduces the system to

$$\begin{cases} 0.8x - 0.1y - 0.6z = 0 \\ -0.4x + 0.5y - 0.1z = 0 \\ -0.4x - 0.4y + 0.7z = 0 \end{cases}$$

Solving for x, y, z, we find that

$$x = \tfrac{31}{36}z \qquad y = \tfrac{32}{36}z$$

where z is the parameter. To get solutions that fall close to \$3000, we set $z = 3600$.* The wages to be paid out are therefore

$$x = \$3100 \qquad y = \$3200 \qquad z = \$3600$$

■

The matrix in Example 1, namely,

$$\begin{bmatrix} 0.2 & 0.1 & 0.6 \\ 0.4 & 0.5 & 0.1 \\ 0.4 & 0.4 & 0.3 \end{bmatrix}$$

is called an **input–output matrix.**

* Other choices for z are, of course, possible. The choice of which value to use is up to the homeowners. No matter what choice is made, the amount paid by each one equals the amount received by each one.

In the general closed model we have an economy consisting of n components. Each component produces an *output* of some goods or services, which, in turn, is completely used up by the n components. The proportionate use of each component's output by the economy makes up the input–output matrix of the economy. The problem is to find suitable pricing levels for each component so that income equals expenditure.

Closed Leontief Model

In general, an input–output matrix for a closed Leontief model is of the form

$$A = [a_{ij}] \qquad i, j = 1, 2, \ldots, n$$

where the a_{ij} represent the fractional amount of goods or services used by i and produced by j. For a closed model the sum of each column equals 1 (this is the condition that all production is consumed internally) and $0 \leq a_{ij} \leq 1$ for all entries (this is the restriction that each entry is a fraction).

If A is the input–output matrix of a closed system with n components and X is a column vector representing the price of each output of the system, then

$$X = AX$$

represents the requirement that income equal expenditure.

For example, the first entry of the matrix equality $X = AX$ requires that

$$x_1 = a_{11} x_1 + a_{12} x_2 + \cdots + a_{1n} x_n$$

The right side represents the price paid by component 1 for the goods it uses, while x_1 represents the income of component 1; we are requiring they be equal.

We can rewrite the equation $X = AX$ as

$$X - AX = \mathbf{0}$$
$$I_n X - AX = \mathbf{0}$$
$$(I_n - A)X = \mathbf{0}$$

This matrix equation, which represents a system of equations in which the right-hand side is always $\mathbf{0}$, is called a **homogeneous system of equations.** It can be shown that if the entries in the input–output matrix A are positive and if the sum of each column of A equals 1, then this system has a one-parameter solution; that is, we can solve for $n - 1$ of the variables in terms of the remaining one, which serves as the parameter. This parameter serves as a "scale factor."

✍ **Now Work Problem 1**

THE OPEN MODEL

For the open model, in addition to internal consumption of goods produced, there is an outside demand for the goods produced. This outside demand may take the form of exportation of goods or may be the goods needed to support consumer demand. Again, however, we make the assumption that whatever is produced is also consumed.

For example, suppose an economy consists of three industries R, S, and T, and suppose each one produces a single product. We assume that a portion of R's production is used by each of the three industries, while the remainder is used up by consumers.

The same is true of the production of S and T. To organize our thoughts, we construct a table that describes the interaction of the use of R, S, and T's production over some fixed period of time. See Table 1.

Table 1

	R	S	T	Consumer	Total
R	50	20	40	70	180
S	20	30	20	90	160
T	30	20	20	50	120

All entries in the table are in appropriate units, say, in dollars. The first row (row R) represents the production in dollars of industry R (input). Out of the total of $180 worth of goods produced, R, S, and T use $50, $20, and $40, respectively, for the production of their goods, while consumers purchase the remaining $70 for their consumption (output). Observe that input equals output since everything produced by R is used up by R, S, T, and consumers.

The second and third rows are interpreted in the same way.

An important observation is that the goal of R's production is to produce $70 worth of goods, since this is the demand of consumers. In order to meet this demand, R must produce a total of $180, since the difference, $110, is required internally by R, S, and T.

Suppose, however, that consumer demand is expected to change. To effect this change, how much should each industry now produce? For example, in Table 1, current demand for R, S, and T can be represented by a **demand vector:**

$$D_0 = \begin{bmatrix} 70 \\ 90 \\ 50 \end{bmatrix}$$

But suppose marketing forecasts predict that in 3 years the demand vector will be

$$D_3 = \begin{bmatrix} 60 \\ 110 \\ 60 \end{bmatrix}$$

Here, the demand for item R has decreased; the demand for item S has significantly increased, and the demand for item T is higher. Given the current total output of R, S, and T at 180, 160, and 120, respectively, what must it be in 3 years to meet this projected demand?

The solution of this type of forecasting problem is derived from the *open Leontief model* in input–output analysis. In using input–output analysis to obtain a solution to such a forecasting problem, we take into account the fact that the output of any one of these industries is affected by changes in the other two, since the total demand for, say, R in 3 years depends not only on consumer demand for R, but also on consumer demand for S and T. That is, the industries are interrelated.

To obtain the solution, we need to determine how much of each of the three products R, S, and T is required to produce 1 unit of R. For example, to obtain 180 units of R requires the use of 50 units of R, 20 units of S, and 30 units of T (the entries in column 1). Forming the ratios, we find that to produce 1 unit of R requires $\frac{50}{180} = 0.278$ of R, $\frac{20}{180} = 0.111$ of S, and $\frac{30}{180} = 0.167$ of T. If we want, say, x units of R, we will require $0.278x$ units of R, $0.111x$ units of S, and $0.167x$ units of T.

Continuing in this way, we can construct the matrix

$$A = \begin{array}{c} \\ R \\ S \\ T \end{array} \begin{array}{ccc} R & S & T \\ \left[\begin{array}{ccc} 0.278 & 0.125 & 0.333 \\ 0.111 & 0.188 & 0.167 \\ 0.167 & 0.125 & 0.167 \end{array}\right] \end{array}$$

Observe that column 1 represents the amounts of R, S, T required for 1 unit of R; column 2 represents the amounts of R, S, and T required for 1 unit of S; and column 3 represents the amounts of R, S, and T required for 1 unit of T. For example, the entry in row 3, column 2 (0.125), represents the amount of T needed to produce 1 unit of S.

As a result of placing the entries this way, if

$$X = \begin{bmatrix} x \\ y \\ z \end{bmatrix}$$

represents the total output required to obtain a given demand, the product AX represents the amount of R, S, and T required for internal consumption. The condition that production = consumption requires that

Internal consumption + Consumer demand = Total output

In terms of the matrix A, the total output X, and the demand vector D, this requirement is equivalent to the equation

$$AX + D = X$$

In this equation we seek to find X for a prescribed demand D. The matrix A is calculated as above for some initial production process.*

EXAMPLE 2   Finding Production Levels to Meet Future Demand: The Open Leontief Model

For the data given in Table 1, find the total output X required to achieve a future demand of

$$D_3 = \begin{bmatrix} 60 \\ 110 \\ 60 \end{bmatrix}$$

---

* The entries in A can be checked by using the requirement that $AX + D = X$, for $D =$ Initial demand and $X =$ Total output. For our example, it must happen that

$$\begin{array}{ccc} \text{Internal consumption} & + \text{Consumer demand} & = \text{Total ouput} \\ AX & D & = X \end{array}$$

$$\begin{bmatrix} 0.278 & 0.125 & 0.333 \\ 0.111 & 0.188 & 0.167 \\ 0.167 & 0.125 & 0.167 \end{bmatrix} \begin{bmatrix} 180 \\ 160 \\ 120 \end{bmatrix} + \begin{bmatrix} 70 \\ 90 \\ 50 \end{bmatrix} = \begin{bmatrix} 180 \\ 160 \\ 120 \end{bmatrix}$$

SOLUTION   We need to solve for X in

$$AX + D_3 = X$$

Simplifying, we have

$$[I_3 - A]X = D_3$$

Solving for X, we have

$$X = [I_3 - A]^{-1} \cdot D_3$$

$$= \begin{bmatrix} 0.722 & -0.125 & -0.333 \\ -0.111 & 0.812 & -0.167 \\ -0.167 & -0.125 & 0.833 \end{bmatrix}^{-1} \begin{bmatrix} 60 \\ 110 \\ 60 \end{bmatrix}$$

$$= \begin{bmatrix} 1.6048 & 0.3568 & 0.7131 \\ 0.2946 & 1.3363 & 0.3857 \\ 0.3660 & 0.2721 & 1.4013 \end{bmatrix} \begin{bmatrix} 60 \\ 110 \\ 60 \end{bmatrix}$$

$$= \begin{bmatrix} 178.322 \\ 187.811 \\ 135.969 \end{bmatrix}$$

Thus the total output of R, S, and T required for the forecast demand D_3 is

$$x = 178.322 \qquad y = 187.811 \qquad z = 135.969$$

■

The general open model can be described as follows: Suppose there are n industries in the economy. Each industry produces some goods or services, which are partially consumed by the n industries, while the rest are used to meet a prescribed current demand. Given the output required of each industry to meet current demand, what should the output of each industry be to meet some different future demand?

Open Leontief Model

The matrix $A = [a_{ij}]$, $i, j = 1, \ldots, n$, of the open model is defined to consist of entries a_{ij}, where a_{ij} is the amount of output of industry j required for one unit of output of industry i. If X is a column vector representing the production of each industry in the system and D is a column vector representing future demand for goods produced in the system, then

$$X = AX + D$$

From the equation above we find

$$[I_n - A]X = D$$

It can be shown that the matrix $I_n - A$ has an inverse, provided each entry in A is positive and the sum of each column in A is less than 1. Under these conditions we may solve for X to get

$$X = [I_n - A]^{-1} \cdot D$$

This form of the solution is particularly useful since it allows us to find X for a variety of demands D by doing one calculation: $[I_n - A]^{-1}$.

We conclude by noting that the use of an input–output matrix to solve forecasting problems assumes that each industry produces a single commodity and that no technological advances take place in the period of time under investigation (in other words, the proportions found in the matrix A are fixed).

EXERCISE 2.7: APPLICATION 1   Answers to odd-numbered problems begin on page AN-12.

In Problems 1–4 find the relative wages of each person for the given closed input–output matrix. In each case take the wages of C to be the parameter and use $z = C$'s wages $= \$30,000$.

$$
\text{1. } \begin{array}{c} \\ A \\ B \\ C \end{array}
\begin{array}{ccc} A & B & C \end{array}
\begin{bmatrix}
\frac{1}{2} & \frac{1}{3} & \frac{1}{4} \\
\frac{1}{4} & \frac{1}{3} & \frac{1}{4} \\
\frac{1}{4} & \frac{1}{3} & \frac{1}{2}
\end{bmatrix}
\qquad
\text{2. } \begin{array}{c} \\ A \\ B \\ C \end{array}
\begin{array}{ccc} A & B & C \end{array}
\begin{bmatrix}
\frac{1}{4} & \frac{2}{3} & \frac{1}{2} \\
\frac{1}{2} & \frac{1}{6} & \frac{1}{4} \\
\frac{1}{4} & \frac{1}{6} & \frac{1}{4}
\end{bmatrix}
$$

$$
\text{3. } \begin{array}{c} \\ A \\ B \\ C \end{array}
\begin{array}{ccc} A & B & C \end{array}
\begin{bmatrix}
0.2 & 0.3 & 0.1 \\
0.6 & 0.4 & 0.2 \\
0.2 & 0.3 & 0.7
\end{bmatrix}
\qquad
\text{4. } \begin{array}{c} \\ A \\ B \\ C \end{array}
\begin{array}{ccc} A & B & C \end{array}
\begin{bmatrix}
0.4 & 0.3 & 0.2 \\
0.2 & 0.3 & 0.3 \\
0.4 & 0.4 & 0.5
\end{bmatrix}
$$

5. For the three industries R, S, and T in the open Leontief model of Table 1 on page 111, compute the total output vector X if the forecast demand vector is

$$
D_2 = \begin{bmatrix} 80 \\ 90 \\ 60 \end{bmatrix}
$$

6. Rework Problem 5 if the forecast demand vector is

$$
D_4 = \begin{bmatrix} 100 \\ 80 \\ 60 \end{bmatrix}
$$

7. Closed Leontief Model   A society consists of four individuals: a farmer, a builder, a tailor, and a rancher (who produces meat products). Of the food produced by the farmer, $\frac{3}{10}$ is used by the farmer, $\frac{2}{10}$ by the builder, $\frac{2}{10}$ by the tailor, and $\frac{3}{10}$ by the rancher. The builder's production is utilized 30% by the farmer, 30% by the builder, 10% by the tailor, and 30% by the rancher. The tailor's production is used in the ratios $\frac{3}{10}$, $\frac{3}{10}$, $\frac{1}{10}$, and $\frac{3}{10}$ by the farmer, builder, tailor, and rancher, respectively. Finally, meat products are used 20% by each of the farmer, builder, and tailor, and

40% by the rancher. What is the relative income of each if the rancher's income is scaled at $\$25,000$?

8. If in Problem 7 the meat production utilization changes so that it is used equally by all four individuals, while everyone else's production utilization remains the same, what are the relative incomes?

9. Open Leontief Model   Suppose the interrelationships between the production of two industries R and S in a given year are given in the table:

	R	S	Current Consumer Demand	Total Output
R	30	40	60	130
S	20	10	40	70

If the forecast demand in 2 years is

$$
D_2 = \begin{bmatrix} 80 \\ 40 \end{bmatrix}
$$

what should the total output X be?

Technology Exercises

In Problems 1–4 use a graphing calculator (or a computer) to solve each open Leontief model. Find the production vector X for given input–output matrix A and demand vector D, using the formula $X = [I_n - A]^{-1} \cdot D$.

$$
\text{1. } A = \begin{bmatrix}
0 & 0.1 & 0.2 & 0.1 \\
0.2 & 0 & 0.4 & 0.3 \\
0.2 & 0.7 & 0 & 0 \\
0.1 & 0.2 & 0 & 0.2
\end{bmatrix};\quad
D = \begin{bmatrix} 3 \\ 0 \\ 8 \\ 1 \end{bmatrix}
$$

$$
\text{2. } A = \begin{bmatrix}
0 & 0 & 0.2 & 0.1 & 0.3 \\
0.2 & 0 & 0.2 & 0.3 & 0 \\
0.2 & 0.2 & 0 & 0 & 0.1 \\
0.1 & 0.2 & 0 & 0.2 & 0.3 \\
0.4 & 0 & 0.1 & 0 & 0.2
\end{bmatrix};\quad
D = \begin{bmatrix} 3 \\ 0 \\ 7 \\ 5 \\ 1 \end{bmatrix}
$$

3. $A = \begin{bmatrix} 0 & 0.2 & 0.1 & 0.3 \\ 0 & 0.2 & 0.3 & 0 \\ 0.2 & 0 & 0 & 0.1 \\ 0 & 0 & 0.2 & 0.3 \end{bmatrix}$; $D = \begin{bmatrix} 5 \\ 7 \\ 2 \\ 0 \end{bmatrix}$

4. $A = \begin{bmatrix} 0 & 0 & 0.02 & 0 & 0.3 \\ 0.2 & 0 & 0 & 0.3 & 0 \\ 0.2 & 0 & 0 & 0 & 0.01 \\ 0.1 & 0.7 & 0 & 0.01 & 0.03 \\ 0.04 & 0.4 & 0 & 0 & 0.2 \end{bmatrix}$; $D = \begin{bmatrix} 7 \\ 0 \\ 6 \\ 8 \\ 2 \end{bmatrix}$

5. Open Leontief Model Suppose the interrelationships between the production of five industries (manufacturing, electric power, petroleum, transportation, and textiles) in a given year are as listed in the table:

	Manu.	**E.P.**	**Petr.**	**Tran.**	**Text.**
Manu.	0.2	0.12	0.15	0.18	0.1
E.P.	0.17	0.11	0	0.19	0.28
Petr.	0.11	0.11	0.12	0.46	0.12
Tran.	0.1	0.14	0.18	0.17	0.19
Text.	0.16	0.18	0.02	0.1	0.3

If the demand vector is given by

$$D = \begin{bmatrix} 100 \\ 100 \\ 200 \\ 100 \\ 100 \end{bmatrix}$$

what should the production vector X be? [*Hint:* Use your graphing calculator to find the inverse $[I_n - A]^{-1}$.]

Application 2: Cryptography

Our second application is to *cryptography,* the art of writing or deciphering secret codes. We begin by giving examples of elementary codes.

EXAMPLE 1   Encoding a Message

A message can be encoded by associating each letter of the alphabet with some other letter of the alphabet according to a prescribed pattern. For example, we might have

A B C D E F G H I J K L M N O P Q R S T U V W X Y Z
↓ ↓
C D E F G H I J K L M N O P Q R S T U V W X Y Z A B

With the above code the word *BOMB* would become DQOD. ∎

EXAMPLE 2   Encoding a Message

Another code may associate numbers with the letters of the alphabet. For example, we might have

A B C D E F G H I J K L M N O P Q R S T U V W X Y Z
↓ ↓
26 25 24 23 22 21 20 19 18 17 16 15 14 13 12 11 10 9 8 7 6 5 4 3 2 1

In this code the word *PEACE* looks like 11 22 26 24 22. ∎

Both the above codes have one important feature in common. The association of letters with the coding symbols is made using a one-to-one correspondence so that no possible ambiguities can arise.

Suppose we want to encode the following message:

BEWARE THE IDES OF MARCH

If we decide to divide the message into pairs of letters, the message becomes

BE WA RE TH EI DE SO FM AR CH

(If there is a letter left over, we arbitrarily assign Z to the last position.) Using the correspondence of letters to numbers given in Example 2, and writing each pair of letters as a column vector, we obtain

$$\begin{bmatrix} B \\ E \end{bmatrix} = \begin{bmatrix} 25 \\ 22 \end{bmatrix} \qquad \begin{bmatrix} W \\ A \end{bmatrix} = \begin{bmatrix} 4 \\ 26 \end{bmatrix} \qquad \begin{bmatrix} R \\ E \end{bmatrix} = \begin{bmatrix} 9 \\ 22 \end{bmatrix} \qquad \begin{bmatrix} T \\ H \end{bmatrix} = \begin{bmatrix} 7 \\ 19 \end{bmatrix} \qquad \text{etc.}$$

Next, we arbitrarily choose a 2×2 matrix A, which we know has an inverse A^{-1} (the reason for this is seen later). Suppose we choose

$$A = \begin{bmatrix} 2 & 3 \\ 1 & 2 \end{bmatrix}$$

Its inverse is

$$A^{-1} = \begin{bmatrix} 2 & -3 \\ -1 & 2 \end{bmatrix}$$

Now, we transform the column vectors representing the message by multiplying each of them on the left by the matrix A:

$$A \begin{bmatrix} B \\ E \end{bmatrix} = A \begin{bmatrix} 25 \\ 22 \end{bmatrix} = \begin{bmatrix} 116 \\ 69 \end{bmatrix}$$

$$A \begin{bmatrix} W \\ A \end{bmatrix} = A \begin{bmatrix} 4 \\ 26 \end{bmatrix} = \begin{bmatrix} 86 \\ 56 \end{bmatrix}$$

$$A \begin{bmatrix} R \\ E \end{bmatrix} = A \begin{bmatrix} 9 \\ 22 \end{bmatrix} = \begin{bmatrix} 84 \\ 53 \end{bmatrix} \qquad \text{etc.}$$

The coded message is

116   69   86   56   84   53   etc.

To decode or unscramble the above message, pair the numbers in 2×1 column vectors. Multiply each of these column vectors by A^{-1} on the left:

$$A^{-1} \begin{bmatrix} 116 \\ 69 \end{bmatrix} = \begin{bmatrix} 25 \\ 22 \end{bmatrix}$$

$$A^{-1} \begin{bmatrix} 86 \\ 56 \end{bmatrix} = \begin{bmatrix} 4 \\ 26 \end{bmatrix}$$

By reassigning letters to these numbers, we obtain the original message.

✍ **Now Work Problem 1**

EXAMPLE 3   Encoding a Message

The message to be encoded is

THE   END   IS   NEAR

We agree to associate numbers to letters as follows:

A B C D E F G H I J K L M N O P Q R S T U V W X Y Z
↓ ↓
1 2 3 4 5 6 7 8 9 10 11 12 13 14 15 16 17 18 19 20 21 22 23 24 25 26

The encoded message is to be formed of triplets of numbers.

SOLUTION   This time we must divide the message into triplets of letters, obtaining

THE   END   ISN   EAR

in order for the encoded message to have triplets of numbers. (If the message required additional letters to complete the triplet, we would have used Z or YZ.)

Now we choose a 3×3 matrix such as

$$A = \begin{bmatrix} 1 & 0 & 0 \\ 3 & 1 & 5 \\ -2 & 0 & 1 \end{bmatrix}$$

Its inverse is

$$A^{-1} = \begin{bmatrix} 1 & 0 & 0 \\ -13 & 1 & -5 \\ 2 & 0 & 1 \end{bmatrix}$$

The encoded message is obtained by multiplying the matrix A times each column vector of the original message:

$$A \begin{bmatrix} T \\ H \\ E \end{bmatrix} = \begin{bmatrix} 1 & 0 & 0 \\ 3 & 1 & 5 \\ -2 & 0 & 1 \end{bmatrix} \begin{bmatrix} 20 \\ 8 \\ 5 \end{bmatrix} = \begin{bmatrix} 20 \\ 93 \\ -35 \end{bmatrix}$$

$$A \begin{bmatrix} E \\ N \\ D \end{bmatrix} = \begin{bmatrix} 1 & 0 & 0 \\ 3 & 1 & 5 \\ -2 & 0 & 1 \end{bmatrix} \begin{bmatrix} 5 \\ 14 \\ 4 \end{bmatrix} = \begin{bmatrix} 5 \\ 49 \\ -6 \end{bmatrix}$$

$$A \begin{bmatrix} I \\ S \\ N \end{bmatrix} = \begin{bmatrix} 1 & 0 & 0 \\ 3 & 1 & 5 \\ -2 & 0 & 1 \end{bmatrix} \begin{bmatrix} 9 \\ 19 \\ 14 \end{bmatrix} = \begin{bmatrix} 9 \\ 116 \\ -4 \end{bmatrix}$$

$$A \begin{bmatrix} E \\ A \\ R \end{bmatrix} = \begin{bmatrix} 1 & 0 & 0 \\ 3 & 1 & 5 \\ -2 & 0 & 1 \end{bmatrix} \begin{bmatrix} 5 \\ 1 \\ 18 \end{bmatrix} = \begin{bmatrix} 5 \\ 106 \\ 8 \end{bmatrix}$$

The coded message is

20 93 −35 5 49 −6 9 116 −4 5 106 8

To decode the message in Example 3, form 3×1 column vectors of the numbers in the coded message and multiply on the left by A^{-1}.

The above are elementary examples of encoding and decoding. Modern-day cryptography uses sophisticated computer-implemented codes that depend on higher-level mathematics.

EXERCISE 2.7: APPLICATION 2   Answers to odd-numbered problems begin on page AN-13.

Problems 1–5 use the correspondence

$$\begin{array}{cccccccccccccccccccccccccc} A & B & C & D & E & F & G & H & I & J & K & L & M & N & O & P & Q & R & S & T & U & V & W & X & Y & Z \\ \downarrow & \downarrow & \downarrow & \downarrow & \downarrow & \downarrow & \downarrow & \downarrow & \downarrow & \downarrow & \downarrow & \downarrow & \downarrow & \downarrow & \downarrow & \downarrow & \downarrow & \downarrow & \downarrow & \downarrow & \downarrow & \downarrow & \downarrow & \downarrow & \downarrow & \downarrow \\ 1 & 2 & 3 & 4 & 5 & 6 & 7 & 8 & 9 & 10 & 11 & 12 & 13 & 14 & 15 & 16 & 17 & 18 & 19 & 20 & 21 & 22 & 23 & 24 & 25 & 26 \end{array}$$

1. Use the matrix

$$A = \begin{bmatrix} 2 & 3 \\ 1 & 2 \end{bmatrix}$$

to decode the following messages:

(a) 51   30   27   16   75   47   19   10   48   26
(b) 70   45   103   62   58   38   102   61   88   57

2. Use the matrix

$$A = \begin{bmatrix} 1 & 0 & 0 \\ 3 & 1 & 5 \\ -2 & 0 & 1 \end{bmatrix}$$

to decode the message

25   195   −29   6   135   9   14   183   −2

3. The matrix A used to encode a message has as its inverse the matrix

$$A^{-1} = \begin{bmatrix} 2 & -3 \\ -1 & 2 \end{bmatrix}$$

Decode the message

11   7   84   51   51   28   66   43   44   29   107   65   64   41

4. Use the matrix

$$A = \begin{bmatrix} 1 & 0 & 0 \\ 3 & 1 & 5 \\ -2 & 0 & 1 \end{bmatrix}$$

to encode the message: SELL THE COMPANY

5. Use the matrices

(I) $A = \begin{bmatrix} 2 & 3 \\ 1 & 2 \end{bmatrix}$ (II) $A = \begin{bmatrix} 1 & 0 & 0 \\ 3 & 1 & 5 \\ -2 & 0 & 1 \end{bmatrix}$

to encode the following messages:

(a) MEET ME AT THE CASBAH
(b) TOMORROW NEVER COMES
(c) THE MISSION IS IMPOSSIBLE

Application 3: Accounting

Consider a firm that has two types of departments, production and service. The production departments produce goods that can be sold in the market, and the service departments provide services to the production departments. A major objective of the cost accounting process is the determination of the full cost of manufactured products on a per unit basis. This requires an allocation of indirect costs, first, from the service department (where they are incurred) to the producing department in which the goods are manufactured and, second, to the specific goods themselves. For example, an accounting department usually provides accounting services for service departments, as well as for the production departments. Thus the indirect costs of service rendered by a service department must be determined in order to correctly assess the production departments. The total costs of a service department consist of its direct costs (salaries, wages, and materials) and its indirect costs (charges for the services it receives from other service departments). The nature of the problem and its solution are illustrated by the following example.

EXAMPLE 1   Finding Total Monthly Costs

Consider a firm with two production departments, P_1 and P_2, and three service departments, S_1, S_2, and S_3. These five departments are listed in the leftmost column of Table 2.

Table 2

Department	Total Costs	Direct Costs, Dollars	Indirect Costs for Services from Departments		
			S_1	S_2	S_3
S_1	x_1	600	$0.25x_1$	$0.15x_2$	$0.15x_3$
S_2	x_2	1100	$0.35x_1$	$0.20x_2$	$0.25x_3$
S_3	x_3	600	$0.10x_1$	$0.10x_2$	$0.35x_3$
P_1	x_4	2100	$0.15x_1$	$0.25x_2$	$0.15x_3$
P_2	x_5	1500	$0.15x_1$	$0.30x_2$	$0.10x_3$
Totals		5900	x_1	x_2	x_3

The total monthly costs of these departments are unknown and are denoted by x_1, x_2, x_3, x_4, x_5. The direct monthly costs of the five departments are shown in the third column of the table. The fourth, fifth, and sixth columns show the allocation of charges for the services of S_1, S_2, and S_3 to the various departments. Since the total cost for each department is its direct costs plus its indirect costs, the first three rows of the table yield the total costs for the three service departments:

$$x_1 = \ \ 600 + 0.25x_1 + 0.15x_2 + 0.15x_3$$
$$x_2 = 1100 + 0.35x_1 + 0.20x_2 + 0.25x_3$$
$$x_3 = \ \ 600 + 0.10x_1 + 0.10x_2 + 0.35x_3$$

Let X, C, and D denote the following matrices:

$$X = \begin{bmatrix} x_1 \\ x_2 \\ x_3 \end{bmatrix} \qquad C = \begin{bmatrix} 0.25 & 0.15 & 0.15 \\ 0.35 & 0.20 & 0.25 \\ 0.10 & 0.10 & 0.35 \end{bmatrix} \qquad D = \begin{bmatrix} 600 \\ 1100 \\ 600 \end{bmatrix}$$

Then the system of equations above can be written in matrix notation as

$$X = D + CX$$

which is equivalent to

$$[I_3 - C]X = D$$

The total costs of the three service departments can be obtained by solving this matrix equation for X:

$$X = [I_3 - C]^{-1}D$$

Now

$$[I_3 - C] = \begin{bmatrix} 0.75 & -0.15 & -0.15 \\ -0.35 & 0.80 & -0.25 \\ -0.10 & -0.10 & 0.65 \end{bmatrix}$$

from which it can be verified that

$$[I_3 - C]^{-1} = \begin{bmatrix} 1.57 & 0.36 & 0.50 \\ 0.79 & 1.49 & 0.76 \\ 0.36 & 0.28 & 1.73 \end{bmatrix}$$

It is significant that the inverse of $[I_3 - C]$ exists, and that all of its entries are nonnegative. Because of this and the fact that the matrix D contains only nonnegative entries, the matrix X will also have only nonnegative entries. This means there is a meaningful solution to the accounting problem:

$$X = [I_3 - C]^{-1}D = \begin{bmatrix} 1638.00 \\ 2569.00 \\ 1562.00 \end{bmatrix}$$

Thus, $x_1 = \$1638.00$, $x_2 = \$2569.00$, and $x_3 = \$1562.00$. All direct and indirect costs can now be determined by substituting these values in Table 2, as shown in Table 3.

From Table 3 we learn that department P_1 pays \$1122.25 for the services it receives from S_1, S_2, S_3, and P_2 pays \$1172.60 for the services it receives from these departments. The procedure we have followed charges the direct costs of the service departments to the production departments, and each production department is charged according to the services it utilizes. Furthermore, the total cost for P_1 and P_2 is \$5894.85, and this figure approximates the sum of the direct costs of the three service departments and the two production departments. The results are consistent with conventional accounting procedure. Discrepancies that occur are due to rounding.

Table 3

Department	Total Costs, Dollars	Direct Costs, Dollars	Indirect Costs for Services from Departments, Dollars		
			S_1	S_2	S_3
S_1	1629.15	600	409.50	385.35	234.30
S_2	2577.60	1100	573.30	513.80	390.50
S_3	1567.40	600	163.80	256.90	546.70
P_1	3222.25	2100	245.70	642.25	234.30
P_2	2672.60	1500	245.70	770.70	156.20

Finally, a comment should be made about the allocation of charges for services as shown in Table 2. How is it determined that 25% of the total cost x_1 of S_1 should be charged to S_1, 35% to S_2, 10% to S_3, 15% to P_1, and 15% to P_2? The services of each department can be measured in some suitable unit, and each department can be charged according to the number of these units of service it receives. If 20% of the accounting items concern a given department, that department is charged 20% of the total cost of the accounting department. When services are not readily measurable, the allocation basis is subjectively determined.

EXERCISE 2.7: APPLICATION 3   Answers to odd-numbered problems begin on page AN-13.

1. Consider the accounting problem described by the data in the table:

Department	Total Costs	Direct Costs, Dollars	Indirect Costs for Services from Departments	
			S_1	S_2
S_1	x_1	2,000	$\frac{1}{9}x_1$	$\frac{3}{9}x_2$
S_2	x_2	1,000	$\frac{3}{9}x_1$	$\frac{1}{9}x_2$
P_1	x_3	2,500	$\frac{1}{9}x_1$	$\frac{2}{9}x_2$
P_2	x_4	1,500	$\frac{3}{9}x_1$	$\frac{1}{9}x_2$
P_3	x_5	3,000	$\frac{1}{9}x_1$	$\frac{2}{9}x_2$
Totals		10,000	x_1	x_2

Determine whether this accounting problem has a solution. If it does, find the total costs. Prepare a table similar to Table 3. Show that the total of the service charges allocated to P_1, P_2, and P_3 is equal to the sum of the direct costs of the service departments S_1 and S_2.

2. Follow the directions of Problem 1 for the accounting problem described by the following data:

Department	Total Costs	Direct Costs, Dollars	Indirect Costs for Services from Departments		
			S_1	S_2	S_3
S_1	x_1	500	$0.20x_1$	$0.10x_2$	$0.10x_3$
S_2	x_2	1000	$0.40x_1$	$0.15x_2$	$0.30x_3$
S_3	x_3	500	$0.10x_1$	$0.05x_2$	$0.30x_3$
P_1	x_4	2000	$0.20x_1$	$0.35x_2$	$0.20x_3$
P_2	x_5	1500	$0.10x_1$	$0.35x_2$	$0.10x_3$
Totals		5500	x_1	x_2	x_3

Application 4: The Method of Least Squares

The method of least squares refers to a technique that is often used in data analysis to find the "best" linear equation that fits a given collection of experimental data. (We shall see a little later just what we mean by the word *best*.) It is a technique employed by statisticians, economists, business forecasters, and those who try to interpret and analyze data. Before we begin our discussion, we will need to define the *transpose* of a matrix.

Transpose

  Let A be a matrix of dimension $m \times n$. The **transpose** of A, written A^T, is the $n \times m$ matrix obtained from A by interchanging the rows and columns of A.

  Thus, the first row of A^T is the first column of A; the second row of A^T is the second column of A; and so on.

EXAMPLE 1   Finding the Transpose of a Matrix

If

$$A = \begin{bmatrix} 1 & 2 & 3 \\ 0 & -1 & 2 \end{bmatrix} \qquad B = \begin{bmatrix} 1 & 1 \\ 0 & 1 \\ 2 & 3 \end{bmatrix} \qquad C = \begin{bmatrix} 1 & 0 & -1 \end{bmatrix}$$

then

$$A^T = \begin{bmatrix} 1 & 0 \\ 2 & -1 \\ 3 & 2 \end{bmatrix} \qquad B^T = \begin{bmatrix} 1 & 0 & 2 \\ 1 & 1 & 3 \end{bmatrix} \qquad C^T = \begin{bmatrix} 1 \\ 0 \\ -1 \end{bmatrix}$$

Note how dimensions are reversed when computing A^T: A has the dimension 2×3, while A^T has the dimension 3×2, and so on.

We begin our discussion of the method of least squares with an example.

Suppose a product has been sold over time at various prices and that we have some data that show the demand for the product (in thousands of units) in terms of its price (in dollars). If we use x to represent price and y to represent demand, the data might look like the information in the table to the left.

Price, x	4	5	9	12
Demand, y	9	8	6	3

Suppose also that we have reason to assume that y and x are *linearly related*. That is, we assume we can write

$$y = mx + b$$

for some, as yet unknown, m and b. In other words, we are assuming that when demand is graphed against price, the resulting graph will be a line. Our belief that y and x are linearly related might be based on past experience or economic theory.

If we plot the data from the above table (see Figure 6), it seems pretty clear that no single line passes through the plotted points. Though no line will pass through all the points, we might still ask: "Is there a line that 'best fits' the points?" That is, we seek a line of the sort in Figure 7.

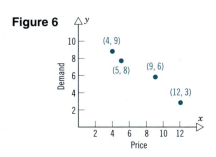

Figure 6

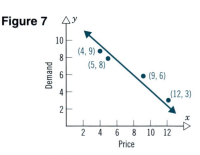

Figure 7

Our Problem

Given a set of noncollinear points, find the line that "best fits" these points.

We will need to clarify our use of the phrase "best fits." First, though, suppose there are four data points, labeled from left to right as (x_1, y_1), (x_2, y_2), and so on. See Figure 8. Now suppose the equation of the line L in Figure 8 is

$$y = mx + b$$

where we do not know what m and b are. If (x_1, y_1) is actually on the line, it would be true that

$$y_1 = mx_1 + b$$

Figure 8

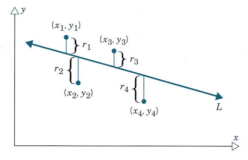

But, since we can't be sure (x_1, y_1) lies on L, the above equation may not be true. That is, there may be a nonzero difference between y_1 and $mx_1 + b$. We designate this difference by r_1. That is,

$$r_1 = y_1 - (mx_1 + b)$$

or

$$y_1 = mx_1 + b + r_1$$

What we have just said about (x_1, y_1) we can repeat for the remaining data points (x_i, y_i), $i = 2, 3$, and 4, obtaining

$$r_i = y_i - (mx_i + b)$$

or, equivalently,

$$y_i = mx_i + b + r_i \qquad i = 1, 2, 3, 4 \tag{1}$$

Geometrically, $|r_i|$ measures the vertical distance between the sought-after line L and the data points.

Recall that our problem is to find a line L that "best fits" the data points (x_i, y_i). Intuitively, we would seek a line L for which the r_i's are all simultaneously small. Since some r_i's may be positive and some may be negative, it will not do to just add them up. To eliminate the signs, we square each r_i. The method of least squares consists of finding a line L for which the sum of the squares of the r_i's is as small as possible. Since finding the line $y = mx + b$ is the same as finding m and b, we restate the least squares problem for this example as follows:

Least Squares Problem

Given the data points (x_i, y_i), $i = 1, 2, 3, 4$, find m and b satisfying

$$y_i = mx_i + b + r_i \qquad i = 1, 2, 3, 4 \tag{2}$$

so that $r_1^2 + r_2^2 + r_3^2 + r_4^2$ is minimized.

A solution to the problem can be compactly expressed using matrix language. Since a derivation requires either calculus or more advanced linear algebra, we omit the proof and simply present the solution:

Solution to Least Squares Problem

The line of best fit

$$y = mx + b \tag{3}$$

to a set of four points (x_1, y_1), (x_2, y_2), (x_3, y_3), (x_4, y_4) is obtained by solving the system of two equations in two unknowns

$$A^T A X = A^T Y \tag{4}$$

for m and b, where

$$A = \begin{bmatrix} x_1 & 1 \\ x_2 & 1 \\ x_3 & 1 \\ x_4 & 1 \end{bmatrix} \qquad X = \begin{bmatrix} m \\ b \end{bmatrix} \qquad Y = \begin{bmatrix} y_1 \\ y_2 \\ y_3 \\ y_4 \end{bmatrix} \tag{5}$$

EXAMPLE 2   Finding the Line of Best Fit

Use least squares to find the line of best fit for the points

$$(4, 9), (5, 8), (9, 6), (12, 3)$$

SOLUTION   We use Equations (5) to set up the matrices A and Y:

$$A = \begin{bmatrix} 4 & 1 \\ 5 & 1 \\ 9 & 1 \\ 12 & 1 \end{bmatrix} \qquad Y = \begin{bmatrix} 9 \\ 8 \\ 6 \\ 3 \end{bmatrix}$$

Equation (4), $A^T A X = A^T Y$, becomes

$$\begin{bmatrix} 4 & 5 & 9 & 12 \\ 1 & 1 & 1 & 1 \end{bmatrix} \begin{bmatrix} 4 & 1 \\ 5 & 1 \\ 9 & 1 \\ 12 & 1 \end{bmatrix} \begin{bmatrix} m \\ b \end{bmatrix} = \begin{bmatrix} 4 & 5 & 9 & 12 \\ 1 & 1 & 1 & 1 \end{bmatrix} \begin{bmatrix} 9 \\ 8 \\ 6 \\ 3 \end{bmatrix}$$

This reduces to a system of two equations in two unknowns:

$$\begin{cases} 266m + 30b = 166 \\ 30m + 4b = 26 \end{cases}$$

The solution, which you can verify, is given by

$$m = -\frac{29}{41} \qquad \text{and} \qquad b = \frac{484}{41}$$

Hence, the least squares solution to the problem is given by the line

$$y = -\frac{29}{41}x + \frac{484}{41}$$

Note, for example, that corresponding to the value $x = 5$, the above line of "best fit" has $y = -\frac{29}{41} \cdot 5 + \frac{484}{41} = 8.27$, while the experimentally observed demand had value 8. Were we to use our computed line to approximate the connection between price and demand, we would predict that a selling price of $x = 6$ would yield a demand of $y = -\frac{29}{41} \cdot 6 + \frac{484}{41} = 7.56$.

THE GENERAL LEAST SQUARES PROBLEM

We presented the method of least squares using a particular example with four data points. It is easy to extend this analysis.

The general least squares problem consists of finding the line of best fit to a set of n data points:

$$(x_1, y_1), (x_2, y_2), \ldots, (x_n, y_n) \tag{6}$$

If we let

$$A = \begin{bmatrix} x_1 & 1 \\ x_2 & 1 \\ \cdot & \cdot \\ \cdot & \cdot \\ \cdot & \cdot \\ x_n & 1 \end{bmatrix} \qquad X = \begin{bmatrix} m \\ b \end{bmatrix} \qquad Y = \begin{bmatrix} y_1 \\ y_2 \\ \cdot \\ \cdot \\ \cdot \\ y_n \end{bmatrix}$$

then the line of best fit to the data points in (6) is

$$y = mx + b$$

where $X = \begin{bmatrix} m \\ b \end{bmatrix}$ is found by solving the system of two equations in two unknowns,

$$A^T A X = A^T Y \tag{7}$$

It can be shown that the system given by Equation (7) always has a unique solution provided the data points do not all lie on the same vertical line.

The least squares techniques presented here along with more elaborate variations are frequently used today by statisticians and researchers.

EXERCISE 2.7: APPLICATION 4 Answers to odd-numbered problems begin on page AN-13.

In Problems 1–6 compute A^T.

1. $A = \begin{bmatrix} 4 & 1 & 2 \\ 3 & 1 & 0 \end{bmatrix}$

2. $A = \begin{bmatrix} 5 & 2 & -1 \\ 1 & 3 & 6 \\ 1 & -1 & 2 \end{bmatrix}$

3. $A = \begin{bmatrix} 1 & 11 \\ 0 & 12 \\ 1 & 4 \end{bmatrix}$

4. $A = \begin{bmatrix} -1 & 6 & 4 \end{bmatrix}$

5. $A = \begin{bmatrix} 8 \\ 6 \\ 3 \end{bmatrix}$

6. $A = \begin{bmatrix} 5 & 3 \\ 3 & 7 \end{bmatrix}$

7. Supply Equation   The following table shows the supply (in thousands of units) of a product at various prices (in dollars):

Price, x	3	5	6	7
Supply, y	10	13	15	16

(a) Find the least squares line of best fit to the above data.

(b) Use the equation of this line to estimate the supply of the product at a price of $8.

8. **Study Time vs. Performance**   Data giving the number of hours a person had studied compared to his or her performance on an exam are given in the table:

Hours Studied x	Exam Score y
0	50
2	74
4	85
6	90
8	92

(a) Find a least squares line of best fit to the above data.
(b) What prediction would this line make for a student who studied 9 hours?

9. **Advertising vs. Sales**   A business would like to determine the relationship between the amount of money spent on advertising and its total weekly sales. Over a period of 5 weeks it gathers the following data:

Amount Spent on Advertising (in thousands) x	Weekly Sales Volume (in thousands) y
10	50
17	61
11	55
18	60
21	70

Find a least squares line of best fit to the above data.

10. **Drug Concentration vs. Time**   The following data show the connection between the number of hours a drug has been in a person's body and its concentration in the body.

Number of Hours	Drug Concentration (parts per million)
2	2.1
4	1.6
6	1.4
8	1.0

(a) Fit a least squares line to the above data.
(b) Use the equation of the line to estimate the drug concentration after 5 hours.

11. A matrix is **symmetric** if $A^T = A$. Which of the following matrices are symmetric?

(a) $\begin{bmatrix} 1 & 1 & 2 \\ 1 & 0 & 1 \\ 3 & 2 & 3 \end{bmatrix}$ (b) $\begin{bmatrix} 0 & 1 & 3 \\ 1 & 4 & 7 \\ 3 & 7 & 5 \end{bmatrix}$

(c) $\begin{bmatrix} 1 & 2 & 3 & 0 \\ 2 & 4 & 5 & 0 \\ 3 & 5 & 1 & 0 \end{bmatrix}$

Need a symmetric matrix be square?

12. Show that the matrix $A^T A$ is always symmetric.

Technology Exercises

Most computer software programs and many graphing calculators have the capability of finding the line of best fit for data. Consult your manual and use the LINear REGression feature to find the line of best fit for the data given in Problems 7–10. Compare the answers obtained using technology with those obtained using the method of least squares.

CHAPTER REVIEW

IMPORTANT TERMS AND CONCEPTS

steps for solving a system of linear equations by substitution 41
rules for obtaining an equivalent system of equations 42

steps for solving a system of linear equations by the method of elimination 43
parameter 46
matrix representation of a system of linear equations 52

row operations 54
steps for solving a system of linear equations using matrices 56
conditions for the reduced row-echelon form of a matrix 56

steps for solving a system of
   m linear equations
     containing *n* variables 73
definition of a matrix 79
equality of matrices 80
addition of matrices 82
scalar multiplication 85
matrix multiplication 91

identity matrix 95
inverse of a matrix 99
steps for finding the inverse
   of a matrix 102
solving a system of *n* linear
   equations containing *n*
   variables using inverses
   104

closed Leontief model 110
open Leontief model 113
cryptography 115
method of least squares 121

TRUE–FALSE ITEMS Answers are on page AN-13.

T_____ F_____ **1.** Matrices of the same dimension can always be added.

T_____ F_____ **2.** Matrices of the same dimension can always be multiplied.

T_____ F_____ **3.** A square matrix will always have an inverse.

T_____ F_____ **4.** The reduced row-echelon form of a matrix A is unique.

T_____ F_____ **5.** If A and B are each of dimension 4×4, then $AB = BA$.

T_____ F_____ **6.** Matrix addition is always defined.

T_____ F_____ **7.** Matrix multiplication is commutative.

FILL IN THE BLANKS Answers are on page AN-13.

1. If matrix A is of dimension 3×4 and matrix B is of dimension 4×2, then AB is of dimension _____ .

2. A system of three linear equations containing three variables has either _____ solution, or no solution, or _____ _____ solutions.

3. If A is a matrix of dimension 3×4, the 3 tells the number of _____ and the 4 tells the number of _____ .

4. If $AB = I$, the identity matrix, then B is called the _____ of A.

5. If B is a 2×3 matrix and BA^2 is defined, then A is a matrix of dimension _____ .

6. If A is a 4×5 matrix and AB^3 is defined, then B is a matrix of dimension _____ .

REVIEW EXERCISES Answers to odd-numbered problems begin on page AN-13.

In Problems 1–16 perform the indicated operations using

$$A = \begin{bmatrix} -2 & 0 & 7 \\ 1 & 8 & 3 \\ 2 & 4 & 21 \end{bmatrix} \quad B = \begin{bmatrix} 1 & 3 & 9 \\ 2 & 7 & 5 \\ 3 & 6 & 8 \end{bmatrix} \quad C = \begin{bmatrix} 0 & 1 & 2 \\ 0 & 5 & 1 \\ 8 & 7 & 9 \end{bmatrix}$$

1. $A + B$

2. $B + A$

3. $3(A + B)$

4. $3A + 3B$

5. $3A + 5B$

6. $6B - 3C$

7. $2(5A)$

8. $\frac{3}{2}A$

9. $2A + \frac{1}{2}B - 3C$

10. $A - 2B + 3C$

11. AB

12. BA

13. $(B - A)C$

14. $BC - AC$

15. $A(BC)$

16. $(BA)C$

In Problems 17–24 find the inverse, if it exists, of each matrix.

17. $\begin{bmatrix} 3 & 0 \\ -2 & 1 \end{bmatrix}$

18. $\begin{bmatrix} 4 & 1 \\ 3 & 1 \end{bmatrix}$

19. $\begin{bmatrix} 1 & 2 & 3 \\ 2 & 4 & 5 \\ 3 & 5 & 6 \end{bmatrix}$

20. $\begin{bmatrix} -1 & 2 & 0 \\ 3 & 2 & -1 \\ 4 & 0 & 3 \end{bmatrix}$ **21.** $\begin{bmatrix} 4 & 3 & -1 \\ 0 & 2 & 2 \\ 3 & -1 & 0 \end{bmatrix}$ **22.** $\begin{bmatrix} -6 & 6 & 2 \\ 13 & 3 & 1 \\ 8 & -8 & 8 \end{bmatrix}$

23. $\begin{bmatrix} 1 & 2 & -3 \\ 4 & 6 & 2 \\ -3 & -6 & 9 \end{bmatrix}$ **24.** $\begin{bmatrix} 9 & 6 & -3 \\ 2 & -6 & 4 \\ -3 & 2 & 1 \end{bmatrix}$

In Problems 25–42 find the solution, if it exists, of each system of linear equations. If the system has infinitely many solutions, list at least three solutions.

25. $\begin{cases} -5x + 2y = -2 \\ -3x + 3y = 4 \end{cases}$ **26.** $\begin{cases} -3x + 2y = 3 \\ -3x - 4y = 4 \end{cases}$ **27.** $\begin{cases} x + 2y + 5z = 6 \\ 3x + 7y + 12z = 23 \\ x + 4y = 25 \end{cases}$

28. $\begin{cases} x + 2y - z = -4 \\ 3x + 7y - 6z = -21 \\ x + 4y - 6z = -17 \end{cases}$ **29.** $\begin{cases} x + 2y + 7z = 2 \\ 3x + 7y + 18z = -1 \\ x + 4y + 2z = -13 \end{cases}$ **30.** $\begin{cases} x + 2y - 7z = -1 \\ 3x + 7y - 24z = 6 \\ x + 4y - 12z = 26 \end{cases}$

31. $\begin{cases} 2x - y + z = 1 \\ x + y - z = 2 \\ 3x - y + z = 0 \end{cases}$ **32.** $\begin{cases} 2x + 3y - z = 5 \\ x - y + z = 1 \\ 3x - 3y + 3z = 3 \end{cases}$ **33.** $\begin{cases} y - 2z = 6 \\ 3x + 2y - z = 2 \\ 4x + 3z = -1 \end{cases}$

34. $\begin{cases} 2x - y + 3z = 5 \\ x + 2z = 0 \\ 3x + 2y + z = -3 \end{cases}$ **35.** $\begin{cases} x - 3y = 5 \\ 3y + z = 0 \\ 2x - y + 2z = 2 \end{cases}$ **36.** $\begin{cases} x - z = 2 \\ 2x - y = 4 \\ x + y + z = 6 \end{cases}$

37. $\begin{cases} 3x + y - 2z = 3 \\ x - 2y + z = 4 \end{cases}$ **38.** $\begin{cases} 2x - y - 3z = 0 \\ x - 2y + z = 4 \end{cases}$ **39.** $\begin{cases} x + 2y - z = 5 \\ 2x - y + 2z = 0 \end{cases}$

40. $\begin{cases} x - y + 2z = 6 \\ 2x + 2y - z = -1 \end{cases}$ **41.** $\begin{cases} 2x - y = 6 \\ x - 2y = 0 \\ 3x - y = 6 \end{cases}$ **42.** $\begin{cases} x - 2y = 0 \\ 2x + y = 5 \\ x - 3y = -3 \end{cases}$

43. What must be true about x, y, z, w, if we require $AB = BA$ for the matrices

$$A = \begin{bmatrix} x & y \\ z & w \end{bmatrix} \quad \text{and} \quad B = \begin{bmatrix} 1 & 1 \\ -1 & 1 \end{bmatrix}$$

44. Let $t = [t_1 \quad t_2]$, with $t_1 + t_2 = 1$, and let $A = \begin{bmatrix} \frac{1}{4} & \frac{3}{4} \\ \frac{2}{3} & \frac{1}{3} \end{bmatrix}$. Find t such that $tA = t$.

45. Mixture Sweet Delight Candies, Inc., sells boxes of candy consisting of creams and caramels. Each box sells for $4 and holds 50 pieces of candy (all pieces are the same size). If the caramels cost $0.05 to produce and the creams cost $0.10 to produce, how many caramels and creams should be in each box for no profit and no loss? Would you increase or decrease the number of caramels in order to obtain a profit?

46. Cookie Orders A cookie company makes three kinds of cookies, oatmeal raisin, chocolate chip, and shortbread, packaged in small, medium, and large boxes. The small box contains 1 dozen oatmeal raisin and 1 dozen chocolate chip; the medium box has 2 dozen oatmeal raisin, 1 dozen chocolate chip, and 1 dozen shortbread; the large box contains 2 dozen oatmeal raisin, 2 dozen chocolate

chip, and 3 dozen shortbread. If you require exactly 15 dozen oatmeal raisin, 10 dozen chocolate chip, and 11 dozen shortbread cookies, how many of each size box should you buy?

47. Mixture Problem A store sells almonds for $6 per pound, cashews for $5 per pound, and peanuts for $2 per pound. One week the manager decides to prepare 100 16-ounce packages of nuts by mixing the peanuts, almonds, and cashews. Each package will be sold for $4. The mixture is to produce the same revenue as selling the nuts separately. Prepare a table that shows some of the possible ways the manager can prepare the mixture.

48. Financial Planning Three retired couples each require an additional annual income of $1800 per year. As their financial consultant, you recommend they invest some money in Treasury bills that yield 6%, some money in corporate bonds that yield 8%, and some money in junk bonds that yield 10%. Prepare a table for each couple showing the various ways their goal can be achieved

(a) If the first couple has $20,000 to invest
(b) If the second couple has $25,000 to invest
(c) If the third couple has $30,000 to invest

49. Financial Planning A retired couple has $40,000 to invest. As their financial consultant, you recommend they invest some money in Treasury bills that yield 6%, some money in corporate bonds that yield 8%, and some money in junk bonds that yield 10%. Prepare a table showing the various ways this couple can achieve the following goals:

(a) They want $2500 per year in income.
(b) They want $3000 per year in income.
(c) They want $3500 per year in income.

MATHEMATICAL QUESTIONS FROM PROFESSIONAL EXAMS

Use the following information to answer Problems 1–4:

Akron, Inc. owns 80% of the capital stock of Benson Company and 70% of the capital stock of Cashin, Inc. Benson Company owns 15% of the capital stock of Cashin, Inc. Cashin, Inc., in turn, owns 25% of the capital stock of Akron, Inc. These ownership interrelationships are illustrated in the following diagram:

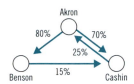

Net income before adjusting for interests in intercompany net income for each corporation follows:

Akron, Inc.	*$190,000*
Benson Co.	*$170,000*
Cashin, Inc.	*$230,000*

Ignore all income tax considerations.

A_e = *Akron's consolidated net income;*
that is, its net income plus its share of the consolidated net income of Benson and Cashin

B_e = *Benson's consolidated net income;*
that is, its net income plus its share of the consolidated net income of Cashin

C_e = *Cashin's consolidated net income;*
that is, its net income plus its share of the consolidated net income of Akron

1. CPA Exam The equation, in a set of simultaneous equations, which computes A_e is

(a) $A_e = .75(190,000 + .8B_e + .7C_e)$
(b) $A_e = 190,000 + .8B_e + .7C_e$
(c) $A_e = .75(190,000) + .8(170,000) + .7(230,000)$
(d) $A_e = .75(190,000) + .8B_e + .7C_e$

2. CPA Exam The equation, in a set of simultaneous equations, which computes B_e is

(a) $B_e = 170,000 + .15C_e - .75A_e$
(b) $B_e = 170,000 + .15C_e$

(c) $B_e = .2(170,000) + .15(230,000)$
(d) $B_e = .2(170,000) + .15C_e$

3. CPA Exam Cashin's minority interest in consolidated net income is

(a) $.15(230,000)$
(b) $230,000 + .25A_e$
(c) $.15(230,000) + .25A_e$
(d) $.15C_e$

4. CPA Exam Benson's minority interest in consolidated net income is

(a) $34,316   (b) $25,500   (c) $45,755   (d) $30,675

Chapter 3

Linear Programming: Geometric Approach

3.1 Linear Inequalities

3.2 A Geometric Approach to Linear Programming Problems

3.3 Applications

Chapter Review

Whenever the analysis of a problem leads to minimizing or maximizing a linear expression in which the variables must obey a collection of linear inequalities, a solution may be obtained using linear programming techniques.

Historically, linear programming problems evolved out of the need to solve problems involving resource allocation by the U.S. Army during World War II. Among those who worked on such problems was George Dantzig, who later gave a general formulation of the linear programming problem and offered a method for solving it, called the *simplex method*. This method is discussed in Chapter 4.

In this chapter we study ways to solve linear programming problems that involve only two variables. As a result, we can use a geometric approach utilizing the graph of a system of linear inequalities to solve the problem.

3.1 LINEAR INEQUALITIES

We have already discussed linear equations (linear equalities) in two variables x and y (Section 1.1). These are equations of the form

$$Ax + By = C \tag{1}$$

where A, B, C are real numbers and A and B are not both zero. If in Equation (1) we replace the equal sign by an inequality symbol, namely, one of the symbols $<, >, \le,$ $\ge$, we obtain a **linear inequality in two variables** x and y. For example, the expressions

$$3x + 2y \ge 4 \qquad 2x - 3y < 0 \qquad 3x + 5y > -8$$

are each linear inequalities in two variables. The first of these is called a **nonstrict inequality** since the inequality symbol $\ge$ is nonstrict; the remaining two linear inequalities are **strict.*

The Graph of a Linear Inequality

The **graph of a linear inequality** in two variables x and y is the set of all points (x, y) for which the inequality is satisfied.

Let's look at an example.

EXAMPLE 1   Graphing a Linear Inequality

Graph the inequality: $2x + 3y \ge 6$

SOLUTION   The inequality $2x + 3y \ge 6$ is equivalent to $2x + 3y > 6$ or $2x + 3y = 6$. So we begin by graphing the line $2x + 3y = 6$, noting that any point on the line must satisfy the inequality $2x + 3y \ge 6$. See Figure 1(a).

Figure 1

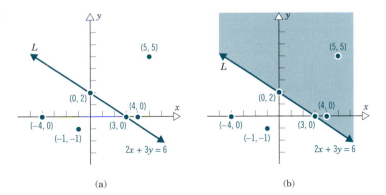

(a)   (b)

Now let's test a few points, such as $(-1, -1)$, $(5, 5)$, $(4, 0)$, $(-4, 0)$, to see if they satisfy the inequality. We do this by substituting the coordinates of each point into the left member of the inequality and determining whether the result is ≥ 6 or < 6.

	$2x$	$+ 3y$		*Conclusion*
$(-1, -1)$:	$2(-1) + 3(-1)$	$= -2 - 3 = -5 < 6$		Not part of graph
$(5, 5)$:	$2(5)$	$+ 3(5)$	$= 25 > 6$	Part of graph
$(4, 0)$:	$2(4)$	$+ 3(0)$	$= 8 > 6$	Part of graph
$(-4, 0)$:	$2(-4) + 3(0)$	$= -8 < 6$		Not part of graph

* A review of inequalities may be found in Appendix A, Section A.1.

Notice that the two points (4, 0) and (5, 5) that are part of the graph both lie on one side of L, while the points $(-4, 0)$ and $(-1, -1)$ (not part of the graph) lie on the other side of L. This is not an accident. The graph of the inequality consists of all points on the same side of L as (4, 0) and (5, 5). The shaded region of Figure 1(b) illustrates the graph of the inequality.

The inequality in Example 1 was *nonstrict,* so the *corresponding line was part of the graph of the inequality.* If the inequality is *strict,* the *corresponding line is not part of the graph of the inequality.* We will indicate the latter by using dashes to graph the line.

Let's outline the procedure for graphing a linear inequality:

Steps for Graphing a Linear Inequality

Step 1 Graph the corresponding linear equation, a line L. If the inequality is nonstrict, graph L using a solid line; if the inequality is strict, graph L using dashes.

Step 2 Select a test point P not on the line L.

Step 3 Substitute the coordinates of the test point P into the given inequality. If the coordinates of this point P satisfy the linear inequality, then all points on the same side of L as the point P satisfy the inequality. If the coordinates of the point P do not satisfy the linear inequality, then all points on the opposite side of L from P satisfy the inequality.

EXAMPLE 2   Graphing a Linear Inequality

Graph the linear inequality: $2x - y < -4$

SOLUTION   The corresponding linear equation is the line

$$L: \quad 2x - y = -4$$

Since the inequality is strict, points on L are not part of the graph of the linear inequality. When we graph L, we use a dashed line to indicate this fact. See Figure 2(a).

We select a point not on the line L to be tested, for example, (0, 0):

$$2(0) - 0 = 0 > -4$$

Figure 2

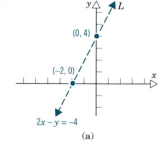

(a)

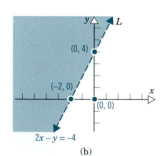

(b)

Since (0, 0) does not satisfy the inequality, all points on the opposite side of L from (0, 0) are on the graph. The graph of $2x - y < -4$ is the shaded region of Figure 2(b). ■

EXAMPLE 3   Graphing Linear Inequalities

Graph: (a) $x \le 3$ (b) $2x \le y$

SOLUTION

(a) The corresponding linear equation is $x = 3$, a vertical line. If we choose (0, 0) as the test point, we find that it satisfies the inequality. Thus all points to the left of, and on, the vertical line are on the graph. See Figure 3(a).

(b) The corresponding linear equation is $2x = y$. Since it passes through (0, 0), we choose the point (0, 2) as the test point. The inequality is satisfied by (0, 2), so it is on the graph. See Figure 3(b).

Figure 3

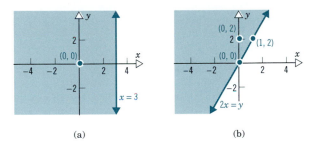

(a)   (b)

The set of points belonging to the graph of a linear inequality [for example, the shaded region in Figure 3(b)] is sometimes called a **half-plane.**

✏ **Now Work Problem 7**

Systems of Linear Inequalities

A **system of linear inequalities** is a collection of two or more linear inequalities. To **graph** a system of two inequalities we locate all points whose coordinates satisfy each of the linear inequalities of the system.

EXAMPLE 4   Determining Whether a Point Belongs to the Graph of a System

Which of the points $P_1 = (7, 5)$, $P_2 = (9, 12)$, $P_3 = (3, 1)$ are part of the graph of the following system?

$$\begin{cases} 10x - y \ge 0 \\ -x + 2y \ge 0 \\ x + y \le 15 \end{cases}$$

SOLUTION For $P_1 = (7, 5)$ to be part of the graph, it must satisfy each of the linear inequalities. Since $10(7) - 5 = 65 \geq 0$, the first inequality is satisfied. Since $-7 + 2(5) = 3 \geq 0$, so is the second. And since $7 + 5 = 12 \leq 15$, so is the third. Thus P_1 is part of the graph.

For $P_2 = (9, 12)$ we have

$$10(9) - 12 = 78 \geq 0 \qquad -9 + 2(12) = 15 \geq 0 \qquad 9 + 12 = 21 \text{ is not} \leq 15$$

Thus P_2 is not part of the graph.

For $P_3 = (3, 1)$ we have

$$10(3) - 1 = 29 \geq 0 \qquad -3 + 2(1) = -1 \text{ is not} \geq 0$$

Thus P_3 is not part of the graph.

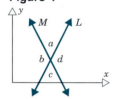 **Now Work Problem 13**

Figure 4

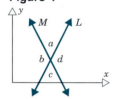

Several possible graphs can result from a system of two linear inequalities in two variables. For example, suppose L and M are the lines corresponding to two linear inequalities, and suppose L and M intersect. See Figure 4. Then the two intersecting lines L and M divide the plane into four regions, a, b, c, and d. One of these regions is the solution of the system.

EXAMPLE 5   Graphing a System of Linear Inequalities

Graph the system: $\begin{cases} 2x - y \leq -4 \\ x + y \geq -1 \end{cases}$

SOLUTION   First we graph each inequality separately. See Figures 5(a) and 5(b).

The solution of the system consists of all points common to these two half-planes. The dark blue shaded region in Figure 6 represents the solution of the system.

Figure 5

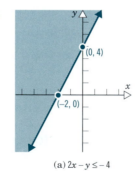

(a) $2x - y \leq -4$ (b) $x + y \geq -1$

Figure 6

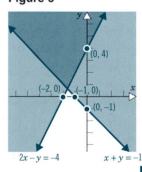

$2x - y = -4$ $x + y = -1$

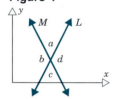 **Now Work Problem 19**

If the lines L and M are parallel, the system of linear inequalities may or may not have a solution. Examples of such situations are given below.

EXAMPLE 6   Graphing a System of Linear Inequalities

Graph the system: $\begin{cases} 2x - y \le -4 \\ 2x - y \le -2 \end{cases}$

SOLUTION   First we graph each inequality separately. See Figures 7(a) and 7(b). The dark blue shaded region in Figure 8 represents the solution of the system.

Figure 7

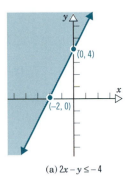

(a) $2x - y \le -4$

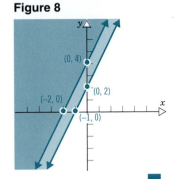

(b) $2x - y \le -2$

Figure 8

Notice that the solution of this system is the same as that of the single linear inequality $2x - y \le -4$.

EXAMPLE 7   Graphing a System of Linear Inequalities

The solution of the system

$$\begin{cases} 2x - y \ge -4 \\ 2x - y \le -2 \end{cases}$$

is the dark blue shaded region in Figure 9.

Figure 9

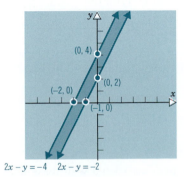

$2x - y = -4$ $2x - y = -2$

EXAMPLE 8   Graphing a System of Linear Inequalities

Figure 10

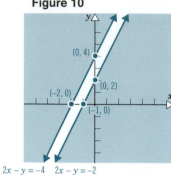

The system

$$\begin{cases} 2x - y \le -4 \\ 2x - y \ge -2 \end{cases}$$

has no solution, as Figure 10 indicates, because the two half-planes have no points in common.

Until now, we have considered systems of only two linear inequalities. The next example is of a system of four linear inequalities. As we shall see, the technique for graphing such systems is the same as that used for graphing systems of two linear inequalities in two variables.

EXAMPLE 9   Graphing a System of Four Linear Inequalities

Graph the system: $\begin{cases} x + y \ge 2 \\ 2x + y \ge 3 \\ x \ge 0 \\ y \ge 0 \end{cases}$

SOLUTION   Again we first graph the four lines:

$$\begin{aligned} L_1: & \quad x + y = 2 \\ L_2: & \quad 2x + y = 3 \\ L_3: & \quad x = 0 \\ L_4: & \quad y = 0 \end{aligned}$$

Figure 11

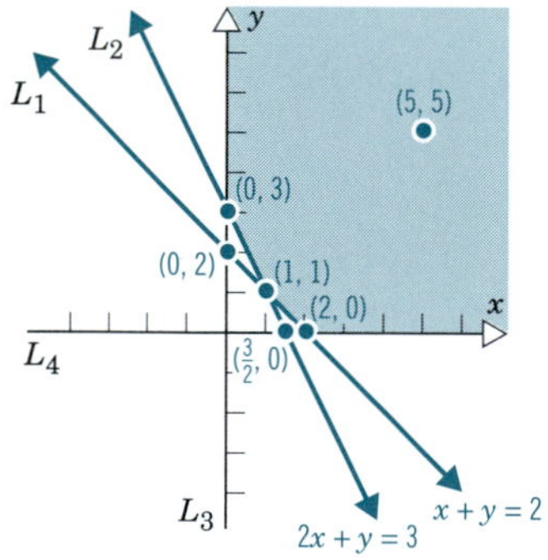

The lines L_1 and L_2 intersect at the point $(1, 1)$. (Do you see why?) The inequalities $x \ge 0$ and $y \ge 0$ indicate that the graph of the system lies in quadrant I. Thus the graph of the system consists of that part of the graph of the inequalities $x + y \ge 2$ and $2x + y \ge 3$ that lies in the first quadrant. Since $(5, 5)$ is a point that satisfies each of these inequalities, we obtain Figure 11.

EXAMPLE 10   Graphing a System of Four Linear Inequalities

Graph the system: $\begin{cases} x + y \le 2 \\ 2x + y \le 3 \\ x \ge 0 \\ y \ge 0 \end{cases}$

SOLUTION   Since the lines associated with these linear inequalities are the same as those of the previous example, we proceed directly to the graph. See Figure 12.

Figure 12

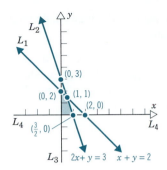

Some Terminology

Compare the graphs of the systems of linear inequalities given in Figures 11 and 12. The graph in Figure 11 is said to be **unbounded** in the sense that it extends infinitely far in some direction. The graph in Figure 12 is **bounded** in the sense that it can be enclosed by some circle of sufficiently large radius. See Figure 13.

Figure 13

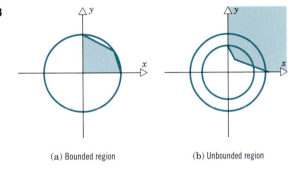

(a) Bounded region   (b) Unbounded region

The boundary of each of the graphs in Figures 11 and 12 consists of line segments. In fact, the graph of any system of linear inequalities will have line segments as boundaries. The point of intersection of two line segments that form the boundary is called a **corner point** of the graph. For example, the graph of the system given in Example 9 has the corner points $(0, 3)$, $(1, 1)$, and $(2, 0)$. See Figure 11. The graph of the system given in Example 10 has the corner points $(0, 2)$, $(0, 0)$, $(\frac{3}{2}, 0)$, $(1, 1)$. See Figure 12.

We shall soon see that the corner points of the graph of a system of linear inequalities play a major role in the procedure for solving linear programming problems.

✏️ **Now Work Problem 27**

Application

EXAMPLE 11   Analyzing a Mixture Problem

Nutt's Nuts has 75 pounds of cashews and 120 pounds of peanuts. These are to be mixed in 1-pound packages as follows: a low-grade mixture that contains 4 ounces of cashews and 12 ounces of peanuts and a high-grade mixture that contains 8 ounces of cashews and 8 ounces of peanuts.

(a) Use x to denote the number of packages of the low-grade mixture and use y to denote the number of packages of the high-grade mixture and write a system of linear inequalities that describes the possible number of each kind of package.

(b) Graph the system and list its corner points.

SOLUTION

(a) We begin by naming the variables:

$$x = \text{Number of packages of low-grade mixture}$$
$$y = \text{Number of packages of high-grade mixture}$$

First, we note that the only meaningful values for x and y are nonnegative values. Thus we must restrict x and y so that

$$x \geq 0 \quad \text{and} \quad y \geq 0$$

Next, we note that there is a limit to the number of pounds of cashews and peanuts available. That is, the total number of pounds of cashews cannot exceed 75 pounds (1200 ounces), and the number of pounds of peanuts cannot exceed 120 pounds (1920 ounces). This means that

$$\begin{pmatrix} \text{Ounces of} \\ \text{cashews} \\ \text{required} \\ \text{for low-grade} \\ \text{mixture} \end{pmatrix} \begin{pmatrix} \text{Number of} \\ \text{packages of} \\ \text{low-grade} \\ \text{mixture} \end{pmatrix} + \begin{pmatrix} \text{Ounces of} \\ \text{cashews} \\ \text{required} \\ \text{for high-} \\ \text{grade} \\ \text{mixture} \end{pmatrix} \begin{pmatrix} \text{Number of} \\ \text{packages} \\ \text{of high-} \\ \text{grade} \\ \text{mixture} \end{pmatrix} \begin{matrix} \text{cannot} \\ \text{exceed} \end{matrix} 1200$$

$$\begin{pmatrix} \text{Ounces of} \\ \text{peanuts} \\ \text{required} \\ \text{for low-grade} \\ \text{mixture} \end{pmatrix} \begin{pmatrix} \text{Number of} \\ \text{packages of} \\ \text{low-grade} \\ \text{mixture} \end{pmatrix} + \begin{pmatrix} \text{Ounces of} \\ \text{peanuts} \\ \text{for high-} \\ \text{grade} \\ \text{mixture} \end{pmatrix} \begin{pmatrix} \text{Number of} \\ \text{packages of} \\ \text{high-grade} \\ \text{mixture} \end{pmatrix} \begin{matrix} \text{cannot} \\ \text{exceed} \end{matrix} 1920$$

In terms of the data given and the variables introduced, we can write these statements compactly as

$$4x + 8y \leq 1200$$
$$12x + 8y \leq 1920$$

The system of linear inequalities that gives the possible values x and y can take on is

$$\begin{cases} 4x + 8y \leq 1200 \\ 12x + 8y \leq 1920 \\ \quad\quad x \geq 0 \\ \quad\quad y \geq 0 \end{cases}$$

(b) The system of linear inequalities given above can be simplified to the equivalent form

$$\begin{cases} x + 2y \le 300 \\ 3x + 2y \le 480 \\ \quad x \ge \quad 0 \\ \quad y \ge \quad 0 \end{cases}$$

The graph of the system is given in Figure 14. The corner points of the graph are the points of intersection of the lines ① and ②, ① and ③, ② and ④, and ③ and ④ as shown in Figure 14. The last three are easy to identify by inspection; the first one requires that we solve the system of equations

$$\begin{cases} x + 2y = 300 \\ 3x + 2y = 480 \end{cases}$$

Subtracting the first equation from the second gives $2x = 180$ or $x = 90$. Using this in the first equation, we find

$$2y = 300 - x \quad \text{or} \quad 2y = 210 \quad \text{or} \quad y = 105$$

Therefore, $(90, 105)$ is a corner point. The four corner points are

$$(0, 0), (0, 150), (160, 0), \text{ and } (90, 105)$$

Figure 14

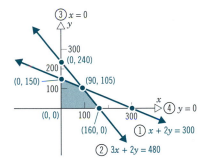

EXERCISE 3.1   Answers to odd-numbered problems begin on page AN-14.

In Problems 1–12 graph each inequality.

1. $x \ge 0$

2. $y \ge 0$

3. $x \le 4$

4. $y \le 6$

5. $y \ge 1$

6. $x \ge 2$

7. $2x + 3y \le 6$

8. $3x + 2y \ge 6$

9. $5x + y \le 10$

10. $x + 2y > 4$

11. $x + 5y \le 5$

12. $3x + y \le 3$

13. Without graphing, determine which of the points $P_1 = (3, 8)$, $P_2 = (12, 9)$, $P_3 = (5, 1)$ are part of the graph of the following system:

$$\begin{cases} -10x + 3y \le \quad 0 \\ \quad -3x + 2y \ge \quad 0 \\ \quad 2x + \quad y \le 15 \end{cases}$$

14. Without graphing, determine which of the points $P_1 = (9, -5)$, $P_2 = (12, -4)$, $P_3 = (4, 1)$ are part of the graph of the following system:

$$\begin{cases} 10x + \quad y \le \quad 0 \\ \quad -x + 2y \ge \quad 0 \\ \quad 4x + \quad y \le 15 \end{cases}$$

15. Without graphing, determine which of the points $P_1 = (5, -8)$, $P_2 = (10, 10)$, $P_3 = (5, 1)$ are part of the graph of the following system:

$$\begin{cases} 10x + 3y \geq 0 \\ 3x + 2y \geq 0 \\ x + y \leq 15 \end{cases}$$

16. Without graphing, determine which of the points $P_1 = (2, 6)$, $P_2 = (12, 4)$, $P_3 = (4, 1)$ are part of the graph of the following system:

$$\begin{cases} -10x + y \leq 0 \\ 2x - 5y \leq 0 \\ x + 3y \leq 15 \end{cases}$$

17. Without graphing, determine which of the points $P_1 = (5, 3)$, $P_2 = (6, 12)$, $P_3 = (6, 1)$ are part of the graph of the following system:

$$\begin{cases} 2y - 10x \geq 0 \\ 2y - x \geq 0 \\ y + 6x \geq 15 \end{cases}$$

18. Without graphing, determine which of the points $P_1 = (1, -4)$, $P_2 = (10, 6)$, $P_3 = (6, -2)$ are part of the graph of the following system:

$$\begin{cases} 3y - 10x \leq 0 \\ 2y + 5x \geq 0 \\ 4y + x \leq 15 \end{cases}$$

In Problems 19–26 determine which region a, b, c, or d represents the graph of the given system of linear inequalities. The regions a, b, c, and d are nonoverlapping regions bounded by the indicated lines.

19. $\begin{cases} 5x - 4y \leq 8 \\ 2x + 5y \leq 23 \end{cases}$

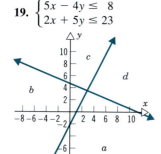

20. $\begin{cases} 4x - 5y \leq 0 \\ 4x + 2y \leq 28 \end{cases}$

21. $\begin{cases} 2x - 3y \leq -3 \\ 4x + 6y \leq 30 \end{cases}$

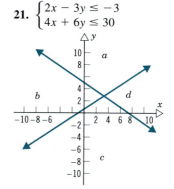

22. $\begin{cases} 6x - 5y \leq 5 \\ 2x + 4y \leq 30 \end{cases}$

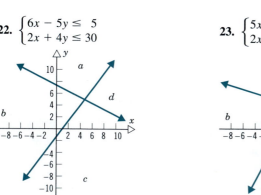

23. $\begin{cases} 5x - 3y \leq 3 \\ 2x + 6y \leq 30 \end{cases}$

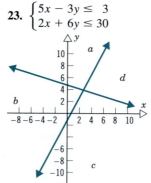

24. $\begin{cases} 5x - 5y \leq 10 \\ 6x + 4y \leq 48 \end{cases}$

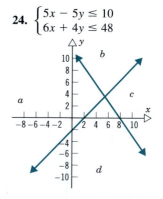

25. $\begin{cases} 5x - 4y \le 0 \\ 2x + 4y \le 28 \end{cases}$

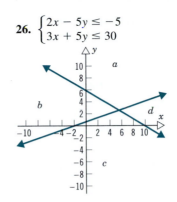

26. $\begin{cases} 2x - 5y \le -5 \\ 3x + 5y \le 30 \end{cases}$

In Problems 27–38 graph each system of linear inequalities. Tell whether the graph is bounded or unbounded and list each corner point of the graph.

27. $\begin{cases} x + y \le 2 \\ x \ge 0 \\ y \ge 0 \end{cases}$

28. $\begin{cases} 2x + 3y \le 6 \\ x \ge 0 \\ y \ge 0 \end{cases}$

29. $\begin{cases} x + y \ge 2 \\ 2x + 3y \le 6 \\ x \ge 0 \\ y \ge 0 \end{cases}$

30. $\begin{cases} x + y \ge 2 \\ 2x + 3y \le 12 \\ 3x + 2y \le 12 \\ x \ge 0 \\ y \ge 0 \end{cases}$

31. $\begin{cases} 2 \le x + y \\ x + y \le 8 \\ 2x + y \le 10 \\ x \ge 0 \\ y \ge 0 \end{cases}$

32. $\begin{cases} 2 \le x + y \\ x + y \le 8 \\ 1 \le x + 2y \\ x \ge 0 \\ y \ge 0 \end{cases}$

33. $\begin{cases} x + y \ge 2 \\ 2x + 3y \le 12 \\ 3x + y \le 12 \\ x \ge 0 \\ y \ge 0 \end{cases}$

34. $\begin{cases} 2 \le x + y \\ x + y \le 10 \\ 2x + y \le 3 \\ x \ge 0 \\ y \ge 0 \end{cases}$

35. $\begin{cases} 1 \le x + 2y \\ x + 2y \le 10 \\ x \ge 0 \\ y \ge 0 \end{cases}$

36. $\begin{cases} 1 \le x + 2y \\ x + 2y \le 10 \\ 2 \le x + y \\ x + y \le 8 \\ x \ge 0 \\ y \ge 0 \end{cases}$

37. $\begin{cases} x + 2y \ge 2 \\ x + y \le 4 \\ 3x + y \le 3 \\ x \ge 0 \\ y \ge 0 \end{cases}$

38. $\begin{cases} 2x + y \ge 2 \\ 3x + 2y \le 6 \\ x + y \ge 2 \\ x \ge 0 \\ y \ge 0 \end{cases}$

39. Rework Example 11 if 60 pounds of cashews and 90 pounds of peanuts are available.

40. Rework Example 11 if the high-grade mixture contains 10 ounces of cashews and 6 ounces of peanuts.

41. Manufacturing Mike's Famous Toy Trucks company manufactures two kinds of toy trucks—a dumpster and a tanker. In the manufacturing process, each dumpster requires 3 hours of grinding and 4 hours of finishing, while each tanker requires 2 hours of grinding and 3 hours of finishing. The company has two grinders and three finishers, each of whom works at most 40 hours per week.

(a) Using x to denote the number of dumpsters and y to denote the number of tankers, write a system of linear inequalities that describes the possible numbers of each truck that can be manufactured.

(b) Graph the system and list its corner points.

42. Manufacturing Repeat Problem 41 if one grinder and two finishers, each of whom works at most 40 hours per week, are available.

43. Financial Planning A retired couple has up to $25,000 to invest. As their financial adviser, you recommend they place at least $15,000 in Treasury bills yielding 6% and at most $10,000 in corporate bonds yielding 9%.

(a) Using x to denote the amount of money invested in Treasury bills and y to denote the amount invested in corporate bonds, write a system of inequalities that describes this situation.
(b) Graph the system and list its corner points.
(c) Interpret the meaning of each corner point in relation to the investments it represents.

44. Financial Planning   Use the information supplied in Problem 43, along with the fact that the couple will invest at least $20,000, to answer Parts (a), (b), and (c).

45. Nutrition   A farmer prepares feed for livestock by combining two types of grain. Each unit of the first grain contains 1 unit of protein and 5 units of iron while each unit of the second grain contains 2 units of protein and 1 unit of iron. Each animal must receive at least 5 units of protein and 16 units of iron each day.

(a) Write a system of linear inequalities that describes the possible amounts of each grain the farmer needs to prepare.
(b) Graph the system and list the corner points.

46. Investment Strategy   Laura wishes to invest up to a total of $40,000 in class AA bonds and stocks. Furthermore, she believes that the amount invested in class AA bonds should be at most one-third of the amount invested in stocks.

(a) Write a system of linear inequalities that describes the possible amount of investments in each security.
(b) Graph the system and list the corner points.

47. Nutrition   To maintain an adequate daily diet, nutritionists recommend the following: at least 85 g of carbohydrate, 70 g of fat, and 50 g of protein. An ounce of food A contains 5 g of carbohydrate, 3 g of fat, and 2 g of protein, while an ounce of food B contains 4 g of carbohydrate, 3 g of fat, and 3 g of protein.

(a) Write a system of linear inequalities that describes the possible quantities of each food.
(b) Graph the system and list the corner points.

48. Transportation   A microwave company has two plants, one on the East Coast and one in the Midwest. It takes 25 hours (packing, transportation, and so on) to transport an order of microwaves from the Eastern plant to its central warehouse and 20 hours from the Midwest plant to its central warehouse. It costs $80 to transport an order from the Eastern plant to the central warehouse and $40 from the Midwestern plant to its central warehouse. There are 1000 work-hours available for packing, transportation, and so on, and $3000 for transportation cost.

(a) Write a system of linear inequalities that describes the transportation system.
(b) Graph the system and list the corner points.

Technology Exercises

Some graphing calculators can graph systems of inequalities and shade the region representing the solution to the system on the screen. Consult your user's manual, and use your graphing calculator to solve the systems of inequalities in Problems 1–6. If your calculator does not graph inequalities, graph the boundary line of each inequality. Use INTERSECT to find the corner points.

1. $\begin{cases} 1 < x + y \\ x \leq 2 \\ y \leq 2 \end{cases}$

2. $\begin{cases} x + 2y \leq 4 \\ x \geq 0 \\ y \geq 0 \end{cases}$

3. $\begin{cases} 2x + 3y \leq 6 \\ x \geq 1 \\ y \geq 0 \end{cases}$

4. $\begin{cases} 3x + 4y \leq 12 \\ x \geq 0 \\ y \geq 0 \end{cases}$

5. $\begin{cases} y - x + 1 \geq 0 \\ y + x \leq 5 \\ x \geq 0 \end{cases}$

6. $\begin{cases} 2x - 3y \geq 6 \\ x \geq 0 \\ y \leq 2 \\ y \geq 0 \end{cases}$

3.2 A GEOMETRIC APPROACH TO LINEAR PROGRAMMING PROBLEMS

To help motivate the character of a linear programming problem, we begin by looking again at Example 11 of the previous section.

Nutt's Nuts has 75 pounds of cashews and 120 pounds of peanuts. These are to be mixed in 1-pound packages as follows: a low-grade mixture that contains 4 ounces of cashews and 12 ounces of peanuts and a high-grade mixture that contains 8 ounces of cashews and 8 ounces of peanuts.

Suppose that in addition to the information given above, we also know what the profit will be on each type of mixture. For example, suppose the profit is $0.25 on each package of the low-grade mixture and is $0.45 on each package of the high-grade mixture. The question of importance to the manager is "How many packages of each type of mixture should be prepared to maximize the profit?"

If P symbolizes the profit, x the number of packages of low-grade mixture, and y the number of high-grade packages, then the question can be restated as "What are the values of x and y so that the expression

$$P = \$0.25x + \$0.45y$$

is a maximum?"

This problem is typical of a **linear programming problem.** It requires that a certain linear expression, in this case the profit, be maximized. This linear expression is called the **objective function.** Furthermore, the problem requires that the maximum profit be achieved under certain restrictions or **constraints,** each of which are linear inequalities involving the variables. The linear programming problem may be restated as

Maximize

$$P = \$0.25x + \$0.45y \quad \text{Objective function}$$

subject to the conditions that

$$x + 2y \leq 300 \quad \text{Cashew constraint}$$
$$3x + 2y \leq 480 \quad \text{Peanut constraint}$$
$$x \geq 0 \quad \text{Nonnegativity constraint}$$
$$y \geq 0 \quad \text{Nonnegativity constraint}$$

In general, every linear programming problem has two components:

1. A linear objective function to be maximized or minimized.
2. A collection of linear inequalities that must be satisfied simultaneously.

Linear Programming Problem

A **linear programming problem** in two variables, x and y, consists of **maximizing** or **minimizing** an **objective function**

$$z = Ax + By$$

where A and B are given real numbers, subject to certain conditions or **constraints** expressible as a system of linear inequalities in x and y.

Let's look at this definition more closely. To maximize (or minimize) the quantity $z = Ax + By$ means to locate the points (x, y) that result in the largest (or smallest) value of z. But not all points (x, y) are eligible. Only the points that obey *all* the constraints are potential solutions. Hence, we refer to such points as **feasible points.**

In a linear programming problem we want to find the feasible point that maximizes (or minimizes) the objective function.

By a **solution to a linear programming problem** we mean a feasible point (x, y), together with the value of the objective function at that point, which maximizes (or minimizes) the objective function. If none of the feasible points maximizes (or minimizes) the objective function, or if there are no feasible points, then the linear programming problem has no solution.

EXAMPLE 1   **Solving a Linear Programming Problem**

Minimize the quantity

$$z = x + 2y$$

subject to the constraints

$$x + y \geq 1$$
$$x \geq 0$$
$$y \geq 0$$

SOLUTION   The objective function to be minimized is $z = x + 2y$. The constraints are the linear inequalities

$$x + y \geq 1$$
$$x \geq 0$$
$$y \geq 0$$

Figure 15

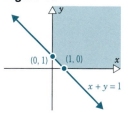

The shaded portion of Figure 15 illustrates the set of feasible points.

To see if there is a smallest z, we graph $z = x + 2y$ for some choice of z, say, $z = 3$. See Figure 16. By moving the line $x + 2y = 3$ parallel to itself, we can observe what happens for different values of z. Since we want a minimum value for z, we try to move $z = x + 2y$ down as far as possible while keeping some part of the line within the set of feasible points. The "best" solution is obtained when the line just touches a corner point of the set of feasible points. If you refer to Figure 16, you will see that the best solution is $x = 1$, $y = 0$, which yields $z = 1$. There is no other feasible point for which z is smaller.

Figure 16

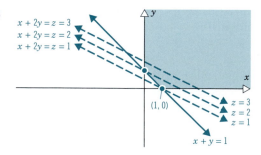

The next example illustrates a linear programming problem that has no solution.

EXAMPLE 2   A Linear Programming Problem without a Solution

Maximize the quantity

$$z = x + 2y$$

subject to the constraints

$$x + y \geq 1$$
$$x \geq 0$$
$$y \geq 0$$

SOLUTION   First, we graph the constraints. The shaded portion of Figure 17 illustrates the set of feasible points.

The graphs of the objective function $z = x + 2y$ for $z = 2$, $z = 8$, and $z = 12$ are also shown in Figure 17. Observe that we continue to get larger values for z by moving the graph of the objective function upward. But there is no feasible point that will make z *largest*. No matter how large a value is assigned to z, there is a feasible point that will give a larger value. Since there is no feasible point that makes z largest, we conclude that this linear programming problem has no solution.

Figure 17

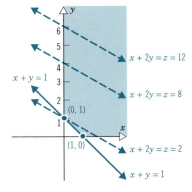

Let's compare the set of feasible points in Examples 1 and 2. In Example 1, which had a solution, the set of feasible points was bounded; in Example 2, which had no solution, it was unbounded. The next result gives some conditions on the set of feasible points that determine when a solution to the linear programming problem exists.

Existence of a Solution

Consider a linear programming problem with the set R of feasible points and objective function $z = Ax + By$.

1. If R is bounded, then z has both a maximum and a minimum value on R.

2. If R is unbounded and $A \geq 0$, $B \geq 0$, and the constraints include $x \geq 0$ and $y \geq 0$, then z has a minimum value on R but not a maximum value (see Example 2).

3. If R is the empty set, then the linear programming problem has no solution and z has neither a maximum nor a minimum value.

In Example 1 we found that the feasible point that minimizes z occurs at a corner point. This is not an unusual situation. If there are feasible points minimizing (or maximizing) the objective function, at least one will be at a corner point of the set of feasible points.

Fundamental Theorem of Linear Programming

If a linear programming problem has a solution, it is located at a corner point of the set of feasible points; if a linear programming problem has multiple solutions, at least one of them is located at a corner point of the set of feasible points. In either case the corresponding value of the objective function is unique.

The result just stated indicates that it is possible for a feasible point that is not a corner point to minimize (or maximize) the objective function. For example, if the slope of the objective function is the same as the slope of one of the boundaries of the set of feasible points and if the two adjacent corner points are solutions, then so are all the points on the line segment joining them. The following example illustrates this situation.

EXAMPLE 3   A Linear Programming Problem with Multiple Solutions

Minimize the quantity

$$z = x + 2y$$

subject to the constraints

$$x + y \geq 1$$
$$2x + 4y \geq 3$$
$$x \geq 0$$
$$y \geq 0$$

SOLUTION   Again we first graph the constraints. The shaded portion of Figure 18 illustrates the set of feasible points.

Figure 18

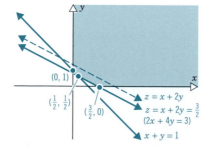

If we graph the objective equation $z = x + 2y$ for some choice of z and move it down, we find that a minimum is reached when $z = \frac{3}{2}$. In fact, any point on the line

$2x + 4y = 3$ between the adjacent corner points $(\frac{1}{2}, \frac{1}{2})$ and $(\frac{3}{2}, 0)$ and including these corner points will minimize the objective function. Of course, the reason any feasible point on $2x + 4y = 3$ minimizes the objective equation $z = x + 2y$ is that these two lines each have slope $-\frac{1}{2}$. Thus this linear programming problem has infinitely many solutions.

✏ **Now Work Problem 1**

Since the objective function attains its maximum or minimum value at the corner points of the set of feasible points, we can outline a procedure for solving a linear programming problem provided that it has a solution.

Steps for Solving a Linear Programming Problem

If a linear programming problem has a solution, follow these steps to find it:

Step 1 Write an expression for the quantity that is to be maximized or minimized (the objective function).

Step 2 Determine all the constraints and graph the set of feasible points.

Step 3 List the corner points of the set of feasible points.

Step 4 Determine the value of the objective function at each corner point.

Step 5 Select the optimal solution, that is, the maximum or minimum value of the objective function.

Let's look at some examples.

EXAMPLE 4   Solving a Linear Programming Problem

Maximize and minimize the objective function

$$z = x + 5y$$

subject to the constraints

$$
\begin{aligned}
x + 4y &\leq 12 \quad ① \\
x &\leq 8 \quad ② \\
x + y &\geq 2 \quad ③ \\
x &\geq 0 \quad ④ \\
y &\geq 0 \quad ⑤
\end{aligned}
$$

SOLUTION   The objective function and the constraints (numbered for convenience) are given (this will not be the case when we do word problems), so we can proceed to graph the constraints. The shaded portion of Figure 19 illustrates the set of feasible points. Since this set is bounded, we know a solution exists.

Figure 19

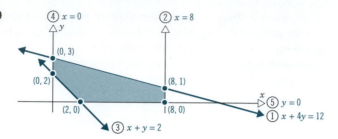

Now we locate the corner points of the set of feasible points at the points of intersection of lines ① and ④, ① and ②, ② and ⑤, ③ and ⑤, and ③ and ④. Using methods discussed earlier, we find that the corner points are

$$(0, 3) \qquad (8, 1) \qquad (8, 0) \qquad (2, 0) \qquad (0, 2)$$

To find the maximum and minimum value of $z = x + 5y$, we set up a table:

Corner Point (x, y)	Value of Objective Function $z = x + 5y$
(0, 3)	$z = 0 + 5(3) = 15$
(8, 1)	$z = 8 + 5(1) = 13$
(8, 0)	$z = 8 + 5(0) = 8$
(2, 0)	$z = 2 + 5(0) = 2$
(0, 2)	$z = 0 + 5(2) = 10$

The maximum value of z is 15, and it occurs at the point $(0, 3)$. The minimum value of z is 2, and it occurs at the point $(2, 0)$.

 **Now Work Problems**

17 and 29

Now let's solve the problem of the cashews and peanuts that we discussed at the start of this section.

EXAMPLE 5   **Maximizing Profit**

Maximize

$$P = 0.25x + 0.45y$$

subject to the constraints

$$x + 2y \le 300 \qquad ①$$
$$3x + 2y \le 480 \qquad ②$$
$$x \ge 0 \qquad ③$$
$$y \ge 0 \qquad ④$$

SOLUTION   Before applying the method of this chapter to solve this problem, let's discuss a solution that might be suggested by intuition. Namely, since the profit is higher for the high-grade mixture, you might think that Nutt's Nuts should prepare as many packages of the high-grade mixture as possible. If this were done, then there would be a total of 150 packages (8 ounces divides into 75 pounds of cashews exactly 150 times) and the total profit would be

$$150(0.45) = \$67.50$$

As we shall see, this is not the best solution to the problem.

To obtain the maximum profit, we use linear programming. The graph of the set of feasible points is given in Figure 20.

Figure 20

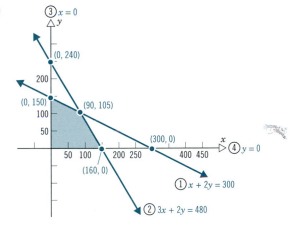

Since this set is bounded, we proceed to locate its corner points. The corner points of the set of feasible points are the points of intersection of lines ③ and ④, ① and ③, ② and ④, and ① and ②:

$$(0, 0) \qquad (0, 150) \qquad (160, 0) \qquad (90, 105)$$

(Notice that the points of intersection of lines ① and ④ and lines ② and ③ are not feasible points.) It remains only to evaluate the objective equation at each corner point:

Corner Point (x, y)	Value of Objective Function $P = (\$0.25)x + (\$0.45)y$
(0, 0)	$P = (0.25)(0) + (0.45)(0) = 0$
(0, 150)	$P = (0.25)(0) + (0.45)(150) = \$67.50$
(160, 0)	$P = (0.25)(160) + (0.45)(0) = \$40.00$
(90, 105)	$P = (0.25)(90) + (0.45)(105) = \$69.75$

Thus a maximum profit is obtained if 90 packages of low-grade mixture and 105 packages of high-grade mixture are made. The maximum profit obtainable under the conditions described is $69.75.

Now Work Problem 49

EXERCISE 3.2 Answers to odd-numbered problems begin on page AN-16.

In Problems 1–10 the given figure illustrates the graph of the set of feasible points of a linear programming problem. Find the maximum and minimum values of each objective function.

1. $z = 2x + 3y$ **2.** $z = 3x + 2y$

3. $z = x + y$ **4.** $z = 3x + 3y$

5. $z = x + 6y$ **6.** $z = 6x + y$

7. $z = 3x + 4y$ **8.** $z = 4x + 3y$

9. $z = 10x + y$ **10.** $z = x + 10y$

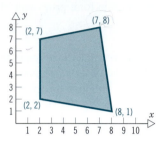

In Problems 11–16 list the corner points for each collection of constraints of a linear programming problem.

11.
$$x \le 13 \quad ①$$
$$4x + 3y \ge 12 \quad ②$$
$$x \ge 0 \quad ③$$
$$y \ge 0 \quad ④$$

12.
$$x \le 8 \quad ①$$
$$2x + 3y \ge 6 \quad ②$$
$$x \ge 0 \quad ③$$
$$y \ge 0 \quad ④$$

13.
$$y \le 10 \quad ①$$
$$x + y \le 15 \quad ②$$
$$x \ge 0 \quad ③$$
$$y \ge 0 \quad ④$$

14.
$$y \le 8 \quad ①$$
$$2x + y \ge 10 \quad ②$$
$$x \ge 0 \quad ③$$
$$y \ge 0 \quad ④$$

15.
$$x \le 10 \quad ①$$
$$y \le 8 \quad ②$$
$$4x + 3y \ge 12 \quad ③$$
$$x \ge 0 \quad ④$$
$$y \ge 0 \quad ⑤$$

16.
$$x \le 9 \quad ①$$
$$y \le 12 \quad ②$$
$$2x + 3y \le 24 \quad ③$$
$$x \ge 0 \quad ④$$
$$y \ge 0 \quad ⑤$$

In Problems 17–24 maximize (if possible) the quantity $z = 5x + 7y$ subject to the given constraints.

17.
$$x + y \le 2$$
$$y \ge 1$$
$$x \ge 0$$
$$y \ge 0$$

18.
$$2x + 3y \le 6$$
$$x \le 2$$
$$x \ge 0$$
$$y \ge 0$$

19.
$$x + y \ge 2$$
$$2x + 3y \le 6$$
$$x \ge 0$$
$$y \ge 0$$

20.
$$x + y \ge 2$$
$$2x + 3y \le 12$$
$$3x + 2y \le 12$$
$$x \ge 0$$
$$y \ge 0$$

21.
$$2 \le x + y$$
$$x + y \le 8$$
$$2x + y \le 10$$
$$x \ge 0$$
$$y \ge 0$$

22.
$$2 \le x + y$$
$$x + y \le 8$$
$$1 \le x + 2y$$
$$x + 2y \le 10$$
$$x \ge 0$$
$$y \ge 0$$

23.
$$x + y \le 10$$
$$x \ge 6$$
$$x \ge 0$$
$$y \ge 0$$

24.
$$x + y \le 8$$
$$y \ge 2$$
$$x \ge 0$$
$$y \ge 0$$

In Problems 25–32 minimize (if possible) the quantity $z = 2x + 3y$ subject to the given constraints.

25.
$$x + y \le 2$$
$$y \le x$$
$$x \ge 0$$
$$y \ge 0$$

26.
$$3x + y \le 3$$
$$y \ge x$$
$$x \ge 0$$
$$y \ge 0$$

27.
$$x + y \ge 2$$
$$x + 3y \le 12$$
$$3x + y \le 12$$
$$x \ge 0$$
$$y \ge 0$$

28.
$$x + y \le 8$$
$$2x + 3y \ge 6$$
$$x + y \ge 2$$
$$x \ge 0$$
$$y \ge 0$$

29.
$$2 \leq x + y$$
$$x + y \leq 10$$
$$2x + 3y \leq 6$$
$$x \geq 0$$
$$y \geq 0$$

30.
$$2y \leq x$$
$$x + 2y \leq 10$$
$$x + 2y \geq 4$$
$$x \geq 0$$
$$y \geq 0$$

31.
$$1 \leq x + 2y$$
$$x + 2y \leq 10$$
$$y \geq 2x$$
$$x + y \leq 8$$
$$x \geq 0$$
$$y \geq 0$$

32.
$$2 \leq 2x + y$$
$$x + y \leq 6$$
$$2x \geq y$$
$$x \geq 0$$
$$y \geq 0$$

In Problems 33–40 find the maximum and minimum values (if possible) of the given objective function subject to the constraints

$$x + y \leq 10$$
$$2x + y \geq 10$$
$$x + 2y \geq 10$$
$$x \geq 0$$
$$y \geq 0$$

33. $z = x + y$

34. $z = 2x + 3y$

35. $z = 5x + 2y$

36. $z = x + 2y$

37. $z = 3x + 4y$

38. $z = 3x + 6y$

39. $z = 10x + y$

40. $z = x + 10y$

41. Find the maximum and minimum values of $z = 18x + 30y$ subject to the constraints $3y + 3x \geq 9$, $-x + 4y \leq 12$, and $4x - y \leq 12$.

42. Find the maximum and minimum values of $z = 20x + 16y$ subject to the constraints $3y + 4x \geq 12$, $-2x + 4y \leq 16$, and $6x - y \leq 18$.

43. Find the maximum and minimum values of $z = 7x + 6y$ subject to the constraints $3y + 2x \geq 6$, $-3x + 4y \leq 8$, and $5x - y \leq 15$.

44. Find the maximum and minimum values of $z = 6x + 3y$ subject to the constraints $2y + 2x \geq 4$, $-x + 5y \leq 10$, and $3x - 3y \leq 6$.

45. Maximize $z = -20x + 30y$ subject to the constraints $0 \leq x \leq 15$, $0 \leq y \leq 10$, $3y + 5x \geq 15$, and $3y - 3x \leq 21$.

46. Maximize $z = -10x + 10y$ subject to the constraints $0 \leq x \leq 15$, $0 \leq y \leq 10$, $y + 6x \geq 6$, and $y - 3x \leq 7$.

47. Maximize $z = -12x + 24y$ subject to the constraints $0 \leq x \leq 15$, $0 \leq y \leq 10$, $3y + 3x \geq 9$, and $2y - 3x \leq 14$.

48. Maximize $z = -20x + 10y$ subject to the constraints $0 \leq x \leq 15$, $0 \leq y \leq 10$, $3y + 4x \geq 12$, and $y - 3x \leq 7$.

49. In Example 5, if the profit on the low-grade mixture is $0.30 per package and the profit on the high-grade mixture is $0.40 per package, how many packages of each mixture should be made for a maximum profit?

3.3 APPLICATIONS

In this section, several situations that lead to linear programming problems are presented.

EXAMPLE 1   Maximizing Profit

Mike's Famous Toy Trucks manufactures two kinds of toy trucks—a standard model and a deluxe model. In the manufacturing process each standard model requires 2 hours of grinding and 2 hours of finishing, and each deluxe model needs 2 hours of grinding and 4 hours of finishing. The company has two grinders and three finishers, each of whom works at most 40 hours per week. Each standard model toy truck brings a profit of $3 and each deluxe model a profit of $4. Assuming that every truck made will be sold, how many of each should be made to maximize profits?

SOLUTION   First, we name the variables:

$$x = \text{Number of standard models made}$$
$$y = \text{Number of deluxe models made}$$

The quantity to be maximized is the profit, which we denote by P:

$$P = \$3x + \$4y$$

This is the objective function. To manufacture one standard model requires 2 grinding hours and to make one deluxe model requires 2 grinding hours. Thus, the number of grinding hours of x standard and y deluxe models is

$$2x + 2y$$

But the total amount of grinding time available is 80 hours per week. This means we have the constraint

$$2x + 2y \leq 80 \qquad \text{Grinding time constraint}$$

Similarly, for the finishing time we have the constraint

$$2x + 4y \leq 120 \qquad \text{Finishing time constraint}$$

By simplifying each of these constraints and adding the nonnegativity constraints $x \geq 0$ and $y \geq 0$, we may list all the constraints for this problem:

$$x + y \leq 40 \quad \textcircled{1}$$
$$x + 2y \leq 60 \quad \textcircled{2}$$
$$x \geq 0 \quad \textcircled{3}$$
$$y \geq 0 \quad \textcircled{4}$$

Figure 21 illustrates the set of feasible points, which is bounded.

Figure 21

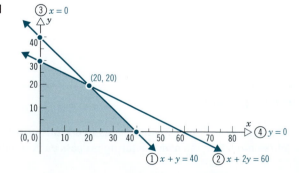

The corner points of the set of feasible points are

$$(0, 0) \qquad (0, 30) \qquad (40, 0) \qquad (20, 20)$$

The table lists the corresponding values of the objective equation:

Corner Point (x, y)	Value of Objective Function $P = \$3x + \$4y$
(0, 0)	$P = 0$
(0, 30)	$P = \$120$
(40, 0)	$P = \$120$
(20, 20)	$P = 3(20) + 4(20) = \$140$

Thus a maximum profit is obtained if 20 standard trucks and 20 deluxe trucks are manufactured. The maximum profit is $140.

EXAMPLE 2   Financial Planning

A retired couple has up to $30,000 to invest in fixed-income securities. Their broker recommends investing in two bonds: one a AAA bond yielding 8%; the other a B$^+$ bond paying 12%. After some consideration, the couple decides to invest at most $12,000 in the B$^+$-rated bond and at least $6000 in the AAA bond. They also want the amount invested in the AAA bond to exceed or equal the amount invested in the B$^+$ bond. What should the broker recommend if the couple (quite naturally) wants to maximize the return on their investment?

SOLUTION   First, we name the variables:

$$x = \text{Amount invested in the AAA bond}$$
$$y = \text{Amount invested in the B}^+ \text{ bond}$$

The quantity to be maximized, the couple's return on investment, which we denote by P, is

$$P = 0.08x + 0.12y$$

This is the objective function. The conditions specified by the problem are

Up to $30,000 available to invest	$x + y \leq 30{,}000$
Invest at most $12,000 in the B$^+$ bond	$y \leq 12{,}000$
Invest at least $6000 in the AAA bond	$x \geq 6000$
Amount in the AAA bond must exceed or equal amount in the B$^+$ bond	$x \geq y$

In addition, we must have the conditions $x \geq 0$ and $y \geq 0$. The total list of constraints is

$$x + y \leq 30{,}000 \quad ①$$
$$y \leq 12{,}000 \quad ②$$
$$x \geq 6000 \quad ③$$
$$x \geq y \quad ④$$
$$x \geq 0 \quad ⑤$$
$$y \geq 0 \quad ⑥$$

Figure 22 illustrates the set of feasible points, which is bounded. The corner points of the set of feasible points are

(6000, 0)   (6000, 6000)   (12,000, 12,000)   (18,000, 12,000)   (30,000, 0)

Figure 22

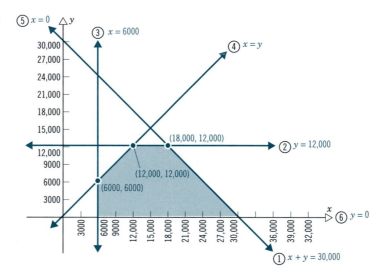

The corresponding return on investment at each corner point is

$$P = 0.08(6000) + 0.12(0) = \$480$$
$$P = 0.08(6000) + 0.12(6000) = 480 + 720 = \$1200$$
$$P = 0.08(12{,}000) + 0.12(12{,}000) = 960 + 1440 = \$2400$$
$$P = 0.08(18{,}000) + 0.12(12{,}000) = 1440 + 1440 = \$2880$$
$$P = 0.08(30{,}000) + 0.12(0) = \$2400$$

Thus the maximum return on investment is $2880, obtained by placing $18,000 in the AAA bond and $12,000 in the B⁺ bond.

✎ Now Work Problem 5

EXAMPLE 3   Land Reclamation*

This example concerns reclaimed land and its allocation into two major uses—agricultural and urban (or nonagricultural). The reclamation of land for urban purposes costs $400 per acre and for agricultural uses, $300. The reclamation agency wishes to minimize the total cost C of reclaiming the land:

$$C = \$400x + \$300y$$

where x = the number of acres of urban land and y = the number of acres of agricultural land. Although this equation can be minimized by setting both x and y at zero, that is, reclaiming nothing, the problem derives from a number of constraints due to three different groups.

The first is an urban group, which insists that at least 4000 acres of land be reclaimed for urban purposes. The second group is concerned with agriculture and says that at least 5000 acres of land must be reclaimed for agricultural uses. Finally, the third group is concerned only with reclamation and is quite uninterested in the use to which the land will be put. The third group, however, says that at least 10,000 acres of land must be reclaimed. The problem and the constraints can, therefore, be written in full as follows:

Minimize

$$C = \$400x + \$300y$$

subject to the constraints

$$x \geq 4000 \quad ①$$
$$y \geq 5000 \quad ②$$
$$x + y \geq 10,000 \quad ③$$
$$x \geq 0 \quad ④$$
$$y \geq 0 \quad ⑤$$

Figure 23 illustrates the set of feasible points, which is not bounded. However, this minimum problem has a solution since the coefficients of x and y in the objective function are positive and the constraints include $x \geq 0$ and $y \geq 0$. See Condition 2 for existence of a solution (p. 145).

Figure 23

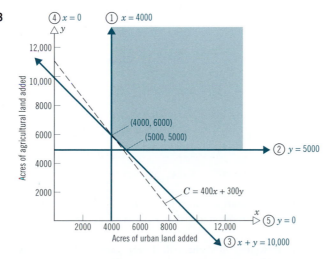

*This example is adapted from Maurice Yeates, *An Introduction to Quantitative Analysis in Economic Geography.* New York: McGraw-Hill.

The combination of urban and agricultural land at the corner point (4000, 6000) reveals that if 4000 acres are devoted to urban purposes and 6000 acres to agricultural purposes, the cost is a minimum and is

$$C = (\$400)(4000) + (\$300)(6000) = \$3,400,000$$

EXAMPLE 4   Pollution Control Model*

The following model is based on a paper by Robert C. Kohn. In this paper a linear programming model is proposed to help determine what air pollution controls should be adopted in an airshed. The basic premise is that air quality goals should be achieved at the least possible cost. Advantages of the model are its simplicity, its emphasis on economic efficiency, and its appropriateness for the kind of data that are already available.

To illustrate the model, consider a hypothetical airshed with a single industry, cement manufacturing. Annual production is 2,500,000 barrels of cement. Although the kilns are equipped with mechanical collectors for air pollution control, they are still emitting 2 pounds of dust for every barrel of cement produced. The industry can be required to replace the mechanical collectors with four-field electrostatic precipitators, which would reduce emissions to 0.5 pound of dust per barrel of cement or with five-field electrostatic precipitators, which would reduce emissions to 0.2 pound per barrel. If the capital and operating costs of the four-field precipitator are $0.14 per barrel of cement produced and of the five-field precipitator are $0.18 per barrel, what control methods should be required of this industry? Assume that, for this hypothetical airshed, it has been determined that particulate emissions (which now total 5,000,000 pounds per year) should be reduced by 4,200,000 pounds per year.

If C represents the cost of control, x the number of barrels of annual cement production subject to the four-field electrostatic precipitator (cost is $0.14 a barrel of cement produced and pollutant reduction is $2 - 0.5 = 1.5$ pounds of particulates per barrel of cement produced), and y the number of barrels of annual cement production subject to the five-field electrostatic precipitator (cost is $0.18 a barrel and pollutant reduction is $2 - 0.2 = 1.8$ pounds per barrel of cement produced), then the problem can be stated as follows:

Minimize

$$C = \$0.14x + \$0.18y$$

subject to

$$x + y \leq 2,500,000 \quad ①$$
$$1.5x + 1.8y \geq 4,200,000 \quad ②$$
$$x \geq 0 \quad ③$$
$$y \geq 0 \quad ④$$

The first equation states that our objective is to minimize air pollution control costs; the first constraint states that barrels of cement production subject to the two control

---

*R. E. Kohn, "A Mathematical Programming Model for Air Pollution Control," *School Science and Mathematics* (June 1969), pp. 487–499.

methods cannot exceed the annual production; the second constraint states that the particulate reduction from the two methods must be greater than or equal to the particulate reduction target; and the last two inequalities mean that we cannot have negative quantities of cement. Figure 24 illustrates a geometric solution to the problem.

Figure 24

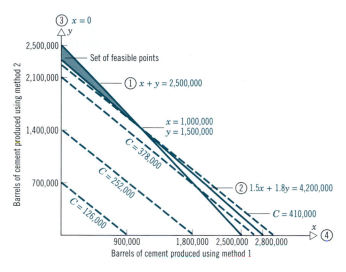

The least costly solution would be to install the four-field precipitator on kilns producing 1,000,000 ($x = 1,000,000$) and the five-field precipitator on kilns producing 1,500,000 ($y = 1,500,000$) barrels of cement at a cost of $C = \$410,000$.

EXERCISE 3.3 Answers to odd-numbered problems begin on page AN-17.

1. **Optimal Land Use** A farmer has 70 acres of land available on which to grow some soybeans and some corn. The cost of cultivation per acre, the workdays needed per acre, and the profit per acre are indicated in the table:

	Soybeans	**Corn**	**Total Available**
Cultivation Cost per Acre	$60	$30	$1800
Days of Work per Acre	3 days	4 days	120 days
Profit per Acre	$300	$150	

As indicated in the last column, the acreage to be cultivated is limited by the amount of money available for cultivation costs and by the number of working days that can be put into this part of the business. Find the number

of acres of each crop that should be planted in order to maximize the profit.

2. **Investment Strategy** An investment broker wants to invest up to $20,000. She can purchase a type A bond yielding a 10% return on the amount invested and she can purchase a type B bond yielding a 15% return on the amount invested. She also wants to invest at least as much in the type A bond as in the type B bond. She will also invest at least $5000 in the type A bond and no more than $8000 in the type B bond. How much should she invest in each type of bond to maximize her return?

3. **Manufacturing** A factory manufactures two products, each requiring the use of three machines. The first machine can be used at most 70 hours; the second machine at most 40 hours; and the third machine at most 90 hours. The first product requires 2 hours on machine 1, 1 hour on machine 2, and 1 hour on machine 3; the second product requires 1 hour each on machines 1 and 2, and 3 hours on machine 3. If the profit is $40 per unit for the first

product and $60 per unit for the second product, how many units of each product should be manufactured to maximize profit?

4. **Diet** A diet is to contain at least 400 units of vitamins, 500 units of minerals, and 1400 calories. Two foods are available: F_1, which costs $0.05 per unit, and F_2, which costs $0.03 per unit. A unit of food F_1 contains 2 units of vitamins, 1 unit of minerals, and 4 calories; a unit of food F_2 contains 1 unit of vitamins, 2 units of minerals, and 4 calories. Find the minimum cost for a diet that consists of a mixture of these two foods and also meets the minimal nutrition requirements.

5. **Investment Strategy** A financial consultant wishes to invest up to a total of $30,000 in two types of securities, one that yields 10% per year and another that yields 8% per year. Furthermore, she believes that the amount invested in the first security should be at most one-third of the amount invested in the second security. What investment program should the consultant pursue in order to maximize income?

6. **Scheduling** Blink appliances has a sale on microwaves and stoves. Each microwave requires 2 hours to unpack and set up, and each stove requires 1 hour. The storeroom space is limited to 50 items. The budget of the store allows only 80 hours of employee time for unpacking and setup. Microwaves sell for $300 each, and stoves sell for $200 each. How many of each should the store order to maximize revenue?

7. **Cost Control** An appliance repair shop has 5 vacuum cleaners, 12 TV sets, and 18 VCRs to be repaired. The store employs two part-time repairmen. One repairman can repair one vacuum cleaner, three TV sets, and three VCRs in 1 week, while the second repairman can repair one vacuum cleaner, two TV sets and six VCRs in 1 week. The first employee is paid $250 a week and the second employee is paid $220 a week. To minimize the cost, how many weeks should each of the two repairmen be employed?

8. **Transportation** An appliance company has a warehouse and two terminals. To minimize shipping costs, the manager must decide how many appliances should be shipped to each terminal. There is a total supply of 1200 units in the warehouse and a demand for 400 units in terminal A and 500 units in terminal B. It costs $12 to ship each unit to terminal A and $16 to ship to terminal B. How many units should be shipped to each terminal in order to minimize cost?

9. **Pollution Control** A chemical plant produces two items A and B. For each item A produced, 2 cubic feet of carbon monoxide and 6 cubic feet of sulfur dioxide are emitted into the atmosphere; to produce item B, 4 cubic feet of carbon monoxide and 3 cubic feet of sulfur dioxide are emitted into the atmosphere. Government pollution standards permit the manufacturer to emit a maximum of 3000 cubic feet of carbon monoxide and 5400 cubic feet of sulfur dioxide per week. The manufacturer can sell all of the items that it produces and make a profit of $1.50 per unit for item A and $1.00 per unit for item B. Determine the number of units of each item to be produced each week to maximize profit without exceeding government standards.

10. **Production Scheduling** A company produces two types of steel. Type 1 requires 2 hours of melting, 4 hours of cutting, and 10 hours of rolling per ton. Type 2 requires 5 hours of melting, 1 hour of cutting, and 5 hours of rolling per ton. Forty hours are available for melting, 20 for cutting, and 60 for rolling. Each ton of Type 1 produces $240 profit, and each ton of Type 2 yields $80 profit. Find the maximum profit and the production schedule that will produce this profit.

11. **Diet** Danny's Chicken Farm is a producer of frying chickens. In order to produce the best fryers possible, the regular chicken feed is supplemented by four vitamins. The minimum amount of each vitamin required per 100 ounces of feed is: vitamin 1, 50 units; vitamin 2, 100 units; vitamin 3, 60 units; vitamin 4, 180 units. Two supplements are available: supplement I costs $0.03 per ounce and contains 5 units of vitamin 1 per ounce, 25 units of vitamin 2 per ounce, 10 units of vitamin 3 per ounce, and 35 units of vitamin 4 per ounce. Supplement II costs $0.04 per ounce and contains 25 units of vitamin 1 per ounce, 10 units of vitamin 2 per ounce, 10 units of vitamin 3 per ounce, and 20 units of vitamin 4 per ounce. How much of each supplement should Danny buy to add to each 100 ounces of feed in order to minimize his cost, but still have the desired vitamin amounts present?

12. **Maximizing Income** J. B. Rug Manufacturers has available 1200 square yards of wool and 1000 square yards of nylon for the manufacture of two grades of carpeting: high-grade, which sells for $500 per roll, and low-grade, which sells for $300 per roll. Twenty square yards of wool and 40 square yards of nylon are used in a roll of high-grade carpet, and 40 square yards of nylon are used in a roll of low-grade carpet. Forty work-hours are required to manufacture each roll of the high-grade carpet, and 20 work-hours are required for each roll of the low-grade carpet, at an average cost of $6 per work-hour. A maximum of 800 work-hours are available. The cost of wool is $5 per square yard and the cost of nylon is $2 per square yard. How many rolls of each type of carpet should be manufactured to maximize income? [*Hint:*

Income = revenue from sale − (production cost for material + labor)]

13. The rug manufacturer in Problem 12 finds that maximum income occurs when no high-grade carpet is produced. If the price of the low-grade carpet is kept at \$300 per roll, in what price range should the high-grade carpet be sold so that income is maximized by selling some rolls of each type of carpet? Assume all other data remain the same.

14. Maximize

$$P = 2x + y + 3z$$

subject to

$$x + 2y + z \le 25 \quad \text{①}$$
$$3x + 2y + 3z \le 30 \quad \text{②}$$
$$x \ge 0 \quad \text{③}$$
$$y \ge 0 \quad \text{④}$$
$$z \ge 0 \quad \text{⑤}$$

[*Hint:* Solve the constraints three at a time, find the feasible points, and test each of them in the objective function. Assume a solution exists.]

Technology Exercises

Some graphing calculators can graph systems of inequalities and shade the region representing the solution to the system on the screen. Consult your user's manual, and use your graphing calculator to solve the systems of inequalities in Problems 1–6. If your calculator does not graph inequalities, graph the boundary line of each inequality.

For each system in Problems 1–6, find the minimum and the maximum of the objective function

$$z = 3.5x + 1.25y$$

within the region representing the solution to the system of inequalities. Use INTERSECT to find the corner points.

1.
$$\begin{cases} y \le x + 1 \\ y + x \le 9 \\ x + y \ge 3 \\ y + 3 \ge x \end{cases}$$

2.
$$\begin{cases} x + y \le 7 \\ y \le 2x + 2 \\ y + \frac{1}{2}x \ge 2 \\ y + 3 \ge x \end{cases}$$

3.
$$\begin{cases} y \le \frac{1}{2}x + 5 \\ y + 2x \le 15 \\ 5x + y \ge 8 \\ y + 1 \ge x \end{cases}$$

4.
$$\begin{cases} y \le 1.3x + 4 \\ y + 1.17x \le 12.33 \\ y \ge 3 \\ x \ge 0 \end{cases}$$

5.
$$\begin{cases} y \le \frac{1}{2}x + 5 \\ y + 2x \le 15 \\ 5x + y \ge 8 \\ y + 5 \ge x \\ y \ge 0 \end{cases}$$

6.
$$\begin{cases} y \le 6 \\ y + 2x \le 6 \\ x + y \ge 0 \\ x \le 4 \end{cases}$$

CHAPTER REVIEW

IMPORTANT TERMS AND CONCEPTS

graph of a linear inequality 131
steps for graphing a linear inequality 132
systems of linear inequalities 133
linear programming problem 143
solution to a linear programming
    problem 144

existence of solutions to a
    linear programming
    problem 145
fundamental theorem of
    linear programming 146

steps for solving a
    linear programming
    problem 147

TRUE–FALSE ITEMS   Answers are on page AN-18.

T_____ F_____ **1.** The graph of a system of linear inequal-ities may be bounded or unbounded.

T_____ F_____ **2.** The graph of the set of constraints of a linear programming problem, under cer-tain conditions, could have a circle for a boundary.

T_____ F_____ **3.** The objective function of a linear pro-gramming problem is always a linear equation involving the variables.

T_____ F_____ **4.** In a linear programming problem, there may be more than one point that maxi-mizes or minimizes the objective func-tion.

T_____ F_____ **5.** Some linear programming problems will have no solution.

T_____ F_____ **6.** If a linear programming problem has a solution, it is located at the center of the set of feasible points.

FILL IN THE BLANKS   Answers are on page AN-18.

1. The graph of a linear inequality in two variables is called a _____ .

2. In a linear programming problem the quantity to be max-imized or minimized is referred to as the _____ function.

3. The points that obey the collection of constraints of a linear programming problem are called _____ points.

4. A linear programming problem will always have a solution if the set of feasible points is _____ .

5. If a linear programming problem has a solution, it is lo-cated at a _____ of the set of feasible points.

REVIEW EXERCISES   Answers to odd-numbered problems begin on page AN-18.

In Problems 1–4 graph each linear inequality.

1. $x + 3y \leq 0$

2. $4x + y \geq 0$

3. $5x + y \geq 10$

4. $2x + 3y \geq 6$

5. Without graphing, determine which of the points $P_1 = (4, -3)$, $P_2 = (2, -6)$, $P_3 = (8, -3)$ are part of the graph of the following system:

$$\begin{cases} 7y + 10x \leq 0 \\ 2y + 9x \geq 0 \\ y + 3x \leq 15 \end{cases}$$

6. Without graphing, determine which of the points $P_1 = (8, 6)$, $P_2 = (2, 5)$, $P_3 = (4, 1)$ are part of the graph of the following system:

$$\begin{cases} y - 10x \leq 0 \\ 2y - 3x \geq 0 \\ y + x \leq 15 \end{cases}$$

In Problems 7–8 determine which region, a, b, c, or d, represents the graph of the given system of linear inequalities.

7. $\begin{cases} 6x - 4y \leq 12 \\ 3x + 2y \leq 18 \end{cases}$

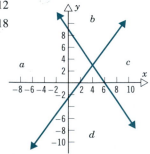

8. $\begin{cases} 6x - 5y \leq 5 \\ 6x + 6y \leq 60 \end{cases}$

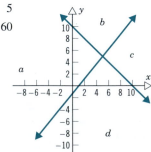

In Problems 9–14 graph each system of linear inequalities. Locate the corner points and tell whether the graph is bounded or unbounded.

9. $\begin{cases} 3x + 2y \le 12 \\ x + y \ge 4 \\ x \ge 0 \quad y \ge 0 \end{cases}$

10. $\begin{cases} x + y \le 8 \\ 2x + y \ge 4 \\ x \ge 0 \quad y \ge 0 \end{cases}$

11. $\begin{cases} x + 2y \ge 4 \\ 3x + y \le 6 \\ x \ge 0 \quad y \ge 0 \end{cases}$

12. $\begin{cases} 2x + y \ge 4 \\ 3x + 2y \ge 6 \\ x \ge 0 \quad y \ge 0 \end{cases}$

13. $\begin{cases} 3x + 2y \ge 6 \\ 3x + 2y \le 12 \\ x + 2y \le 8 \\ x \ge 0 \quad y \ge 0 \end{cases}$

14. $\begin{cases} x + 2y \ge 2 \\ x + 2y \le 10 \\ 2x + y \le 10 \\ x \ge 0 \quad y \ge 0 \end{cases}$

In Problems 15–22 use the constraints below to solve each linear programming problem.

$$x + 2y \le 40$$
$$2x + y \le 40$$
$$x + y \ge 10$$
$$x \ge 0 \quad y \ge 0$$

15. Maximize $z = x + y$

16. Maximize $z = 2x + 3y$

17. Minimize $z = 5x + 2y$

18. Minimize $z = 3x + 2y$

19. Maximize $z = 2x + y$

20. Maximize $z = x + 2y$

21. Minimize $z = 2x + 5y$

22. Minimize $z = x + y$

In Problems 23–26 maximize and minimize (if possible) the quantity $z = 15x + 20y$ subject to the given constraints.

23.
$$x \le 5$$
$$y \le 8$$
$$3x + 4y \ge 12$$
$$x \ge 0 \quad y \ge 0$$

24.
$$x \le 6$$
$$y \le 6$$
$$3x + 2y \ge 6$$
$$x \ge 0 \quad y \ge 0$$

25.
$$2x + 3y \le 22$$
$$x \le 5$$
$$y \le 6$$
$$x \ge 0 \quad y \ge 0$$

26.
$$x + 2y \le 20$$
$$x + 10y \ge 36$$
$$5x + 2y \ge 36$$
$$x \ge 0 \quad y \ge 0$$

In Problems 27–32 solve each linear programming problem.

27. Maximize

$$z = 2x + 3y$$

subject to the constraints

$$0 \le x \le 9$$
$$0 \le y \le 8$$
$$x + y \ge 3$$

28. Maximize

$$z = 4x + y$$

subject to the constraints

$$0 \le x \le 7$$
$$0 \le y \le 8$$
$$x + y \ge 2$$

29. Maximize

$$z = x + 2y$$

subject to the constraints

$$0 \le x \le 8$$
$$0 \le y \le 8$$
$$x + y \ge 1$$
$$y \le 2x$$

30. Maximize

$$z = 3x + 4y$$

subject to the constraints

$$0 \le x \le 8$$
$$0 \le y \le 7$$
$$x + 2y \ge 2$$
$$y \le 2x$$

31. Minimize

$$z = 3x + 2y$$

subject to the constraints

$$0 \le x \le 10$$
$$0 \le y \le 8$$
$$x + 2y \ge 8$$
$$y \ge x$$

32. Minimize

$$z = 2x + 5y$$

subject to the constraints

$$0 \le x \le 10$$
$$0 \le y \le 12$$
$$2x + y \ge 10$$
$$y \le x$$

33. Nutrition Katy needs at least 60 units of carbohydrates, 45 units of protein, and 30 units of fat each month. From each pound of food A, she receives 5 units of carbohydrates, 3 of protein, and 4 of fat. Food B contains 2 units of carbohydrates, 2 units of protein, and 1 unit of fat per pound. If food A costs $1.30 per pound and food B costs $0.80 per pound, how many pounds of each food should Katy buy each month to keep costs at a minimum?

34. Production Scheduling A company sells two types of shoes. The first uses 2 units of leather and 2 units of synthetic material and yields a profit of $8 per pair. The second type requires 5 units of leather and 1 unit of synthetic material and gives a profit of $10 per pair. If there are 40 units of leather and 16 units of synthetic material available, how many pairs of each type of shoe should be sold to maximize profit? What is the maximum profit?

35. Maximizing Profit A ski manufacturer makes two types of skis: downhill and cross-country. Using the information given in the table below, how many of each type of ski should be made for a maximum profit to be achieved? What is the maximum profit?

	Downhill	Cross-Country	Maximum Time Available
Manufacturing Time per Ski	2 hours	1 hour	40 hours
Finishing Time per Ski	1 hour	1 hour	32 hours
Profit per Ski	$70	$50	

36. Rework Problem 35 if the manufacturing unit has a maximum of 48 hours available.

37. Maximizing Profit A company makes two explosives: type I and type II. Due to storage problems, a maximum of 100 pounds of type I and 150 pounds of type II can be mixed and packaged each week. One pound of type I takes 60 hours to mix and 70 hours to package; 1 pound of type II takes 40 hours to mix and 40 hours to package. The mixing department has at most 7200 work-hours available each week, and packaging has at most 7800 workhours available. If the profit for 1 pound of type I is $60 and for 1 pound of type II is $40, what is the maximum profit possible each week?

38. Mixture A company makes two kinds of animal food, A and B, which contain two food supplements. It takes 2 pounds of the first supplement and one pound of the second to make a dozen cans of food A, and 4 pounds of the first supplement and 5 pounds of the second to make a dozen cans of food B. On a certain day 80 pounds of the first supplement and 70 pounds of the second is available. Maximize company profits if the profit on a dozen cans of food A is $3.00 and the profit on a dozen cans of food B is $10.00.

MATHEMATICAL QUESTIONS FROM PROFESSIONAL EXAMS

Use the following information to answer Problems 1–3:

CPA Exam *The Random Company manufactures two products, Zeta and Beta. Each product must pass through two processing operations. All materials are introduced at the start of process 1. There are no work-in-process inventories. Random may produce either one product exclusively or various combinations of both products subject to the following constraints:*

	Process No. 1	Process No. 2	Contribution Margin per Unit
Hours required to produce one unit of			
Zeta	1 hour	1 hour	$4.00
Beta	2 hours	3 hours	5.25
Total capacity in hours per day	1000 hours	1275 hours	

A shortage of technical labor has limited Beta production to 400 units per day. There are no constraints on the production of Zeta other than the hour constraints in the above schedule. Assume that all relationships between capacity and production are linear, and that all of the above data and relationships are deterministic rather than probabilistic.

1. Given the objective to maximize total contribution margin, what is the production constraint for process 1?

(a) Zeta + Beta ≤ 1000
(b) Zeta + 2 Beta ≤ 1000
(c) Zeta + Beta ≥ 1000
(d) Zeta + 2 Beta ≥ 1000

2. Given the objective to maximize total contribution margin, what is the labor constraint for production of Beta?

(a) Beta ≤ 400   (b) Beta ≥ 400
(c) Beta ≤ 425   (d) Beta ≥ 425

3. What is the objective function of the data presented?

(a) Zeta + 2 Beta = $9.25
(b) ($4.00)Zeta + 3($5.25)Beta = Total contribution margin
(c) ($4.00)Zeta + ($5.25)Beta = Total contribution margin
(d) 2($4.00)Zeta + 3($5.25)Beta = Total contribution margin

4. CPA Exam   Williamson Manufacturing intends to produce two products, X and Y. Product X requires 6 hours of time on machine 1 and 12 hours of time on machine 2. Product Y requires 4 hours of time on machine 1 and no time on machine 2. Both machines are available for 24 hours. Assuming that the objective function of the total contribution margin is $2X + $1Y$, what product mix will produce the maximum profit?

(a) No units of product X and 6 units of product Y.
(b) 1 unit of product X and 4 units of product Y.
(c) 2 units of product X and 3 units of product Y.
(d) 4 units of product X and no units of product Y.

5. CPA Exam   Quepea Company manufactures two products, Q and P, in a small building with limited capacity. The selling price, cost data, and production time are given below:

	Product Q	Product P
Selling price per unit	$20	$17
Variable costs of producing and selling a unit	$12	$13
Hours to produce a unit	3	1

Based on this information, the profit maximization objective function for a linear programming solution may be stated as

(a) Maximize $20Q + $17P$.
(b) Maximize $12Q + $13P$.
(c) Maximize $3Q + $1P$.
(d) Maximize $8Q + $4P$.

6. CPA Exam   Patsy, Inc., manufactures two products, X and Y. Each product must be processed in each of three departments: machining, assembling, and finishing. The hours needed to produce one unit of product per department and the maximum possible hours per department follow:

Department	Production Hours per Unit X	Y	Maximum Capacity in Hours
Machining	2	1	420
Assembling	2	2	500
Finishing	2	3	600

Other restrictions follow:

$$X \geq 50 \qquad Y \geq 50$$

The objective function is to maximize profits where profit = $4X + $2Y$. Given the objective and constraints, what is the most profitable number of units of X and Y, respectively, to manufacture?

(a) 150 and 100   (b) 165 and 90
(c) 170 and 80   (d) 200 and 50

7. CPA Exam   Milford Company manufactures two models, medium and large. The contribution margin expected is $12 for the medium model and $20 for the large model. The medium model is processed 2 hours in the machining department and 4 hours in the polishing department. The large model is processed 3 hours in the machining department and 6 hours in the polishing department. How would the formula for determining the maximization of total contribution margin be expressed?

(a) $5X + 10Y$ (b) $6X + 9Y$
(c) $12X + 20Y$ (d) $12X(2 + 4) + 20Y(3 + 6)$

8. CPA Exam   Hale Company manufactures products A and B, each of which requires two processes, polishing and grinding. The contribution margin is $3 for product A and $4 for product B. The illustration shows the maximum number of units of each product that may be processed in the two departments.

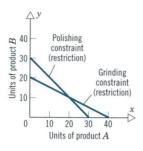

Considering the constraints (restrictions) on processing, which combination of products A and B maximizes the total contribution margin?

(a) 0 units of A and 20 units of B.
(b) 20 units of A and 10 units of B.
(c) 30 units of A and 0 units of B.
(d) 40 units of A and 0 units of B.

9. CPA Exam   Johnson, Inc., manufactures product X and product Y, which are processed as follows:

	Machine A	Machine B
Product X	6 hours	4 hours
Product Y	9 hours	5 hours

The contribution margin is $12 for product X and $7 for product Y. The available time daily for processing the two products is 120 hours for machine A and 80 hours for machine B. How would the restriction (constraint) for machine B be expressed?

(a) $4X + 5Y$ (b) $4X + 5Y \leq 80$
(c) $6X + 9Y \leq 120$ (d) $12X + 7Y$

10. CMA Exam   A small company makes only two products, with the following two production constraints representing two machines and their maximum availability:

$$2X + 3Y \leq 18$$
$$2X + Y \leq 10$$

where $X =$ Units of the first product
          $Y =$ Units of the second product

If the profit equation is $Z = \$4X + \$2Y$, the maximum possible profit is

(a) $20   (b) $21   (c) $18   (d) $24
(e) Some profit other than those given above

CMA Exam

Questions 11–13 are based on the Jarten Company, which manufactures and sells two products. Demand for the two products has grown to such a level that Jarten can no longer meet the demand with its facilities. The company can work a total of 600,000 direct labor-hours annually using three shifts. A total of 200,000 hours of machine time is available annually. The company plans to use linear programming to determine a production schedule that will maximize its net return.

The company spends $2,000,000 in advertising and promotion and incurs $1,000,000 for general and administrative costs. The unit sale price for model A is $27.50; model B sells for $75.00 each. The unit manufacturing requirements and unit cost data are as shown below. Overhead is assigned on a machine-hour (MH) basis.

	Model A		*Model B*	
Raw material		$ 3		$ 7
Direct labor	1 DLH @ $8	8	1.5 DLH @ $8	12
Variable overhead	0.5 MH @ $12	6	2.0 MH @ $12	24
Fixed overhead	0.5 MH @ $4	2	2.0 MH @ $4	8
		$19		$51

11. The objective function that would maximize Jarten's net income is

(a) $10.50A + 32.00B$
(b) $8.50A + 24.00B$
(c) $27.50A + 75.00B$
(d) $19.00A + 51.00B$
(e) $17.00A + 43.00B$

12. The constraint function for the direct labor is

(a) $1A + 1.5B \leq 200,000$ (b) $8A + 12B \leq 600,000$
(c) $8A + 12B \leq 200,000$
(d) $1A + 1.5B \leq 4,800,000$
(e) $1A + 1.5B \leq 600,000$

13. The constraint function for the machine capacity is

(a) $6A + 24B \leq 200,000$

(b) $(1/0.5)A + (1.5/2.0)B \leq 800,000$

(c) $0.5A + 2B \leq 200,000$

(d) $(0.5 + 0.5)A + (2 + 2)B \leq 200,000$

(e) $(0.5 \times 1) + (1.5 \times 2.00) \leq (200,000 \times 600,000)$

14. CPA Exam Boaz Company manufactures two models, medium (X) and large (Y). The contribution margin expected is $24 for the medium model and $40 for the large model. The medium model is processed 2 hours in the machining department and 4 hours in the polishing department. The large model is processed 3 hours in the machining department and 6 hours in the polishing department. If total contribution margin is to be maximized, using linear programming, how would the objective function be expressed?

(a) $24X(2 + 4) + 40Y(3 + 6)$

(b) $24X + 40Y$

(c) $6X + 9Y$

(d) $5X + 10Y$

Chapter 4

Linear Programming: Simplex Method

4.1 The Simplex Tableau; Pivoting

4.2 The Simplex Method: Solving Maximum Problems in Standard Form

4.3 Solving Minimum Problems in Standard Form; The Duality Principle

4.4 The Simplex Method with Mixed Constraints

Chapter Review

In Chapter 3 we described a geometric method (using graphs) for solving linear programming problems. Unfortunately, this method is useful only when there are no more than two variables and the number of constraints is small.

If we have a large number of either variables or constraints, it is still true that if an optimal solution exists, it will be found at a corner point of the set of feasible points. In fact, we could find these corner points by writing all the equations corresponding to the inequalities of the problem and then proceeding to solve all possible combinations of these equations. We would, of course, have to discard any solutions that are not feasible (because they do not satisfy one or more of the constraints). Then we could evaluate the objective function at the remaining corner points.

Just how difficult is this procedure? Well, if there were just 4 variables and 7 constraints, we would have to solve all possible combinations of 4 equations chosen from a set of 7 equations—that would be 35 solutions in all. Each of these solutions would then have to be tested for feasibility. So even for this relatively small number of variables and constraints, the work would be quite tedious. In the real world of applications, it is fairly common to encounter problems with *hundreds,* even *thousands,* of variables and constraints. Of course, such problems must be solved by computer. Even so, choosing a more efficient problem-solving strategy than the geometric method might reduce the computer's running time from hours to seconds, or, for very large problems, from years to hours.

A more systematic approach would involve choosing a solution at one corner point of the feasible set, then moving from there to another corner point at which the objective function has a better value, and continuing in this way until the best possible value is found. One very efficient and popular way of doing this is the subject of the present chapter: the *simplex method.*

A discussion of LINDO, a software package that closely mimics the simplex method, may be found in Appendix B.

4.1 THE SIMPLEX TABLEAU; PIVOTING

Standard Form of a Maximum Problem

A linear programming problem in which the objective function is to be maximized is referred to as a **maximum linear programming problem.** Such problems are said to be in **standard form** provided two conditions are met:

Standard Form of a Maximum Linear Programming Problem

Condition 1 All the variables are nonnegative.

Condition 2 All other constraints are written as a linear expression that is less than or equal to a positive constant.

EXAMPLE 1 **Determining Whether a Maximum Linear Programming Problem Is in Standard Form**

Determine which of the following maximum linear programming problems are in standard form.*

(a) Maximize

$$z = 5x_1 + 4x_2$$

subject to the constraints

$$3x_1 + 4x_2 \leq 120$$
$$4x_1 + 3x_2 \leq 20$$
$$x_1 \geq 0, x_2 \geq 0$$

(b) Maximize

$$z = 8x_1 + 2x_2 + 3x_3$$

subject to the constraints

$$4x_1 + 8x_2 \qquad \leq 120$$
$$3x_2 + 4x_3 \leq 120$$
$$x_1 \geq 0, x_2 \geq 0$$

---

*Due to the nature of solving linear programming problems using the simplex method, we shall find it convenient to use subscripted variables throughout this chapter.

(c) Maximize

$$z = 6x_1 - 8x_2 + x_3$$

subject to the constraints

$$\begin{aligned} 3x_1 + x_2 &\leq 10 \\ 4x_1 - x_2 &\leq 5 \\ x_1 - x_2 - x_3 &\geq -3 \\ x_1 \geq 0, \ x_2 \geq 0, \ x_3 &\geq 0 \end{aligned}$$

(d) Maximize

$$z = 8x_1 + x_2$$

subject to the constraints

$$\begin{aligned} 3x_1 + 4x_2 &\geq 2 \\ x_1 + x_2 &\leq 6 \\ x_1 \geq 0, \ x_2 &\geq 0 \end{aligned}$$

SOLUTION

(a) This is a maximum problem containing two variables x_1 and x_2. Since both variables are nonnegative and since the other constraints

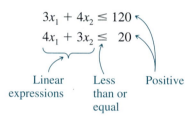

$$3x_1 + 4x_2 \leq 120$$
$$4x_1 + 3x_2 \leq 20$$

Linear   Less   Positive
expressions   than or
                equal

are each written as linear expressions less than or equal to a positive constant, we conclude the maximum problem is in standard form.

(b) This is a maximum problem containing three variables x_1, x_2, and x_3. Since the variable x_3 is not given as nonnegative, the maximum problem is not in standard form.

(c) This is a maximum problem containing three variables x_1, x_2, and x_3. Each variable is nonnegative. The set of constraints

$$\begin{aligned} 3x_1 + x_2 &\leq 10 \\ 4x_1 - x_2 &\leq 5 \\ x_1 - x_2 - x_3 &\geq -3 \end{aligned}$$

contains $x_1 - x_2 - x_3 \geq -3$, which is not a linear expression that is less than or equal to a positive constant. Thus the maximum problem is not in standard form. Notice, however, that by multiplying this constraint by -1, we get

$$-x_1 + x_2 + x_3 \leq 3$$

which is in the desired form. Thus, although the maximum problem as stated is not in standard form, it can easily be modified to conform to the requirements of the standard form.

(d) The maximum problem contains two variables x_1 and x_2, each of which is non-negative. Of the other constraints, the first one, $3x_1 + 4x_2 \geq 2$ does not conform. Thus the maximum problem is not in standard form. Notice that we cannot modify this problem to place it in standard form. Even though multiplying by -1 will change the $\geq$ to $\leq$, in so doing the 2 will change to -2.

 **Now Work Problem 1**

Slack Variables and the Simplex Tableau

In order to apply the simplex method to a maximum problem, we need to first

1. Introduce *slack variables.*
2. Construct the *initial simplex tableau.*

We will show how these steps are done by working with a specific maximum problem in standard form. (This problem is the same as Example 1, Section 3.3, page 151.)
    Maximize

$$P = 3x_1 + 4x_2$$

subject to the constraints

$$2x_1 + 4x_2 \leq 120$$
$$2x_1 + 2x_2 \leq 80$$
$$x_1 \geq 0, \quad x_2 \geq 0$$

First, observe that this maximum problem is in standard form.
    Next, recall that when we say that $2x_1 + 4x_2 \leq 120$, we mean that there is a number greater than or equal to 0, which we designate by s_1, such that

$$2x_1 + 4x_2 + s_1 = 120$$

This number s_1 is a variable. It must be nonnegative since it is the difference between 120 and a number that is less than or equal to 120. We call it a **slack variable** since it "takes up the slack" between the left and right sides of the inequality.
    Similarly, for the constraint $2x_1 + 2x_2 \leq 80$, we introduce the slack variable s_2:

$$2x_1 + 2x_2 + s_2 = 80, \quad s_2 \geq 0$$

Finally, we write the objective function $P = 3x_1 + 4x_2$ as

$$P - 3x_1 - 4x_2 = 0$$

In effect, we have now replaced our original system of constraints and the objective function by a system of three equations containing five variables, P, x_1, x_2, s_1, and s_2:

$$
\left.
\begin{aligned}
2x_1 + 4x_2 + s_1 \qquad &= 120 \\
2x_1 + 2x_2 \qquad + s_2 &= 80
\end{aligned}
\right\} \quad \text{Constraints}
$$

$$P - 3x_1 - 4x_2 \qquad\qquad = 0 \qquad \text{Objective function}$$

where
$$x_1 \geq 0 \qquad x_2 \geq 0 \qquad s_1 \geq 0 \qquad s_2 \geq 0$$

To solve the maximum problem is to find the particular solution (P, x_1, x_2, s_1, s_2) that gives the largest possible value for P. The augmented matrix for this system is given below:

$$
\begin{array}{ccccc}
P & x_1 & x_2 & s_1 & s_2 \\
\end{array}
$$
$$
\left[
\begin{array}{ccccc|c}
0 & 2 & 4 & 1 & 0 & 120 \\
0 & 2 & 2 & 0 & 1 & 80 \\
1 & -3 & -4 & 0 & 0 & 0 \\
\end{array}
\right]
$$

If we write the augmented matrix in the form given next, we have the **initial simplex tableau** for the maximum problem:

$$
\begin{array}{c|ccccc|c}
\text{BV} & P & x_1 & x_2 & s_1 & s_2 & \text{RHS} \\
\hline
s_1 & 0 & 2 & 4 & 1 & 0 & 120 \\
s_2 & 0 & 2 & 2 & 0 & 1 & 80 \\
\hline
P & 1 & -3 & -4 & 0 & 0 & 0 \\
\end{array}
\tag{1}
$$

The bottom row of the initial simplex tableau represents the objective function and is called the **objective row.** The rows above it represent the constraints. We separate the objective row from these rows with a horizontal rule. Notice that we have written the symbol for each variable above the column in which its coefficients appear. The notation **BV** stands for **basic variables.** These are the variables that have a coefficient of 1 and 0 elsewhere in their column. The notation **RHS** stands for **right-hand side,** that is, the numbers to the right of the equal sign in each equation.

So far, we have seen this much of the simplex method:

For a maximum problem in standard form:

1. The constraints are changed from inequalities to equations by the introduction of additional variables—one for each constraint and all nonnegative—called **slack variables.**

2. These equations, together with one that describes the objective function, are placed in the **initial simplex tableau.**

EXAMPLE 2   Setting Up the Initial Simplex Tableau

The following maximum problems are in standard form. For each one introduce slack variables and set up the initial simplex tableau.

(a) Maximize

$$P = 3x_1 + 2x_2 + x_3$$

subject to the constraints

$$3x_1 + x_2 + x_3 \leq 30$$
$$5x_1 + 2x_2 + x_3 \leq 24$$
$$x_1 + x_2 + 4x_3 \leq 20$$
$$x_1 \geq 0 \qquad x_2 \geq 0 \qquad x_3 \geq 0$$

(b) Maximize

$$P = x_1 + 4x_2 + 3x_3 + x_4$$

subject to the constraints

$$2x_1 + x_2 \qquad\qquad \leq 10$$
$$3x_1 + x_2 + x_3 + 2x_4 \leq 18$$
$$x_1 + x_2 + x_3 + x_4 \leq 14$$
$$x_1 \geq 0 \qquad x_2 \geq 0 \qquad x_3 \geq 0 \qquad x_4 \geq 0$$

SOLUTION

(a) We write the objective function in the form

$$P - 3x_1 - 2x_2 - x_3 = 0$$

For each constraint we introduce a nonnegative slack variable to obtain the following equations:

$$3x_1 + x_2 + x_3 + s_1 \qquad\qquad = 30$$
$$5x_1 + 2x_2 + x_3 \qquad + s_2 \qquad = 24$$
$$x_1 + x_2 + 4x_3 \qquad\qquad + s_3 = 20$$
$$x_1 \geq 0 \qquad x_2 \geq 0 \qquad x_3 \geq 0$$
$$s_1 \geq 0 \qquad s_2 \geq 0 \qquad s_3 \geq 0$$

These equations, together with the objective function P, give the initial simplex tableau:

BV	P	x_1	x_2	x_3	s_1	s_2	s_3	RHS
s_1	0	3	1	1	1	0	0	30
s_2	0	5	2	1	0	1	0	24
s_3	0	1	1	4	0	0	1	20
P	1	−3	−2	−1	0	0	0	0

(b) We write the objective function in the form

$$P - x_1 - 4x_2 - 3x_3 - x_4 = 0$$

For each constraint we introduce a nonnegative slack variable to obtain the equations

$$2x_1 + x_2 \qquad\qquad + s_1 \qquad\qquad = 10$$
$$3x_1 + x_2 + x_3 + 2x_4 \qquad + s_2 \qquad = 18$$
$$x_1 + x_2 + x_3 + x_4 \qquad\qquad + s_3 = 14$$
$$x_1 \geq 0 \qquad x_2 \geq 0 \qquad x_3 \geq 0 \qquad x_4 \geq 0$$
$$s_1 \geq 0 \qquad s_2 \geq 0 \qquad s_3 \geq 0$$

These equations, together with the objective function P, give the initial simplex tableau:

BV	P	x_1	x_2	x_3	x_4	s_1	s_2	s_3	RHS
s_1	0	2	1	0	0	1	0	0	10
s_2	0	3	1	1	2	0	1	0	18
s_3	0	1	1	1	1	0	0	1	14
P	1	−1	−4	−3	−1	0	0	0	0

Notice that in each initial simplex tableaux an identity matrix appears under the columns headed by P and the slack variables. Notice too that the right-hand column (RHS) will always contain nonnegative constants.

 **Now Work Problem 17**

The Pivot Operation

Before going any further in our discussion of the simplex method, we need to discuss the matrix operation known as *pivoting*. The first thing one does in a pivot operation is to choose a *pivot element*. However, for now the pivot element will be specified in advance. The method of selecting pivot elements in the simplex tableau will be shown in the next section.

> **Pivoting**
>
> To **pivot** a matrix about a given element, called the **pivot element,** is to apply certain row operations so that the pivot element is replaced by a 1 and all other entries in the same column, called the **pivot column,** become 0s.

> **Steps for Pivoting**
>
> **Step 1** In the *pivot row* (where the pivot element appears), divide each entry by the *pivot element* (we assume it is not 0). (This causes the pivot element to become 1.)
>
> **Step 2** Obtain 0s elsewhere in the *pivot column* by performing row operations using the pivot row.

The steps for pivoting utilize two variations of the three row operations for matrices, namely:

> **Row Operations Used in Pivoting**
>
> **Step 1** Replace the pivot row by a positive multiple of that same row.
>
> **Step 2** Replace a row by the sum of that row and a multiple of the pivot row.

> ***Warning!*** Notice that Step 2 requires row operations that must involve the pivot row.

We continue with the initial simplex tableau given in Display (1), page 170, to illustrate the pivot operation.

EXAMPLE 3   Performing a Pivot Operation

Perform a pivot operation on the initial simplex tableau given in Display (1) and repeated below in Display (2), where the pivot element is circled, and the pivot row and pivot column are marked by arrows:

$$
\begin{array}{c|ccccc|c}
\text{BV} & P & x_1 & x_2 & s_1 & s_2 & \text{RHS} \\
\hline
\rightarrow s_1 & 0 & 2 & ④ & 1 & 0 & 120 \\
s_2 & 0 & 2 & 2 & 0 & 1 & 80 \\
\hline
P & 1 & -3 & -4 & 0 & 0 & 0
\end{array}
\tag{2}
$$

SOLUTION   In this tableau the pivot column is column x_2 and the pivot row is row s_1. Step 1 of the pivoting procedure tells us to divide the pivot row by 4, so we use the row operation

$$R_1 = \tfrac{1}{4}r_1$$

$$
\begin{array}{c|ccccc|c}
 & P & x_1 & x_2 & s_1 & s_2 & \text{RHS} \\
\hline
 & 0 & \tfrac{1}{2} & ① & \tfrac{1}{4} & 0 & 30 \\
 & 0 & 2 & 2 & 0 & 1 & 80 \\
\hline
 & 1 & -3 & -4 & 0 & 0 & 0
\end{array}
$$

For Step 2 the pivot row is row 1. To obtain 0s elsewhere in the pivot column, we multiply row 1 by -2 and add it to row 2; then we multiply row 1 by 4 and add it to row 3. The row operations specified are

$$R_2 = r_2 + (-2)r_1 \qquad R_3 = r_3 + 4r_1$$

The new tableau looks like this:

$$
\begin{array}{c|ccccc|c}
\text{BV} & P & x_1 & x_2 & s_1 & s_2 & \text{RHS} \\
\hline
x_2 & 0 & \tfrac{1}{2} & 1 & \tfrac{1}{4} & 0 & 30 \\
s_2 & 0 & 1 & 0 & -\tfrac{1}{2} & 1 & 20 \\
\hline
P & 1 & -1 & 0 & 1 & 0 & 120
\end{array}
\tag{3}
$$

This completes the pivot operation since the pivot column (column x_2) has a 1 in the pivot row x_2 and has 0's everywhere else.

∎

This process should look familiar. It is similar to the one used in Chapter 2 to obtain the reduced row-echelon form of a matrix.

Just what has the pivot operation done? To see, we look again at the initial simplex tableau, in Display (2). Observe that the entries in columns P, s_1, and s_2 form an identity matrix (I_3, to be exact). This makes it easy to solve for P, s_1, and s_2 using the other variables as parameters:

$$
\begin{array}{llll}
2x_1 + 4x_2 + s_1 & = 120 & \text{or} & s_1 = -2x_1 - 4x_2 + 120 \\
2x_1 + 2x_2 \quad + s_2 & = 80 & \text{or} & s_2 = -2x_1 - 2x_2 + 80 \\
P - 3x_1 - 4x_2 & = 0 & \text{or} & P = \quad 3x_1 + 4x_2
\end{array}
$$

The variables P, s_1, and s_2 are the original basic variables (BV) listed in the tableau.

After pivoting, we obtain the tableau given in Display (3). Notice that in this form, it is easy to solve for P, x_2, and s_2 in terms of x_1 and s_1.

$$
\begin{array}{rl}
x_2 = & -\tfrac{1}{2}x_1 - \tfrac{1}{4}s_1 + 30 \\
s_2 = & -x_1 + \tfrac{1}{2}s_1 + 20 \\
P = & x_1 - s_1 + 120
\end{array}
\tag{4}
$$

The variables P, x_2, and s_2 are the new basic variables of the tableau. The variables x_1 and s_1 are the **nonbasic variables.** Thus the result of pivoting is that x_2 becomes a basic variable, while s_1 becomes a nonbasic variable.

Notice in Equations (4) that if we let the value of the nonbasic variables x_1 and s_1 equal 0, then the basic variables P, x_2, and s_2 equal the entries across from them in the right-hand side (RHS) of the tableau in Display (3). Thus for the tableau in Display (3) the current value of the objective function is $P = 120$, obtained for $x_1 = 0$, $s_1 = 0$. The values of x_2 and s_2 are $x_2 = 30$, $s_2 = 20$. Because $P = x_1 - s_1 + 120$ and $x_1 \geq 0$, the value of P can be increased beyond 120 when $x_1 > 0$ and $s_1 = 0$. So we have not maximized P yet.

We summarize this discussion below.

Analyzing a Tableau

To obtain the current values of the objective function and the basic variables in a tableau, follow these steps:

Step 1 From the tableau write the equations corresponding to each row.

Step 2 Solve the bottom equation for P and the remaining equations for the basic variables.

Step 3 Set each nonbasic variable equal to zero to obtain the current values of P and the basic variables.

Now Work Problem 29

EXAMPLE 4   Performing a Pivot Operation; Analyzing a Tableau

Perform another pivot operation on the tableau given in Display (3). Use the circled pivot element in the tableau below. Then analyze the new tableau.

$$
\begin{array}{c|ccccc|c}
\text{BV} & P & x_1 & x_2 & s_1 & s_2 & \text{RHS} \\
\hline
x_2 & 0 & \frac{1}{2} & 1 & \frac{1}{4} & 0 & 30 \\
\rightarrow s_2 & 0 & \textcircled{1} & 0 & -\frac{1}{2} & 1 & 20 \\
\hline
P & 1 & -1 & 0 & 1 & 0 & 120
\end{array}
$$

SOLUTION   Since the pivot element happens to be a 1 in this case, we skip Step 1. For Step 2 we perform the row operations

$$
R_1 = r_1 + (-\tfrac{1}{2})r_2 \qquad R_3 = r_3 + r_2
$$

The result is

$$
\begin{array}{c|ccccc|c}
\text{BV} & P & x_1 & x_2 & s_1 & s_2 & \text{RHS} \\
\hline
x_2 & 0 & 0 & 1 & \frac{1}{2} & -\frac{1}{2} & 20 \\
x_1 & 0 & 1 & 0 & -\frac{1}{2} & 1 & 20 \\
\hline
P & 1 & 0 & 0 & \frac{1}{2} & 1 & 140
\end{array} \tag{5}
$$

In the tableau given in Display (5), the new basic variables are P, x_2, and x_1. The variables s_1 and s_2 are the nonbasic variables. The result of pivoting caused x_1 to become a basic variable and s_2 to become a nonbasic variable. Finally, the equations represented by Display (5) can be written as

$$
\begin{aligned}
x_2 &= -\tfrac{1}{2}s_1 + \tfrac{1}{2}s_2 + 20 \\
x_1 &= \tfrac{1}{2}s_1 - s_2 + 20 \\
P &= -\tfrac{1}{2}s_1 - s_2 + 140
\end{aligned}
$$

If we let the nonbasic variables s_1 and s_2 equal 0, then $P = 140$, $x_2 = 20$, and $x_1 = 20$. Thus the current values of the basic variables are $P = 140$, $x_2 = 20$, and $x_1 = 20$. The pivot process has improved the value of P.

Because $P = -\tfrac{1}{2}s_1 - s_2 + 140$ and $s_1 \geq 0$ and $s_2 \geq 0$, the value of P cannot increase beyond 140 (any values of s_1 and s_2, other than 0, reduce the value of P). So we have maximized P.

We'll stop here to do some problems before continuing our discussion of the simplex method.

EXERCISE 4.1   Answers to odd-numbered problems begin on page AN-19.

In Problems 1–10 determine which maximum linear programming problems are in standard form. Do not attempt to solve them!

1. Maximize

$$P = 2x_1 + x_2$$

subject to the constraints

$$x_1 + x_2 \le 5$$
$$2x_1 + 3x_2 \le 2$$
$$x_1 \ge 0 \qquad x_2 \ge 0$$

2. Maximize

$$P = 3x_1 + 4x_2$$

subject to the constraints

$$3x_1 + x_2 \le 6$$
$$x_1 + 4x_2 \le 74$$
$$x_1 \ge 0 \qquad x_2 \ge 0$$

3. Maximize

$$P = 3x_1 + x_2 + x_3$$

subject to the constraints

$$x_1 + x_2 + x_3 \le 6$$
$$2x_1 + 3x_2 + 4x_3 \le 10$$
$$x_1 \ge 0$$

4. Maximize

$$P = 2x_1 + x_2 + 4x_3$$

subject to the constraints

$$2x_1 + x_2 + x_3 \le 10$$
$$x_2 \ge 0$$

5. Maximize

$$P = 3x_1 + x_2 + x_3$$

subject to the constraints

$$x_1 + x_2 + x_3 \le 8$$
$$2x_1 + x_2 + 4x_3 \ge 6$$
$$x_1 \ge 0 \qquad x_2 \ge 0$$

6. Maximize

$$P = 2x_1 + x_2 + 4x_3$$

subject to the constraints

$$2x_1 + x_2 + x_3 \le -1$$
$$x_1 \ge 0 \qquad x_2 \ge 0$$

7. Maximize

$$P = 2x_1 + x_2$$

subject to the constraints

$$x_1 + x_2 \ge -6$$
$$2x_1 + x_2 \le 4$$
$$x_1 \ge 0 \qquad x_2 \ge 0$$

8. Maximize

$$P = 3x_1 + x_2$$

subject to the constraints

$$x_1 + 3x_2 \le 4$$
$$2x_1 - x_2 \ge 1$$
$$x_1 \ge 0 \qquad x_2 \ge 0$$

9. Maximize

$$P = 2x_1 + x_2 + 3x_3$$

subject to the constraints

$$x_1 + x_2 - x_3 \le 10$$
$$x_2 + x_3 \le 4$$
$$x_1 \ge 0 \qquad x_2 \ge 0 \qquad x_3 \ge 0$$

10. Maximize

$$P = 2x_1 + 2x_2 + 3x_3$$

subject to the constraints

$$x_1 - x_2 + x_3 \le 6$$
$$x_1 \le 4$$
$$x_1 \ge 0 \qquad x_2 \ge 0 \qquad x_3 \ge 0$$

In Problems 11–16 each maximum problem is not in standard form. Determine if the problem can be modified so as to be in standard form. If it can, write the modified version.

11. Maximize

$$P = x_1 + x_2$$

subject to the constraints

$$3x_1 - 4x_2 \le -6$$
$$x_1 + x_2 \le 4$$
$$x_1 \ge 0 \qquad x_2 \ge 0$$

12. Maximize

$$P = 2x_1 + 3x_2$$

subject to the constraints

$$-4x_1 + 2x_2 \ge -8$$
$$x_1 - x_2 \le 6$$
$$x_1 \ge 0 \qquad x_2 \ge 0$$

13. Maximize

$$P = x_1 + x_2 + x_3$$

subject to the constraints

$$x_1 + x_2 + x_3 \le 6$$
$$4x_1 + 3x_2 \qquad \ge 12$$
$$x_1 \ge 0 \qquad x_2 \ge 0 \qquad x_3 \ge 0$$

14. Maximize

$$P = 2x_1 + x_2 + 3x_3$$

subject to the constraints

$$x_1 + x_2 + x_3 \ge -8$$
$$x_1 - x_2 \qquad \le -6$$
$$x_1 \ge 0 \qquad x_2 \ge 0 \qquad x_3 \ge 0$$

15. Maximize

$$P = 2x_1 + x_2 + 3x_3$$

subject to the constraints

$$-x_1 + x_2 + x_3 \ge -6$$
$$2x_1 - 3x_2 \qquad \ge -12$$
$$x_3 \le 2$$
$$x_1 \ge 0 \qquad x_2 \ge 0 \qquad x_3 \ge 0$$

16. Maximize

$$P = x_1 + x_2 + x_3$$

subject to the constraints

$$2x_1 - x_2 + 3x_3 \le 8$$
$$x_1 - x_2 \qquad \ge 6$$
$$x_3 \le 4$$
$$x_1 \ge 0 \qquad x_2 \ge 0 \qquad x_3 \ge 0$$

In Problems 17–24 each maximum problem is in standard form. For each one introduce slack variables and set up the initial simplex tableau.

17. Maximize

$$P = 2x_1 + x_2 + 3x_3$$

subject to the constraints

$$5x_1 + 2x_2 + x_3 \le 20$$
$$6x_1 + x_2 + 4x_3 \le 24$$
$$x_1 + x_2 + 4x_3 \le 16$$
$$x_1 \ge 0 \qquad x_2 \ge 0 \qquad x_3 \ge 0$$

18. Maximize

$$P = 3x_1 + 2x_2 + x_3$$

subject to the constraints

$$3x_1 + 2x_2 - x_3 \le 10$$
$$x_1 - x_2 + 3x_3 \le 12$$
$$2x_1 + x_2 + x_3 \le 6$$
$$x_1 \ge 0 \qquad x_2 \ge 0 \qquad x_3 \ge 0$$

19. Maximize

$$P = 3x_1 + 5x_2$$

subject to the constraints

$$2.2x_1 - 1.8x_2 \le 5$$
$$0.8x_1 + 1.2x_2 \le 2.5$$
$$x_1 + x_2 \le 0.1$$
$$x_1 \ge 0 \qquad x_2 \ge 0$$

20. Maximize

$$P = 2x_1 + 3x_2$$

subject to the constraints

$$1.2x_1 - 2.1x_2 \le 0.5$$
$$0.3x_1 + 0.4x_2 \le 1.5$$
$$x_1 + x_2 \le 0.7$$
$$x_1 \ge 0 \qquad x_2 \ge 0$$

21. Maximize

$$P = 2x_1 + 3x_2 + x_3$$

subject to the constraints

$$x_1 + x_2 + x_3 \leq 50$$
$$3x_1 + 2x_2 + x_3 \leq 10$$
$$x_1 \geq 0 \qquad x_2 \geq 0 \qquad x_3 \geq 0$$

22. Maximize

$$P = x_1 + 4x_2 + 2x_3$$

subject to the constraints

$$3x_1 + x_2 + x_3 \leq 10$$
$$x_1 + x_2 + 3x_3 \leq 5$$
$$x_1 \geq 0 \qquad x_2 \geq 0 \qquad x_3 \geq 0$$

23. Maximize

$$P = 3x_1 + 4x_2 + 2x_3$$

subject to the constraints

$$3x_1 + x_2 + 4x_3 \leq 5$$
$$x_1 + x_2 \qquad \leq 5$$
$$2x_1 - x_2 + x_3 \leq 6$$
$$x_1 \geq 0 \qquad x_2 \geq 0 \qquad x_3 \geq 0$$

24. Maximize

$$P = 2x_1 + x_2 + 3x_3$$

subject to the constraints

$$2x_1 + x_2 + x_3 \leq 2$$
$$x_1 - x_2 \qquad \leq 4$$
$$2x_1 + x_2 - x_3 \leq 5$$
$$x_1 \geq 0 \qquad x_2 \geq 0 \qquad x_3 \geq 0$$

In Problems 25–28 each maximum problem can be modified so as to be in standard form. Write the modified version and, for each one, introduce slack variables and set up the initial simplex tableau.

25. Maximize

$$P = x_1 + 2x_2 + 5x_3$$

subject to the constraints

$$-x_1 + 2x_2 + 3x_3 \geq -10$$
$$3x_1 + x_2 - x_3 \geq -12$$
$$x_1 \geq 0 \qquad x_2 \geq 0 \qquad x_3 \geq 0$$

26. Maximize

$$P = 2x_1 + 4x_2 + x_3$$

subject to the constraints

$$-2x_1 - 3x_2 + x_3 \geq -8$$
$$-3x_1 + x_2 - 2x_3 \geq -12$$
$$2x_1 + x_2 + x_3 \leq 10$$
$$x_1 \geq 0 \qquad x_2 \geq 0 \qquad x_3 \geq 0$$

27. Maximize

$$P = 2x_1 + 3x_2 + x_3 + 6x_4$$

subject to the constraints

$$-x_1 + x_2 + 2x_3 + x_4 \leq 10$$
$$x_1 - x_2 + x_3 - x_4 \geq -8$$
$$x_1 + x_2 + x_3 + x_4 \leq 9$$
$$x_1 \geq 0 \qquad x_2 \geq 0 \qquad x_3 \geq 0 \qquad x_4 \geq 0$$

28. Maximize

$$P = x_1 + 5x_2 + 3x_3 + 6x_4$$

subject to the constraints

$$x_1 - x_2 + 2x_3 - 2x_4 \geq -8$$
$$-x_1 + x_2 + x_3 - x_4 \geq -10$$
$$x_1 + x_2 + x_3 + x_4 \leq 12$$
$$x_1 \geq 0 \qquad x_2 \geq 0 \qquad x_3 \geq 0 \qquad x_4 \geq 0$$

In Problems 29–33 perform a pivot operation on each tableau. The pivot element is circled. Using the new tableau obtained, write the corresponding system of equations. Indicate the current values of the objective function and the basic variables.

29.

BV	P	x_1	x_2	s_1	s_2	RHS
s_1	0	1	②	1	0	300
s_2	0	3	2	0	1	480
P	1	−1	−2	0	0	0

30.

BV	P	x_1	x_2	s_1	s_2	RHS
s_1	0	1	4	1	0	100
s_2	0	②	5	0	1	50
P	1	−2	−1	0	0	0

31.

BV	P	x_1	x_2	x_3	s_1	s_2	s_3	RHS
s_1	0	1	2	4	1	0	0	24
s_2	0	2	−1	1	0	1	0	32
s_3	0	3	2	④	0	0	1	18
P	1	−1	−2	−3	0	0	0	0

32.

BV	P	x_1	x_2	x_3	s_1	s_2	s_3	RHS
s_1	0	1	2	1	1	0	0	6
s_2	0	2	3	1	0	1	0	12
s_3	0	1	−2	③	0	0	1	0
P	1	−1	−2	−3	0	0	0	0

33.

BV	P	x_1	x_2	x_3	x_4	s_1	s_2	s_3	s_4	RHS
s_1	0	−3	0	1	0	1	0	0	0	20
s_2	0	2	0	0	①	0	1	0	0	24
s_3	0	0	−3	1	0	0	0	1	0	28
s_4	0	0	−3	0	1	0	0	0	1	24
P	1	−1	−2	−3	−4	0	0	0	0	0

Technology Exercises

In Problems 1–4 use the elementary row operations on your graphing calculator to perform a pivot operation on each tableau. The pivot element is circled.

1.

BV	P	x_1	x_2	s_1	s_2	RHS
s_1	0	3	2	1	0	200
s_2	0	1	③	0	1	150
P	1	−2	−3	0	0	0

2.

BV	P	x_1	x_2	s_1	s_2	RHS
s_1	0	1	3	1	0	300
s_2	0	③	4	0	1	480
P	1	−5	−2	0	0	0

3.

BV	P	x_1	x_2	s_1	s_2	RHS
s_1	0	1	④	1	0	100
s_2	0	2	5	0	1	135
P	1	−2	−3	0	0	0

4.

BV	P	x_1	x_2	s_1	s_2	RHS
s_1	0	③.25	2.12	1	0	125
s_2	0	1.50	3.35	0	1	75
P	1	−2.75	−1.45	0	0	0

4.2 THE SIMPLEX METHOD: SOLVING MAXIMUM PROBLEMS IN STANDARD FORM

We are now ready to state the details of the simplex method for solving a maximum linear programming problem. This method requires that the problem be in standard form and that the problem be placed in an initial simplex tableau with slack variables.

The Simplex Method for Solving a Maximum Problem in Standard Form

Step 1 Begin with the initial simplex tableau of a maximum problem in standard form.

Step 2

Step 3

Step 4

Step 5

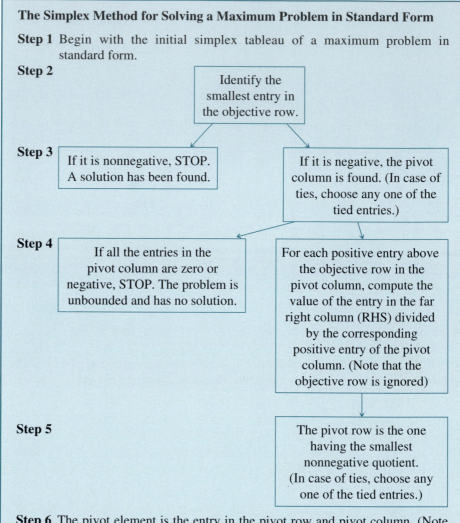

Step 6 The pivot element is the entry in the pivot row and pivot column. (Note that the pivot element can never be in the objective row.) Pivot and repeat the process from Step 2 until a STOP is obtained.

Let's go through the process.

EXAMPLE 1   Using the Simplex Method

Maximize

$$P = 3x_1 + 4x_2$$

subject to the constraints

$$2x_1 + 4x_2 \leq 120$$
$$2x_1 + 2x_2 \leq 80$$
$$x_1 \geq 0 \qquad x_2 \geq 0$$

SOLUTION

Step 1 This is a maximum problem in standard form. To obtain the initial simplex tableau, we proceed as follows: The objective function is written in the form

$$P - 3x_1 - 4x_2 = 0$$

After introducing slack variables s_1 and s_2, the constraints take the form

$$\begin{aligned}
2x_1 + 4x_2 + s_1 \quad\quad &= 120 \\
2x_1 + 2x_2 \quad\quad + s_2 &= 80 \\
x_1 \geq 0 \quad\quad x_2 &\geq 0 \\
s_1 \geq 0 \quad\quad s_2 &\geq 0
\end{aligned}$$

The initial simplex tableau is

BV	P	x_1	x_2	s_1	s_2	RHS
s_1	0	2	4	1	0	120
s_2	0	2	2	0	1	80
P	1	−3	−4	0	0	0

Step 2 The smallest entry in the objective row is −4.

Step 3 Since −4 is negative, the pivot column is column x_2.

Step 4 For each positive entry above the objective row in the pivot column, form the quotient of the corresponding RHS entry by the positive entry.

BV	Positive entry, x_2	RHS	Quotient
s_1	4	120	$120 \div 4 = 30$
s_2	2	80	$80 \div 2 = 40$

Step 5 The smallest nonnegative value is 30, so the pivot row is row s_1. The tableau below shows the pivot element (circled) and the current values.

BV	P	x_1	x_2	s_1	s_2	RHS	Current values
→ s_1	0	2	④	1	0	120	$s_1 = 120$
s_2	0	2	2	0	1	80	$s_2 = 80$
P	1	−3	−4	0	0	0	$P = 0$

(1)

Step 6 The pivot element is 4. Next, we pivot by using the row operations:

1. $R_1 = \frac{1}{4}r_1$
2. $R_2 = r_2 + (-2)r_1, \quad R_3 = r_3 + 4r_1$

After pivoting, we have this tableau:

BV	P	x_1	x_2	s_1	s_2	RHS	Current values
x_2	0	$\frac{1}{2}$	1	$\frac{1}{4}$	0	30	$x_2 = 30$
s_2	0	1	0	$-\frac{1}{2}$	1	20	$s_2 = 20$
P	1	−1	0	1	0	120	$P = 120$

(2)

Notice that x_2 is now a basic variable, replacing s_1. The objective row

$$P = x_1 - s_1 + 120$$

has a current value of 120. Because $P = x_1 - s_1 + 120$ and $x_1 \geq 0$, the value of P can be increased beyond 120, when $x_1 > 0$ and $s_1 = 0$. So we have not maximized P yet.

We continue with the simplex method at Step 2.

Step 2 The smallest entry in the objective row is -1.

Step 3 Since -1 is negative, the pivot column is column x_1.

Step 4 For each positive entry above the objective row in the pivot column, form the quotient of the corresponding RHS entry by the positive entry.

BV	Positive entry, x_2	RHS	Quotient
x_2	$\frac{1}{2}$	30	$30 \div \frac{1}{2} = 60$
s_2	1	20	$20 \div 1 = 20$

Step 5 The smallest nonnegative value is 20, so the pivot row is row s_2. The tableau below shows the pivot element (circled).

BV	P	x_1	x_2	s_1	s_2	RHS	Current values
x_2	0	$\frac{1}{2}$	1	$\frac{1}{4}$	0	30	$x_2 = 30$
$\rightarrow s_2$	0	①	0	$-\frac{1}{2}$	1	20	$s_2 = 20$
P	1	-1	0	1	0	120	$P = 120$

Step 6 The pivot element is 1. Next, we pivot by using the row operations:

$$R_1 = r_1 + (-\tfrac{1}{2})r_2 \qquad R_3 = r_3 + r_2$$

The result is this tableau:

BV	P	x_1	x_2	s_1	s_2	RHS	Current values
x_2	0	0	1	$\frac{1}{2}$	$-\frac{1}{2}$	20	$x_2 = 20$
x_1	0	1	0	$-\frac{1}{2}$	1	20	$x_1 = 20$
P	1	0	0	$\frac{1}{2}$	1	140	$P = 140$

(3)

We continue with the simplex method at Step 2.

Step 2 The smallest entry in the objective row is 0.

Step 3 Since it is nonnegative, we STOP. To see why we STOP, we write the equation from the objective row, namely,

$$P = -\tfrac{1}{2}s_1 - s_2 + 140$$

Since $s_1 \geq 0$ and $s_2 \geq 0$, any positive value of s_1 or s_2 would make the value of P less than 140. By choosing $s_1 = 0$ and $s_2 = 0$, we obtain the largest possible value for P, namely, 140. If we write the equations from the second and third rows, substituting 0 for s_1 and s_2, we have

$$x_2 = -\tfrac{1}{2}s_1 + \tfrac{1}{2}s_2 + 20 = 20$$
$$x_1 = \tfrac{1}{2}s_1 - s_2 + 20 = 20$$

In other words, we have found the optimal solution,

$$P = 140$$

the maximum, which occurs at

$$x_1 = 20 \qquad x_2 = 20 \qquad s_1 = 0 \qquad s_2 = 0$$

The tableau in Display (3) is called the **final tableau** because with this tableau an optimal solution is found.

To summarize, each tableau obtained by applying the simplex method provides information about the current status of the solution:

1. The RHS entry in the objective row gives the current value of the objective function.
2. The remaining entries in the RHS column give the current values of the corresponding basic variables.

EXAMPLE 2   Analyzing Tableaus

Determine whether each tableau

1. is a final tableau. If it is, give the solution.
2. requires additional pivoting. If so, identify the pivot element.
3. indicates no solution.

(a)

BV	P	x_1	x_2	s_1	s_2	RHS
s_1	0	0	0	1	1	40
x_1	0	1	-1	0	1	20
P	1	0	-2	0	1	20

(b)

BV	P	x_1	x_2	s_1	s_2	RHS
x_1	0	1	0	2	-1	40
x_2	0	0	1	-1	1	20
P	1	0	0	1	2	220

SOLUTION

(a) The smallest entry in the objective row is -2, which is negative. The pivot column is column x_2. Since all the entries in the pivot column are zero or negative, the problem is unbounded and has no solution.

(b) The objective row contains no negative entries, so this is a final tableau. The solution is

$$P = 220 \qquad \text{when} \qquad x_1 = 40 \quad x_2 = 20 \quad s_1 = 0 \quad s_2 = 0$$

Now Work Problem 1

EXAMPLE 3   Solving a Maximum Linear Programming Problem Using the Simplex Method

Maximize

$$P = 6x_1 + 8x_2 + x_3$$

subject to the constraints

$$3x_1 + 5x_2 + 3x_3 \leq 20$$
$$x_1 + 3x_2 + 2x_3 \leq 9$$
$$6x_1 + 2x_2 + 5x_3 \leq 30$$
$$x_1 \geq 0 \qquad x_2 \geq 0 \qquad x_3 \geq 0$$

SOLUTION   Note that the problem is in standard form. By introducing slack variables s_1, s_2, and s_3, the constraints take the form

$$3x_1 + 5x_2 + 3x_3 + s_1 \qquad\qquad = 20$$
$$x_1 + 3x_2 + 2x_3 \qquad + s_2 \qquad = 9$$
$$6x_1 + 2x_2 + 5x_3 \qquad\qquad + s_3 = 30$$
$$x_1 \geq 0 \qquad x_2 \geq 0 \qquad x_3 \geq 0$$
$$s_1 \geq 0 \qquad s_2 \geq 0 \qquad s_3 \geq 0$$

Since

$$P - 6x_1 - 8x_2 - x_3 = 0$$

the initial simplex tableau is

BV	P	x_1	$x_2 \downarrow$	x_3	s_1	s_2	s_3	RHS	Current values	
s_1	0	3	5	3	1	0	0	20	$s_1 = 20$	$20 \div 5 = 4$
$\rightarrow s_2$	0	1	③	2	0	1	0	9	$s_2 = 9$	$9 \div 3 = 3$
s_3	0	6	2	5	0	0	1	30	$s_3 = 30$	$30 \div 2 = 15$
P	1	-6	-8	-1	0	0	0	0	$P = 0$	

The pivot column is found by locating the column containing the smallest entry in the objective row (-8 in column x_2). The pivot row is obtained by dividing each entry in the RHS column by the corresponding entry in the pivot column and selecting the smallest nonnegative quotient. Thus the pivot row is row s_2. The pivot element is 3, which is circled. After pivoting, the new tableau is

BV	P	$x_1 \downarrow$	x_2	x_3	s_1	s_2	s_3	RHS	Current values	
$\rightarrow s_1$	0	$\left(\frac{4}{3}\right)$	0	$-\frac{1}{3}$	1	$-\frac{5}{3}$	0	5	$s_1 = 5$	$5 \div \frac{4}{3} = 3.75$
x_2	0	$\frac{1}{3}$	1	$\frac{2}{3}$	0	$\frac{1}{3}$	0	3	$x_2 = 3$	$3 \div \frac{1}{3} = 9$
s_3	0	$\frac{16}{3}$	0	$\frac{11}{3}$	0	$-\frac{2}{3}$	1	24	$s_3 = 24$	$24 \div \frac{16}{3} = 4.5$
P	1	$-\frac{10}{3}$	0	$\frac{13}{3}$	0	$\frac{8}{3}$	0	24	$P = 24$	

The value of P has improved to 24, but the negative entry, $-\frac{10}{3}$, in the objective row indicates further improvement is possible. We determine the next pivot element to be $\frac{4}{3}$. After pivoting, we obtain the tableau

BV	P	x_2	x_2	x_3	s_1	$s_2 \downarrow$	s_3	RHS	Current values	
x_1	0	1	0	$-\frac{1}{4}$	$\frac{3}{4}$	$-\frac{5}{4}$	0	$\frac{15}{4}$	$x_1 = \frac{15}{4}$	
x_2	0	0	1	$\frac{3}{4}$	$-\frac{1}{4}$	$\frac{3}{4}$	0	$\frac{7}{4}$	$x_2 = \frac{7}{4}$	$\frac{7}{4} \div \frac{3}{4} = \frac{7}{3}$
$\rightarrow s_3$	0	0	0	5	-4	⑥	1	4	$s_3 = 4$	$4 \div 6 = \frac{2}{3}$
P	1	0	0	$\frac{7}{2}$	$\frac{5}{2}$	$-\frac{3}{2}$	0	$\frac{73}{2}$	$P = \frac{73}{2}$	

The value of P has improved to $\frac{73}{2}$, but, since we still observe a negative entry in the objective row, we pivot again. (Remember, in finding the pivot row, we ignore the objective row and any rows in which the pivot column contains a negative number or zero—in this case, the row x_1, containing $-\frac{5}{4}$.) The new tableau is

BV	P	x_1	x_2	x_3	s_1	s_2	s_3	RHS	Current values
x_1	0	1	0	$\frac{19}{24}$	$-\frac{1}{12}$	0	$\frac{5}{24}$	$\frac{55}{12}$	$x_1 = \frac{55}{12}$
x_2	0	0	1	$\frac{1}{8}$	$\frac{1}{4}$	0	$-\frac{1}{8}$	$\frac{5}{4}$	$x_2 = \frac{5}{4}$
s_2	0	0	0	$\frac{5}{6}$	$-\frac{2}{3}$	1	$\frac{1}{6}$	$\frac{2}{3}$	$s_2 = \frac{2}{3}$
P	1	0	0	$\frac{19}{4}$	$\frac{3}{2}$	0	$\frac{1}{4}$	$\frac{75}{2}$	$P = \frac{75}{2}$

This is a final tableau since all the entries in the objective row are nonnegative. The objective (bottom) row yields the equation

$$P = -\tfrac{19}{4}x_3 - \tfrac{3}{2}s_1 - \tfrac{1}{4}s_3 + \tfrac{75}{2}$$

Thus P is a maximum when $x_3 = 0$, $s_1 = 0$, and $s_3 = 0$, giving $P = \frac{75}{2}$. From the rows x_1, x_2, and s_2 of the final tableau we have the equations

$$x_1 = -\tfrac{19}{24}x_3 + \tfrac{1}{12}s_1 - \tfrac{5}{24}s_3 + \tfrac{55}{12}$$
$$x_2 = -\tfrac{1}{8}x_3 - \tfrac{1}{4}s_1 + \tfrac{1}{8}s_3 + \tfrac{5}{4}$$
$$s_2 = -\tfrac{5}{6}x_3 + \tfrac{2}{3}s_1 - \tfrac{1}{6}s_3 + \tfrac{2}{3}$$

Using $x_3 = 0$, $s_1 = 0$, and $s_3 = 0$ in these equations, we find

$$x_1 = \tfrac{55}{12} \qquad x_2 = \tfrac{5}{4} \qquad s_2 = \tfrac{2}{3} \qquad\qquad (4)$$

Thus the optimal solution is $P = \frac{75}{2}$, obtained when $x_1 = \frac{55}{12}$, $x_2 = \frac{5}{4}$, and $x_3 = 0$. ∎

Note that the solution may also be found by looking at the current values of the final tableau.

✍ **Now Work Problem 9**

EXAMPLE 4 **Solving a Maximum Linear Programming Problem Using the Simplex Method**

Maximize

$$P = 4x_1 + 4x_2 + 3x_3$$

subject to the constraints

$$2x_1 + 3x_2 + x_3 \le 6$$
$$x_1 + 2x_2 + 3x_3 \le 6$$
$$x_1 + x_2 + x_3 \le 5$$
$$x_1 \ge 0 \qquad x_2 \ge 0 \qquad x_3 \ge 0$$

SOLUTION   The problem is in standard form. We introduce slack variables s_1, s_2, s_3, and write the constraints as

$$2x_1 + 3x_2 + x_3 + s_1 \qquad\qquad = 6$$
$$x_1 + 2x_2 + 3x_3 \qquad + s_2 \qquad = 6$$
$$x_1 + x_2 + x_3 \qquad\qquad + s_3 = 5$$
$$x_1 \geq 0 \qquad x_2 \geq 0 \qquad x_3 \geq 0$$
$$s_1 \geq 0 \qquad s_2 \geq 0 \qquad s_3 \geq 0$$

Since $P - 4x_1 - 4x_2 - 3x_3 = 0$, the initial simplex tableau is

BV	P	x_1	x_2	x_3	s_1	s_2	s_3	RHS	
s_1	0	②	3	1	1	0	0	6	$6 \div 2 = 3$
s_2	0	1	2	3	0	1	0	6	$6 \div 1 = 6$
s_3	0	1	1	1	0	0	1	5	$5 \div 1 = 5$
P	1	-4	-4	-3	0	0	0	0	

Since -4 is the smallest negative entry in the objective row, we have a tie for the pivot column between the columns x_1 and x_2. We choose (arbitrarily) as pivot column the column x_1. The pivot row is row s_1. (Do you see why?) The pivot element is 2, which is circled. After pivoting, we obtain the following tableau:

BV	P	x_1	x_2	x_3	s_1	s_2	s_3	RHS	Current values	
x_1	0	1	$\frac{3}{2}$	$\frac{1}{2}$	$\frac{1}{2}$	0	0	3	$x_1 = 3$	$3 \div \frac{1}{2} = 6$
s_2	0	0	$\frac{1}{2}$	$\frac{5}{2}$	$-\frac{1}{2}$	1	0	3	$s_2 = 3$	$3 \div \frac{5}{2} = \frac{6}{5}$
s_3	0	0	$-\frac{1}{2}$	$\frac{1}{2}$	$-\frac{1}{2}$	0	1	2	$s_3 = 2$	$2 \div \frac{1}{2} = 4$
P	1	0	2	-1	2	0	0	12	$P = 12$	

The pivot element is $\frac{5}{2}$, which is circled. After pivoting, we obtain

BV	P	x_1	x_2	x_3	s_1	s_2	s_3	RHS	Current values
x_1	0	1	$\frac{7}{5}$	0	$\frac{3}{5}$	$-\frac{1}{5}$	0	$\frac{12}{5}$	$x_1 = \frac{12}{5}$
x_3	0	0	$\frac{1}{5}$	1	$-\frac{1}{5}$	$\frac{2}{5}$	0	$\frac{6}{5}$	$x_3 = \frac{6}{5}$
s_3	0	0	$-\frac{3}{5}$	0	$-\frac{2}{5}$	$-\frac{1}{5}$	1	$\frac{7}{5}$	$s_3 = \frac{7}{5}$
P	1	0	$\frac{11}{5}$	0	$\frac{9}{5}$	$\frac{2}{5}$	0	$\frac{66}{5}$	$P = \frac{66}{5}$

This is a final tableau. The solution is $P = \frac{66}{5}$, obtained when $x_1 = \frac{12}{5}$, $x_2 = 0$, $x_3 = \frac{6}{5}$.

The reader is encouraged to solve Example 4 again, this time choosing column x_2 as the pivot column for the first pivot.

EXAMPLE 5   Maximizing Profit

Mike's Famous Toy Trucks specializes in making four kinds of toy trucks: a delivery truck, a dump truck, a garbage truck, and a gasoline truck. Three machines—a metal casting machine, a paint spray machine, and a packaging machine—are used in the production of these trucks. The time, in hours, each machine works to make each type of truck and the profit for each truck are given in Table 1 on page 187. The maximum time available per week for each machine is given: metal casting, 4000 hours; paint spray, 1800 hours; and packaging, 1000 hours. How many of each type truck should be produced to maximize profit? Assume that every truck made is sold.

Table 1

	Delivery Truck	Dump Truck	Garbage Truck	Gasoline Truck	Maximum Time
Metal Casting	2 hours	2.5 hours	2 hours	2 hours	4000 hours
Paint Spray	1 hour	1.5 hours	1 hour	2 hours	1800 hours
Packaging	0.5 hour	0.5 hour	1 hour	1 hour	1000 hours
Profit	$0.50	$1.00	$1.50	$2.00	

SOLUTION   Let x_1, x_2, x_3, and x_4 denote the number of delivery trucks, dump trucks, garbage trucks, and gasoline trucks, respectively, to be made. If P denotes the profit to be maximized, we have this problem:

Maximize

$$P = 0.5x_1 + x_2 + 1.5x_3 + 2x_4$$

subject to the constraints

$$2x_1 + 2.5x_2 + 2x_3 + 2x_4 \leq 4000$$
$$x_1 + 1.5x_2 + x_3 + 2x_4 \leq 1800$$
$$0.5x_1 + 0.5x_2 + x_3 + x_4 \leq 1000$$
$$x_1 \geq 0 \qquad x_2 \geq 0 \qquad x_3 \geq 0 \qquad x_4 \geq 0$$

Since this problem is in standard form, we introduce slack variables s_1, s_2, and s_3, write the initial simplex tableau, and solve:

BV	P	x_1	x_2	x_3	x_4	s_1	s_2	s_3	RHS	Current values
s_1	0	2	2.5	2	2	1	0	0	4000	$s_1 = 4000$
s_2	0	1	1.5	1	②	0	1	0	1800	$s_2 = 1800$
s_3	0	0.5	0.5	1	1	0	0	1	1000	$s_3 = 1000$
P	1	−0.5	−1	−1.5	−2	0	0	0	0	$P = 0$

BV	P	x_1	x_2	x_3	x_4	s_1	s_2	s_3	RHS	Current values
s_1	0	1	1	1	0	1	−1	0	2200	$s_1 = 2200$
x_4	0	0.5	0.75	0.5	1	0	0.5	0	900	$x_4 = 900$
s_3	0	0	−0.25	⓪.5	0	0	−0.5	1	100	$s_3 = 100$
P	1	0.5	0.5	−0.5	0	0	1	0	1800	$P = 1800$

BV	P	x_1	x_2	x_3	x_4	s_1	s_2	s_3	RHS	Current values
s_1	0	1	1.5	0	0	1	0	−2	2000	$s_1 = 2000$
x_4	0	0.5	1	0	1	0	1	−1	800	$x_4 = 800$
x_3	0	0	−0.5	1	0	0	−1	2	200	$x_3 = 200$
P	1	0.5	0.25	0	0	0	0.5	1	1900	$P = 1900$

This is a final tableau. The maximum profit is $P = \$1900$, and it is attained for

$$x_1 = 0 \qquad x_2 = 0 \qquad x_3 = 200 \qquad x_4 = 800$$

The practical considerations of the situation described in Example 5 are that delivery trucks and dump trucks are too costly to produce or too little profit is being gained from their sale. Since the slack variable s_1 has a value of 2000 for maximum P and, since s_1 represents the number of hours the metal casting machine is idle, it may be possible to release this machine for other duties. Also note that both the paint spray and packaging machines are operating at full capacity. This means that to increase productivity, more paint spray and packaging capacity is required.

✍ **Now Work Problem 25**

Analyzing the Simplex Method

To justify some of the steps of the simplex method, we return to Example 1 and analyze more carefully what we did. Look back to Display (1).

The reason we choose the most negative entry in the objective row is that it is the negative of the *largest* coefficient in the objective function:

$$P = 3x_1 + 4x_2$$

If we were to set $x_1 = x_2 = 0$, we would obtain $P = 0$ as a first approximation for the profit P. Of course, this is not a very good approximation; it can easily be improved by increasing either x_1 or x_2. But the profit per unit of x_2 is $4, while the profit per unit of x_1 is only $3. Thus it is more effective to increase x_2 than x_1. But what is the largest amount by which x_2 can be increased?

We can answer this question by referring back to Display (2), the tableau that resulted after the first pivot. The corresponding equations are

$$x_2 = -\tfrac{1}{2}x_1 - \tfrac{1}{4}s_1 + \; 30$$
$$s_2 = - \; x_1 + \tfrac{1}{2}s_1 + \; 20$$
$$P = \quad x_1 - \quad s_1 + 120$$

This suggests that x_2 can be as large as 30, if we take both x_1 and s_1 to be 0, in which case P will be 120.

So we chose the pivot column for Display (1) because we wanted to increase x_2. But why did we choose the pivot row? Let's see what would have happened if instead we had chosen the row s_2 as the pivot row. The result would have been

BV	P	x_1	x_2	s_1	s_2	RHS
s_1	0	−2	0	1	−2	−40
x_2	0	1	1	0	$\frac{1}{2}$	40
P	1	1	0	0	2	160

However, this is not acceptable because of the negative number in the last column. (In effect, this matrix tells us that we could get P to be as large as 160, by setting $x_1 = 0$ and $s_2 = 0$. Then $x_2 = 40$ and $s_1 = -40$; but this is not a feasible point because s_1 is supposed to be greater than or equal to 0.)

Each iteration of the simplex method consists in choosing an "entering" variable from the nonbasic ones and a "leaving" variable from the basic ones, using a selective criterion so that the value of the objective function is not decreased (sometimes the value may remain unchanged). The iterative procedure stops when no variable can enter from the nonbasic variables.

So the reasoning behind the simplex method for standard maximum problems is not too complicated, and the process of "moving to a better solution" is made quite easy simply by following the rules. Briefly, the pivoting strategy works like this:

> The choice of the pivot column forces us to pivot the variable that apparently improves the value of the objective function most effectively.
>
> The choice of the pivot row prevents us from making this variable *too large* to be feasible.

Geometry of the Simplex Method

The maximum value (provided it exists) of the objective function will occur at one of the corner points of the feasible region. The simplex method is designed to move from corner point to corner point of the feasible region, at each stage improving the value of the objective function until a solution is found. More precisely, the geometry behind the simplex method is outlined below.

1. A given tableau corresponds to a corner point of the feasible region.
2. The operation of pivoting moves us to an adjacent corner point, where the objective function has a value at least as large as it did at the previous corner point.
3. The process continues until the final tableau is reached—which produces a corner point that maximizes the objective function.

Though drawings depicting this process can be rendered in only two or three dimensions (that is, when the objective function has two or three variables in it), the same interpretation can be shown to hold regardless of the number of variables involved. Let's look at an example.

EXAMPLE 6   **The Geometry of the Simplex Method**

Maximize

$$P = 3x_1 + 5x_2$$

subject to the constraints

$$x_1 + x_2 \leq 60$$
$$x_1 + 2x_2 \leq 80$$
$$x_1 \geq 0 \qquad x_2 \geq 0$$

Figure 1

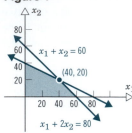

SOLUTION   The feasible region is shown in Figure 1.

Below, we apply the simplex method, indicating the corner point corresponding to each tableau and the value of the objective function there. You should supply the details.

Tableau							Corner Point (x_1, x_2)	Value of $P = 3x_1 + 5x_2$ at the Corner Point

BV	P	x_1	x_2	s_1	s_2	RHS		
s_1	0	1	1	1	0	60		
s_2	0	1	2	0	1	80	(0, 0)	0
P	1	-3	-5	0	0	0		

(Pivot)

BV	P	x_1	x_2	s_1	s_2	RHS		
s_1	0	$\frac{1}{2}$	0	1	$-\frac{1}{2}$	20		
x_2	0	$\frac{1}{2}$	1	0	$\frac{1}{2}$	40	(0, 40)	200
P	1	$-\frac{1}{2}$	0	0	$\frac{5}{2}$	200		

(Pivot)

BV	P	x_1	x_2	s_1	s_2	RHS		
x_1	0	1	0	2	-1	40		
x_2	0	0	1	-1	1	20	(40, 20)	220
P	1	0	0	1	2	220		

(Final tableau)   (Maximum value)

The Unbounded Case

So far in our discussion, it has always been possible to continue to choose pivot elements until the problem has been solved. But it may turn out that all the entries in a column of a tableau are 0 or negative at some stage. If this happens, it means that the problem is *unbounded* and a maximum solution does not exist.

For example, consider the tableau

BV	P	x_1	x_2	s_1	s_2	RHS
s_1	0	-1	1	1	0	2
s_2	0	1	-1	0	1	2
P	1	-1	-1	0	0	0

When the smallest negative entries in the objective (bottom) row are equal, you may choose either column as the pivot column. Suppose we arbitrarily choose column x_1 as the pivot column. After pivoting, the tableau becomes

BV	P	x_1	x_2	s_1	s_2	RHS
s_1	0	0	0	1	1	4
x_1	0	1	-1	0	1	2
P	1	0	-2	0	1	2

Figure 2

Now the only negative entry in the objective (bottom) row is in column x_2, and it is impossible to choose a pivot element in that column. This implies that the objective function is unbounded. Indeed, it is easy to see that if the only constraints are $-x_1 + x_2 \leq 2$, $x_1 - x_2 \leq 2$, $x_1 \geq 0$, and $x_2 \geq 0$, then $P = x_1 + x_2$ has no maximum. The feasible region is shown in Figure 2.

Summary of the Simplex Method

The general procedure for solving a maximum linear programming problem in standard form using the simplex method can be outlined as follows:

1. The maximum problem is stated in standard form as

Maximize

$$P = c_1 x_1 + c_2 x_2 + \cdots + c_n x_n$$

subject to the constraints

$$a_{11} x_1 + a_{12} x_2 + \cdots + a_{1n} x_n \leq b_1$$
$$a_{21} x_1 + a_{22} x_2 + \cdots + a_{2n} x_n \leq b_2$$
$$\vdots$$
$$a_{m1} x_1 + a_{m2} x_2 + \cdots + a_{mn} x_n \leq b_m$$
$$x_1 \geq 0, x_2 \geq 0, \ldots, x_n \geq 0$$

in which $b_1 > 0, b_2 > 0, \ldots, b_m > 0$.

2. Introduce slack variables $s_1, s_2, \ldots, s_m$ so that the constraints take the form of equalities:

$$a_{11} x_1 + a_{12} x_2 + \cdots + a_{1n} x_n + s_1 = b_1$$
$$a_{21} x_1 + a_{22} x_2 + \cdots + a_{2n} x_n + s_2 = b_2$$
$$\vdots$$
$$a_{m1} x_1 + a_{m2} x_2 + \cdots + a_{mn} x_n + s_m = b_m$$
$$x_1 \geq 0, x_2 \geq 0, \ldots, x_n \geq 0$$
$$s_1 \geq 0, s_2 \geq 0, \ldots, s_m \geq 0$$

3. Write the objective function in the form

$$P - c_1 x_1 - c_2 x_2 - \cdots - c_n x_n = 0$$

4. Set up the initial simplex tableau

BV	P	x_1	$x_2 \cdots x_n$	s_1	$s_2 \cdots s_m$	RHS
s_1	0	a_{11}	$a_{12} \cdots a_{1n}$	1	$0 \cdots 0$	b_1
s_2	0	a_{21}	$a_{22} \cdots a_{2n}$	0	$1 \cdots 0$	b_2
$\vdots$	$\vdots$	$\vdots$	$\vdots$	$\vdots$	$\vdots$	$\vdots$
s_m	0	a_{m1}	$a_{m2} \cdots a_{mn}$	0	$0 \cdots 1$	b_m
P	1	$-c_1$	$-c_2 \cdots -c_n$	0	$0 \cdots 0$	0

5. Pivot until
   (a) All the entries in the objective row are nonnegative. This is a final tableau from which a solution can be read.

   Or until

   (b) The pivot column is a column whose entries are negative or zero. In this case the problem is unbounded and there is no solution.

The flowchart in Figure 3 illustrates the steps to be used in solving standard maximum linear programming problems.

Figure 3

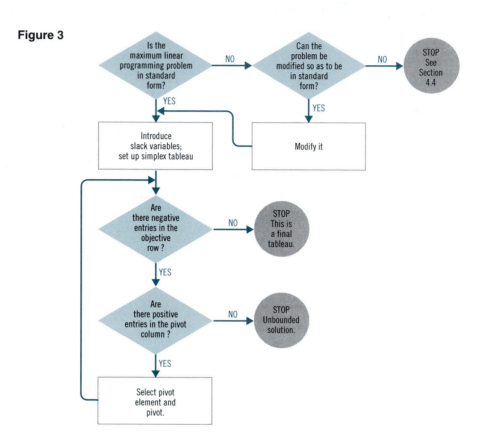

EXERCISE 4.2   Answers to odd-numbered problems begin on page AN-21.

In Problems 1–8 determine which of the following statements is true about each tableau:

(a) It is the final tableau.

(b) It requires additional pivoting.

(c) It indicates no solution to the problem.

If the answer is (a), write down the solution; if the answer is (b), indicate the pivot element.

1.

BV	P	x_1	x_2	s_1	s_2	RHS
s_1	0	1	0	1	$-\frac{1}{2}$	20
x_2	0	$\frac{1}{2}$	1	0	$\frac{1}{4}$	30
P	1	-1	0	0	1	120

2.

BV	P	x_1	x_2	s_1	s_2	RHS
x_1	0	1	0	1	$-\frac{1}{2}$	20
x_2	0	0	1	$-\frac{1}{2}$	$\frac{1}{2}$	20
P	1	0	0	1	$\frac{1}{2}$	140

3.

BV	P	x_1	x_2	s_1	s_2	RHS
s_1	0	0	$\frac{1}{14}$	1	$-\frac{1}{7}$	$\frac{186}{21}$
x_1	0	1	$\frac{12}{7}$	0	$\frac{4}{7}$	$\frac{32}{7}$
P	1	0	$\frac{12}{7}$	0	$\frac{32}{7}$	$\frac{256}{7}$

4.

BV	P	x_1	x_2	s_1	s_2	RHS
s_1	0	$\frac{1}{4}$	$\frac{1}{2}$	1	0	10
s_2	0	$\frac{7}{4}$	3	0	1	8
P	1	-8	-12	0	0	0

5.

BV	P	x_1	x_2	s_1	s_2	RHS
x_1	0	1	-2	0	4	24
s_1	0	0	-2	1	4	36
P	1	5	-10	12	4	20

6.

BV	P	x_1	x_2	s_1	s_2	RHS
s_1	0	1	3	1	0	30
s_2	0	2	1	0	1	12
P	1	-2	-5	0	0	0

7.

BV	P	x_1	x_2	s_1	s_2	s_3	RHS
x_2	0	0	1	-2	0	1	6
s_2	0	0	0	1	1	4	6
x_1	0	1	0	1	0	-1	1
P	1	0	0	-10	0	-5	110

8.

BV	P	x_1	x_2	s_1	s_2	s_3	RHS
x_2	0	2	1	0	0	-1	8
s_2	0	-1	0	0	1	5	5
s_1	0	1	0	1	0	-1	1
P	1	10	0	0	0	-15	120

In Problems 9–24 use the simplex method to solve each maximum linear programming problem.

9. Maximize

$$P = 5x_1 + 7x_2$$

subject to

$$2x_1 + 3x_2 \leq 12$$
$$3x_1 + x_2 \leq 12$$
$$x_1 \geq 0 \qquad x_2 \geq 0$$

10. Maximize

$$P = x_1 + 5x_2$$

subject to

$$2x_1 + x_2 \leq 10$$
$$x_1 + 2x_2 \leq 10$$
$$x_1 \geq 0 \qquad x_2 \geq 0$$

11. Maximize

$$P = 5x_1 + 7x_2$$

subject to

$$x_1 + 2x_2 \leq 2$$
$$2x_1 + x_2 \leq 2$$
$$x_1 \geq 0 \qquad x_2 \geq 0$$

12. Maximize

$$P = 5x_1 + 4x_2$$

subject to

$$x_1 + x_2 \leq 2$$
$$2x_1 + 3x_2 \leq 6$$
$$x_1 \geq 0 \qquad x_2 \geq 0$$

13. Maximize

$$P = 3x_1 + x_2$$

subject to

$$x_1 + x_2 \leq 2$$
$$2x_1 + 3x_2 \leq 12$$
$$3x_1 + x_2 \leq 12$$
$$x_1 \geq 0 \qquad x_2 \geq 0$$

14. Maximize

$$P = 3x_1 + 5x_2$$

subject to

$$2x_1 + x_2 \leq 4$$
$$x_1 + 2x_2 \leq 6$$
$$x_1 \geq 0 \qquad x_2 \geq 0$$

15. Maximize

$$P = 2x_1 + x_2 + x_3$$

subject to

$$-2x_1 + x_2 - 2x_3 \leq 4$$
$$x_1 - 2x_2 + x_3 \leq 2$$
$$x_1 \geq 0 \quad x_2 \geq 0 \quad x_3 \geq 0$$

16. Maximize

$$P = 4x_1 + 2x_2 + 5x_3$$

subject to

$$x_1 + 3x_2 + 2x_3 \leq 30$$
$$2x_1 + x_2 + 3x_3 \leq 12$$
$$x_1 \geq 0 \quad x_2 \geq 0 \quad x_3 \geq 0$$

17. Maximize

$$P = 2x_1 + x_2 + 3x_3$$

subject to

$$x_1 + 2x_2 + x_3 \leq 25$$
$$3x_1 + 2x_2 + 3x_3 \leq 30$$
$$x_1 \geq 0 \quad x_2 \geq 0 \quad x_3 \geq 0$$

18. Maximize

$$P = 6x_1 + 3x_2 + 2x_3$$

subject to

$$2x_1 + 2x_2 + 3x_3 \leq 30$$
$$2x_1 + 2x_2 + x_3 \leq 12$$
$$x_1 \geq 0 \quad x \geq 0 \quad x_3 \geq 0$$

19. Maximize

$$P = 2x_1 + 4x_2 + x_3 + x_4$$

subject to

$$2x_1 + x_2 + 2x_3 + 3x_4 \leq 12$$
$$2x_2 + x_3 + 2x_4 \leq 20$$
$$2x_1 + x_2 + 4x_3 \leq 16$$
$$x_1 \geq 0 \quad x_2 \geq 0 \quad x_3 \geq 0 \quad x_4 \geq 0$$

20. Maximize

$$P = 2x_1 + 4x_2 + x_3$$

subject to

$$-x_1 + 2x_2 + 3x_3 \leq 6$$
$$-x_1 + 4x_2 + 5x_3 \leq 5$$
$$x_1 + 5x_2 + 7x_3 \leq 7$$
$$x_1 \geq 0 \quad x_2 \geq 0 \quad x \geq 0$$

21. Maximize

$$P = 2x_1 + x_2 + x_3$$

subject to

$$x_1 + 2x_2 + 4x_3 \leq 20$$
$$2x_1 + 4x_2 + 4x_3 \leq 60$$
$$3x_1 + 4x_2 + x_3 \leq 90$$
$$x_1 \geq 0 \quad x_2 \geq 0 \quad x_3 \geq 0$$

22. Maximize

$$P = x_1 + 2x_2 + 4x_3$$

subject to

$$8x_1 + 5x_2 - 4x_3 \leq 30$$
$$-2x_1 + 6x_2 + x_3 \leq 5$$
$$-2x_1 + 2x_2 + x_3 \leq 15$$
$$x_1 \geq 0 \quad x_2 \geq 0 \quad x_3 \geq 0$$

23. Maximize

$$P = x_1 + 2x_2 + 4x_3 - x_4$$

subject to

$$5x_1 + 4x_3 + 6x_4 \leq 20$$
$$4x_1 + 2x_2 + 2x_3 + 8x_4 \leq 40$$
$$x_1 \geq 0 \quad x_2 \geq 0 \quad x_3 \geq 0 \quad x_4 \geq 0$$

24. Maximize

$$P = x_1 + 2x_2 - x_3 + 3x_4$$

subject to

$$2x_1 + 4x_2 + 5x_3 + 6x_4 \leq 24$$
$$4x_1 + 4x_2 + 2x_3 + 2x_4 \leq 4$$
$$x_1 \geq 0 \quad x_2 \geq 0 \quad x_3 \geq 0 \quad x_4 \geq 0$$

25. Process Utilization   A jean manufacturer makes three types of jeans, each of which goes through three manufacturing phases—cutting, sewing, and finishing. The number of minutes each type of product requires in each of the three phases is given below:

Jean	Cutting	Sewing	Finishing
I	8	12	4
II	12	18	8
III	18	24	12

There are 5200 minutes of cutting time, 6000 minutes of sewing time, and 2200 minutes of finishing time each day. The company can sell all the jeans it makes and make a profit of $3 on each Jean I, $4.50 on each Jean II, and $6 on each Jean III. Determine the number of jeans in each category that should be made each day to maximize profits.

26. Process Utilization   A company manufactures three types of toys A, B, and C. Each requires rubber, plastic, and aluminum as listed below:

Toy	Rubber	Plastic	Aluminum
A	2	2	4
B	1	2	2
C	1	2	4

The company has available 600 units of rubber, 800 units of plastic, and 1400 units of aluminum. The company makes a profit of $4, $3, and $2 on toys A, B, and C, respectively. Assuming all toys manufactured can be sold, determine a production order so that profit is maximum.

27. Scheduling   Products A, B, and C are sold door-to-door. Product A costs $3 per unit, takes 10 minutes to sell (on the average), and costs $0.50 to deliver to the customer. Product B costs $5, takes 15 minutes to sell, and is left with the customer at the time of sale. Product C costs $4, takes 12 minutes to sell, and costs $1.00 to deliver. During any week a salesperson is allowed to draw up to $500 worth of A, B, and C (at cost) and is allowed delivery expenses not to exceed $75. If a salesperson's selling time is not expected to exceed 30 hours (1800 minutes) in a week, and if the salesperson's profit (net after all expenses) is $1 each on a unit of A or B and $2 on a unit of C, what combination of sales of A, B, and C will lead to maximum profit and what is this maximum profit?

28. Resource Allocations   Suppose that a large hospital classifies its surgical operations into three categories according to their length and charges a fee of $600, $900, and $1200, respectively, for each of the categories. The average time of the operations in the three categories is 30 minutes, 1 hour, and 2 hours, respectively; and the hospital has four operating rooms, each of which can be used for 10 hours per day. If the total number of operations cannot exceed 60, how many of each type should the hospital schedule to maximize its revenues?

29. Mixture   The Lee refinery blends high and low octane gasoline into three intermediate grades: regular, premium, and super premium. The regular grade consists of 60%

high octane and 40% low octane, the premium consists of 70% high octane and 30% low octane, and the super premium consists of 80% high octane and 20% low octane. The company has available 140,000 gallons of high octane and 120,000 gallons of low octane, but can mix only 225,000 gallons. Regular gas sells for $1.20 per gallon, premium sells for $1.30 per gallon, and super premium sells for $1.40 per gallon. How many gallons of each grade should the company mix in order to maximize revenues?

30. Mixture Repeat Problem 29 under the assumption that the combined total number of gallons produced by the refinery cannot exceed 200,000 gallons.

31. Investment A financial consultant has at most $90,000 to invest in stocks, corporate bonds, and municipal bonds. The average yields for stocks, corporate bonds, and municipal bonds is 10%, 8%, and 6%, respectively. Determine how much she should invest in each security to maximize the return on her investments, if she has decided that her investment in stocks should not exceed half her funds, and that twice her investment in corporate bonds should not exceed her investment in municipal bonds.

32. Investment Repeat Problem 31 under the assumption that no more than $25,000 can be invested in stocks.

33. Crop Planning A farmer has at most 200 acres of farmland suitable for cultivating crops A, B, and C. The costs for cultivating crops A, B, and C are $40, $50, and $30 per acre, respectively. The farmer has a maximum of $18,000 available for land cultivation. Crops A, B, and C require 20, 30, and 15 hours per acre of labor, respectively, and there is a maximum of 4200 hours of labor available. If the farmer expects to make a profit of $70, $90, and $50 per acre on crops A, B, and C, respectively, how many acres of each crop should he plant in order to maximize his profit?

34. Crop Planning Repeat Problem 33 if the farmer modified his allocations as follows:

	Cost of Cultivating			Maximum
	Crop A	**Crop B**	**Crop C**	**Available**
Cost	$30	$40	$20	$12,000
Hours	10	20	18	3,600
Profit	$50	$60	$40	

35. Mixture Problem Nutt's Nut Company has 500 pounds of peanuts, 100 pounds of pecans, and 50 pounds of cashews on hand. They package three types of 5-pound cans of nuts: can I contains 3 pounds peanuts, 1 pound pecans, and 1 pound cashews; can II contains 4 pounds peanuts,

$\frac{1}{2}$ pound pecans, and $\frac{1}{2}$ pound cashews; and can III contains 5 pounds peanuts. The selling price is $28 for can I, $24 for can II, and $20 for can III. How many cans of each kind should be made to maximize revenue?

36. Maximizing Profit One of the methods used by the Alexander Company to separate copper, lead, and zinc from ores is the flotation separation process. This process consists of three steps: oiling, mixing, and separation. These steps must be applied for 2, 2, and 1 hour, respectively, to produce 1 unit of copper; 2, 3, and 1 hour, respectively, to produce 1 unit of lead; and 1, 1, and 3 hours, respectively, to produce 1 unit of zinc. The oiling and separation phases of the process can be in operation for a maximum of 10 hours a day, while the mixing phase can be in operation for a maximum of 11 hours a day. The Alexander Company makes a profit of $45 per unit of copper, $30 per unit of lead, and $35 per unit of zinc. The demand for these metals is unlimited. How many units of each metal should be produced daily by use of the flotation process to achieve the highest profit?

37. Maximizing Profit A wood cabinet manufacturer produces cabinets for television consoles, stereo systems, and radios, each of which must be assembled, decorated, and crated. Each television console requires 3 hours to assemble, 5 hours to decorate, and 0.1 hour to crate and returns a profit of $10. Each stereo system requires 10 hours to assemble, 8 hours to decorate, and 0.6 hour to crate and returns a profit of $25. Each radio requires 1 hour to assemble, 1 hour to decorate, and 0.1 hour to crate and returns a profit of $3. The manufacturer has 30,000, 40,000, and 120 hours available weekly for assembling, decorating, and crating, respectively. How many units of each product should be manufactured to maximize profit?

38. Maximizing Profit The finishing process in the manufacture of cocktail tables and end tables requires sanding, staining, and varnishing. The time in minutes required for each finishing process is given below:

	Sanding	*Staining*	*Varnishing*
End table	8	10	4
Cocktail table	4	4	8

The equipment required for each process is used on one table at a time and is available for 6 hours each day. If the profit on each cocktail table is $20 and on each end table is $15, how many of each should be manufactured each day in order to maximize profit?

39. Maximizing Profit A large TV manufacturer has warehouse facilities for storing its 25-inch color TVs in Chicago, New York, and Denver. Each month the city of Atlanta is shipped at most four hundred 25-inch TVs. The cost of transporting each TV to Atlanta from Chicago,

New York, and Denver averages $20, $20, and $40, respectively, while the cost of labor required for packing averages $6, $8, and $4, respectively. Suppose $10,000 is allocated each month for transportation costs and $3000 is allocated for labor costs. If the profit on each TV made in Chicago is $50, in New York is $80, and in Denver is $40, how should monthly shipping arrangements be scheduled to maximize profit?

Technology Exercises

A graphing calculator can be programmed to perform pivoting using elementary row operations. In Problems 1–4 use your graphing calculator to solve each maximum linear programming problem by the simplex method.

1. Maximize

$$P = 5x_1 + 3x_2$$

subject to

$$2x_1 + 3x_2 \le 125$$
$$3x_1 + x_2 \le 75$$
$$x_1 \ge 0 \quad x_2 \ge 0$$

2. Maximize

$$P = x_1 + 5x_2$$

subject to

$$x_1 + 3x_2 \le 50$$
$$3x_1 + x_2 \le 75$$
$$x_1 \ge 0 \quad x_2 \ge 0$$

3. Maximize

$$P = 5x_1 + 7x_2$$

subject to

$$x_1 + 2x_2 \le 2$$
$$2x_1 + x_2 \le 2$$
$$x_1 \ge 0 \quad x_2 \ge 0$$

4. Maximize

$$P = 3.25x_1 + 4.25x_2$$

subject to

$$1.25x_1 + 1.75x_2 \le 2.50$$
$$3.75x_1 + 1.50x_2 \le 7.50$$
$$x_1 \ge 0 \quad x_2 \ge 0$$

4.3 SOLVING MINIMUM PROBLEMS IN STANDARD FORM; THE DUALITY PRINCIPLE*

Standard Form of a Minimum Problem

A linear programming problem in which the objective function is to be minimized is referred to as a **minimum linear programming problem.** Such problems are said to be in **standard form** provided the following three conditions are met:

Standard Form of a Minimum Linear Programming Problem

Condition 1 All the variables are nonnegative.

Condition 2 All other constraints are written as linear expressions that are greater than or equal to a constant.

Condition 3 The objective function must be expressed as a linear expression with nonnegative coefficients.

---

*The solution of general minimum problems is discussed in Section 4.4. If you plan to cover Section 4.4, this section may be omitted without loss of continuity.

EXAMPLE 1   **Recognizing Minimum Linear Programming Problems in Standard Form**

Determine which of the following minimum problems are in standard form.

SOLUTION

(a) Minimize

$$C = 2x_1 + 3x_2$$

subject to the constraints

$$x_1 + 3x_2 \geq 24$$
$$2x_1 + x_2 \geq 18$$
$$x_1 \geq 0 \qquad x_2 \geq 0$$

(a) Since all three conditions are met, this minimum problem is in standard form.

(b) Minimize

$$C = 3x_1 - x_2 + 4x_3$$

subject to the constraints

$$3x_1 + x_2 + x_3 \geq 12$$
$$x_1 + x_2 + x_3 \geq 8$$
$$x_1 \geq 0 \qquad x_2 \geq 0 \qquad x_3 \geq 0$$

(b) Conditions 1 and 2 are met, but Condition 3 is not, since the coefficient of x_2 in the objective function is negative. Thus, this minimum problem is not in standard form.

(c) Minimize

$$C = 2x_1 + x_2 + x_3$$

subject to the constraints

$$x_1 - 3x_2 + x_3 \leq 12$$
$$x_1 + x_2 + x_3 \geq 1$$
$$x_1 \geq 0 \qquad x_2 \geq 0 \qquad x_3 \geq 0$$

(c) Conditions 1 and 3 are met, but Condition 2 is not, since the first constraint

$$x_1 - 3x_2 + x_3 \leq 12$$

is not written with a $\geq$ sign. Thus, the minimum problem as stated is not in standard form. Notice, however, that by multiplying by -1, we can write this constraint as

$$-x_1 + 3x_2 - x_3 \geq -12$$

Written in this way, the minimum problem is in standard form.

(d) Minimize

$$C = 2x_1 + x_2 + 3x_3$$

subject to the constraints

$$-x_1 + 2x_2 + x_3 \geq -2$$
$$x_1 + x_2 + x_3 \geq 6$$
$$x_1 \geq 0 \qquad x_2 \geq 0 \qquad x_3 \geq 0$$

(d) Conditions 1, 2, and 3 are each met, so this minimum problem is in standard form.

Now Work Problem 1

The Duality Principle

One technique for solving a minimum problem in standard form was developed by John von Neumann and others. The solution (if it exists) is found by solving a related maximum problem, called the **dual problem.** The next example illustrates how to obtain the dual problem.

EXAMPLE 2   Obtaining the Dual Problem

Obtain the dual problem of the following minimum problem:

Minimize

$$C = 300x_1 + 480x_2$$

subject to the constraints

$$x_1 + 3x_2 \geq 0.25$$
$$2x_1 + 2x_2 \geq 0.45$$
$$x_1 \geq 0 \qquad x_2 \geq 0$$

SOLUTION   First notice that the minimum problem is in standard form. We begin by writing down a matrix that represents the constraints and the objective function:

$$
\begin{array}{cc}
x_1 & x_2 \\
\end{array}
$$

$$
\left[
\begin{array}{cc|c}
1 & 3 & 0.25 \\
2 & 2 & 0.45 \\
300 & 480 & 0
\end{array}
\right]
\begin{array}{l}
\text{Constraint: } x_1 + 3x_2 \geq 0.25 \\
\text{Constraint: } 2x_1 + 2x_2 \geq 0.45 \\
\text{Objective function: } C = 300x_1 + 480x_2
\end{array}
$$

Now form the matrix that has as columns the rows of the above matrix by taking column 1 above and writing it as row 1 below, taking column 2 above and writing it as row 2 below, and taking column 3 above and writing it as row 3 below.

$$
\left[
\begin{array}{cc|c}
1 & 2 & 300 \\
3 & 2 & 480 \\
0.25 & 0.45 & 0
\end{array}
\right]
$$

This matrix is called the **transpose** of the first matrix.

From this matrix, create the following maximum problem:

Maximize

$$P = 0.25y_1 + 0.45y_2$$

subject to the conditions

$$y_1 + 2y_2 \leq 300$$
$$3y_1 + 2y_2 \leq 480$$
$$y_1 \geq 0 \qquad y_2 \geq 0$$

This maximum problem is the dual of the given minimum problem.

Notice that the dual of a minimum problem in standard form is a maximum problem in standard form so it can be solved by using techniques discussed in the previous section. The significance of this is expressed in the following principle:

Von Neumann Duality Principle

Suppose a minimum linear programming problem in standard form has a solution. The minimum value of the objective function of the minimum problem in standard form equals the maximum value of the objective function of the dual problem, a maximum problem in standard form.

So, one way to solve a minimum problem in standard form is to form the dual problem and solve it. Another way to solve minimum problems (even those not in standard form) is given in Section 4.4.

The steps to use for obtaining the dual problem are listed below.

Steps for Obtaining the Dual Problem

Step 1 Write the minimum problem in standard form.

Step 2 Construct a matrix that represents the constraints and the objective function.

Step 3 Interchange the rows and columns to form the matrix of the dual problem.

Step 4 Translate this matrix into a maximum problem in standard form.

EXAMPLE 3   Obtaining the Dual Problem

Find the dual of the following minimum problem:

Minimize

$$C = 2x_1 + 3x_2$$

subject to

$$2x_1 + x_2 \geq 6$$
$$x_1 + 2x_2 \geq 4$$
$$x_1 + x_2 \geq 5$$
$$x_1 \geq 0 \qquad x_2 \geq 0$$

SOLUTION   Observe that the minimum problem is in standard form. The matrix that represents the constraints and the objective function is

$$
\begin{bmatrix}
2 & 1 & | & 6 \\
1 & 2 & | & 4 \\
1 & 1 & | & 5 \\
2 & 3 & | & 0
\end{bmatrix}
\quad
\begin{array}{l}
\text{Constraint: } 2x_1 + x_2 \geq 6 \\
\text{Constraint: } x_1 + 2x_2 \geq 4 \\
\text{Constraint: } x_1 + x_2 \geq 5 \\
\text{Objective function: } C = 2x_1 + 3x_2
\end{array}
$$

Interchanging rows and columns, we obtain the matrix

$$\begin{bmatrix} 2 & 1 & 1 & | & 2 \\ 1 & 2 & 1 & | & 3 \\ 6 & 4 & 5 & | & 0 \end{bmatrix}$$

Constraint: $2y_1 + y_2 + y_3 \leq 2$
Constraint: $y_1 + 2y_2 + y_3 \leq 3$
Objective function: $P = 6y_1 + 4y_2 + 5y_3$

This matrix represents the following maximum problem:

Maximize

$$P = 6y_1 + 4y_2 + 5y_3$$

subject to

$$2y_1 + y_2 + y_3 \leq 2$$
$$y_1 + 2y_2 + y_3 \leq 3$$
$$y_1 \geq 0 \qquad y_2 \geq 0 \qquad y_3 \geq 0$$

This maximum problem is in standard form and is the dual problem of the minimum problem.

Some observations about Example 3:

1. The variables (x_1, x_2) of the minimum problem are different from the variables of its dual problem (y_1, y_2, y_3).
2. The minimum problem has three constraints and two variables, while the dual problem has two constraints and three variables. (In general, if a minimum problem has m constraints and n variables, its dual problem will have n constraints and m variables.)
3. The inequalities defining the constraints are $\geq$ for the minimum problem and $\leq$ for the maximum problem.
4. Since the coefficients in the minimal objective function are positive, the dual problem has nonnegative numbers to the right of the $\leq$ signs.
5. We follow the custom of denoting an objective function by C (for *Cost*) if it is to be minimized and P (for *Profit*) if it is to be maximized.

Now Work Problem 7

EXAMPLE 4　Solving a Minimum Problem by Solving the Dual Problem

Solve the maximum problem of Example 3 by the simplex method and thereby obtain the solution for the minimum problem.

SOLUTION　We introduce slack variables s_1 and s_2 to obtain

$$2y_1 + y_2 + y_3 + s_1 \qquad = 2, \quad s_1 \geq 0$$
$$y_1 + 2y_2 + y_3 \qquad + s_2 = 3, \quad s_2 \geq 0$$

The initial simplex tableau is

BV	P	y_1	y_2	y_3	s_1	s_2	RHS
s_1	0	②	1	1	1	0	2
s_2	0	1	2	1	0	1	3
P	1	-6	-4	-5	0	0	0

The pivot element, 2, is circled. After pivoting, we obtain this tableau:

BV	P	y_1	y_2	y_3	s_1	s_2	RHS	Current value
y_1	0	1	$\frac{1}{2}$	$\left(\frac{1}{2}\right)$	$\frac{1}{2}$	0	1	$y_1 = 1$
s_2	0	0	$\frac{3}{2}$	$\frac{1}{2}$	$-\frac{1}{2}$	1	2	$s_2 = 2$
P	1	0	-1	-2	3	0	6	$P = 6$

The pivot element, $\frac{1}{2}$, is circled. After pivoting, we obtain this tableau:

BV	P	y_1	y_2	y_3	s_1	s_2	RHS	Current value
y_3	0	2	1	1	1	0	2	$y_3 = 2$
s_2	0	-1	1	0	-1	1	1	$s_2 = 1$
P	1	4	1	0	5	0	10	$P = 10$

This is a final tableau so an optimal solution has been found. We read from it that the solution to the maximum problem is

$$P = 10 \qquad y_1 = 0 \qquad y_2 = 0 \qquad y_3 = 2$$

The duality principle states that the minimum value of the objective function in the original problem is the same as the maximum value in the dual; that is,

$$C = 10$$

But which values of x_1 and x_2 will yield this minimum value? As it turns out, the value of x_1 is found in the objective row in column s_1 ($x_1 = 5$) and x_2 is found in the objective row in column s_2 ($x_2 = 0$). As a consequence, the solution to the minimum problem can be read from the right end of the objective row of the final tableau of the maximum problem:

$$x_1 = 5 \qquad x_2 = 0 \qquad C = 10$$

We summarize how to solve a minimum linear programming problem below.

Solving a Minimum Problem in Standard Form Using the Dual Problem

Step 1 Write the dual (maximum) problem.

Step 2 Solve this maximum problem by the simplex method.

Step 3 The minimum value of the objective function (C) will appear in the lower right corner of the final tableau; it is equal to the maximum value of the dual objective function (P). The values of the variables that give rise to the minimum value are located in the objective row in the slack variable columns.

EXAMPLE 5   Solving a Minimum Linear Programming Problem Using the Dual Problem

Minimize

$$C = 6x_1 + 8x_2 + x_3$$

subject to

$$3x_1 + 5x_2 + 3x_3 \geq 20$$
$$x_1 + 3x_2 + 2x_3 \geq 9$$
$$6x_1 + 2x_2 + 5x_3 \geq 30$$
$$x_1 \geq 0 \qquad x_2 \geq 0 \qquad x_3 \geq 0$$

SOLUTION   This minimum problem is in standard form. The matrix representing this problem is

$$\begin{bmatrix} 3 & 5 & 3 & | & 20 \\ 1 & 3 & 2 & | & 9 \\ 6 & 2 & 5 & | & 30 \\ 6 & 8 & 1 & | & 0 \end{bmatrix}$$
Constraint: $3x_1 + 5x_2 + 3x_3 \geq 20$
Constraint: $x_1 + 3x_2 + 2x_3 \geq 9$
Constraint: $6x_1 + 2x_2 + 5x_3 \geq 30$
Objective function: $C = 6x_1 + 8x_2 + x_3$

We interchange rows and columns to get

$$\begin{bmatrix} 3 & 1 & 6 & | & 6 \\ 5 & 3 & 2 & | & 8 \\ 3 & 2 & 5 & | & 1 \\ 20 & 9 & 30 & | & 0 \end{bmatrix}$$
Constraint: $3y_1 + y_2 + 6y_3 \leq 6$
Constraint: $5y_1 + 3y_2 + 2y_3 \leq 8$
Constraint: $3y_1 + 2y_2 + 5y_3 \leq 1$
Objective function: $P = 20y_1 + 9y_2 + 30y_3$

The dual problem is:

Maximize

$$P = 20y_1 + 9y_2 + 30y_3$$

subject to

$$3y_1 + y_2 + 6y_3 \leq 6$$
$$5y_1 + 3y_2 + 2y_3 \leq 8$$
$$3y_1 + 2y_2 + 5y_3 \leq 1$$
$$y_1 \geq 0 \qquad y_2 \geq 0 \qquad y_3 \geq 0$$

We introduce slack variables s_1, s_2, and s_3. The initial tableau for this problem is

BV	P	y_1	y_2	y_3	s_1	s_2	s_3	RHS
s_1	0	3	1	6	1	0	0	6
s_2	0	5	3	2	0	1	0	8
s_3	0	3	2	5	0	0	1	1
P	1	-20	-9	-30	0	0	0	0

The final tableau (as you should verify) is

BV	P	y_1	y_2	y_3	s_1	s_2	s_3	RHS	Current value
s_1	0	0	-1	1	1	0	-1	5	$s_1 = 5$
s_2	0	0	$-\frac{1}{3}$	$-\frac{19}{3}$	0	1	$-\frac{5}{3}$	$\frac{19}{3}$	$s_2 = \frac{19}{3}$
y_1	0	1	$\frac{2}{3}$	$\frac{5}{3}$	0	0	$\frac{1}{3}$	$\frac{1}{3}$	$y_1 = \frac{1}{3}$
P	1	0	$\frac{13}{3}$	$\frac{10}{3}$	0	0	$\frac{20}{3}$	$\frac{20}{3}$	$P = \frac{20}{3}$

The solution to the maximum problem is

$$P = \tfrac{20}{3} \qquad y_1 = \tfrac{1}{3} \qquad y_2 = 0 \qquad y_3 = 0$$

For the minimum problem, the values of x_1, x_2, and x_3 are read as the entries in the objective row in the columns under s_1, s_2, and s_3, respectively. Hence, the solution to the minimum problem is

$$x_1 = 0 \qquad x_2 = 0 \qquad x_3 = \tfrac{20}{3}$$

and the minimum value is $C = \tfrac{20}{3}$.

EXERCISE 4.3   Answers to odd-numbered problems begin on page AN-21.

In Problems 1–6 determine which of the given minimum problems are in standard form.

1. Minimize

$$C = 2x_1 + 3x_2$$

subject to the constraints

$$4x_1 - x_2 \geq 2$$
$$x_1 + x_2 \geq 1$$
$$x_1 \geq 0 \qquad x_2 \geq 0$$

2. Minimize

$$C = 3x_1 + 5x_2$$

subject to the constraints

$$3x_1 - x_2 \geq 4$$
$$x_1 - 2x_2 \geq 3$$
$$x_1 \geq 0 \qquad x_2 \geq 0$$

3. Minimize

$$C = 2x_1 - x_2$$

subject to the constraints

$$2x_1 - x_2 \geq 1$$
$$-2x_1 \geq -3$$
$$x_1 \geq 0 \qquad x_2 \geq 0$$

4. Minimize

$$C = 2x_1 + 3x_2$$

subject to the constraints

$$x_1 - x_2 \leq 3$$
$$2x_1 + 3x_2 \geq 4$$
$$x_1 \geq 0 \qquad x_2 \geq 0$$

5. Minimize

$$C = 3x_1 + 7x_2 + x_3$$

subject to the constraints

$$x_1 + x_3 \leq 6$$
$$2x_1 + x_2 \geq 4$$
$$x_1 \geq 0 \qquad x_2 \geq 0 \qquad x_3 \geq 0$$

6. Minimize

$$C = x_1 - x_2 + x_3$$

subject to the constraints

$$x_1 + x_2 \geq 6$$
$$2x_1 - x_3 \geq 4$$
$$x_1 \geq 0 \qquad x_2 \geq 0 \qquad x_3 \geq 0$$

In Problems 7–12 write the dual problem for each minimum linear programming problem.

7. Minimize

$$C = 2x_1 + 3x_2$$

subject to

$$x_1 + x_2 \geq 2$$
$$2x_1 + 3x_2 \geq 6$$
$$x_1 \geq 0 \qquad x_2 \geq 0$$

8. Minimize

$$C = 3x_1 + 4x_2$$

subject to

$$2x_1 + x_2 \geq 2$$
$$2x_1 + x_2 \geq 6$$
$$x_1 \geq 0 \qquad x_2 \geq 0$$

9. Minimize

$$C = 3x_1 + x_2 + x_3$$

subject to

$$x_1 + x_2 + x_3 \geq 5$$
$$2x_1 + x_2 \geq 4$$
$$x_1 \geq 0 \qquad x_2 \geq 0 \qquad x_3 \geq 0$$

10. Minimize

$$C = 2x_1 + x_2 + x_3$$

subject to

$$2x_1 + x_2 + x_3 \geq 4$$
$$x_1 + 2x_2 + x_3 \geq 6$$
$$x_1 \geq 0 \qquad x_2 \geq 0 \qquad x_3 \geq 0$$

11. Minimize

$$C = 3x_1 + 4x_2 + x_3 + 2x_4$$

subject to

$$x_1 + x_2 + x_3 + 2x_4 \geq 60$$
$$3x_1 + 2x_2 + x_3 + 2x_4 \geq 90$$
$$x_1 \geq 0 \quad x_2 \geq 0 \quad x_3 \geq 0 \quad x_4 \geq 0$$

12. Minimize

$$C = 2x_1 + x_2 + 4x_3 + x_4$$

subject to

$$2x_1 + x_2 + x_3 + x_4 \geq 80$$
$$2x_1 + 3x_2 + x_3 + 2x_4 \geq 100$$
$$x_1 \geq 0 \quad x_2 \geq 0 \quad x_3 \geq 0 \quad x_4 \geq 0$$

In Problems 13–20 solve each minimum linear programming problem.

13. Minimize

$$C = 6x_1 + 3x_2$$

subject to

$$x_1 + x_2 \geq 2$$
$$2x_1 + 6x_2 \geq 6$$
$$x_1 \geq 0 \qquad x_2 \geq 0$$

14. Minimize

$$C = 3x_1 + 4x_2$$

subject to

$$x_1 + x_2 \geq 3$$
$$2x_1 + x_2 \geq 4$$
$$x_1 \geq 0 \qquad x_2 \geq 0$$

15. Minimize

$$C = 6x_1 + 3x_2$$

subject to

$$x_1 + x_2 \geq 4$$
$$3x_1 + 4x_2 \geq 12$$
$$x_1 \geq 0 \qquad x_2 \geq 0$$

16. Minimize

$$C = 2x_1 + 3x_2 + 4x_3$$

subject to

$$x_1 - 2x_2 - 3x_3 \geq -2$$
$$x_1 + x_2 + x_3 \geq 2$$
$$2x_1 + x_3 \geq 3$$
$$x_1 \geq 0 \qquad x_2 \geq 0 \qquad x_3 \geq 0$$

17. Minimize

$$C = x_1 + 2x_2 + x_3$$

subject to

$$x_1 - 3x_2 + 4x_3 \geq 12$$
$$3x_1 + x_2 + 2x_3 \geq 10$$
$$x_1 - x_2 - x_3 \geq -8$$
$$x_1 \geq 0 \qquad x_2 \geq 0 \qquad x_3 \geq 0$$

18. Minimize

$$C = x_1 + 2x_2 + 4x_3$$

subject to

$$x_1 - x_2 + 3x_3 \geq 4$$
$$2x_1 + 2x_2 - 3x_3 \geq 6$$
$$-x_1 + 2x_2 + 3x_3 \geq 2$$
$$x_1 \geq 0 \qquad x_2 \geq 0 \qquad x_3 \geq 0$$

19. Minimize

$$C = x_1 + 4x_2 + 2x_3 + 4x_4$$

subject to

$$x_1 + x_3 \geq 1$$
$$x_2 + x_4 \geq 1$$
$$-x_1 - x_2 - x_3 - x_4 \geq -3$$
$$x_1 \geq 0 \quad x_2 \geq 0 \quad x_3 \geq 0 \quad x_4 \geq 0$$

20. Minimize

$$C = x_1 + 2x_2 + 3x_3 + 4x_4$$

subject to

$$x_1 + x_3 \geq 1$$
$$x_2 + x_4 \geq 1$$
$$-x_1 - x_2 - x_3 - x_4 \geq -3$$
$$x_1 \geq 0 \quad x_2 \geq 0 \quad x_3 \geq 0 \quad x_4 \geq 0$$

21. Diet Preparation Mr. Jones needs to supplement his diet with at least 50 mg calcium and 8 mg iron daily. The minerals are available in two types of vitamin pills, P and Q. Pill P contains 5 mg calcium and 2 mg iron, while Pill Q contains 10 mg calcium and 1 mg iron. If each P pill costs 3 cents and each Q pill costs 4 cents, how could Mr. Jones minimize the cost of adding the minerals to his diet? What would the daily minimum cost be?

22. Production Schedule A company owns two mines. Mine A produces 1 ton of high-grade ore, 3 tons of medium-grade ore, and 5 tons of low-grade ore each day.

Mine B produces 2 tons of each grade ore per day. The company needs at least 80 tons of high-grade ore, at least 160 tons of medium-grade ore, and at least 200 tons of low-grade ore. How many days should each mine be operated to minimize costs if it costs $2000 per day to operate each mine?

23. Production Schedule Argus Company makes three products: A, B, and C. Each unit of A costs $4, each unit of B costs $2, and each unit of C costs $1 to produce. Argus must produce at least 20 As, 30 Bs, and 40 Cs, and cannot produce fewer than 200 total units of As, Bs, and Cs combined. Minimize Argus's costs.

24. **Diet Planning**   A health clinic dietician is planning a meal consisting of three foods whose ingredients are summarized as follows:

	One Unit of		
	Food I	**Food II**	**Food III**
Units of protein	10	15	20
Units of carbohydrates	1	2	1
Units of iron	4	8	1
Calories	80	120	100

The dietician wishes to determine the number of units of each food to use to create a balanced meal containing at least 40 units of protein, 6 units of carbohydrates, and 12 units of iron, with as few calories as possible. How many units of each food should be used in order to minimize calories?

25. **Menu Planning**   Fresh Starts Catering offers the following lunch menu:

Menu		
Lunch #1	Soup, salad, sandwich	$6.20
Lunch #2	Salad, pasta	$7.40
Lunch #3	Salad, sandwich, pasta	$9.10

Mrs. Mintz and her friends would like to order 4 bowls of soup, 9 salads, 6 sandwiches, and 5 orders of pasta and keep the cost as low as possible. Compose their order.

26. **Inventory Control**   A department store stocks three brands of toys: A, B, and C. Each unit of brand A occupies 1 square foot of shelf space, each unit of brand B occupies 2 square feet, and each unit of brand C occupies 3 square feet. The store has 120 square feet available for storage. Surveys show that the store should have on hand at least 12 units of brand A and at least 30 units of A and B combined. Each unit of brand A costs the store $8, each unit of brand B $6, and each unit of brand C $10. Minimize the cost to the store.

Technology Exercises

A graphing calculator can be programmed to perform pivoting using elementary row operations. In Problems 1–4 use your graphing calculator to solve each minimum linear programming problem by the simplex method.

1. Minimize

$$C = 5x_1 + 3x_2$$

subject to

$$2x_1 + 3x_2 \geq 125$$
$$3x_1 + x_2 \geq 75$$
$$x_1 \geq 0 \qquad x_2 \geq 0$$

2. Minimize

$$C = x_1 + 5x_2$$

subject to

$$x_1 + 3x_2 \geq 53$$
$$3x_1 + x_2 \geq 74$$
$$x_1 \geq 0 \qquad x_2 \geq 0$$

3. Minimize

$$C = 5x_1 + 7x_2$$

subject to

$$x_1 + 2x_2 \geq 2$$
$$2x_1 + x_2 \geq 2$$
$$x_1 \geq 0 \qquad x_2 \geq 0$$

4. Minimize

$$C = 3.25x_1 + 4.25x_2$$

subject to

$$1.25x_1 + 1.75x_2 \geq 2.50$$
$$3.75x_1 + 1.50x_2 \geq 7.50$$
$$x_1 \geq 0 \qquad x_2 \geq 0$$

4.4 THE SIMPLEX METHOD WITH MIXED CONSTRAINTS

Thus far, we have developed the simplex method only for solving linear programming problems in standard form. In this section we develop the simplex method for linear programming problems that cannot be written in standard form.

The Simplex Method with Mixed Constraints

Recall that for a maximum problem in standard form each constraint must be of the form

$$a_1x_1 + a_2x_2 + \cdots + a_nx_n \leq b_1 \qquad b_1 > 0$$

That is, each is a linear expression *less than or equal to a positive constant.* When the constraints are of any other form (greater than or equal to, or equal to) we have what are called **mixed constraints.** The following example illustrates the simplex method for solving problems with mixed constraints.

EXAMPLE 1   **Solving a Maximum Problem with Mixed Constraints**

Maximize

$$P = 20x_1 + 15x_2$$

subject to the constraints

$$
\begin{aligned}
x_1 + x_2 &\geq 7 \\
9x_1 + 5x_2 &\leq 45 \\
2x_1 + x_2 &\geq 8 \\
x_1 \geq 0 \qquad x_2 &\geq 0
\end{aligned}
$$

SOLUTION   We first observe this is a maximum problem that is not in standard form. Second, it cannot be modified so as to be in standard form.

> **Step 1** Write each constraint, except the nonnegative constraints, as an inequality with the variables on the left side of a $\leq$ sign.

For Step 1, we merely multiply the first and third inequality by -1. The result is that the constraints become

$$
\begin{aligned}
-x_1 - x_2 &\leq -7 \\
9x_1 + 5x_2 &\leq 45 \\
-2x_1 - x_2 &\leq -8 \\
x_1 \geq 0 \qquad x_2 &\geq 0
\end{aligned}
$$

> **Step 2** Introduce nonnegative slack variables on the left side of each inequality to form an equality.

For Step 2, we introduce the slack variables s_1, s_2, s_3 to obtain

$$
\begin{aligned}
-x_1 - x_2 + s_1 \qquad\qquad &= -7 \\
9x_1 + 5x_2 \qquad + s_2 \qquad &= 45 \\
-2x_1 - x_2 \qquad\qquad + s_3 &= -8
\end{aligned}
$$

$$x_1 \geq 0 \qquad x_2 \geq 0 \qquad s_1 \geq 0 \qquad s_2 \geq 0 \qquad s_3 \geq 0$$

Step 3 Set up the initial simplex tableau.

BV	P	x_1	x_2	s_1	s_2	s_3	RHS	Current value
s_1	0	-1	-1	1	0	0	-7	$s_1 = -7$
s_2	0	9	5	0	1	0	45	$s_2 = 45$
s_3	0	-2	-1	0	0	1	-8	$s_3 = -8$
P	1	-20	-15	0	0	0	0	$P = 0$

(1)

This initial tableau represents the solution $x_1 = 0$, $x_2 = 0$, $s_1 = -7$, $s_2 = 45$, $s_3 = -8$. This is not a feasible point. The two negative entries in the right-hand column violate the nonnegativity requirement. Whenever this occurs, the simplex algorithm requires an *alternative pivoting strategy*.

Alternative Pivoting Strategy

Step 4 Whenever negative entries occur in the right-hand column RHS of the constraint equations, the pivot element is selected as follows:

Pivot row: Identify the negative entries on the RHS and their corresponding basic variables (BV). (Ignore the objective row.) If any of the basic variables is an x-variable, choose the one with the smallest subscript. Otherwise choose the slack variable with the smallest subscript. The basic variable BV chosen identifies the pivot row. Because the objective row is ignored, it can never be the pivot row.

Pivot column: Go from left to right along the pivot row until a negative entry is found. (Ignore the RHS.) This entry identifies the pivot column and is the pivot element. If there are no negative entries in the pivot row except the one in the RHS column, then the problem has no solution.

Step 5 Pivot.

1. If, in the new tableau, negative entries appear on the RHS of the constraint equations, repeat Step 4.

2. If, in the new tableau, only nonnegative entries appear on the RHS of the constraint equations, then the tableau represents a maximum problem in standard form so the method outlined on page 180 of Section 4.2 is followed.*

---

* It can be shown that the alternative pivoting strategy will always lead to a tableau that represents a maximum problem in standard form. See Alan Sultan, *Linear Programming,* Academic Press, 1993.

To continue with the example, we notice there are negative entries in the RHS column of Display (1). We follow Step 4.

Step 4 The RHS column has negative entries in the rows corresponding to the basic variables s_1 and s_3. Neither of these is an x-variable so row s_1, the row with slack variable having the smallest subscript, is the pivot row. Going across from left to right along row s_1, the first negative entry we find is -1 in column x_1. The pivot column is x_1 and -1 is the pivot element.

Step 5 Pivot.

BV	P	x_1	x_2	s_1	s_2	s_3	RHS
s_1	0	(−1)	−1	1	0	0	−7
s_2	0	9	5	0	1	0	45
s_3	0	−2	−1	0	0	1	−8
P	1	−20	−15	0	0	0	0

Pivot →

BV	P	x_1	x_2	s_1	s_2	s_3	RHS
x_1	0	1	1	−1	0	0	7
s_2	0	0	−4	9	1	0	−18
s_3	0	0	1	−2	0	1	6
P	1	0	5	−20	0	0	140

The new tableau has a negative entry, -18, in the RHS column, so we repeat Step 4.

Step 4 The pivot row corresponds to the slack variable s_2. Going across, the first negative entry is -4, so the pivot column is column x_2. The pivot element is -4.

Step 5 Pivot.

BV	P	x_1	x_2	s_1	s_2	s_3	RHS
x_1	0	1	1	−1	0	0	7
s_2	0	0	(−4)	9	1	0	−18
s_3	0	0	1	−2	0	1	6
P	1	0	5	−20	0	0	140

Pivot →

BV	P	x_1	x_2	s_1	s_2	s_3	RHS
x_1	0	1	0	$\frac{5}{4}$	$\frac{1}{4}$	0	$\frac{5}{2}$
x_2	0	0	1	$-\frac{9}{4}$	$-\frac{1}{4}$	0	$\frac{9}{2}$
s_3	0	0	0	$\frac{1}{4}$	$\frac{1}{4}$	1	$\frac{3}{2}$
P	1	0	0	$-\frac{35}{4}$	$\frac{5}{4}$	0	$\frac{235}{2}$

Since the new tableau has only nonnegative entries in the RHS column, it represents a maximum problem in standard form. Since the tableau is not a final tableau (the objective row contains a negative entry), we use the standard pivoting strategy of Section 4.2. The pivot column is column s_1. Form the quotients:

$$\tfrac{5}{2} \div \tfrac{5}{4} = 2 \qquad \tfrac{3}{2} \div \tfrac{1}{4} = 6$$

The smaller of these is 2, so the pivot row is row x_1. The pivot element is $\frac{5}{4}$.

BV	P	x_1	x_2	s_1	s_2	s_3	RHS
x_1	0	1	0	$\left(\frac{5}{4}\right)$	$\frac{1}{4}$	0	$\frac{5}{2}$
x_2	0	0	1	$-\frac{9}{4}$	$-\frac{1}{4}$	0	$\frac{9}{2}$
s_3	0	0	0	$\frac{1}{4}$	$\frac{1}{4}$	1	$\frac{3}{2}$
P	1	0	0	$-\frac{35}{4}$	$\frac{5}{4}$	0	$\frac{235}{2}$

Pivot →

BV	P	x_1	x_2	s_1	s_2	s_3	RHS
s_1	0	$\frac{4}{5}$	0	1	$\frac{1}{5}$	0	2
x_2	0	$\frac{9}{5}$	1	0	$\frac{1}{5}$	0	9
s_3	0	$-\frac{1}{5}$	0	0	$\frac{1}{5}$	1	1
P	1	7	0	0	3	0	135

This is a final tableau. The maximum value of P is 135, and it is achieved when $x_1 = 0$, $x_2 = 9$, $s_1 = 2$, $s_2 = 0$, $s_3 = 1$.

✍ **Now Work Problem 1**

The Minimum Problem

In general, a minimum problem can be changed to a maximum problem by using the fact that minimizing z is the same as maximizing $P = -z$. The following example illustrates this.

EXAMPLE 2 **Solving a Minimum Linear Programming Problem**

Minimize

$$z = 5x_1 + 6x_2$$

subject to the constraints

$$x_1 + x_2 \leq 10$$
$$x_1 + 2x_2 \geq 12$$
$$2x_1 + x_2 \geq 12$$
$$x_1 \geq 3$$
$$x_1 \geq 0 \qquad x_2 \geq 0$$

SOLUTION We change the problem from minimizing $z = 5x_1 + 6x_2$ to maximizing $P = -z = -5x_1 - 6x_2$ and follow the steps for a mixed-constraint problem.

Step 1 Write each constraint with $\leq$.

$$x_1 + x_2 \leq 10$$
$$-x_1 - 2x_2 \leq -12$$
$$-2x_1 - x_2 \leq -12$$
$$-x_1 \leq -3$$

Step 2 Introduce nonnegative slack variables to form equalities:

$$x_1 + x_2 + s_1 \qquad\qquad = 10$$
$$-x_1 - 2x_2 \qquad + s_2 \qquad\qquad = -12$$
$$-2x_1 - x_2 \qquad\qquad + s_3 \qquad = -12$$
$$-x_1 \qquad\qquad\qquad + s_4 = -3$$
$$x_1 \geq 0 \qquad x_2 \geq 0 \qquad s_1 \geq 0 \qquad s_2 \geq 0 \qquad s_3 \geq 0 \qquad s_4 \geq 0$$

The objective function is: $P = -5x_1 - 6x_2$.

Step 3 Set up the initial simplex tableau

BV	P	x_1	x_2	s_1	s_2	s_3	s_4	RHS
s_1	0	1	1	1	0	0	0	10
s_2	0	-1	-2	0	1	0	0	-12
s_3	0	-2	-1	0	0	1	0	-12
s_4	0	-1	0	0	0	0	1	-3
P	1	5	6	0	0	0	0	0

Because of the negative entries in the RHS column, we follow the alternative pivoting strategy given in Step 4.

Step 4 The pivot row is s_2. The pivot column is x_1. The pivot element is -1.

Step 5 Pivot.

BV	P	x_1	x_2	s_1	s_2	s_3	s_4	RHS
s_1	0	1	1	1	0	0	0	10
s_2	0	(-1)	-2	0	1	0	0	-12
s_3	0	-2	-1	0	0	1	0	-12
s_4	0	-1	0	0	0	0	1	-3
P	1	5	6	0	0	0	0	0

Pivot →

BV	P	x_1	x_2	s_1	s_2	s_3	s_4	RHS
s_1	0	0	-1	1	1	0	0	-2
x_1	0	1	2	0	-1	0	0	12
s_3	0	0	3	0	-2	1	0	12
s_4	0	0	2	0	-1	0	1	9
P	1	0	-4	0	5	0	0	-60

The new tableau has a negative entry, -2, in the RHS column. Repeat Step 4.

Step 4 The pivot row is s_1; the pivot column is x_2; the pivot element is -1.

Step 5 Pivot.

BV	P	x_1	x_2	s_1	s_2	s_3	s_4	RHS
s_1	0	0	(-1)	1	1	0	0	-2
x_1	0	1	2	0	-1	0	0	12
s_3	0	0	3	0	-2	1	0	12
s_4	0	0	2	0	-1	0	1	9
P	1	0	-4	0	5	0	0	-60

Pivot →

BV	P	x_1	x_2	s_1	s_2	s_3	s_4	RHS
x_2	0	0	1	-1	-1	0	0	2
x_1	0	1	0	2	1	0	0	8
s_3	0	0	0	3	1	1	0	6
s_4	0	0	0	2	1	0	1	5
P	1	0	0	-4	1	0	0	-52

Since the new tableau has only nonnegative entries in the RHS column (remember, the objective row is ignored), it represents a maximum problem in standard form. Since the tableau is not a final tableau (the objective row contains a negative entry), we use the standard pivoting strategy of Section 4.2. The pivot column is column s_1. Form the quotients:

$$8 \div 2 = 4, \qquad 6 \div 3 = 2, \qquad 5 \div 2 = 2.5$$

The smallest of these is 2. The pivot row is row s_3. The pivot element is 3.

BV	P	x_1	x_2	s_1	s_2	s_3	s_4	RHS
x_2	0	0	1	-1	-1	0	0	1
x_1	0	1	0	2	1	0	0	8
s_3	0	0	0	(3)	1	1	0	6
s_4	0	0	0	2	1	0	1	5
P	1	0	0	-4	1	0	0	-52

Pivot →

BV	P	x_1	x_2	s_1	s_2	s_3	s_4	RHS
x_2	0	0	1	0	$-\frac{2}{3}$	$\frac{1}{3}$	0	4
x_1	0	1	0	0	$\frac{1}{3}$	$-\frac{2}{3}$	0	4
s_1	0	0	0	1	$\frac{1}{3}$	$\frac{1}{3}$	0	2
s_4	0	0	0	0	$\frac{1}{3}$	$-\frac{2}{3}$	1	1
P	1	0	0	0	$\frac{7}{3}$	$\frac{4}{3}$	0	-44

This is a final tableau. The maximum value of P is -44, so the minimum value of z is 44. This occurs when $x_1 = 4$, $x_2 = 4$, $s_1 = 2$, $s_2 = 0$, $s_3 = 0$, $s_4 = 1$. ■

Thus, to solve a minimum linear programming problem, change it to a maximum linear programming problem as follows:

Steps for Solving a Minimum Problem

Step 1 If z is to be minimized, let $P = -z$.

Step 2 Solve the linear programming problem: Maximize P subject to the same constraints as the minimum problem.

Step 3 Use the principle that

$$\text{Minimum of } z = -\text{Maximum of } P$$

 **Now Work Problem 5**

Equality Constraints

So far, all our constraints used $\leq$ or $\geq$. What can be done if one of the constraints is an equality? One method is to replace the $=$ constraint with the two constraints $\leq$ and $\geq$. The next example illustrates this method.

EXAMPLE 3 **Solving a Minimum Linear Programming Problem with Equality Constraints**

Minimize

$$z = 7x_1 + 5x_2 + 6x_3$$

subject to the constraints

$$x_1 + x_2 + x_3 = 10$$
$$x_1 + 2x_2 + 3x_3 \leq 19$$
$$2x_1 + 3x_2 \qquad \geq 21$$
$$x_1 \geq 0 \qquad x_2 \geq 0 \qquad x_3 \geq 0$$

SOLUTION We wish to maximize $P = -z = -7x_1 - 5x_2 - 6x_3$ subject to the constraints

$$x_1 + x_2 + x_3 \leq 10$$
$$x_1 + x_2 + x_3 \geq 10$$
$$x_1 + 2x_2 + 3x_3 \leq 19$$
$$2x_1 + 3x_2 \qquad \geq 21$$
$$x_1 \geq 0 \qquad x_2 \geq 0 \qquad x_3 \geq 0$$

Step 1 Rewrite the constraints with $\leq$:

$$x_1 + x_2 + x_3 \leq 10$$
$$-x_1 - x_2 - x_3 \leq -10$$
$$x_1 + 2x_2 + 3x_3 \leq 19$$
$$-2x_1 - 3x_2 \qquad \leq -21$$

Step 2 Introduce nonnegative slack variables:

$$
\begin{aligned}
x_1 + x_2 + x_3 + s_1 &= 10 \\
-x_1 - x_2 - x_3 \qquad\quad + s_2 &= -10 \\
x_1 + 2x_2 + 3x_3 \qquad\qquad\quad + s_3 &= 19 \\
-2x_1 - 3x_2 \qquad\qquad\qquad\qquad\qquad + s_4 &= -21
\end{aligned}
$$

$$x_1 \geq 0 \qquad x_2 \geq 0 \qquad x_3 \geq 0 \qquad s_1 \geq 0 \qquad s_2 \geq 0 \qquad s_3 \geq 0 \qquad s_4 \geq 0$$

Step 3 Set up the initial simplex tableau

BV	P	x_1	x_2	x_3	s_1	s_2	s_3	s_4	RHS
s_1	0	1	1	1	1	0	0	0	10
s_2	0	-1	-1	-1	0	1	0	0	-10
s_3	0	1	2	3	0	0	1	0	19
s_4	0	-2	-3	0	0	0	0	1	-21
P	1	7	5	6	0	0	0	0	0

Because of the negative entries in the RHS column, we follow the alternative pivoting strategy.

Step 4 The pivot row is row s_2; the pivot column is column x_1. The pivot element is -1.

Step 5 Pivot.

BV	P	x_1	x_2	x_3	s_1	s_2	s_3	s_4	RHS
s_1	0	1	1	1	1	0	0	0	10
s_2	0	$\boxed{-1}$	-1	-1	0	1	0	0	-10
s_3	0	1	2	3	0	0	1	0	19
s_4	0	-2	-3	0	0	0	0	1	-21
P	1	7	5	6	0	0	0	0	0

	BV	P	x_1	x_2	x_3	s_1	s_2	s_3	s_4	RHS
	s_1	0	0	0	0	1	1	0	0	0
Pivot	x_1	0	1	1	1	0	-1	0	0	10
	s_3	0	0	1	2	0	1	1	0	9
	s_4	0	0	-1	2	0	-2	0	1	-1
	P	1	0	-2	-1	0	7	0	0	-70

The new tableau has a negative entry, -1, in the RHS column, so we repeat Step 4.

Step 4 The pivot row is row s_4. Going across, the first negative entry is -1, so the pivot column is column x_2. The pivot element is -1.

BV	P	x_1	x_2	x_3	s_1	s_2	s_3	s_4	RHS
s_1	0	0	0	0	1	1	0	0	0
x_1	0	1	1	1	0	-1	0	0	10
s_3	0	0	1	2	0	1	1	0	9
s_4	0	0	$\boxed{-1}$	2	0	-2	0	1	-1
P	1	0	-2	-1	0	7	0	0	-70

BV	P	x_1	x_2	x_3	s_1	s_2	s_3	s_4	RHS
s_1	0	0	0	0	1	1	0	0	0
Pivot x_1	0	1	0	3	0	-3	0	1	9
s_3	0	0	0	4	0	-1	1	1	8
x_2	0	0	1	-2	0	2	0	-1	1
P	1	0	0	-5	0	11	0	-2	-68

Since the new tableau has only nonnegative entries in the RHS column (remember, the objective row is ignored), it represents a maximum problem in standard form. Since the tableau is not a final tableau (the objective row contains negative entries), we use the standard pivoting strategy of Section 4.2. The pivot column is column x_3. Form the nonnegative quotients:

$$9 \div 3 = 3, \qquad 8 \div 4 = 2$$

The smaller of these is 2, so the pivot row is row s_3. The pivot element is 4.

BV	P	x_1	x_2	x_3	s_1	s_2	s_3	s_4	RHS
s_1	0	0	0	0	1	1	0	0	0
x_1	0	1	0	3	0	-3	0	1	9
s_3	0	0	0	④	0	-1	1	1	8
x_2	0	0	1	-2	0	2	0	-1	1
P	1	0	0	-5	0	11	0	-2	-68

BV	P	x_1	x_2	x_3	s_1	s_2	s_3	s_4	RHS
s_1	0	0	0	0	1	1	0	0	0
Pivot x_1	0	1	0	0	0	$-\frac{9}{4}$	$-\frac{3}{4}$	$\frac{1}{4}$	3
x_3	0	0	0	1	0	$-\frac{1}{4}$	$\frac{1}{4}$	$\frac{1}{4}$	2
x_2	0	0	1	0	0	$\frac{3}{2}$	$\frac{1}{2}$	$-\frac{1}{2}$	5
P	1	0	0	0	0	$\frac{39}{4}$	$\frac{5}{4}$	$-\frac{3}{4}$	-58

This is not a final tableau. The pivot column is column s_4. Form the quotients:

$$3 \div \tfrac{1}{4} = 12, \qquad 2 \div \tfrac{1}{4} = 8$$

The smaller of these is 8. The pivot row is x_3; the pivot element is $\frac{1}{4}$.

BV	P	x_1	x_2	x_3	s_1	s_2	s_3	s_4	RHS
s_1	0	0	0	0	1	1	0	0	0
x_1	0	1	0	0	0	$-\frac{9}{4}$	$-\frac{3}{4}$	$\frac{1}{4}$	3
x_3	0	0	0	1	0	$-\frac{1}{4}$	$\frac{1}{4}$	⑴⁄₄	2
x_2	0	0	1	0	0	$\frac{3}{2}$	$\frac{1}{2}$	$-\frac{1}{2}$	5
P	1	0	0	0	0	$\frac{39}{4}$	$\frac{5}{4}$	$-\frac{3}{4}$	-58

BV	P	x_1	x_2	x_3	s_1	s_2	s_3	s_4	RHS
s_1	0	0	0	0	1	1	0	0	0
Pivot x_1	0	1	0	-1	0	-2	-1	0	1
s_4	0	0	0	4	0	-1	1	1	8
x_2	0	0	1	2	0	1	1	0	9
P	1	0	0	3	0	9	2	0	-52

This is a final tableau. The maximum value of P is -52, so the minimum value of z is 52. This occurs when $x_1 = 1$, $x_2 = 9$, $x_3 = 0$, $s_1 = 0$, $s_2 = 0$, $s_3 = 0$, $s_4 = 8$.

✏ **Now Work Problem 7**

EXAMPLE 4   Minimizing Cost

The Red Tomato Company operates two plants for canning its tomatoes and has two warehouses for storing the finished products until they are purchased by retailers. The schedule shown in the table represents the per case shipping costs from plant to warehouse.

		Warehouse	
		A	B
Plant	I	$0.25	$0.18
	II	$0.25	$0.14

Each week plant I can produce at most 450 cases and plant II can produce at most 350 cases of tomatoes. Also, each week warehouse A requires at least 300 cases and warehouse B requires at least 500 cases. If we represent the number of cases shipped from plant I to warehouse A by x_1, from plant I to warehouse B by x_2, and so on, the above data can be represented by the following table:

		Warehouse		Maximum Available
		A	B	
Plant	I	x_1	x_2	450
	II	x_3	x_4	350
Minimum Demand		300	500	

The company wants to arrange its shipments from the plants to the warehouses so that the requirements of the warehouses are met and shipping costs are kept at a minimum. How should the company proceed?

SOLUTION   The linear programming problem is stated as follows:

Minimize the cost equation

$$C = 0.25x_1 + 0.18x_2 + 0.25x_3 + 0.14x_4$$

subject to

$$x_1 + x_2 \leq 450$$
$$x_3 + x_4 \leq 350$$
$$x_1 + x_3 \geq 300$$
$$x_2 + x_4 \geq 500$$
$$x_1 \geq 0 \quad x_2 \geq 0 \quad x_3 \geq 0 \quad x_4 \geq 0$$

Thus we shall maximize

$$P = -C = -0.25x_1 - 0.18x_2 - 0.25x_3 - 0.14x_4$$

subject to the same constraints.

Step 1 Write each constraint with $\leq$.

$$
\begin{aligned}
x_1 + x_2 &\leq 450 \\
x_3 + x_4 &\leq 350 \\
-x_1 \quad - x_3 &\leq -300 \\
-x_2 \quad - x_4 &\leq -500
\end{aligned}
$$

Step 2 Introduce nonnegative slack variables to form equalities:

$$
\begin{aligned}
x_1 + x_2 \quad\quad\quad + s_1 &= 450 \\
x_3 + x_4 \quad + s_2 &= 350 \\
-x_1 \quad - x_3 \quad\quad + s_3 &= -300 \\
-x_2 \quad - x_4 \quad\quad\quad + s_4 &= -500
\end{aligned}
$$

$$x_1 \geq 0 \quad x_2 \geq 0 \quad x_3 \geq 0 \quad x_4 \geq 0 \quad s_1 \geq 0 \quad s_2 \geq 0 \quad s_3 \geq 0 \quad s_4 \geq 0$$

Step 3 Set up the initial simplex tableau:

BV	P	x_1	x_2	x_3	x_4	s_1	s_2	s_3	s_4	RHS
s_1	0	1	1	0	0	1	0	0	0	450
s_2	0	0	0	1	1	0	1	0	0	350
s_3	0	−1	0	−1	0	0	0	1	0	−300
s_4	0	0	−1	0	−1	0	0	0	1	−500
P	1	0.25	0.18	0.25	0.14	0	0	0	0	0

Because of the negative entries in the RHS column, we follow the alternative pivoting strategy.

Step 4 The pivot row is row s_3; the pivot column is column x_1. The pivot element is -1.

Step 5 Pivot.

BV	P	x_1	x_2	x_3	x_4	s_1	s_2	s_3	s_4	RHS
s_1	0	1	1	0	0	1	0	0	0	450
s_2	0	0	0	1	1	0	1	0	0	350
s_3	0	⊝1	0	−1	0	0	0	1	0	−300
s_4	0	0	−1	0	−1	0	0	0	1	−500
P	1	0.25	0.18	0.25	0.14	0	0	0	0	0

	BV	P	x_1	x_2	x_3	x_4	s_1	s_2	s_3	s_4	RHS
	s_1	0	0	1	−1	0	1	0	1	0	150
Pivot ⟶	s_2	0	0	0	1	1	0	1	0	0	350
	x_1	0	1	0	1	0	0	0	−1	0	300
	s_4	0	0	−1	0	−1	0	0	0	1	−500
	P	1	0	0.18	0	0.14	0	0	0.25	0	−75

The new tableau has a negative entry, -500, in the RHS column. Repeat Step 4.

Step 4 The pivot row is row s_4; the pivot column is column x_2; the pivot element is -1.

Step 5 Pivot.

BV	P	x_1	x_2	x_3	x_4	s_1	s_2	s_3	s_4	RHS
s_1	0	0	1	-1	0	1	0	1	0	150
s_2	0	0	0	1	1	0	1	0	0	350
x_1	0	1	0	1	0	0	0	-1	0	300
s_4	0	0	(-1)	0	-1	0	0	0	1	-500
P	1	0	0.18	0	0.14	0	0	0.25	0	-75

Pivot →

BV	P	x_1	x_2	x_3	x_4	s_1	s_2	s_3	s_4	RHS
s_1	0	0	0	-1	-1	1	0	1	1	-350
s_2	0	0	0	1	1	0	1	0	0	350
x_1	0	1	0	1	0	0	0	-1	0	300
x_2	0	0	1	0	1	0	0	0	-1	500
P	1	0	0	0	-0.04	0	0	0.25	0.18	-165

The new tableau has a negative entry, -350, in the RHS column, so we repeat Step 4.

Step 4 The pivot row is row s_1. Going across, the first negative entry is -1, so the pivot column is column x_3. The pivot element is -1.

Step 5 Pivot.

BV	P	x_1	x_2	x_3	x_4	s_1	s_2	s_3	s_4	RHS
s_1	0	0	0	(-1)	-1	1	0	1	1	-350
s_2	0	0	0	1	1	0	1	0	0	350
x_1	0	1	0	1	0	0	0	-1	0	300
x_2	0	0	1	0	1	0	0	0	-1	500
P	1	0	0	0	-0.04	0	0	0.25	0.18	-165

Pivot →

BV	P	x_1	x_2	x_3	x_4	s_1	s_2	s_3	s_4	RHS
x_3	0	0	0	1	1	-1	0	-1	-1	350
s_2	0	0	0	0	0	1	1	1	1	0
x_1	0	1	0	0	-1	1	0	0	1	-50
x_2	0	0	1	0	1	0	0	0	-1	500
P	1	0	0	0	-0.04	0	0	0.25	0.18	-165

The new tableau has a negative entry, -50, in the RHS column. Repeat Step 4.

Step 4 The pivot row is row x_1; the pivot column is column x_4; the pivot element is -1.

Step 5 Pivot.

BV	P	x_1	x_2	x_3	x_4	s_1	s_2	s_3	s_4	RHS
x_3	0	0	0	1	1	-1	0	-1	-1	350
s_2	0	0	0	0	0	1	1	1	1	0
x_1	0	1	0	0	(-1)	1	0	0	1	-50
x_2	0	0	1	0	1	0	0	0	-1	500
P	1	0	0	0	-0.04	0	0	0.25	0.18	-165

Pivot $\longrightarrow$

BV	P	x_1	x_2	x_3	x_4	s_1	s_2	s_3	s_4	RHS
x_3	0	1	0	1	0	0	0	-1	0	300
s_2	0	0	0	0	0	1	1	1	1	0
x_4	0	-1	0	0	1	-1	0	0	-1	50
x_2	0	1	1	0	0	1	0	0	0	450
P	1	-0.04	0	0	0	-0.04	0	0.25	0.14	-163

Since the new tableau has only nonnegative entries in the RHS column (remember, the objective row is ignored), it represents a maximum problem in standard form. Since the tableau is not a final tableau (the objective row contains a negative entry), we use the standard pivoting strategy of Section 4.2. The pivot column is column s_1 (column x_1 could also be chosen). Form the quotients:

$$0 \div 1 = 0 \qquad 450 \div 1 = 450$$

The smaller of these is 0. The pivot row is row s_2. The pivot element is 1.

BV	P	x_1	x_2	x_3	x_4	s_1	s_2	s_3	s_4	RHS
x_3	0	1	0	1	0	0	0	-1	0	300
s_2	0	0	0	0	0	(1)	1	1	1	0
x_4	0	-1	0	0	1	-1	0	0	-1	50
x_2	0	1	1	0	0	1	0	0	0	450
P	1	-0.04	0	0	0	-0.04	0	0.25	0.14	-163

Pivot $\longrightarrow$

BV	P	x_1	x_2	x_3	x_4	s_1	s_2	s_3	s_4	RHS
x_3	0	1	0	1	0	0	0	-1	0	300
s_1	0	0	0	0	0	1	1	1	1	0
x_4	0	-1	0	0	1	0	1	1	0	50
x_2	0	1	1	0	0	0	-1	-1	-1	450
P	1	-0.04	0	0	0	0	0.04	0.29	0.18	-163

The tableau is not a final tableau since the objective row contains a negative entry. We use the standard pivoting strategy of Section 4.2. The pivot column is column x_1. Form the quotients:

$$300 \div 1 = 300 \qquad 450 \div 1 = 450$$

The smaller of these is 300. The pivot row is row x_3, the pivot element is 1.

BV	P	x_1	x_2	x_3	x_4	s_1	s_2	s_3	s_4	RHS
x_3	0	①	0	1	0	0	0	-1	0	300
s_1	0	0	0	0	0	1	1	1	1	0
x_4	0	-1	0	0	1	0	1	1	0	50
x_2	0	1	1	0	0	0	-1	-1	-1	450
P	1	-0.04	0	0	0	0	0.04	0.29	0.18	-163

BV	P	x_1	x_2	x_3	x_4	s_1	s_2	s_3	s_4	RHS
x_1	0	1	0	1	0	0	0	-1	0	300
Pivot $\longrightarrow$ s_1	0	0	0	0	0	1	1	1	1	0
x_4	0	0	0	1	1	0	1	0	0	350
x_2	0	0	1	-1	0	0	-1	0	-1	150
P	1	0	0	0.04	0	0	0.04	0.25	0.18	-151

This is a final tableau. The maximum value of P is -151, so the minimum cost C is \$151. This occurs when $x_1 = 300$, $x_2 = 150$, $x_3 = 0$, $x_4 = 350$. This means plant I should deliver 300 cases to warehouse A and 150 cases to warehouse B; and plant II should deliver 350 cases to warehouse B to keep costs at the minimum (\$151). ∎

EXERCISE 4.4   Answers to odd-numbered problems begin on page AN-21.

In Problems 1–8 use the mixed-constraint method to solve each linear programming problem.

1. Maximize

$$P = 3x_1 + 4x_2$$

subject to the constraints

$$\begin{aligned} x_1 + x_2 &\le 12 \\ 5x_1 + 2x_2 &\ge 36 \\ 7x_1 + 4x_2 &\ge 14 \\ x_1 \ge 0 \qquad x_2 &\ge 0 \end{aligned}$$

2. Maximize

$$P = 5x_1 + 2x_2$$

subject to the constraints

$$\begin{aligned} x_1 + x_2 &\ge 11 \\ 2x_1 + 3x_2 &\ge 24 \\ x_1 + 3x_2 &\le 18 \\ x_1 \ge 0 \qquad x_2 &\ge 0 \end{aligned}$$

3. Maximize

$$P = 3x_1 + 2x_2 - x_3$$

subject to the constraints

$$\begin{aligned} x_1 + 3x_2 + x_3 &\le 9 \\ 2x_1 + 3x_2 - x_3 &\ge 2 \\ 3x_1 - 2x_2 + x_3 &\ge 5 \\ x_1 \le 0 \quad x_2 \ge 0 \quad x_2 &\ge 0 \end{aligned}$$

4. Maximize

$$P = 3x_1 + 2x_2 - x_3$$

subject to the constraints

$$\begin{aligned} 2x_1 - x_2 - x_3 &\le 2 \\ x_1 + 2x_2 + x_3 &\ge 2 \\ x_1 - 3x_2 - 2x_3 &\le -5 \\ x_1 \ge 0 \quad x_2 \ge 0 \quad x_3 &\ge 0 \end{aligned}$$

5. Minimize

$$z = 6x_1 + 8x_2 + x_3$$

subject to the constraints

$$3x_1 + 5x_2 + 3x_3 \geq 20$$
$$x_1 + 3x_2 + 2x_3 \geq 9$$
$$6x_1 + 2x_2 + 5x_3 \geq 30$$
$$x_1 + x_2 + x_3 \leq 10$$
$$x_1 \geq 0 \quad x_2 \geq 0 \quad x_3 \geq 0$$

6. Minimize

$$z = 2x_1 + x_2 + x_3$$

subject to the constraints

$$3x_1 - x_2 - 4x_3 \leq -12$$
$$x_1 + 3x_2 + 2x_3 \geq 10$$
$$x_1 - x_2 + x_3 \leq 8$$
$$x_1 \geq 0 \quad x_2 \geq 0 \quad x_3 \geq 0$$

7. Maximize

$$P = 3x_1 + 2x_2$$

subject to the constraints

$$2x_1 + x_2 \leq 4$$
$$x_1 + x_2 = 3$$
$$x_1 \geq 0 \quad x_2 \geq 0$$

8. Maximize

$$P = 45x_1 + 27x_2 + 18x_3 + 36x_4$$

subject to the constraints

$$5x_1 + x_2 + x_3 + 8x_4 = 30$$
$$2x_1 + 4x_2 + 3x_3 + 2x_4 = 30$$
$$x_1 \geq 0 \quad x_2 \geq 0 \quad x_3 \geq 0 \quad x_4 \geq 0$$

9. **Shipping** Private Motors, Inc., has two plants, M1 and M2, which manufacture engines; the company also has two assembly plants, A1 and A2, which assemble the cars. M1 can produce at most 600 engines per week. M2 can produce at most 400 engines per week. A1 needs at least 500 engines per week and A2 needs at least 300 engines per week. Following is a table of charges to ship engines to assembly plants.

	A1	**A2**
M1	$400	$100
M2	$200	$300

How many engines should be shipped each week from each engine plant to each assembly plant? [*Hint:* Consider four variables: x_1 = number of units shipped from M1 to A1, x_2 = number of units shipped from M1 to A2, x_3 = number of units shipped from M2 to A1, and x_4 = number of units shipped from M2 to A2.]

10. **Minimizing Materials** Quality Oak Tables, Inc., has an individual who does all its finishing work, and it wishes to use him in this capacity at least 6 hours each day. The assembly area can be used at most 8 hours each day. The company has three models of oak tables, T1, T2, T3. T1 requires 1 hour for assembly, 2 hours for finishing, and 9 board feet of oak. T2 requires 1 hour for assembly, 1 hour for finishing, and 9 board feet of oak. T3 requires 2 hours for assembly, 1 hour for finishing, and 3 board feet of oak. If we wish to minimize the board feet of oak used, how many of each model should be made?

11. **Mixture** Minimize the cost of preparing the following mixture, which is made up of three foods, I, II, III.

Food I costs $2 per unit, food II costs $1 per unit, and food III costs $3 per unit. Each unit of food I contains 2 ounces of protein and 4 ounces of carbohydrate; each unit of food II has 3 ounces of protein and 2 ounces of carbohydrate; and each unit of food III has 4 ounces of protein and 2 ounces of carbohydrate. The mixture must contain at least 20 ounces of protein and 15 ounces of carbohydrate.

12. **Advertising** A local appliance store has decided on an advertising campaign utilizing newspaper and radio. Each dollar spent on newspaper advertising is expected to reach 50 people in the "Under $25,000" and 40 in the "Over $25,000" bracket. Each dollar spent on radio advertising is expected to reach 70 people in the "Under $25,000" and 20 people in the "Over $25,000" bracket. If the store wants to reach at least 100,000 people in the "Under $25,000" and at least 120,000 in the "Over $25,000" bracket, how should it proceed so that the cost of advertising is minimized?

13. **Shipping Schedule** A television manufacturer must fill orders from two retailers. The first retailer, R_1, has ordered 55 television sets, while the second retailer, R_2, has ordered 75 sets. The manufacturer has the television sets stored in two warehouses, W_1 and W_2. There are 100 sets in W_1 and 120 sets in W_2. The shipping costs per television set are: $8 from W_1 to R_1; $12 from W_1 to R_2; $13 from W_2 to R_1; $7 from W_2 to R_2. Find the number of television sets to be shipped from each warehouse to each retailer if the total shipping cost is to be a minimum. What is this minimum cost?

14. **Shipping Schedule** A motorcycle manufacturer must fill orders from two dealers. The first dealer, D_1, has or-

dered 20 motorcycles, while the second dealer, D_2, has ordered 30 motorcycles. The manufacturer has the motorcycles stored in two warehouses, W_1 and W_2. There are 40 motorcycles in W_1 and 15 in W_2. The shipping costs per motorcycle are as follows: $15 from W_1 to D_1; $13 from W_1 to D_2; $14 from W_2 to D_1; $16 from W_2 to D_2. Under these conditions, find the number of motorcycles to be shipped from each warehouse to each dealer if the total shipping cost is to be held to a minimum. What is this minimum cost?

15. Production Control   RCA manufacturing received an order for a machine. The machine is to weigh 150 pounds. The two raw materials used to produce the machine are A, with a cost of $4 per unit, and B, with a cost of $8 per unit. At least 14 units of B and no more than 20 units of A must be used. Each unit of A weighs 5 pounds; each unit of B weighs 10 pounds. How much of each type of raw material should be used for each machine if we wish to minimize cost?

CHAPTER REVIEW

IMPORTANT TERMS AND CONCEPTS

standard form of a maximum problem 167
initial simplex tableau 170
pivoting 172
the simplex method 180

flowchart to solve a maximum linear programming problem 180
standard form of a minimum problem 196
duality principle 199

steps for obtaining the dual problem 199
steps for solving a minimum problem in standard form 201
simplex method with mixed constraints 206–207
alternative pivoting strategy 207
steps for solving a minimum problem 211

TRUE–FALSE ITEMS   Answers are on page AN-22.

T_____ F_____ **1.** For a maximum problem in standard form, each of the constraints, with the exception of the nonnegativity constraints, is written with a $\leq$ symbol.

T_____ F_____ **2.** For a maximum problem in standard form, the slack variables are sometimes negative.

T_____ F_____ **3.** Once the pivot element is identified in a tableau, the pivot operation causes the pivot element to become a 1 and causes the remaining entries in the pivot column to become 0s.

T_____ F_____ **4.** The pivot element is sometimes in the objective row.

T_____ F_____ **5.** One way to solve a minimum problem is to first solve its dual, which is a maximum problem.

T_____ F_____ **6.** Another way to solve a minimum problem is to solve the maximum problem whose objective function is the negative of the minimum problem's objective function.

FILL IN THE BLANKS   Answers are on page AN-22.

1. The constraints of a maximum problem in standard form are changed from an inequality to an equation by introducing _____ _____ .

2. For a maximum problem in standard form, the pivot _____ is located by selecting the most negative entry in the objective row.

3. For a minimum problem to be in standard form all the constraints must be written with _____ signs.

4. The _____ _____ _____ principle states that the optimal solution of a minimum linear programming problem, if it exists, has the same value as the optimal solution of the maximum problem, which is its dual.

REVIEW EXERCISES   Answers to odd-numbered problems begin on page AN-22.

In Problems 1–9 use the simplex method.

1. Maximize

$$P = 100x_1 + 200x_2 + 50x_3$$

subject to the constraints

$$5x_1 + 5x_2 + 10x_3 \leq 1000$$
$$10x_1 + 8x_2 + 5x_3 \leq 2000$$
$$10x_1 + 5x_2 \qquad \leq 500$$
$$x_1 \geq 0 \qquad x_2 \geq 0 \qquad x_3 \geq 0$$

2. Maximize

$$P = x_1 + 2x_2 + x_3$$

subject to the constraints

$$3x_1 + x_2 + x_3 \leq 3$$
$$x_1 - 10x_2 - 4x_3 \leq 20$$
$$x_1 \geq 0 \qquad x_2 \geq 0 \qquad x_3 \geq 0$$

3. Maximize

$$P = 40x_1 + 60x_2 + 50x_3$$

subject to the constraints

$$2x_1 + 2x_2 + x_3 \leq 8$$
$$x_1 - 4x_2 + 3x_3 \leq 12$$
$$x_1 \geq 0 \qquad x_2 \geq 0 \qquad x_3 \geq 0$$

4. Maximize

$$P = 2x_1 + 8x_2 + 10x_3 + x_4$$

subject to the constraints

$$x_1 + 2x_2 + x_3 + x_4 \leq 50$$
$$3x_1 + x_2 + 2x_3 + x_4 \leq 100$$
$$x_1 \geq 0 \quad x_2 \geq 0 \quad x_3 \geq 0 \quad x_4 \geq 0$$

5. Minimize

$$C = 2x_1 + x_2$$

subject to the constraints

$$2x_1 + 2x_2 \geq 8$$
$$x_1 - x_2 \geq 2$$
$$x_1 \geq 0 \qquad x_2 \geq 0$$

6. Minimize

$$C = 4x_1 + 2x_2$$

subject to the constraints

$$x_1 + 2x_2 \geq 4$$
$$x_1 + 4x_2 \geq 6$$
$$x_1 \geq 0 \qquad x_2 \geq 0$$

7. Minimize

$$C = 5x_1 + 4x_2 + 3x_3$$

subject to the contraints

$$x_1 + x_2 + x_3 \geq 100$$
$$2x_1 + x_2 \qquad \geq 50$$
$$x_1 \geq 0 \qquad x_2 \geq 0 \qquad x_3 \geq 0$$

8. Minimize

$$C = 2x_1 + x_2 + 3x_3 + x_4$$

subject to the constraints

$$x_1 + x_2 + x_3 + x_4 \geq 50$$
$$3x_1 + x_2 + 2x_3 + x_4 \geq 100$$
$$x_1 \geq 0 \quad x_2 \geq 0 \quad x_3 \geq 0 \quad x_4 \geq 0$$

9. Maximize

$$P = 300x_1 + 200x_2 + 450x_3$$

subject to the constraints

$$4x_1 + 3x_2 + 5x_3 \leq 140$$
$$x_1 + x_2 + x_3 = 30$$
$$x_1 \geq 0 \qquad x_2 \geq 0 \qquad x_3 \geq 0$$

10. **Mixture**   A brewery manufactures three types of beer—lite, regular, and dark. Each vat of lite beer requires 6 bags of barley, 1 bag of sugar, and 1 bag of hops. Each vat of regular beer requires 4 bags of barley, 3 bags of sugar, and 1 bag of hops. Each vat of dark beer requires 2 bags of barley, 2 bags of sugar, and 4 bags of hops. Each day the brewery has 800 bags of barley, 600 bags of sugar, and 300 bags of hops available. The brewery realizes a profit of $10 per vat of lite beer, $20 per vat of regular beer, and $30 per vat of dark beer. How many vats of lite, regular, and dark beer should be brewed in order to maximize profits? What is the maximum profit?

11. **Management**   The manager of a supermarket meat department finds that there are 160 pounds of round steak, 600 pounds of chuck steak, and 300 pounds of pork in stock on Saturday morning. From experience, the manager knows that half of these quantities can be sold as straight cuts. The remaining meat will have to be ground into hamburger patties and picnic patties for which there is a large weekend demand. Each pound of hamburger patties contains 20% ground round and 60% ground chuck. Each pound of picnic patties contains 30% ground pork and 50% ground chuck. The remainder of each prod-

uct consists of an inexpensive nonmeat filler that the store has in unlimited quantities. How many pounds of each product should be made if the objective is to maximize the amount of meat used to make the patties?

12. **Scheduling**   An automobile manufacturer must fill orders from two dealers. The first dealer, D_1, has ordered 40 cars, while the second dealer, D_2, has ordered 25 cars. The manufacturer has the cars stored in two locations, W_1 and W_2. There are 30 cars in W_1 and 50 cars in W_2. The shipping costs per car are as follows: $180 from W_1 to D_1; $150 from W_1 to D_2; $160 from W_2 to D_1; $170 from W_2 to D_2. Under these conditions, how many cars should be shipped from each storage location to each dealer so as to minimize the total shipping costs? What is this minimum cost?

13. **Optimal Land Use**   A farmer has 1000 acres of land on which corn, wheat, or soybeans can be grown. Each acre of corn costs $100 for preparation, requires 7 days of labor, and yields a profit of $30. An acre of wheat costs $120 to prepare, requires 10 days of labor, and yields $40 profit. An acre of soybeans costs $70 to prepare, requires 8 days of labor, and yields $40 profit. If the farmer has

$10,000 for preparation and can count on enough workers to supply 8000 days of labor, how many acres should be devoted to each crop to maximize profits?

14. **Minimizing Cost**   The ACE Meat Market makes up a combination package of ground beef and ground pork for meat loaf. The ground beef is 75% lean (75% beef, 25%

fat) and costs the market 70¢ per pound. The ground pork is 60% lean (60% pork, 40% fat) and costs the market 50¢ per pound. If the meat loaf is to be at least 70% lean, how much ground beef and ground pork should be mixed to keep cost at a minimum?

Mathematical Questions from Professional Exams

Use the following information to answer Problems 1–4.

CPA Exam   The Ball Company manufactures three types of lamps, which are labeled A, B, and C. Each lamp is processed in two departments—I and II. Total available work-hours per day for departments I and II are 400 and 600, respectively. No additional labor is available. Time requirements and profit per unit for each lamp type are as follows:

	A	B	C
Work-hours required in department I	2	3	1
Work-hours required in department II	4	2	3
Profit per unit (sales price less all variable costs)	$5	$4	$3

The company has assigned you, as the accounting member of its profit planning committee, to determine the number of types of A, B, and C lamps that it should produce in order to maximize its total profit from the sale of lamps. The following questions relate to a linear programming model that your group has developed.

1. The coefficients of the objective function would be

   (a) 4, 2, 3   (b) 2, 3, 1   (c) 5, 4, 3   (d) 400, 600

2. The constraints in the model would be

   (a) 2, 3, 1   (b) 5, 4, 3   (c) 4, 2, 3   (d) 400, 600

3. The constraint imposed by the available work-hours in department I could be expressed as

   (a) $4X_1 + 2X_2 + 3X_3 \leq 400$
   (b) $4X_1 + 2X_2 + 3X_3 \geq 400$
   (c) $2X_1 + 3X_2 + 1X_3 \leq 400$
   (d) $2X_1 + 3X_2 + 1X_3 \geq 400$

4. The most types of lamps that would be included in the optimal solution would be

   (a) 2   (b) 1   (c) 3   (d) 0

5. **CPA Exam**   In a system of equations for a linear programming model, what can be done to equalize an inequality such as $3X + 2Y \leq 15$?

   (a) Nothing.
   (b) Add a slack variable.
   (c) Add a tableau.
   (d) Multiply each element by -1.

Use the following information to answer Problems 6 and 7.

CPA Exam   The Golden Hawk Manufacturing Company wants to maximize the profits on products A, B, and C. The contribution margin for each product follows:

Product	Contribution Margin
A	$2
B	$5
C	$4

The production requirements and departmental capacities, by departments, are as follows:

Department	Production Requirements by Product (Hours)			Department	Departmental Capacity (Total Hours)
	A	B	C	Assembling	30,000
Assembling	2	3	2	Painting	38,000
Painting	1	2	2	Finishing	28,000
Finishing	2	3	1		

6. What is the profit maximization formula for the Golden Hawk Company?

(a) $\$2A + \$5B + \$4C = X$ (where X = Profit)
(b) $5A + 8B + 5C \leq 96,000$
(c) $\$2A + \$5B + \$4C \leq X$ (where X = Profit)
(d) $\$2A + \$5B + \$4C = 96,000$

7. What is the constraint for the painting department of the Golden Hawk Company?

(a) $1A + 2B + 2C \geq 38,000$
(b) $\$2A + \$5B + \$4C \geq 38,000$
(c) $1A + 2B + 2C \leq 38,000$
(d) $2A + 3B + 2C \leq 30,000$

8. CPA Exam   Watch Corporation manufactures products A, B, and C. The daily production requirements are shown at the right.

What is Watch's objective function in determining the daily production of each unit?

(a) $A + B + C \leq \$60$
(b) $\$3A + \$6B + \$7C = \$60$
(c) $A + B + C \leq$ Profit
(d) $\$10A + \$20B + \$30C =$ Profit

Product	Profit per Unit	Hours Required per Unit per Department		
		Machining	Plating	Polishing
A	$10	1	1	1
B	$20	3	1	2
C	$30	2	3	2
Total Hours per Day per Department		16	12	6

CPA Exam   Problems 9–11 are based on a company that uses a linear programming model to schedule the production of three products. The per-unit selling prices, variable costs, and labor time required to produce these products are presented below. Total labor time available is 200 hours.

Product	Selling Price	Variable Cost	Labor (Hours)
A	$4.00	$1.00	2
B	$2.00	$ .50	2
C	$3.50	$1.50	3

9. The objective function to maximize the company's gross profit (Z) is

(a) $4A + 2B + 3.5C = Z$
(b) $2A + 2B + 3C = Z$
(c) $5A + 2.5B + 5C = Z$
(d) $3A + 1.5B + 2C = Z$
(e) $A + B + C = Z$

10. The constraint of labor time available is represented by

(a) $2A + 2B + 3C \leq 200$
(b) $2A + 2B + 3C \geq 200$
(c) $A + B + C \geq 200$

(d) $4A + 2B + 3.5C = 200$
(e) $A/2 + B/2 + C/3 = 200$

11. A linear programming model produces an optimal solution by

(a) Ignoring resource constraints.
(b) Minimizing production costs.
(c) Minimizing both variable production costs and labor costs.
(d) Maximizing the objective function subject to resource constraints.
(e) Finding the point at which various resource constraints intersect.

Chapter 5

Finance

5.1 Interest

5.2 Compound Interest

5.3 Annuities; Sinking Funds

5.4 Present Value of an Annuity; Amortization

5.5 Applications: Leasing; Capital Expenditure; Bonds

Chapter Review

In this chapter we discuss several types of problems from the field of finance, such as compound interest, annuities, sinking funds, and mortgage payments.

5.1 INTEREST

Interest is money paid for the use of money. The total amount of money borrowed, whether by an individual from a bank in the form of a loan or by a bank from an individual in the form of a savings account, is called the **principal.**

The **rate of interest** is the amount charged for the use of the principal for a given length of time, usually on a yearly, or *per annum,* basis. Rates of interest are usually expressed as percents: 10% per annum, 14% per annum, $7\frac{1}{2}$% per annum, and so on.

The word **percent** means "per hundred." The familiar symbol % thus means to divide by one hundred. For example,

$$1\% = \frac{1}{100} = 0.01 \qquad 12\% = \frac{12}{100} = 0.12 \qquad 0.3\% = \frac{0.3}{100} = \frac{3}{1000} = 0.003$$

By reversing these ideas, we can write decimals as percents.

$$0.35 = \frac{35}{100} = 35\% \qquad 1.25 = \frac{125}{100} = 125\% \qquad 0.005 = \frac{5}{1000} = \frac{0.5}{100}$$
$$= 0.5\% = \tfrac{1}{2}\%$$

 Now Work Problems 1

and 9

EXAMPLE 1 Working with Percents

(a) Find 12% of 80.
(b) What percent of 40 is 18?
(c) 15% of what number is 8?

SOLUTION

(a) The English word "of" translates to "multiply" when percents are involved. Thus

$$12\% \text{ of } 80 = 12\% \text{ times } 80 = (0.12) \cdot (80) = 9.6$$

(b) Let x represent the unknown percent. Then

$$x\% \text{ of } 40 = 18$$

$$\frac{x}{100} \cdot 40 = 18 \qquad x\% = \frac{x}{100}$$

$$40x = 1800 \qquad \text{Multiply both sides by 100}$$

$$x = \frac{1800}{40} \qquad \text{Divide both sides by 40}$$

$$x = 45$$

Thus 45% of 40 is 18.

(c) Let x represent the number. Then

$$15\% \text{ of } x = 8$$

$$0.15x = 8 \qquad 15\% = 0.15$$

$$x = \frac{8}{0.15} \qquad \text{Divide both sides by 0.15}$$

$$x = 53.33 \qquad \text{Use a calculator}$$

 Now Work Problems   Thus 15% of 53.33 is 8.

17, 23, and 27

EXAMPLE 2 Computing a State Income Tax

A resident of Illinois has base income, after adjustment for deductions, of $18,000. The state income tax on this base income is 3%. What tax is due?

SOLUTION   We must find 3% of $18,000. We convert 3% to its decimal equivalent and then multiply by $18,000.

$$3\% \text{ of } \$18,000 = (0.03)(\$18,000) = \$540$$

The state income tax is $540.00.

Simple Interest

The easiest type of interest to deal with is called *simple interest.*

Simple Interest

Simple interest is interest computed on the principal for the entire period it is borrowed.

Simple Interest Formula

If a principal P is borrowed at a simple interest rate of $r\%$ per annum (where r is expressed as a decimal) for a period of t years, the interest charge I is

$$I = Prt$$

The **amount** A owed at the end of t years is the sum of the principal P borrowed and the interest I charged: That is,

$$A = P + I = P + Prt = P(1 + rt) \tag{1}$$

EXAMPLE 3   Computing Interest and the Amount Due

A loan of \$250 is made for 9 months at a simple interest rate of 10% per annum. What is the interest charge? What amount is due after 9 months?

SOLUTION   The actual period the money is borrowed for is 9 months, which is $\frac{9}{12} = \frac{3}{4}$ of a year. The interest charge is the product of the amount borrowed, \$250, the annual rate of interest, 0.10, and the length of time in years, $\frac{3}{4}$. Thus

$$I = Prt$$

$$\text{Interest charge} = (\$250)(0.10)(\tfrac{3}{4}) = \$18.75$$

The amount A due after 9 months is

$$A = P + I = 250 + 18.75 = \$268.75$$

 Now Work Problem 31

EXAMPLE 4   Computing the Rate of Interest

A person borrows \$1000 for a period of 6 months. What simple interest rate is being charged if the amount A that must be repaid after 6 months is \$1045?

SOLUTION   The principal P is \$1000, the period is $\frac{1}{2}$ year (6 months), and the amount A owed after 6 months is \$1045. We substitute the known values of A, P, and t in Equation (1), and then solve for r.

$$A = P + Prt$$
$$1045 = 1000 + 1000r(\tfrac{1}{2}) \quad \text{$A = 1045$; $P = 1000$; $t = \tfrac{1}{2}$}$$
$$45 = 500r \qquad \text{Subtract 1000 from each side}$$
$$\text{and simplify}$$
$$r = \frac{45}{500} = 0.09 \qquad \text{Solve for r.}$$

The per annum rate of interest is 9%.

 **Now Work Problem 37**

EXAMPLE 5   Computing the Amount Due

A bank borrows $1,000,000 for 1 month at a simple interest rate of 9% per annum. How much must the bank pay back at the end of 1 month?

SOLUTION   The principal P is $1,000,000, the period t is $\tfrac{1}{12}$ year, and the rate r is 0.09.

$$A = P(1 + rt)$$
$$A = 1,000,000\,[1 + 0.09(\tfrac{1}{12})]$$
$$= 1,000,000(1.0075) \qquad \text{Use a calculator}$$
$$= \$1,007,500$$

At the end of 1 month, the bank must pay back $1,007,500.

Discounted Loans

If a lender deducts the interest from the amount of the loan at the time the loan is made, the loan is said to be **discounted.** The interest deducted from the amount of the loan is the **discount.** The amount the borrower receives is called the **proceeds.**

Discounted Loans

Let r be the per annum rate of interest, t the time in years, and A the amount of the loan. Then the proceeds P is given by

$$P = A - Art = A(1 - rt) \qquad (2)$$

where Art is the discount, the interest deducted from the amount of the loan.

EXAMPLE 6   Computing the Proceeds of a Discounted Loan

A borrower signs a note for a discounted loan and agrees to pay $1000 in 9 months at a 10% rate of interest. How much does this borrower receive?

SOLUTION   The amount of the loan is $A = 1000$. The rate of interest is $r = 10\% = 0.10$. The time is $t = 9$ months $= \frac{9}{12}$ year. The discount is

$$Art = \$1000(0.10)(\tfrac{9}{12}) = \$75$$

The discount is deducted from the loan amount of \$1000, so that the proceeds, the amount the borrower receives, is

$$P = A - Art = 1000 - 75 = \$925$$

 Now Work Problem 43

EXAMPLE 7   Computing the Simple Interest on a Discounted Loan

What simple rate of interest is the borrower in Example 6 paying on the \$925 that was borrowed for 9 months?

SOLUTION   The principal P is \$925, t is $\frac{9}{12} = \frac{3}{4}$ of a year, and the amount A is \$1000. If r is the simple rate of interest, then, from (1),

$$A = P + Prt$$
$$1000 = 925 + 925r(\tfrac{3}{4})$$
$$75 = 693.75r$$
$$r = \frac{75}{693.75} = 0.108108$$

The simple rate of interest is 10.81%.

EXAMPLE 8   Finding the Amount of a Discounted Loan

You wish to borrow \$10,000 for 3 months. If the person you are borrowing from offers a discounted loan at 8%, how much must you repay at the end of 3 months?

SOLUTION   The principal P you borrow (the proceeds) is \$10,000, r is 0.08, and the time t is $\frac{3}{12} = \frac{1}{4}$ year. From Equation (2), the amount A you repay obeys

$$P = A(1 - rt)$$
$$10,000 = A[1 - 0.08\,(\tfrac{1}{4})]$$
$$10,000 = 0.98A$$
$$A = \frac{10,000}{0.98} = \$10,204.08$$

You will repay \$10,204.08 for this loan.

Treasury Bills

Treasury bills (T-bills) are short-term securities issued by the Federal Reserve. The bills do not specify a rate of interest. They are sold at public auction with financial institutions making competitive bids. For example, a financial institution may bid $982,400 for a 3-month $1 million treasury bill. At the end of 3 months the financial institution receives $1 million, which includes the interest earned and the cost of the T-bill. This bidding process is an example of a discounted loan.

EXAMPLE 9   Bidding on Treasury Bills

How much should a bank bid to earn 7.65% simple interest on a 6-month $1 million treasury bill?

SOLUTION   The maturity value, the amount to be repaid to the bank by the government, is $A = \$1,000,000$. The rate of interest is $r = 7.65\% = 0.0765$. The time is $t = 6$ months $= \frac{1}{2}$ year. The proceeds P to the government are

$$P = A(1 - rt) = \$1,000,000[1 - 0.0765\left(\tfrac{1}{2}\right)]$$
$$= 1,000,000(0.96175)$$
$$= 961,750$$

The bank should bid $961,750.

EXERCISE 5.1   Answers to odd-numbered problems begin on page AN-22.

In Problems 1–8 write each decimal as a percent.

1. 0.60   **2.** 0.40   **3.** 1.1   **4.** 1.2

5. 0.06   **6.** 0.07   **7.** 0.0025   **8.** 0.0015

In Problems 9–16 write each percent as a decimal.

9. 25%   **10.** 15%   **11.** 100%   **12.** 300%

13. 6.5%   **14.** 4.3%   **15.** 73.4%   **16.** 92%

In Problems 17–30 calculate the indicated quantity.

17. 15% of 1000   **18.** 20% of 500   **19.** 18% of 100

20. 10% of 50   **21.** 210% of 50   **22.** 135% of 1000

23. What percent of 80 is 4?   **24.** What percent of 60 is 5?

25. What percent of 5 is 8?   **26.** What percent of 25 is 45?

27. 8% of what number is 20?   **28.** 12% of what number is 25?

29. 15% of what number is 50?   **30.** 18% of what number is 40?

In Problems 31–36 find the interest due on each loan.

31. $1000 is borrowed for 3 months at 4% simple interest.   **32.** $100 is borrowed for 6 months at 8% simple interest.

33. $500 is borrowed for 9 months at 12% simple interest.   **34.** $800 is borrowed for 8 months at 5% simple interest.

35. $1000 is borrowed for 18 months at 10% simple interest.

36. $100 is borrowed for 24 months at 12% simple interest.

In Problems 37–42 find the simple interest rate for each loan.

37. $1000 is borrowed; the amount owed after 6 months is $1050.

38. $500 is borrowed; the amount owed after 8 months is $600.

39. $300 is borrowed; the amount owed after 12 months is $400.

40. $600 is borrowed; the amount owed after 9 months is $660.

41. $900 is borrowed; the amount owed after 10 months is $1000.

42. $800 is borrowed; the amount owed after 3 months is $900.

In Problems 43–46 find the proceeds for each discounted loan.

43. $1200 repaid in 6 months at 10%.

44. $500 repaid in 8 months at 9%.

45. $2000 repaid in 24 months at 8%.

46. $1500 repaid in 18 months at 10%.

In Problems 47–50 find the amount you must repay for each discounted loan.

47. You borrow $1200 for 6 months at 10%.

49. You borrow $2000 for 24 months at 8%.

48. You borrow $500 for 8 months at 9%.

50. You borrow $1500 for 18 months at 10%.

51. Buying a Stereo Madalyn wants to buy a $500 stereo set in 9 months. How much should she invest at 8% simple interest to have the money then?

52. Interest on a Loan Mike borrows $10,000 for a period of 3 years at a simple interest rate of 10%. Determine the interest due on the loan.

53. Term of a Loan Tami borrowed $600 at 8% simple interest. The amount of interest paid was $156. What was the length of the loan?

54. Equipment Loan The owner of a restaurant would like to borrow $12,000 from a bank to buy some equipment. The bank will give the owner a discounted loan at an 11% rate of interest for 9 months. What maturity value should be used so that the owner will receive $12,000?

55. Comparing Loans You need to borrow $1000 right now, but can repay the loan in 6 months. Since you want to pay as little interest as possible, which type of loan should you take: a discounted loan at 9% per annum or a simple interest loan at 10% per annum?

56. Comparing Loans You need to borrow $5000 right now, but can repay the loan in 9 months. Since you want to pay as little interest as possible, which type of loan should you take: a discounted loan at 8% per annum or a simple interest loan at 8.5% per annum?

57. Comparing Loans You need to borrow $4000 right now, but can repay the loan in 1 year. Since you want to pay as little interest as possible, which type of loan should you take: a discounted loan at 6% per annum or a simple interest loan at 6.3% per annum?

58. Comparing Loans You need to borrow $5000 right now, but can repay the loan in 18 months. Since you want to pay as little interest as possible, which type of loan should you take: a discounted loan at 5.3% per annum or a simple interest loan at 5.6% per annum?

59. Comparing Loans Ruth would like to borrow $2000 for one year from a bank. She is given a choice of a simple interest loan at 12.3% or a discounted loan at 12.1%. What should she do?

60. Refer to Problem 59. If Ruth only needs to borrow the $2000 for 3 months, what should she do?

61. Bidding on Treasury Bills A bank wants to earn 8.5% simple interest on a 3-month $1 million treasury bill. How much should they bid?

62. Interest on Treasury Bills A bank paid $979,000 for a 3-month $1 million treasury bill. What simple interest was earned?

63. Bidding on Treasury Bills How much should a bank bid on a 6-month $3 million treasury bill to earn 7.715% simple interest?

5.2 COMPOUND INTEREST

Compound Interest Formula

In working with problems involving interest we use the term **payment period** as follows:

Annually	Once per year
Semiannually	Twice per year
Quarterly	4 times per year
Monthly	12 times per year
Daily	365 times per year*

If the interest due at the end of each payment period is added to the principal, so that the interest computed for the next payment period is based on this new amount of the old principal plus interest, then the interest is said to have been **compounded.** That is, **compound interest** is interest paid on the initial principal and previously earned interest.

EXAMPLE 1   Computing Compound Interest

A bank pays 6% per annum compounded quarterly. If $200 is placed in a savings account and the quarterly interest is left in the account, how much money is in the account after 1 year?

SOLUTION   At the end of the first quarter (3 months) the interest earned is

$$I = Prt = (\$200)(0.06)(\tfrac{1}{4}) = \$3.00$$

The new principal is $P + I = \$203$. The interest on this principal at the end of the second quarter is

$$I = (\$203)(0.06)(\tfrac{1}{4}) = \$3.05$$

The interest at the end of the third quarter on the principal of $203 + \$3.05 = \$206.05$ is

$$I = (\$206.05)(0.06)(\tfrac{1}{4}) = \$3.09$$

The interest at the end of the fourth quarter on the principal of $206.05 + \$3.09 = \$209.14$ is

$$I = (\$209.14)(0.06)(\tfrac{1}{4}) = \$3.14$$

Thus after 1 year the total in the savings account is $\$209.14 + \$3.14 = \$212.28$.
These results are shown in Figure 1.

Figure 1

* Some banks use 360 times per year.

Let's develop a formula for computing the amount when interest is compounded. Suppose r is the per annum rate of interest compounded each payment period. Then the rate of interest per payment period is

$$i = \frac{\text{Per annum rate of interest}}{\text{Number of payment periods}}$$

For example, if the annual rate of interest is 10% and the compounding is monthly, then there are 12 payment periods per year and

$$i = \frac{0.10}{12} = 0.00833$$

If 18% is the annual rate compounded daily (365 payment periods), then

$$i = \frac{0.18}{365} = 0.000493$$

If P is the principal and i is the interest rate per payment period, then the amount A_1 at the end of the first payment period is

$$A_1 = P + Pi = P(1 + i)$$

At the end of the second payment period, and subsequent ones, the amounts are

$$A_2 = A_1 + A_1 i = A_1(1 + i) = P(1 + i)(1 + i) = P(1 + i)^2$$
$$A_3 = A_2 + A_2 i = A_2(1 + i) = P(1 + i)^2(1 + i) = P(1 + i)^3$$
$$\cdot$$
$$\cdot$$
$$\cdot$$
$$A_n = A_{n-1} + A_{n-1} i = A_{n-1}(1 + i) = P(1 + i)^{n-1}(1 + i) = P(1 + i)^n$$

Compound Interest Formula

The amount A_n accrued on a principal P after n payment periods at i interest per payment period is

$$A_n = P(1 + i)^n$$

In working with the compound interest formula, we use a calculator with a $\boxed{y^x}$ key. To use this key, enter the value of y, press $\boxed{y^x}$, enter the value of x, and press $\boxed{=}$.

EXAMPLE 2   Working with the Compound Interest Formula

If $1000 is invested at an annual rate of interest of 10%, what is the amount after 5 years if the compounding takes place

(a) Annually?   (b) Monthly?   (c) Daily?

How much interest is earned in each case?

SOLUTION   The principal is $P = \$1000$.

(a) We use a calculator. For annual compounding, $i = 0.10$ and $n = 5$. The amount A is

$$A = P(1 + i)^n = (\$1000)(1 + 0.10)^5 = (\$1000)(1.61051) = \$1610.51$$

The interest earned is

$$A - P = \$1610.51 - \$1000.00 = \$610.51$$

(b) For monthly compounding, there are $5 \cdot 12 = 60$ payment periods over 5 years. The interest rate per payment period is $i = 0.10/12$. Using a calculator, the amount A is

$$A = P(1 + i)^n = (\$1000)\left(1 + \frac{0.10}{12}\right)^{60} = (\$1000)(1.64531) = \$1645.31$$

The interest earned is

$$A - P = \$1645.31 - \$1000.00 = \$645.31$$

(c) For daily compounding, there are $5 \cdot 365 = 1825$ payment periods over 5 years. The interest rate per payment period is $i = 0.10/365$. Using a calculator, the amount A is

$$A = P(1 + i)^n = (\$1000)\left(1 + \frac{0.10}{365}\right)^{1825} = (\$1000)(1.64861) = \$1648.61$$

The interest earned is

$$A - P = \$1648.61 - \$1000.00 = \$648.61$$

The results of Example 2 are summarized in Table 1.

Table 1

Per Annum Rate	Compounding Method	Interest Rate per Payment Period	Initial Principal	Amount after 5 Years	Interest Earned
10%	Annual	0.10	$1000	$1610.51	$610.51
10%	Monthly	0.00833	$1000	$1645.31	$645.31
10%	Daily	0.000274	$1000	$1648.61	$648.61

 Now Work Problem 1

EXAMPLE 3   **Comparing Compound Rates of Interest with Simple Interest**

(a) If $100 is invested at an annual rate of interest of 10% compounded monthly, what is the interest earned after 1 year?

(b) What simple rate of interest is required to obtain this amount of interest?

SOLUTION   We begin with a principal of $100 and proceed to compute the amount after 1 year at 10% compounded monthly. For monthly compounding at 10%, there are 12 payment periods, and the interest rate per period is $i = 0.10/12$. Using a calculator, the amount A is

$$A = P(1 + i)^n = \$100\left(1 + \frac{0.10}{12}\right)^{12} = \$110.47$$

The interest earned is $\$110.47 - \$100 = \$10.47$.

(b) The simple interest rate required to earn interest of $10.47 on a principal of $100 after one year is

$$I = Prt$$
$$10.47 = (100)r(1)$$
$$r = .1047$$

Thus, a simple interest rate of 10.47% is required to obtain interest equal to that obtained using 10% compounded monthly.

Let's look at the effect of various compounding periods on a principal of $100 after 1 year using a rate of interest of 8% per annum.

Annual compounding:	$A = \$100(1 + 0.08)^1$	$= \$108.00$
Semiannual compounding:	$A = \$100(1 + 0.04)^2$	$= \$108.16$
Quarterly compounding(s)	$A = \$100\left(1 + \frac{.08}{4}\right)^4$	$= \$108.24$
Monthly compounding:	$A = \$100\left(1 + \frac{0.08}{12}\right)^{12}$	$= \$108.30$
Daily compounding:	$A = \$100\left(1 + \frac{0.08}{365}\right)^{365}$	$= \$108.33$

With semiannual compounding the amount $108.16 could have been obtained with a simple interest of 8.16%. We describe this by saying that the *effective rate of interest* of 8% compounded semiannually is 8.16%.

In general, the **effective rate of interest** is the equivalent annual simple rate of interest that yields the same amount as compounding does after one year.

Table 2 summarizes the calculations given above for 8%.

Table 2

Rate/Compounding Period	Effective Rate of Interest
8% Compounded Semiannually	8.16%
8% Compounded Quarterly	8.24%
8% Compounded Monthly	8.3%
8% Compounded Daily	8.33%

Now Work Problem 17

EXAMPLE 4   Comparing Certificates of Deposit

Three local banks offer the following 1-year Certificates of Deposit (CDs):

(a) Simple interest of 5.2% per annum
(b) 5% per annum compounded monthly
(c) $4\frac{3}{4}$% per annum compounded daily

Which CD results in the most interest?

SOLUTION   To compare the three CDs, we compute the amount $1000 (or any other amount could also be used) will grow to in each case.

(a) At simple interest of 5.2%, $1000 will grow to

$$A = P + Prt = \$1000 + \$1000(0.052)(1) = \$1052.00$$

(b) There are 12 payment periods and the rate of interest per payment period is $i = 0.05/12$. The amount A that $1000 will grow to is

$$A = P(1 + i)^n = \$1000\left(1 + \frac{0.05}{12}\right)^{12} = \$1051.16$$

(c) There are 365 payment periods and the rate of interest per payment period is $i = 0.0475/365$. The amount A that $1000 will grow to is

$$A = P(1 + i)^n = \$1000\left(1 + \frac{0.0475}{365}\right)^{365} = \$1048.64$$

The CD offering 5.2% simple interest results in the most interest.

 **Now Work Problem 31**

Present Value

The compound interest formula states that a principal P earning an interest rate per payment period i will, after n payment periods, be worth the amount A, where

$$A = P(1 + i)^n$$

If we solve for P, we obtain

$$P = \frac{A}{(1 + i)^n} = A(1 + i)^{-n}$$

In this formula P is called the **present value** of the amount A due at the end of n interest periods at i interest per payment period. In other words, P is the amount that must be invested for n interest periods at i interest per payment period in order to accumulate the amount A.

The compound interest formula and the present value formula can be used to solve many different kinds of problems. The examples below illustrate some of these applications.

EXAMPLE 5 Computing the Present Value of $10,000

How much money should be invested at 8% per annum so that after 2 years the amount will be $10,000 when the interest is compounded

(a) Annually?   (b) Monthly?   (c) Daily?

SOLUTION In this problem we want to find the principal P when we know that the amount A after 2 years is going to be $10,000. That is, we want to find the present value of $10,000.

(a) Since compounding is once per year for 2 years, $n = 2$. Using a calculator with a $\boxed{y^x}$ key, the present value P of $10,000 is

$$P = A(1 + i)^{-n} = 10{,}000(1 + 0.08)^{-2} = 10{,}000(0.8573388) = \$8573.39$$

(b) Since compounding is 12 times per year for 2 years, $n = 24$. The present value P of $10,000 is

$$P = A(1 + i)^{-n} = 10{,}000\left(1 + \frac{0.08}{12}\right)^{-24} = 10{,}000(0.852596) = \$8525.96$$

(c) Since compounding is 365 times per year for 2 years, $n = 730$. The present value P of $10,000 is

$$P = A(1 + i)^{-n} = 10{,}000\left(1 + \frac{0.08}{365}\right)^{-730} = 10{,}000(0.8521587) = \$8521.59$$

 **Now Work Problem 7**

EXAMPLE 6 Finding the Rate of Interest to Double an Investment

What annual rate of interest compounded annually should you seek if you want to double your investment in 5 years?

SOLUTION If P is the principal and we want P to double, the amount A will be $2P$. We use the compound interest formula with $n = 5$ to find i:

$$A = P(1 + i)^n$$
$$2P = P(1 + i)^5 \qquad\qquad A = 2P, n = 5$$
$$2 = (1 + i)^5$$
$$\sqrt[5]{2} = 1 + i \qquad\qquad \text{Take the 5th root of each side}$$
$$i = \sqrt[5]{2} - 1 = 1.148698 - 1 = 0.148698 \qquad \text{Solve for } i$$

$$\sqrt[5]{2} = 2^{1/5} = 2^{0.2}$$
Use the $\boxed{y^x}$ key on your calculator

The annual rate of interest needed to double the principal in 5 years is 14.87%.

EXAMPLE 7*   **Finding the Time Required to Double/Triple an Investment**

(a) How long will it take for an investment to double in value if it earns 5% compounded monthly?

(b) How long will it take to triple at this rate?

SOLUTION

(a) If P is the initial investment and we want P to double, the amount A will be $2P$. We use the compound interest formula with $i = 0.05/12$. Then

$$A = P(1 + i)^n$$

$$2P = P\left(1 + \frac{0.05}{12}\right)^n \qquad A = 2P, i = \frac{0.05}{12}$$

$$2 = (1.0041667)^n \qquad \text{Apply the definition of a logarithm}$$

$$n = \log_{1.0041667} 2 = \frac{\log_{10} 2}{\log_{10} 1.0041667} = 166.7 \text{ months}$$

$$\underset{\substack{\uparrow \\ \text{Change-of-} \\ \text{base formula}}}{} \qquad \underset{\substack{\uparrow \\ \text{Payment period} \\ \text{measured in months}}}{}$$

It will take about 13 years 11 months to double the investment.

(b) To triple, we have

$$A = P(1 + i)^n$$

$$3P = P\left(1 + \frac{0.05}{12}\right)^n \qquad A = 3P, i = \frac{0.05}{12}$$

$$3 = (1.0041667)^n \qquad \text{Apply the definition of a logarithm}$$

$$n = \log_{1.0041667} 3 = \frac{\log_{10} 3}{\log_{10} 1.0041667} = 264.2 \text{ months}$$

It will take about 22 years to triple the investment.

■

EXERCISE 5.2   Answers to odd-numbered problems begin on page AN-22.

In Problems 1–6 find the amount.

1. $1000 is invested at 8% compounded monthly for 36 months.

2. $100 is invested at 6% compounded monthly for 20 months.

3. $500 is invested at 9% compounded annually for 3 years.

4. $200 is invested at 10% compounded annually for 10 years.

5. $800 is invested at 12% compounded daily for 200 days.

6. $400 is invested at 7% compounded daily for 180 days.

In Problems 7–10 find the principal needed now to get each amount.

7. To get $100 in 6 months at 10% compounded monthly

8. To get $500 in 1 year at 12% compounded annually

9. To get $500 in 1 year at 9% compounded daily

10. To get $800 in 2 years at 5% compounded monthly

---

* Requires a knowledge of logarithms, especially the Change-of-Base Formula. See Appendix A for a review.

11. If $1000 is invested at 9% compounded

    (a) Annually   (b) Semiannually

    (c) Quarterly   (d) Monthly

what is the amount after 3 years? How much interest is earned?

12. If $2000 is invested at 12% compounded

    (a) Annually   (b) Semiannually

    (c) Quarterly   (d) Monthly

what is the amount after 5 years? How much interest is earned?

13. If $1000 is invested at 12% compounded quarterly, what is the amount after

    (a) 2 years?   (b) 3 years?   (c) 4 years?

14. If $2000 is invested at 8% compounded quarterly, what is the amount after

    (a) 2 years?   (b) 3 years?   (c) 4 years?

15. If a bank pays 6% compounded semiannually, how much should be deposited now to have $5000

    (a) 4 years later?   (b) 8 years later?

16. If a bank pays 8% compounded quarterly, how much should be deposited now to have $10,000

    (a) 5 years later?   (b) 10 years later?

17. Find the effective rate of interest for

    (a) 8% compounded semiannually

    (b) 12% compounded monthly

18. Find the effective rate of interest for

    (a) 6% compounded monthly

    (b) 14% compounded semiannually

19. What annual rate of interest compounded annually is required to double an investment in 3 years?

20. What annual rate of interest compounded annually is required to double an investment in 10 years?

21. Approximately how long will it take to triple an investment at 10% compounded annually?

22. Approximately how long will it take to triple an investment at 9% compounded annually?

23. Mr. Nielsen wants to borrow $1000 for 2 years. He is given the choice of (a) a simple interest loan of 12% or (b) a loan at 10% compounded monthly. Which loan results in less interest due?

24. Rework Problem 23 if the simple interest loan is 15% and the other loan is at 14% compounded daily.

25. What principal is needed now to get $1000 1 year from today and $1000 2 years from today at 9% compounded annually?

26. Repeat Problem 25 using 9% compounded daily.

27. Find the effective rate of interest for $5\frac{1}{4}$% compounded quarterly.

28. Repeat Problem 27 using 6% compounded quarterly.

29. What interest rate compounded quarterly will give an effective interest rate of 7%?

30. Repeat Problem 29 using 10%.

In Problems 31–34 which of the two rates would yield the larger amount in 1 year?
Hint: Start with a principal of $10,000 in each instance.

31. 6% compounded quarterly or $6\frac{1}{4}$% compounded annually

32. 9% compounded quarterly or $9\frac{1}{4}$% compounded annually

33. 9% compounded monthly or 8.8% compounded daily

34. 8% compounded semiannually or 7.9% compounded daily

35. Future Price of a Home If the price of homes rises an average of 5% per year for the next 4 years, what will be the selling price of a home that is selling for $90,000 today 4 years from today? Express your answer rounded to the nearest hundred dollars.

36. Amount Due on a Charge Card A department store charges 1.25% per month on the unpaid balance for customers with charge accounts (interest is compounded monthly). A customer charges $200 and does not pay her bill for 6 months. What is the bill at that time?

37. Amount Due on a Charge Card A major credit card company has a finance charge of 1.5% per month on the outstanding indebtedness. Caryl charged $600 and did not pay her bill for 6 months. What is the bill at that time?

38. Buying a Car Laura wishes to have $8000 available to buy a car in 3 years. How much should she invest in a savings account now so that she will have enough if the bank pays 8% interest compounded quarterly?

39. Down Payment on a House Tami and Todd will need $40,000 for a down payment on a house in 4 years. How much should they invest in a savings account now so that they will be able to do this? The bank pays 8% compounded quarterly.

40. **Saving for College**   A newborn child receives a $3000 gift toward a college education. How much will the $3000 be worth in 17 years if it is invested at 10% compounded quarterly?

41. **Gifting**   A child's grandparents have opened a $6000 savings account for the child on the day of her birth. The account pays 8% compounded semiannually. The child will be allowed to withdraw the money when she reaches the age of 25. What will the account be worth at that time?

42. **Future Price of a House**   What will a $90,000 house cost 5 years from now if the inflation rate over that period averages 5% compounded annually? Express your answer rounded to the nearest hundred dollars.

43. **Population Increases**   A town increased in population 2% per year for 8 years. If the population was 17,000 at the beginning, what is the size of the population at the end of 8 years?

44. **Investing Earnings**   Omega Company can invest its earnings at (a) 7% per year compounded monthly, at (b) 7.15% per year compounded semiannually, or at (c) 7.20% per year compounded annually. Which rate should Omega choose?

45. **Deciding on a Stock Purchase**   Jack is considering buying 1000 shares of a stock that sells at $15 per share. The stock pays no dividends. From the history of the stock, Jack is certain that he will be able to sell it 4 years from now at $20 per share. Jack's goal is not to make any investment unless it returns at least 7% compounded quarterly. Should Jack buy the stock?

46. Repeat Problem 45 if Jack requires a return of at least 14% compounded quarterly.

47. **Value of an IRA**   An Individual Retirement Account (IRA) has $2000 in it, and the owner decides not to add any more money to the account other than the interest earned at 9% compounded quarterly. How much will be in the account 25 years from the day the account was opened?

48. **Return on Investment**   If Jack sold a stock for $35,281.50 (net) that cost him $22,485.75 three years ago, what annual compound rate of return did Jack make on his investment?

For Problems 49–52 zero coupon bonds are used. A **zero coupon bond** *is a bond that is sold now at a discount and will pay its face value at some time in the future when it matures; no interest payments are made.*

49. **Saving for College**   Tami's grandparents are considering buying a $40,000 face value zero coupon bond at birth so that she will have enough money for her college education 17 years later. If money is worth 8% compounded annually, what should they pay for the bond?

50. **Price of a Bond**   How much should a $10,000 face value zero coupon bond, maturing in 10 years, be sold for now if its rate of return is to be 8% compounded annually?

51. **Rate of Return of a Bond**   If you pay $12,485.52 for a $25,000 face value zero coupon bond that matures in 8 years, what is your annual compound rate of return?

52. **Effective Rates of Interest**   A bank advertises that it pays interest on saving accounts at the rate of 6.25% compounded daily.

   (a) Find the effective rate if the bank uses 360 days in determining the daily rate.
   (b) What if 365 days is used?

Problems 53 and 54 require logarithms.

53. **Length of Investment**   How many years will it take for an initial investment of $10,000 to grow to $25,000? Assume a rate of interest of 6% compounded daily.

54. **Length of Investment**   How many years will it take for an initial investment of $25,000 to grow to $80,000? Assume a rate of interest of 7% compounded daily.

Use the following discussion for Problems 55–62. **Inflation** *erodes the purchasing power of money. For example, suppose there is an annual rate of inflation of 3%. Then $1000 worth of purchasing power now will be worth only $970 in one year. In general, for an annual rate of inflation of r%, the amount A that $P will purchase after n years is*

$$A = P(1 - r)^n$$

where r is a decimal.

55. Suppose the inflation rate is 3%. After 2 years, how much will $1000 purchase?

56. Suppose the inflation rate is 4%. After 2 years, how much will $1000 purchase?

57. Suppose the inflation rate is 3%. After 5 years, how much will $1000 purchase?

58. Suppose the inflation rate is 4%. After 5 years, how much will $1000 purchase?

Problems 59–62 require logarithms.

59. Suppose the inflation rate is 3%. How long is it until purchasing power is halved?

60. Suppose the inflation rate is 4%. How long is it until purchasing power is halved?

61. Suppose the inflation rate is 6%. How long is it until purchasing power is halved?

62. Suppose the inflation rate is 9%. How long is it until purchasing power is halved?

Technology Exercises

A business or financial calculator is preprogrammed to use the compound interest formula. If you have one, consult your manual and use your calculator to check your answers to Problems 9, 11(c), 13, 19, 20, 53, and 54. A graphing calculator can be programmed to do what a financial calculator is preprogrammed to do.

1. Write a program that will calculate the amount after n years if a principal P is invested at $r\%$ per annum compounded quarterly. Use it to verify your answers to Problems 11(c) and 13 above.

2. Write a program that will calculate the principal needed now to get the amount A in n years at $r\%$ per annum compounded daily. Use it to verify your answer to Problem 9 above.

3. Write a program that will calculate the rate of interest required to double an investment in n years. Use it to verify your answers to Problems 19 and 20 (previous page).

4. Write a program that will calculate the number of years required for an initial investment of x dollars to grow to y dollars at $r\%$ per annum compounded daily. Use it to verify your answer to Problems 53 and 54 (previous page).

5.3 ANNUITIES; SINKING FUNDS

Annuity

In the previous sections we saw how to compute the future value of an investment when a fixed amount of money is deposited in an account that pays interest compounded periodically. Often, however, people and financial institutions do not deposit money and then sit back and watch it grow. Rather, money is invested in small amounts at periodic intervals. Examples of such investments are annual life insurance premiums, monthly deposits in a bank, installment loan payments, and dollar averaging in the stock market with 401(k) or 403(b) accounts.

An **annuity** is a sequence of equal periodic deposits. When the deposits are made at the same time the interest is credited, the annuity is termed **ordinary.** We shall concern ourselves only with ordinary annuities in this book.

The payment period can be annual, semiannual, quarterly, monthly, or any other fixed length of time.

Amount of an Annuity

The **amount of an annuity** is the sum of all deposits made plus all interest accumulated.

EXAMPLE 1   Finding the Amount of an Annuity

Find the amount of an annuity after 5 deposits if each deposit is equal to $100 and is made on an annual basis at an interest rate of 10% per annum compounded annually.

SOLUTION   After 5 deposits the first $100 deposit will have accumulated interest compounded annually at 10% for 4 years. Its value A_1 after 4 years is

$$A_1 = \$100(1 + 0.10)^4 = \$100(1.4641) = \$146.41$$

The second deposit of $100, made 1 year after the first deposit, will accumulate interest compounded at 10% for 3 years. Its value A_2 after the fifth deposit is

$$A_2 = \$100(1 + 0.10)^3 = \$100(1.331) = \$133.10$$

Similarly, the third, fourth, and fifth deposits will have the values

$$A_3 = \$100(1 + 0.10)^2 = \$100(1.21) = \$121.00$$
$$A_4 = \$100(1 + 0.10)^1 = \$100(1.10) = \$110.00$$
$$A_5 = \$100$$

The amount of the annuity after 5 deposits is

$$A_1 + A_2 + A_3 + A_4 + A_5 = \$146.41 + \$133.10 + \$121.00$$
$$+ \$110.00 + \$100.00$$
$$= \$610.51$$

Figure 2

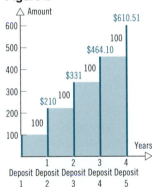

Figure 2 illustrates the growth of the annuity described in Example 1.

To develop a formula for the amount of an annuity, suppose $\$P$ is the deposit for an annuity at an interest rate of i percent per payment period (in decimal form) and with a term of n payment periods. Since deposits are made at the end of each period, the first deposit will accumulate the value A_1 compounded over $n - 1$ periods at i percent per payment period. The second deposit will accumulate the value A_2 compounded over $n - 2$ periods at i percent per payment period, and so on. Thus

$$A_1 = \$P(1 + i)^{n-1}, \quad A_2 = \$P(1 + i)^{n-2}, \quad \ldots, \quad A_n = \$P(1 + i)^0 = \$P$$

The total amount A of the annuity after n payment periods is

$$A = A_1 + \cdots + A_n = P(1 + i)^{n-1} + P(1 + i)^{n-2} + \cdots + P(1 + i) + P$$
$$= P[1 + (1 + i) + \cdots + (1 + i)^{n-1}]$$

The expression in brackets is the sum of a geometric sequence* with n terms and common ratio $1 + i$. As a result,

$$1 + (1 + i) + \cdots + (1 + i)^{n-1} = \frac{1 - (1 + i)^n}{1 - (1 + i)} = \frac{1 - (1 + i)^n}{-i} = \frac{(1 + i)^n - 1}{i}$$

---

* The sum of the first n terms of the geometric sequence with common ratio r is

$$1 + r + r^2 + \cdots + r^{n-1} = \frac{1 - r^n}{1 - r}$$

See Appendix A for a more detailed discussion.

We have established the following result.

Amount of an Annuity

If P represents the deposit in dollars made at each payment period for an annuity at i percent interest per payment period, the amount A of the annuity after n payment periods is

$$A = P\frac{(1 + i)^n - 1}{i}$$

Table I, Amount of an Annuity, in the back of the book, gives values for

$$\frac{(1 + i)^n - 1}{i}$$

for per annum rates of 8%, 10%, and 12% compounded annually and monthly. For most of our work, a calculator will be used.

EXAMPLE 2   Finding the Amount of an Annuity

Find the amount of an annuity if a deposit of $100 per year is made for 5 years at 10% compounded annually. How much interest is earned?

SOLUTION   The deposit is $P = \$100$. The number of payment periods is $n = 5$ years, and the interest per payment period is $i = 0.10$. The amount A after 5 years is

$$A = 100\left[\frac{(1 + 0.10)^5 - 1}{0.10}\right] = \$100(6.1051) = \$610.51$$

The interest accrued is the amount after 5 years less the 5 annual payments of $100 each:

$$\text{Interest accrued} = A - 500 = 610.51 - 500 = \$110.51$$

Now Work Problem 1

EXAMPLE 3   Finding the Amount of an Annuity

Mary decides to put aside $100 every month in an insurance fund that pays 8% compounded monthly. After making 8 deposits, how much money does Mary have?

SOLUTION   This is an annuity with $P = \$100$, $n = 8$, and $i = 0.08/12$. The amount A after 8 deposits is

$$A = 100\left[\frac{\left(1 + \dfrac{0.08}{12}\right)^8 - 1}{\dfrac{0.08}{12}}\right] = \$100(8.1892) = \$818.92$$

Mary has $818.92 after making 8 deposits.

EXAMPLE 4 Saving for College

To save for his son's college education, Mr. Graff decides to put $50 aside every month in a credit union account paying 10% interest compounded monthly. If he begins this savings program when his son is 3 years old, how much will he have saved by the time his son is 18 years old?

SOLUTION When his son is 18 years old, Mr. Graff will have made his 180th payment (15 years $\times$ 12 payments per year). This is an annuity with $P = \$50$, $n = 180$, and $i = 0.10/12$, The amount A saved is

$$A = 50 \left[\frac{\left(1 + \dfrac{0.10}{12}\right)^{180} - 1}{\dfrac{0.10}{12}} \right] = \$50(414.4703) = \$20{,}723.52$$

Mr. Graff will have $20,723.52 for his son's college education.

EXAMPLE 5 Funding an IRA

Joe, at age 35, decides to invest in an IRA. He will put aside $2000 per year for the next 30 years. How much will he have at age 65 if his rate of return is assumed to be 10% per annum?

SOLUTION This is an annuity with $P = \$2000$, $n = 30$, and $i = 0.10$. The amount A in Joe's IRA after 30 years is

$$A = 2000 \left[\frac{(1 + 0.10)^{30} - 1}{0.10} \right] = \$2000(164.49402) = \$328{,}988.05$$

Joe will have $328,988.05 in his IRA when he's 65.

EXAMPLE 6 Funding an IRA

If, in Example 5, Joe had begun his IRA at age 25, instead of 35, what would his IRA be worth at age 65?

SOLUTION

$$A = 2000 \left[\frac{(1 + 0.10)^{40} - 1}{0.10} \right] = \$2000(442.59256) = \$885{,}185.11$$

Joe will have $885,185.11 in his IRA when he's 65.

EXAMPLE 7*   Finding the Time it Takes to Reach a Certain Savings Goal

How long does it take to save $500,000 if you place $500 per month in an account paying 6% compounded monthly?

SOLUTION   This is an annuity in which $A = 500,000$, $P = 500$, and $i = 0.06/12$. We begin with the Amount of an Annuity formula:

$$A = P \frac{(1 + i)^n - 1}{i}$$

$$500,000 = 500 \frac{\left(1 + \dfrac{0.06}{12}\right)^n - 1}{\dfrac{0.06}{12}} \qquad A = 500,000; \; P = 500; \; i = \frac{0.06}{12}$$

$$5 = \left(1 + \frac{0.06}{12}\right)^n - 1 \qquad \text{Simplify}$$

$$6 = (1.005)^n \qquad\qquad \text{Apply the definition of a logarithm}$$

$$n = \log_{1.005} 6 = \frac{\log_{10} 6}{\log_{10} 1.005} = 359.25 \text{ months}$$

$\underset{\text{Change-of-base formula}}{\uparrow} \qquad \underset{\substack{\text{Payment period} \\ \text{in months}}}{\uparrow}$

It takes $359\tfrac{1}{4}$ months (almost 30 years) to save the $500,000.

■

Sinking Fund

A person with a debt may decide to accumulate sufficient funds to pay off the debt by agreeing to set aside enough money each month (or quarter, or year) so that when the debt becomes payable, the money set aside each month plus the interest earned will equal the debt. The fund created by such a plan is called a **sinking fund.**

We shall limit our discussion of sinking funds to those in which equal payments are made at equal time intervals at the end of each period.

EXAMPLE 8   Funding a Bond Obligation

The Board of Education received permission to issue $4,000,000 in bonds to build a new high school. The board is required to make payments every 6 months into a sinking fund paying 8% compounded semiannually. At the end of 12 years the bond obligation will be retired. What should each payment be?

SOLUTION   This is an example of a sinking fund. The payment P required twice a year to accumulate $4,000,000 in 12 years (24 payments at a rate of interest of $i =$

---

* This example requires a knowledge of logarithms, especially the Change-of-Base Formula. A review of logarithms appears in Appendix A.

0.08/2 per payment period) obeys the Amount of an Annuity formula

$$A = P\,\frac{(1 + i)^n - 1}{i}$$

$$4,000,000 = P\,\frac{\left(1 + \dfrac{0.08}{2}\right)^{24} - 1}{\dfrac{0.08}{2}} \qquad A = 4,000,000;\ n = 24;\ i = \frac{0.08}{2}$$

$$4,000,000 = P(39.082604) \qquad \text{Simplify}$$

$$P = \$102,347.33$$

The Board will need to make a payment of $102,347.33 every 6 months to redeem the bonds in 12 years.

■

EXAMPLE 9   Funding a Loan

A woman borrows $3000, which will be paid back to the lender in one payment at the end of 5 years. She agrees to pay interest monthly at an annual rate of 12%. At the same time she sets up a sinking fund in order to repay the loan at the end of 5 years. She decides to make monthly deposits into her sinking fund, which earns 8% interest compounded monthly.

(a) What is the monthly sinking fund deposit?
(b) Construct a table that shows how the sinking fund grows over time.
(c) How much does she need each month to be able to pay the interest on the loan and make the sinking fund deposit?

SOLUTION

(a) The sinking fund deposit is the value of P in the Amount of an Annuity formula

$$A = P\,\frac{(1 + i)^n - 1}{i}$$

where A equals the amount to be accumulated, namely, $A = \$3000$, $n = 60$ (5 years of monthly deposits), and $i = 0.08/12$. The sinking fund deposit is therefore

$$P = 3000 \left[\frac{\left(1 + \dfrac{0.08}{12}\right)^{60} - 1}{\dfrac{0.08}{12}}\right]^{-1} = 3000(0.0136097) = \$40.83$$

The monthly sinking fund deposit is $40.83.

(b) Table 3 (p. 246) shows the growth of the sinking fund over time. For example, the total in the account after payment number 12 is obtained by using the Amount of an Annuity formula for a monthly payment of $40.83 made for 12 months at 8% compounded monthly:

$$\text{Total} = \$40.83 \left[\frac{\left(1 + \frac{0.08}{12}\right)^{12} - 1}{\frac{0.08}{12}} \right] = \$40.83(12.449926) = \$508.33$$

The deposit for payment number 60, the final payment, is only $40.77 because a deposit of $40.83 results in a total of $3000.06.

Table 3

Payment Number	Sinking Fund Deposit, $	Cumulative Deposits	Accumulated Interest, $	Total, $
1	40.83	40.83	0	40.83
12	40.83	489.96	18.37	508.33
24	40.83	979.92	78.93	1058.85
36	40.83	1469.88	185.19	1655.07
48	40.83	1959.84	340.93	2300.77
60	40.77	2449.74	550.26	3000.00

(c) The monthly interest payment due on the loan of $3000 at 12% interest is found using the simple interest formula.

$$I = 3000(0.12)\left(\tfrac{1}{12}\right) = \$30$$

Thus the woman needs to be able to pay $40.83 + $30 = $70.83 each month. ■

Now Work Problem 11

EXAMPLE 10   Depletion Investment

A gold mine is expected to yield an annual net return of $200,000 for the next 10 years, after which it will be worthless. An investor wants an annual return on his investment of 18%. If he can establish a sinking fund earning 10% annually, how much should he be willing to pay for the mine?

SOLUTION   Let p denote the purchase price. Then $0.18p$ represents an 18% annual return on investment. The annual sinking fund contribution needed to recover the purchase price p in 10 years is

$$p \left[\frac{(1 + i)^n - 1}{i} \right]^{-1}$$

where $n = 10$ and $i = 0.10$. The investor should be willing to pay an amount p so that

$$\left(\begin{array}{c}\text{Annual return}\\\text{on investment}\end{array}\right) + \left(\begin{array}{c}\text{Annual sinking}\\\text{fund requirement}\end{array}\right) = \text{Annual return}$$

$$0.18p + p\left[\frac{(1 + 0.10)^{10} - 1}{0.10}\right]^{-1} = \$200{,}000$$

$$0.18p + (0.0627453)p = \$200{,}000$$

$$0.2427453p = \$200{,}000$$

$$p = \$823{,}909$$

A purchase price of $823,909 will achieve the investor's goals.

EXERCISE 5.3   Answers to odd-numbered problems begin on page AN-22.

In Problems 1–10 find the amount of each annuity.

1. The deposit is $100 annually for 10 years at 10% compounded annually.

2. The deposit is $200 monthly for 1 year at 5% compounded monthly.

3. The deposit is $400 monthly for 1 year at 12% compounded monthly.

4. The deposit is $1000 annually for 5 years at 10% compounded annually.

5. The deposit is $200 monthly for 3 years at 6% compounded monthly.

6. The deposit is $2000 semiannually for 20 years at 5% compounded semiannually.

7. The deposit is $100 monthly for 5 years at 6% compounded monthly.

8. The deposit is $1000 quarterly for 2 years at 4% compounded quarterly.

9. The deposit is $9000 annually for 10 years at 5% compounded annually.

10. The deposit is $5000 annually for 20 years at 4% compounded annually.

In Problems 11–20 find the payment required for each sinking fund.

11. The amount required is $10,000 after 5 years at 5% compounded monthly. What is the monthly payment?

12. The amount required is $5000 after 180 days at 4% compounded daily. What is the daily payment?

13. The amount required is $20,000 after $2\frac{1}{2}$ years at 6% compounded quarterly. What is the quarterly payment?

14. The amount required is $50,000 after 10 years at 7% compounded semiannually. What is the semiannual payment?

15. The amount required is $25,000 after 6 months at $5\frac{1}{2}\%$ compounded monthly. What is the monthly payment?

16. The amount required is $100,000 after 25 years at $4\frac{1}{4}\%$ compounded annually. What is the annual payment?

17. The amount required is $5000 after 2 years at 4% compounded monthly. What is the monthly payment?

18. The amount required is $5000 after 2 years at 4% compounded semiannually. What is the semiannual payment?

19. The amount required is $9000 after 4 years at 5% compounded annually. What is the annual payment?

20. The amount required is $9000 after 2 years at 5% compounded quarterly. What is the quarterly payment?

21. Market Value of a Mutual Fund   Caryl invests $2500 a year in a mutual fund for 15 years. If the market value of the fund increases on the average 7% per year, what will be the value of the fund at the end of the 15th year?

22. Saving for a Car   Laura wants to invest an amount every 3 months so that she will have $12,000 in 3 years to buy a new car. The account pays 8% componded quarterly. How much should she deposit each quarter?

23. **Value of an Annuity**　Todd and Tami pay $300 every 3 months for 6 years into an ordinary annuity paying 8% compounded quarterly. What is the value of the annuity at the end of 6 years?

24. **Saving for a House**　In 4 years Colleen and Bill would like to have $30,000 for a down payment on a house. How much should they deposit each month into an account paying 9% compounded monthly?

25. **Funding a Pension**　Dan wishes to have $350,000 in a pension fund 20 years from now. How much should he deposit each month in an account paying 9% compounded monthly?

26. **Funding a Keogh Plan**　Pat has a Keogh retirement plan (this type of plan is tax-deferred until money is withdrawn). If deposits of $7500 are made each year into an account paying 8% compounded annually, how much will be in the account after 25 years?

27. **Sinking Fund Payment**　A company establishes a sinking fund to provide for the payment of a $100,000 debt maturing in 4 years. Contributions to the fund are to be made each year. Find the amount of each annual deposit if interest is 8% per annum.

28. **Paying Off Bonds**　A state has $5,000,000 worth of bonds that are due in 20 years. A sinking fund is established to pay off the debt. If the state can earn 10% annually on its money, what is the annual sinking fund deposit needed?

29. **Depletion Investment**　An investor wants to know the amount she should pay for an oil well expected to yield an annual return of $30,000 for the next 30 years, after which the well will be dry. Find the amount she should pay to yield a 14% annual return if a sinking fund earns 10% annually.

30. **Time Needed for a Million Dollars**　If you deposit $10,000 every year in an account paying 8% compounded annually, how long will it take to accumulate $1,000,000?

31. **Bond Payments**　A city has issued bonds to finance a new library. The bonds have a total face value of $1,000,000 and are payable in 10 years. A sinking fund has been opened to retire the bonds. If the interest rate on the fund is 8% compounded quarterly, what will the quarterly payments be?

32. **Value of an IRA**

    (a) Laura invested $2000 per year in an IRA each year for 10 years earning 8% compounded annually. At the end of 10 years she ceased the IRA payments, but continued to invest her accumulated amount at 8% compounded annually, for the next 30 years. What was the value of her IRA investment at the end of 10 years? What was the value of her investment at the end of the next 30 years?

    (b) Tami started her IRA investment in the 11th year and invested $2000 per year for the next 30 years at 8% compounded annually. What was the value of her investment at the end of 30 years?

    (c) Who had more money at the end of the period?

33. **Managing a Condo**　The Crown Colony Condo Association is required by law to set aside funds to replace its roof. The current cost to replace the roof is $100,000 and it will need to be replaced in 20 years. The cost of a roof is expected to increase at the rate of 3% per year. The Condo can invest in Treasuries yielding 6% paid semiannually.

    (a) What will the roof cost in 20 years?

    (b) If the Condo invests in the Treasuries, what semiannual payment is required to have the funds to replace the roof in 20 years?

34. **Managing a Condo**　The Crown Colony Condo Association is required by law to set aside funds to replace its common-area carpet. The current cost to replace the carpet is $20,000 and it will need to be replaced in 6 years. The cost of carpet is expected to increase at the rate of 2% per year. The Condo can invest in Treasuries yielding 5% paid semiannually.

    (a) What will the carpet cost in 6 years?

    (b) If the Condo invests in the Treasuries, what semiannual payment is required to have the funds to replace the carpet in 6 years?

Problems 35 and 36 require logarithms.

35. How many years will it take to save $1,000,000 if you place $600 per month in an account that earns 7% compounded monthly?

36. How many years will it take to save $1,000,000 if you place $1000 per month in an account that earns 6% compounded monthly?

Technology Exercises

A business or financial calculator is preprogrammed to use the annuity formula. If you have one, consult your manual and use your calculator to check your answers to Problems 1, 4, 11, 15, 35, and 36 above. A graphing calculator can be programmed to do what a financial calculator is preprogrammed to do.

1. Write a program that will calculate the amount of an annuity after n years if a principal P is invested at $r\%$ per annum compounded annually. Use it to verify your answers to Problems 1 and 4 above.

2. Write a program that will calculate the monthly payment needed to get the amount A after n years at $r\%$ per annum

compounded monthly. Use it to verify your answers to Problems 11 and 15 above.

3. Write a program that will calculate the number of months required for a monthly investment of x dollars to grow to y dollars at $r\%$ per annum compounded monthly. Use it to verify your answers to Problems 35 and 36 above.

5.4 PRESENT VALUE OF AN ANNUITY; AMORTIZATION

In Section 5.2 we defined present value (as it relates to the compound interest formula) as the amount of money needed now to obtain an amount A in the future. A similar idea is used for periodic withdrawals.

Suppose you want to withdraw $10,000 per year each year for the next 5 years from a retirement account that earns 10% compounded annually. How much money is required initially in this account for this to happen? In fact, the amount is the present value of each of the $10,000 withdrawals. This leads to the following definition.

> **Present Value of an Annuity**
>
> The **present value** of an annuity is the sum of the present values of the payments.

In other words, the present value of an annuity is the amount of money needed now so that if it is invested at i percent, n equal dollar amounts can be withdrawn without any money left over.

EXAMPLE 1   Finding the Present Value of an Annuity

Compute the amount of money required to pay out $10,000 per year for 5 years at 10% compounded annually.

SOLUTION   For the first $10,000 withdrawal, the present value V_1 (the dollars needed now to do this) is

$$V_1 = \$10,000(1 + 0.10)^{-1} = \$10,000(0.9090909) = \$9090.91$$

For the second $10,000 withdrawal, the present value V_2 is

$$V_2 = \$10,000(1 + 0.10)^{-2} = \$10,000(0.826446) = \$8264.46$$

Similarly,

$$V_3 = \$10,000(1 + 0.10)^{-3} = \$10,000(0.7513148) = \$7513.15$$

$$V_4 = \$10,000(1 + 0.10)^{-4} = \$10,000(0.6830134) = \$6830.13$$

$$V_5 = \$10,000(1 + 0.10)^{-5} = \$10,000(0.6209213) = \$6209.21$$

The present value V for 5 withdrawals of $10,000 each is

$$
\begin{aligned}
V &= V_1 + V_2 + V_3 + V_4 + V_5 \\
&= \$9090.91 + \$8264.46 + \$7513.15 + \$6830.13 + \$6209.21 \\
&= \$37,907.86
\end{aligned}
$$

Thus a person would need \$37,907.86 now invested at 10% per annum in order to withdraw \$10,000 per year for the next 5 years.

Table 4(a) summarizes these results. Table 4(b) lists the amount at the beginning of each year.

Table 4(a)

Withdrawal	Present Value
1st	$\$10{,}000(1.10)^{-1} = \$\ 9{,}090.91$
2nd	$\$10{,}000(1.10)^{-2} = \$\ 8{,}264.46$
3rd	$\$10{,}000(1.10)^{-3} = \$\ 7{,}513.15$
4th	$\$10{,}000(1.10)^{-4} = \$\ 6{,}830.13$
5th	$\$10{,}000(1.10)^{-5} = \$\ 6{,}209.21$
Total	\$37,907.86

Table 4(b)

Year	Amount at the Beginning of the Year	Add Interest	Subtract Withdrawal
1	37,907.86	3,790.79	10,000.00
2	31,698.65	3,169.87	10,000.00
3	24,868.52	2,486.85	10,000.00
4	17,355.37	1,735.54	10,000.00
5	9,090.91	909.09	10,000.00
6	0		

To derive a formula for present value, suppose \$$P$ is the withdrawal per payment period for an annuity at an interest rate of i percent per payment period (in decimal form) and with a term of n payment periods. Then the present value V_1 of the first withdrawal is

$$V_1 = P(1 + i)^{-1}$$

The present value V_2 for the second withdrawal is

$$V_2 = P(1 + i)^{-2}$$

The present value V_n for the nth withdrawal is

$$V_n = P(1 + i)^{-n}$$

The total present value V of the annuity is

$$V = V_1 + \cdots + V_n = P(1 + i)^{-1} + \cdots + P(1 + i)^{-n}$$
$$= P(1 + i)^{-n}[1 + (1 + i) + \cdots + (1 + i)^{n-1}]$$

The expression in brackets is the sum of the first n terms of a geometric sequence, whose ratio is $1 + i$. As a result,

$$V = P(1 + i)^{-n}\,\frac{1 - (1 + i)^n}{1 - (1 + i)} = P\,\frac{(1 + i)^{-n} - 1}{-i} = P\,\frac{1 - (1 + i)^{-n}}{i}$$

Present Value of an Annuity

If P represents the withdrawal per payment period, the present value V of an annuity at a rate i per payment period for n payment periods is

$$V = P\,\frac{1 - (1 + i)^{-n}}{i}$$

Table II, Present Value of an Annuity, in the back of the book, gives values for

$$\frac{1 - (1 + i)^{-n}}{i}$$

for per annum rates of 8%, 10%, and 12% compounded annually and monthly. For most of our work, a calculator will be used.

 **Now Work Problem 1**

EXAMPLE 2   Finding the Cost of a Car

A man agrees to pay $300 per month for 48 months to pay off a car loan. If interest of 12% per annum is charged monthly, how much did the car originally cost? How much interest was paid?

SOLUTION   This is the same as asking for the present value V of an annuity of $300 per month at 12% for 48 months. The original cost of the car is

$$V = 300\left[\frac{1 - \left(1 + \dfrac{0.12}{12}\right)^{-48}}{\dfrac{0.12}{12}}\right] = \$300(37.9739595) = \$11{,}392.19$$

The total payment is ($300)(48) = $14,400. Thus the interest paid is

$$\$14{,}400 - \$11{,}392.19 = \$3007.81$$

■

Amortization

We can look at Example 2 differently. What it also says is that if the man pays $300 per month for 48 months with an interest of 12% compounded monthly, then the car is his. In other words, he *amortized* the debt in 48 equal monthly payments. (The Latin word *mort* means "death." Paying off a loan is regarded as "killing" it.) A loan with a fixed rate of interest is said to be **amortized** if both principal and interest are paid by a sequence of equal payments made over equal periods of time.

When a loan of V dollars is amortized at a rate of interest i per payment period over n payment periods, the customary question is, "What is the payment P?" In other words, in amortization problems, we want to find the amount of payment P that, after n payment periods at the rate of interest i per payment period, gives us a present value equal to the amount of the loan. Thus, we need to find P in the formula

$$V = P\left[\frac{1 - (1 + i)^{-n}}{i}\right]$$

Thus

$$P = V\left[\frac{1 - (1 + i)^{-n}}{i}\right]^{-1}$$

Amortization

The payment P required to pay off a loan of V dollars borrowed for n payment periods at a rate of interest i per payment period is

$$P = V\left[\frac{1 - (1 + i)^{-n}}{i}\right]^{-1}$$

Table II, Present Value of an Annuity, in the back of the book, gives values for

$$\left[\frac{1 - (1 + i)^{-n}}{i}\right]^{-1}$$

for per annum rates of 8%, 10%, and 12% compounded annually and monthly. For most of our work, a calculator will be used.

EXAMPLE 3   Finding the Payment for an Amortized Loan

What monthly payment is necessary to pay off a loan of $800 at 10% per annum

(a) In 2 years?   (b) In 3 years?
(c) What total amount is paid out for each loan?

SOLUTION

(a) For the 2-year loan $V = \$800$, $n = 24$, and $i = 0.10/12$. The monthly payment P is

$$P = 800\left[\frac{1 - \left(1 + \dfrac{0.10}{12}\right)^{-24}}{\dfrac{0.10}{12}}\right]^{-1} = \$800(0.04614493) = \$36.92$$

(b) For the 3-year loan $V = \$800$, $n = 36$, and $i = 0.10/12$. The monthly payment P is

$$P = 800\left[\frac{1 - \left(1 + \dfrac{0.10}{12}\right)^{-36}}{\dfrac{0.10}{12}}\right]^{-1} = \$800(0.03226719) = \$25.81$$

(c) For the 2-year loan, the total amount paid out is $(36.92)(24) = \$886.08$; for the 3-year loan, the total amount paid out is $(\$25.81)(36) = \$929.16$.

Now Work Problem 9

EXAMPLE 4 Mortgage Payments

Mr. and Mrs. Corey have just purchased a $70,000 house and have made a down payment of $15,000. They can amortize the balance ($70,000 − $15,000 = $55,000) at 9% for 25 years.

(a) What are the monthly payments?
(b) What is their total interest payment?
(c) After 20 years, what equity do they have in their house (that is, what is the sum of the down payment and the amount paid on the loan)?

SOLUTION

(a) The monthly payment P needed to pay off the loan of $55,000 at 9% for 25 years (300 months) is

$$P = \$55,000 \left[\frac{1 - \left(1 + \dfrac{0.09}{12}\right)^{-300}}{\dfrac{0.09}{12}} \right]^{-1} = \$55,000(0.008392) = \$461.56$$

(b) The total paid out for the loan is

$$(\$461.56)(300) = \$138,468.00$$

Thus the interest on this loan amounts to

$$\$138,468 - \$55,000 = \$83,468.00$$

(c) After 20 years (240 months) there remains 5 years (or 60 months) of payments. The present value of the loan is the present value of a monthly payment of $461.56 for 60 months at 9%, namely,

$$V = \$461.56 \left[\frac{1 - \left(1 + \dfrac{0.09}{12}\right)^{-60}}{\dfrac{0.09}{12}} \right]$$

$$= (461.56)(48.17337) = \$22,234.90$$

The amount paid on the loan is

$$\begin{pmatrix} \text{Original loan} \\ \text{amount} \end{pmatrix} - \begin{pmatrix} \text{Present} \\ \text{value} \end{pmatrix} = \$55,000 - \$22,234.90 = \$32,765.10$$

Thus the equity after 20 years is

$$\begin{pmatrix} \text{Down} \\ \text{payment} \end{pmatrix} + \begin{pmatrix} \text{Amount paid} \\ \text{on loan} \end{pmatrix} = \$15,000 + \$32,765.10 = \$47,765.10$$

Table 5 gives a partial schedule of payments for the loan in Example 4. It is interesting to observe how slowly the amount paid on the loan increases early in the payment schedule, with very little of the payment used to reduce principal, and how quickly the amount paid on the loan increases during the last 5 years.

Table 5

Payment Number	Monthly Payment	Principal	Interest	Amount Paid on Loan
1	$461.56	$49.06	$412.50	$49.06
60	$461.56	$75.94	$385.62	$3,699.94
120	$461.56	$119.27	$342.29	$9,493.23
180	$461.56	$186.85	$274.71	$18,563.67
240	$461.56	$292.56	$169.00	$32,765.10
300	$461.56	$458.16	$3.40	$55,000.00

✍ **Now Work Problem 11**

EXAMPLE 5　Inheritance

When Mr. Nicholson died, he left an inheritance of $15,000 for his family to be paid to them over a 10-year period in equal amounts at the end of each year. If the $15,000 is invested at 10% per annum, what is the annual payout to the family?

SOLUTION　This example asks what annual payment is needed at 10% for 10 years to disperse $15,000. That is, we can think of the $15,000 as a loan amortized at 10% for 10 years. The payment needed to pay off the loan is the yearly amount Mr. Nicholson's family will receive. Thus the yearly payout P is

$$P = \$15,000 \left[\frac{1 - (1 + 0.10)^{-10}}{0.10} \right]^{-1}$$

$$= \$15,000(0.16274539) = \$2441.18$$

EXAMPLE 6　Determining Retirement Income

Joan is 20 years away from retiring and starts saving $100 a month in an account paying 6% compounded monthly. When she retires, she wishes to withdraw a fixed amount each month for 25 years. What will this fixed amount be?

SOLUTION　After 20 years the amount accumulated in her account is the amount of an annuity with a monthly payment of $100 and an interest rate of $i = 0.06/12$. Thus

$$A = 100 \left[\frac{\left(1 + \dfrac{0.06}{12}\right)^{240} - 1}{\dfrac{0.06}{12}} \right]$$

$$= 100 \cdot (462.0408951) = \$46,204.09$$

The amount P she can withdraw for 300 payments (25 years) at 6% compounded monthly is

$$P = 46{,}204.09 \left[\frac{1 - \left(1 + \dfrac{0.06}{12}\right)^{-300}}{\dfrac{0.06}{12}} \right]^{-1} = \$297.69$$

EXERCISE 5.4   Answers to odd-numbered problems begin on page AN-22.

In Problems 1–6 find the present value of each annuity.

1. The withdrawal is to be $500 per month for 36 months at 10% compounded monthly.

2. The withdrawal is to be $1000 per year for 3 years at 8% compounded annually.

3. The withdrawal is to be $100 per month for 9 months at 12% compounded monthly.

4. The withdrawal is to be $400 per month for 18 months at 5% compounded monthly.

5. The withdrawal is to be $10,000 per year for 20 years at 10% compounded annually.

6. The withdrawal is to be $2000 per month for 3 years at 4% compounded monthly.

7. **Value of a IRA**   A husband and wife contribute $4000 per year to an IRA paying 10% compounded annually for 20 years. What is the value of their IRA? How much can they withdraw each year for 25 years at 10% compounded annually?

8. Rework Problem 7 if the interest rate is 8%.

9. **Loan Payments**   What monthly payment is needed to pay off a loan of $10,000 amortized at 12% compounded monthly for 2 years?

10. **Loan Payments**   What monthly payment is needed to pay off a loan of $500 amortized at 12% compounded monthly for 2 years?

11. In Example 4 if Mr. and Mrs. Corey amortize the $55,000 loan at 10% for 20 years, what is their monthly payment?

12. In Example 4 if Mr. and Mrs. Corey amortize their $55,000 loan at 12% for 15 years, what is their monthly payment?

13. In Example 5 if Mr. Nicholson left $15,000 to be paid over 20 years in equal yearly payments and if this amount were invested at 12%, what would the annual payout be?

14. Joan has a sum of $30,000 that she invests at 10% compounded monthly. What equal monthly payments can she receive over a 10-year period? Over a 20-year period?

15. **Planning Retirement**   Mr. Doody, at age 65, can expect to live for 20 years. If he can invest at 10% per annum compounded monthly, how much does he need now to guarantee himself $250 every month for the next 20 years?

16. **Planning Retirement**   Sharon, at age 65, can expect to live for 25 years. If she can invest at 10% per annum compounded monthly, how much does she need now to guarantee herself $300 every month for the next 25 years?

17. **House Mortgage**   A couple wishes to purchase a house for $120,000 with a down payment of $25,000. They can amortize the balance either at 8% for 20 years or at 9% for 25 years. Which monthly payment is greater? For which loan is the total interest paid greater? After 10 years, which loan provides the greater equity?

18. **House Mortgage**   A couple has decided to purchase a $100,000 house using a down payment of $20,000. They can amortize the balance at 8% for 25 years. (a) What is their monthly payment? (b) What is the total interest paid? (c) What is their equity after 5 years? (d) What is the equity after 20 years?

19. **Planning Retirement**   John is 45 years old and wants to retire at 65. He wishes to make monthly deposits in an account paying 9% compounded monthly so when he retires he can withdraw $300 a month for 30 years. How much should John deposit each month?

20. **Cost of a Lottery**   The grand prize in an Illinois lottery is $6,000,000, which will be paid out in 20 equal annual payments of $300,000 each. Assume the first payment of $300,000 is made, leaving the state with the obligation to pay out $5,700,000 in 19 equal yearly payments of $300,000 each. How much does the state need to deposit in an account paying 6% compounded annually to achieve this?

21. **College Expenses** Dan works during the summer to help with expenses at school the following year. He is able to save $100 each week for 12 weeks, and he invests it at 6% compounded weekly.

    (a) How much has he saved after 12 weeks?
    (b) When school starts, Dan will begin to withdraw equal amounts from this account each week. What is the most Dan can withdraw each week for 34 weeks?

22. Repeat Problem 21 if the rate of interest is 8%. What if it is only 4%?

23. **Analyzing a Town House Purchase** Mike and Yola have just purchased a town house for $76,000. They obtain financing with the following terms: a 20% down payment and the balance to be amortized over 30 years at 9%.

    (a) What is their down payment?
    (b) What is the loan amount?
    (c) How much is their monthly payment on the loan?
    (d) How much total interest do they pay over the life of the loan?
    (e) If they pay an additional $100 each month toward the loan, when will the loan be paid?
    (f) With the $100 additional monthly payment, how much total interest is paid over the life of the loan?

24. **House Mortgage** Mr. Smith obtained a 25-year mortgage on a house. The monthly payments are $749.19 (principal and interest) and are based on a 7% interest rate. How much did Mr. Smith borrow? How much interest will be paid?

25. **Car Payments** A car costs $12,000. You put 20% down and amortize the rest with equal monthly payments over a 3-year period at 15% to be compounded monthly. What will the monthly payment be?

26. **Cost of Furniture** Jay pays $320 per month for 36 months for furniture, making no down payment. If the interest charged is 0.5% per month on the unpaid balance, what was the original cost of the furniture? How much interest did he pay?

27. **Paying for Restaurant Equipment** A restaurant owner buys equipment costing $20,000. If the owner pays 10% down and amortizes the rest with equal monthly payments over 4 years at 12% compounded monthly, what will be the monthly payments? How much interest is paid?

28. **Refinancing a Mortgage** A house that was bought 8 years ago for $50,000 is now worth $100,000. Originally the house was financed by paying 20% down with the rest financed through a 25-year mortgage at 10.5% interest. The owner (after making 96 equal monthly payments) is in need of cash, and would like to refinance the house. The finance company is willing to loan 80% of the new value of the house amortized over 25 years with the same interest rate. How much cash will the owner receive after paying the balance of the original loan?

29. **Comparing Mortgages** A home buyer is purchasing a $140,000 house. The down payment will be 20% of the price of the house, and the remainder will be financed by a 30-year mortgage at a rate of 9.8% interest compounded monthly. What will the monthly payment be? Compare the monthly payments and the total amounts of interest paid if a 15-year mortgage is chosen instead of a 30-year mortgage.

30. **Mortgage Payments** Mr. and Mrs. Hoch are interested in building a house that will cost $180,000. They intend to use the $60,000 equity in their present house as a down payment on the new one and will finance the rest with a 25-year mortgage at an interest rate of 10.2% compounded monthly. How large will their monthly payment be on the new house?

Problems 31 and 32 require logarithms.

31. **IRA Withdrawals** How long will it take to exhaust an IRA of $100,000 if you withdraw $2000 every month? Assume a rate of interest of 5% compounded monthly.

32. **IRA Withdrawals** How long will it take to exhaust an IRA of $200,000 if you withdraw $3000 every month? Assume a rate of interest of 4% compounded monthly.

Technology Exercises

A business or financial calculator is preprogrammed to use the present value of an annuity formula. If you have one, consult your manual and use your calculator to check your answers to Problems 2, 5, 9, 11, 31, and 32 above. A graphing calculator can be programmed to do what a financial calculator is preprogrammed to do.

1. Write a program that will calculate the present value of an annuity after n years if P dollars is withdrawn each year at $r\%$ per annum compounded annually. Use it to verify your answers to Problems 2 and 5 above.

2. Write a program that will calculate the monthly payment needed to pay off a loan of V dollars borrowed for n months at $r\%$ per annum compounded monthly. Use it to verify your answers to Problems 9 and 11 above.

3. Write a program that will calculate the number of months it will take to exhaust A dollars if you withdraw x dollars every month. Assume a rate of interest of $r\%$ per annum compounded monthly. Use it to verify your answers to Problems 31 and 32 above.

5.5 APPLICATIONS: LEASING; CAPITAL EXPENDITURE; BONDS

Leasing

EXAMPLE 1   Lease or Purchase

A corporation may obtain a particular machine either by leasing it for 4 years (the useful life) at an annual rent of $1000 or by purchasing the machine for $3000.

(a) Which alternative is preferable if the corporation can invest money at 10% per annum?
(b) What if it can invest at 14% per annum?

SOLUTION

(a) Suppose the corporation may invest money at 10% per annum. The present value of an annuity of $1000 for 4 years at 10% equals $3169.87, which exceeds the purchase price. Therefore, purchase is preferable.
(b) Suppose the corporation may invest at 14% per annum. The present value of an annuity of $1000 for 4 years at 14% equals $2913.71, which is less than the purchase price. Hence, leasing is preferable.

 **Now Work Problem 1**

Capital Expenditure

EXAMPLE 2   Selecting Equipment

A corporation is faced with a choice between two machines, both of which are designed to improve operations by saving on labor costs. Machine A costs $8000 and will generate an annual labor savings of $2000. Machine B costs $6000 and will save $1800 annually. However, machine A has a useful life of 7 years while machine B has a useful life of only 5 years. Assuming that the time value of money (the investment opportunity rate) of the corporation is 10% per annum, which machine is preferable? (Assume annual compounding and that the savings is realized at the end of each year.)

SOLUTION   Machine A costs $8000 and has a life of 7 years. Since an annuity of $1 for 7 years at 10% interest has a present value of $4.87, the cost of machine A may be considered the present value of an annuity:

$$\frac{\$8000}{4.87} = \$1642.71$$

The $1642.71 may be termed the *equivalent annual cost* of machine A. Similarly, the equivalent annual cost of machine B may be calculated by reference to the present value of an annuity of 5 years, namely,

$$\frac{\$6000}{3.79} = \$1583.11$$

Thus the net annual savings of each machine is as follows:

	A	B
Labor savings	$2000.00	$1800.00
Equivalent annual cost	1642.71	1583.11
Net savings	$ 357.29	$ 216.89

Therefore, machine A is preferable.

Now Work Problem 3

Bonds

We begin our third application with some definitions of terms concerning corporate bonds. A calculator will be needed for the example that follows.

Face Amount (Face Value or Par Value)

The **face amount** or **denomination** of a bond (normally $1000) is the amount paid to the bondholder at maturity. It is also the amount usually paid by the bond-holder when the bond is originally issued.

Nominal Interest (Coupon Rate)

The contractual periodic interest payments on a bond.

Nominal interest is normally quoted as an annual percentage of the face amount. Nominal interest payments are conventionally made semiannually, so semiannual periods are used for compound interest calculations. For example, if a bond has a face amount of $1000 and a coupon rate of 8%, then every 6 months the owner of the bond would receive $(1000)(0.08)(\frac{1}{2}) = \$40$.

But, because of market conditions, such as the current prime rate of interest, or the discount rate set by the Federal Reserve Board, or changes in the credit rating of the company issuing the bond, the price of a bond will fluctuate. When the bond price is higher than the face amount, it is trading at a **premium;** when it is lower, it is trading at a **discount.** Thus, for example, a bond with a face amount of $1000 and a coupon rate of 8% may trade in the marketplace at a price of $1100, which means the **true yield** is less than 8%.

To obtain the **true interest rate** of a bond, we view the bond as a combination of an annuity of semiannual interest payments plus a single future amount payable at maturity. The price of a bond is therefore the sum of the present value of the annuity of semiannual interest payments plus the present value of the single future payment at maturity. This present value is calculated by discounting at the true interest rate and assuming semiannual discounting periods.

EXAMPLE 3   Pricing Bonds

A bond has a face amount of $1000 and matures in 10 years. The nominal interest rate is 8.5%. What is the price of the bond to yield a true interest rate of 8%?

SOLUTION

Step 1 Calculate the amount of each semiannual interest payment:

$$(\$1000)(\tfrac{1}{2})(0.085) = \$42.50$$

Step 2 Calculate the present value of the annuity of semiannual payments:

$$V = 1{,}000 \left[\frac{(1 +)^{20} - 1}{i} \right]$$

Amount of each payment from Step 1	$ 42.50
Number of payments (2 × 10 years): 20	
True interest rate per period	
(half of stated true rate): 4%	
Factor from formula for V (use a calculator)	13.5903
Present value of interest payments	$ 577.59

Step 3 Calculate the present value of the amount payable at maturity:

Amount payable at maturity	$ 1000
Number of semiannual compounding	
periods before maturity: 20	
True interest rate per period: 4%	
Factor from formula for $(1 + i)^{-n}$ (use a calculator)	0.45639
Present value of maturity value	$ 456.39

Step 4 Determine the price of the bond:

Present value of interest payments	$ 577.59
Present value of maturity payment	456.39
Price of bond	$1033.98

EXERCISE 5.5   Answers to odd-numbered problems begin on page AN-23.

1. **Leasing Problem**   A corporation may obtain a machine either by leasing it for 5 years (the useful life) at an annual rent of $2000 or by purchasing the machine for $8100. If the corporation can borrow money at 10% per annum, which alternative is preferable?

2. If the corporation in Problem 1 can borrow money at 14% per annum, which alternative is preferable?

3. **Capital Expenditure Analysis**   Machine A costs $10,000 and has a useful life of 8 years, and machine B costs $8000 and has a useful life of 6 years. Suppose machine A generates an annual labor savings of $2000 while machine B generates an annual labor savings of $1800. Assuming the time value of money (investment opportunity rate) is 10% per annum, which machine is preferable?

4. In Problem 3 if the time value of money is 14% per annum, which machine is preferable?

5. Corporate Bonds   A bond has a face amount of $1000 and matures in 15 years. The nominal interest rate is 9%.

What is the price of the bond that will yield an effective interest rate of 8%?

6. For the bond in Problem 5 what is the price of the bond to yield an effective interest rate of 10%?

CHAPTER REVIEW

IMPORTANT TERMS AND CONCEPTS

simple interest formula 226
discounted loans 227
compound interest formula 232

present value 235
amount of an annuity 240
sinking fund 244

present value of an annuity 249
amortization 251

IMPORTANT FORMULAS

Simple Interest Formula

$$I = Prt$$

Discounted Loans

$$P = A - Art$$

Compound Interest Formula

$$A_n = P(1 + i)^n$$

Amount of an Annuity

$$A = P\frac{(1 + i)^n - 1}{i}$$

Present Value of an Annuity; Amortization

$$V = P\frac{1 - (1 + i)^{-n}}{i}$$

TRUE–FALSE ITEMS   Answers are on page AN-23.

T_____F_____ **1.** Simple interest is interest computed on the principal for the entire period it is borrowed.

T_____F_____ **2.** The amount A accrued on a principal P after five payment periods at i interest per payment period is

$$A = P(1 + i)^5$$

T_____F_____ **3.** The effective rate of interest for 10% compounded daily is about 9.58%.

T_____F_____ **4.** The present value of an annuity is the sum of all the present values of the payment less any interest.

FILL IN THE BLANKS   Answers are on page AN-23.

1. For a discounted loan, the amount the borrower receives is called the _____ .

2. In the formula $P = A(1 + i)^{-n}$, P is called the _____ _____ of the amount A due at the end of n interest periods at i interest per payment period.

3. The amount of an _____ is the sum of all deposits made plus all interest accumulated.

4. A loan with a fixed rate of interest is _____ if both principal and interest are paid by a sequence of equal payments over equal periods of time.

REVIEW EXERCISES   Answers to odd-numbered problems begin on page AN-23.

1. Find the interest (I) and amount (A) if $400 is borrowed for 9 months at 12% simple interest.

2. Dan borrows $500 at 9% per annum simple interest for 1 year and 2 months. What is the interest charged, and what is the amount due?

3. Find the amount of an investment of $100 after 2 years and 3 months at 10% compounded monthly.

4. Mike places $200 in a savings account that pays 4% per annum compounded monthly. How much is in his account after 9 months?

5. **Choosing a Car Loan**   A car dealer offers Mike the choice of two loans:

   (a) $3000 for 3 years at 12% per annum simple interest
   (b) $3000 for 3 years at 10% per annum compounded monthly

   Which loan costs Mike the least?

6. A mutual fund pays 9% per annum compounded monthly. How much should I invest now so that 2 years from now I will have $100 in the account?

7. **Saving for a Bicycle**   Katy wants to buy a bicycle that costs $75 and will purchase it in 6 months. How much should she put in her savings account for this if she can get 10% per annum compounded monthly?

8. **Saving for a Car**   Mike decides he needs $500 1 year from now to buy a used car. If he can invest at 8% compounded monthly, how much should he save each month to buy the car?

9. **Saving for a House**   Mr. and Mrs. Corey are newlyweds and want to purchase a home, but they need a down payment of $10,000. If they want to buy their home in 2 years, how much should they save each month in their savings account that pays 3% per annum compounded monthly?

10. **True Cost of a Car**   Mike has just purchased a used car and will make equal payments of $50 per month for 18 months at 12% per annum charged monthly. How much did the car actually cost? Assume no down payment.

11. **House Mortgage**   Mr. and Mrs. Ostedt have just purchased an $80,000 home and made a 25% down payment. The balance can be amortized at 10% for 25 years.

    (a) What are the monthly payments?
    (b) How much interest will be paid?
    (c) What is their equity after 5 years?

12. **Inheritance Payouts**   An inheritance of $25,000 is to be paid in equal amounts over a 5-year period at the end of each year. If the $25,000 can be invested at 10% per annum, what is the annual payment?

13. **House Mortgage**   A mortgage of $125,000 is to be amortized at 9% per annum for 25 years. What are the monthly payments? What is the equity after 10 years?

14. **Paying Off Construction Bonds**   A state has $8,000,000 worth of construction bonds that are due in 25 years. What annual sinking fund deposit is needed if the state can earn 10% per annum on its money?

15. **Depletion Problem**   How much should Mr. Graff pay for a gold mine expected to yield an annual return of $20,000 and to have a life expectancy of 20 years, if he wants to have a 15% annual return on his investment and he can set up a sinking fund that earns 10% a year?

16. **Retirement Income**   Mr. Doody, at age 70, is expected to live for 15 years. If he can invest at 12% per annum compounded monthly, how much does he need now to guarantee himself $300 every month for the next 15 years?

17. **Depletion Problem**   An oil well is expected to yield an annual net return of $25,000 for the next 15 years, after which it will run dry. An investor wants a return on his investment of 20%. He can establish a sinking fund earning 10% annually. How much should he pay for the oil well?

18. Hal deposited $100 a month in an account paying 9% per annum compounded monthly for 25 years. What is the largest amount he may withdraw monthly for 35 years?

19. **Saving for College**   Mr. Jones wants to save for his son's college education. If he deposits $500 every 6 months at 6% compounded semiannually, how much will he have on hand at the end of 8 years?

20. How much money should be invested at 8% compounded quarterly in order to have $20,000 in 6 years?

21. Bill borrows $1000 at 5% compounded annually. He is able to establish a sinking fund to pay off the debt and interest in 7 years. The sinking fund will earn 8% compounded quarterly. What should be the size of the quarterly payments into the fund?

22. A man has $50,000 invested at 12% compounded quarterly at the time he retires. If he wishes to withdraw money every 3 months for the next 7 years, what will be the size of his withdrawals?

23. How large should monthly payments be to amortize a loan of $3000 borrowed at 12% compounded monthly for 2 years?

24. **Paying off School Bonds**   A school board issues bonds in the amount of $20,000,000 to be retired in 25 years. How much must be paid into a sinking fund at 6% compounded annually to pay off the total amount due?

25. What effective rate of interest corresponds to a nominal rate of 9% compounded monthly?

26. If an amount was borrowed 5 years ago at 6% compounded quarterly, and $6000 is owed now, what was the original amount borrowed?

27. **Trust Fund Payouts**   John is the beneficiary of a trust fund set up for him by his grandparents. If the trust fund amounts to $20,000 earning 8% compounded semiannually and he is to receive the money in equal semiannual installments for the next 15 years, how much will he receive each 6 months?

28. For 20 years a person has had $4000 on deposit. At 6% annual interest compounded monthly, how much will have been saved at the end of the 20 years?

29. **Retirement Funds**   An employee gets paid at the end of each month and $60 is withheld from her paycheck for a retirement fund. The fund pays 1% per month (equivalent to 12% annually compounded monthly). What amount will be in the fund at the end of 30 months?

30. $6000 is borrowed at 10% compounded semiannually. The amount is to be paid back in 5 years. If a sinking fund is established to repay the loan and interest in 5 years, and the fund earns 8% compounded quarterly, how much will have to be paid into the fund every 3 months?

31. **Buying a Car**   A student borrowed $4000 from a credit union toward purchasing a car. The interest rate on such a loan is 14% compounded quarterly, with payments due every quarter. The student wants to pay off the loan in 4 years. Find the quarterly payment.

MATHEMATICAL QUESTIONS FROM PROFESSIONAL EXAMS

1. **CPA Exam**   Which of the following should be used to calculate the amount of the equal periodic payments that could be equivalent to an outlay of $3000 at the time of the last payment?

   (a) Amount of 1
   (b) Amount of an annuity of 1
   (c) Present value of an annuity of 1
   (d) Present value of 1

2. **CPA Exam**   A businessman wants to withdraw $3000 (including principal) from an investment fund at the end of each year for 5 years. How should he compute his required initial investment at the beginning of the first year if the fund earns 6% compounded annually?

   (a) $3000 times the amount of an annuity of $1 at 6% at the end of each year for 5 years
   (b) $3000 divided by the amount of an annuity of $1 at 6% at the end of each year for 5 years
   (c) $3000 times the present value of an annuity of $1 at 6% at the end of each year for 5 years
   (d) $3000 divided by the present value of an annuity of $1 at 6% at the end of each year for 5 years

3. **CPA Exam**   A businesswoman wants to invest a certain sum of money at the end of each year for 5 years. The investment will earn 6% compounded annually. At the end of 5 years, she will need a total of $30,000 accumulated. How should she compute the required annual investment?

   (a) $30,000 times the amount of an annuity of $1 at 6% at the end of each year for 5 years
   (b) $30,000 divided by the amount of an annuity of $1 at 6% at the end of each year for 5 years
   (c) $30,000 times the present value of an annuity of $1 at 6% at the end of each year for 5 years
   (d) $30,000 divided by the present value of an annuity of $1 at 6% at the end of each year for 5 years

4. **CPA Exam**   Shaid Corporation issued $2,000,000 of 6%, 10-year convertible bonds on June 1, 1993, at 98 plus accrued interest. The bonds were dated April 1, 1993, with interest payable April 1 and October 1. Bond discount is amortized semiannually on a straight-line basis.

   On April 1, 1994, $500,000 of these bonds were converted into 500 shares of $20 par value common stock. Accrued interest was paid in cash at the time of conversion.

   What was the effective interest rate on the bonds when they were issued?

   (a) 6%   (b) Above 6%   (c) Below 6%
   (d) Cannot be determined from the information given.

CPA Exam

Items 5–7 apply to the appropriate use of present value tables. Given below are the present value factors for $1.00 discounted at 8% for 1 to 5 periods. Each of the following items is based on 8% interest compounded annually from day of deposit to day of withdrawal.

Periods	Present Value of $1 Discounted at 8% per Period
1	0.926
2	0.857
3	0.794
4	0.735
5	0.681

5. What amount should be deposited in a bank today to grow to $1000 3 years from today?

(a) $\dfrac{\$1000}{0.794}$ (b) $\$1000 \times 0.926 \times 3$

(c) ($1000 × 0.926) + ($1000 × 0.857) + ($1000 × 0.794)

(d) $1000 × 0.794

6. What amount should an individual have in her bank account today before withdrawal if she needs $2000 each year for 4 years with the first withdrawal to be made today and each subsequent withdrawal at 1-year intervals? (She is to have exactly a zero balance in her bank account after the fourth withdrawal.)

(a) $2000 + ($2000 × 0.926) + ($2000 × 0.857) + ($2000 × 0.794)

(b) $\dfrac{\$2000}{0.735} \times 4$

(c) ($2000 × 0.926) + ($2000 × 0.857) + ($2000 × 0.794) + ($2000 × 0.735)

(d) $\dfrac{\$2000}{0.926} \times 4$

7. If an individual put $3000 in a savings account today, what amount of cash would be available 2 years from today?

(a) $3000 × 0.857   (b) $3000 × 0.857 × 2

(c) $\dfrac{\$3000}{0.857}$ (d) $\dfrac{\$3000}{0.926} \times 2$

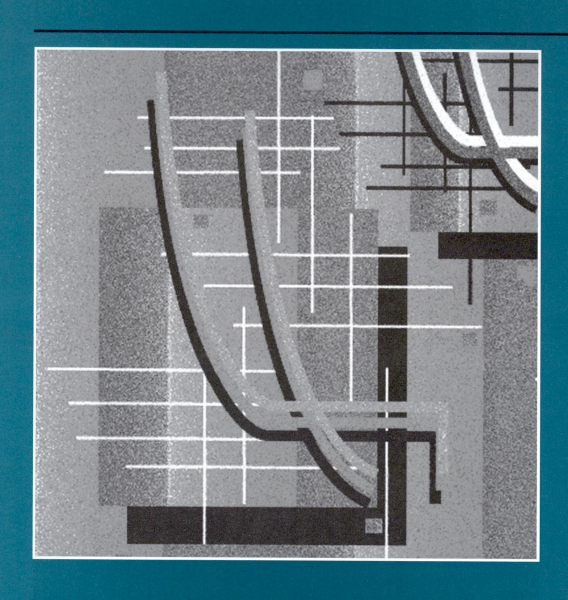

PART TWO

Probability

Chapter 6   Sets; Counting Techniques
Chapter 7   Probability
Chapter 8   Additional Probability Topics
Chapter 9   Statistics
Chapter 10   Markov Chains; Games

Sets; Counting Techniques

6.1 Sets

6.2 The Number of Elements in a Set

6.3 The Multiplication Principle

6.4 Permutations

6.5 Combinations

6.6 More Counting Problems

6.7 The Binomial Theorem

Chapter Review

This chapter discusses techniques and formulas for counting the number of objects in a set, a part of the branch of mathematics called *combinatorics.* These formulas are used in computer science to analyze algorithms and recursive functions and to study stacks and queues. They are also used to determine *probabilities,* the likelihood that a certain outcome of a random experiment will occur, which is the subject of Chapter 7. The final section explores the binomial theorem and some of its uses.

6.1 SETS

Set Properties and Set Notation

When we want to treat a collection of distinct objects as a whole, we use the idea of a **set.** For example, the set of **digits** consists of the collection of numbers 0, 1, 2, 3,

4, 5, 6, 7, 8, and 9. If we use the symbol D to denote the set of digits, then we can write

$$D = \{0, 1, 2, 3, 4, 5, 6, 7, 8, 9\}$$

In this notation the braces { } are used to enclose the objects, or **elements,** in the set. This method of denoting a set is called the **roster method.** A second way to denote a set is to use **set-builder notation,** where the set D of digits is written as

$$D = \{\quad x \quad | \quad x \text{ is a digit}\}$$

Read as "D is the set of all x such that x is a digit."

EXAMPLE 1   Examples of Sets

(a) $E = \{x | x \text{ is an even digit}\} = \{0, 2, 4, 6, 8\}$
(b) $O = \{x | x \text{ is an odd digit}\} = \{1, 3, 5, 7, 9\}$

EXAMPLE 2   Identifying the Elements of a Set

(a) Let E denote the set that consists of all possible outcomes resulting from tossing a coin three times. If we let H denote "heads" and T denote "tails," then the set E can be written as

$$E = \{TTT, TTH, THT, THH, HHH, HTH, HHT, HTT\}$$

where, for instance, THT means the first toss resulted in tails, the second toss in heads, and the third toss in tails.
(b) Let F denote the set consisting of all possible arrangements of the digits without repetition. Some typical elements of F are

$$1478906532 \qquad 4875326019 \qquad 3214569870$$

The number of elements in F is very large, so listing all of them is impractical. Later in this chapter we will study a technique to compute the number of elements in F.

The elements of a set are not repeated. That is, we do not write $\{3, 2, 2\}$ but rather write $\{3, 2\}$. Finally, the order in which the elements of a set are listed does not make any difference. Thus the three sets

$$\{3, 2, 4\} \qquad \{2, 3, 4\} \qquad \{4, 3, 2\}$$

are different listings of the same set. The *elements* of a set distinguish the set—not the *order* in which the elements are written.

A set that has no elements is called the **empty set** or **null set** and is denoted by the symbol $\varnothing$.

Equality of Sets

　　Let A and B be two sets. We say that A **is equal to** B, written as

$$A = B$$

if and only if A and B have the same elements. If two sets A and B are not equal, we write

$$A \neq B$$

Subset

　　Let A and B be two sets. We say that A **is a subset of** B or that A **is contained in** B, written as

$$A \subseteq B$$

if and only if every element of A is also an element of B. If a set A is not a subset of a set B, we write

$$A \nsubseteq B$$

　　Thus, $A \subseteq B$ if and only if whenever x is in A, then x is in B for all x. This latter way of interpreting the meaning of $A \subseteq B$ is useful for obtaining various laws that sets obey. For example, it follows that for any set A, $A \subseteq A$; that is, every set is a subset of itself. (Do you see why? Whenever x is in A, then x is in A!)

　　When we say that A is a subset of B, it is equivalent to saying "there are no elements in set A that are not also elements in set B." In particular, suppose $A = \varnothing$. Since the empty set $\varnothing$ has no elements, there is no element of the set $\varnothing$ that is not also in B. That is,

$$\varnothing \subseteq B, \qquad \text{for any set } B$$

Proper Subset

　　Let A and B be two sets. We say that A **is a proper subset of** B or that A **is properly contained in** B, written as

$$A \subset B$$

if and only if every element in the set A is also in the set B, but there is at least one element in set B that is *not* in set A.

　　Notice that "A is a proper subset of B" means that there are *no* elements of A that are not also elements of B, but there is at least one element of B that is not in A. Thus, if B is any nonempty set, that is, any set having at least one element, then

$$\varnothing \subset B$$

　　If a set A is *not* a proper subset of a set B, we write

$$A \not\subset B$$

From the definition it follows that $A \not\subset A$; that is, a set is never a proper subset of itself.

The following example illustrates some uses of the three relationships, $=$, $\subseteq$, and $\subset$, just defined.

EXAMPLE 3   Finding Relationships Between Sets

Consider three sets A, B, and C given by

$$A = \{1, 2, 3\} \qquad B = \{1, 2, 3, 4, 5\} \qquad C = \{3, 2, 1\}$$

Some of the relationships between pairs of these sets are

(a) $A = C$ (b) $A \subseteq B$ (c) $A \subseteq C$ (d) $A \subset B$ (e) $C \subseteq A$ (f) $\varnothing \subseteq A$ ∎

In comparing the two definitions of *subset* and *proper subset,* you should notice that if a set A is a subset of a set B, then either A is a proper subset of B or else A equals B. That is,

$$A \subseteq B \qquad \text{if and only if either} \qquad A \subset B \text{ or } A = B$$

Also, if A is a proper subset of B, we can infer that A is a subset of B, but A does not equal B. That is,

$$A \subset B \qquad \text{if and only if} \qquad A \subseteq B \text{ and } A \neq B$$

The distinction that is made between *subset* and *proper subset* is rather subtle but quite important.

We can think of the relationship $\subset$ as a refinement of $\subseteq$. On the other hand, the relationship $\subseteq$ is an extension of $\subset$, in the sense that $\subseteq$ may include equality whereas with $\subset$, equality cannot be included.

Now Work Problems 1 and 3

In applications the elements that may be considered are usually limited to some specific all-encompassing set. For example, in discussing students eligible to graduate from Midwestern University, the discussion would be limited to students enrolled at the university.

Universal Set

The **universal set** U is defined as the set consisting of all elements under consideration.

Thus if A is any set and if U is the universal set, then every element in A must be in U (since U consists of all elements under consideration). Hence, we may write

$$A \subseteq U$$

for *any* set A.

It is convenient to represent a set as the interior of a circle. Two or more sets may be depicted as circles enclosed in a rectangle, which represents the universal set. The circles may or may not overlap, depending on the situation. Such diagrams of sets are called **Venn diagrams.** See Figure 1.

Figure 1

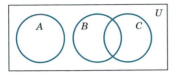

Operations on Sets

Next, we introduce operations that may be performed on sets.

Union of Two Sets

Let A and B be any two sets. The **union** of A with B, written as

$$A \cup B$$

and read as "A union B" or as "A or B," is defined to be the set consisting of those elements either in A or in B or in both A and B. That is,

$$A \cup B = \{x \mid x \text{ is in } A \text{ or } x \text{ is in } B\}$$

Figure 2

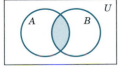

Warning! In English, the word *or* has two meanings: the *inclusive or* "A or B" means A or B or both. The *exclusive or* "A or B" means A or B but *not* both. Thus A or B in mathematics means use the inclusive or from English.

In the Venn diagram in Figure 2 the shaded area corresponds to $A \cup B$.

Intersection of Two Sets

Let A and B be any two sets. The **intersection** of A with B, written as

$$A \cap B$$

and read as "A intersect B" or as "A and B," is defined as the set consisting of those elements that are in both A and B. That is,

$$A \cap B = \{x \mid x \text{ is in } A \text{ and } x \text{ is in } B\}$$

Figure 3

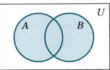

In other words, to find the intersection of two sets A and B means to find the elements *common* to A *and* B. In the Venn diagram in Figure 3 the shaded region is $A \cap B$.

For any set A it follows that

$$A \cup \varnothing = A \qquad A \cap \varnothing = \varnothing$$

EXAMPLE 4   Finding the Union and Intersection of Two Sets

Use the sets

$$A = \{1, 3, 5\} \qquad B = \{3, 4, 5, 6\} \qquad C = \{6, 7\}$$

to find   (a) $A \cup B$ (b) $A \cap B$ (c) $A \cap C$

SOLUTION

(a) $A \cup B = \{1, 3, 5\} \cup \{3, 4, 5, 6\} = \{1, 3, 4, 5, 6\}$
(b) $A \cap B = \{1, 3, 5\} \cap \{3, 4, 5, 6\} = \{3, 5\}$
(c) $A \cap C = \{1, 3, 5\} \cap \{6, 7\} = \varnothing$

EXAMPLE 5   Interpreting the Intersection of Two Sets

Let T be the set of all taxpayers and let S be the set of all people over 65 years of age. Describe $T \cap S$.

SOLUTION $T \cap S$ is the set of all taxpayers who are also over 65 years of age.

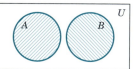
**Now Work Problems
11 and 13**

Disjoint Sets

If two sets A and B have no elements in common, that is, if

$$A \cap B = \varnothing$$

then A and B are called **disjoint sets.**

Figure 4

Two disjoint sets A and B are illustrated in the Venn diagram in Figure 4. Notice that the areas corresponding to A and B do not overlap anywhere because $A \cap B$ is empty.

EXAMPLE 6   Tossing a Die*

Suppose that a die is tossed. What is the universal set? Let A be the set of outcomes in which an even number turns up; let B be the set of outcomes in which an odd number shows. Find A and B. Find $A \cap B$.

SOLUTION   The universal set is $U = \{1, 2, 3, 4, 5, 6\}$. It follows that

$$A = \{2, 4, 6\} \qquad B = \{1, 3, 5\}$$

* A **die** (*plural* **dice**) is a cube with 1, 2, 3, 4, 5, or 6 dots showing on the six faces.

We note that A and B have no elements in common since an even number and an odd number cannot occur simultaneously. Therefore, $A \cap B = \varnothing$ and the sets A and B are disjoint sets.

Suppose we consider all the employees of some company as our universal set U. Let A be the subset of employees who smoke. Then all the nonsmokers will make up the subset of U that is called the *complement* of the set of smokers.

> **Complement of a Set**
>
>   Let A be any set. The **complement** of A, written as
> $$\overline{A} \qquad \text{(or } A')$$
> is defined as the set consisting of elements in the universe U that are not in A. Thus
> $$\overline{A} = \{x \mid x \text{ is not in } A\}$$

Figure 5

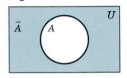

The shaded region in Figure 5 illustrates the complement, $\overline{A}$.
For any set A it follows that
$$A \cup \overline{A} = U \qquad A \cap \overline{A} = \varnothing \qquad \overline{\overline{A}} = A$$

EXAMPLE 7   Working with the Complement of a Set

Use the sets
$$U = \{a, b, c, d, e, f\} \qquad A = \{a, b, c\} \qquad B = \{a, c, f\}$$
to list the elements of the following sets:

(a) $\overline{A}$ (b) $\overline{B}$ (c) $\overline{A \cup B}$

(d) $\overline{A} \cap \overline{B}$ (e) $\overline{A \cap B}$ (f) $\overline{A} \cup \overline{B}$

SOLUTION

(a) $\overline{A}$ consists of all the elements in U that are not in A:
$$\overline{A} = \{d, e, f\}.$$

(b) Similarly,
$$\overline{B} = \{b, d, e\}.$$

(c) To determine $\overline{A \cup B}$, we first determine the elements in $A \cup B$:
$$A \cup B = \{a, b, c, f\}$$
The complement of the set $A \cup B$ is then
$$\overline{A \cup B} = \{d, e\}$$

(d) From Parts (a) and (b) we find that
$$\overline{A} \cap \overline{B} = \{d, e\}$$

(e) As in Part (c) we first determine the elements in $A \cap B$:

$$A \cap B = \{a, c\}$$

Then

$$\overline{A \cap B} = \{b, d, e, f\}$$

(f) From Parts (a) and (b) we find that

$$\overline{A} \cup \overline{B} = \{b, d, e, f\}$$

Now Work Problems

19(d) and (e)

The answers to Parts (c) and (d) in Example 7 are the same, and so are the results from Parts (e) and (f). This is no coincidence. There are two important properties involving intersections and unions of complements of sets. They are known as *De Morgan's properties:*

De Morgan's Properties

Let A and B be any two sets. Then

     (a) $\overline{A \cup B} = \overline{A} \cap \overline{B}$ **(b)** $\overline{A \cap B} = \overline{A} \cup \overline{B}$

De Morgan's properties state that all we need to do to form the complement of a union (or intersection) of sets is to form the complements of the individual sets and then change the union symbol to an intersection (or the intersection to a union). We shall employ Venn diagrams to verify De Morgan's properties.

(a) First we draw two diagrams, as shown in Figure 6.

Figure 6

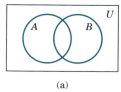 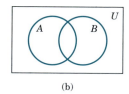

      (a)   (b)

We will use the diagram on the left for $\overline{A} \cap \overline{B}$ and the one on the right for $\overline{A \cup B}$.

Figure 7 illustrates the completed Venn diagrams of these sets.

Figure 7

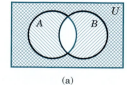 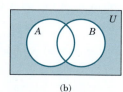

      (a)   (b)

Thus in Figure 7(a) $\overline{A} \cap \overline{B}$ is represented by the cross-hatched region, and in Figure 7(b) $\overline{A \cup B}$ is represented by the shaded region. Since these regions correspond, this illustrates that the two sets $\overline{A} \cap \overline{B}$ and $\overline{A \cup B}$ are equal.

(b) This verification is left to you. See Problem 24, part (c).

EXAMPLE 8   Using Venn Diagrams to Illustrate a Set

Use a Venn diagram to illustrate the set $(A \cup B) \cap C$.

SOLUTION   First we construct Figure 8(a). Then we hatch $A \cup B$ and C as in Figure 8(b). The cross-hatched region of Figure 8(b) is the set $(A \cup B) \cap C$.

Figure 8

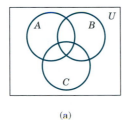

(a)

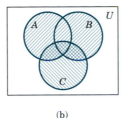

(b)

EXAMPLE 9   Using a Venn Diagram to Illustrate the Equality of Two Sets

Use a Venn diagram to illustrate the following equality:

$$A \cup B = (A \cap \overline{B}) \cup (A \cap B) \cup (\overline{A} \cap B)$$

SOLUTION   First we construct Figure 9. Now hatch the regions $A \cap \overline{B}$, $A \cap B$, and $\overline{A} \cap B$, as shown in Figure 10. The union of the three regions in Figure 10 is the set $A \cup B$. See Figure 11.

Figure 9

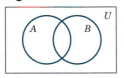

Figure 10

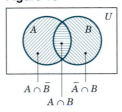

$A \cap \overline{B}$ $\quad$ $\overline{A} \cap B$

$A \cap B$

Figure 11

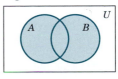

 **Now Work Problems**

23(a) and (g)

EXERCISE 6.1   Answers to odd-numbered problems begin on page AN-23.

In Problems 1–10 tell whether the given statement is true or false.

1. $\{1, 3, 2\} = \{2, 1, 3\}$

2. $\{2, 3\} \subseteq \{1, 2, 3\}$

3. $\{6, 7, 9\} \subseteq \{1, 6, 9\}$

4. $\{4, 8, 2\} = \{2, 4, 6, 8\}$

5. $\{6, 7, 9\} \subset \{1, 6, 9\}$

6. $\{4, 8, 2\} \subset \{2, 4, 6, 8\}$

7. $\{1, 2\} \cap \{2, 3, 4\} = \{2\}$

8. $\{2, 3\} \cap \{2, 3, 4\} = \{2, 3, 4\}$

9. $\{4, 5\} \cap \{1, 2, 3, 4\} \subseteq \{4\}$

10. $\{1, 4\} \cup \{2, 3\} \subseteq \{1, 2, 3, 4\}$

In Problems 11–18 write each expression as a single set.

11. $\{1, 2, 3\} \cap \{2, 3, 4, 5\}$

12. $\{0, 1, 3\} \cup \{2, 3, 4\}$

13. $\{1, 2, 3\} \cup \{2, 3, 4, 5\}$

14. $\{0, 1, 2\} \cap \{2, 3, 4\}$

15. $\{2, 4, 6, 8\} \cap \{1, 3, 5, 7\}$

16. $\{2, 4, 6\} \cup \{1, 3, 5\}$

17. $\{a, b, e\} \cup \{d, e, f, q\}$

18. $\{a, e, m\} \cup \{p, o, m\}$

19. If $U = $ Universal set $= \{0, 1, 2, 3, 4, 5, 6, 7, 8, 9\}$ and if $A = \{0, 1, 5, 7\}$, $B = \{2, 3, 5, 8\}$, $C = \{5, 6, 9\}$, find

    (a) $A \cup B$ (b) $B \cap C$

    (c) $A \cap B$ (d) $\overline{A \cap B}$

    (e) $\overline{A} \cap \overline{B}$ (f) $A \cup (B \cap A)$

    (g) $(C \cap A) \cap (\overline{A})$ (h) $(A \cap B) \cup (B \cap C)$

20. If $U = $ Universal set $= \{1, 2, 3, 4, 5\}$ and if $A = \{3, 5\}$, $B = \{1, 2, 3\}$, $C = \{2, 3, 4\}$, find

    (a) $\overline{A} \cap \overline{C}$ (b) $(A \cup B) \cap C$

    (c) $A \cup (B \cap C)$ (d) $(A \cup B) \cap (A \cup C)$

    (e) $\overline{A} \cap \overline{C}$ (f) $\overline{A \cup B}$

    (g) $\overline{A} \cap \overline{B}$ (h) $(A \cap B) \cup C$

21. Let $U = \{$All letters of the alphabet$\}$, $A = \{b, c, d\}$, and $B = \{c, e, f, g\}$. List the elements of the sets:

    (a) $A \cup B$ (b) $A \cap B$

    (c) $\overline{A} \cap \overline{B}$ (d) $\overline{A} \cup \overline{B}$

22. Let $U = \{a, b, c, d, e, f\}$, $A = \{b, c\}$, and $B = \{c, d, e\}$. List the elements of the sets:

(a) $A \cup B$ (b) $A \cap B$

(c) $\overline{A}$ (d) $\overline{B}$

(e) $\overline{A \cap B}$ (f) $\overline{A \cup B}$

23. Use Venn diagrams to illustrate the following sets:

(a) $\overline{A} \cap B$ (b) $(\overline{A} \cap \overline{B}) \cup C$

(c) $A \cap (A \cup B)$ (d) $A \cup (A \cap B)$

(e) $(A \cup B) \cap (A \cup C)$

(f) $A \cup (B \cap C)$

(g) $A = (A \cap B) \cup (A \cap \overline{B})$

(h) $B = (A \cap B) \cup (\overline{A} \cap B)$

24. Use Venn diagrams to illustrate the following properties:

(a) $A \cap (B \cup C) = (A \cap B) \cup (A \cap C)$
    (Distributive property)

(b) $A \cap (A \cup B) = A$ (Absorption property)

(c) $\overline{A \cap B} = \overline{A} \cup \overline{B}$ (De Morgan's property)

(d) $(A \cup B) \cup C = A \cup (B \cup C)$
    (Associative property)

In Problems 25–30 use

$A = \{x | x$ is a customer of IBM$\}$

$B = \{x | x$ is a secretary employed by IBM$\}$

$C = \{x | x$ is a computer operator at IBM$\}$

$D = \{x | x$ is a stockholder of IBM$\}$

$E = \{x | x$ is a member of the Board of Directors of IBM$\}$

to describe each set in words.

25. $A \cap E$ **26.** $B \cap D$ **27.** $A \cup D$

28. $C \cap E$ **29.** $\overline{A} \cap D$ **30.** $A \cup \overline{D}$

In Problems 31–36 use

$U = \{$All college students$\}$

$M = \{$All male students$\}$

$S = \{$All students who smoke$\}$

$F = \{$All Freshmen$\}$

to describe each set in words.

31. $M \cap S$ **32.** $M \cup S$ **33.** $\overline{M} \cup \overline{F}$

34. $\overline{M} \cap \overline{S}$ **35.** $F \cap S \cap M$ **36.** $F \cup S \cup M$

37. List all the subsets of $\{a, b, c\}$.

38. List all the subsets of $\{a, b, c, d\}$.

6.2 THE NUMBER OF ELEMENTS IN A SET

When we count objects, what we are actually doing is taking each object to be counted and matching each of these objects exactly once to the counting numbers 1, 2, 3, and so on, until *no* objects remain. Even before numbers had names and symbols assigned to them, this method of counting was used. Prehistoric peoples used rocks to determine how many cattle did not return from pasture. As each cow left, a rock was placed aside. As each cow returned, a rock was removed from the pile. If rocks remained after all the cows returned, it was then known that some cows were missing.

If A is any set, we will denote by $c(A)$ the number of elements in A. Thus, for example, for the set L of letters in the alphabet,

$$L = \{a, b, c, d, e, f, \ldots , x, y, z\}$$

we write $c(L) = 26$ and say "the number of elements in L is 26."

Also, for the set

$$N = \{1, 2, 3, 4, 5\}$$

we write $c(N) = 5$.

The empty set $\varnothing$ has no elements, and we write

$$c(\varnothing) = 0$$

If the number of elements in a set is zero or a positive integer, we say that the set is **finite.** Otherwise, the set is said to be **infinite.** The area of mathematics that deals with the study of finite sets is called **finite mathematics.**

EXAMPLE 1   Analyzing Survey Data

A survey of a group of people indicated there were 25 with brown eyes and 15 with black hair. If 10 people had both brown eyes and black hair and 23 people had neither, how many people were interviewed?

SOLUTION   Let A denote the set of people with brown eyes and B the set of people with black hair. Then the data given tell us

$$c(A) = 25 \qquad c(B) = 15 \qquad c(A \cap B) = 10$$

The number of people with either brown eyes or black hair cannot be $c(A) + c(B)$, since those with both would be counted twice. The correct procedure then would be to subtract those with both. That is,

$$c(A \cup B) = c(A) + c(B) - c(A \cap B) = 25 + 15 - 10 = 30$$

The sum of people found either in A or in B and those found neither in A nor in B is the total interviewed. Thus the number of people interviewed is

$$30 + 23 = 53$$

Figure 12 illustrates how a Venn diagram can be used to represent this situation.

Figure 12

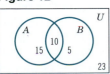

In Example 1 we discovered the following important relationship:

> **Counting Formula**
>
> Let A and B be two finite sets. Then
>
> $$c(A \cup B) = c(A) + c(B) - c(A \cap B) \tag{1}$$

EXAMPLE 2   Using the Counting Formula

Let $A = \{a, b, c, d, e\}$, $B = \{a, e, g, u, w, z\}$. Find $c(A)$, $c(B)$, $c(A \cap B)$, and $c(A \cup B)$.

SOLUTION $c(A) = 5$ and $c(B) = 6$. To find $c(A \cap B)$, we note that $A \cap B = \{a, e\}$ so that $c(A \cap B) = 2$. Since $A \cup B = \{a, b, c, d, e, g, u, w, z\}$, we have $c(A \cup B) = 9$. This checks with the formula

$$c(A \cup B) = c(A) + c(B) - c(A \cap B) = 5 + 6 - 2 = 9$$

 **Now Work Problem 7**   ■

Applications

EXAMPLE 3   Analyzing a Consumer Survey

In a survey of 75 consumers, 12 indicated they were going to buy a new car, 18 said they were going to buy a new refrigerator, and 24 said they were going to buy a new stove. Of these, 6 were going to buy both a car and a refrigerator, 4 were going to buy a car and a stove, and 10 were going to buy a stove and refrigerator. One person indicated he was going to buy all three items.

(a) How many were going to buy none of these items?
(b) How many were going to buy only a car?
(c) How many were going to buy only a stove?
(d) How many were going to buy only a refrigerator?
(e) How many were going to buy a stove and refrigerator, but not a car?

SOLUTION   We denote the sets of people buying cars, refrigerators, and stoves by C, R, and S, respectively. Then we know from the data given that

$$c(C) = 12 \qquad c(R) = 18 \qquad c(S) = 24$$
$$c(C \cap R) = 6 \qquad c(C \cap S) = 4 \qquad c(S \cap R) = 10$$
$$c(C \cap R \cap S) = 1$$

We use the information given above in the reverse order and put it into a Venn diagram. Thus, beginning with the fact that $c(C \cap R \cap S) = 1$, we place a 1 in that set, as shown in Figure 13(a). Now $c(C \cap R) = 6$, $c(C \cap S) = 4$, and $c(S \cap R) = 10$. Thus we place $6 - 1 = 5$ in the proper region (giving a total of 6 in the set $C \cap R$). Similarly, we place 3 and 9 in the proper regions for the sets $C \cap S$ and $S \cap R$. See Figure 13(b). Now $c(C) = 12$ and 9 of these 12 are already accounted for. Also, $c(R) = 18$ with 15

Figure 13

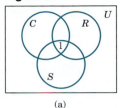
(a)

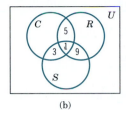
(b)

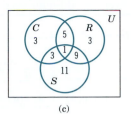
(c)

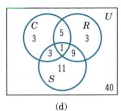
(d)

accounted for and $c(S) = 24$ with 13 accounted for. See Figure 13(c). Finally, the number in $C \cup R \cup S$ is the total of 75 less those accounted for in C, R, and S, namely, $3 + 5 + 1 + 3 + 3 + 9 + 11 = 35$. Thus

$$c\overline{(C \cup R \cup S)} = 75 - 35 = 40$$

See Figure 13(d). From this figure, we can see that (a) 40 were going to buy none of the items; (b) 3 were going to buy only a car; (c) 11 were going to buy only a stove; (d) 3 were going to buy only a refrigerator; and (e) 9 were going to buy a stove and refrigerator, but not a car.

 **Now Work Problem 17**

EXAMPLE 4   Analyzing Data

In a survey of 10,281 people restricted to those who were either black or male or over 18 years of age, the following data were obtained:

Black:	3490	Black males:	1745	Black male over 18:	239
Male:	5822	Over 18 and male:	859		
Over 18:	4722	Over 18 and black:	1341		

The data are inconsistent. Why?

SOLUTION   We denote the set of people who were black by B, male by M, and over 18 by H. Then we know that

$$c(B) = 3490 \qquad c(M) = 5822 \qquad c(H) = 4722$$
$$c(B \cap M) = 1745 \qquad c(H \cap M) = 859 \qquad c(H \cap B) = 1341$$
$$c(H \cap M \cap B) = 239$$

Figure 14

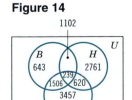

Since $H \cap M \cap B \neq \varnothing$, we use the Venn diagram shown in Figure 14. This means that

$$239 + 1102 + 620 + 1506 + 3457 + 643 + 2761 = 10{,}328$$

people were interviewed. However, it is given that only 10,281 were interviewed. This means the data are inconsistent.

EXERCISE 6.2   Answers to odd-numbered problems begin on page AN-24.

In Problems 1–6 use the sets A = {1, 2, 3, 4, 5, 6} and B = {2, 4, 6, 8} to find the number of elements in each set.

1. A **2.** B **3.** $A \cap B$ **4.** $A \cup B$ **5.** $(A \cap B) \cup A$ **6.** $(B \cap A) \cup B$

7. Find $c(A \cup B)$, given that $c(A) = 4$, $c(B) = 3$, and $c(A \cap B) = 2$.

8. Find $c(A \cup B)$, given that $c(A) = 14$, $c(B) = 11$, and $c(A \cap B) = 6$.

9. Find $c(A \cap B)$, given that $c(A) = 5$, $c(B) = 4$, and $c(A \cup B) = 7$.

10. Find $c(A \cap B)$, given that $c(A) = 8$, $c(B) = 9$, and $c(A \cup B) = 16$.

11. Find $c(A)$, given that $c(B) = 8$, $c(A \cap B) = 4$, and $c(A \cup B) = 14$.

12. Find $c(B)$, given that $c(A) = 10$, $c(A \cap B) = 5$, and $c(A \cup B) = 29$.

13. Motors, Inc., manufactured 325 cars with automatic transmissions, 216 with power steering, and 89 with both these options. How many cars were manufactured if every car has at least one option?

14. Suppose that out of 1500 first-year students at a certain college, 350 are taking history, 300 are taking mathematics, and 270 are taking both history and mathematics. How many first-year students are taking history or mathematics?

In Problems 15–24 use the data in the figure to answer each question.

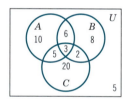

15. How many elements are in set A?

16. How many elements are in set B?

17. How many elements are in A or B?

18. How many elements are in B or C?

19. How many elements are in A but not B?

20. How many elements are in B but not C?

21. How many elements are in A or B or C?

22. How many elements are in neither A nor B nor C?

23. How many elements are in A and B and C?

24. How many elements are in U?

25. Voting Patterns   Suppose the influence of religion and age on voting preference is given by the table on the right.

Find

(a) The number of voters who are Catholic or Republican or both.

(b) The number of voters who are Catholic or over 54 or both.

(c) The number voting Democratic below 35 or over 54.

	Age		
	Below 35	**35–54**	**Over 54**
Protestant Voting Republican	82	152	111
Protestant Voting Democratic	42	33	15
Catholic Voting Republican	27	33	7
Catholic Voting Democratic	44	47	33

26. The Venn diagram illustrates the number of seniors (S), female students (F), and students on the dean's list (D) at a small western college. Describe each number in terms of the sets S, F, or D.

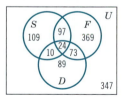

27. At a small midwestern college:

31	female seniors were on the dean's list
62	women were on the dean's list who were not seniors
45	male seniors were on the dean's list
87	female seniors were not on the dean's list
96	male seniors were not on the dean's list
275	women were not seniors and were not on the dean's list
89	men were on the dean's list who were not seniors
227	men were not seniors and were not on the dean's list

(a) How many were seniors?
(b) How many were women?
(c) How many were on the dean's list?
(d) How many were seniors on the dean's list?
(e) How many were female seniors?
(f) How many were women on the dean's list?
(g) How many were students at the college?

28. Survey Analysis In a survey of 75 college students, it was found that of the three weekly news magazines *Time, Newsweek,* and *U.S. News and World Report:*

23	read *Time*
18	read *Newsweek*
14	read *U.S. News and World Report*
10	read *Time* and *Newsweek*
9	read *Time* and *U.S. News and World Report*
8	read *Newsweek* and *U.S. News and World Report*
5	read all three

(a) How many read none of these three magazines?
(b) How many read *Time* alone?

(c) How many read *Newsweek* alone?
(d) How many read *U.S. News and World Report* alone?
(e) How many read neither *Time* nor *Newsweek?*
(f) How many read *Time* or *Newsweek* or both?

29. Car Sales Of the cars sold during the month of July, 90 had air conditioning, 100 had automatic transmissions, and 75 had power steering. Five cars had all three of these extras. Twenty cars had none of these extras. Twenty cars had only air conditioning; 60 cars had only automatic transmissions; and 30 cars had only power steering. Ten cars had both automatic transmission and power steering.

(a) How many cars had both power steering and air conditioning?
(b) How many had both automatic transmission and air conditioning?
(c) How many had neither power steering nor automatic transmission?
(d) How many cars were sold in July?
(e) How many had automatic transmission or air conditioning or both?

30. Incorrect Information A staff member at a large engineering school was presenting data to show that the students there received a liberal education as well as a scientific one. "Look at our record," she said. "Out of our senior class of 500 students, 281 are taking English, 196 are taking English and history, 87 are taking history and a foreign language, 143 are taking a foreign language and English, and 36 are taking all of these." She was fired. Why?

31. Blood Classification Blood is classified as being either Rh-positive or Rh-negative and according to type. If blood contains an A antigen, it is type A; if it has a B antigen, it is type B; if it has both A and B antigens, it is type AB; and if it has neither antigen, it is type O. Use a Venn diagram to illustrate these possibilities. How many different possibilities are there?

32. Survey Analysis A survey of 52 families from a suburb of Chicago indicated that there was a total of 241 children below the age of 18. Of these, 109 were male; 132 were below the age of 11; 143 had played Little League; 69 males were below the age of 11; 45 females under 11 had played Little League; and 30 males under 11 had played Little League.

(a) How many children over 11 and under 18 had played Little League?
(b) How many females under 11 did not play Little League?

33. Survey Analysis   Of 100 personal computer users surveyed: 27 use IBM; 35 use Apple; 35 use AT&T; 10 use both IBM and Apple; 10 use both IBM and AT&T; 10 use both Apple and AT&T; 3 use all three; and 30 use another computer brand. How many people exclusively use one of the three brands mentioned, that is, only IBM or only Apple or only AT&T?

34. List all the subsets of {*a, b, c*}. How many are there?

35. List all the subsets of {*a, b, c, d*}. How many are there?

6.3 THE MULTIPLICATION PRINCIPLE

In this section we introduce a general principle of counting, called the *Multiplication Principle*. We begin with two examples.

EXAMPLE 1   **Counting the Number of Ways a Certain Trip Can Be Taken**

In traveling from New York to Los Angeles, Mr. Doody wishes to stop over in Chicago. If he has five different routes to choose from in driving from New York to Chicago and has three routes to choose from in driving from Chicago to Los Angeles, in how many ways can Mr. Doody travel from New York to Los Angeles?

SOLUTION   The task of traveling from New York to Los Angeles is composed of two consecutive operations:

Choose a route from New York to Chicago Task 1	Choose a route from Chicago to Los Angeles Task 2

In Figure 15 we see after the five routes from New York to Chicago there are three routes from Chicago to Los Angeles.

Figure 15

These different routes can be enumerated as

1*A*, 1*B*, 1*C*   2*A*, 2*B*, 2*C*   3*A*, 3*B*, 3*C*   4*A*, 4*B*, 4*C*   5*A*, 5*B*, 5*C*

Thus, in all, there are 5 · 3 = 15 different routes.

Notice that the total number of ways the trip can be taken is simply the product of the number of ways of doing Task 1 with the number of ways of doing Task 2.

The different routes in Example 1 can also be depicted in a **tree diagram.** See Figure 16.

Figure 16

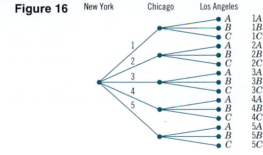

 **Now Work Problem 1**

EXAMPLE 2   Counting the Number of Ways Four Offices Can Be Filled

In a city election there are four candidates for mayor, three candidates for vice-mayor, six candidates for treasurer, and two for secretary. In how many ways can these four offices be filled?

SOLUTION   The task of filling an office can be divided into four consecutive operations:

| Select a mayor | Select a vice-mayor | Select a treasurer | Select a secretary |

Corresponding to each of the four possible mayors, there are three vice-mayors. These two offices can be filled in $4 \cdot 3 = 12$ different ways. Also, corresponding to each of these 12 possibilities, we have six different choices for treasurer—giving $12 \cdot 6 = 72$ different possibilities. Finally, to each of these 72 possibilities there correspond two choices for secretary. Thus, in all, these offices can be filled in $4 \cdot 3 \cdot 6 \cdot 2 = 144$ different ways. A partial illustration is given by the tree diagram in Figure 17.

Figure 17

Mayor   Vice-mayor   Treasurer   Secretary

The examples just solved demonstrate a general type of counting problem, which can be solved by the Multiplication Principle.

> **The Multiplication Principle**
>
> If we can perform a first task in p different ways, a second task in q different ways, a third task in r different ways, . . . , then the total act of performing the first task, followed by performing the second task, and so on, can be done in $p \cdot q \cdot r \cdot$ . . . different ways.

EXAMPLE 3   Combination Locks

A particular type of combination lock has 10 numbers on it.

(a) How many sequences of four numbers can be formed to open the lock?
(b) How many sequences can be formed if no number is repeated?

SOLUTION

(a) Each of the four numbers can be chosen in 10 ways. Therefore, there are

$$10 \cdot 10 \cdot 10 \cdot 10 = 10,000$$

different sequences.

(b) If no number can be repeated, then there are 10 choices for the first number, only 9 for the second number, 8 for the third number, and 7 for the fourth number. By the Multiplication Principle, there are

$$10 \cdot 9 \cdot 8 \cdot 7 = 5040$$

different sequences.

EXAMPLE 4   Batting Orders in Baseball

(a) How many possible batting orders can the manager of a baseball team construct from nine available players?
(b) If the manager adheres to the rule that the pitcher always bats last and the star homerun hitter is always fourth (cleanup), then how many batting orders are possible from nine given players?

SOLUTION

(a) In the first slot any of the nine players can be chosen. In the second slot any one of the remaining eight can be chosen, and so on, so that there are

$$9 \cdot 8 \cdot 7 \cdot 6 \cdot 5 \cdot 4 \cdot 3 \cdot 2 \cdot 1 = 362,880$$

possible batting orders.

(b) Since two of the players have designated spots in the batting order, there are seven remaining positions to be filled. The first position can be filled in any one of seven ways, the second in any of six ways, the third in any of five ways, the fourth in four ways, the fifth in three ways, the sixth in two ways, and the seventh in one way, so that here there are

$$7 \cdot 6 \cdot 5 \cdot 4 \cdot 3 \cdot 2 \cdot 1 = 5040$$

different batting orders of nine players after two have been designated.

 **Now Work Problem 9**

EXAMPLE 5   License Plates in Maryland

License plates in the state of Maryland consist of three letters of the alphabet followed by three digits.

(a) The Maryland system will allow how many possible license plates?
(b) Of these, how many will have all their digits distinct?
(c) How many will have distinct digits and distinct letters?

SOLUTION

(a) There are six positions on the plate to be filled, the first three by letters and the last three by digits. The positions 1, 2, and 3 can be filled in any one of 26 ways, while the remaining positions can each be filled in any of 10 ways. The total number of plates, by the Multiplication Principle, is then

$$26 \cdot 26 \cdot 26 \cdot 10 \cdot 10 \cdot 10 = 17,576,000$$

(b) Here, the tasks involved in filling the digit positions are slightly different. The first digit can be any one of 10, but the second digit can be only any one of 9 (we cannot duplicate the first digit); there are only 8 choices for the third digit (we cannot duplicate either the first or the second). Thus there are

$$26 \cdot 26 \cdot 26 \cdot 10 \cdot 9 \cdot 8 = 12,654,720$$

plates with no repeated digit.

(c) If the letters and digits are each to be distinct, then the total number of possible license plates is

$$26 \cdot 25 \cdot 24 \cdot 10 \cdot 9 \cdot 8 = 11,232,000$$

EXERCISE 6.3   Answers to odd-numbered problems begin on page AN-24.

1. There are 2 roads between towns A and B. There are 4 roads between towns B and C. How many different routes may one travel between towns A and C?

2. A woman has 4 blouses and 5 skirts. How many different outfits can she wear?

3. XYZ Company wants to build a complex consisting of a factory, office building, and warehouse. If the building contractor has 3 different kinds of factories, 2 different office buildings, and 4 different warehouses, how many models must be built to show all possibilities to XYZ Company?

4. Cars, Inc., has 3 different car models and 8 color schemes. If you are one of the dealers, how many cars must you display to show each possibility?

5. A man has 3 pairs of shoes, 8 pairs of socks, 4 pairs of slacks, and 9 sweaters. How many outfits can he wear?

6. There are 18 teachers in a math department. A student is asked to indicate her favorite and her least favorite. In how many ways is this possible?

7. A house has 3 outside doors and 12 windows. In how many ways can a person enter the house through a window and exit through a door?

8. A woman has 4 pairs of gloves. In how many ways can she select a right-hand glove and a left-hand glove that do not match?

9. A corporation has a board of directors consisting of 10 members. The board must select from its members a chairman, vice-chairman, and a secretary. In how many ways can this be done?

10. How many license plates consisting of 2 letters followed by 2 digits are possible?

11. A restaurant offers 3 different salads, 8 different main courses, 10 different desserts, and 4 different drinks. How many different lunches—each consisting of a salad, a main course, a dessert, and a drink—are possible?

12. Five different mathematics books are to be arranged on a student's desk. How many arrangements are possible?

13. How many ways can 6 people be seated in a row of 6 seats? 8 people be seated in a row of 8 seats?

14. How many different arrangements can be formed using the 5 letters of the word CLIPS?

15. How many 4-letter code words are possible from the first 6 letters of the alphabet with no letters repeated? How many when letters are allowed to repeat?

16. Find the number of 7-digit telephone numbers

   (a) With no repeated digits (lead 0 is allowed)
   (b) With no repeated digits (lead 0 not allowed)
   (c) With repeated digits allowed including a lead 0

17. How many ways are there to rank 7 candidates who apply for a job?

18. (a) How many different ways are there to arrange the 7 letters in the word PROBLEM?
   (b) If we insist that the letter P comes first, how many ways are there?
   (c) If we insist that the letter P comes first and the letter M last, how many ways are there?

19. **Testing**   On a math test there are 10 multiple-choice questions with 4 possible answers and 15 true–false questions. In how many possible ways can the 25 questions be answered?

20. Using the digits 1, 2, 3, and 4, how many different 4-digit numbers can be formed?

21. **License Plate Possibilities**   How many different license plate numbers can be made using 2 letters followed by 4 digits, if

   (a) Letters and digits may be repeated?
   (b) Letters may be repeated, but digits are not repeated?
   (c) Neither letters nor digits may be repeated?

22. **Security**   A system has 7 switches, each of which may be either open or closed. The state of the system is described by indicating for each switch whether it is open or closed. How many different states of the system are there?

23. **Product Choice**   An automobile manufacturer produces 3 different models. Models A and B can come in any of 3 body styles; model C can come in only 2 body styles. Each car also comes in either black or green. How many distinguishable car types are there?

24. **Telephone Numbers**   How many 7-digit numbers can be formed if the first digit cannot be 0 or 9 and if the last digit is greater than or equal to 2 and less than or equal to 3? Repeated digits are allowed.

25. **Home Choices**   A contractor constructs homes with 5 different choices of exterior finish, 3 different roof arrangements, and 4 different window designs. How many different types of homes can be built?

26. **License Plate Possibilities**   A license plate consists of 1 letter, excluding O and I, followed by a 4-digit number that cannot have a 0 in the lead position. How many different plates are possible?

27. **Bytes**   Using only the digits 0 and 1, how many different numbers consisting of 8 digits can be formed?

28. **Stock Portfolios**   As a financial planner, you are asked to select one stock from each of the following groups: 8

DOW stocks, 15 NASDAQ stocks, and 4 global stocks. How many different portfolios are possible?

29. **Combination Locks**   A combination lock has 50 numbers on it. To open it, you turn counterclockwise to a number, then rotate clockwise to a second number, and then counterclockwise to the third number. How many different lock combinations are there?

30. **Opinion Polls**   An opinion poll is to be conducted among college students. Eight multiple-choice questions, each with 3 possible answers, will be asked. In how many different ways can a student complete the poll, if exactly one response is given to each question?

31. **Path Selection in a Maze**   The maze on the right is constructed so that a rat must pass through a series of oneway doors. How many different paths are there from start to finish?

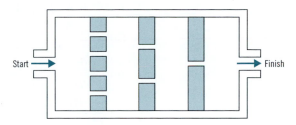

32. How many 3-letter code words are possible using the first 10 letters of the alphabet if

   (a) No letter can be repeated?
   (b) Letters can be repeated?
   (c) Adjacent letters cannot be the same?

6.4 PERMUTATIONS

In the next two sections we use the Multiplication Principle to discuss two general types of counting problems, called *permutations* and *combinations*. These concepts arise often in applications, especially in probability.

Factorial

Before discussing permutations, we introduce a useful shorthand notation—the *factorial symbol.*

Factorial

   The symbol $n!$, read as **"n factorial,"** is defined as

$$0! = 1$$
$$1! = 1$$
$$2! = 2 \cdot 1 \qquad = 2$$
$$3! = 3 \cdot 2 \cdot 1 \quad = 6$$
$$4! = 4 \cdot 3 \cdot 2 \cdot 1 = 24$$

and, in general, for $n \geq 1$ an integer,

$$n! = n \cdot (n - 1) \cdot (n - 2) \cdot \ \ldots \ \cdot 3 \cdot 2 \cdot 1$$

Thus to compute $n!$, we find the product of all consecutive integers from n down to 1, inclusive, or from 1 up to n, inclusive. Remember that by definition, $0! = 1$.

A formula we shall find useful is

$$(n + 1)! = (n + 1) \cdot n!$$

EXAMPLE 1   Evaluating Expressions Containing Factorials

(a) $4! = 4 \cdot 3 \cdot 2 \cdot 1 = 24$

(b) $\dfrac{5!}{4!} = \dfrac{5 \cdot \cancel{4!}}{\cancel{4!}} = 5$

(c) $\dfrac{52!}{5!47!} = \dfrac{52 \cdot 51 \cdot 50 \cdot 49 \cdot 48 \cdot \cancel{47!}}{5 \cdot 4 \cdot 3 \cdot 2 \cdot 1 \cdot \cancel{47!}} = 2{,}598{,}960$

(d) $\dfrac{7!}{(7-5)!5!} = \dfrac{7!}{2!5!} = \dfrac{7 \cdot 6 \cdot \cancel{5!}}{2!\cancel{5!}} = \dfrac{7 \cdot 6}{2} = 21$

(e) $\dfrac{50 \cdot 49 \cdot 48 \cdot 47 \cdot 46}{50!} = \dfrac{50 \cdot 49 \cdot 48 \cdot 47 \cdot 46}{50 \cdot 49 \cdot 48 \cdot 47 \cdot 46 \cdot 45!} = \dfrac{1}{45!}$

✎ **Now Work Problem 1**

Factorials grow very quickly. Compare the following:

$$5! = 120$$
$$10! = 3{,}628{,}800$$
$$15! = 1{,}307{,}674{,}368{,}000$$

In fact, if your calculator has a factorial key, you will find that $69! = 1.71 \cdot 10^{98}$, while 70! produces an error message—indicating you have exceeded the range of the calculator. Because of this, it is important to "cancel out" factorials whenever possible so that "out of range" errors may be avoided. Thus to calculate 100!/95!, we write

$$\frac{100!}{95!} = \frac{100 \cdot 99 \cdot 98 \cdot 97 \cdot 96 \cdot \cancel{95!}}{\cancel{95!}} = 9.03 \cdot 10^{9}$$

Permutations

We start with an example.

EXAMPLE 2   Forming Different Words

Figure 18

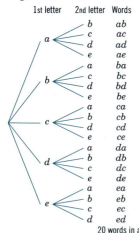

1st letter   2nd letter   Words

20 words in all

Suppose we are setting up a code of two-letter "words" and have five different letters, a, b, c, d, e, from which to choose. If the code must not repeat any letter and if such words as ab and ba are considered different, how many different words can be formed?

SOLUTION   We solve the problem by using the Multiplication Principle. In selecting a first letter, we have five choices. Since whatever letter is chosen cannot be repeated, we have four choices available for the second letter. In all, then, there are $5 \cdot 4 = 20$ words of two letters that can be formed. See Figure 18 for a tree diagram of this solution.

A way of rephrasing the question posed in Example 2 would be to ask: How many ordered arrangements using two distinct letters can be formed from the five letters a through e?

In general, we could ask: Given n distinct objects, how many ordered arrangements can be formed using r of the objects where $r \le n$?

> **Permutation**
>
>   Ordered arrangements of distinct objects are called **permutations.** An r **permutation of a set of** n **distinct objects** is an ordered arrangement using r of the n objects. $P(n, r)$ is defined to be the *number* of r permutations of a set of n distinct objects.

$P(n, r)$ is also referred to as the number of **permutations of n different objects taken r at a time.**

Thus $P(n, r)$ is the number of ordered arrangements that can be formed using r objects chosen from a set of n distinct objects.

In computing $P(n, r)$, we want to find the number of possible different arrangements of r objects that are chosen from n distinct objects in which no item is repeated and order is important. Thus the symbol $P(5, 2)$, means to rearrange two items chosen from five items. Since there are five choices for the first item and four for the second (no repetitions allowed), we find that

$$P(5, 2) = \underbrace{5 \cdot 4}_{2 \text{ factors}} = 20$$

Other examples are

$$P(7, 3) = \underbrace{7 \cdot 6 \cdot 5}_{3 \text{ factors}} = 210 \qquad P(5, 5) = \underbrace{5 \cdot 4 \cdot 3 \cdot 2 \cdot 1}_{5 \text{ factors}} = 5! = 120$$

✐ **Now Work Problem 13**   To find a formula for $P(n, r)$, we note that the first entry can be filled by any one of the n possibilities, the second by any one of the remaining $(n - 1)$ possibilities, the third by any one of the now remaining $(n - 2)$ possibilities, and so on. Since there are r positions to be filled, the number of possibilities is

$$P(n, r) = \underbrace{n(n - 1)(n - 2) \cdot \ldots}_{r \text{ factors}}$$

To obtain the last factor in the expression for $P(n, r)$, we observe the following pattern:

> First factor is n
>
> Second factor is $n - 1$
>
> Third factor is $n - 2$
>
> .
>
> .
>
> .
>
> rth factor is $n - (r - 1) = n - r + 1$

Thus

$$P(n, r) = n(n - 1) \cdot \ldots \cdot (n - r + 1)$$

Multiplying the right side by 1 in the form of $(n - r)!/(n - r)!$, we obtain

$$P(n, r) = n(n - 1)(n - 2) \cdot \ldots \cdot (n - r + 1) \frac{(n - r)!}{(n - r)!}$$

Since $n(n - 1)(n - 2) \cdot \ldots \cdot (n - r + 1)(n - r)! = n!$, we obtain the following alternate formula for $P(n, r)$:

$$P(n, r) = \frac{n!}{(n - r)!}$$

Permutation Formula

The number of different arrangements using r objects chosen from n objects in which

1. The n objects are all different
2. No object is repeated in an arrangement
3. Order is important

is given by the formula*

$$P(n, r) = n(n - 1) \cdot \ldots \cdot (n - r + 1) = \frac{n!}{(n - r)!}$$

Here is a list of short problems with their solutions given in $P(n, r)$ notation.

Problem	Solution
Find the number of ways of choosing five people from a group of 10 and arranging them in a line.	$P(10, 5)$
Find the number of six-letter "words" that can be formed with no letter repeated.	$P(26, 6)$
Find the number of seven-digit telephone numbers, with no repeated digit (allow 0 for a first digit).	$P(10, 7)$
Find the number of ways of arranging eight people in a line.	$P(8, 8)$

Note that in all of the above examples, *order is important.*

EXAMPLE 3   Arranging Books on a Shelf

You own eight different mathematics books. How many ways can a shelf arrangement of five of the mathematics books be formed?

SOLUTION   We are seeking the number of arrangements using five of the eight books. The answer is given by

$$P(8, 5) = \frac{8!}{(8 - 5)!} = \frac{8!}{3!} = \frac{8 \cdot 7 \cdot 6 \cdot 5 \cdot 4 \cdot \cancel{3!}}{\cancel{3!}} = 8 \cdot 7 \cdot 6 \cdot 5 \cdot 4 = 6720$$

✏ **Now Work Problem 25**

---

* Some calculators have a key for computing $P(n, r)$. If yours has such a key, consult your manual to find out how to use it.

EXAMPLE 4   Answering Test Questions

A student has six questions on an examination and is allowed to answer the questions in any order. In how many different orders could the student answer these questions?

SOLUTION   The student wants the number of ordered arrangements of the six questions using all six of them. The number is given by

$$P(6, 6) = \frac{6!}{(6 - 6)!} = \frac{6!}{0!} = \frac{6!}{1} = 720$$

Example 4 leads us to formulate the next result.

> The number of permutations (arrangements) of n different objects using all n of them is given by
>
> $$P(n, n) = n!$$

For example, in a class of n students, there are $n!$ ways of positioning all the students in a line.

 Now Work Problem 33

EXAMPLE 5   Arranging Books on a Shelf

You own eight mathematics books and six computer science books and wish to fill seven positions on a shelf. If the first four positions are to be occupied by math books and the last three by computer science books, in how many ways can this be done?

SOLUTION   We think of the problem as consisting of two tasks. Task 1 is to fill the first four positions with four of the eight mathematics books. This can be done in $P(8, 4)$ ways. Task 2 is to fill the remaining three positions with three of six computer books. This can be done in $P(6, 3)$ ways. By the Multiplication Principle the seven positions can be filled in

$$P(8, 4) \cdot P(6, 3) = \frac{8!}{4!} \cdot \frac{6!}{3!} = 8 \cdot 7 \cdot 6 \cdot 5 \cdot 6 \cdot 5 \cdot 4 = 201{,}600 \text{ ways}$$

EXERCISE 6.4   Answers to odd-numbered problems begin on page AN-24.

In Problems 1–20 evaluate each expression.

1. $\dfrac{5!}{2!}$

2. $\dfrac{8!}{2!}$

3. $\dfrac{6!}{3!}$

4. $\dfrac{9!}{3!}$

5. $\dfrac{10!}{8!}$

6. $\dfrac{11!}{9!}$

7. $\dfrac{9!}{8!}$

8. $\dfrac{10!}{9!}$

9. $\dfrac{8!}{2!6!}$

10. $\dfrac{9!}{3!6!}$

11. $P(7, 2)$

12. $P(8, 1)$

13. $P(8, 7)$

14. $P(6, 6)$

15. $P(6, 0)$

16. $P(6, 4)$

17. $\dfrac{8!}{(8 - 3)!3!}$

18. $\dfrac{9!}{(9 - 5)!5!}$

19. $\dfrac{6!}{(6 - 6)!6!}$

20. $\dfrac{7!}{(0 - 0)!7!}$

21. (a) How many different ways are there to arrange the 6 letters in the word SUNDAY?
    (b) If we insist that the letter S come first, how many ways are there?
    (c) If we insist that the letter S come first and the letter Y be last, how many ways are there?

22. **Ranking Candidates**   How many ways are there to rank 8 candidates who apply for a job?

23. From a pool of 10 job applicants, a list ranking the top 4 must be made. How many such lists are possible?

24. **Forming Words**   How many different 5-letter "words" (sequences of letters) can be formed from the standard alphabet if repeated letters are not allowed? If repeated letters are allowed?

25. A station wagon has 9 seats. In how many different ways can 5 people be seated in it?

26. **Arranging Books**   There are 5 different French books and 5 different Spanish books. How many ways are there to arrange them on a shelf if
    (a) Books of the same language must be grouped together, French on the left, Spanish on the right?
    (b) French and Spanish books must alternate in the grouping, beginning with a French book?

27. **Distributing Books**   In how many ways can 8 different books be distributed to 12 children if no child gets more than one book?

28. A computer must assign each of 4 outputs to one of 8 different printers. In how many ways can it do this provided no printer gets more than one output?

29. **Lottery Tickets**   From 1500 lottery tickets that are sold, 3 tickets are to be selected for first, second, and third prizes. How many possible outcomes are there?

30. **Personnel Assignment**   A salesperson is needed in each of 7 different sales territories. If 10 equally qualified persons apply for the jobs, in how many ways can the jobs be filled?

31. **Choosing Officers**   A club has 15 members. In how many ways can 4 officers consisting of a president, vice-president, secretary, and treasurer be chosen?

32. **Psychology Testing**   In an ESP experiment a person is asked to select and arrange 3 cards from a set of 6 cards labeled A, B, C, D, E, and F. Without seeing the card, a second person is asked to give the arrangement. Determine the number of possible responses by the second person if he simply guesses.

33. How many ways are there to arrange 5 people in a line?

34. How many ways are there to seat 4 people in a 6-passenger automobile?

6.5   COMBINATIONS

Permutations focus on the order in which objects are arranged. However, in many cases, order is not important. For example, in a draw poker hand, the order in which you receive the cards is not relevant—all that matters is what cards are received. That is, with poker hands, we are concerned only with what *combination* of cards we have—not the particular order of the cards.

Combinations

The following example illustrates the distinction between selections in which order is important and those for which order is not important.

EXAMPLE 1   Arranging Letters

Figure 19

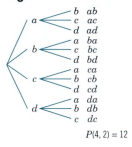

$P(4, 2) = 12$

From the four letters a, b, c, d, choose two without repeating any letter

(a) If order is important   (b) If order is not important

SOLUTION

(a) If order is important, there are $P(4, 2) = 4 \cdot 3 = 12$ possible selections, namely,

$$ab \quad ac \quad ad \quad ba \quad bc \quad bd \quad ca \quad cb \quad cd \quad da \quad db \quad dc$$

See Figure 19.

(b) If order is not important, only 6 of the 12 selections found in Part (a) are listed, namely,

$$ab \qquad ac \qquad ad \qquad bc \qquad bd \qquad cd$$

Notice that the number of ordered selections, 12, is $2! = 2$ times the number of unordered selections, 6. The reason is that each unordered selection consists of two letters that allow for $2!$ rearrangements. For example, the selection ab in the unordered list gives rise to ab and ba in the ordered list. ■

EXAMPLE 2   Arranging Letters

Figure 20

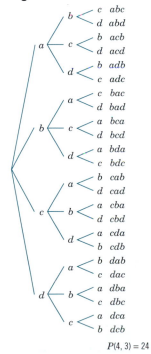

$P(4, 3) = 24$

From the four letters a, b, c, d, choose three without repeating any letter

(a) If order is important   (b) If order is not important

SOLUTION

(a) If order is important, there are $P(4, 3) = 4 \cdot 3 \cdot 2 = 24$ possible selections, namely,

$$abc \quad abd \quad acb \quad acd \quad adb \quad adc \quad bac \quad bad \quad bca \quad bcd \quad bda \quad bdc$$
$$cab \quad cad \quad cba \quad cbd \quad cda \quad cdb \quad dab \quad dac \quad dba \quad dbc \quad dca \quad dcb$$

See Figure 20.

(b) If order is not important, only 4 of the 24 selections found in Part (a) are listed, namely,

$$abc \qquad abd \qquad acd \qquad bcd$$

Notice that the number of ordered selections, 24, is $3! = 6$ times the number of unordered selections, 4. The reason is that each unordered selection consists of three letters that allow for $3!$ rearrangements. For example, the selection abc in the unordered list gives rise to abc, acb, bac, bca, cab, cba in the ordered list. ■

Unordered selections are called *combinations*.

> **Combinations**
>
> Unordered selections of distinct objects are called **combinations.** An r **combination of a set of n distinct objects** is an unordered selection using r of the n objects. $C(n, r)$ is defined to be the *number* of ways of choosing r distinct objects from a set of n distinct objects, $0 \le r \le n$, without regard to the order of the selection.

$C(n, r)$ is also referred to as the number of **combinations of n objects taken r at a time.**

To obtain a formula for $C(n, r)$, we first observe that each unordered selection of r objects will give rise to $r!$ ordered selections. Thus the number of ordered selections, $P(n, r)$, is $r!$ times the number of unordered selections, $C(n, r)$. That is,

$$P(n, r) = r!C(n, r)$$

Using the previously developed formula for $P(n, r)$, we find that

$$C(n, r) = \frac{P(n, r)}{r!} = \frac{n!}{r!(n - r)!}$$

Combination Formula

The number of different selections of r objects chosen from n objects in which

1. The n objects are all different
2. No object is repeated
3. Order is not important

is given by the formula*

$$C(n, r) = \frac{n!}{r!(n - r)!}$$

EXAMPLE 3   Evaluating $C(n, r)$

Compute:

(a) $C(50, 2)$ (b) $C(7, 5)$ (c) $C(7, 7)$ (d) $C(7, 0)$

SOLUTION

(a) $C(50, 2) = \dfrac{50!}{2!(50 - 2)!} = \dfrac{50!}{2!48!} = \dfrac{50 \cdot 49 \cdot \cancel{48!}}{2 \cdot \cancel{48!}} = 1225$

(b) $C(7, 5) = \dfrac{7!}{5!(7 - 5)!} = \dfrac{7!}{5!2!} = \dfrac{7 \cdot 6 \cdot \cancel{5!}}{\cancel{5!} \cdot 2} = 21$

*Some calculators have a key for computing $C(n, r)$. If yours has such a key, consult your manual to find out how to use it.

(c) $C(7, 7) = \dfrac{7!}{7!(7 - 7)!} = \dfrac{\cancel{7!}}{\cancel{7!}0!} = 1$

(d) $C(7, 0) = \dfrac{7!}{0!(7 - 0)!} = \dfrac{\cancel{7!}}{0!\cancel{7!}} = 1$

 **Now Work Problem 1**

In general,

$$C(n, 0) = \dfrac{n!}{0!(n - 0)!} = \dfrac{n!}{n!} = 1, \qquad C(n, n) = \dfrac{n!}{n!(n - n)!} = \dfrac{1}{0!} = 1$$

EXAMPLE 4 Forming Committees

From five faculty members a committee of two is to be formed. In how many ways can this be done?

SOLUTION The formation of committees is an example of a combination. The five faculty members are all different. The members of the committee are distinct. Order is not important. (On committees it is membership, not the order of selection, that is important.) Thus, for the situation described, we can form

$$C(5, 2) = 10$$
$$\qquad\quad \;= \dfrac{5!}{2!3!}$$

different committees.

Here are some other problems that are examples of combinations. The solutions are given in $C(n, r)$ notation.

Problem	Solution
Find the number of ways of selecting four people from a group of six	$C(6, 4)$
Find the number of committees of six that can be formed from the U.S. Senate (100 members)	$C(100, 6)$
Find the number of ways of selecting five courses from a catalog containing 200	$C(200, 5)$

EXAMPLE 5 Playing Cards

From a deck of 52 cards a hand of 5 cards is dealt. How many different hands are possible?

SOLUTION   Such a hand is an unordered selection of 5 cards from a deck of 52. So the number of different hands is

$$C(52, 5) = \frac{52!}{5!47!} = \frac{52 \cdot 51 \cdot 50 \cdot 49 \cdot 48 \cdot \cancel{47!}}{5 \cdot 4 \cdot 3 \cdot 2 \cdot 1 \cdot \cancel{47!}} = 2{,}598{,}960$$

■

EXAMPLE 6   Six-bit Strings

A bit is a 0 or a 1. A six-bit string is a sequence of length six consisting of 0s and 1s. How many six-bit strings contain

(a) Exactly one 1?   (b) Exactly two 1s?

SOLUTION

(a) To form a six-bit string having one 1, we only need to specify where the single 1 is located (the other positions are 0s). The location for the 1 can be chosen in

$$C(6, 1) = 6 \text{ ways}$$

(b) Here, we must choose two of the six positions to contain 1s. Hence, there are

$$C(6, 2) = 15 \text{ such strings}$$

■

EXAMPLE 7   Forming Committees

From five faculty members and four students a committee of four is to be chosen that includes two students and two faculty members. In how many ways can this be done?

SOLUTION   The faculty members can be chosen in $C(5, 2)$ ways. The students can be chosen in $C(4, 2)$ ways. By the Multiplication Principle there are then

$$C(5, 2) \cdot C(4, 2) = \frac{5!}{2!3!} \cdot \frac{4!}{2!2!} = 10 \cdot 6 = 60 \text{ different ways}$$

 Now Work Problem 17

■

EXAMPLE 8   Forming Committees

From six women and four men a committee of three is to be formed. The committee must include at least two women. In how many ways can this be done?

SOLUTION   A committee of three that includes at least two women will contain either exactly two women and one man, or exactly three women and zero men. Since no committee can contain exactly two women and simultaneously exactly three women, once we have counted the number of ways a committee of exactly two women and the number of ways a committee of exactly three women can be formed, their sum will give the number of ways exactly two women or exactly three women are on the committee. (Refer to Equation (1) on page 278, noting that the sets are disjoint.)

Following the solution to Example 7, a committee of exactly two women and one man can be formed from six women, four men in $C(6, 2) \cdot C(4, 1)$ ways, while a committee of exactly three women, zero men can be formed in $C(6, 3) \cdot C(4, 0)$ ways. Thus a committee of three consisting of at least two women can be formed in

$$C(6, 2) \cdot C(4, 1) + C(6, 3) \cdot C(4, 0) = \frac{6!}{4!2!} \cdot \frac{4!}{3!1!} + \frac{6!}{3!3!} \cdot \frac{4!}{4!0!}$$

$$= 15 \cdot 4 + 20 \cdot 1 = 60 + 20 = 80 \text{ ways}$$

Pascal's Triangle

Sometimes the notation $\binom{n}{r}$, read as "from n choose r," is used in place of $C(n, r)$. $\binom{n}{r}$ is called a **binomial coefficient** because of its connection with the binomial theorem (discussed in Section 6.7). A triangular display of $\binom{n}{r}$ for $n = 0$ to $n = 6$ is given in Figure 21. This triangular display is called **Pascal's triangle.**

For example, $\binom{5}{2} = 10$ is found in the row marked $n = 5$ and on the diagonal marked $r = 2$.

In the Pascal triangle successive entries can be obtained by adding the two nearest entries in the row above it. The shaded triangles in Figure 21 illustrate this. For example, $10 + 5 = 15$, etc.

Figure 21

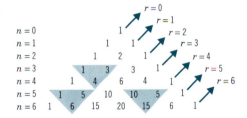

The Pascal triangle, as the figure indicates, is symmetric. Thus when n is even, the largest entry occurs in the middle, and corresponding entries on either side are equal. When n is odd, there are two equal middle entries with corresponding equal entries on either side.

The reasons behind these properties of Pascal's triangle as well as other properties of binomial coefficients are discussed in Section 6.7.

EXERCISE 6.5   Answers to odd-numbered problems begin on page AN-25.

In Problems 1–8 find the value of each expression.

1. $C(6, 4)$ **2.** $C(5, 4)$ **3.** $C(7, 2)$ **4.** $C(8, 7)$ **5.** $\binom{5}{1}$

6. $\binom{8}{1}$ **7.** $\binom{8}{6}$ **8.** $\binom{8}{4}$

9. In how many ways can a committee of 5 senators be selected from a group of 8 senators?

10. In how many ways can a committee of 5 representatives be selected from a group of 9 representatives?

11. A math department is allowed to tenure 4 of 17 eligible teachers. In how many ways can the selection for tenure be made?

12. How many different hands are possible in a bridge game? (A bridge hand consists of 13 cards dealt from a deck of 52 cards.)

13. There are 20 students in the Math Club. In how many ways can a subcommittee of 3 members be formed?

14. How many different relay teams of 4 persons can be chosen from a group of 10 runners?

15. **Basketball Teams**   The starting lineup of a basketball team consists of 2 guards, 2 forwards, and 1 center. A basketball team has 6 players who play guard. How many different starting lineups are possible, assuming the remaining 3 positions are filled and it is not possible to distinguish a left guard from a right guard?

16. **Basketball Teams**   On a basketball team of 12 players, 2 play only center, 3 play only guard, and the rest play forward (5 players on a team: 2 forwards, 2 guards, and 1 center). How many different starting lineups are possible, assuming it is not possible to distinguish left and right guards and left and right forwards?

17. The Student Affairs Committee has 3 faculty members, 2 administration members, and 5 students on it. In how many ways can a subcommittee of 1 faculty, 1 administrator, and 2 students be formed?

18. **Stock Trading**   Of 1520 stocks traded in 1 day on the New York Stock Exchange, 841 advanced, 434 declined, and the remainder were unchanged. In how many ways can this happen?

19. **Football Teams**   How many different ways can an offensive football team be formed from a squad that consists of 20 linemen, 3 quarterbacks, 8 halfbacks, and 4 fullbacks? This football team must have 1 quarterback, 2 halfbacks, 1 fullback, and 7 linemen.

20. **Baseball Teams**   How many different ways can a baseball team be made up from a squad of 25 players, if 8 players are used only as pitchers and the remaining players can be placed at any position except pitcher (9 players on a team)?

21. A little girl has 1 penny, 1 nickel, 1 dime, 1 quarter, and 1 half dollar in her purse. If she pulls out 3 coins, how many different sums are possible?

22. How many eight-bit strings contain exactly two 1s? Exactly three 1s?

23. **Lottery Tickets**   A state of Maryland million dollar lottery ticket consists of 6 distinct numbers chosen from the range 00 through 99. The order in which the numbers appear on the ticket is irrelevant. How many distinct lottery tickets can be issued?

24. How many 5-card poker hands contain all spades? (A deck of 52 cards contains 13 spades.)

25. How many committees of 5 can be formed from members of the U.S. Senate?

26. A test has 3 parts. In part 1 a student must do 3 of 5 questions, in part 2 a student must choose 2 of 4 questions, and in part 3 a student must pick 3 of 6 questions. In how many different ways can a student complete the test?

In Problems 27–35 use the Multiplication Principle, permutations, or combinations, as appropriate.

27. **Test Panel Selection**   A sample of 8 persons is selected for a test from a group containing 40 smokers and 15 nonsmokers. In how many ways can the 8 persons be selected?

28. **Resource Allocation**   A trucking company has 8 trucks and 6 drivers available when requests for 4 trucks are received. How many different ways are there of selecting the trucks and the drivers to meet these requests?

29. **Group Selection**   From a group of 5 people we are required to select a different person to participate in each of 3 different tests. In how many ways can the selections be made?

30. **Congressional Committees**   In the U.S. Congress a conference committee is to be composed of 5 senators and 4 representatives. In how many ways can this be done? (There are 435 representatives and 100 senators.)

31. **Quality Control**   A box contains 24 light bulbs. The quality control engineer will pick a sample of 4 light bulbs for inspection. How many different samples are there?

32. **Investment Selection**   An investor is going to invest $21,000 in 3 stocks from a list of 12 prepared by his broker. How many different investments are possible if

   (a) $7000 is to be invested in each stock?
   (b) $10,000 is to be invested in one stock, $6000 in another, and $5000 in the third?
   (c) $8000 is to be invested in each of 2 stocks and $5000 in a third stock?

33. Rating   A sportswriter makes a preseason guess of the top 15 university basketball teams (in order) from among 50 major university teams. How many different possibilities are there?

34. Packaging   A manufacturer produces 8 different items. He packages assortments of equal parts of 3 different items. How many different assortments can be packaged?

35. The digits 0 through 9 are written on 10 cards. Four different cards are drawn, and a 4-digit number is formed. How many different 4-digit numbers can be formed in this way?

6.6 MORE COUNTING PROBLEMS

In this section we consider some counting problems that will be useful in our discussion of probability.

Examples

The first example deals with a coin-tossing experiment in which a coin is tossed a fixed number of times. There are exactly two possible outcomes on each trial or toss (heads, H, or tails, T). For instance, in tossing a coin three times, one possible outcome is HTH—heads on the first toss, tails on the second toss, and heads on the third toss.

EXAMPLE 1   Tossing a Coin Four Times

Suppose an experiment consists of tossing a coin four times.

(a) How many different outcomes are possible?
(b) How many different outcomes have exactly 3 tails?
(c) How many outcomes have at most 2 tails?
(d) How many outcomes have at least 1 tail?

SOLUTION

(a) Each outcome of the experiment consists of a sequence of four letters H or T, where the first letter records the result of the first toss, the second letter the result of the second toss, and so forth. Thus the process can be visualized as the tree diagram in Figure 22.

Figure 22

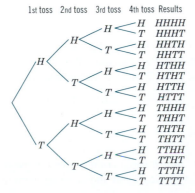

| 1st toss | 2nd toss | 3rd toss | 4th toss | Results |

We can also visualize the process as filling an empty box at each toss with either an H or a T.

1st toss	T			
2nd toss	T	H		
3rd toss	T	H	H	
4th toss	T	H	H	H

Since each box can be filled in two ways, by the Multiplication Principle the sequence of four boxes can be filled in

$$\underbrace{2 \cdot 2 \cdot 2 \cdot 2}_{4 \text{ factors}} = 2^4 = 16 \text{ ways}$$

Thus there are $2^4 = 16$ different outcomes.

(b) Any sequence that contains exactly 3 Ts must contain 1 H. A particular outcome is determined as soon as we decide where to place the Ts in the four boxes. The three boxes to receive the Ts can be selected from the four boxes in $C(4, 3)$ different ways. So the number of outcomes with exactly 3 tails is

$$C(4, 3) = \frac{4!}{3!1!} = 4$$

(c) The outcomes with at most 2 tails correspond to the sequences with 0, 1, or 2 Ts:

0 T Only one outcome is possible, namely, $HHHH$.

1 T These outcomes, which include $THHH$ and $HTHH$, are determined by selecting one box out of four in which to place the single T. This can be done in $C(4, 1) = 4$ ways.

2 Ts   These outcomes, which include $THHT$ and $HTTH$, are determined by selecting two boxes out of four in which to place the two Ts. This can be done in $C(4, 2) = 6$ ways.

Thus the number of outcomes with at most 2 tails is just the sum of all these results, which is $1 + 4 + 6 = 11$.

(d) The outcomes with at least 1 tail are the results with 1, 2, 3, or 4 tails. The total number of such outcomes is

$$C(4, 1) + C(4, 2) + C(4, 3) + C(4, 4)$$

But there is a simpler way of obtaining the answer. If we start with the total number of outcomes obtained in Part (a) and subtract the number of outcomes of at most 0 tails, namely $C(4, 0) = 1$, we get the number of outcomes with at least 1 tail:

$$16 - 1 = 15$$

Part (d) of Example 1 illustrates a counting technique that is often useful: Count the outcomes that are not favorable to you and subtract from the total number of outcomes.

This "backdoor" approach can be easier at times than a direct attack. We use it again in the next example.

✐ Now Work Problem 1

EXAMPLE 2   Telephone Numbers with a Repeated Digit

Find the number of seven-digit telephone numbers that have at least one repeated digit. Leading 0s are allowed.

SOLUTION   A direct application of the Multiplication Principle shows that there are 10^7 possible seven-digit telephone numbers. The number of telephone numbers that have *no* repeated digits is

$$P(10, 7) = \frac{10!}{3!} = 604{,}800$$

Hence, the number that have at least one repeated digit is

$$10^7 - \frac{10!}{3!} = 9{,}395{,}200$$

EXAMPLE 3   Selecting Balls from an Urn

An urn contains 8 white balls and 4 red balls. Four balls are selected. In how many ways can the 4 balls be drawn from the total of 12 balls

(a) If 3 balls are white and 1 is red?   (b) If all 4 balls are white?
(c) If all 4 balls are red?

SOLUTION

(a) The desired answer involves two tasks: first, the selection of 3 white balls from 8; and second, the selection of 1 red ball from 4:

Select 3 white balls task 1	Select 1 red ball task 2

The first task can be performed in $C(8, 3)$ ways; the second task can be performed in $C(4, 1)$ ways. By the Multiplication Principle, the answer is

$$C(8, 3) \cdot C(4, 1) = \frac{8!}{3!5!} \cdot \frac{4!}{1!3!} = 224$$

(b) Since all 4 balls must be selected from the 8 that are white, the answer is

$$C(8, 4) = \frac{8!}{4!4!} = \frac{8 \cdot 7 \cdot 6 \cdot 5}{4 \cdot 3 \cdot 2 \cdot 1} = 70 \text{ ways}$$

(c) Since the 4 red balls must be selected from 4 red balls, the answer is 1 way.

✐ Now Work Problem 3

Permutations with Repetition

Our previous discussion of permutations required that the objects we were rearranging be distinct. We now examine what happens when repetitions are allowed. The following example shows that allowing repetition of some objects introduces modifications.

EXAMPLE 4   Forming Words

How many three-letter words (real or imaginary) can be formed from the letters in the word

(a) MAD?   (b) DAD?

SOLUTION

(a) The three distinct letters in MAD can be rearranged in

$$P(3, 3) = 3! = 6 \text{ ways}$$

(b) Straightforward listing shows that there are only three ways of rearranging the letters in the word DAD:

DAD, DDA, and ADD

■

The word DAD in Example 4 contains 2 Ds, and it is this duplication that results in fewer rearrangements for DAD than for MAD. In the next example we describe a way of dealing with the problem of duplication.

EXAMPLE 5   Forming Words

How many distinct "words" can be formed using all the letters of the six-letter word

M A M M A L ?

SOLUTION   Any such word will have 6 letters formed from 3 Ms, 2 As, and 1 L. To form a word, think of six blank positions that will have to be filled by the above letters.

$$\overline{} \ \overline{} \ \overline{} \ \overline{} \ \overline{} \ \overline{}$$
$$1 \ 2 \ 3 \ 4 \ 5 \ 6$$

We separate the construction of a word into three tasks.

Task 1 Choose 3 of the positions for the Ms.
Task 2 Choose 2 of the remaining positions for the As.
Task 3 Choose the remaining position for the L.

Doing this sequence of tasks will result in a word and, conversely, every rearrangement of MAMMAL can be interpreted as resulting from this sequence of tasks.

Task 1 can be done in $C(6, 3)$ ways. There are now three positions left for the 2 As, so Task 2 can be done in $C(3, 2)$ ways. Five blanks have been filled, so that the L must go in the remaining blank. That is, Task 3 can be done in $C(1, 1)$ way. The Multiplication Principle says that the number of rearrangements is

$$C(6, 3) \cdot C(3, 2) \cdot C(1, 1) = \frac{6!}{3!3!} \cdot \frac{3!}{2!1!} \cdot \frac{1!}{1!}$$

$$= \frac{6!}{3!2!1!}$$

The form of the answer in Example 5 is suggestive. Had the letters in MAMMAL been distinct, there would have been 6! rearrangements possible. The presence of 3 Ms, 2 As, and 1 L yielded the denominator above. The very same reasoning used in Example 5 can be used to derive the following general result.

Permutation with Repetition

The number of distinct permutations of n objects, of which n_1 are of one kind, n_2 of a second kind, . . . , n_k of a kth kind, is

$$\frac{n!}{n_1! \cdot n_2! \cdot \ \ldots \ \cdot n_k!}$$

where $n_1 + n_2 + \cdots + n_k = n$.

EXAMPLE 6 Arranging Flags

How many different vertical arrangements are possible for 10 flags, if 2 are white, 3 are red, and 5 are blue?

SOLUTION Here we want the different arrangements of 10 objects, which are not all different. Following the result above, we have

$$\frac{10!}{2!3!5!} = \frac{10 \cdot 9 \cdot 8 \cdot 7 \cdot \cancel{6} \cdot \cancel{5!}}{2 \cdot \cancel{3} \cdot \cancel{2} \cdot \cancel{5!}} = 2520 \text{ different arrangements}$$ ∎

EXAMPLE 7 Forming Words

How many different 11-letter words (real or imaginary) can be formed from the word below?

MISSISSIPPI

SOLUTION Here we want the number of distinct 11-letter words with 4 Is, 4 Ss, 2 Ps, and 1 M, so that the total number of 11-letter words is

$$\frac{11!}{1!2!4!4!} = \frac{39{,}916{,}800}{1152} = 34{,}650$$

 **Now Work Problem 9**

The ideas above can be adapted to problems involving assignments of objects to locations.

EXAMPLE 8   Assigning Rooms

A sorority house has three bedrooms and 10 students. One bedroom has three beds, the second has two beds, and the third has five beds. In how many different ways can the students be assigned rooms?

SOLUTION Designate the bedrooms as A, B, and C. We think of the 10 students as standing in a row and we hand each student a letter corresponding to her assigned bedroom. An assignment of rooms can then be visualized as a sequence of length 10 (the number of students) containing 3 As, 2 Bs, and 5 Cs (the capacity of the rooms). For example, the sequence

$$A \quad B \quad B \quad C \quad \ldots$$

would have the first student in room A, students 2 and 3 in room B, and so on. There are

$$\frac{10!}{3!2!5!} = 2520$$

such sequences and, hence, room assignments.

EXERCISE 6.6   Answers to odd-numbered problems begin on page AN-25.

1. An experiment consists of tossing a coin 10 times.

(a) How many different outcomes are possible?
(b) How many different outcomes have exactly 4 heads?
(c) How many different outcomes have at most 2 heads?
(d) How many different outcomes have at least 3 heads?

2. An experiment consists of tossing a coin six times.

(a) How many different outcomes are possible?
(b) How many different outcomes have exactly 3 heads?
(c) How many different outcomes have at least 2 heads?
(d) How many different outcomes have 4 heads or 5 heads?

3. An urn contains 7 white balls and 3 red balls. Three balls are selected. In how many ways can the 3 balls be drawn from the total of 10 balls

(a) If 2 balls are white and 1 is red?
(b) If all 3 balls are white?
(c) If all 3 balls are red?

4. An urn contains 15 red balls and 10 white balls. Five balls are selected. In how many ways can the 5 balls be drawn from the total of 25 balls

(a) If all balls are red?

(b) If 3 balls are red and 2 are white?

(c) If at least 4 are red balls?

5. In the World Series the American League team (A) and the National League team (N) play until one team wins four games. If the sequence of winners is designated by letters (for example, *NAAAA* means the National League team won the first game and the American League team won the next four), how many different sequences are possible?

6. How many different ways can 3 red, 4 yellow, and 5 blue bulbs be arranged in a string of Christmas tree lights with 12 sockets?

7. In how many ways can 3 apple trees, 4 peach trees, and 2 plum trees be arranged along a fence line if one does not distinguish between trees of the same kind?

8. How many different 9-letter words (real or imaginary) can be formed from the letters in the word ECONOMICS?

9. How many different 11-letter words (real or imaginary) can be formed from the letters in the word MATHE-MATICS?

10. The U.S. Senate has 100 members. Suppose it is desired to place each senator on exactly 1 of 7 possible committees. The first committee has 22 members, the second has 13, the third has 10, the fourth has 5, the fifth has 16, and the sixth and seventh have 17 apiece. In how many ways can these committees be formed?

11. In how many ways can 12 children be placed on 3 distinct teams of 3, 5, and 4 members?

12. A group of 9 people is going to be formed into committees of 4, 3, and 2 people. How many committees can be formed if

(a) A person can serve on any number of committees?

(b) No person can serve on more than one committee?

13. A group consists of 5 men and 8 women. A committee of 4 is to be formed from this group, and policy dictates that at least 1 woman be on this committee.

(a) How many committees can be formed that contain exactly 1 man?

(b) How many committees can be formed that contain exactly 2 women?

(c) How many committees can be formed that contain at least 1 man?

14. How many distinct seven-digit telephone numbers can be formed

(a) If the digits 1, 1, 2, 2, 5, 5, 5 are used?

(b) If it is required that the first digit be a 5?

15. In how many ways can 30 diplomats be assigned to 5 countries, with each country receiving an equal number of diplomats?

16. How many rearrangements of the letters in the word SUCCESS have the U before the E?

17. An experiment consists of tossing a coin 8 times. How many outcomes have more heads than tails?

18. Eight couples (husband and wife) are present at a meeting where a committee of 3 is to be chosen. How many ways can this be done so that the committee

(a) Contains a couple?   (b) Contains no couple?

19. In how many ways can a committee of 4 be selected from 6 men and 8 women if the committee must contain at least 2 women?

20. A man wants to invite 1 or more of his 4 friends to dinner. In how many ways can he do this?

21. How many eight-bit strings contain an even number of 1s? An odd number of 1s?

22. An office manager must locate 12 secretaries into three offices that hold, respectively, 6, 4, and 2 secretaries. In how many ways can the three groups be chosen to occupy the three offices?

6.7 THE BINOMIAL THEOREM

The *binomial theorem* deals with the problem of expanding an expression of the form $(x + y)^n$, where n is a positive integer.

Expressions such as $(x + y)^2$ and $(x + y)^3$ are not too difficult to expand. For example,

$$(x + y)^2 = x^2 + 2xy + y^2$$
$$(x + y)^3 = (x + y)^2(x + y) = (x^2 + 2xy + y^2)(x + y) = x^3 + 3x^2y + 3xy^2 + y^3$$

However, expanding expressions such as $(x + y)^6$ or $(x + y)^8$ by the normal process of multiplication would be tedious and time consuming. It is here that the binomial theorem is especially useful.

Recall that

$$C(n, r) = \binom{n}{r} = \frac{n!}{r!(n - r)!}$$

Since $\binom{2}{0} = 1$, $\binom{2}{1} = 2$, and $\binom{2}{2} = 1$, we can write the expression

$$(x + y)^2 = x^2 + 2xy + y^2$$

in the form

$$(x + y)^2 = \binom{2}{0} x^2 + \binom{2}{1} xy + \binom{2}{2} y^2$$

Since $\binom{3}{0} = 1$, $\binom{3}{1} = 3$, $\binom{3}{2} = 3$, and $\binom{3}{3} = 1$, the expansion of $(x + y)^3$ can be written as

$$(x + y)^3 = x^3 + 3x^2y + 3xy^2 + y^3 = \binom{3}{0} x^3 + \binom{3}{1} x^2y + \binom{3}{2} xy^2 + \binom{3}{3} y^3$$

Similarly, the expansion of $(x + y)^4$ can be written as

$$(x + y)^4 = x^4 + 4x^3y + 6x^2y^2 + 4xy^3 + y^4$$
$$= \binom{4}{0} x^4 + \binom{4}{1} x^3y + \binom{4}{2} x^2y^2 + \binom{4}{3} xy^3 + \binom{4}{4} y^4$$

The binomial theorem generalizes this pattern.

Binomial Theorem

If n is a positive integer,

$$(x + y)^n = \binom{n}{0} x^n + \binom{n}{1} x^{n-1}y + \binom{n}{2} x^{n-2}y^2$$
$$+ \cdots + \binom{n}{k} x^{n-k}y^k + \cdots + \binom{n}{n} y^n \tag{1}$$

Observe that the powers of x begin at n and decrease by 1, while the powers of y begin with 0 and increase by 1. Also, the coefficient of the term involving y^k is always $\binom{n}{k}$.

Let's get some practice using the binomial theorem.

EXAMPLE 1   Expanding a Binomial

Expand $(x + y)^6$ using the binomial theorem.

SOLUTION

$$(x + y)^6 = \binom{6}{0} x^6 + \binom{6}{1} x^5y + \binom{6}{2} x^4y^2 + \binom{6}{3} x^3y^3$$

$$+ \binom{6}{4} x^2y^4 + \binom{6}{5} xy^5 + \binom{6}{6} y^6$$

$$= x^6 + 6x^5y + 15x^4y^2 + 20x^3y^3 + 15x^2y^4 + 6xy^5 + y^6$$

Note that the coefficients in the expansion of $(x + y)^6$ are the entries in the Pascal triangle for $n = 6$. (See Figure 21, page 297.)

 **Now Work Problem 1**

EXAMPLE 2   Finding a Particular Coefficient

Find the coefficient of x^3y^4 in the expansion of $(x + y)^7$.

SOLUTION   The expansion of $(x + y)^7$ is

$$(x + y)^7 = \binom{7}{0} x^7 + \binom{7}{1} x^6y + \binom{7}{2} x^5y^2 + \binom{7}{3} x^4y^3 + \binom{7}{4} x^3y^4$$

$$+ \binom{7}{5} x^2y^5 + \binom{7}{6} xy^6 + \binom{7}{7} y^7$$

Thus the coefficient of x^3y^4 is

$$\binom{7}{4} = \frac{7 \cdot 6 \cdot 5}{3 \cdot 2 \cdot 1} = 35$$

 Now Work Problem 7

EXAMPLE 3   Expanding a Binomial

Expand $(x + 2y)^4$ using the binomial theorem.

SOLUTION   Here, we let "$2y$" play the role of "y" in the binomial theorem. We then get

$$(x + 2y)^4 = \binom{4}{0} x^4 + \binom{4}{1} x^3(2y) + \binom{4}{2} x^2(2y)^2$$

$$+ \binom{4}{3} x(2y)^3 + \binom{4}{4} (2y)^4$$

$$= x^4 + 8x^3y + 24x^2y^2 + 32xy^3 + 16y^4$$

 **Now Work Problem 3**

To explain why the binomial theorem is true, we take a close look at what happens when we compute $(x + y)^3$. Think of $(x + y)^3$ as the product of three factors, namely,

$$(x + y)^3 = (x + y)(x + y)(x + y)$$

Were we to multiply these three factors together without any attempt at simplification or collecting of terms, we would get

$$(x + y)\,(x + y)\,(x + y) = xxx + xyx + yxx + yyx + xxy + xyy + yxy + yyy$$

Factor   Factor   Factor
   1   2   3

Notice that the terms on the right yield all possible products that can be formed by picking either an x or y from each of the three factors on the left. Thus, for example,

$$xyx \qquad \text{results from choosing an } x \text{ from factor 1,}$$
a y from factor 2, and an x from factor 3

Now, the number of terms on the right that will simplify to, say, xy^2, will be those terms that resulted from choosing y's from two of the factors and an x from the remaining factor. How many such terms are there? There are as many as there are ways of choosing two of the three factors to contribute ys—that is, there are $C(3, 2) = \binom{3}{2}$ such terms. This is why the coefficient of xy^2 in the expansion of $(x + y)^3$ is $\binom{3}{2}$.

In general, if we think of $(x + y)^n$ as the product of n factors,

$$(x + y)^n = \underbrace{(x + y) \cdot (x + y) \cdot \ \ldots \ \cdot (x + y)}_{n \text{ factors}}$$

then, upon multiplying out and simplifying, there will be as many terms of the form $x^{n-k}\,y^k$ as there are ways of choosing k of the n factors to contribute ys (and the remaining $n - k$ to contribute x's). There are $C(n, k)$ ways of making this choice. So the coefficient of $x^{n-k}\,y^k$ is thus $\binom{n}{k}$, and this is the assertion of the binomial theorem.

Binomial Identities

Binomial coefficients have some interesting properties. For example, it turns out that $\binom{n}{k}$ and $\binom{n}{n-k}$ are equal. We can explain this equality as follows: Suppose we wanted to pick a team of k players from n people. Then choosing those k who will play is the same as choosing those $n - k$ who will not. So the number of ways of choosing the players equals the number of ways of choosing the nonplayers. The players can be chosen in $C(n, k) = \binom{n}{k}$ ways, while those to be left out can be chosen in $\binom{n}{n-k}$ ways, and the equality follows.

EXAMPLE 4   Proving a Property of Binomial Coefficients

Show that

$$\binom{n}{k} = \binom{n}{n - k}$$

SOLUTION By definition,

$$\binom{n}{k} = \frac{n!}{k!(n - k)!}$$

while

$$\binom{n}{n-k} = \frac{n!}{(n-k)![n-(n-k)]!}$$

Since $n - (n-k) = k$, a comparison of the expressions shows that they are equal. ■

Thus

$$\binom{5}{3} = \binom{5}{2} \qquad \binom{10}{2} = \binom{10}{8}$$

and so on. This identity accounts for the symmetry in the rows of Pascal's triangle.

EXAMPLE 5   Proving a Property of Binomial Coefficients

Show that

$$\binom{n}{k} = \binom{n-1}{k} + \binom{n-1}{k-1}$$

SOLUTION   We could expand both sides of the above identity using the definition of binomial coefficients and, after some algebra, demonstrate the equality. But we choose the route of posing a problem that we solve two different ways. Equating the two solutions will prove the identity.

A committee of k is to be chosen from n people. The total number of ways this can be done is $C(n, k) = \binom{n}{k}$.

We now count the total number of committees a different way. Assume that one of the n people is Jennifer. We compute

(1) those committees not containing Jennifer

and

(2) those committees containing Jennifer

The number of committees of type 1 is $\binom{n-1}{k}$ since the k people must be chosen from the $n-1$ people who are not Jennifer. The number of committees of type 2 will correspond to the number of ways we can choose the $k-1$ people other than Jennifer to be on the committee. So the number of committees of type 2 is given by $\binom{n-1}{k-1}$. Since the number of committees of type 1 plus the number of committees of type 2 equals the total number of committees, our identity follows. ■

For example, $\binom{8}{5} = \binom{7}{5} + \binom{7}{4}$. It is precisely this identity that explains the reason why an entry in Pascal's triangle can be obtained by adding the nearest two entries in the row above it. Due to its recursive character, the identity in Example 5 is sometimes used in computer programs that evaluate binomial coefficients.

Now Work Problem 21

EXAMPLE 6   Establishing a Relationship of Specific Binomial Coefficients

Show that

$$\binom{6}{3} = \binom{2}{2} + \binom{3}{2} + \binom{4}{2} + \binom{5}{2}$$

SOLUTION   We make repeated use of the identity in Example 5. So

$$\binom{6}{3} = \binom{5}{3} + \binom{5}{2}$$

$$= \left[\binom{4}{3} + \binom{4}{2}\right] + \binom{5}{2} \qquad \text{Apply the identity to } \binom{5}{3}.$$

$$= \left[\binom{3}{3} + \binom{3}{2}\right] + \binom{4}{2} + \binom{5}{2} \qquad \text{Apply the identity to } \binom{4}{3}.$$

$$= \binom{2}{2} + \binom{3}{2} + \binom{4}{2} + \binom{5}{2} \qquad \text{Since } \binom{2}{2} = \binom{3}{3} = 1$$

■

EXAMPLE 7   Proving a Property of Binomial Coefficients

Show that

$$\binom{n}{0} + \binom{n}{1} + \binom{n}{2} + \cdots + \binom{n}{n} = 2^n$$

SOLUTION   We make use of the binomial theorem. Since the binomial theorem is valid for all x and y, we may set $x = y = 1$ in Equation (1). This gives

$$2^n = (1 + 1)^n = \binom{n}{0} + \binom{n}{1} + \binom{n}{2} + \cdots + \binom{n}{n}$$

■

This shows, for example, that the sum of the elements in the row marked $n = 6$ of Pascal's triangle is $2^6 = 64$. The result in Example 7 can be used to find the number of subsets of a set with n elements. $\binom{n}{0}$ gives the number of subsets with 0 elements; $\binom{n}{1}$ the number of subsets with 1 element; $\binom{n}{2}$ the number of subsets with 2 elements; and so on. The sum $\binom{n}{0} + \binom{n}{1} + \cdots + \binom{n}{n}$ is thus the total number of subsets of a set with n elements. The result in Example 7 can then be rephrased as follows:

A set with n elements has 2^n subsets.

Thus a set with 5 elements has $2^5 = 32$ subsets.

📖 **Now Work Problem 11**

EXAMPLE 8   Proving a Property of Binomial Coefficients

Show that

$$\binom{n}{0} - \binom{n}{1} + \binom{n}{2} - \cdots + (-1)^n \binom{n}{n} = 0$$

(The last term will be preceded by a plus or minus sign depending on whether n is even or odd.)

SOLUTION   We again make use of the binomial theorem. This time we let $x = 1$ and $y = -1$ in Equation (1). This produces

$$(1 - 1)^n = \binom{n}{0} + \binom{n}{1}(-1) + \binom{n}{2}(-1)^2 + \binom{n}{3}(-1)^3$$

$$+ \cdots + \binom{n}{n}(-1)^n$$

$$0 = \binom{n}{0} - \binom{n}{1} + \binom{n}{2} - \cdots + (-1)^n \binom{n}{n}$$

We mention an interpretation of this identity by examining the instance where $n = 5$. The identity gives

$$\binom{5}{0} - \binom{5}{1} + \binom{5}{2} - \binom{5}{3} + \binom{5}{4} - \binom{5}{5} = 0$$

Rearranging some terms yields

$$\binom{5}{0} + \binom{5}{2} + \binom{5}{4} = \binom{5}{1} + \binom{5}{3} + \binom{5}{5}$$

This says that a set with 5 elements has as many subsets containing an even number of elements as it has subsets containing an odd number of elements.

EXERCISE 6.7   Answers to odd-numbered problems begin on page AN-25.

In Problems 1–6 use the binomial theorem to expand each expression.

1. $(x + y)^5$ **2.** $(x + y)^4$ **3.** $(x + 3y)^3$ **4.** $(2x + y)^3$ **5.** $(2x - y)^4$ **6.** $(x - y)^4$

7. What is the coefficient of x^2y^3 in the expansion of $(x + y)^5$?

8. What is the coefficient of x^2y^6 in the expansion of $(x + y)^8$?

9. What is the coefficient of x^8 in the expansion of $(x + 3)^{10}$?

10. What is the coefficient of x^3 in the expansion of $(x + 2)^5$?

11. How many different subsets can be chosen from a set with 5 elements?

12. How many different subsets can be chosen from a set with 50 elements?

13. How many nonempty subsets does a set with 10 elements have?

14. How many subsets with an even number of elements does a set with 10 elements have?

15. How many subsets with an odd number of elements does a set with 10 elements have?

16. Show that

$$\binom{8}{5} = \binom{4}{4} + \binom{5}{4} + \binom{6}{4} + \binom{7}{4}$$

17. Show that

$$\binom{10}{7} = \binom{6}{6} + \binom{7}{6} + \binom{8}{6} + \binom{9}{6}$$

18. Show that

$$\binom{7}{1} + \binom{7}{3} + \binom{7}{5} + \binom{7}{7} = 2^6$$

19. Replace $\binom{11}{6} + \binom{11}{5}$ by a single binomial coefficient.

20. Replace $\binom{8}{8} + \binom{9}{8} + \binom{10}{8}$ by a single binomial coefficient.

21. Show that

$$k\binom{n}{k} = n\binom{n-1}{k-1}$$

CHAPTER REVIEW

IMPORTANT TERMS AND CONCEPTS

set 267	disjoint sets 272	permutation 289
subset 269	complement of a set 273	combination 294
proper subset 269	De Morgan's properties 274	binomial coefficient 297
universal set 270	counting formula 278	Pascal's triangle 297
Venn diagram 271	Multiplication Principle 284	permutation with repetition 303
union of sets 271	factorial 287	binomial theorem 306
intersection of sets 271		

IMPORTANT FORMULAS

Counting Formula

$$c(A \cup B) = c(A) + c(B) - c(A \cap B)$$

Permutation Formula

$$P(n, r) = n(n - 1) \cdot \ \ldots \ \cdot (n - r + 1) = \frac{n!}{(n - r)!}$$

Combination Formula

$$C(n, r) = \binom{n}{r} = \frac{n!}{r!(n - r)!} = \frac{P(n, r)}{r!}$$

Binomial Theorem

$$(x + y)^n = \binom{n}{0}x^n + \binom{n}{1}x^{n-1}y + \binom{n}{2}x^{n-2}y^2$$
$$+ \cdots + \binom{n}{k}x^{n-k}y^k + \cdots + \binom{n}{n}y^n$$

TRUE–FALSE ITEMS   Answers are on page AN-25.

T_____F_____ **1.** If $A \cup B = A \cap B$, then $A = B$.

T_____F_____ **2.** If A and B are disjoint sets, then $c(A \cup B) = c(A) + c(B)$.

T_____F_____ **3.** The number of permutations of 4 different objects taken 4 at a time is 12.

T_____F_____ **4.** $C(5, 3) = 20$

T_____F_____ **5.** $5! = 120$

T_____F_____ **6.** $\dfrac{7!}{6!} = \dfrac{7}{6}$

T_____F_____ **7.** In the binomial expansion of $(x + 1)^7$, the coefficient of x^4 is 4.

FILL IN THE BLANKS   Answers are on page AN-25.

1. Two sets that have no elements in common are called _____ .

2. The number of different selections of r objects from n objects in which (a) the n objects are different, (b) no object is repeated more than once in a selection, and (c) order is important is called a _____ .

3. If in 2 above, condition (c) is replaced by "order is not important," we have a _____ .

4. A triangular display of the number of combinations is called the _____ triangle.

5. The numbers $\binom{n}{r}$ are sometimes called _____ _____ .

6. To expand an expression such as $(x + y)^n$, n a positive integer, we can use the _____ _____ .

7. The coefficient of x^3 in the expansion of $(x + 2)^5$ is _____ .

REVIEW EXERCISES   Answers to odd-numbered problems begin on page AN-25.

In Problems 1–16 replace the asterisk by any of the symbol(s) $\subset$, $\subseteq$, $=$ that result in a true statement. If none result in a true statement, write "None of these." More than one answer may be possible.

1. $\{0\} * \varnothing$

2. $\{0\} * \{1, 0, 3\}$

3. $\{5, 6\} \cap \{2, 6\} * \{8\}$

4. $\{2, 3\} \cup \{3, 4\} * \{3\}$

5. $\{8, 9\} * \{9, 10, 11\}$

6. $\{1\} * \{1, 3, 5\} \cup \{3, 4\}$

7. $\{5\} * \{0, 5\}$

8. $\varnothing * \{1, 2, 3\}$

9. $\varnothing * \{1, 2\} \cap \{3, 4, 5\}$

10. $\{2, 3\} * \{3, 4\}$

11. $\{1, 2\} * \{1\} \cup \{3\}$

12. $\{5\} * \{1\} \cup \{2, 3\}$

13. $\{4, 5\} \cap \{5, 6\} * \{4, 5\}$

14. $\{6, 8\} * \{8, 9, 10\}$

15. $\{6, 7, 8\} \cap \{8\} * \{6\}$

16. $\{4\} * \{6, 8\} \cap \{4, 8\}$

17. For the sets

$$A = \{1, 3, 5, 6, 8\} \quad B = \{2, 3, 6, 7\} \quad C = \{6, 8, 9\}$$

find

(a) $(A \cap B) \cup C$ (b) $(A \cap B) \cap C$

(c) $(A \cup B) \cap B$

18. For the sets U = universal set = $\{1, 2, 3, 4, 5, 6, 7\}$ and

$$A = \{1, 3, 5, 6\} \quad B = \{2, 3, 6, 7\} \quad C = \{4, 6, 7\}$$

find:

(a) $\overline{A \cap B}$ (b) $(B \cap C) \cap A$ (c) $\overline{B} \cup \overline{A}$

19. If A and B are sets and if $c(A) = 24$, $c(A \cup B) = 33$, $c(B) = 12$, find $c(A \cap B)$.

20. **Car Options** During June, Colleen's Motors sold 75 cars with air conditioning, 95 with power steering, and 100 with automatic transmissions. Twenty cars had all three options, 10 cars had none of these options, and 10 cars were sold that had only air conditioning. In addition, 50 cars had both automatic transmissions and power steering, and 60 cars had both automatic transmissions and air conditioning.

(a) How many cars were sold in June?

(b) How many cars had only power steering?

21. **Student Survey** In a survey of 125 college students, it was found that of three newspapers, the *Wall Street Journal, New York Times,* and *Chicago Tribune:*

60	read the *Chicago Tribune*
40	read the *New York Times*
15	read the *Wall Street Journal*
25	read the *Chicago Tribune* and *New York Times*
8	read the *New York Times* and *Wall Street Journal*
3	read the *Chicago Tribune* and *Wall Street Journal*
1	read all three

(a) How many read none of these papers?

(b) How many read only the *Chicago Tribune?*

(c) How many read neither the *Chicago Tribune* nor the *New York Times?*

22. If U = universal set = {1, 2, 3, 4, 5} and B = {1, 4, 5}, find all sets A for which $A \cap B$ = {1}.

23. Compute $P(6, 3)$.

24. Compute $C(6, 2)$.

25. In how many different ways can a committee of 3 people be formed from a group of 5 people?

26. In how many different ways can 4 people line up?

27. In how many different ways can 3 books be placed on a shelf?

28. In how many different ways can 3 people be seated in 4 chairs?

29. How many house styles are possible if a contractor offers 3 choices of roof designs, 4 choices of window designs, and 6 choices of brick?

30. How many different answers are possible in a true–false test consisting of 10 questions?

31. You are to set up a code of 2-digit words using the digits 1, 2, 3, 4 without using any digit more than once. What is the maximum number of words in such a language? If all words of the form *ab* and *ba* are the same, how many words are possible?

32. You are to set up a code of 3-digit words using the digits 1, 2, 3, 4, 5, 6 without using any digit more than once in the same word. What is the maximum number of words in such a language? If the words 124, 142, etc., designate the same word, how many different words are possible?

33. A small town consists of a north side and a south side. The north side has 16 houses and the south side has 10 houses. A pollster is asked to visit 4 houses on the north side and 3 on the south side. In how many ways can this be done?

34. **Program Selection** A ceremony is to include 7 speeches and 6 musical selections.

    (a) How many programs are possible?
    (b) How many programs are possible if speeches and musical selections are to be alternated?

35. There are 7 boys and 6 girls willing to serve on a committee. How many 7-member committees are possible if a committee is to contain:

    (a) 3 boys and 4 girls?
    (b) At least one member of each sex?

36. Juan's Ice Cream Parlor offers 31 different flavors to choose from and specializes in double dip cones.

    (a) How many different cones are there to choose from if you may select the same flavor for each dip?
    (b) How many different cones are there to choose from if you cannot repeat any flavor? Assume that a cone with vanilla on top of chocolate is different from a cone with chocolate on top of vanilla.
    (c) How many different cones are there if you consider any cone having chocolate on top and vanilla on the bottom the same as having vanilla on top and chocolate on the bottom?

37. A person has 4 history, 5 English, and 6 mathematics books. How many ways can the books be arranged on a shelf if books on the same subject must be together?

38. Five people are to line up for a group photograph. If two of them refuse to stand next to each other, in how many ways can the photograph be taken?

39. In how many ways can a committee of 8 boys and 5 girls be formed if there are 10 boys and 11 girls eligible to serve on the committee?

40. A football squad has 7 linemen, 11 linebackers, and 9 safeties. How many different teams composed of 5 linemen, 3 linebackers, and 3 safeties can be formed?

41. In how many ways can we choose three words, one each from five 3-letter words, six 4-letter words, and eight 5-letter words?

42. A newborn child can be given 1, 2, or 3 names. In how many ways can a child be named if we can choose from 100 names?

43. In how many ways can 5 girls and 3 boys be divided into 2 teams of 4 if each team is to include at least 1 boy?

44. A meeting is to be addressed by 5 speakers, A, B, C, D, E. In how many ways can the speakers be ordered if B must not precede A?

45. What is the answer to Problem 44 if B is to speak immediately after A?

46. **License Plate Numbers** An automobile license number contains 1 or 2 letters followed by a 4-digit number. Compute the maximum number of different licenses.

47. There are 5 rotten plums in a crate of 25 plums. How many samples of 4 of the 25 plums contain

    (a) Only good plums?
    (b) Three good plums and 1 rotten plum?
    (c) One or more rotten plums?

48. An admissions test given by a university contains 10 true–false questions. Eight or more of the questions must be answered correctly in order to be admitted.

   (a) How many different ways can the answer sheet be filled out?
   (b) How many different ways can the answer sheet be filled out so that 8 or more questions are answered correctly?

49. The figure below indicates the locations of two houses, A and B, in a city, where the lines represent streets. A person at A wishes to reach B and can travel in only two directions, to the right and up. How many different paths are there from A to B?

50. A car driver picks up a passenger at point A (see the figure below) whose destination is point B. After completing the trip, the driver is to proceed to the garage at point C. If the cab must travel to the right or up, how many different routes are there from A to C?

51. Expand $(x + 2)^4$.

52. Expand $(x - 1)^5$.

53. What is the coefficient of x^3 in the expansion of $(x + 2)^7$?

54. What is the coefficient of x^4 in the expansion of $(2x + 1)^6$?

Chapter 7

Probability

7.1 Sample Spaces and Assignment of Probabilities

7.2 Properties of the Probability of an Event

7.3 Probability Problems Using Counting Techniques

7.4 Conditional Probability

7.5 Independent Events

Chapter Review

Probability theory is a part of mathematics that is useful for discovering and investigating the *regular* features of *random events*. Although it is not really possible to give a precise and simple definition of what is meant by the words *random* and *regular,* the explanation and the examples given below will help you understand these concepts.

Certain phenomena in the real world may be considered *chance phenomena.* These phenomena do not always produce the same observed outcome, and the outcome of any given observation of the phenomena may not be predictable. But they have a long-range behavior known as *statistical regularity.* Some examples of such cases, called *random events,* follow.

Tossing a fair coin gives a result that is either a head or a tail. For any one throw, we cannot predict the result, although it is obvious that it is determined by definite causes (such as the initial velocity of the coin, the initial angle of throw, and the surface on which the coin rests). Even though some of these causes can be controlled, we cannot predetermine the result of any particular toss. Thus the result of tossing a coin is a *random event.*

Although we cannot predict the result of any particular toss of the coin, if we perform a long sequence of tosses, we expect that the number of heads is approximately equal to the number of tails. That is, it seems *reasonable* to say that in any toss of this fair coin, a head or a tail is *equally likely* to occur. As a result, we might *assign a probability* of $\frac{1}{2}$ for obtaining a head (or tail) on a particular toss.

In throwing an ordinary die, we cannot predict the result with certainty. Thus the result of throwing a die is a *random event.* We do know that one of the faces 1, 2, 3, 4, 5, or 6 will occur.

The appearance of any particular face of the die is an *outcome.* If we perform a long series of tosses, any face is as *likely* to occur as any other, provided the die is fair. Here we might *assign a probability* of $\frac{1}{6}$ for obtaining a particular face.

The sex of a newborn baby is either male or female. This, too, is an example of a *random event.*

Our intuition tells us that a boy baby and a girl baby are *equally likely* to occur. If we follow this reasoning, we might *assign a probability* of $\frac{1}{2}$ to having a girl baby. However, if we consult the data about births in the United States found in Table 1, we see that it might be more accurate to assign a probability of .488 to having a girl baby.

Table 1*

Year of Birth	Number of Births (in thousands)		Total Number of Births (in thousands)	Ratio of Births	
	Boys b	Girls g	$b + g$	$\dfrac{b}{b+g}$	$\dfrac{g}{b+g}$
1995	1,996	1,903	3,899	.512	.488
1994	2,023	1,930	3,953	.512	.488
1993	2,049	1,951	4,000	.512	.488
1992	2,082	1,983	4,065	.512	.488
1991	2,102	2,009	4,111	.511	.489
1990	2,129	2,029	4,158	.512	.488
1989	2,069	1,921	3,990	.519	.481
1988	2,002	1,907	3,909	.512	.488
1987	1,951	1,858	3,809	.512	.488
1986	1,925	1,832	3,757	.512	.488

* *Source:* U.S. Department of Health and Human Services, *Monthly Vital Statistics Report,* December 1998.

These examples demonstrate that in studying a sequence of random experiments it is not possible to forecast individual results. These are subject to irregular, random fluctuations that cannot be exactly predicted. However, if the number of observations is large—that is, if we deal with a *mass phenomenon*—some regularity appears. This leads to the study of probability.

7.1 SAMPLE SPACES AND ASSIGNMENT OF PROBABILITIES

In studying probability we are concerned with experiments, real or conceptual, and their outcomes. In this study we try to formulate in a precise manner a mathematical theory that closely resembles the experiment in question. The first stage of development of such a mathematical theory is the building of what is termed a *probability model.* This model is then used to analyze and predict outcomes of the experiment. The purpose of this section is to learn how a probability model can be constructed.

Sample Spaces

We begin by writing the associated *sample space* of an experiment; that is, we write all outcomes that can occur as a result of the experiment.

EXAMPLE 1   Finding a Sample Space

If the experiment consists of flipping a coin, we would agree that the only possible outcomes are heads (H) and tails (T). Therefore, a sample space for the experiment is the set $\{H, T\}$.

■

EXAMPLE 2   Finding a Sample Space

Consider an experiment in which, for the sake of simplicity, one die is green and the other is red. When the dice are rolled, the set of outcomes consists of all the different ways that the dice may come to rest, that is, the set of all *possibilities* that can occur as a result of the experiment.

An application of the Multiplication Principle reveals that the number of outcomes of this experiment is 36 since there are 6 possible ways for the green die to come up and 6 ways for the red die to come up.

We can use a tree diagram to obtain a list of the 36 outcomes. See Figure 1. Figure 2 gives a graphical representation of the 36 outcomes.

Figure 1

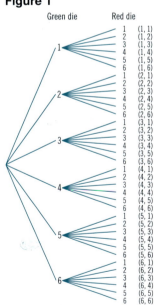

Figure 2

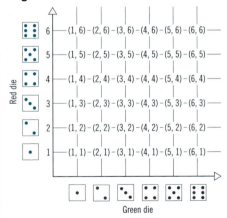

■

Sample Space; Outcome

A **sample space** S, associated with a real or conceptual experiment, is the set of all possibilities that can occur as a result of the experiment. Each element of a sample space S is called an **outcome**.

We list below some experiments and their sample spaces.

Experiment	*Sample Space*
(a) A spinner is marked from 1 to 8. An experiment consists of spinning the dial once.	$\{1, 2, 3, 4, 5, 6, 7, 8\}$
(b) An experiment consists of tossing two coins, a penny and a nickel, and observing whether the coins match (M) or do not match (D).	$\{M, D\}$
(c) An experiment consists of tossing two coins, a penny and a nickel, and observing the number of heads that appear.	$\{0, 1, 2\}$
(d) An experiment consists of tossing two coins, a penny and a nickel, and observing whether each coin falls heads (H) or tails (T).	$\{HH, HT, TH, TT\}$
(e) An experiment consists of tossing three coins and observing whether the coins fall heads (H) or tails (T).	$\{HHH, HHT, HTH, HTT,$ $THH, THT, TTH, TTT\}$
(f) An experiment consists of selecting three manufactured parts from the production process and observing whether they are acceptable (A) or defective (D).	$\{AAA, AAD, ADA, ADD,$ $DAA, DAD, DDA, DDD\}$

✎ **Now Work Problem 1**

The sample space of an experiment plays the same role in probability as the universal set does in set theory for all questions concerning the experiment.

In this chapter we confine our attention to those cases for which the sample space is finite, that is, to those situations in which it is possible to have only a finite number of outcomes.

Notice that in our definition we say *a* sample space, rather than *the* sample space, since an experiment can be described in many different ways. In general, it is a safe guide to include as much detail as possible in the description of the outcomes of the experiment in order to answer all pertinent questions concerning the result of the experiment.

EXAMPLE 3   Finding a Sample Space

Consider the set of all different types of families with three children. Describe a sample space for the experiment of selecting one family from the set of all possible three-child families.

SOLUTION   One way of describing a sample space is by denoting the number of girls in the family. The only possibilities are members of the set

$$\{0, 1, 2, 3\}$$

That is, a three-child family can have 0, 1, 2, or 3 girls. This sample space has four outcomes.

Another way of describing a sample space for this experiment is by first defining *B* and *G* as "boy" and "girl," respectively. Then the sample space would be given as

$$\{BBB, BBG, BGB, BGG, GBB, GBG, GGB, GGG\}$$

where *BBB* means first child is a boy, second child is a boy, third child is a boy, and so on. This experiment can be depicted by the tree diagram in Figure 3. The experiment has $2 \cdot 2 \cdot 2 = 8$ possible outcomes, as the Multiplication Principle indicates.

Figure 3

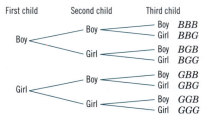

Assignment of Probabilities

We are now ready to give a definition of the probability of an outcome of a sample space.

Probability of an Outcome

Suppose the sample space *S* of an experiment has *n* outcomes so that

$$S = \{e_1, e_2, \ldots, e_n\}$$

To each outcome of *S*, we assign a real number, $P(e)$, called the **probability of the outcome** *e*, so that

$$P(e_1) > 0, P(e_2) > 0, \ldots, P(e_n) > 0 \qquad (1)$$

and

$$P(e_1) + P(e_2) + \ldots + P(e_n) = 1 \qquad (2)$$

Condition (1) states that each probability assignment must be nonnegative. Condition (2) states that the sum of all the probability assignments must equal one.

EXAMPLE 4   Acceptable Probability Assignments

Let a die be thrown. A sample space *S* is then

$$S = \{1, 2, 3, 4, 5, 6\}$$

There are six outcomes in *S*: 1, 2, 3, 4, 5, 6.

One acceptable assignment of probabilities is

$$P(1) = \tfrac{1}{6} \qquad P(2) = \tfrac{1}{6} \qquad P(3) = \tfrac{1}{6} \qquad P(4) = \tfrac{1}{6} \qquad P(5) = \tfrac{1}{6} \qquad P(6) = \tfrac{1}{6}$$

This choice is consistent with the definition since the probability assigned each outcome is nonnegative and their sum is 1. This assignment is made when the die is **fair.**

Another assignment that is consistent with the definition is

$$P(1) = 0 \qquad P(2) = 0 \qquad P(3) = \tfrac{1}{3} \qquad P(4) = \tfrac{2}{3} \qquad P(5) = 0 \qquad P(6) = 0$$

This assignment, although unnatural, is made when the die is "loaded" in such a way that only a 3 or a 4 can occur and the 4 is twice as likely as the 3 to occur.

Many other assignments can also be made that are consistent with the definition.

 **Now Work Problem 25**

EXAMPLE 5   Finding Probability Assignments

A coin is tossed. The coin is weighted so that heads (H) is 5 times more likely to occur than tails (T). What probability should we assign to heads? To tails?

SOLUTION   A sample space S for this experiment is

$$S = \{T, H\}$$

Let x denote the probability that tails occurs. Then

$$P(T) = x \qquad \text{and} \qquad P(H) = 5x$$

Since the sum of all the probability assignments must equal 1, we have

$$\begin{aligned} P(H) + P(T) = 5x + x &= 1 \\ 6x &= 1 \\ x &= \tfrac{1}{6} \end{aligned}$$

Thus we assign the probabilities

$$P(H) = \tfrac{5}{6} \qquad P(T) = \tfrac{1}{6}$$

 **Now Work Problem 29**

The sample space and the assignment of probabilities to each outcome of an experiment constitutes a *probability model* for the experiment.

> **Constructing a Probability Model**
>
> To construct a probability model requires two steps:
>
> **Step 1** Find a sample space. List all the possible outcomes of the experiment, or, if this is not easy to do, determine the number of outcomes of the experiment.
>
> **Step 2** Assign to each outcome e a probability $P(e)$ so that
> (a) $P(e) \geq 0$
> (b) The sum of all the probabilities assigned to the outcomes equals 1.

EXAMPLE 6   Constructing a Probability Model

A fair coin is tossed. If it comes up heads (H), then a fair die is rolled. If it comes up tails (T), then the coin is tossed once more. Construct a probability model for this experiment.

SOLUTION   We begin by constructing a tree diagram (Figure 4) that reflects the experiment.

Figure 4

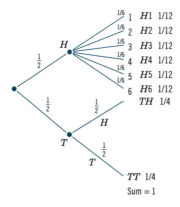

Based on the tree diagram, we can list the possible outcomes of the experiment. Thus a sample space for the experiment is

$$S = \{H1, H2, H3, H4, H5, H6, TH, TT\}$$

where $H1$ indicates heads for the coin and then 1 for the die, and so on.

Next, since the coin is fair, the probability is $\frac{1}{2}$ for a head H and $\frac{1}{2}$ for a tail T. Also, since the die is fair, the probability for any face to occur is $\frac{1}{6}$. Refer again to Figure 4. Based on all this, we assign the probabilities to each outcome in S follows:

$$P(H1) = P(H2) = P(H3) = P(H4) = P(H5) = P(H6) = \tfrac{1}{12}$$
$$P(TH) = P(TT) = \tfrac{1}{4}$$

The above discussion constitutes a probability model, or **stochastic model,** for the experiment.

◼

Notice that the probability assignments of Example 6 are obtained by multiplying along the branches of the tree. We'll have more to say about this later.

Events and Simple Events

> **Event; Simple Event**
>
> An **event** is any subset of the sample space. If an event has exactly one element, that is, consists of only one outcome, it is called a **simple event.**

For example, consider the experiment discussed in Example 3: three-child families. A sample space S for this experiment is

$$S = \{BBB, BBG, BGB, BGG, GBB, GBG, GGB, GGG\}$$

The event E that the family consists of exactly two boys is

$$E = \{BBG, BGB, GBB\}$$

Every event can be written as the union of simple events.

For example, the event E is the union of the three simple events $\{BBG\}$, $\{BGB\}$, $\{GBB\}$. That is,

$$E = \{BBG\} \cup \{BGB\} \cup \{GBB\}$$

Since a sample space S is also an event, we can express a sample space as the union of simple events. Thus if the sample space S consists of n outcomes,

$$S = \{e_1, e_2, \ldots, e_n\}$$

then

$$S = \{e_1\} \cup \{e_2\} \cup \cdots \cup \{e_n\}$$

Suppose probabilities have been assigned to each outcome of S. "What is the probability of an event E of S?" Let S be a sample space and let E be any event of S. It is clear that either $E = \varnothing$ or E is a simple event or E is the union of two or more simple events.

Probability of an Event

If $E = \varnothing$, the event E is **impossible.** We define the **probability of** $\varnothing$ as

$$P(\varnothing) = 0$$

If $E = \{e\}$ is a simple event, then $P(E) = P(e)$; that is, $P(E)$ equals the probability assigned to the outcome e.

If E is the union of r simple events $\{e_1\}$, $\{e_2\}$, $\ldots$, $\{e_r\}$, we define the **probability of E** to be the sum of the probabilities assigned to each simple event in E. That is,

$$P(E) = P(e_1) + P(e_2) + \cdots + P(e_r) \tag{3}$$

In particular, if the sample space S is given by

$$S = \{e_1, e_2, \ldots, e_n\}$$

we must have

$$P(S) = P(e_1) + \cdots + P(e_n) = 1$$

Thus the probability of S, the sample space, is 1.

EXAMPLE 7   Finding the Probability of an Event

In an experiment with two fair dice, find the probability of each of the following events:

(a) The sum of the faces is 3.   (b) The sum of the faces is 7.
(c) The sum of the faces is 7 or 3.   (d) The sum of the faces is 7 and 3.

SOLUTION   The tree diagram shown in Figure 5 illustrates the experiment. Since the dice are fair, we assign a probability of $\frac{1}{6}$ along each branch. The probability assigned to each outcome in S is therefore $\frac{1}{36}$.

Figure 5

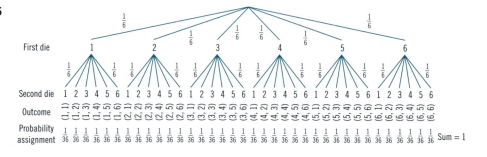

(a) The sum of the faces is 3 if and only if the outcome is an element of the event $A = \{(1, 2), (2, 1)\}$. Thus

$$P(A) = \tfrac{1}{36} + \tfrac{1}{36} = \tfrac{2}{36} = \tfrac{1}{18}$$

(b) The sum of the faces is 7 if and only if the outcome is an element of the event $B = \{(1, 6), (2, 5), (3, 4), (4, 3), (5, 2), (6, 1)\}$. Thus

$$P(B) = \tfrac{1}{36} + \tfrac{1}{36} + \tfrac{1}{36} + \tfrac{1}{36} + \tfrac{1}{36} + \tfrac{1}{36} = \tfrac{6}{36} = \tfrac{1}{6}$$

(c) The sum of the faces is 7 or 3 if and only if the outcome is an element of $A \cup B$, where A and B are the sets given in Parts (a) and (b). Then,

$$A \cup B = \{(2, 1), (1, 2), (1, 6), (2, 5), (3, 4), (4, 3), (5, 2), (6, 1)\}$$

Thus,

$$P(A \cup B) = \tfrac{1}{36} + \tfrac{1}{36} + \tfrac{1}{36} + \tfrac{1}{36} + \tfrac{1}{36} + \tfrac{1}{36} + \tfrac{1}{36} + \tfrac{1}{36} = \tfrac{8}{36} = \tfrac{2}{9}$$

(d) The sum of the faces is 3 and 7 if and only if the outcome is an element of $A \cap B$. Since $A \cap B = \varnothing$, the event is impossible. That is, $P(A \cap B) = 0$.   ∎

 **Now Work Problem 33**

EXAMPLE 8   Finding the Probability of an Event

Let two coins be tossed. A sample space S is

$$S = \{HH, TH, HT, TT\}$$

Let E be the event that both coins show heads or both show tails. Compute the probability for this event E using the following two assignments of probabilities:

(a) $P(HH) = P(TH) = P(HT) = P(TT) = \frac{1}{4}$

(b) $P(HH) = \frac{1}{9}$, $P(TH) = \frac{2}{9}$, $P(HT) = \frac{2}{9}$, $P(TT) = \frac{4}{9}$

SOLUTION   The event E is $\{HH, TT\}$.

(a) $P(E) = P(HH) + P(TT) = \frac{1}{4} + \frac{1}{4} = \frac{1}{2}$

(b) $P(E) = P(HH) + P(TT) = \frac{1}{9} + \frac{4}{9} = \frac{5}{9}$

The fact that we obtained different probabilities for the same event in Example 8 is not unexpected since it results from our original assignment of probabilities to the outcomes of the experiment. Any assignment that conforms to the restrictions given in (1) and (2) is mathematically correct. The question of which assignment to make is not a mathematical question, but is one that depends on the real world situation to which the theory is applied. In this example the coins were fair in case (a) and were loaded in case (b).

EXERCISE 7.1   Answers to odd-numbered problems begin on page AN-26.

1. A nickel and a dime are tossed. List the elements of the sample space

(a) If we are interested in whether the dime falls heads (H) or tails (T).

(b) If we are interested only in the number of heads that appear on a single toss of the two coins.

(c) If we are interested in whether the coins match (M) or do not match (D).

2. A card is selected from a regular deck of cards.* List the elements of the sample space

(a) If we are interested in the color of the card

(b) If we are interested in the suit of the card

In Problems 3–8 describe a sample space associated with each experiment. List the outcomes of each sample space. In each experiment we are interested in whether the coin falls heads (H) or tails (T).

3. Tossing 2 coins

4. Tossing 1 coin twice

5. Tossing a coin 3 times

6. Tossing 3 coins

7. Tossing a coin 2 times and then a die

8. Tossing two coins and then a die

In Problems 9–16 use the pictured spinners to list the outcomes of a sample space associated with each experiment.

---

* A regular deck of cards has 52 cards. There are four suits of 13 cards each. The suits are called *clubs* (black), *diamonds* (red), *hearts* (red), and *spades* (black). In each suit the 13 cards are labeled A (ace), 2, 3, 4, 5, 6, 7, 8, 9, 10, J (jack), Q (queen), and K (king).

Spinner 1   Spinner 2   Spinner 3

9. First spinner 1 is spun and then spinner 2 is spun.

10. First spinner 2 is spun and then spinner 3 is spun.

11. Spinner 2 is spun twice.

12. Spinner 3 is spun twice.

13. Spinner 2 is spun twice and then spinner 3 is spun.

14. Spinner 3 is spun once and then spinner 2 is spun twice.

15. Spinners 1, 2, and 3 are each spun once in this order.

16. Spinners 3, 2, and 1 are each spun once in this order.

In Problems 17–24 find the number of outcomes of a sample space associated with each experiment.

17. Tossing a coin 4 times

18. Tossing a coin 5 times

19. Tossing 3 dice

20. Tossing 2 dice and then a coin

21. Selecting 2 cards (without replacement) from a regular deck of cards (Assume order is not important.)

22. Selecting 3 cards (without replacement) from a regular deck of 52 cards (Assume order is not important.)

23. Picking two letters from the alphabet (repetitions allowed; assume order is important)

24. Picking two letters from the alphabet (no repetitions allowed; assume order is important)

In Problems 25–28 consider the experiment of tossing a coin twice. The table lists possible assignments of probabilities for this experiment:

		Sample Space			
		HH	**HT**	**TH**	**TT**
Assignments	1	$\frac{1}{4}$	$\frac{1}{4}$	$\frac{1}{4}$	$\frac{1}{4}$
	2	0	0	0	1
	3	$\frac{3}{16}$	$\frac{5}{16}$	$\frac{5}{16}$	$\frac{3}{16}$
	4	$\frac{1}{2}$	$\frac{1}{2}$	$-\frac{1}{2}$	$\frac{1}{2}$
	5	$\frac{1}{8}$	$\frac{1}{4}$	$\frac{1}{4}$	$\frac{1}{8}$
	6	$\frac{1}{9}$	$\frac{2}{9}$	$\frac{2}{9}$	$\frac{4}{9}$

Using this table, answer the following four questions.

25. Which of the assignments of probabilities are consistent with the definition of the probability of an outcome?

26. Which of the assignments of probabilities should be used if the coin is known to be fair?

27. Which of the assignments of probabilities should be used if the coin is known to always come up tails?

28. Which of the assignments of probabilities should be used if tails is twice as likely as heads to occur?

29. A coin is weighted so that heads is three times as likely as tails to occur. What probability should we assign to heads? to tails?

30. A coin is weighted so that tails is twice as likely as heads to occur. What probability should we assign to heads? to tails?

31. A die is weighted so that an odd-numbered face is twice as likely as an even-numbered face. What probability should we assign to each face?

32. A die is weighted so that a 6 cannot appear. If the other faces each have the same probability, what probability should we assign to each face?

In Problems 33–38 the experiment consists of tossing 2 fair dice. Construct a probability model for this experiment and find the probability of each event.

33. $A = \{(1, 4), (4, 1)\}$

34. $B = \{(1, 5), (2, 4), (3, 3), (4, 2), (5, 1)\}$

35. $C = \{(1, 4), (2, 4), (3, 4), (4, 4)\}$

36. $D = \{(1, 2), (2, 1), (2, 4), (4, 2), (3, 6), (6, 3)\}$

37. $E = \{(1, 1), (2, 2), (3, 3), (4, 4), (5, 5), (6, 6)\}$

38. $F = \{(1, 3), (2, 2), (3, 1)\}$

In Problems 39–48 the experiment consists of tossing a fair die and then a fair coin. Construct a probability model for this experiment. Then find the probability of each event.

39. The coin comes up heads.

40. The coin comes up tails.

41. The die comes up 4.

42. The die comes up 1.

43. The die does not come up 4.

44. The die does not come up 1.

45. The die comes up 5 or 6.

46. The die comes up 1 or 2.

47. The die comes up 3, 4, or 5 and the coin comes up heads.

48. The coin comes up tails and the die comes up a number less than 4.

In Problems 49–52 assign valid probabilities to the outcomes of each random experiment.

49. Tossing a fair coin twice

50. Tossing a fair coin 3 times

51. Tossing a fair die and then a fair coin

52. Tossing a fair die and then 2 fair coins

In Problems 53–58 the random experiment consists of tossing a fair coin 4 times.

53. List the outcomes or the sample space and assign probabilities to each one.

54. Write the elements of the event, "The first 2 tosses are heads."

55. Write the elements of the event, "The last 3 tosses are tails."

56. Write the elements of the event, "Exactly 3 tosses come up tails."

57. Write the elements of the event, "The number of heads exceeds 1 but is fewer than 4."

58. Write the elements of the event, "The first 2 tosses are heads and the second 2 are tails."

59. In a T-maze a mouse may turn to the right (*R*) and receive a mild shock, or to the left (*L*) and get a piece of cheese. Its behavior in making such "choices" is studied by psychologists. Suppose a mouse runs a T-maze 3 times. List the set of all possible outcomes and assign valid probabilities to each outcome under the assumption that the first two times the maze is run the mouse chooses equally between left and right, but on the third run, the mouse is twice as likely to choose cheese. Find the probability of each of the events listed.

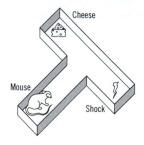

(a) E: Run to the right 2 consecutive times
(b) F: Never run to the right
(c) G: Run to the left on the first trial
(d) H: Run to the right on the second trial

60. Let $S = \{e_1, e_2, e_3, e_4, e_5, e_6, e_7\}$ be a given sample space. Let the probabilities assigned to each outcome be given as follows:

$$P(e_1) =\ \ P(e_2) = P(e_6)$$
$$P(e_3) = 2P(e_4) = \tfrac{1}{2}P(e_1)$$
$$P(e_5) = \tfrac{1}{2}P(e_7) = \tfrac{1}{4}P(e_1)$$

(a) Find $P(e_1)$, $P(e_2)$, $P(e_3)$, $P(e_4)$, $P(e_5)$, $P(e_6)$, $P(e_7)$.
(b) If $A = \{e_1, e_2\}$, $B = \{e_2, e_3, e_4\}$, $C = \{e_5, e_6, e_7\}$, and $D = \{e_1, e_5, e_6\}$, find $P(A)$, $P(B)$, $P(C)$, $P(D)$, $P(A \cup B)$, $P(A \cap D)$, $P(D \cap B)$, and $P(A \cap \bar{B})$.

61. Three cars, C_1, C_2, C_3, are in a race. If the probability of C_1 winning is p, that is, $P(C_1) = p$, and $P(C_1) = \tfrac{1}{2}P(C_2)$ and $P(C_3) = \tfrac{1}{2}P(C_2)$, find $P(C_1)$, $P(C_2)$, and $P(C_3)$.

62. Consider an experiment with a loaded die such that the probability of any of the faces appearing in a toss is equal to that face times the probability that a 1 will occur. That is, $P(6) = 6 \cdot P(1)$, and so on.

(a) Describe the sample space.
(b) Find $P(1)$, $P(2)$, $P(3)$, $P(4)$, $P(5)$, and $P(6)$.
(c) Let the events A, B, and C be described as

A: Even-numbered face
B: Odd-numbered face
C: Prime number on face (2, 3, 5 are prime)

Find $P(A)$, $P(B)$, $P(C)$, $P(A \cup B)$, and $P(A \cup \bar{C})$.

7.2 PROPERTIES OF THE PROBABILITY OF AN EVENT

In this section we investigate properties of the probability of an event. Sets and Venn diagrams form the basis for much of what we shall do.

Mutually Exclusive Events

> **Mutually Exclusive Events**
>
> Two or more events of a sample space S are said to be **mutually exclusive** if and only if they have no outcomes in common.

If we treat mutually exclusive events as sets, they are disjoint.

The following result gives us a way of computing probabilities for mutually exclusive events.

> **Probability of E or F if E and F Are Mutually Exclusive**
>
> Let E and F be two events of a sample space S. If E and F are mutually exclusive, that is, if $E \cap F = \emptyset$, then the probability of E or F is the sum of their probabilities. That is,
>
> $$P(E \cup F) = P(E) + P(F) \qquad \text{if } E \cap F = \emptyset \tag{1}$$

Since E and F can be written as a union of simple events in which no simple event of E appears in F and no simple event of F appears in E, the result follows.

EXAMPLE 1   Finding the Probability When (1) Applies

In the experiment of tossing two fair dice, what is the probability of obtaining either a sum of 7 or a sum of 11?

SOLUTION   Each simple event of the sample space is assigned the probability $\frac{1}{36}$. Let E and F be the events

$$E: \quad \text{Sum is 7} \qquad F: \quad \text{Sum is 11}$$

Since the dice are fair, and a sum of 7 can occur in six different ways, we have

$$P(E) = P(\text{Sum is 7}) = \tfrac{1}{36} + \tfrac{1}{36} + \tfrac{1}{36} + \tfrac{1}{36} + \tfrac{1}{36} + \tfrac{1}{36} = \tfrac{6}{36}$$

Similarly, since a sum of 11 can occur in two different ways, we have

$$P(F) = \tfrac{2}{36}$$

The two events E and F are mutually exclusive. Therefore, by (1), the probability that the sum is 7 or 11 is

$$P(E \cup F) = P(E) + P(F) = \tfrac{6}{36} + \tfrac{2}{36} = \tfrac{8}{36} = \tfrac{2}{9}$$

 **Now Work Problem 17**

■

Additive Rule

The following result, called the **Additive Rule,** provides a technique for finding the probability of the union of two events whether they are mutually exclusive or not.

Additive Rule

For any two events E and F of a sample space S,

$$P(E \cup F) = P(E) + P(F) - P(E \cap F) \tag{2}$$

This result concerning probability is closely related to the counting formula discussed in the previous chapter (page 278). A proof is outlined in Problem 57.

EXAMPLE 2   Using the Additive Rule

If $P(E) = .30$, $P(F) = .20$, and $P(E \cup F) = .40$, find $P(E \cap F)$.

SOLUTION   By the Additive Rule we know that

$$P(E \cup F) = P(E) + P(F) - P(E \cap F)$$

Since $P(E) = .30$, $P(F) = .20$, and $P(E \cup F) = .40$, we have

$$.40 = .30 + .20 - P(E \cap F)$$

Thus

$$P(E \cap F) = .30 + .20 - .40 = .10$$

■

EXAMPLE 3　Using the Additive Rule

Consider the two events

E:　A shopper spends at least \$40 for food

F:　A shopper spends at least \$15 for meat

Based on recent studies, we might assign

$$P(E) = .56 \qquad P(F) = .63$$

Suppose the probability that a shopper spends at least \$40 for food and \$15 for meat is .33. What is the probability that a shopper spends at least \$40 for food or at least \$15 for meat?

SOLUTION　From the information given, we conclude that $P(E \cap F) = .33$. Since we are looking for the probability of $E \cup F$, we use the Additive Rule and find that

$$P(E \cup F) = P(E) + P(F) - P(E \cap F)$$
$$= .56 + .63 - .33 = .86$$

■

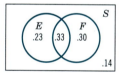 **Now Work Problem 5**

Figure 6

Often a Venn diagram is helpful in solving probability problems. A Venn diagram depicting the information of Example 3 is given in Figure 6. To obtain the diagram, we begin with the fact that $P(E \cap F) = .33$. Since $P(E) = .56$ and $P(F) = .63$, we fill in E with $.56 - .33 = .23$ and we fill in F with $.63 - .33 = .30$. Since $P(S) = 1$, we complete Figure 6 by entering $1 - (.23 + .33 + .30) = .14$. Now it is easy to see that the probability of E, but not F, is .23. The probability of neither E nor F is .14.

EXAMPLE 4　Using a Venn Diagram

In an experiment with two fair dice, consider the events

E:　The sum of the faces is 8

F:　Doubles are thrown

What is the probability of obtaining E, but not F?

SOLUTION　We write down the elements of E and F.

$$E = \{(2, 6), (3, 5), (4, 4), (5, 3), (6, 2)\}$$
$$F = \{(1, 1), (2, 2), (3, 3), (4, 4), (5, 5), (6, 6)\}$$
$$E \cap F = \{(4, 4)\}$$

Then, the probabilities of each of these events is

$$P(E) = \tfrac{5}{36} \qquad P(F) = \tfrac{6}{36} \qquad P(E \cap F) = \tfrac{1}{36}$$

Figure 7

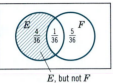

E, but not *F*

Now construct a Venn diagram. See Figure 7.
The probability of E, but not F, is $\tfrac{4}{36} = \tfrac{1}{9}$.

We could also have solved Example 4 by actually finding the set E, but not F, namely:

$$\{(2, 6), (3, 5), (5, 3), (6, 2)\}$$

Since this set has four simple events, the probability is

$$\tfrac{1}{36} + \tfrac{1}{36} + \tfrac{1}{36} + \tfrac{1}{36} = \tfrac{4}{36} = \tfrac{1}{9}$$

as before.

The probability of any outcome of a sample space S is nonnegative. Furthermore, since any event E of S is the union of outcomes in S, and since $P(S) = 1$, it follows that

$$0 \leq P(E) \leq 1$$

Properties of the Probability of an Event

To summarize, the probability of an event E of a sample space S has the following three properties:

(I) $\mathbf{0 \leq P(E) \leq 1}$ for every event E of S

(II) $\mathbf{P(\varnothing) = 0}$ **and** $\mathbf{P(S) = 1}$

(III) $\mathbf{P(E \cup F) = P(E) + P(F) - P(E \cap F)}$ for any two events E and F of S

Complement of an Event

Let E be an event of a sample space S. The complement of E is the event "Not E" in S. The next result gives a relationship between their probabilities.

Probability of the Complement of an Event

Let E be an event of a sample space S. Then

$$P(\overline{E}) = 1 - P(E) \tag{3}$$

where $\overline{E}$ is the complement of E.

Proof   We know that

$$S = E \cup \overline{E} \qquad E \cap \overline{E} = \varnothing$$

Since E and $\bar{E}$ are mutually exclusive,

$$P(S) = P(E) + P(\bar{E})$$

Since $P(S) = 1$ (Property II), it follows that

$$1 = P(E) + P(\bar{E})$$
$$P(\bar{E}) = 1 - P(E)$$

■

This result gives us a tool for finding the probability that an event does not occur if we know the probability that it does occur. Thus the probability $P(\bar{E})$ that E does not occur is obtained by subtracting from 1 the probability $P(E)$ that E does occur. We will see shortly that it is sometimes easier to find $P(E)$ by finding $P(\bar{E})$ and using (3), than it is to proceed directly.

EXAMPLE 5   Using Formula (3)

A study of people over 40 with an MBA degree shows that it is reasonable to assign a probability of .756 that such a person will have annual earnings in excess of $80,000. The probability that such a person will earn $80,000 or less is then

 **Now Work Problem 1**

$$1 - .756 = .244$$

■

EXAMPLE 6   Using Formula (3)

In an experiment of tossing two fair dice, find:

(a) The probability that the sum of the faces is less than or equal to 3
(b) The probability that the sum of the faces is greater than 3

SOLUTION

(a) The number of outcomes in the event E, "the sum of the faces is less than or equal to 3," is 3. The number of outcomes of the sample space S is 36. Thus, because the dice are fair,

$$P(E) = \tfrac{3}{36} = \tfrac{1}{12}$$

(b) The event "the sum of the faces is greater than 3" is the complement of the event E defined in part (a). Since we seek $P(\bar{E})$, we can use the result of part (a) and Formula (3):

$$P(\bar{E}) = 1 - P(E) = 1 - \tfrac{1}{12} = \tfrac{11}{12}$$

That is, the probability that the sum of the faces is greater than 3 is $\tfrac{11}{12}$.

■

Compare the work required to do part (a) first and then apply Formula (3) (as we did in Example 6) to solving part (b) by listing the elements of the set and then computing the probability. We'll see that in many cases the approach of finding $P(E)$ first to get $P(\bar{E})$ is not only easier, but sometimes is the only choice available.

Equally Likely Outcomes

When the same probability is assigned to each outcome of a sample space, the outcomes are termed *equally likely outcomes.*

Equally likely outcomes often occur when items are selected randomly. For example, in randomly selecting 1 person from a group of 10, the probability of selecting a particular individual is $\frac{1}{10}$. If a card is chosen randomly from a deck of 52 cards, the probability of drawing a particular card is $\frac{1}{52}$. In general, if a sample space S has n equally likely outcomes, the probability assigned to each outcome is $\frac{1}{n}$.

Let a sample space S be given by

$$S = \{e_1, e_2, \ldots, e_n\}$$

Suppose each of the outcomes $e_1, \ldots, e_n$ is equally likely to occur. If E contains m of these n outcomes, then

$$P(E) = \frac{m}{n} \qquad (4)$$

A proof of Equation (4) is outlined in Problem 58.

For example, if a person is chosen randomly from a group of 100 people, 60 female and 40 male, the probability a male is chosen is $\frac{40}{100}$.

This leads to the following formulation:

Probability of an Event E in a Sample Space with Equally Likely Outcomes

If the sample space S of an experiment has n equally likely outcomes, and the event E in S occurs m times, then the probability of event E, written as $P(E)$, is m/n. That is,

$$P(E) = \frac{\text{Number of possible ways the event } E \text{ can take place}}{\text{Number of outcomes in } S} = \frac{c(E)}{c(S)} \qquad (5)$$

Thus, to compute the probability of an event E in which the outcomes are equally likely, count the number $c(E)$ of outcomes in E, and divide by the total number $c(S)$ of outcomes in the sample space.

EXAMPLE 7   Equally Likely Outcomes

A jar contains 10 marbles; 5 are solid color, 4 are speckled, and 1 is clear.

(a) If one marble is picked at random, what is the probability it is speckled?

(b) If one marble is picked at random, what is the probability it is clear or a solid color?

SOLUTION   The experiment is an example of one in which the outcomes are equally likely. That is, no one marble is more likely to be picked than another. If S is the sample space, then there are 10 possible outcomes in S, so $c(S) = 10$.

(a) Define the event E: A speckled marble is picked. There are 4 ways E can occur. Thus

$$P(E) = \frac{c(E)}{c(S)} = \frac{4}{10} = .4$$

(b) Define the events F: A clear marble is picked and G: A solid color marble is picked. Then there is 1 way for F to occur and 5 ways for G to occur. Thus

$$P(F) = \frac{c(F)}{c(S)} = \frac{1}{10} \qquad P(G) = \frac{c(G)}{c(S)} = \frac{5}{10}$$

We seek $P(F \cup G)$. Since F and G are mutually exclusive,

$$P(F \cup G) = P(F) + P(G) = \tfrac{1}{10} + \tfrac{5}{10} = \tfrac{6}{10} = .6$$

 **Now Work Problem 23**

Odds

In many instances the probability of an event may be expressed as *odds* — either *odds for* an event or *odds against* an event.

If E is an event:

The **odds for** E are $\qquad \dfrac{P(E)}{P(\bar{E})} \qquad$ or $\qquad P(E)$ to $P(\bar{E})$

The **odds against** E are $\qquad \dfrac{P(\bar{E})}{P(E)} \qquad$ or $\qquad P(\bar{E})$ to $P(E)$

EXAMPLE 8   **Computing Odds Given a Probability**

Suppose the probability of the event

$$E: \quad \text{It will rain}$$

is .3. The odds for rain are

$$\frac{.3}{.7} \qquad \text{or} \qquad 3 \text{ to } 7$$

The odds against rain are

$$\frac{.7}{.3} \qquad \text{or} \qquad 7 \text{ to } 3$$

To obtain the probability of the event E when either the odds for E or the odds against E are known, we use the following formulas:

If the odds for E are a to b, then

$$P(E) = \frac{a}{a + b} \tag{6}$$

If the odds against E are a to b, then

$$P(E) = \frac{b}{a + b}$$

The proof of Equation (6) is outlined in Problem 59.

EXAMPLE 9   Computing a Probability from Odds

(a) The odds for a Republican victory in the next Presidential election are 7 to 5. What is the probability that a Republican victory occurs?
(b) The odds against the Chicago Cubs winning the league pennant are 200 to 1. What is the probability that the Cubs win the pennant?

SOLUTION

(a) The event E is "A Republican victory occurs." The odds for E are 7 to 5. Thus,

$$P(E) = \frac{7}{7 + 5} = \frac{7}{12} \approx .583$$

(b) The event F is "The Cubs win the pennant." The odds against F are 200 to 1. Thus

$$P(F) = \frac{1}{200 + 1} = \frac{1}{201} \approx .00498$$

 Now Work Problem 41

EXERCISE 7.2   Answers to odd-numbered problems begin on page AN-26.

In Problems 1–6 find the probability of the indicated event if $P(A) = .25$ and $P(B) = .40$.

1. $P(\overline{A})$

2. $P(\overline{B})$

3. $P(A \cup B)$ if A, B are mutually exclusive

4. $P(A \cap B)$ if A, B are mutually exclusive

5. $P(A \cup B)$ if $P(A \cap B) = .15$

6. $P(A \cap B)$ if $P(A \cup B) = .55$

In Problems 7–16 a card is drawn at random from a regular deck of 52 cards. Calculate the probability of each event.

7. The ace of hearts is drawn.

8. An ace is drawn.

9. A spade is drawn.

10. A red card is drawn.

11. A picture card (J, Q, K) is drawn.

12. A number card (A, 2, 3, 4, 5, 6, 7, 8, 9, 10) is drawn.

13. A card with a number less than 6 is drawn (count A as 1).

14. A card with a value of 10 or higher is drawn.

15. A card that is not an ace is drawn.

16. A card that is either a queen or king of any suit is drawn.

17. In tossing 2 fair dice, are the events "Sum is 2" and "Sum is 12" mutually exclusive? What is the probability of obtaining either a 2 or a 12?

18. In tossing 2 fair dice, are the events "Sum is 6" and "Sum is 8" mutually exclusive? What is the probability of obtaining either a 6 or an 8?

In Problems 19–26 a ball is picked at random from a box containing 3 white, 5 red, 8 blue, and 7 green balls. Find the probability of each event.

19. White ball is picked.

20. Blue ball is picked.

21. Green ball is picked.

22. Red ball is picked.

23. White or red ball is picked.

24. Green or blue ball is picked.

25. Neither red nor green ball is picked.

26. Red or white or blue ball is picked.

27. In a throw of 2 fair dice, what is the probability that the number on one die is double the number on the other?

28. In a throw of 2 fair dice, what is the probability that one die gives a 5 and the other die a number less than 5?

29. The Chicago Bears football team has a probability of winning of .65 and of tying of .05. What is their probability of losing?

30. The Chicago Black Hawks hockey team has a probability of winning of .6 and a probability of losing of .25. What is the probability of a tie?

31. Jenny is taking courses in both mathematics and English. She estimates her probability of passing mathematics at .4 and English at .6, and she estimates her probability of passing at least one of them at .8. What is her probability of passing both courses?

32. After midterm exams, Jenny (see Problem 31) reassesses her probability of passing mathematics to .7. She feels her probability of passing at least one of these courses is still .8, but she has a probability of only .1 of passing both courses. If her probability of passing English is less than .4, she will drop English. Should she drop English? Why?

33. Let A and B be events of a sample space S and let $P(A) = .5$, $P(B) = .4$, and $P(A \cup B) = .2$. Find the probabilities of each of the following events:

(a) A or B
(b) A but not B
(c) B but not A
(d) Neither A nor B

34. If A and B represent two mutually exclusive events such that $P(A) = .35$ and $P(B) = .60$, find each of the following probabilities:

(a) $P(A \cup B)$
(b) $P(\overline{A \cup B})$
(c) $P(\overline{B})$
(d) $P(\overline{A})$
(e) $P(A \cap B)$

35. At the Milex tune-up and brake repair shop, the manager has found that a car will require a tune-up with a probability of .6, a brake job with a probability of .1, and both with a probability of .02.

(a) What is the probability that a car requires either a tune-up or a brake job?
(b) What is the probability that a car requires a tune-up but not a brake job?
(c) What is the probability that a car requires neither type of repair?

36. A factory needs two raw materials, say, E and F. The probability of not having an adequate supply of material E is .06, whereas the probability of not having an adequate supply of material F is .04. A study shows that the probability of a shortage of both E and F is .02. What is the probability of the factory being short of either material E or F?

37. In a survey of the number of TV sets in a house, the following probability table was constructed:

Number of TV sets	0	1	2	3	4 or more
Probability	.05	.24	.33	.21	.17

Find the probability of a house having

(a) 1 or 2 TV sets
(b) 1 or more TV sets
(c) 3 or fewer TV sets
(d) 3 or more TV sets
(e) Fewer than 2 TV sets
(f) Not even 1 TV set
(g) 1, 2, or 3 TV sets
(h) 2 or more TV sets

38. Through observation it has been determined that the probability for a given number of people waiting in line at a particular checkout register of a supermarket is as shown in the table:

Number waiting in line	0	1	2	3	4 or more
Probability	.10	.15	.20	.24	.31

Find the probability of

(a) At most 2 people in line
(b) At least 2 people in line
(c) At least 1 person in line

39. From a sales force of 150 people, 1 person will be chosen to attend a special sales meeting. If 52 are single, 72 are college graduates, and, of the 52 who are single, $\frac{3}{4}$ are college graduates, what is the probability that a salesperson selected at random will be neither single nor a college graduate?

40. In an election two amendments were proposed. The results indicate that of 850 people eligible to cast a ballot, 480 voted in favor of Amendment I, 390 voted for Amendment II, 120 voted for both, and 100 approved of neither. If an eligible voter is selected at random (that is, any one is as likely to be chosen as another), compute the following probabilities:

(a) The voter is in favor of I, but not II.
(b) The voter is in favor of II, but not I.

In Problems 41–46 determine the probability of E for the given odds.

41. 3 to 1 for E

42. 4 to 1 against E

43. 7 to 5 against E

44. 2 to 9 for E

45. 1 to 1 for E (even)

46. 50 to 1 for E

In Problems 47–50 determine the odds for and against each event for the given probability.

47. $P(E) = .6$

48. $P(H) = \frac{1}{4}$

49. $P(F) = \frac{3}{4}$

50. $P(G) = .1$

51. If two fair dice are thrown, what are the odds of obtaining a sum of 7? a sum of 11? a sum of 7 or 11?

52. If the odds for event A are 1 to 5 and the odds for event B are 1 to 3, what are the odds for the event A or B, assuming the event A and B is impossible?

53. The probability of a person getting a job interview is .54; what are the odds against getting the interview?

54. If the probability of war is .6, what are the odds against war?

55. In a track contest the odds that A will win are 1 to 2, and the odds that B will win are 2 to 3. Find the probability and the odds that A or B wins the race, assuming a tie is impossible.

56. It has been estimated that in 70% of the fatal accidents involving two cars, at least one of the drivers is drunk. If you hear of a two-car fatal accident, what odds should you give a friend for the event that at least one of the drivers was drunk?

57. Prove the Additive Rule, Equation (2) on page 329. [*Hint:* From Example 9 in Section 6.1 (page 275) we have

$$E \cup F = (E \cap \bar{F}) \cup (E \cap F) \cup (\bar{E} \cap F)$$

Since $E \cap \bar{F}, E \cap F$, and $\bar{E} \cap F$ are pairwise disjoint, we can write

$$P(E \cup F) = P(E \cap \bar{F}) + P(E \cap F) + P(\bar{E} \cap F) \quad \text{(a)}$$

We may write the sets E and F in the form

$$E = (E \cap F) \cup (E \cap \bar{F})$$
$$F = (E \cap F) \cup (\bar{E} \cap F)$$

Since $E \cap F$ and $E \cap \bar{F}$ are disjoint and $E \cap F$ and $\bar{E} \cap F$ are disjoint, we have

$$P(E) = P(E \cap F) + P(E \cap \bar{F})$$
$$P(F) = P(E \cap F) + P(\bar{E} \cap F) \quad \text{(b)}$$

Now combine (a) and (b).]

58. Prove Equation (4) on page 333.
[*Hint:* If $S = \{e_1, e_2, \ldots, e_n\}$, then

$$S = \{e_1\} \cup \{e_2\} \cup \cdots \cup \{e_n\}$$

and

$$P(S) = P(e_1) + P(e_2) + \cdots + P(e_n) = 1$$

If $P(e_1) = P(e_2) = \cdots = P(e_n)$, show that the common value is $1/n$. Next, if

$$E = \{e_1, e_2, \ldots, e_m\}$$

then

$$E = \{e_1\} \cup \{e_2\} \cup \cdots \cup \{e_m\}$$

Use the definition of probability of an event to calculate $P(E) = m/n$.]

59. Prove Equation (6) on page 335.

[*Hint:* If the odds for E are a to b, then, by the definition of odds,

$$\frac{P(E)}{P(\bar{E})} = \frac{a}{b}$$

But $P(\bar{E}) = 1 - P(E)$. So

$$\frac{P(E)}{1 - P(E)} = \frac{a}{b}$$

Now solve for $P(E)$.]

60. Generalize the Additive Rule by showing the probability of the occurrence of at least one of the three events A, B, C is given by

$$P(A \cup B \cup C) = P(A) + P(B) + P(C)$$
$$- P(A \cap B) - P(A \cap C)$$
$$- P(B \cap C) + P(A \cap B \cap C)$$

Technology Exercises

Most graphing calculators have a random number function (usually RAND or RND) generating numbers between 0 and 1. Check your user's manual to see how to use this function.

Sometimes experiments are simulated using a random number function instead of actually performing the experiment. In Problems 1–6 use a random number function to simulate each experiment.

1. Tossing a Fair Coin   Consider an experiment of tossing a fair coin. Simulate the experiment using a random number function on your calculator, considering a toss to be tails (T) if the result is less than 0.5, and considering a toss to be heads (H) if the result is greater than or equal to 0.5. [*Note:* Most calculators repeat the action of the last entry if you simply press the ENTER, or EXE, key again.] Repeat the experiment 10 times. Using these 10 outcomes of the experiment you can estimate the probabilities $P(H)$ and $P(T)$ by the ratios

$$P(H) \approx \frac{\text{Number of times } H \text{ occurred}}{10}$$

$$P(T) \approx \frac{\text{Number of times } T \text{ occurred}}{10}$$

What are the actual probabilities? How close are the results of the experiment to the actual values?

2. Urn and Balls   Consider an experiment of choosing a ball from an urn containing 18 red and 12 white balls. Simulate the experiment using a random number function on your calculator, considering a selection to be a red ball (R) if the result is less than 0.6, and considering a toss to be a white ball (W) if the result is greater than or equal to 0.6. [*Note:* Most calculators repeat the action of the last entry if you simply press the ENTER, or EXE, key again.] Repeat the experiment 10 times. Using these 10 outcomes of the experiment you can estimate the probabilities $P(R)$ and $P(W)$ by the ratios

$$P(R) \approx \frac{\text{Number of times } R \text{ occurred}}{10}$$

$$P(W) \approx \frac{\text{Number of times } W \text{ occurred}}{10}$$

What are the actual probabilities? How close are the results of the experiment to the actual values?

3. Tossing a Loaded Coin   Consider an experiment of tossing a loaded coin. Simulate the experiment using a random number function on your calculator, considering a toss to be tails (T) if the result is less than 0.25, and considering a toss to be heads (H) if the result is greater than or equal to 0.25. [*Note:* Most calculators repeat the action of the last entry if you simply press the ENTER, or EXE, key again.] Repeat the experiment 20 times. Using these 20 outcomes of the experiment you can estimate the probabilities $P(H)$ and $P(T)$ by the ratios

$$P(H) \approx \frac{\text{Number of times } H \text{ occurred}}{20}$$

$$P(T) \approx \frac{\text{Number of times } T \text{ occurred}}{20}$$

What are the actual probabilities? How close are the results of the experiment to the actual values?

4. Urn and Balls   Consider an experiment of choosing a ball from an urn containing 15 red and 35 white balls. Simulate the experiment using a random number function on your calculator, considering a selection to be a red ball (R) if the result is less than 0.3, and considering a toss to be a white ball (W) if the result is greater than or equal to 0.3. [*Note:* Most calculators repeat the action of the last entry if you simply press the ENTER, or EXE, key again.] Repeat the experiment 10 times. Using these 10 outcomes of the experiment you can estimate the probabilities $P(R)$ and $P(W)$ by the ratios

$$P(R) \approx \frac{\text{Number of times } R \text{ occurred}}{10}$$

$$P(W) \approx \frac{\text{Number of times } W \text{ occurred}}{10}$$

What are the actual probabilities? How close are the results of the experiment to the actual values?

5. **Jar and Marbles** Consider an experiment of choosing a marble from a jar containing 5 red, 2 yellow, and 8 white marbles. Simulate the experiment using a random number function on your calculator, considering a selection to be a red marble (R) if the result is less than or equal to 0.33, a yellow marble (Y) if the result is between 0.33 and 0.47, and a white marble (W) if the result is greater than or equal to 0.47. [*Note:* Most calculators repeat the action of the last entry if you simply press the ENTER, or EXE, key again.] Repeat the experiment 10 times. Using these 10 outcomes of the experiment you can estimate the probabilities $P(R)$, $P(Y)$, and $P(W)$ by the ratios

$$P(R) \approx \frac{\text{Number of times } R \text{ occurred}}{10}$$

$$P(Y) \approx \frac{\text{Number of times } Y \text{ occurred}}{10}$$

$$P(W) \approx \frac{\text{Number of times } W \text{ occurred}}{10}$$

What are the actual probabilities? How close are the results of the experiment to the actual values?

6. **Poker Game** The probabilities in a game of poker that Adam, Beatrice, or Cathy win are .22, .60, .18, respectively. Simulate the game using a random number function on your calculator, considering that Adam won (A) if the result is less than or equal to 0.22, that Beatrice won (B) if the result is between 0.22 and 0.82, and that Cathy won (C) if the result is greater than or equal to 0.82. [*Note:* Most calculators repeat the action of the last entry if you simply press the ENTER, or EXE, key again.] Repeat the experiment 20 times; that is, simulate 20 games. Using these 20 outcomes of the experiment you can estimate the probabilities $P(A)$, $P(B)$, and $P(C)$ by the ratios

$$P(A) \approx \frac{\text{Number of times } A \text{ won}}{20}$$

$$P(B) \approx \frac{\text{Number of times } B \text{ won}}{20}$$

$$P(C) \approx \frac{\text{Number of times } C \text{ won}}{20}$$

What are the actual values? How close are the results of the experiment to the actual values?

7.3 PROBABILITY PROBLEMS USING COUNTING TECHNIQUES

In this section we see how counting techniques can be effectively used to find probabilities.

EXAMPLE 1 Finding Probabilities Using Counting Techniques

From a box containing four white, three yellow, and one green ball, two balls are drawn one at a time without replacing the first before the second is drawn. Find the probability that one white and one yellow ball are drawn.

SOLUTION The number of ways that 2 balls can be drawn from 8 balls is $C(8, 2) = 28$. We define the events E and F as

E: A white ball is drawn

F: A yellow ball is drawn

The number of ways a white ball can be drawn is $C(4, 1) = 4$ and the number of ways a yellow ball can be drawn is $C(3, 1) = 3$. By the Multiplication Principle, the number of ways a white and yellow ball can be drawn is

$$C(4, 1) \cdot C(3, 1) = 4 \cdot 3 = 12$$

Since the outcomes are equally likely, the probability of drawing a white ball and a yellow ball is

$$P(E \cap F) = \frac{C(4, 1) \cdot C(3, 1)}{C(8, 2)} = \frac{12}{28} = \frac{3}{7}$$

■

EXAMPLE 2   Finding Probabilities Using Counting Techniques

A box contains 12 light bulbs, of which 5 are defective. All bulbs look alike and have equal probability of being chosen. Three lightbulbs are selected and placed in a box.

(a) What is the probability that all 3 are defective?
(b) What is the probability that exactly 2 are defective?
(c) What is the probability that at least 2 are defective?

SOLUTION

(a) The number of elements in the sample space S is equal to the number of combinations of 12 light bulbs taken 3 at a time, namely,

$$C(12, 3) = \frac{12!}{3!9!} = 220$$

Define E as the event, "3 bulbs are defective." Then E can occur in $C(5, 3)$ ways, that is, the number of ways in which 3 defective bulbs can be chosen from 5 defectives ones. The probability $P(E)$ is

$$P(E) = \frac{C(5, 3)}{C(12, 3)} = \frac{\dfrac{5!}{3!2!}}{220} = \frac{10}{220} = .04545$$

(b) Define F as the event, "2 bulbs are defective." To obtain 2 defective bulbs when 3 are chosen requires that we select 2 defective bulbs from the 5 available defective ones and 1 good bulb from the 7 good ones. By the Multiplication Principle this can be done in the following number of ways:

$$C(5, 2) \cdot C(7, 1) = \frac{5!}{2!3!} \cdot \frac{7!}{1!6!} = 10 \cdot 7 = 70$$

Number of ways   Number of ways
to select 2 defectives   to select 1 good bulb
from 5 defectives   from 7 good ones

The probability of selecting exactly 2 defective bulbs is therefore

$$P(F) = P(\text{Exactly 2 defectives}) = \frac{C(5, 2) \cdot C(7, 1)}{C(12, 3)} = \frac{70}{220} = .31818$$

(c) Define G as the event, "At least 2 are defective." The event G is equivalent to asking for the probability of selecting either exactly 2 or exactly 3 defective bulbs. Since these events are mutually exclusive, the sum of their probabilities will give the probability of G. Therefore,

$$P(G) = P(\text{Exactly 2 defectives}) + P(\text{Exactly 3 defectives})$$
$$= P(F) + P(E) = .31818 + .04545 = .36363$$

✍ **Now Work Problem 9**

■

EXAMPLE 3   **Finding Probabilities Using Counting Techniques**

A fair coin is tossed 10 times.

(a) What is the probability of obtaining exactly 5 heads and 5 tails?
(b) What is the probability of obtaining between 4 and 6 heads, inclusive?

SOLUTION   The number of elements in the sample space S is found by using the Multiplication Principle. Each toss results in a head (H) or a tail (T). Since the coin is tossed 10 times, we have

$$c(S) = \underbrace{2 \cdot 2 \cdot \ \ldots \ \cdot 2}_{10 \text{ twos}} = 2^{10}$$

The outcomes are equally likely since the coin is fair.

(a) Any sequence that contains 5 heads and 5 tails is determined once the position of the 5 heads (or 5 tails) is known. The number of ways we can position 5 heads in a sequence of 10 slots is $C(10, 5)$. The probability of the event E: Exactly 5 heads, 5 tails is

$$P(E) = \frac{c(E)}{c(S)} = \frac{C(10, 5)}{2^{10}} = .2461$$

(b) Let F be the event: Between 4 and 6 heads, inclusive. To obtain between 4 and 6 heads is equivalent to the event: Exactly 4 heads or exactly 5 heads or exactly 6 heads. Since these are mutually exclusive (for example, it is impossible to obtain exactly 4 heads and exactly 5 heads when tossing a coin 10 times), we have

$$P(F) = P(4 \text{ heads or 5 heads or 6 heads})$$
$$= P(4 \text{ heads}) + P(5 \text{ heads}) + P(6 \text{ heads})$$

Proceeding as we did in Part (a), we find

$$P(F) = \frac{C(10, 4)}{2^{10}} + \frac{C(10, 5)}{2^{10}} + \frac{C(10, 6)}{2^{10}} = .2051 + .2461 + .2051$$
$$= .6563$$

 **Now Work Problem 5**

EXAMPLE 4   **Finding Probabilities Using Counting Techniques**

What is the probability that a four-digit telephone extension has one or more repeated digits? Assume no one digit is more likely than another to be used.

SOLUTION   By the Multiplication Principle, there are $10^4 = 10,000$ distinct four-digit telephone extensions, so the number of outcomes in the sample space is

$$c(S) = 10,000$$

We wish to find the probability that a four-digit telephone extension has one or more repeated digits. Define the event E as follows:

E: The extension has one or more repeated digits

Since it is difficult to count the outcomes of this event, we first compute the probability of the complement, namely,

$$\overline{E}: \quad \text{No repeated digits in four-digit extension}$$

To find $c(\overline{E})$, we use the Multiplication Principle. The number of four-digit extensions that have *no* repeated digits is

$$c(\overline{E}) = 10 \cdot 9 \cdot 8 \cdot 7 = 5040$$

Hence,

$$P(\overline{E}) = \frac{c(\overline{E})}{c(S)} = \frac{5040}{10,000} = .504$$

Therefore, the probability of one or more repeated digits is

$$P(E) = 1 - P(\overline{E}) = 1 - .504 = .496$$

Now Work Problem 1

The next example is very similar to the one just completed.

EXAMPLE 5   Birthday Problem

An interesting problem, called the **birthday problem,** is to find the probability that in a group of r people there are at least two people who have the same birthday (the same month and day of the year).

SOLUTION   We assume a person is as likely to be born on one day as another.

We first determine the number of outcomes in the sample space S. There are 365 possibilities for each person's birthday (we exclude February 29 for simplicity). Since there are r people in the group, there are 365^r possibilities for the birthdays. [For one person in the group, there are 365 days on which his or her birthday can fall; for two people, there are $(365)(365) = 365^2$ pairs of days; and, in general, using the Multiplication Principle, for r people there are 365^r possibilities.] Thus,

$$c(S) = 365^r$$

We wish to find the probability of the event E:

$$E: \quad \text{at least two people have the same birthday.}$$

It is difficult to count the elements in this set; it is much easier to count the elements of the complement:

$$\overline{E}: \quad \text{No two people have the same birthday}$$

We proceed to find $c(\overline{E})$ as follows:

Choose one person at random. There are 365 possibilities for his or her birthday. Choose a second person. There are 364 possibilities for this birthday, if no two people are to have the same birthday. Choose a third person. There are 363 possibilities left for this birthday. Finally, we arrive at the rth person. There are $365 - (r - 1)$ possibilities left for this birthday. By the Multiplication Principle, the total number of possibilities is

$$c(\overline{E}) = 365 \cdot 364 \cdot 363 \cdot \ \ldots \ \cdot (365 - r + 1).$$

Hence, the probability of event $\overline{E}$ is

$$P(\overline{E}) = \frac{c(\overline{E})}{c(S)} = \frac{365 \cdot 364 \cdot 363 \cdot \ \ldots \ \cdot (365 - r + 1)}{365^r}$$

The probability that two or more people have the same birthday is then

$$P(E) = 1 - P(\overline{E})$$

For example, in a group of eight people, the probability that two or more will have the same birthday is

$$
\begin{aligned}
P(E) &= 1 - \frac{365 \cdot 364 \cdot 363 \cdot 362 \cdot 361 \cdot 360 \cdot 359 \cdot 358}{365^8} \\
&= 1 - .93 \\
&= .07
\end{aligned}
$$

The table below gives the probabilities for two or more people having the same birthday. Notice that the probability is better than $\frac{1}{2}$ for any group of 23 or more people.

	Number of People															
	5	**10**	**15**	**20**	**21**	**22**	**23**	**24**	**25**	**30**	**40**	**50**	**60**	**70**	**80**	**90**
Probability That 2 or More Have Same Birthday	.027	.117	.253	.411	.444	.476	.507	.538	.569	.706	.891	.970	.994	.99916	.99991	.99999

 Now Work Problem 13

EXAMPLE 6 **Finding Probabilities Using Counting Techniques**

If the letters in the word MISTER are randomly scrambled, what is the probability that the resulting rearrangement has the I preceding the E?

SOLUTION The 6 letters in the word MISTER can be rearranged in $P(6, 6) = 6!$ ways. We need to count the number of arrangements in which the letter I precedes the letter E. We can view the problem as consisting of two tasks.

Task 1 is to choose two of the six positions for the I and E, with I preceeding E. There are $P(6, 2)$ ways of arranging I and E in six positions, $\frac{1}{2}$ of which have I before E. Thus, Task 1 can be done in $\dfrac{P(6, 2)}{2}$ ways.

Task 2 would be to arrange the letters MSTR in the remaining four positions. Task 2 can be done in 4! ways.

Hence, by the Multiplication Principle, the number of arrangements with the I preceding the E is

$$\frac{P(6, 2)}{2} \cdot 4! = \frac{6 \cdot 5}{2} \cdot 4! = 15 \cdot 4!$$

So the desired probability is

$$\frac{15 \cdot 4!}{6!} = \frac{15}{6 \cdot 5} = \frac{15}{30} = \frac{1}{2}$$

■

The answer should be intuitively plausible—we would expect one half of the arrangements to have I before E and the other half to have E before I.

EXERCISE 7.3   Answers to odd-numbered problems begin on page AN-27.

1. What is the probability that a seven-digit phone number has one or more repeated digits?

2. What is the probability that a seven-digit phone number contains the number 7?

3. Five letters, with repetition allowed, are selected from the alphabet. What is the probability that none is repeated?

4. Four letters, with repetition allowed, are selected from the alphabet. What is the probability that none of them is a vowel (a, e, i, o, u)?

5. A fair coin is tossed 5 times.

(a) Find the probability that exactly 3 heads appear.
(b) Find the probability that no heads appear.

6. A fair coin is tossed 6 times.

(a) Find the probability that exactly 1 tail appears.
(b) Find the probability that no more than 1 tail appears.

7. A pair of fair dice are tossed 3 times.

(a) Find the probability that the sum of seven appears 3 times.
(b) Find the probability that a sum of 7 or 11 appears at least twice.

8. A pair of fair dice are tossed 5 times.

(a) Find the probability that the sum is never 2.
(b) Find the probability that the sum is never 7.

9. Through a mix-up on the production line, 6 defective refrigerators were shipped out with 44 good ones. If 5 are selected at random, what is the probability that all 5 are defective? What is the probability that at least 2 of them are defective?

10. In a shipment of 50 transformers 10 are known to be defective. If 30 transformers are picked at random, what is the probability that all 30 are nondefective? Assume that all transformers look alike and have an equal probability of being chosen.

11. What is the probability that, in a group of 3 people, at least 2 were born in the same month (disregard day and year)?

12. What is the probability that, in a group of 6 people, at least 2 were born in the same month (disregard day and year)?

13. A box contains 100 slips of paper numbered from 1 to 100. If 3 slips are drawn in succession with replacement, what is the probability that at least 2 of them have the same number?

14. If, in Problem 13, 10 slips are drawn with replacement, what is the probability that at least 2 of them have the same number?

15. Use the idea behind Example 5 to find the approximate probability that 2 or more U.S. senators have the same birthday. (There are 100 Senators.)

16. Follow the directions of Problem 15 for the House of Representatives. (There are more than 365 representatives.)

17. If the five letters in the word VOWEL are rearranged, what is the probability the L will precede the E?

18. If the four letters in the word MATH are rearranged, what is the probability the word begins with the letters TH?

19. If the five letters in the word VOWEL are rearranged, what is the probability the word will begin with L?

20. If the seven letters in the word DEFAULT are rearranged, what is the probability the word will end in E and begin with T?

21. A spinner has 26 equally spaced wedges, each labeled consecutively, 1, 2, 3, . . . , 26. What is the probability, from one spin, of landing on an even integer or on any of the last 19 integers?

22. A rare coins dealer has 37 distinctly valued silver dollars in a bag, one of which is valued at more than $10,000. The winner of a certain contest is given the opportunity to reach into the bag, while blindfolded, and pull out 4 of the coins. What is the probability that one of the 4 coins is the one valued at more than $10,000?

23. A person is dealt 3 cards from a regular deck of 52 cards. What is the probability that they are all hearts, all diamonds, or all spades?

24. Five cards are dealt at random from a regular deck of 52 playing cards. Find the probability that

(a) All are hearts.   (b) Exactly 4 are spades.

(c) Exactly 2 are clubs.

25. In a game of bridge find the probability that a hand of 13 cards consists of 5 spades, 4 hearts, 3 diamonds, and 1 club.

26. Find the probability of obtaining each of the following poker hands:

(a) Royal flush (10, J, Q, K, A all of the same suit)

(b) Straight flush (5 cards in sequence in a single suit, but not a royal flush)

(c) Four of a kind (4 cards of the same face value)

(d) Full house (one pair and one triple of the same face values)

(e) Flush (5 nonconsecutive cards each of the same suit)

(f) Straight (5 consecutive cards, not all of the same suit)

27. Elevator Problem   An elevator starts with 5 passengers and stops at 8 floors. Find the probability that no 2 passengers leave at the same floor. Assume that all arrangements of discharging the passengers have the same probability.

7.4 CONDITIONAL PROBABILITY

Whenever we compute the probability of an event, we do it relative to the entire sample space. Thus when we ask for the probability $P(E)$ of the event E, this probability $P(E)$ represents an appraisal of the likelihood that a chance experiment will produce an outcome in the set E relative to the sample space S.

However, sometimes we want to compute the probability of an event E of a sample space relative to another event F of the same sample space. That is, if we have *prior* information that the outcome must be in a set F, this information should be used to reappraise the likelihood that the outcome will also be in E. This reappraised probability is denoted by $P(E|F)$, and is read as the *conditional probability of E given F*. It represents the answer to the question, "How probable is E, given that F has occurred?"

Let's work some examples to illustrate conditional probability before giving the definition.

EXAMPLE 1   Example of Conditional Probability

Figure 8

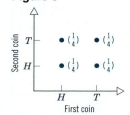

Consider the experiment of flipping two fair coins. As we have previously seen, the sample space S is

$$S = \{HH, HT, TH, TT\}$$

Figure 8 illustrates the sample space and, for convenience, the probability of each outcome.

Suppose the experiment is performed by another person and we have no knowledge

Figure 9

of the result, but we are informed that at least one tail was tossed. This information means the outcome *HH* could not have occurred. But the remaining outcomes *HT*, *TH*, *TT* are still possible. See Figure 9. How does this alter the probabilities of the remaining outcomes?

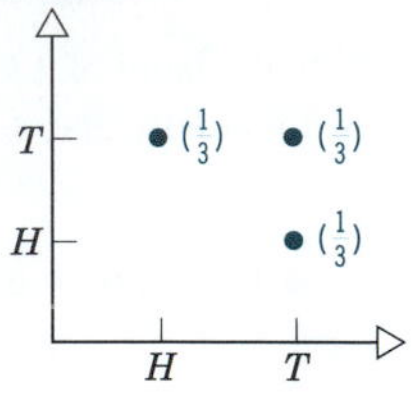

For instance, we might be interested in calculating the probability of the event {*TT*}. The three outcomes *TH*, *HT*, *TT* were each assigned the probability $\frac{1}{4}$ *before* we knew the information that at least one tail occurred, so it is not reasonable to assign them this same probability now. Since only three equally likely outcomes are now possible, we assign to each of them the probability $\frac{1}{3}$.

EXAMPLE 2   Example of Conditional Probability

Suppose a population of 1000 people includes 70 accountants and 520 females. There are 40 females who are accountants. A person is chosen at random, and we are told the person is female. The probability the person is an accountant, given that the person is female, is $\frac{40}{520}$. The ratio $\frac{40}{520} = \frac{1}{13}$ represents the *conditional probability* of the event *E* (accountant) assuming the event *F* (the person chosen is female).

We shall use the symbol $P(E|F)$, read "the probability of *E* given *F*," to denote conditional probability. Thus, for Example 2, if we let *E* be the event "A person chosen at random is an accountant" and we let *F* be the event "A person chosen at random is female," we would write

$$P(E|F) = \frac{40}{520} = \frac{1}{13}$$

Figure 10

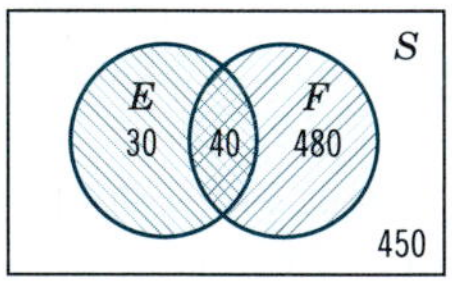

Figure 10 illustrates that in computing $P(E|F)$ in Example 2, we form the ratio of the numbers of those entries in *E* and in *F* with the numbers that are in *F*. Since $c(E \cap F) = 40$ and $c(F) = 520$, then

$$P(E|F) = \frac{c(E \cap F)}{c(F)} = \frac{40}{520} = \frac{1}{13}$$

Notice that

$$P(E|F) = \frac{\dfrac{c(E \cap F)}{c(S)}}{\dfrac{c(F)}{c(S)}} = \frac{P(E \cap F)}{P(F)}$$

With this result in mind we define conditional probability as follows:

Conditional Probability

Let *E* and *F* be events of a sample space *S* and suppose $P(F) > 0$. The **conditional probability of the event E, assuming the event F,** denoted by $P(E|F)$, is defined as

$$P(E|F) = \frac{P(E \cap F)}{P(F)} \tag{1}$$

For sample spaces involving equally likely outcomes, this formula can be proved. Suppose E and F are two events for a particular experiment. Assume that the sample space S for this experiment has n equally likely outcomes. Suppose event F has m outcomes, while $E \cap F$ has k outcomes $(k \le m)$. Since the outcomes are equally likely, we have

$$P(F) = \frac{c(F)}{c(S)} = \frac{m}{n}$$

and

$$P(E \cap F) = \frac{c(E \cap F)}{c(S)} = \frac{k}{n}$$

We wish to compute $P(E|F)$, the probability that E occurs given that F has occurred. Since we assume F has occurred, look only at the m outcomes in F. Of these m outcomes, there are k outcomes where E also occurs since $E \cap F$ has k outcomes. See Figure 11. Thus,

Figure 11

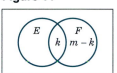

$$P(E|F) = \frac{k}{m}$$

Divide numerator and denominator by n to get

$$P(E|F) = \frac{\dfrac{k}{n}}{\dfrac{m}{n}} = \frac{P(E \cap F)}{P(F)}$$

EXAMPLE 3   Finding a Conditional Probability

Consider a three-child family for which the sample space S is

$$S = \{BBB,\ BBG,\ BGB,\ BGG,\ GBB,\ GBG,\ GGB,\ GGG\}$$

We assume that each outcome is equally likely, so that each is assigned a probability of $\frac{1}{8}$. Let E be the event, "The family has exactly two boys" and let F be the event "The first child is a boy." What is the probability that the family has two boys, given that the first child is a boy?

SOLUTION   We want to find $P(E|F)$. The events E and F are

$$E = \{BBG,\ BGB,\ GBB\} \qquad F = \{BBB,\ BBG,\ BGB,\ BGG\}$$

Since $E \cap F = \{BBG,\ BGB\}$, we have

$$P(E \cap F) = \tfrac{1}{4} \qquad P(F) = \tfrac{1}{2}$$

$$P(E|F) = \frac{P(E \cap F)}{P(F)} = \frac{\frac{1}{4}}{\frac{1}{2}} = \frac{1}{2}$$

Figure 12

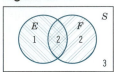

Since the sample space S of Example 3 consists of equally likely outcomes, we can also compute $P(E|F)$ using the Venn diagram in Figure 12. Then

$$P(E|F) = \frac{c(E \cap F)}{c(F)} = \frac{2}{4} = \frac{1}{2}$$

Let's analyze the situation in Example 3 more carefully. See Figure 12. The event E is "the family has exactly 2 boys." We have computed the probability of event E, knowing event F has occurred. This means we are computing a probability relative to a *new sample space* F. That is, F is treated as the universal set and we should consider only that part of E that is included in F, namely, $E \cap F$. Thus if E is any subset of the sample space S, then $P(E|F)$ provides the reappraisal of the likelihood that an outcome of the experiment will be in the set E if we have prior information that it must be in the set F.

✍ Now Work Problem 3

EXAMPLE 4   Finding Probabilities

In a certain experiment the events E and F have the characteristics

$$P(E) = .7 \qquad P(F) = .8 \qquad P(E \cap F) = .6$$

Find

(a) $P(E \cup F)$ (b) $P(E|F)$ (c) $P(F|E)$ (d) $P(\overline{E}|\overline{F})$

SOLUTION We can visualize the situation by using a Venn diagram. See Figure 13.

Figure 13

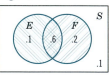

(a) To find $P(E \cup F)$, we use the Additive Rule.

$$P(E \cup F) = P(E) + P(F) - P(E \cap F) = .7 + .8 - .6 = .9$$

(b) To find the conditional probability $P(E|F)$, we use (1).

$$P(E|F) = \frac{P(E \cap F)}{P(F)} = \frac{.6}{.8} = .75$$

(c) To find the conditional probability $P(F|E)$, we use a variation of (1).

$$P(F|E) = \frac{P(F \cap E)}{P(E)} = \frac{P(E \cap F)}{P(E)} = \frac{.6}{.7} = .857$$

(d) To find the conditional probability $P(\overline{E}|\overline{F})$, we use a variation of (1).

$$P(\overline{E}|\overline{F}) = \frac{P(\overline{E} \cap \overline{F})}{P(\overline{F})}$$

Since $\overline{E} \cap \overline{F} = \overline{E \cup F}$ (De Morgan's property), we have

✍ Now Work Problems

23 and 25

$$P(\overline{E}|\overline{F}) = \frac{P(\overline{E \cup F})}{P(\overline{F})} = \frac{1 - P(E \cup F)}{1 - P(F)} = \frac{1 - .9}{1 - .8} = \frac{.1}{.2} = .50$$

Product Rule

If in Formula (1) we multiply both sides of the equation by $P(F)$, we obtain the following useful relationship, which is referred to as the **Product Rule:**

Product Rule

$$P(E \cap F) = P(F) \cdot P(E|F) \qquad (2)$$

The next example illustrates how the Product Rule is used to compute the probability of an event that is itself a sequence of two events.

EXAMPLE 5   Using the Product Rule

Two cards are drawn at random (without replacement) from a regular deck of 52 cards. What is the probability that the first card is a diamond and the second is red?

SOLUTION   We seek the probability of an event that is a sequence of two events, namely,

$$A: \qquad \text{The first card is a diamond}$$
$$B: \qquad \text{The second card is red}$$

Since there are 52 cards in the deck, of which 13 are diamonds, it follows that

$$P(A) = \tfrac{13}{52} = \tfrac{1}{4}$$

If A occurred, it means that there are only 51 cards left in the deck, of which 25 are red, so

$$P(B|A) = \tfrac{25}{51}$$

By the Product Rule,

$$P(B \cap A) = P(A) \cdot P(B|A) = \tfrac{1}{4} \cdot \tfrac{25}{51} = \tfrac{25}{204}$$

A tree diagram is helpful for problems like that of Example 5. See Figure 14 on page 350. The branch leading to A: Diamond has probability $\tfrac{13}{52} = \tfrac{1}{4}$; the branch leading to Heart also has probability $\tfrac{13}{52} = \tfrac{1}{4}$; and the branch leading to Club or Spade has probability $\tfrac{26}{52} = \tfrac{1}{2}$. These must add up to 1 since no other possibilities (branches) are possible. The branch from A to Red is the conditional probability of drawing a red card on the second draw after a diamond on the first draw, $P(\text{Red}|\text{Diamond})$, which is $\tfrac{25}{51}$. The branch from A to Black is the conditional probability $P(\text{Black}|\text{Diamond})$, which is $\tfrac{26}{51}$ (26 black cards and 51 total cards remain after a diamond on the first draw). The remaining entries are obtained similarly. Notice that the probability the first card is a diamond and the second a red corresponds to tracing the top branch of the tree. The product rule then tells us to multiply the branch probabilities.

Figure 14

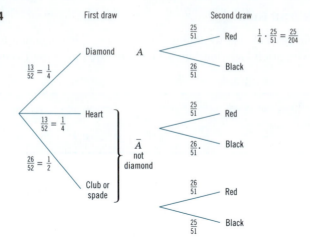

A further advantage of using a tree diagram is that it enables us to easily answer other questions about the experiment. For example, based on Figure 14, we see that

(a) Probability the first card is a heart and the second is black equals $\qquad$ $\frac{1}{4} \cdot \frac{26}{51}$

(b) Probability the first card is black and the second is red equals $\qquad$ $\frac{1}{2} \cdot \frac{26}{51}$

(c) Probability the first card is black and the second is black equals $\qquad$ $\frac{1}{2} \cdot \frac{25}{51}$

EXAMPLE 6   Using a Tree Diagram and the Product Rule

From a box containing four white, three yellow, and one green ball, two balls are drawn one at a time without replacing the first before the second is drawn. Use a tree diagram to find the probability that one white and one yellow ball are drawn.

SOLUTION   To fix our ideas, we define the events

W: White ball drawn Y: Yellow ball drawn G: Green ball drawn

The tree diagram for this experiment is given in Figure 15.

Figure 15

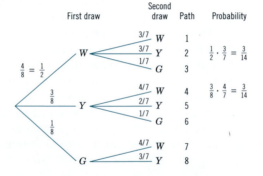

The event of drawing one white ball and one yellow ball can occur in two ways: drawing a white ball first and then a yellow ball (path 2 of the tree diagram in Figure 15), or drawing a yellow ball first and then a white ball (path 4).

Consider path 2. Since four of the eight balls are white, on the first draw we have

$$P(W) = P(W \text{ on 1st}) = \tfrac{4}{8} = \tfrac{1}{2}$$

Since one white ball has been removed, leaving seven balls in the box, of which three are yellow, on the second draw we have

$$P(Y|W) = P(Y \text{ on 2nd} | W \text{ on 1st}) = \tfrac{3}{7}$$

Thus for path 2 we have

$$P(W) \cdot P(Y|W) = P(W \text{ on 1st}) \cdot P(Y \text{ on 2nd} | W \text{ on 1st}) = \tfrac{1}{2} \cdot \tfrac{3}{7} = \tfrac{3}{14}$$

Similarly, for path 4 we have

$$P(Y) \cdot P(W|Y) = P(Y \text{ on 1st}) \cdot P(W \text{ on 2nd} | Y \text{ on 1st}) = \tfrac{3}{8} \cdot \tfrac{4}{7} = \tfrac{3}{14}$$

Since the two events are mutually exclusive, the probability of drawing one white ball and one yellow ball is the sum of these two probabilities:

$$\tfrac{3}{14} + \tfrac{3}{14} = \tfrac{6}{14} = \tfrac{3}{7}$$

■

Compare the solution to Example 6 given here with the solution to Example 1 of Section 7.3 (page 339). Which of the two solutions do you prefer?

 **Now Work Problem 33**

EXAMPLE 7   Using a Tree Diagram to Find Conditional Probabilities

Motors, Inc., has two plants to manufacture cars. Plant I manufactures 80% of the cars and plant II manufactures 20%. At plant I, 85 out of every 100 cars are rated standard quality or better. At plant II, only 65 out of every 100 cars are rated standard quality or better. We would like to answer the following questions:

(a) What is the probability that a customer obtains a standard quality car if he buys a car from Motors, Inc.?
(b) What is the probability that the car came from plant I if it is known that the car is of standard quality?

SOLUTION   We begin with a tree diagram. See Figure 16.

Figure 16

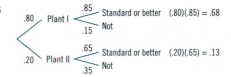

We define the following events:

I: Car came from plant I   II: Car came from plant II

A: Car is of standard quality

(a) There are two ways a standard car can be obtained: either it is standard and came from plant I, or else it is standard and came from plant II. By the Product Rule,

$$P(A \cap I) = P(I) \cdot P(A|I) = (.8)(.85) = .68$$
$$P(A \cap II) = P(II) \cdot P(A|II) = (.2)(.65) = .13$$

Since $A \cap I$ and $A \cap II$ are mutually exclusive, we have

$$P(A) = P(A \cap I) + P(A \cap II) = .68 + .13 = .81$$

(b) To compute $P(I|A)$, we use the definition of conditional probability.

$$P(I|A) = \frac{P(I \cap A)}{P(A)} = \frac{.68}{.81} = .8395$$

EXERCISE 7.4   Answers to odd-numbered problems begin on page AN-27.

In Problems 1–8 use the Venn diagram below to find each probability.

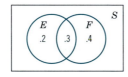

1. $P(E)$ **2.** $P(F)$ **3.** $P(E|F)$ **4.** $P(F|E)$

5. $P(E \cap F)$ **6.** $P(E \cup F)$ **7.** $P(\overline{E})$ **8.** $P(\overline{F})$

9. If E and F are events with $P(E) = .2$, $P(F) = .4$, and $P(E \cap F) = .1$, find the probability of E given F. Find $P(F|E)$.

10. If E and F are events with $P(E) = .5$, $P(F) = .6$, and $P(E \cap F) = .3$, find the probability of E given F. Find $P(F|E)$.

11. If E and F are events with $P(E \cap F) = .2$ and $P(E|F) = .4$, find $P(F)$.

12. If E and F are events with $P(E \cap F) = .2$ and $P(E|F) = .6$, find $P(F)$.

13. If E and F are events with $P(F) = \frac{5}{13}$ and $P(E|F) = \frac{4}{5}$, find $P(E \cap F)$.

14. If E and F are events with $P(F) = .38$ and $P(E|F) = .46$, find $P(E \cap F)$.

15. If E and F are events with $P(E \cap F) = \frac{1}{3}$, $P(E|F) = \frac{1}{2}$, and $P(F|E) = \frac{2}{3}$, find
(a) $P(E)$ (b) $P(F)$

16. If E and F are events with $P(E \cap F) = .1$, $P(E|F) = .25$, and $P(F|E) = .125$, find
(a) $P(E)$ (b) $P(F)$

In Problems 17–22 find each probability by referring to the following tree diagram:

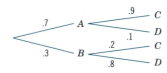

17. $P(C)$ **18.** $P(D)$ **19.** $P(C|A)$ **20.** $P(D|A)$ **21.** $P(C|B)$ **22.** $P(D|B)$

In Problems 23–28 E and F are events in a sample space S for which

$$P(E) = .5 \qquad P(F) = .4 \qquad P(E \cup F) = .8$$

Find each probability

23. $P(E \cap F)$ **24.** $P(E|F)$ **25.** $P(F|E)$

26. $P(\overline{E}|F)$ **27.** $P(E|\overline{F})$ **28.** $P(\overline{E}|\overline{F})$

29. For a 3-child family, find the probability of exactly 2 girls, given that the first child is a girl.

30. For a 3-child family, find the probability of exactly 1 girl, given that the first child is a boy.

31. A fair coin is tossed 4 successive times. Find the probability of obtaining 4 heads. Does the probability change if we are told that the second throw resulted in a head?

32. A pair of fair dice is thrown and we are told that at least one of them shows a 2. If we know this, what is the probability that the total is 7?

33. Two cards are drawn at random (without replacement) from a regular deck of 52 cards. What is the probability that the first card is a heart and the second is red?

34. If 2 cards are drawn from a regular deck of 52 cards without replacement, what is the probability that the second card is a queen?

35. From a box containing 3 white, 2 green, and 1 yellow ball, 2 balls are drawn at a time without replacing the first before the second is drawn. Find the probability that 1 white and 1 yellow ball are drawn.

36. A box contains 2 red, 4 green, 1 black, and 8 yellow marbles. If 2 marbles are selected without replacement, what is the probability that 1 is red and 1 is green?

37. A card is drawn at random from a regular deck of 52 cards. What is the probability that

(a) The card is a red ace?
(b) The card is a red ace if it is known an ace was picked?
(c) The card is a red ace if it is known a red card was picked?

38. A card is drawn at random from a regular deck of 52 cards. What is the probability that

(a) The card is a black jack?
(b) The card is a black jack if it is known a jack was picked?
(c) The card is a black jack if it is known a black card was picked?

39. In a small town it is known that 20% of the families have no children, 30% have 1 child, 20% have 2 children, 16% have 3 children, 8% have 4 children, and 6% have 5 or more children. Find the probability that a family has more than 2 children if it is known that it has at least 1 child.

40. A sequence of 2 cards is drawn from a regular deck of 52 cards (without replacement). What is the probability that the first card is red and the second is black?

In Problems 41–52 use the table to obtain probabilities for events in a sample space S.

	E	F	G	Totals
H	.10	.06	.08	.24
I	.30	.14	.32	.76
Totals	.40	.20	.40	1.00

For Problems 41–48, read each probability directly from the table:

41. $P(E)$ **42.** $P(G)$ **43.** $P(H)$ **44.** $P(I)$

45. $P(E \cap H)$ **46.** $P(E \cap I)$ **47.** $P(G \cap H)$ **48.** $P(G \cap I)$

For Problems 49–52 use Equation (1) and the appropriate values from the above table to compute the conditional probability.

49. $P(E|H)$ **50.** $P(E|I)$ **51.** $P(G|H)$ **52.** $P(G|I)$

53. A recent poll of residents in a certain community revealed the following information about voting preferences:

	Democrat	Republican	Independent
Male	50	40	30
Female	60	30	25

Events $M, F, D, R,$ and I are defined as follows:

M: Resident is male

F: Resident is female

D: Resident is a Democrat

R: Resident is a Republican

I: Resident is an Independent

Find

(a) $P(F|I)$ (b) $P(R|F)$ (c) $P(M|D)$
(d) $P(D|M)$ (e) $P(M|R \cup I)$ (f) $P(I|M)$

54. Let E be the event, "A person is an executive" and let F be the event "A person earns over \$65,000 per year." State in words what is expressed by each of the following probabilities:

(a) $P(E|F)$ (b) $P(F|E)$
(c) $P(\overline{E}|F)$ (d) $P(\overline{E}|F)$

55. The following table summarizes the graduating class of a midwestern university:

	Arts and Sciences *A*	Education *E*	Business *B*	Total
Male, *M*	342	424	682	1448
Female, *F*	324	102	144	570
Total	666	526	826	2018

A student is selected at random from the graduating class. Find the probability that the student

(a) Is male
(b) Is receiving an arts and sciences degree
(c) Is a female receiving a business degree
(d) Is a female, given that the student is receiving an education degree
(e) Is receiving an arts and sciences degree, given that the student is a male
(f) Is a female, given that the student is receiving an arts and sciences degree or an education degree

(g) Is not receiving a business degree and is male
(h) Is female, given that an education degree is not received

56. The following data are the result of a survey conducted by a marketing company to determine beer preferences.

	Do Not Drink Beer *N*	Light Beer *L*	Regular Beer *R*	Total
Female, *F*	224	420	622	1266
Male, *M*	196	512	484	1192
Total	420	932	1106	2458

A student is selected at random from this class. Find the probability that

(a) The student does not drink beer.
(b) The student is a female.
(c) The student is a female who prefers regular beer.
(d) The student prefers regular beer, given that the student is male.
(e) The student is male, given that the student prefers regular beer.
(f) The student is female, given that the student prefers regular beer or does not drink beer.

57. In a sample survey it is found that 35% of the men and 70% of the women weigh less than 160 pounds. Assume that 50% of the sample are men. If a person is selected at random and this person weighs less than 160 pounds, what is the probability that this person is a woman?

58. If the probability that a married man will vote in a given election is .50, the probability that a married woman will vote in the election is .60, and the probability that a woman will vote in the election, given that her husband votes, is .90, find

(a) The probability that a husband and wife will both vote in the election.
(b) The probability that a married man will vote in the election, given that at least one member of the married couple will vote.

59. Of the first-year students in a certain college, it is known that 40% attended private secondary schools and 60% attended public schools. The registrar reports that 30% of all students who attended private schools maintain an A average in their first year at college and that 24% of all first-year students had an A average. At the end of the year, one student is chosen at random from the class. If the student has an A average, what is the conditional probability that the student attended a private school? [*Hint:* Use a tree diagram.]

60. In a rural area in the north, registered Republicans outnumber registered Democrats by 3 to 1. In a recent election all Democrats voted for the Democratic candidate and enough Republicans also voted for the Democratic candidate so that the Democrat won by a ratio of 5 to 4. If a voter is selected at random, what is the probability he or she is Republican? What is the probability a voter is Republican, if it is known that he or she voted for the Democratic candidate?

61. "Temp Help" uses a preemployment test to screen applicants for the job of programmer. The test is passed by 70% of the applicants. Among those who pass the test, 85% complete training successfully. In an experiment a

random sample of applicants who do not pass the test is also employed. Training is successfully completed by only 40% of this group. If no preemployment test is used, what percentage of applicants would you expect to complete the training successfully?

62. If E and F are two events with $P(E) > 0$ and $P(F) > 0$, show that

$$P(F) \cdot P(E|F) = P(E) \cdot P(F|E)$$

63. Show that $P(E|E) = 1$ when $P(E) \neq 0$.

64. Show that $P(E|F) + P(\overline{E}|F) = 1$.

65. If S is the sample space, show that $P(E|S) = P(E)$.

66. If $P(E) > 0$ and $P(E|F) = P(E)$, show that $P(F|E) = P(F)$.

7.5 INDEPENDENT EVENTS

One of the most important concepts in probability is that of independence. In this section we define what is meant by two events being *independent*. First, however, we provide an intuitive idea of the meaning of independent events.

EXAMPLE 1   Example of Independent Events

Consider a group of 36 students. Define the events E and F as

E: Student has blue eyes F: Student is female

With regard to these two characteristics, suppose it is found that the 36 students are distributed as follows:

	Blue Eyes E	Not Blue Eyes $\overline{E}$	Totals
Female, F	12	12	24
Male, $\overline{F}$	6	6	12
Totals	18	18	36

What is $P(E|F)$?

SOLUTION   If we choose a student at random, the following probabilities can be obtained from the table:

$$P(E) = \tfrac{18}{36} = \tfrac{1}{2} \qquad P(F) = \tfrac{24}{36} = \tfrac{2}{3}$$
$$P(E \cap F) = \tfrac{12}{36} = \tfrac{1}{3}$$

Then we find that

$$P(E|F) = \frac{P(E \cap F)}{P(F)} = \frac{\frac{1}{3}}{\frac{2}{3}} = \frac{1}{2} = P(E)$$

∎

In Example 1 the probability of E given F equals the probability of E. This situation can be described by saying that the information that the event F has occurred does not affect the probability of the event E. If this is the case, we say that E *is independent of* F.

E Is Independent of F

Let E and F be two events of a sample space S with $P(F) > 0$. **The event E is independent of the event F** if and only if

$$P(E|F) = P(E)$$

Theorem

Let E, F be events for which $P(E) > 0$ and $P(F) > 0$. If E is independent of F, then F is independent of E.

The proof of this result is outlined in Problem 41.

Independent Events

If two events E and F have positive probabilities and if the event E is independent of F, then F is also independent of E. In this case E and F are called **independent events.**

We have the following result concerning independent events:

Test for Independence

Two events E and F of a sample space S are independent events if and only if

$$P(E \cap F) = P(E) \cdot P(F) \tag{1}$$

That is, the probability of E *and* F is equal to the product of the probability of E and the probability of F.

The proof of this result is outlined in Problem 42.

The Test for Independence can be used to determine whether two events are independent.

EXAMPLE 2   Testing for Independent Events

Suppoose $P(E) = \frac{1}{4}$, $P(F) = \frac{2}{3}$, and $P(E \cap F) = \frac{1}{6}$. Show that E and F are independent.

SOLUTION   We see if Equation (1) holds.

$$P(E) \cdot P(F) = \tfrac{1}{4} \cdot \tfrac{2}{3} = \tfrac{1}{6}$$
$$P(E \cap F) = \tfrac{1}{6}$$

Since Equation (1) holds, by the Test for Independence E and F are independent. ∎

If E and F are two independent events, the Test for Independence can also be used to find $P(E \cap F)$, the probability of E and F.

Warning! Two events E and F must be independent in order to use (1) to find $P(E \cap F)$.

EXAMPLE 3   Using Equation (1) with Independent Events

Suppose E and F are independent events with $P(E) = .4$ and $P(F) = .3$. Find $P(E \cap F)$.

SOLUTION   Since E and F are independent, we may use Equation (1) to find $P(E \cap F)$.

$$
\begin{aligned}
P(E \cap F) &= P(E) \cdot P(F) \quad \text{By (1)}\\
&= (.4) \cdot (.3)\\
&= .12
\end{aligned}
$$

 Now Work Problem 1

EXAMPLE 4   Showing Events are Independent

Suppose a red die and a green die are thrown. Let event E be "Throw a 5 with the red die," and let event F be "Throw a 6 with the green die." Show that E and F are independent events.

SOLUTION   In this experiment the events E and F are

$$E = \{(5, 1), (5, 2), (5, 3), (5, 4), (5, 5), (5, 6)\}$$
$$F = \{(1, 6), (2, 6), (3, 6), (4, 6), (5, 6), (6, 6)\}$$

Then

$$P(E) = \tfrac{6}{36} = \tfrac{1}{6} \qquad P(F) = \tfrac{6}{36} = \tfrac{1}{6}$$

Also, the event E and F is

$$E \cap F = \{(5, 6)\}$$

so that

$$P(E \cap F) = \tfrac{1}{36}$$

Since $P(E) \cdot P(F) = \tfrac{1}{6} \cdot \tfrac{1}{6} = \tfrac{1}{36} = P(E \cap F)$, E and F are independent events.

 **Now Work Problem 5**

EXAMPLE 5   Testing for Independent Events

In a T-maze a mouse may run to the right, R, or to the left, L. Suppose its behavior in making such "choices" is random so that R and L are equally likely outcomes. Suppose a mouse is put through the T-maze three times. Define events E, G, and H as

E: Run to the right 2 consecutive times

G: Run to the left on first trial

H: Run to the right on second trial

(a) Show that E and G are not independent.
(b) Show that G and H are independent.

SOLUTION

(a) The events E and G are

$$E = \{RRL, LRR, RRR\}$$
$$G = \{LLL, LLR, LRL, LRR\}$$

The sample space S has eight elements, so that

$$P(E) = \tfrac{3}{8} \qquad P(G) = \tfrac{1}{2}$$

Also, the event E and G is

$$E \cap G = \{LRR\}$$

so that

$$P(E \cap G) = \tfrac{1}{8}$$

Since $P(E \cap G) \neq P(E) \cdot P(G)$, the events E and G are not independent.
(b) The event H is

$$H = \{RRL, RRR, LRL, LRR\}$$

so that

$$P(H) = \tfrac{1}{2}$$

The event G and H and its probability are

$$G \cap H = \{LRL, LRR\} \qquad P(G \cap H) = \tfrac{1}{4}$$

Since $P(G \cap H) = P(G) \cdot P(H)$, the events G and H are independent.

Example 5 illustrates that the question of whether two events are independent can be answered simply by determining whether Equation (1) is satisfied. Although we may often suspect two events E and F as being independent, our intuition must be checked by computing $P(E)$, $P(F)$, and $P(E \cap F)$ and determining whether $P(E \cap F) = P(E) \cdot P(F)$.

✍ Now Work Problem 13

For some probability models, an assumption of independence is made. In such instances when two events are independent, Equation (1) may be used to compute the probability that both events occur. The following example illustrates such a situation.

EXAMPLE 6   Planting Seeds

In a group of seeds, $\frac{1}{4}$ of which should produce white plants, the best germination that can be obtained is 75%. If one seed is planted, what is the probability that it will grow into a white plant?

SOLUTION   Let G and W be the events

$\quad\quad$ G: The plant will grow

$\quad\quad$ W: The seed will produce a white plant

Assume that W and G are independent events.

Then the probability that the plant grows and is white, namely, $P(W \cap G)$ is

$$P(W \cap G) = P(W) \cdot P(G) = \tfrac{1}{4} \cdot \tfrac{3}{4} = \tfrac{3}{16}$$

A white plant will grow 3 out of 16 times.

■

There is a danger that mutually exclusive events and independent events may be confused. A source of this confusion is the common expression, "Events are independent if they have nothing to do with each other." This expression provides a description of independence when applied to everyday events; but when it is applied to sets, it suggests nonoverlapping. Nonoverlapping sets are mutually exclusive but in general are not independent. See Problem 36.

Independence for More than Two Events

The concept of independence can be applied to more than two events:

Independence

A set $E_1, E_2, \ldots, E_n$ of n events is called **independent** if the occurrence of one or more of them does not change the probability of any of the others. It can be shown that, for such events,

$$P(E_1 \cap E_2 \cap \cdots \cap E_n) = P(E_1) \cdot P(E_2) \cdot \ldots \cdot P(E_n) \tag{2}$$

EXAMPLE 7 **Using Equation (2)**

A new skin cream can cure skin infection 90% of the time. If five randomly selected people with skin infections use this cream, assuming independence, what is the probability that

(a) All five are cured?　　　　　　(b) All five still have the infection?

SOLUTION

(a) Let

E_1:　First person does not have infection

E_2:　Second person does not have infection

E_3:　Third person does not have infection

E_4:　Fourth person does not have infection

E_5:　Fifth person does not have infection

Then, since events E_1, E_2, E_3, E_4, E_5, are given to be independent,

$$
\begin{aligned}
P(\text{All 5 are cured}) &= P(E_1 \cap E_2 \cap E_3 \cap E_4 \cap E_5) \\
&= P(E_1) \cdot P(E_2) \cdot P(E_3) \cdot P(E_4) \cdot P(E_5) \\
&= (.9)^5 \\
&= .59
\end{aligned}
$$

(b) Let

$\overline{E}_i$:　ith person has an infection, $i = 1, 2, 3, 4, 5$

Then

$$
\begin{aligned}
P(\overline{E}_i) &= 1 - .9 = .1 \\
P(\text{All 5 are infected}) &= P(\overline{E}_1 \cap \overline{E}_2 \cap \overline{E}_3 \cap \overline{E}_4 \cap \overline{E}_5) \\
&= P(\overline{E}_1) \cdot P(\overline{E}_2) \cdot P(\overline{E}_3) \cdot P(\overline{E}_4) \cdot P(\overline{E}_5) \\
&= (.1)^5 \\
&= .00001
\end{aligned}
$$

■

EXERCISE 7.5　Answers to odd-numbered problems begin on page AN-27.

1. If E and F are independent events and if $P(E) = .4$ and $P(F) = .6$, find $P(E \cap F)$.

2. If E and F are independent events and if $P(E) = .6$ and $P(E \cap F) = .2$, find $P(F)$.

3. If E and F are independent events, find $P(F)$ if $P(E) = .2$ and $P(E \cup F) = .3$.

4. If E and F are independent events, find $P(E)$ if $P(F) = .3$ and $P(E \cup F) = .6$.

5. Suppose E and F are two events such that $P(E) = \frac{4}{21}$, $P(F) = \frac{7}{12}$, and $P(E \cap F) = \frac{2}{9}$. Are E and F independent?

6. If E and F are two events such that $P(E) = .25$, $P(F) = .36$, and $P(E \cap F) = .09$. then are E and F independent?

7. If E and F are two independent events with $P(E) = .2$, and $P(F) = .4$, find

　(a) $P(E|F)$　　　　　　　(b) $P(F|E)$
　(c) $P(E \cap F)$　　　　　　(d) $P(E \cup F)$

8. If E and F are independent events with $P(E) = .3$ and $P(F) = .5$, find

　(a) $P(E|F)$　　　　　　　(b) $P(F|E)$
　(c) $P(E \cap F)$　　　　　　(d) $P(E \cup F)$

9. If E, F, and G are three independent events with $P(E) = \frac{2}{3}$, $P(F) = \frac{3}{7}$, and $P(G) = \frac{2}{21}$, then find $P(E \cap F \cap G)$.

10. If E_1, E_2, E_3, and E_4 are four independent events with $P(E_1) = .6$, $P(E_2) = .3$, $P(E_3) = .5$, and $P(E_4) = .4$, find $P(E_1 \cap E_2 \cap E_3 \cap E_4)$.

11. If $P(E) = .3$, $P(F) = .2$, and $P(E \cup F) = .4$, what is $P(E|F)$? Are E and F independent?

12. If $P(E) = .4$, $P(F) = .6$, and $P(E \cup F) = .7$, what is $P(E|F)$? Are E and F independent?

13. A fair die is rolled. Let E be the event "1, 2, or 3 is rolled" and let F be the event "3, 4, or 5 is rolled." Are E and F independent?

14. A loaded die is rolled. The probabilities for this die are $P(1) = P(2) = P(4) = P(5) = \frac{1}{8}$ and $P(3) = P(6) = \frac{1}{4}$. Are the events defined in Problem 13 independent in this case?

15. For a 3-child family let E be the event "The family has at most 1 boy" and let F be the event "The family has children of each sex." Are E and F independent events?

16. For a 2-child family are the events E and F as defined in Problem 15 independent?

17. A first card is drawn at random from a regular deck of 52 cards and is then put back into the deck. A second card is drawn. What is the probability that

   (a) The first card is a club?
   (b) The second card is a heart, given that the first is a club?
   (c) The first card is a club and the second is a heart?

18. For the situation described in Problem 17 what is the probability that

   (a) The first card is an ace?
   (b) The second card is a king, given that the first card is an ace?
   (c) The first card is an ace and the second is a king?

19. In the T-maze of Example 5 are the two events E and H independent?

20. A fair coin is tossed twice. Define the events E and F to be

   E: A head turns up on the first throw of a fair coin
   F: A tail turns up on the second throw of a fair coin

   Show that E and F are independent events.

21. A die is loaded so that

   $$P(1) = P(2) = P(3) = \tfrac{1}{4} \qquad P(4) = P(5) = P(6) = \tfrac{1}{12}$$

   If $A = \{1, 2\}$, $B = \{2, 3\}$, $C = \{1, 3\}$, show that any pair of these events is independent.

22. Determine whether the events E, F, and G defined below for the experiment of tossing 2 fair dice are independent.

   E: The first die shows a 6
   F: The second die shows a 3
   G: The sum of the 2 dice is 7

23. **Cardiovascular Disease** Records show that a child of parents with heart disease has a probability of $\frac{3}{4}$ of inheriting the disease. Assuming independence, what is the probability that, for a couple with heart disease that have two children:

   (a) Both children have heart disease.
   (b) Neither child has heart disease.
   (c) Exactly one child has heart disease.

24. **Hitting a Target** A marksman hits a target with probability $\frac{4}{5}$. Assuming independence for successive firings, find the probabilities of getting

   (a) One miss followed by two hits
   (b) Two misses and one hit (in any order)

25. A coin is loaded so that tails is three times as likely as heads. If the coin is tossed three times, find the probability of getting

   (a) All tails
   (b) Two heads and one tail (in any order)

26. **Sex of Newborns** The probability of a newborn baby being a girl is .49. Assuming that the sex of one baby is independent of the sex of all other babies—that is, the events are independent—what is the probability that all four babies born in a certain hospital on one day are girls?

27. **Recovery Rate** The recovery rate from a flu is .9. If 4 people have this flu, what is the probability (assume independence) that

   (a) All will recover?   (b) Exactly 2 will recover?
   (c) At least 2 will recover?

28. **Germination** In a group of seeds, $\frac{1}{3}$ of which should produce violets, the best germination that can be obtained is 60%. If one seed is planted, what is the probability that it will grow into violets?

29. A box has 10 marbles in it, 6 red and 4 white. Suppose we draw a marble from the box, replace it, and then draw another. Find the probability that

   (a) Both marbles are red.
   (b) Just one of the two marbles is red.

30. **Survey** In a survey of 100 people, categorized as drinkers or nondrinkers, with or without a liver ailment, the

following data were obtained:

	Liver Ailment F	No Liver Ailment $\overline{F}$
Drinkers, E	52	18
Nondrinkers, $\overline{E}$	8	22

(a) Are the events E and F independent?
(b) Are the events $\overline{E}$ and $\overline{F}$ independent?
(c) Are the events E and $\overline{F}$ independent?

31. Insurance   By examining the past driving records of 840 randomly selected drivers over a period of 1 year, the following data were obtained.

	Under 25 U	Over 25 $\overline{U}$	Totals
Accident, A	40	5	45
No Accident, $\overline{A}$	285	510	795
Totals	325	515	840

(a) What is the probability of a driver having an accident, given that the person is under 25?
(b) What is the probability of a driver having an accident, given that the person is over 25?
(c) Are events U and A independent?
(d) Are events U and $\overline{A}$ independent?
(e) Are events $\overline{U}$ and A independent?
(f) Are events $\overline{U}$ and $\overline{A}$ independent?

32. Voting Patterns   The following data show the number of voters in a sample of 1000 from a large city, categorized by religion and their voting preference.

	Democrat D	Republican R	Independent	Totals
Catholic, C	160	150	90	400
Protestant, P	220	220	60	500
Jewish, J	20	30	50	100
Totals	400	400	200	1000

(a) Find the probability a person is a Democrat.
(b) Find the probability a person is a Catholic.
(c) Find the probability a person is Catholic, knowing the person is a Democrat.
(d) Are the events R and D independent?
(e) Are the events P and R independent?

33. Election   A candidate for office believes that $\frac{2}{3}$ of registered voters in her district will vote for her in the next election. If two registered voters are independently selected at random, what is the probability that

(a) Both of them will vote for her in the next election?
(b) Neither will vote for her in the next election?
(c) Exactly one of them will vote for her in the next election?

34. A woman has 10 keys but only 1 fits her door. She tries them successively (without replacement). Find the probability that a key fits in exactly 5 tries.

35. Chevalier de Mere's Problem   Which of the following random events do you think is more likely to occur?

(a) To obtain a 1 on at least one die in a simultaneous throw of four fair dice.
(b) To obtain at least one pair of 1s in a series of 24 throws of a pair of fair dice.
   [*Hint:*   Part (a): P(No 1s are obtained) $= 5^4/6^4 = \frac{625}{1296} = .4823$. Part (b): The probability of not obtaining a double 1 on any given toss is $\frac{35}{36}$. Thus P(No double 1s are obtained) $= (\frac{35}{36})^{24} = .509$.]

36. Give an example of two events that are

(a) Independent, but not mutually exclusive
(b) Not independent, but mutually exclusive
(c) Not independent and not mutually exclusive

37. Show that whenever two events are both independent and mutually exclusive, then at least one of them is impossible.

38. Let E be any event. If F is an impossible event, show that E and F are independent.

39. Show that if E and F are independent events, so are $\overline{E}$ and $\overline{F}$. [*Hint:*   Use De Morgan's properties.]

40. Show that if E and F are independent events and if $P(E) \neq 0$, $P(F) \neq 0$, then E and F are not mutually exclusive.

41. Suppose $P(E) > 0$, $P(F) > 0$ and E is independent of F. Show that F is independent of E.
[*Hint:*   First we note that

$$P(F) \cdot P(E|F) = P(F) \cdot \frac{P(E \cap F)}{P(F)} = P(E \cap F)$$

and

$$P(E) \cdot P(F|E) = P(E) \cdot \frac{P(F \cap E)}{P(E)} = P(E \cap F)$$

Then $P(F) \cdot P(E|F) = P(E) \cdot P(F|E)$. Now use the fact that E is independent of F—that is, $P(E|F) = P(E)$—to show that F is independent of E—that is, $P(F|E) = P(F)$.]

42. Prove Equation (1) on page 356.
[*Hint:* If E and F are independent events, then

$$P(E|F) = \frac{P(E \cap F)}{P(F)} \qquad \text{and} \qquad P(E|F) = P(E)$$

Now show that $P(E \cap F) = P(E) \cdot P(F)$. Conversely, suppose $P(E \cap F) = P(E) \cdot P(F)$. Use the fact that

$$P(E|F) = \frac{P(E \cap F)}{P(F)}$$

to show that $P(E|F) = P(E)$.]

CHAPTER REVIEW

IMPORTANT TERMS AND CONCEPTS

sample space 318
outcome 318
probability of an outcome 320
constructing a probability model 321
event; simple event 322
probability of an event 323
mutually exclusive events 328
probability of E or F, if E, F are
   mutually exclusive 328

Additive Rule 329
properties of the probability
   of an event 331
probability of the complement
   of an event 331
probability of an event
   with equally likely
   outcomes 333

odds 334
conditional probability 346
Product Rule 349
independent events 356
Test for Independence 356
independence of more than
   two events 359

IMPORTANT FORMULAS

Mutually Exclusive Events

$$P(E \cup F) = P(E) + P(F)$$

Additive Rule

$$P(E \cup F) = P(E) + P(F) - P(E \cap F)$$

Complementary Events

$$P(\overline{E}) = 1 - P(E)$$

Equally Likely Outcomes

$$P(E) = \frac{c(E)}{c(S)}$$

Odds for E are a to b

$$P(E) = \frac{a}{a + b}$$

Conditional Probability

$$P(E|F) = \frac{P(E \cap F)}{P(F)}$$

Product Rule

$$P(E \cap F) = P(F) \cdot P(E|F)$$

E Is Independent of F

$$P(E|F) = P(E)$$

Test for Independence E, F are independent if and only if

$$P(E \cap F) = P(E) \cdot P(F)$$

TRUE–FALSE ITEMS   Answers are on page AN-27.

T_____ F_____ **1.** If the odds for an event E are 2 to 1, then $P(E) = \frac{2}{3}$.

T_____ F_____ **2.** The conditional probability of E given F is

$$P(E|F) = \frac{P(E \cap F)}{P(E)}$$

T_____ F_____ **3.** If two events in a sample space have no outcomes in common, they are said to be independent.

T_____ F_____ **4.** $P(E|F) = P(F|E)$ for any events E and F.

T_____ F_____ **5.** $P(E) + P(\overline{E}) = 1$

T_____ F_____ **6.** If $P(E) = .4$ and $P(F) = .3$, then $P(E \cup F)$ must be .7.

T_____ F_____ **7.** If $P(E \cap F) = 0$, then E and F are said to be mutually exclusive.

T_____ F_____ **8.** If $P(E \cup F) = .7$, $P(E) = .4$, and $P(F) = .3$, then $P(E \cap F) = 0$.

FILL IN THE BLANKS   Answers are on page AN-27.

1. If $S = \{a, b, c, d\}$ is a sample space, the outcomes are equally likely, and $E = \{a, b\}$, then $P(\overline{E}) = $ _____ .

2. If a coin is tossed five times, the number of outcomes in the sample space of this experiment is _____ .

3. If an event is certain to occur, then its probability is _____ . If an event is impossible, its probability is _____ .

4. If $P(\overline{E}) = .2$, then $P(E) = $ _____ .

5. If $P(E) = .6$, the odds _____ E are 3 to 2.

6. When each outcome in a sample space is assigned the same probability, the outcomes are termed _____ _____ .

7. If two events in a sample space have no outcomes in common, they are said to be _____ _____ .

REVIEW EXERCISES   Answers to odd-numbered problems begin on page AN-27.

1. A survey of families with 2 children is made, and the gender of the children is recorded. Describe the sample space and draw a tree diagram of this experiment.

2. A fair coin is tossed three times.
(a) Construct a probability model corresponding to this experiment.
(b) Find the probabilities of the following events:
  (i) The first toss is tails.
  (ii) The first toss is heads.
  (iii) Either the first toss is tails or the third toss is heads.
  (iv) At least one of the tosses is heads.
  (v) There are at least 2 tails.
  (vi) No tosses are heads.

3. A jar contains 3 white marbles, 2 yellow marbles, 4 red marbles, and 5 blue marbles. Two marbles are picked at random. What is the probability that
(a) Both are blue?   (b) Exactly 1 is blue?
(c) At least 1 is blue?

4. Let A and B be events with $P(A) = .3$, $P(B) = .5$, and $P(A \cap B) = .2$. Find the probability that

(a) A or B happens.   (b) A does not happen.
(c) Neither A nor B happens.
(d) Either A does not happen or B does not happen.

5. A loaded die is rolled 400 times, and the following outcomes are recorded:

Face	1	2	3	4	5	6
No. Times Showing	32	45	84	74	92	73

Estimate the probability of rolling a

(a) 3   (b) 5   (c) 6

6. A survey of a group of criminals shows that 65% came from low-income families, 40% from broken homes, and 30% came from low-income and broken families. Define

  E: Criminal came from low income
  F: Criminal came from broken home

A criminal is selected at random.

(a) Find the probability that the criminal is not from a low-income family.

(b) Find the probability that the criminal comes from a broken home or a low-income family.

(c) Are E and F mutually exclusive?

7. If E and F are events with $P(E \cup F) = \frac{5}{8}$, $P(E \cap F) = \frac{1}{3}$, and $P(E) = \frac{1}{2}$, find

(a) $P(\overline{E})$ (b) $P(F)$ (c) $P(\overline{F})$

8. If E and F represent mutually exclusive events, $P(E) = .30$, and $P(F) = .45$, find each of the probabilities:

(a) $P(\overline{E})$ (b) $P(\overline{F})$
(c) $P(E \cap F)$ (d) $P(E \cup F)$
(e) $P(\overline{E} \cap F)$ (f) $P(\overline{E} \cup F)$
(g) $P(\overline{E} \cup \overline{F})$ (h) $P(\overline{E} \cap \overline{F})$

9. Consider the experiment of spinning the spinner shown in the figure 3 times. (Assume the spinner cannot fall on a line.)

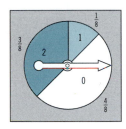

(a) Are all outcomes equally likely?
(b) If not, which of the outcomes has the highest probability?
(c) Let F be the event, "Each digit will occur exactly once." Find $P(F)$.

10. What are the odds in favor of a 5 when a fair die is thrown?

11. A bettor is willing to give 7 to 6 odds that the Bears will win the NFL title. What is the probability of the Bears winning?

12. A biased coin is such that the probability of heads (H) is $\frac{1}{4}$ and the probability of tails (T) is $\frac{3}{4}$. Show that in flipping this coin twice the events E and F defined below are independent.

   E: A head turns up in the first throw

   F: A tail turns up in the second throw

13. The records of Midwestern University show that in one semester, 38% of the students failed mathematics, 27% of the students failed physics, and 9% of the students failed mathematics and physics. A student is selected at random.

(a) If a student failed physics, what is the probability that he or she failed mathematics?

(b) If a student failed mathematics, what is the probability that he or she failed physics?

(c) What is the probability that he or she failed mathematics or physics?

14. A pair of fair dice is thrown 3 times. What is the probability that on the first toss the sum of the 2 dice is even, on the second toss the sum is less than 6, and on the third toss the sum is 7?

15. In a certain population of people, 25% are blue-eyed and 75% are brown-eyed. Also, 10% of the blue-eyed people are left-handed and 5% of the brown-eyed people are left-handed.

(a) What is the probability that a person chosen at random is blue-eyed and left-handed?

(b) What is the probability that a person chosen at random is left-handed?

(c) What is the probability that a person is blue-eyed, given that the person is left-handed?

16. Score Distribution   Two forms of a standardized math exam were given to 100 students. The following are the results.

Score	Form A	Form B	Total
Over 80%	8	12	20
Under 80%	32	48	80
Totals	40	60	100

(a) What is the probability that a student who scored over 80% took form A?

(b) What is the probability that a student who took form A scored over 80%?

(c) Show that the events "scored over 80%" and "took form A" are independent.

(d) Are the events "scored over 80%" and "took form B" independent?

17. ACT Scores   The following data compare ACT scores of students with their performance in the classroom (based on a 4.0 grade point average (GPA)).

GPA	Below 21	22–27	Above 28	Total
3.6–4	8	56	104	168
3.0–3.5	47	70	30	147
Below 3	47	34	4	85
Totals	102	160	138	400

A graduating student is selected at random. Find the probability that

(a) The student scored above 28.

(b) The student's GPA is 3.6–4.

(c) The student scored above 28 with GPA of 3.6–4.

(d) The student's ACT score was in the 22–27 range.

(e) The student had a 3.0–3.5 GPA.

(f) Show that "ACT above 28" and "GPA below 3" are not independent.

18. Three envelopes are addressed for 3 secret letters written in invisible ink. A secretary randomly places each of the letters in an envelope and mails them. What is the probability that at least 1 person receives the correct letter?

Mathematical Questions from Professional Exams

1. Actuary Exam—Part I If P and Q are events having positive probability in the same space S such that $P \cap Q = \varnothing$, then all of the following pairs are independent EXCEPT

(a) $\varnothing$ and P (b) P and Q

(c) P and S (d) P and $P \cap Q$

(e) $\varnothing$ and the complement of P

2. Actuary Exam—Part I A box contains 12 varieties of candy and exactly 2 pieces of each variety. If 12 pieces of candy are selected at random, what is the probability that a given variety is represented?

(a) $\dfrac{2^{12}}{(12!)^2}$ (b) $\dfrac{2^{12}}{24!}$ (c) $\dfrac{2^{12}}{\binom{24}{12}}$ (d) $\dfrac{11}{46}$ (e) $\dfrac{35}{46}$

3. Actuary Exam—Part II What is the probability that a 3-card hand drawn at random and without replacement from a regular deck consists entirely of black cards?

(a) $\dfrac{1}{17}$ (b) $\dfrac{2}{17}$ (c) $\dfrac{1}{8}$ (d) $\dfrac{3}{17}$ (e) $\dfrac{4}{17}$

4. Actuary Exam—Part II Events S and T are independent with $\Pr(S) < \Pr(T)$, $\Pr(S \cap T) = \frac{6}{25}$, and $\Pr(S|T) + \Pr(T|S) = 1$. What is $\Pr(S)$?

(a) $\dfrac{1}{25}$ (b) $\dfrac{1}{5}$ (c) $\dfrac{5}{25}$ (d) $\dfrac{2}{5}$ (e) $\dfrac{3}{5}$

5. Actuary Exam—Part II What is the least number of independent times that an unbiased die must be thrown to make the probability that all throws do not give the same result greater than .999?

(a) 3   (b) 4   (c) 5   (d) 6   (e) 7

6. Actuary Exam—Part II In a group of 20,000 men and 10,000 women, 6% of the men and 3% of the women

19. Jones lives at O (see the figure). He owns 5 gas stations located 4 blocks away (dots). Each afternoon he checks on one of his gas stations. He starts at O. At each intersection he flips a fair coin. If it shows heads, he will head north (N); otherwise, he will head toward the east (E). What is the probability that he will end up at gas station G before coming to one of the other stations?

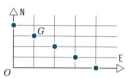

have a certain affliction. What is the probability that an afflicted member of the group is a man?

(a) $\dfrac{3}{5}$ (b) $\dfrac{2}{3}$ (c) $\dfrac{3}{4}$ (d) $\dfrac{4}{5}$ (e) $\dfrac{8}{9}$

7. Actuary Exam—Part II An unbiased die is thrown 2 independent times. Given that the first throw resulted in an even number, what is the probability that the sum obtained is 8?

(a) $\dfrac{5}{36}$ (b) $\dfrac{1}{6}$ (c) $\dfrac{4}{21}$ (d) $\dfrac{7}{36}$ (e) $\dfrac{1}{3}$

8. Actuary Exam—Part II If the events S and T have equal probability and are independent with $\Pr(S \cap T) = p > 0$, then $\Pr(S) =$

(a) $\sqrt{p}$ (b) p^2 (c) $\dfrac{p}{2}$ (d) p (e) $2p$

9. Actuary Exam—Part II The probability that both S and T occur, the probability that S occurs and T does not, and the probability that T occurs and S does not are all equal to p. What is the probability that either S or T occurs?

(a) p (b) $2p$ (c) $3p$ (d) $3p^2$ (e) p^3

10. Actuary Exam—Part II What is the probability that a bridge hand contains 1 card of each denomination (i.e., 1 ace, 1 king, 1 queen, . . . , 1 three, 1 two)?

(a) $\dfrac{13!}{13^{13}}$ (b) $\dfrac{4^{13}}{\binom{52}{13}}$ (c) $\dfrac{\binom{52}{4}}{\binom{52}{13}}$

(d) $\left(\dfrac{1}{13}\right)^{13}$ (e) $\dfrac{13^4}{\binom{52}{13}}$

Chapter 8

Additional Probability Topics

8.1 Bayes' Formula

8.2 The Binomial Probability Model

8.3 Expected Value

8.4 Applications

8.5 Random Variables

Chapter Review

This chapter deals with various applications of probability. Bayes' formula involves sample spaces that can be divided into mutually exclusive events. Applications include the analysis of testing. How easy is it for a nonqualified person to actually be admitted to medical school? What is the likelihood a person without cancer gets a false positive test result? The binomial probability model investigates experiments that are repetitive. Applications include quality control techniques, such as product testing and the evaluation of vaccines. Expected value uses probability to make decisions, such as which oil field a driller should bid on or whether insurance should be purchased. The section on applications provides examples in the field of operations research and a discussion of the digital transmission of data. The chapter closes with a brief discussion of random variables.

8.1 BAYES' FORMULA

In this section we consider experiments with sample spaces that will be divided or partitioned into two (or more) mutually exclusive events. This study involves a further application of conditional probabilities and leads us to the famous Bayes' formula, named after Thomas Bayes, who first published it in 1763.

We begin by considering the following example.

EXAMPLE 1   Introduction to Bayes' Theorem

Given two urns, suppose urn I contains four black and seven white balls. Urn II contains three black, one white, and four yellow balls. We select an urn at random and then draw a ball. What is the probability that we obtain a black ball?

SOLUTION   Let U_I and U_{II} stand for the events "Urn I is chosen" and "Urn II is chosen," respectively. Similarly, let B, W, Y stand for the event that "a black," "a white," or "a yellow" ball is chosen, respectively.

SOLUTION A   Example 1 can be solved using the tree diagram shown in Figure 1.

Figure 1

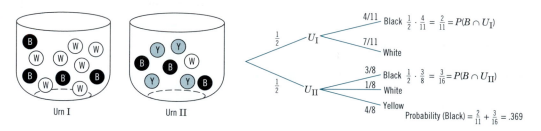

Then

$$P(B) = \tfrac{2}{11} + \tfrac{3}{16} = \tfrac{65}{176} = .369$$

SOLUTION B   Let's see how we can solve Example 1 without using a tree diagram. First we observe that

$$P(U_I) = P(U_{II}) = \tfrac{1}{2}$$
$$P(B|U_I) = \tfrac{4}{11} \qquad P(B|U_{II}) = \tfrac{3}{8}$$

The event B can be written as

$$B = (B \cap U_I) \cup (B \cap U_{II})$$

Since $B \cap U_I$ and $B \cap U_{II}$ are disjoint, we add their probabilities. Then

$$P(B) = P(B \cap U_I) + P(B \cap U_{II}) \tag{1}$$

Using the Product Rule, we have

$$P(B \cap U_I) = P(U_I) \cdot P(B|U_I) \qquad P(B \cap U_{II}) = P(U_{II}) \cdot P(B|U_{II}) \tag{2}$$

Combining (1) and (2), we have

$$P(B) = P(U_I) \cdot P(B|U_I) + P(U_{II}) \cdot P(B|U_{II})$$
$$= \tfrac{1}{2} \cdot \tfrac{4}{11} + \tfrac{1}{2} \cdot \tfrac{3}{8} = \tfrac{2}{11} + \tfrac{3}{16} = .369$$

Partitions

The preceding discussion leads to the following generalization. Suppose A_1 and A_2 are two nonempty, mutually exclusive events of a sample space S and the union of A_1 and A_2 is S; that is,

$$A_1 \neq \varnothing \qquad A_2 \neq \varnothing \qquad A_1 \cap A_2 = \varnothing \qquad S = A_1 \cup A_2$$

Figure 2

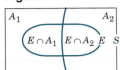

In this case we say that A_1 and A_2 form a **partition** of S. See Figure 2. Now if we let E be any event in S, we may write the set E in the form

$$E = (E \cap A_1) \cup (E \cap A_2)$$

The sets $E \cap A_1$ and $E \cap A_2$ are disjoint since

$$(E \cap A_1) \cap (E \cap A_2) = (E \cap E) \cap (A_1 \cap A_2) = E \cap \varnothing = \varnothing$$

Using the Product Rule, the probability of E is therefore

$$\begin{aligned} P(E) &= P(E \cap A_1) + P(E \cap A_2) \\ &= P(A_1) \cdot P(E|A_1) + P(A_2) \cdot P(E|A_2) \end{aligned} \tag{3}$$

Formula (3) is used to find the probability of an event E of a sample space when the sample space is partitioned into two sets A_1 and A_2. See Figure 3 for a tree diagram depicting Formula (3). You may find it easier to construct Figure 3 to obtain Formula (3) than to memorize it.

Figure 3

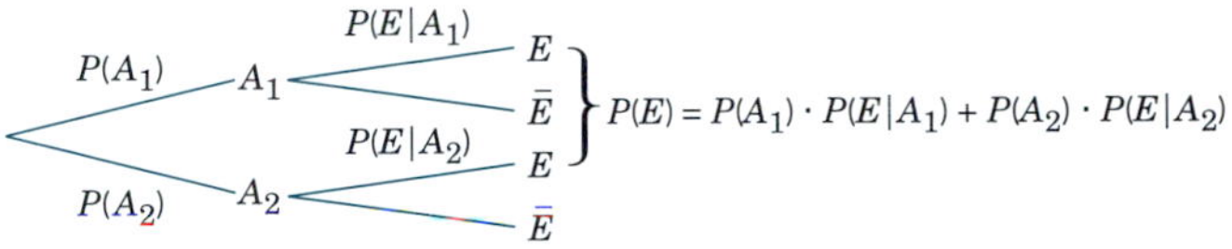

$$\begin{array}{l} P(A_1) \quad A_1 \overset{P(E|A_1)}{\underset{P(E|A_2)}{\diagdown}} \begin{array}{l} E \\ \bar{E} \end{array} \\ P(A_2) \quad A_2 \diagdown \begin{array}{l} E \\ \bar{E} \end{array} \end{array} \left.\right\} P(E) = P(A_1) \cdot P(E|A_1) + P(A_2) \cdot P(E|A_2)$$

EXAMPLE 2   Admissions Tests for Medical School

Of the applicants to a medical school, 80% are eligible to enter and 20% are not. To aid in the selection process, an admissions test is administered that is designed so that an eligible candidate will pass 90% of the time, while an ineligible candidate will pass only 30% of the time. What is the probability that an applicant for admission will pass the admissions test?

Figure 4

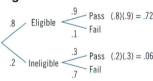

SOLUTION A   Figure 4 provides a tree diagram solution.

SOLUTION B   The sample space S consists of the applicants for admission, and S can be partitioned into the following two events:

$$A_1: \quad \text{Eligible applicant} \qquad A_2: \quad \text{Ineligible applicant}$$

These two events are disjoint, and their union is S. See Figure 5.

The event E is

$$E: \quad \text{Applicant passes admissions test}$$

Figure 5

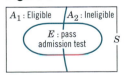

Now

$$\begin{array}{cc} P(A_1) = .8 & P(A_2) = .2 \\ P(E|A_1) = .9 & P(E|A_2) = .3 \end{array}$$

Using Formula (3), we have

$$P(E) = P(A_1) \cdot P(E|A_1) + P(A_2) \cdot P(E|A_2) = (.8)(.9) + (.2)(.3) = .78$$

Thus the probability that an applicant will pass the admissions test is .78.

Now Work Problem 15

If we partition a sample space S into three sets A_1, A_2, and A_3 so that

$$S = A_1 \cup A_2 \cup A_3$$

$$A_1 \cap A_2 = \varnothing \qquad A_2 \cap A_3 = \varnothing \qquad A_1 \cap A_3 = \varnothing$$

$$A_1 \neq \varnothing \qquad A_2 \neq \varnothing \qquad A_3 \neq \varnothing$$

we may write any set E in S in the form

$$E = (E \cap A_1) \cup (E \cap A_2) \cup (E \cap A_3)$$

Figure 6

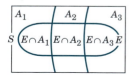

See Figure 6.

Since $E \cap A_1$, $E \cap A_2$, and $E \cap A_3$ are disjoint, the probability of event E is

$$\begin{aligned} P(E) &= P(E \cap A_1) + P(E \cap A_2) + P(E \cap A_3) \\ &= P(A_1) \cdot P(E|A_1) + P(A_2) \cdot P(E|A_2) + P(A_3) \cdot P(E|A_3) \end{aligned} \qquad (4)$$

Formula (4) is used to find the probability of an event E of a sample space when the sample space is partitioned into three sets A_1, A_2, and A_3. See Figure 7 for a tree diagram depicting Formula (4). Again, you may find it easier to construct Figure 7 to obtain Formula (4) than to memorize it.

Figure 7

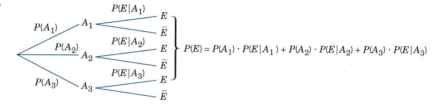

EXAMPLE 3   Determining Quality Control

Three machines, I, II, and III, manufacture .4, .5, and .1 of the total production in a plant, respectively. The percentage of defective items produced by I, II, and III is 2%, 4%, and 1%, respectively. For an item chosen at random, what is the probability that it is defective?

SOLUTION A   Figure 8 gives a tree diagram solution.

Figure 8

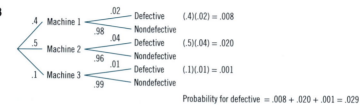

SOLUTION B   The sample space S is partitioned into three events A_1, A_2, and A_3 defined as follows:

A_1: Item produced by machine I

A_2: Item produced by machine II

A_3: Item produced by machine III

Clearly, the events A_1, A_2, and A_3 are mutually exclusive, and their union is S. Define the event E in S to be

E: Item is defective

Now

$$P(A_1) = .4 \qquad P(A_2) = .5 \qquad P(A_3) = .1$$
$$P(E|A_1) = .02 \qquad P(E|A_2) = .04 \qquad P(E|A_3) = .01$$

Thus, using Formula (4), we see that

$$P(E) = (.4)(.02) + (.5)(.04) + (.1)(.01)$$
$$= .008 + .020 + .001 = .029$$

 **Now Work Problem 17**

To generalize Formulas (3) and (4) to a sample space S partitioned into n subsets, we require the following definition:

Partition

A sample space S is **partitioned** into n subsets $A_1, A_2, \ldots, A_n$, provided:

(a) The intersection of any two of the subsets is empty.
(b) Each subset is nonempty.
(c) $A_1 \cup A_2 \cup \cdots \cup A_n = S$

Let S be a sample space and let $A_1, A_2, A_3, \ldots, A_n$ be n events that form a partition of the set S. If E is any event in S, then

$$E = (E \cap A_1) \cup (E \cap A_2) \cup \cdots \cup (E \cap A_n)$$

Since $E \cap A_1, E \cap A_2, \ldots, E \cap A_n$ are mutually exclusive events, we have

$$P(E) = P(E \cap A_1) + P(E \cap A_2) + \cdots + P(E \cap A_n) \tag{5}$$

In (5) replace $P(E \cap A_1), P(E \cap A_2), \ldots, P(E \cap A_n)$ using the Product Rule. Then we obtain the formula

$$P(E) = P(A_1) \cdot P(E|A_1) + P(A_2) \cdot P(E|A_2) + \cdots + P(A_n) \cdot P(E|A_n) \tag{6}$$

EXAMPLE 4   Admissions Tests for Medical School

In Example 2 suppose an applicant passes the admissions test. What is the probability that he or she was among those eligible; that is, what is the probability $P(A_1|E)$?

SOLUTION By the definition of conditional probability,

$$P(A_1|E) = \frac{P(A_1 \cap E)}{P(E)} = \frac{P(A_1) \cdot P(E|A_1)}{P(E)}$$

But $P(E)$ is given by (6) when $n = 2$, or by (3). Thus

$$P(A_1|E) = \frac{P(A_1) \cdot P(E|A_1)}{P(A_1) \cdot P(E|A_1) + P(A_2) \cdot P(E|A_2)} \qquad (7)$$

Using the information supplied in Example 2, we find

$$P(A_1|E) = \frac{(.8)(.9)}{.78} = \frac{.72}{.78} = .923$$

The admissions test is a reasonably effective device. Less than 8% of the students passing the test are ineligible.

Now Work Problem 19

Bayes' Formula

Equation (7) is a special case of **Bayes' formula** when the sample space is partitioned into two sets A_1 and A_2. The general formula is given below.

Bayes' Formula

Let S be a sample space partitioned into n events, $A_1, \ldots, A_n$. Let E be any event of S for which $P(E) > 0$. The probability of the event $A_j (j = 1, 2, \ldots, n)$, given the event E, is

$$P(A_j|E) = \frac{P(A_j) \cdot P(E|A_j)}{P(E)} \qquad (8)$$

$$= \frac{P(A_j) \cdot P(E|A_j)}{P(A_1) \cdot P(E|A_1) + P(A_2) \cdot P(E|A_2) + \cdots + P(A_n) \cdot P(E|A_n)}$$

The proof is left as an exercise (see Problem 42).

The following example illustrates a use for Bayes' formula when the sample space is partitioned into three events.

EXAMPLE 5 Quality Control: Source of Defective Cars

Motors, Inc., has three plants. Plant I produces 35% of the car output, plant II produces 20%, and plant III produces the remaining 45%. One percent of the output of plant I is defective, as is 1.8% of the output of plant II, and 2% of the output of plant III. The annual total output of Motors, Inc., is 1,000,000 cars. A car is chosen at random from the annual output and it is found to be defective. What is the probability that it came from plant I? Plant II? Plant III?

SOLUTION   To answer these questions, we first define the following events:

$$E: \quad \text{Car is defective}$$
$$A_1: \quad \text{Car produced by plant I}$$
$$A_2: \quad \text{Car produced by plant II}$$
$$A_3: \quad \text{Car produced by plant III}$$

Now, $P(A_1|E)$ indicates the probability that a car is produced by plant I, given that it was defective; $P(A_2|E)$ and $P(A_3|E)$ are similarly defined. To find these probabilities, we first need to find $P(E)$.

From the data given in the problem we can determine the following:

$$\begin{array}{ll} P(A_1) = .35 & P(E|A_1) = .010 \\ P(A_2) = .20 & P(E|A_2) = .018 \\ P(A_3) = .45 & P(E|A_3) = .020 \end{array} \qquad (9)$$

Now,

$A_1 \cap E$ is the event "Produced by plant I and is defective"

$A_2 \cap E$ is the event "Produced by plant II and is defective"

$A_3 \cap E$ is the event "Produced by plant III and is defective"

From the Product Rule we find

$$P(A_1 \cap E) = P(A_1) \cdot P(E|A_1) = (.35)(.010) = .0035$$
$$P(A_2 \cap E) = P(A_2) \cdot P(E|A_2) = (.20)(.018) = .0036$$
$$P(A_3 \cap E) = P(A_3) \cdot P(E|A_3) = (.45)(.020) = .0090$$

Since $E = (A_1 \cap E) \cup (A_2 \cap E) \cup (A_3 \cap E)$, we have

$$\begin{aligned} P(E) &= P(A_1 \cap E) + P(A_2 \cap E) + P(A_3 \cap E) \\ &= .0035 + .0036 + .0090 \\ &= .0161 \end{aligned}$$

Thus the probability that a defective car is chosen is .0161. See Figure 9.

Figure 9

Probability of defective = .0161

Given that the car chosen is defective, the probability that it came from plant I is $P(A_1|E)$, from plant II is $P(A_2|E)$, and from plant III is $P(A_3|E)$. To compute these probabilities, we use Bayes' formula:

$$P(A_1|E) = \frac{P(A_1) \cdot P(E|A_1)}{P(A_1) \cdot P(E|A_1) + P(A_2) \cdot P(E|A_2) + P(A_3) \cdot P(E|A_3)}$$

$$= \frac{P(A_1) \cdot P(E|A_1)}{P(E)} = \frac{(.35)(.01)}{.0161} = .217$$

$$P(A_2|E) = \frac{P(A_2) \cdot P(E|A_2)}{P(E)} = \frac{.0036}{.0161} = .224$$

(10)

$$P(A_3|E) = \frac{P(A_3) \cdot P(E|A_3)}{P(E)} = \frac{.0090}{.0161} = .559$$

✍ **Now Work Problem 24**

A Priori, A Posteriori Probabilities

In Bayes' formula the probabilities $P(A_j)$ are referred to as *a priori* probabilities, while the $P(A_j|E)$ are called *a posteriori* probabilities. We use Example 5 to explain the reason for this terminology. Knowing nothing else about a car, the probability that it was produced by plant I is given by $P(A_1)$. Thus $P(A_1)$ can be regarded as a "before the fact," or a priori, probability. With the additional information that the car is defective, we reassess the likelihood of whether it came from plant I and compute $P(A_1|E)$. Thus $P(A_1|E)$ can be viewed as an "after the fact," or a posteriori, probability.

Note that $P(A_1) = .35$, while $P(A_1|E) = .217$. So the knowledge that a car is defective decreases the chances that it came from plant I.

EXAMPLE 6   Testing for Cancer

The residents of a community are examined for cancer. The examination results are classified as positive $(+)$ if a malignancy is suspected, and as negative $(-)$ if there are no indications of a malignancy. If a person has cancer, the probability of a positive result from the examination is .98. If a person does not have cancer, the probability of a positive result is .15. If 5% of the community has cancer, what is the probability of a person not having cancer if the examination is positive?

SOLUTION   Define the following events:

$$A_1: \quad \text{Person has cancer}$$
$$A_2: \quad \text{Person does not have cancer}$$
$$E: \quad \text{Examination is positive}$$

We want to know the probability of a person not having cancer if it is known that the examination is positive; that is, we wish to find $P(A_2|E)$. Now

$$P(A_1) = .05 \qquad P(A_2) = .95$$
$$P(E|A_1) = .98 \qquad P(E|A_2) = .15$$

Using Bayes' formula, we get

$$P(A_2|E) = \frac{P(A_2) \cdot P(E|A_2)}{P(A_1) \cdot P(E|A_1) + P(A_2) \cdot P(E|A_2)}$$

$$= \frac{(.95)(.15)}{(.05)(.98) + (.95)(.15)} = .744$$

Thus even if the examination is positive, the person examined is more likely not to have cancer than to have cancer. The reason the test is designed this way is that it is better for a healthy person to be examined more thoroughly than for someone with cancer to go undetected. Simply stated, the test is useful because of the high probability (.98) that a person with cancer will not go undetected.

 **Now Work Problem 29**

EXAMPLE 7   Car Repair Diagnosis

The manager of a car repair shop knows from past experience that when a call is received from a person whose car will not start, the probabilities for various troubles (assuming no two can occur simultaneously) are as follows:

Event	Trouble	Probability
A_1	Flooded	.3
A_2	Battery cable loose	.2
A_3	Points bad	.1
A_4	Out of gas	.3
A_5	Something else	.1

The manager also knows that if the person will hold the gas pedal down and try to start the car, the probability that it will start (E) is

$$P(E|A_1) = .9 \qquad P(E|A_2) = 0 \qquad P(E|A_3) = .2$$

$$P(E|A_4) = 0 \qquad P(E|A_5) = .2$$

(a) If a person has called and is instructed to "hold the pedal down . . . ," what is the probability that the car will start?

(b) If the car does start after holding the pedal down, what is the probability that the car was flooded?

SOLUTION

(a) We need to compute $P(E)$. Using Formula (6) for $n = 5$ (the sample space is partitioned into five disjoint sets), we have

$$\begin{aligned} P(E) &= P(A_1) \cdot P(E|A_1) + P(A_2) \cdot P(E|A_2) + P(A_3) \cdot P(E|A_3) \\ &\quad + P(A_4) \cdot P(E|A_4) + P(A_5) \cdot P(E|A_5) \\ &= (.3)(.9) + (.2)(0) + (.1)(.2) + (.3)(0) + (.1)(.2) \\ &= .27 + .02 + .02 = .31 \end{aligned}$$

See Figure 10.

Figure 10

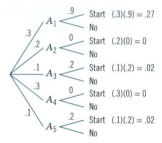

Probability of starting = .31

(b) We use Bayes' formula to compute the a posteriori probability $P(A_1|E)$:

$$P(A_1|E) = \frac{P(A_1) \cdot P(E|A_1)}{P(E)} = \frac{(.3)(.9)}{.31} = .87$$

Thus the probability that the car was flooded, after it is known that holding down the pedal started the car, is .87.

EXERCISE 8.1   Answers to odd-numbered problems begin on page AN-28.

In Problems 1–14 find the indicated probabilities by referring to the tree diagram below and using Bayes' formula.

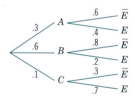

1. $P(E|A)$

2. $P(\bar{E}|A)$

3. $P(E|B)$

4. $P(\bar{E}|B)$

5. $P(E|C)$

6. $P(\bar{E}|C)$

7. $P(E)$

8. $P(\bar{E})$

9. $P(A|E)$

10. $P(B|\bar{E})$

11. $P(C|E)$

12. $P(A|\bar{E})$

13. $P(B|E)$

14. $P(C|\bar{E})$

15. Events A_1 and A_2 form a partition of a sample space S with $P(A_1) = .4$ and $P(A_2) = .6$. If E is an event in S with $P(E|A_1) = .03$ and $P(E|A_2) = .02$, compute $P(E)$.

16. Events A_1 and A_2 form a partition of a sample space S with $P(A_1) = .3$ and $P(A_2) = .7$. If E is an event in S with $P(E|A_1) = .04$ and $P(E|A_2) = .01$, compute $P(E)$.

17. Events A_1, A_2, and A_3 form a partition of a sample space S with $P(A_1) = .6$, $P(A_2) = .2$, and $P(A_3) = .2$. If E is an event in S with $P(E|A_1) = .01$, $P(E|A_2) = .03$, and $P(E|A_3) = .02$, compute $P(E)$.

18. Events A_1, A_2, and A_3 form a partition of a sample space S with $P(A_1) = .3$, $P(A_2) = .2$, and $P(A_3) = .5$. If E is an

event in S with $P(E|A_1) = .01$, $P(E|A_2) = .02$, and $P(E|A_3) = .02$, compute $P(E)$.

19. Use the information in Problem 15 to find $P(A_1|E)$ and $P(A_2|E)$.

20. Use the information in Problem 16 to find $P(A_1|E)$ and $P(A_2|E)$.

21. Use the information in Problem 17 to find $P(A_1|E)$, $P(A_2|E)$, and $P(A_3|E)$.

22. Use the information in Problem 18 to find $P(A_1|E)$, $P(A_2|E)$, and $P(A_3|E)$.

23. In Example 3 (page 370) suppose it is known that a defective item was produced. Find the probability that it came from machine I. From machine II. From machine III.

24. In Example 5 (page 372) suppose $P(A_1) = P(A_2) = P(A_3) = \frac{1}{3}$. Find $P(A_2|E)$ and $P(A_3|E)$.

25. In Example 7 (page 375), compute the a posteriori probabilities $P(A_2|E)$, $P(A_3|E)$, $P(A_4|E)$, and $P(A_5|E)$.

26. In Example 6 (page 374) compute $P(A_1|E)$.

27. Three jars contain colored balls as follows:

Jar	Red, R	White, W	Blue, B
I	5	6	5
II	3	4	9
III	7	5	4

One jar is chosen at random and a ball is withdrawn. The ball is red. What is the probability that it came from jar I? From jar II? From jar III? [*Hint:* Define the events E: Ball selected is red; U_I: jar I selected; U_{II}: jar II selected; and U_{III}: jar III selected.] Determine $P(U_I|E)$, $P(U_{II}|E)$, and $P(U_{III}|E)$ by using Bayes' formula.

28. **Car Production**   Cars are being produced by two factories, but factory I produces twice as many cars as factory II in a given time. Factory I is known to produce 2% defectives and factory II produces 1% defectives. A car is examined and found to be defective. What are the a priori and a posteriori probabilities that the car was produced by factory I?

29. **Color-Blind**   In a human population 51% are male and 49% are female. 5% of the males and 0.3% of the females are color-blind. If a person randomly chosen from the population is found to be color-blind, what is the probability that the person is a male?

30. **Medical Diagnosis**   In a certain small town 16% of the population developed lung cancer. If 45% of the population are smokers, and 85% of those developing lung cancer are smokers, what is the probability that a smoker in this population will develop lung cancer?

31. **Voting Pattern**   In Cook County 55% of the registered voters are Democrats, 30% are Republicans, and 15% are independents. During a recent election, 35% of the Democrats voted, 65% of the Republicans voted, and 75% of the independents voted. What is the probability that someone who voted is a Democrat? Republican? Independent?

32. **Quality Control**   A computer manufacturer has three assembly plants. Records show that 2% of the sets shipped from plant A turn out to be defective, as compared to 3% of those that come from plant B and 4% of those that come from plant C. In all, 30% of the manu-

facturer's total production comes from plant A, 50% from plant B, and 20% from plant C. If a customer finds that his computer is defective, what is the probability it came from plant B?

33. **Oil Drilling**   An oil well is to be drilled in a certain location. The soil there is either rock (probability .53), clay (probability .21), or sand. If it is rock, a geological test gives a positive result with 35% accuracy; if it is clay, this test gives a positive result with 48% accuracy; and if it is sand, the test gives a positive result with 75% accuracy. Given that the test is positive, what is the probability that the soil is rock? What is the probability that the soil is clay? What is the probability that the soil is sand?

34. **Oil Drilling**   A geologist is using seismographs to test for oil. It is found that if oil is present, the test gives a positive result 95% of the time, and if oil is not present, the test gives a positive result 2% of the time. Oil is actually present in 1% of the cases tested. If the test shows positive, what is the probability that oil is present?

35. **Political Polls**   In conducting a political poll, a pollster divides the United States into four sections: Northeast (N), containing 40% of the population; South (S), containing 10% of the population; Midwest (M), containing 25% of the population; and West (W), containing 25% of the population. From the poll it is found that in the next election 40% of the people in the Northeast say they will vote for Republicans, in the South 56% will vote Republican, in the Midwest 48% will vote Republican, and in the West 52% will vote Republican. What is the probability that a person chosen at random will vote Republican? Assuming a person votes Republican, what is the probability that he or she is from the Northeast?

36. **TB Screening**   Suppose that if a person with tuberculosis is given a TB screening, the probability that his or her condition will be detected is .90. If a person without tuberculosis is given a TB screening, the probability that he or she will be diagnosed incorrectly as having tuberculosis is .3. Suppose, further, that 11% of the adult residents of a certain city have tuberculosis. If one of these adults is diagnosed as having tuberculosis based on the screening, what is the probability that he or she actually has tuberculosis? Interpret your result.

37. **Detective Columbo**   An absent-minded nurse is to give Mr. Brown a pill each day. The probability that the nurse forgets to administer the pill is $\frac{2}{3}$. If he receives the pill, the probability that Mr. Brown will die is $\frac{1}{3}$. If he does not get his pill, the probability that he will die is $\frac{3}{4}$. Mr. Brown died. What is the probability that the nurse forgot to give Mr. Brown the pill?

38. Marketing To introduce a new beer, a company conducted a survey. It divided the United States into four regions, eastern, northern, southern, and western. The company estimates that 35% of the potential customers for the beer are in the eastern region, 30% are in the northern region, 20% are in the southern region, and 15% are in the western region. The survey indicates that 50% of the potential customers in the eastern region, 40% of the potential customers in the northern region, 36% of the potential customers in the southern region, and 42% of those in the western region will buy the beer. If a potential customer chosen at random indicates that he or she will buy the beer, what is the probability that the customer is from the southern region?

39. College Majors Data collected by the Office of Admissions of a large midwestern university indicate the following choices made by the members of the freshman class regarding their majors:

Major	Percentage of Freshmen Choosing Major	Female (in percent)	Male (in percent)
Engineering	26	40	60
Business	30	35	65
Education	9	80	20
Social science	12	52	48
Natural science	12	56	44
Humanities	9	65	35
Other	2	51	49

What is the probability that a female student selected at random from the freshman class is majoring in engineering?

40. Testing for HIV An article in the *New York Times* some time ago reported that college students are beginning to routinely ask to be tested for the AIDS virus. The standard test for the HIV virus is the Elias test, which tests for the presence of HIV antibodies. It is estimated that this test has a 99.8% sensitivity and a 99.8% specificity. A 99.8% sensitivity means that, in a large-scale screening test, for every 1000 people tested who have the virus we can expect 998 people to test positive and 2 to have a false negative test. A 99.8% specificity means that, in a large-scale screening test, for every 1000 people tested who do not have the virus we can expect 998 people to have a negative test and 2 to have a false positive test.

(a) The *New York Times* article remarks that it is estimated that about 2 in every 1000 college students have the HIV virus. Assume that a large group of randomly chosen college students, say 100,000, are tested by the Elias test. If a student tests positive, what is the chance that this student has the HIV virus?

(b) What would this probability be for a population at high risk, where 5% of the population has the HIV virus?

(c) Suppose Jack tested positive on an Elias test. Another Elias test* is performed and the results are positive again. Assuming that the tests are independent, what is the probability that Jack has the HIV virus?

41. Medical Test A scientist designed a medical test for a certain disease. Among 100 patients who have the disease, the test will show the presence of the disease in 97 cases out of 100, and will fail to show the presence of the disease in the remaining 3 cases out of 100. Among those who do not have the disease, the test will erroneously show the presence of the disease in 4 cases out of 100, and will show that there is no disease in the remaining 96 cases out of 100.

(a) What is the probability that a patient who tested positive on this test actually has the disease, if it is estimated that 20% of the population has the disease?

(b) What is the probability that a patient who tested positive on this test actually has the disease, if it is estimated that 4% of the population has the disease?

(c) What is the probability that a patient who took the test twice and tested positive both times actually has the disease, if it is estimated that 4% of the population has the disease?

42. Prove Bayes' formula (8).

43. Show that $P(E|F) = 1$ if F is a subset of E and $P(F) \neq 0$.

---

* Actually, in practice, if a person tests positive on an Elias test, then two more Elias tests are carried out. If either is positive, then one more confirmatory test, called the Western blot test, is carried out. If this is positive, the person is assumed to have the HIV virus.

8.2 THE BINOMIAL PROBABILITY MODEL

Bernoulli Trials

In this section we study experiments that can be analyzed by using a probability model called the *binomial probability model.* The model was first studied by J. Bernoulli about 1700 and, for this reason, the model is sometimes referred to as a *Bernoulli trial.*

The **binomial probability model** is a sequence of trials, each of which consists of repetition of a single experiment. We assume the outcome of one experiment does not affect the outcome of any other one; that is, we assume the trials to be independent. Furthermore, we assume that there are only two possible outcomes for each trial and label them A, for *success,* and F, for *failure.* We denote the probability of success by $p = P(A)$; $p = P(A)$ remains the same from trial to trial. In addition, since there are only two outcomes in each trial, the probability of failure, denoted by q, must be $1 - p$, so

$$q = 1 - p = P(F)$$

Thus $p + q = 1$.

Any random experiment for which the binomial probability model is appropriate is called a *Bernoulli trial.*

Bernoulli Trial

Random experiments are called **Bernoulli trials** if

(a) The same experiment is repeated several times.
(b) There are only two possible outcomes, success and failure, on each trial.
(c) The repeated trials are independent.
(d) The probability of each outcome remains the same for each trial.

Many real world situations have the characteristics of the binomial probability model. For example, in repeatedly running a subject through a T-maze, we may label a turn to the left by A and a turn to the right by F. The assumption of independence of each trial is equivalent to presuming the subject has no memory.

In opinion polls one person's response is independent of any other person's response, and we may designate the answer "Yes" by an A and any other answer ("No" or "Don't know") by an F.

In testing TVs, we have a sequence of independent trials (each test of a particular TV is a trial), and we label a nondefective TV with an A and a defective one with an F.

In determining whether 9 out of 12 persons will recover from a tropical disease, we assume that each of the 12 persons has the same chance of recovery from the disease and that their recoveries are independent (they are not treated by the same doctor or in the same hospital). We may designate "recovery" by A and "nonrecovery" by F.

Next we consider an experiment that will lead us to formulate a general expression for the probability of obtaining k successes in a sequence of n Bernoulli trials ($k \leq n$).

The experiment consists of tossing a coin three times. We define success as heads

(H) and failure as tails (T). The coin may be fair or loaded, so we let p be the probability of heads and q be the probability of tails.

The tree diagram in Figure 11 lists all the possible outcomes and their respective probabilities.

Figure 11

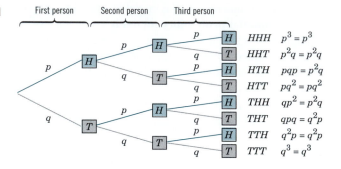

The outcomes are

$$HHH, \quad HHT, \quad HTH, \quad HTT, \quad THH, \quad THT, \quad TTH, \quad TTT$$

Here, as before, *HHT* means that the first two tosses are heads and the third is tails.

The outcome *HHH*, in which all three tosses are heads, has a probability $ppp = p^3$ since each Bernoulli trial has a probability p of resulting in heads.

If we wish to calculate the probability that exactly two heads appear, we consider only the outcomes

$$HHT, HTH, THH$$

The outcome *HHT* has probability $ppq = p^2q$ since the two heads each have probability p and the tail has probability q. In the same way the other two outcomes, *HTH* and *THH*, in which there are two heads and one tail, also have probabilities p^2q. Therefore, the probability that exactly two heads appear is equal to the sum of the probabilities of the three outcomes and, hence, is given by $3p^2q$.

In a similar way we compute the following probabilities:

$$P \text{ (Exactly 1 head and 2 tails)} = 3pq^2$$
$$P \text{ (All 3 heads)} = p^3$$
$$P \text{ (All 3 tails)} = q^3$$

Binomial Probabilities

When considering tosses in excess of three, it would be extremely tedious to solve problems of this type using a tree diagram. This is why it is desirable to find a general formula.

Suppose the probability of a success in a Bernoulli trial is p, and suppose we wish to find the probability of exactly k successes in n repeated trials. One possible outcome is

$$\underbrace{AAA\cdots A}_{k \text{ successes}} \cdot \underbrace{FF\cdots F}_{n - k \text{ failures}} \tag{1}$$

where k successes come first, followed by $n - k$ failures. The probability of this outcome is

$$\underbrace{ppp \cdots p}_{k \text{ factors}} \cdot \underbrace{qq \cdots q}_{n - k \text{ factors}} = p^k q^{n-k}$$

The k successes could also be obtained by rearranging the letters A and F in (1) above. Then the number of such sequences must equal the number of ways of choosing k of the n trials to contain successes—namely, $C(n, k) = \binom{n}{k}$. If we multiply this number by the probability of obtaining any one such sequence, we arrive at the following general result:

Formula for $b(n, k; p)$

In a Bernoulli trial the probability of exactly k successes in n trials is given by

$$b(n, k; p) = \binom{n}{k} p^k \cdot q^{n-k} = \frac{n!}{k!(n-k)!} p^k \cdot q^{n-k} \qquad (2)$$

where p is the probability of success and $q = 1 - p$.

The symbol $b(n, k; p)$, which represents the probability of exactly k successes in n trials, is called a **binomial probability.**

 **Now Work Problem 1**

EXAMPLE 1   **Example of a Bernoulli Trial; Finding a Binomial Probability**

A common example of a Bernoulli trial is the experiment of tossing a fair coin.

1. There are exactly two possible mutually exclusive outcomes on each trial or toss (heads or tails).
2. The probability of a particular outcome (say, heads) remains constant from trial to trial (toss to toss).
3. The outcome on any trial (toss) is independent of the outcome on any other trial (toss).

Find the probability of obtaining exactly one tail in six tosses of a fair coin.

SOLUTION   Let T denote the outcome "Tail shows" and let H denote the outcome "Head shows." Using Formula (2), in which $k = 1$ (one success), $n = 6$ (the number of trials), and $p = \frac{1}{2} = P(T)$ (the probability of success), we obtain

$$P(\text{Exactly 1 success}) = b\left(6, 1; \frac{1}{2}\right) = \binom{6}{1}\left(\frac{1}{2}\right)^1\left(\frac{1}{2}\right)^{6-1} = \frac{6}{64} = .0938$$

 **Now Work Problem 15**

EXAMPLE 2   **Baseball**

A baseball pitcher gives up a hit on the average of once every fifth pitch. If nine pitches are thrown, what is the probability that

(a) Exactly three pitches result in hits?
(b) No pitch results in a hit?
(c) Eight or more pitches result in hits?
(d) No more than seven pitches result in hits?

SOLUTION In this example $p = P(\text{Success}) = P(\text{Allowing a hit}) = \frac{1}{5} = .2$. The number of trials is the number of pitches so $n = 9$. Finally, k is the number of pitches that result in hits.

(a) If exactly three pitches result in hits, then $k = 3$. Using $n = 9$, $p = .2$, and $q = 1 - p = .8$, we find

$$P(\text{Exactly 3 hits}) = b(9, 3; .2) = \binom{9}{3} (.2)^3 (.8)^6 = .1762$$

(b) If none of the nine pitches resulted in a hit, $k = 0$.

$$P(\text{Exactly 0 hits}) = b(9, 0; .2) = \binom{9}{0} (.2)^0 (.8)^9 = .1342$$

(c) Nine pitches are thrown, so the event "eight or more pitches result in hits" is equivalent to the events: "exactly 8 result in hits" or "exactly 9 result in hits." Since these events are mutually exclusive, we have

$$P(\text{At least 8 hits}) = P(\text{Exactly 8 hits}) + P(\text{Exactly 9 hits})$$
$$= b(9, 8; .2) + b(9, 9; .2) = .0000189$$

(d) "No more than seven pitches result in hits" means 0, 1, 2, 3, 4, 5, 6, or 7 pitches result in a hit. We could add $b(9, 0; .2)$, $b(9, 1; .2)$, and so on, but it is easier to use the formula $P(E) = 1 - P(\bar{E})$. The complement of "No more than 7" is "at least 8." We use the result of part (c) to find:

$$P(\text{No more than 7 hits}) = 1 - P(\text{At least 8})$$
$$= 1 - [b(9, 8; .2) + b(9, 9; .2)] = .999981$$

■

Look again at part (c) in Example 2. When we seek the probability of *at least k* successes (or *at most k* successes), it is necessary to find the individual probabilities and, because the events are mutually exclusive, add them. The next example looks once more at this type of situation.

EXAMPLE 3 **Quality Control**

A machine produces light bulbs to meet certain specifications, and 80% of the bulbs produced meet these specifications. A sample of six bulbs is taken from the machine's production and placed in a box. What is the probability that at least three of them fail to meet the specifications?

SOLUTION In this example the number of trials is $n = 6$. We are looking for the probability of the event

E: At least 3 fail to meet specifications

Since the experiment consists of choosing 6 bulbs, the event E is the union of the mutually exclusive events: "Exactly 3 fail," "Exactly 4 fail," "Exactly 5 fail," and "Exactly 6 fail." We use Formula (2) for $n = 6$ and $k = 3, 4, 5,$ and 6. Since the probability of a bulb failing to meet specifications is .20, we have

$$P(\text{Exactly 3 fail}) = b(6, 3; .20) = .0819$$
$$P(\text{Exactly 4 fail}) = b(6, 4; .20) = .0154$$
$$P(\text{Exactly 5 fail}) = b(6, 5; .20) = .0015$$
$$P(\text{Exactly 6 fail}) = b(6, 6; .20) = .0001$$

Therefore,

$$P(\text{At least 3 fail}) = P(E) = .0819 + .0154 + .0015 + .0001 = .0989$$

Another way of getting this answer is to compute the probability of the complementary event

$$\overline{E}: \quad \text{Fewer than 3 fail}$$

Then

$$P(\overline{E}) = P(\text{Exactly 2 fail}) + P(\text{Exactly 1 fail}) + P(\text{Exactly 0 fail})$$
$$= b(6, 2; .20) + b(6, 1; .20) + b(6, 0; .20)$$
$$= .2458 + .3932 + .2621 = .9011$$

As a result,

$$P(\text{At least 3 fail}) = 1 - P(\overline{E}) = 1 - .9011 = .0989$$

as before.

 **Now Work Problem 35**

EXAMPLE 4   Product Testing

A man claims to be able to distinguish between two kinds of wine with 90% accuracy and presents his claim to an agency interested in promoting the consumption of one of the two kinds of wine. The following experiment is conducted to check his claim. The man is to taste the two types of wine and distinguish between them. This is to be done nine times with a 3-minute break after each taste. It is agreed that if the man is correct at least six out of the nine times, he will be hired.

The main questions to be asked are, on the one hand, whether the above procedure gives sufficient protection to the hiring agency against a person guessing and, on the other hand, whether the man is given a sufficient chance to be hired if he is really a wine connoisseur.

SOLUTION   The number of trials is $n = 9$. To answer the first question, let's assume that the man is guessing. Then in each trial he has a probability of $\frac{1}{2}$ of identifying the wine correctly. Let k be the number of correct identifications. Let's compute the binomial probability for $k = 6, 7, 8, 9$, to find the likelihood of the man being hired (at least 6 correct) while guessing ($p = \frac{1}{2}$):

$$b(9, 6; \tfrac{1}{2}) + b(9, 7; \tfrac{1}{2}) + b(9, 8; \tfrac{1}{2}) + b(9, 9; \tfrac{1}{2}) = .1641 + .0703 + .0176 + .0020$$
$$= .2540$$

Thus there is a likelihood of .254 that he will pass if he is just guessing.

To answer the second question in the case where the claim is true ($p = .90$), we need to find the sum of the probabilities $b(9, k; .90)$ for $k = 6, 7, 8, 9$:

$$b(9, 6; .90) + b(9, 7; .90) + b(9, 8; .90) + b(9, 9; .90) = .0446 + .1722 + .3874 + .3874$$
$$= .9916$$

Notice that the test in Example 4 is fair to the man since it practically assures him the position if his claim is true. However, the company may not like the test because 25% of the time a person who guesses will pass the test.

 **Now Work Problem 43**

Application

EXAMPLE 5   Testing a Serum or Vaccine

Suppose that the normal rate of infection of a certain disease in cattle is 25%. To test a newly discovered serum, healthy animals are injected with it. How can we evaluate the result of the experiment?

SOLUTION   For an absolutely worthless serum, the probability that exactly k of n test animals remain free from infection equals $b(n, k; .75)$. For $k = n = 10$, this probability is about $b(10, 10; .75) = .056$. Thus, if out of 10 test animals none is infected, this may be taken as an indication that the serum has had an effect, although it is not conclusive proof. Notice that, without serum, the probability that out of 17 animals at most 1 catches the infection is $b(17, 0; .25) + b(17, 1; .25) = .0501$. Therefore, there is *stronger evidence* in favor of the serum if out of 17 test animals at most 1 gets infected than if out of 10 all remain healthy. For $n = 23$ the probability of at most 2 animals catching the infection is about .0492 and, thus, at most 2 failures out of 23 is again better evidence for the serum than at most 1 out of 17 or 0 out of 10.

EXERCISE 8.2   Answers to odd-numbered problems begin on page AN-28.

In Problems 1–14 use Formula (2), page 381, and a calculator to compute each binomial probability.

1. $b(7, 4; .20)$ **2.** $b(8, 5; .30)$ **3.** $b(15, 8; .80)$

4. $b(8, 5; .70)$ **5.** $b(15, 10; \tfrac{1}{2})$ **6.** $b(12, 6; .90)$

7. $b(15, 3; .3) + b(15, 2; .3) + b(15, 1; .3) + b(15, 0; .3)$ **8.** $b(8, 6; .4) + b(8, 7; .4) + b(8, 8; .4)$

9. $n = 3$, $k = 2$, $p = \tfrac{1}{3}$ **10.** $n = 3$, $k = 1$, $p = \tfrac{1}{3}$ **11.** $n = 3$, $k = 0$, $p = \tfrac{1}{6}$

12. $n = 3$, $k = 3$, $p = \tfrac{1}{6}$ **13.** $n = 5$, $k = 3$, $p = \tfrac{2}{3}$ **14.** $n = 5$, $k = 0$, $p = \tfrac{2}{3}$

15. Find the probability of obtaining exactly 6 successes in 10 trials when the probability of success is .3.

16. Find the probability of obtaining exactly 5 successes in 9 trials when the probability of success is .2.

17. Find the probability of obtaining exactly 9 successes in 12 trials when the probability of success is .8.

18. Find the probability of obtaining exactly 8 successes in 15 trials when the probability of success is .75.

19. Find the probability of obtaining at least 5 successes in 8 trials when the probability of success is .30.

20. Find the probability of obtaining at most 3 successes in 7 trials when the probability of success is .20.

In Problems 21–26 a fair coin is tossed 8 times.

21. What is the probability of obtaining exactly 1 head?

22. What is the probability of obtaining exactly 2 heads?

23. What is the probability of obtaining at least 5 tails?

24. What is the probability of obtaining at most 2 tails?

25. What is the probability of obtaining exactly 2 heads if it is known that at least 1 head appeared?

26. What is the probability of obtaining exactly 3 heads if it is known that at least 1 head appeared?

27. In five rolls of two fair dice, what is the probability of obtaining a sum of 7 exactly twice?

28. In seven rolls of two fair dice, what is the probability of obtaining a sum of 11 exactly three times?

29. **Quality Control** Suppose that 5% of the items produced by a factory are defective. If 8 items are chosen at random, what is the probability that

    (a) Exactly 1 is defective?
    (b) Exactly 2 are defective?
    (c) At least 1 is defective?
    (d) Fewer than 3 are defective?

30. **Opinion Poll** Suppose that 60% of the voters intend to vote for a conservative candidate. What is the probability that a survey polling 8 people reveals that 3 or fewer intend to vote for a conservative candidate?

31. **Family Structure** What is the probability that a family with exactly 6 children will have 3 boys and 3 girls?

32. **Family Structure** What is the probability that in a family of 7 children:

    (a) 4 are girls?
    (b) At least 2 are girls?
    (c) At least 2 and not more than 4 are girls?

33. An experiment is performed 4 times, with 2 possible outcomes, F (failure) and S (success), with probabilities $\frac{1}{4}$ and $\frac{3}{4}$, respectively.

    (a) Draw a tree diagram describing the experiment.
    (b) Calculate the probability of exactly 2 successes and 2 failures by using the tree diagram from Part (a).
    (c) Verify your answer to Part (b) by using Formula (2).

34. **Batting Averages** For a baseball player with a .250 batting average, what is the probability that the player will have at least 2 hits in 4 times at bat? What is the probability of at least 1 hit in 4 times at bat?

35. **Target Shooting** If the probability of hitting a target is $\frac{2}{3}$ and 10 shots are fired independently, what is the probability of the target being hit at least twice?

36. **Quality Control** A television manufacturer tests a random sample of 15 picture tubes to determine whether any are defective. The probability that a picture tube is defective has been found from past experience to be .03.

    (a) What is the probability that there are no defective tubes in the sample?
    (b) What is the probability that more than 2 of the tubes are defective?

37. **True–False Tests** In a 15-item true–false examination, what is the probability that a student who guesses on each question will get at least 10 correct answers? If another student has .8 probability of correctly answering each question, what is the probability that this student will answer at least 12 questions correctly?

38. **Screening Employees** To screen prospective employees, a company gives a 10-question multiple-choice test. Each question has 4 possible answers, of which 1 is correct. The chance of answering the questions correctly by just guessing is $\frac{1}{4}$ or 25%. Find the probability of answering, by chance:

    (a) Exactly 3 questions correctly
    (b) No questions correctly
    (c) At least 8 questions correctly
    (d) No more than 7 questions correctly

39. **Opinion Poll** Mr. Austin and Ms. Moran are running for public office. A survey conducted just before the day of election indicates that 60% of the voters prefer Mr. Austin and 40% prefer Ms. Moran. If 8 people are chosen at random and asked their preference, find the probability that all 8 people will express a preference for Ms. Moran.

40. Working Habits   If 30% of the workers at a large factory bring their lunch each day, what is the probability that in a randomly selected sample of 8 workers

(a) Exactly 2 bring their lunch each day?
(b) At least 2 bring their lunch each day?
(c) No one brings lunch?
(d) No more than 3 bring lunch each day?

41. Heart Attack   Approximately 23% of North American deaths are due to heart attacks. What is the probability that 4 of the next 10 unrelated deaths reported in a certain community will be due to heart attacks?

42. Opinion Polls   Suppose that 60% of the voters intend to vote for a conservative candidate. What is the probability that a survey polling 8 people reveals that 3 or fewer intend to vote for a conservative candidate?

43. Product Testing   A supposed coffee connoisseur claims she can distinguish between a cup of instant coffee and a cup of drip coffee 80% of the time. You give her 6 cups of coffee and tell her that you will grant her claim if she correctly identifies at least 5 of the 6 cups.

(a) What are her chances of having her claim granted if she is in fact only guessing?
(b) What are her chances of having her claim rejected when in fact she really does have the ability she claims?

44. Opinion Poll   Opinion polls based on small samples often yield misleading results. Suppose 65% of the people in a city are opposed to a bond issue and the others favor it. If 7 people are asked for their opinion, what is the probability that a majority of them will favor the bond issue?

Technology Exercises

Most graphing calculators have a random number function (usually RAND or RND) generating numbers between 0 and 1. Check your user's manual to see how to use this function on your graphing calculator.

Sometimes experiments are simulated using a random number function instead of actually performing the experiment. In Problems 1–4 use a random number function to simulate each experiment.

1. Tossing a Fair Coin   Consider the experiment of tossing a coin 4 times, counting the number of heads occurring in these 4 tosses. Simulate the experiment using a random number function on your calculator, considering a toss to be tails (*T*) if the result is less than 0.5, and considering a toss to be heads (*H*) if the result is greater than or equal to 0.5. Record the number of heads in 4 tosses. [*Note:* Most calculators repeat the action of the last entry if you simply press the ENTER, or EXE, key again.] Repeat the experiment 10 times, thus obtaining a sequence of 10 numbers. Using these 10 numbers you can estimate the probability of *k* heads, *P*(*k*), for each *k* = 0, 1, 2, 3, 4, by the ratio

$$\frac{\text{Number of times } k \text{ appears in your sequence}}{10}$$

Enter your estimates in the table to the right. Calculate the actual probabilities using the binomial probability formula, and enter these numbers in the table. How close are your numbers to the actual values?

k	Your Estimate of $P(k)$	Actual Value of $P(k)$
0		
1		
2		
3		
4		

2. Tossing a Loaded Coin   Consider the experiment of tossing a loaded coin 4 times, counting the number of heads occurring in these 4 tosses. Simulate the experiment using a random number function on your calculator, considering a toss to be tails (*T*) if the result is less than 0.80, and considering a toss to be heads (*H*) if the result is greater than or equal to 0.80. Record the number of heads in 4 tosses. Repeat the experiment 10 times, thus obtaining a sequence of 10 numbers. Using these 10 numbers you can estimate the probability of *k* heads, *P*(*k*), for each

$k = 0, 1, 2, 3, 4$, by the ratio

$$\frac{\text{Number of times } k \text{ appears in your sequence}}{10}$$

Enter your estimates in the table below. Calculate the actual probabilities using the binomial probability formula, and enter these numbers in the table. How close are your numbers to the actual values? Why is this coin loaded?

k	Your Estimate of $P(k)$	Actual Value of $P(k)$
0		
1		
2		
3		
4		

3. **Tossing a Fair Coin**   Consider the experiment of tossing a fair coin 8 times, counting the number of heads occurring in these 8 tosses. Simulate the experiment using a random number function on your calculator, considering a toss to be tails (T) if the result is less than 0.50, and considering a toss to be heads (H) if the result is greater than or equal to 0.50. Record the number of heads in 8 tosses. Repeat

the experiment 10 times, thus obtaining a sequence of 10 numbers. Using these 10 numbers you can estimate the probability of 3 heads, $P(3)$, by the ratio

$$\frac{\text{Number of times 3 appears in your sequence}}{10}$$

Calculate the actual probability using the binomial probability formula. How close is your estimate to the actual value?

4. **Tossing a Loaded Coin**   Consider the experiment of tossing a loaded coin 8 times, counting the number of heads occurring in these 8 tosses. Simulate the experiment using a random number function on your calculator, considering a toss to be tails (T) if the result is less than 0.80, and considering a toss to be heads (H) if the result is greater than or equal to 0.80. Record the number of heads in 8 tosses. Repeat the experiment 10 times, thus obtaining a sequence of 10 numbers. Using these 10 numbers you can estimate the probability of 3 heads, $P(3)$, by the ratio

$$\frac{\text{Number of times 3 appears in your sequence}}{10}$$

Calculate the actual probability using the binomial probability formula. How close is your estimate to the actual value?

8.3 EXPECTED VALUE

An important concept, which uses probability, is *expected value*.

EXAMPLE 1   Examples of Expected Value

(a) Suppose that 1000 tickets are sold for a raffle that has the following prizes: one $300 prize, two $100 prizes, and one hundred $1 prizes. Then, of the 1000 tickets, 1 ticket has a cash value of $300, 2 are worth $100, and 100 are worth $1, while the remaining are worth $0. The *expected* (average) *value* of a ticket is then

$$E = \frac{\$300 + (\$100 + \$100) + \overbrace{(\$1 + \$1 + \cdots + \$1)}^{100 \text{ times}} + \overbrace{(\$0 + \$0 + \cdots + \$0)}^{897 \text{ times}}}{1000}$$

$$= \$300 \cdot \frac{1}{1000} + \$100 \cdot \frac{2}{1000} + \$1 \cdot \frac{100}{1000} + \$0 \cdot \frac{897}{1000}$$

$$= \frac{\$600}{1000} = \$0.60$$

Thus if the raffle is to be nonprofit to all, the charge for each ticket should be $0.60.

The situation above can also be viewed as follows: if we entered such a raffle many times, $\frac{1}{1000}$ of the time we would win $300, $\frac{2}{1000}$ of the time we would win $100, and so on, with our winnings in the long run averaging $0.60 per ticket.

(b) Suppose that you are to receive $3.00 each time you obtain two heads when flipping a coin two times and $0 otherwise. Then the *expected value* is

$$E = \$3.00 \cdot \tfrac{1}{4} + \$0 \cdot \tfrac{3}{4} = \$0.75$$

This means, if the game is to be fair, that you should be willing to pay $0.75 each time you toss the coins.

(c) A game consists of flipping a single coin. If a head shows, the player loses $1; but if a tail shows, the player wins $2. Thus half the time the player loses $1 and the other half the player wins $2. The expected value E of the game is

$$E = \$2 \cdot \tfrac{1}{2} + (-\$1) \cdot \tfrac{1}{2} = \$0.50$$

The player is expected to win an average of $0.50 on each play.

In each of the above examples, we arrive at the expected value E by multiplying the amount earned for a given result of the toss times the probability for that toss to occur, and adding all the products.

Look back at Example 1(a). In the expression for the expected value in the raffle problem, the term $\$300 \cdot \frac{1}{1000}$ is the pairing of the value $300 with its corresponding probability $\frac{1}{1000}$, namely, the probability of picking a $300 ticket. Likewise, the probability of getting a $1 ticket is $\frac{100}{1000}$ and this produces the term $\$1 \cdot \frac{100}{1000}$ in the expression for the expected value, E. So the expression for E is obtained by multiplying each ticket value by the probability of its occurrence and adding the results.

In Example 1(b), the term $\$3.00 \cdot \frac{1}{4}$ is the pairing of the value $3.00 with its corresponding probability $\frac{1}{4}$, namely, the probability of obtaining HH. Likewise, the probability of getting HT, TH, and TT is $\frac{3}{4}$ with payoff 0, and this produces $\$0 \cdot \frac{3}{4}$ in the expression for the expected value.

Finally, in Example 1(c) the term $\$2 \cdot \frac{1}{2}$ is the pairing of the value $2.00 with its corresponding probability $\frac{1}{2}$, namely, the probability of obtaining T. Likewise, the probability of getting H is $\frac{1}{2}$ with payoff $-\$1$.

This leads to the following definition.

Expected Value

Let S be a sample space and let $A_1, A_2, \ldots, A_n$ be n events of S that form a partition of S. Let $p_1, p_2, \ldots, p_n$ be the probabilities of the events $A_1, A_2, \ldots, A_n$, respectively. If each event $A_1, A_2, \ldots, A_n$ is assigned the payoff $m_1, m_2, \ldots, m_n$, respectively, the **expected value** E corresponding to these payoffs is

$$E = m_1 \cdot p_1 + m_2 \cdot p_2 + \cdots + m_n \cdot p_n$$

The term *expected value* should not be interpreted as a value that actually occurs in the experiment. In Example 1(a) there was no raffle ticket costing $.60. Rather, this number represents the average value of a raffle ticket.

In gambling, E is interpreted as the average winnings expected for the player in the long run. If E is positive, we say that the game is **favorable** to the player; if $E = 0$, we say the game is **fair**; and if E is negative, we say the game is **unfavorable** to the player.

When the payoff assigned to an outcome of an experiment is positive, it can be interpreted as a profit, winnings, gain, etc. When it is negative, it represents losses, penalties, deficits, etc.

The following steps outline the general procedure involved in determining expected value.

Steps for Computing Expected Value

Step 1 Partition S into n events $A_1, A_2, \ldots, A_n$.

Step 2 Determine the probability $p_1, p_2, \ldots, p_n$ of each event $A_1, A_2, \ldots, A_n$. Since these events constitute a partition of S, it follows that
$$p_1 + p_2 + \cdots + p_n = 1.$$

Step 3 Assign payoff values $m_1, m_2, \ldots, m_n$ to each event $A_1, A_2, \ldots, A_n$.

Step 4 Calculate $E = m_1 \cdot p_1 + m_2 \cdot p_2 + \cdots + m_n \cdot p_n$.

 **Now Work Problem 1**

EXAMPLE 2   Computing Expected Value

Consider the experiment of rolling a fair die. The player recovers an amount of dollars equal to the number of dots on the face that turns up, except when face 5 or 6 turns up, in which case the player will lose $5 or $6, respectively. What is the expected value of the game?

SOLUTION

Step 1 The sample space is $S = \{1, 2, 3, 4, 5, 6\}$. Each of these 6 outcomes constitutes an event that forms a partition of S.

Step 2 Since each outcome (event) is equally likely, the probability of each one is $\frac{1}{6}$.

Step 3 Since the player wins $1 for the outcome 1, $2 for the outcome 2, $3 for a 3, and $4 for a 4 and the player loses $5 for a 5 or equivalently wins $-\$5$ for a 5 and loses $6 for a 6 or equivalently wins $-\$6$ for a 6, we assign payoffs (the winnings of the player) as follows:

event	1	2	3	4	5	6
payoff (winnings)	$1	$2	$3	$4	$-\$5$	$-\$6$

Step 4 The expected value of the game is
$$E = \$1 \cdot \tfrac{1}{6} + \$2 \cdot \tfrac{1}{6} + \$3 \cdot \tfrac{1}{6} + \$4 \cdot \tfrac{1}{6} + (-\$5) \cdot \tfrac{1}{6} + (-\$6) \cdot \tfrac{1}{6}$$
$$= -\$\tfrac{1}{6} = -16.7\cent$$

The player would expect to lose an average of 16.7¢ on each throw.

EXAMPLE 3   Computing Expected Value

What is the expected number of heads in tossing a fair coin three times?

SOLUTION

Step 1 The sample space is

$$S = \{HHH, HHT, HTH, HTT, THH, THT, TTH, TTT\}$$

Since we are interested in the number of heads and since a coin tossed three times will result in 0, 1, 2, or 3 heads, we partition S as follows:

$$A_1 = \{TTT\} \quad A_2 = \{HTT, THT, TTH\} \quad A_3 = \{HHT, HTH, THH\} \quad A_4 = \{HHH\}$$
$$\underset{\text{0 heads}}{\qquad} \quad \underset{\text{1 head}}{\qquad} \quad \underset{\text{2 heads}}{\qquad} \quad \underset{\text{3 heads}}{\qquad}$$

Step 2 The probability of each of these events is

$$P(A_1) = \tfrac{1}{8}, \qquad P(A_2) = \tfrac{3}{8} \qquad P(A_3) = \tfrac{3}{8} \qquad P(A_4) = \tfrac{1}{8}$$

Step 3 Since we are interested in the expected number of heads, we assign payoffs of 0, 1, 2, and 3 respectively, to A_1, A_2, A_3, and A_4.

Step 4 The expected number of heads can now be found by multiplying each payoff by its corresponding probability and finding the sum of these values.

$$\text{Expected number of heads} = E = 0 \cdot \tfrac{1}{8} + 1 \cdot \tfrac{3}{8} + 2 \cdot \tfrac{3}{8} + 3 \cdot \tfrac{1}{8} = \tfrac{3}{2}$$
$$= 1.5$$

Thus, on the average, tossing a coin three times will result in 1.5 heads.

 **Now Work Problem 11**

EXAMPLE 4   Bidding for Oil Wells

An oil company may bid for only one of two contracts for oil drilling in two different locations. If oil is discovered at location I, the profit to the company will be $3,000,000. If no oil is found, the company's loss will be $250,000. If oil is discovered at location II, the profit will be $4,000,000. If no oil is found, the loss will be $500,000. The probability of discovering oil at location I is .7, and at location II it is .6. Which field should the company bid on; that is, for which location is expected profit highest?

SOLUTION In the first field the company expects to discover oil .7 of the time at a profit of $3,000,000. Thus it would not discover oil .3 of the time at a loss of $250,000. The expected profit E_{I} is therefore

$$E_{\text{I}} = (\$3,000,000)(.7) + (-\$250,000)(.3) = \$2,025,000$$

Similarly, for the second field, the expected profit E_{II} is

$$E_{\text{II}} = (\$4,000,000)(.6) + (-\$500,000)(.4) = \$2,200,000$$

Since the expected profit for the second field exceeds that for the first, the oil company should bid on the second field.

Now Work Problem 19

EXAMPLE 5   Evaluating Insurance

Mr. Richmond is producing an outdoor concert. He estimates that he will make $300,000 if it does not rain and make $60,000 if it does rain. The weather bureau predicts that the chance of rain is .34 for the day of the concert.

(a) What are Mr. Richmond's expected earnings from the concert?
(b) An insurance company is willing to insure the concert for $150,000 against rain for a premium of $30,000. If he buys this policy, what are his expected earnings from the concert?
(c) Based on the expected earnings, should Mr. Richmond buy an insurance policy?

SOLUTION   The sample space consists of two outcomes: R: Rain or N: No rain. The probability of rain is $P(R) = .34$; the probability of no rain is $P(N) = .66$.

(a) If it rains, the earnings are $60,000; if there is no rain, the earnings are $300,000. The expected earnings E are

$$E = 60,000 \cdot P(R) + 300,000 \cdot P(N)$$
$$= 60,000\,(.34) + 300,000\,(.66)$$
$$= \$218,400$$

(b) With insurance, if it rains, the earnings are

$$150,000 + 60,000 - 30,000 = 180,000$$

Insurance payout   Concert profit   Cost of insurance

With insurance, if it does not rain, the earnings are

$$0 + 300,000 - 30,000 = 270,000$$

No insurance payout   Concert profit   Cost of insurance

With insurance, the expected earnings E are

$$E = 180,000 \cdot P(R) + 270,000 \cdot P(N)$$
$$= 180,000\,(.34) + 270,000\,(.66)$$
$$= \$239,400$$

(c) Since the expected earnings are higher with insurance, it would be better to obtain the insurance.

EXAMPLE 6   Quality Control

A laboratory contains 10 electron microscopes, of which 2 are defective. If all microscopes are equally likely to be chosen and if 4 are chosen, what is the expected number of defective microscopes?

SOLUTION   The sample of 4 microscopes can contain 0, 1, or 2 defective microscopes.

The probability p_0 that none in the sample is defective is

$$p_0 = \frac{C(2, 0) \cdot C(8, 4)}{C(10, 4)} = \frac{1}{3}$$

Similarly, the probabilities p_1 and p_2 for 1 and 2 defective microscopes are

$$p_1 = \frac{C(2, 1) \cdot C(8, 3)}{C(10, 4)} = \frac{8}{15} \quad \text{and} \quad p_2 = \frac{C(2, 2) \cdot C(8, 2)}{C(10, 4)} = \frac{2}{15}$$

Since we are interested in determining the expected number of defective microscopes, we assign a payoff of 0 to the outcome "0 defectives are selected," a payoff of 1 to the outcome "1 defective is chosen," and a payoff of 2 to the outcome "2 defectives are chosen." The expected value E is then

$$E = 0 \cdot p_0 + 1 \cdot p_1 + 2 \cdot p_2 = \tfrac{8}{15} + \tfrac{4}{15} = \tfrac{4}{5}$$

Of course, we cannot have $\tfrac{4}{5}$ of a defective microscope. However, we can interpret this to mean that in the long run such a sample will average just under 1 defective microscope.

We point out that $\tfrac{4}{5}$ is a reasonable answer for the expected number of defective microscopes since $\tfrac{1}{5}$ of the microscopes in the laboratory are defective and we are selecting a random sample consisting of 4 of these microscopes.

Expected Value for Bernoulli Trials

In 100 tosses of a coin, what is the expected number of heads? If a student guesses at random on a true–false exam with 50 questions, what is her expected grade? These are specific instances of the following more general question:

In n trials of a Bernoulli process, what is the expected number of successes?

We now compute this expected value. As before, p denotes the probability of success on any individual trial.

If $n = 1$ (one trial), then the expected number of successes is

$$E = 1 \cdot p + 0 \cdot (1 - p) = p$$

If $n = 2$ (two trials), then either 0, 1, or 2 successes can occur. The expected number E of successes is

$$E = 2 \cdot p^2 + 1 \cdot 2p(1 - p) + 0 \cdot (1 - p)^2 = 2p$$

If $n = 3$ (three trials), then either 0, 1, 2, or 3 successes can occur. The expected number E of successes is

$$\begin{aligned} E &= 3 \cdot p^3 + 2 \cdot 3p^2(1 - p) + 1 \cdot 3p(1 - p)^2 + 0 \cdot (1 - p)^3 \\ &= 3p^3 + 6p^2 - 6p^3 + 3p - 6p^2 + 3p^3 \\ &= 3p \end{aligned}$$

This would seem to suggest that with n trials the expected value E would be given by $E = np$. This is indeed the case and we have the following result:

> **Expected Value for Bernoulli Trials**
>
> In a Bernoulli process with n trials the expected number of successes is
>
> $$E = np$$
>
> where p is the probability of success on any single trial.

A derivation of this result is included a little later in this section. The intuitive idea behind the result is fairly simple. Thinking of probabilities as percentages, if success results p percent of the time, then out of n attempts, p percent of them, namely, np, should be successful.

EXAMPLE 7   Expected Value in a Bernoulli Trial

In flipping a fair coin five times, what is the expected number of tails?

SOLUTION   The six events 0 tails, 1 tail, 2 tails, 3 tails, 4 tails, 5 tails are events that constitute a partition of the sample space. The respective probability of each of these events is

$$\binom{5}{0}\left(\frac{1}{2}\right)^5 \quad \binom{5}{1}\left(\frac{1}{2}\right)^5 \quad \binom{5}{2}\left(\frac{1}{2}\right)^5 \quad \binom{5}{3}\left(\frac{1}{2}\right)^5 \quad \binom{5}{4}\left(\frac{1}{2}\right)^5 \quad \binom{5}{5}\left(\frac{1}{2}\right)^5$$

If we assign the payoffs 0, 1, 2, 3, 4, 5 respectively to each event, then the expected number of tails is

$$E = 0 \cdot \binom{5}{0}\left(\frac{1}{2}\right)^5 + 1 \cdot \binom{5}{1}\left(\frac{1}{2}\right)^5 + 2 \cdot \binom{5}{2}\left(\frac{1}{2}\right)^5$$

$$+ 3 \cdot \binom{5}{3}\left(\frac{1}{2}\right)^5 + 4 \cdot \binom{5}{4}\left(\frac{1}{2}\right)^5 + 5 \cdot \binom{5}{5}\left(\frac{1}{2}\right)^5 = \frac{5}{2}$$

Using the result $E = np$ is much easier. For $n = 5$ and $p = \frac{1}{2}$, we obtain $E = (5)(\frac{1}{2}) = \frac{5}{2}$.   ■

EXAMPLE 8   Expected Value in a Bernoulli Trial

A multiple-choice test contains 100 questions, each with four choices. If a person guesses, what is the expected number of correct answers?

SOLUTION   This is an example of a Bernoulli trial. The probability for success (a correct answer) when guessing is $p = \frac{1}{4}$. Since there are $n = 100$ questions, the expected number of correct answers is

$$E = np = (100)(\tfrac{1}{4}) = 25$$ ■

✍ **Now Work Problem 21**

Derivation of $E = np$

The derivation is an exercise in handling binomial coefficients and using the binomial theorem. We will make use of the following identity:

$$k \binom{n}{k} = n \binom{n-1}{k-1} \tag{1}$$

Equation (1) can be established by expanding both sides using the definition of a binomial coefficient. (See Problem 21, Exercise 6.7, page 312.)

Recall that the probability of obtaining exactly k successes in n trials is given by $b(n, k; p) = \binom{n}{k}p^k q^{n-k}$. Thus the expected number of successes is

$$E = 0 \cdot \binom{n}{0} p^0 q^n + 1 \cdot \binom{n}{1} p^1 q^{n-1} + 2 \cdot \binom{n}{2} p^2 q^{n-2}$$

$$+ \cdots + k \underbrace{\binom{n}{k} p^k q^{n-k}}_{} + \cdots + n \binom{n}{n} p^n q^0$$

No. of   Corresponding
successes   probability

Using Equation (1) above,

$$E = n \binom{n-1}{0} pq^{n-1} + n \binom{n-1}{1} p^2 q^{n-2} + \cdots$$

$$+ n \binom{n-1}{k-1} p^k q^{n-k} + \cdots + n \binom{n-1}{n-1} p^n$$

Now factor out an n and a p from each of the terms on the right to get

$$E = np \left[\binom{n-1}{0} q^{n-1} + \binom{n-1}{1} pq^{n-2} + \cdots \right.$$

$$\left. + \binom{n-1}{k-1} p^{k-1} q^{(n-1)-(k-1)} + \cdots + \binom{n-1}{n-1} p^{n-1} \right]$$

The expression in brackets is $(q + p)^{n-1}$. To see why, use the binomial theorem. Thus

$$E = np(q + p)^{n-1}$$

Since $p + q = 1$, the result $E = np$ follows.

EXERCISE 8.3   Answers to odd-numbered problems begin on page AN-29.

---

1. For the data given below, compute the expected value.

Outcome	e_1	e_2	e_3	e_4
Probability	.4	.2	.1	.3
Payoff	2	3	−2	0

2. For the data below, compute the expected value.

Outcome	e_1	e_2	e_3	e_4
Probability	$\frac{1}{3}$	$\frac{1}{6}$	$\frac{1}{4}$	$\frac{1}{4}$
Payoff	1	0	4	−2

3. Attendance at a football game in a certain city results in the following pattern. If it is extremely cold, the attendance will be 30,000; if it is cold, it will be 40,000; if it is moderate, 60,000; and if it is warm, 80,000. If the probabilities for extremely cold, cold, moderate, and warm are .08, .42, .42, and .08, respectively, how many fans are expected to attend the game?

4. A player rolls a fair die and receives a number of dollars equal to the number of dots appearing on the face of the die. What is the least the player should expect to pay in order to play the game?

5. Mary will win $8 if she draws an ace from a set of 10 different cards from ace to 10. How much should she pay for one draw?

6. Thirteen playing cards, ace through king, are placed randomly with faces down on a table. The prize for guessing correctly the value of any given card is $1. What would be a fair price to pay for a guess?

7. David gets $10 if he throws a double on a single throw of a pair of dice. How much should he pay for a throw?

8. You pay $1 to toss 2 coins. If you toss 2 heads, you get $2 (including your $1); if you toss only 1 head, you get back your $1; and if you toss no heads, you lose your $1. Is this a fair game to play?

9. **Raffles** In a raffle 1000 tickets are being sold at $1.00 each. The first prize is $100, and there are 3 second prizes of $50 each. By how much does the price of a ticket exceed its expected value?

10. **Raffles** In a raffle 1000 tickets are being sold at $1.00 each. The first prize is $100. There are 2 second prizes of $50 each, and 5 third prizes of $10 each (there are 8 prizes in all). Jenny buys 1 ticket. How much more than the expected value of the ticket does she pay?

11. A fair coin is tossed 3 times, and a player wins $3 if 3 tails occur, wins $2 if 2 tails occur, and loses $3 if no tails occur. If 1 tail occurs, no one wins.

   (a) What is the expected value of the game?
   (b) Is the game fair?
   (c) If the answer to part (b) is "No," how much should the player win or lose for a toss of exactly 1 tail to make the game fair?

12. Colleen bets $1 on a 2-digit number. She wins $75 if she draws her number from the set of all 2-digit numbers, {00, 01, 02, . . . , 99}; otherwise, she loses her $1.

   (a) Is this game fair to the player?
   (b) How much is Colleen expected to lose in a game?

13. Two teams have played each other 14 times. Team A won 9 games, and team B won 5 games. They will play again next week. Bob offers to bet $6 on team A while you bet $4 on team B. The winner gets the $10. Is the bet fair to you in view of the past records of the two teams? Explain your answer.

14. A department store wants to sell 11 purses that cost the store $40 each and 32 purses that cost the store $10 each. If all purses are wrapped in 43 identical boxes and if each customer picks a box randomly, find

   (a) Each customer's expectation.
   (b) The department store's expected profit if it charges $15 for each box.

15. Sarah draws a card from a deck of 52 cards. She receives 40¢ for a heart, 50¢ for an ace, and 90¢ for the ace of hearts. If the cost of a draw is 15¢, should she play the game? Explain.

16. **Family Size** The following data give information about family size in the United States for a household containing a husband and wife where the husband is in the 30–34 age bracket:

Number of Children	0	1	2	3
Proportion of Families	10.2%	15.9%	31.8%	42.1%

A family is chosen at random. Find the expected number of children in the family.

17. Assume that the odds for a certain race horse to win are 7 to 5. If a bettor wins $5 when the horse wins, how much should he bet to make the game fair?

18. **Roulette** In roulette there are 38 equally likely possibilities: the numbers 1–36, 0, and 00 (double zero). See the figure on page 396. What is the expected value for a gambler who bets $1 on number 15 if she wins $35 each time the number 15 turns up and loses $1 if any other number turns up? If the gambler plays the number 15 for 200 consecutive times, what is the total expected gain?

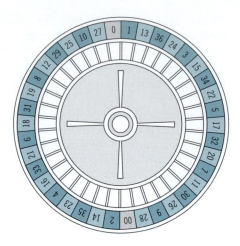

19. Site Selection   A company operating a chain of super-markets plans to open a new store in 1 of 2 locations. They conducted a survey of the 2 locations and estimated that the first location will show an annual profit of $15,000 if it is successful and a $3000 loss otherwise. For the second location, the estimated annual profit is $20,000 if successful and a $6000 loss otherwise. The probability of success at each location is $\frac{1}{2}$. What location should the management decide on in order to maximize its expected profit?

20. For Problem 19 assume the probability of success at the first location is $\frac{2}{3}$ and at the second location is $\frac{1}{3}$. What location should be chosen?

21. Find the number of times the face 5 is expected to occur in a sequence of 2000 throws of a fair die.

22. What is the expected number of tails that will turn up if a fair coin is tossed 582 times?

23. A certain kind of light bulb has been found to have a .02 probability of being defective. A shop owner receives 500 light bulbs of this kind. How many of these bulbs are expected to be defective?

24. A student enrolled in a math course has a .9 probability of passing the course. In a class of 20 students, how many would you expect to fail the math course?

25. Drug Reaction   A doctor has found that the probability that a patient who is given a certain drug will have un-

favorable reactions to the drug is .002. If a group of 500 patients is going to be given the drug, how many of them does the doctor expect to have unfavorable reactions?

26. A true–false test consisting of 30 questions is scored by subtracting the number of wrong answers from the number of right ones. Find the expected number of correct answers of a student who just guesses on each question. What will the test score be?

27. A coin is weighted so that $P(H) = \frac{1}{4}$ and $P(T) = \frac{3}{4}$. Find the expected number of tosses of the coin required in order to obtain either a head or 4 tails.

28. A box contains 3 defective bulbs and 9 good bulbs. If 5 bulbs are drawn from the box without replacement, what is the expected number of defective bulbs?

29. Aircraft Use   An airline must decide which of two aircraft it will use on a flight from New York to Los Angeles. Aircraft A has a seating capacity of 200, while aircraft B has a capacity of 300. Previous experience has allowed the airline to estimate the number of passengers on the flight as follows:

Number of Passengers	150	180	200	250	300
Probability	.2	.3	.2	.2	.1

Regardless of aircraft used, the average cost of a ticket is $500, but there are different operating costs attached to each aircraft. There is a fixed cost (fuel, crew, etc.) of $16,000 attached to using aircraft A, while aircraft B has a fixed cost of $18,000. There is also a per passenger cost (meals, luggage, added fuel) of $200 for aircraft A and $230 for aircraft B. Which aircraft should the airline schedule so that it maximizes its profit on the flight?

30. Prove that if the numerical values assigned to the outcomes of an experiment that has expected value E are all multiplied by the constant k, then the expected value of the new experiment is $k \cdot E$. Similarly, if to all the numerical values we add the same constant k, prove that the expected value of the new experiment is $E + k$.

8.4 APPLICATIONS

The field of **operations research,** the science of making optimal or best decisions, has experienced remarkable growth and development since the 1940s. Let's work some examples from operations research that utilize expected value.

EXAMPLE 1   Car Rentals

A national car rental agency rents cars for $16 per day (gasoline and mileage are additional expenses to the customer). The daily cost per car (for example, lease costs and overhead) is $6 per day. The daily profit to the company is $10 per car if the car is rented, and the company incurs a daily loss of $6 per car if the car is not rented. The daily profit depends on two factors: the demand for cars and the number of cars the company has available to rent. Previous rental records show that the daily demand is as given in the table:

Number of Customers	8	9	10	11	12
Probability	.10	.10	.30	.30	.20

Find the expected number of customers and determine the optimal number of cars the company should have available for rental. (This is the number that yields the largest expected profit.)

SOLUTION   The expected number of customers is

$$8(.1) + 9(.1) + 10(.3) + 11(.3) + 12(.2) = 10.4$$

If 10.4 customers are expected, how many cars should be on hand? Surely, the number should probably not exceed 11 since fewer than 11 customers are expected. However, the number may not be the integer closest to 10.4 since costs play a major role in the determination of profit. We need to compute the expected profit for each possible number of cars. The largest expected profit will tell us how many cars to have on hand.

For example, if there are 10 cars available, the expected profit for 8, 9, or 10 customers is

$$68(.1) + 84(.1) + 100(.8) = \$95.20$$

We obtain the entry 68(.1) by noting that the 10 cars cost the company $60, and 8 cars rented with probability .10 bring in $128, for a profit for $68. Similarly, we obtain the entry 84(.1) by noting that the 10 cars cost the company $60, and 9 cars rented with probability .10 bring in $144, for a profit of $84. The entry 100(.8) is obtained since for 10 or more customers (probability .3 + .3 + .2 = .8) the profit is

$$10 \times \$16 - \$60 = \$100.$$

The table lists the expected profit for 8 to 12 available cars. Clearly, the optimal stock size is 11 cars, since this number of cars maximizes expected profit.

Number of Available Cars	8	9	10	11	12
Expected Profit	$80.00	$88.40	$95.20	$97.20	$94.40

✍ **Now Work Problem 1**

EXAMPLE 2   Quality Control

Figure 12

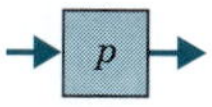

Figure 13

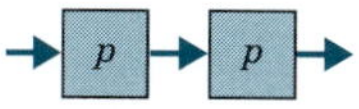

A factory produces electronic components, and each component must be tested. If the component is good, it will allow the passage of current; if the component is defective, it will block the passage of current. Let p denote the probability that a component is good. See Figure 12. With this system of testing, a large number of components requires an equal number of tests. This increases the production cost of the electronic components since it requires one test per component. To reduce the number of tests, a quality control engineer proposes, instead, a new testing procedure: Connect the components pairwise in series, as shown in Figure 13.

If the current passes two components in series, then both components are good and only one test is required. The probability that two components are good is p^2. If the current does not pass, the components must be sent individually to the quality control department, where each component is tested separately. In this case three tests are required. The probability that three tests are needed is $1 - p^2$ (1 minus probability of success p^2). The expected number of tests for a pair of components is

$$E = 1 \cdot p^2 + 3 \cdot (1 - p^2) = p^2 + 3 - 3p^2 = 3 - 2p^2$$

The number of tests saved for a pair is

$$2 - (3 - 2p^2) = 2p^2 - 1$$

The number of tests saved per component is

$$\frac{2p^2 - 1}{2} = p^2 - \frac{1}{2} \text{ tests}$$

The greater the probability p that the component is good, the greater the saving. For example, if p is almost 1, we have a saving of almost $1 - \frac{1}{2}$ or $\frac{1}{2}$, which is 50% of the original number of tests needed. Of course, if p is small, say, less than .7, we do not save anything since $(.7)^2 - \frac{1}{2}$ is less than 0, and we are wasting tests.   ■

If the reliability of the components manufactured in Example 2 is very high, it might even be advisable to make larger groups. Suppose three components are connected in series. See Figure 14.

Figure 14

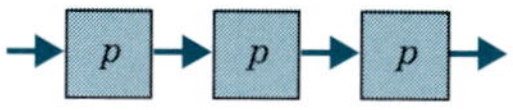

For individual testing we need three tests. For group testing we have

    1   test needed with probability p^3

    4   tests needed with probability $1 - p^3$

The expected number of tests is

$$E = 1 \cdot p^3 + 4 \cdot (1 - p^3) = 4 - 3p^3 \text{ tests}$$

The number of tests saved per component is

$$\frac{3p^3 - 1}{3} = p^3 - \frac{1}{3} \text{ tests}$$

In a similar way, we can show that if the components are arranged in groups of four connected in series, then the number of tests saved per component is

$$p^4 - \frac{1}{4} \text{ tests}$$

In general, for groups of n, the number of tests saved per component is

$$p^n - \frac{1}{n} \text{ tests}$$

Notice from the above formula that as n, the group size, gets very large, the number of tests saved per component gets very, very small.

To determine the optimal group size for $p = .9$, we refer to Table 1. From the table we can see that the optimal group size is four, resulting in a substantial saving of approximately 41%.

Table 1

Group Size	Expected Tests Saved per Component $p = .9$	Percent Saving
2	$p^2 - \frac{1}{2} = .81 - .50 = .31$	31
3	$p^3 - \frac{1}{3} = .729 - .333 = .396$	39.6
4	$p^4 - \frac{1}{4} = .6561 - .25 = .4061$	40.61
5	$p^5 - \frac{1}{5} = .59049 - .2 = .39049$	39.05
6	$p^6 - \frac{1}{6} = .531 - .167 = .364$	36.4
7	$p^7 - \frac{1}{7} = .478 - .143 = .335$	33.5
8	$p^8 - \frac{1}{8} = .430 - .125 = .305$	30.5

We also note that larger group sizes do not increase savings.

 Now Work Problem 3

EXAMPLE 3   Optimizing Hiring

A $75,000 oil detector is lowered under the sea to detect oil fields, and it becomes detached from the ship. If the instrument is not found within 24 hours, it will crack under the pressure of the sea. It is assumed that a scuba diver will find it with probability .85, but it costs $500 to hire each diver. How many scuba divers should be hired?

SOLUTION   Let's assume that x scuba divers are hired. The probability that they will fail to discover the instrument is $.15^x$. Thus the instrument will be found with probability $1 - .15^x$.

The expected gain from hiring the scuba divers is

$$\$75,000(1 - .15^x) = \$75,000 - \$75,000(.15^x)$$

while the cost for hiring them is

$$\$500 \cdot x$$

Thus the expected net gain, denoted by $E(x)$, is

$$E(x) = \$75{,}000 - \$75{,}000(.15^x) - \$500x$$

The problem is then to choose x so that $E(x)$ is maximum.
We begin by evaluating $E(x)$ for various values of x:

$$E(1) = \$75{,}000 - \$75{,}000(.15^1) - \$500(1) = \$63{,}250.00$$
$$E(2) = \$75{,}000 - \$75{,}000(.15^2) - \$500(2) = \$72{,}312.50$$
$$E(3) = \$75{,}000 - \$75{,}000(.15^3) - \$500(3) = \$73{,}246.88$$
$$E(4) = \$75{,}000 - \$75{,}000(.15^4) - \$500(4) = \$72{,}962.03$$
$$E(5) = \$75{,}000 - \$75{,}000(.15^5) - \$500(5) = \$72{,}494.30$$
$$E(6) = \$75{,}000 - \$75{,}000(.15^6) - \$500(6) = \$71{,}999.15$$

Thus the expected net gain is optimal when three divers are hired. Note that hiring additional divers does not necessarily increase expected net gain. In fact, the expected net gain declines if more than three divers are hired.

 **Now Work Problem 5**

Error Correction in the Digital Electronic Transmission of Data

Electronically transmitted data, be it from computer to computer or from a satellite to a ground station, is normally in the form of strings of 0s and 1s—that is, in binary form. Bursts of noise or faults in relays, for example, may at times garble the transmission and produce errors so that the message received is not the same as the one originally sent. For example,

001	⊢─◁◁◁→	101
Message	Noisy	Message
sent	channel	received

is a transmission where the message received is in error since the initial 0 has been changed to a 1.

A naive way of trying to protect against such error would be to repeat the message. So instead of transmitting 001, we would send 001001. Then, were the same error to creep in as it did before, the received message would be

$$101001$$

The receiver would certainly know an error had occurred since the last half of the message is not a duplicate of the first half. But she would have no way of recovering the original message since she would not know where the error happened. For example, she would not be able to distinguish between the two messages

$$001001 \quad \text{and} \quad 101101$$

There are more sophisticated ways of coding binary data with redundancy that not only allow the detection of errors but simultaneously permit their location and correction so that the original message can be recovered. These are referred to as *error-correcting codes* and are commonly used today in computer-implemented transmissions. One such is the (7, 4) Hamming code named after Richard Hamming, a former researcher at AT&T Laboratories. It is a code of length seven, meaning that an individual message is a string consisting of seven items, each of which is either a 0 or 1. (The four refers to the fact that the first four elements in the string can be freely chosen by the sender, while the remaining three are determined by a fixed rule and constitute the redundancy that gives the code its error correction capability.) The (7, 4) Hamming code is capable of locating and correcting a single error. That is, if during transmission a 1 has been changed to a 0 or vice versa in one of the seven locations, the Hamming code is capable of detecting and correcting this. While we will not explain how or why the Hamming code works, we will analyze the benefit obtained by its use.

By an error we mean that an individual 1 has been changed to a 0 or that a 0 has been changed to a 1. We assume that the probability of an error happening remains constant during transmission, and we designate this probability by q. Thus $p = 1 - q$ is the probability that an individual symbol remains unchanged. This is summarized in the diagram. (In practice, values of q are normally small and values of p close to 1 since we would normally be using a relatively reliable channel.)

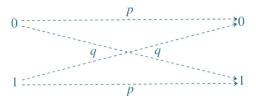

We also assume that errors occur randomly and independently. In short, we can think of the transmission of a binary string of length seven as a Bernoulli trial, with failure corresponding to a symbol being received in error.

For a message of length seven, if *no* coding were used, the probability that the receiver would get the correct message would be

$$b(7, 7; p) = p^7$$

since none of the seven symbols could have been altered. For $p = .98$, this gives $(.98)^7 = .8681$.

Using the Hamming code, the receiver will get the correct message even if one error has occurred. Hence, the corresponding probability of correct reception would be

$$\underbrace{b(7, 7; p)}_{\text{No errors}} + \underbrace{b(7, 6; p)}_{\text{1 error}} = p^7 + 7p^6q$$

For $p = .98$ this now gives $.8681 + .1240 = .9921$, which shows a considerable improvement.

There are codes in use that correct more than a single error. One such code, known by the initials of its originators as a BCH code, is a code of length 15 (messages are binary strings of length 15) that corrects up to two errors. Sending a message of length 15 with no attempt at coding would result in a probability of $b(15, 15; p) = p^{15}$ of the correct message being received. For $p = .98$ this gives $b(15, 15; .98) = .7386$. Using the BCH code that can correct two or fewer errors, the probability that a message will be correctly received becomes

$$\underbrace{b(15, 15; p)}_{\text{No errors}} + \underbrace{b(15, 14; p)}_{\text{1 error}} + \underbrace{b(15, 13; p)}_{\text{2 errors}}$$

Evaluating this for $p = .98$ we get

$$.7386 + .2261 + .0323 = .9970$$

Codes that can correct a high number of errors are clearly very desirable. Yet a basic result in the theory of codes states that as the error-correcting capability of a code increases, so, of necessity, must its length. But lengthier codes require more time for transmission and are clumsier to use. Thus, here, speed and correctness are at odds.

EXERCISE 8.4   Answers to odd-numbered problems begin on page AN-29.

1. **Market Assessment**   A car agency has fixed costs of $10 per car per day and the revenue for each car rented is $30 per day. The daily demand is given in the table:

Number of Customers	7	8	9	10	11
Probability	.10	.20	.40	.20	.10

Find the expected number of customers. Determine the optimal number of cars the company should have on hand each day. What is the expected profit in this case?

2. In Example 2 suppose $p = .8$. Show that the optimal group size is 3.

3. In Example 2 suppose $p = .95$. Show that the optimal group size is 5.

4. In Example 2 suppose $p = .99$. Compute savings for group sizes 10, 11, and 12, and thus show that 11 is the optimal group size. Determine the percent saving.

5. In Example 3 suppose the probability of the scuba divers discovering the instrument is .95. Find
   (a) An equation expressing the net expected gain.
   (b) The number x of scuba divers that maximizes the net gain.

6. In Example 2 compute the expected number of tests saved per component if on the first test the current does not pass through 2 components in series, but on the second test the current does pass through 1 of them. A third test is not made (since the other component is obviously defective).

7. **Hamming Code**   There is a Hamming code of length 15 that corrects a single error. Assuming $p = .98$, find the probability that a message transmitted using this code will be correctly received.

8. **Golay Code**   There is a binary code of length 23 (called the Golay code) that can correct up to 3 errors. With $p = .98$, find the probability that a message transmitted using the Golay code will be correctly received.

8.5 RANDOM VARIABLES

When we perform an experiment, we are often interested not in a particular outcome, but rather in some number associated with that outcome. For example, in tossing a coin three times we may be interested in the number of heads obtained. Similarly, the gambler throwing a pair of dice in a crap game is interested in the sum of the faces rather than the particular number on each face.

In each of these examples we are interested in numbers that are associated with experimental outcomes. This process of assigning a number to each outcome is called *random variable* assignment.

Random Variable

A **random variable** is a rule that assigns a number to each outcome of an experiment.

Consider the experiment of tossing a fair coin three times. Table 2 shows the outcomes in the sample space and the number of heads associated with each outcome.

Table 3 shows the probability of the events 0 heads, 1 head, 2 heads, and 3 heads for this experiment.

Table 2

Sample Space	Number of Heads
e_1: HHH	3
e_2: HHT	2
e_3: HTH	2
e_4: THH	2
e_5: HTT	1
e_6: THT	1
e_7: TTH	1
e_8: TTT	0

Table 3

Number of Heads Obtained in Three Flips of a Coin	Probability
0	$\frac{1}{8}$
1	$\frac{3}{8}$
2	$\frac{3}{8}$
3	$\frac{1}{8}$

The role of the random variable is to transform the original sample space {*HHH, HHT, HTH, HTT, THH, THT, TTH, TTT* } into a new sample space that consists of the number of heads that occur: {0, 1, 2, 3}. If X denotes the random variable, then

$$X(e_1) = X(HHH) = 3 \qquad X(e_2) = X(HHT) = 2 \qquad X(e_3) = X(HTH) = 2$$
$$X(e_4) = X(THH) = 2 \qquad X(e_5) = X(HTT) = 1 \qquad X(e_6) = X(THT) = 1$$
$$X(e_7) = X(TTH) = 1 \qquad X(e_8) = X(TTT) = 0$$

Based on Table 3, we may write

$$\text{Probability}(X = 0) = \tfrac{1}{8} \qquad \text{Probability}(X = 1) = \tfrac{3}{8}$$
$$\text{Probability}(X = 2) = \tfrac{3}{8} \qquad \text{Probability}(X = 3) = \tfrac{1}{8}$$

Probability Distribution

The discussion thus far illustrates two features of a random variable:

1. The values the random variable can assume.
2. The probabilities associated with each of these values.

Because values assumed by a random variable can be used to symbolize all outcomes associated with a given experiment, the probability assigned to a simple event can now be assigned as the likelihood that the random variable takes on the corresponding value.

If a random variable X has the values

$$x_1, x_2, x_3, \ldots, x_n \tag{1}$$

then the rule given by

$$p(x) = P(X = x)$$

where x assumes the values in (1) is called the **probability distribution** of X, or simply, the distribution of X.

It follows that $p(x)$ is a probability distribution of X if it satisfies the following two conditions:

1. $0 \leq p(x) \leq 1$ for $x = x_1, x = x_2, \ldots, x = x_n$

2. $p(x_1) + p(x_2) + \cdots + p(x_n) = 1$

where $x_1, x_2, \ldots, x_n$ are the values of X.

Suppose the probability distribution for the random variable X is given as

x_i	x_1	x_2	$\ldots$	x_n
p_i	p_1	p_2	$\ldots$	p_n

where $p_i = p(x_i)$. Then the **expected value of X,** denoted by $E(X)$, is

$$E(X) = x_1 p_1 + x_2 p_2 + \cdots + x_n p_n$$

Consider again the experiment of tossing a fair coin three times and denote the probability distribution by

$$p(x) \qquad \text{where } x = 0, 1, 2, \text{ or } 3$$

For instance, $p(3)$ is the probability of getting exactly three heads, that is,

$$p(3) = P(X = 3) = \tfrac{1}{8}$$

In a similar way, we define $p(0)$, $p(1)$, and $p(2)$.

Probability distributions may also be represented graphically, as shown in Figure 15. The graph of a probability distribution is often called a *histogram.*

Figure 15

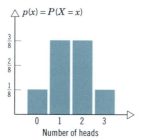

Binomial Probability Using Probability Distribution

We can now state the formula developed for the binomial probability model in terms of its probability distribution.

The binomial probability that assigns probabilities to the number of successes in n trials is an example of a probability distribution. For this distribution X is the random variable whose value for any outcome of the experiment is the number of successes obtained. For this distribution we may write

$$P(X = k) = b(n, k; p) = \binom{n}{k} p^k q^{n-k}$$

where $P(X = k)$ denotes the probability that the random variable equals k, that is, that exactly k successes are obtained.

EXERCISE 8.5   Answers to odd-numbered problems begin on page AN-29.

In Problems 1−6 list the values of the given random variable X together with the probability distributions.

1. A fair coin is tossed two times and X is the random variable whose value for an element in the sample space is the number of heads obtained.

2. A fair die is tossed once. The random variable X is the number showing on the top face.

3. The random variable X is the number of female children in a family with 3 children. (Assume the probability of a female birth is $\frac{1}{2}$).

4. A job applicant takes a 3-question true−false examination and guesses on each question. Let X be the number of right answers minus the number of wrong answers.

5. An urn contains 4 red balls and 6 white balls. Three balls are drawn with replacement. The random variable X is the number of red balls.

6. A couple getting married will have 3 children, and the random variable X denotes the number of boys they will have.

7. For the data given below, compute the expected value.

Outcome	e_1	e_2	e_3	e_4
Probability	.4	.2	.1	.3
x_i	2	3	−2	0

8. For the data below, compute the expected value.

Outcome	e_1	e_2	e_3	e_4
Probability	$\frac{1}{3}$	$\frac{1}{6}$	$\frac{1}{4}$	$\frac{1}{4}$
x_i	1	0	4	−2

Technology Exercises

Most graphing calculators have a random number function (usually RAND or RND) generating numbers between 0 and 1. Check your user's manual to see how to use this function on your graphing calculator.

Sometimes experiments are simulated using a random number function instead of actually performing the experiment. In Problems 1−6 use a random number function to simulate each experiment.

1. **Rolling a Fair Die**   Consider the experiment of rolling a die, and let the random variable X denote the number showing on the top face. Simulate the experiment using a random number function on your calculator, considering a roll to have the outcome k if the value of the random number function is between $(k − 1) \cdot 0.167$ and $k \cdot 0.167$. Record the outcome. Repeat the experiment 50 times, thus obtaining a sequence of 50 numbers. [*Note:* Most calculators repeat the action of the last entry if you simply press the ENTER, or EXE, key again.] Using these 50 numbers you can estimate the probability $P(X = k)$, for $k = 1, 2, 3, 4, 5, 6$, by the ratio

$$\frac{\text{Number of times } k \text{ appears in your sequence}}{50}$$

Enter your estimates in the table. Calculate the actual probabilities, and enter these numbers in the table. How close are your numbers to the actual values?

k	Your Estimate of $P(X = k)$	Actual Value of $P(X = k)$
1		
2		
3		
4		
5		
6		

2. Use a random number function to select a value for the random variable X. Repeat this experiment 50 times. Count the number of times the random variable X is between 0.6 and 0.9. Calculate the ratio

$$R = \frac{\text{Number of times the random variable } X \text{ is between 0.6 and 0.9}}{50}$$

What value of R did you obtain? Calculate the actual probability $P(0.6 \le X < 0.9)$.

3. Use a random number function to select a value for the random variable X. Repeat this experiment 50 times. Count the number of times the random variable X is between 0.1 and 0.3. Calculate the ratio

$$R = \frac{\text{Number of times the random variable } X \text{ is between 0.1 and 0.3}}{50}$$

What value of R did you obtain? Calculate the actual probability $P(0.1 \le X < 0.3)$.

4. **Rolling an Octahedron** Consider the experiment of rolling an octahedron (a regular polyhedron with all 8 faces congruent), and let the random variable X denote the number showing on the top face. Simulate the experiment using a random number function on your calculator, considering a roll to have the outcome k if the value of the random number function is between $(k - 1)/8$ and $k/8$, for $k = 1, 2, 3, 4, 5, \ldots, 8$. Record the outcome. Repeat the experiment 50 times, thus obtaining a sequence of 50 numbers. Using

k	Your Estimate of $P(X = k)$	Actual Value of $P(X = k)$
1		
2		
3		
4		
5		
6		
7		
8		

these 50 numbers you can estimate the probability $P(X = k)$, for $k = 1, 2, 3, 4, 5, 6, 7, 8$, by the ratio

$$\frac{\text{Number of times } k \text{ appears in your sequence}}{50}$$

Enter your estimates in the table shown. Calculate the actual probabilities, and enter these numbers in the table. How close are your numbers to the actual values?

5. **Rolling a Dodecahedron** Consider the experiment of rolling a dodecahedron (a regular polyhedron with all 12 faces congruent), and let the random variable X denote the number showing on the top face. Simulate the experiment using a random number function on your calculator, considering a roll to have the outcome k if the value of the random number function is between $(k - 1)/12$ and $k/12$, for $k = 1, 2, 3, 4, 5, \ldots, 12$. Record the outcome. Repeat the experiment 50 times, thus obtaining a sequence of 50 numbers. Using these 50 numbers you can estimate the probability $P(X = 2)$ by the ratio

$$\frac{\text{Number of times 2 appears in your sequence}}{50}$$

Calculate the actual probability, $P(X = 2)$, and compare these values. How close is your estimate to the actual value?

6. **Rolling an Icosahedron** Consider the experiment of rolling an icosahedron (a regular polyhedron with all 20 faces congruent), and let the random variable X denote the number showing on the top face. Simulate the experiment using a random number function on your calculator, considering a roll to have the outcome k if the value of the random number function is between $(k - 1)/20$ and $k/20$, for $k = 1, 2, 3, 4, 5, \ldots, 20$. Record the outcome. Repeat the experiment 50 times, thus obtaining a sequence of 50 numbers. Using these 50 numbers you can estimate the probability $P(X = 5)$ by the ratio

$$\frac{\text{Number of times 5 appears in your sequence}}{50}$$

Calculate the actual probability, $P(X = 5)$, and compare these values. How close is your estimate to the actual value?

CHAPTER REVIEW

IMPORTANT TERMS AND CONCEPTS

partition of a sample space 371
Bayes' formula 372
a priori/a posteriori probabilities 374

Bernoulli trials 379
formula for $b(n, k; p)$ 381
expected value 388

steps for computing expected value 389
random variable 402
probability distribution 403

IMPORTANT FORMULAS

Probability of an Event E in a Partitioned Sample Space

$$P(E) = P(A_1) \cdot P(E|A_1) + P(A_2) \cdot P(E|A_2)$$
$$+ P(A_3) \cdot P(E|A_3) + \cdots + P(A_n) \cdot P(E|A_n)$$

Binomial Probability Formula

$$b(n, k; p) = \binom{n}{k} p^k q^{n-k}, \quad q = 1 - p$$

Bayes' Formula

$$P(A_j|E) = \frac{P(A_j) \cdot P(E|A_j)}{P(E)}$$

Expected Value for Bernoulli Trials

$$E = np$$

TRUE–FALSE ITEMS   Answers are on page AN-29.

T_____ F_____ **1.** $b(3, 2; \frac{1}{2}) = 3 \cdot (\frac{1}{2})^3$

T_____ F_____ **2.** $P(A_1|E) = P(A_1)P(E)$

T_____ F_____ **3.** The expected value of an experiment is never negative.

T_____ F_____ **4.** Bayes' formula is useful for computing a posteriori probability.

T_____ F_____ **5.** $b(n, k; p)$ gives the probability of exactly n successes in k trials.

T_____ F_____ **6.** In flipping a fair coin 10 times, the expected number of heads is 5.

FILL IN THE BLANKS   Answers are on page AN-29.

1. The formula

$$P(A_1|E) = \frac{P(A_1) \cdot P(E|A_1)}{P(E)}$$

is called _____ _____ .

2. Random experiments are called Bernoulli trials if

(a) The same experiment is repeated several times.
(b) There are only two possible outcomes, success and failure.
(c) The repeated trials are _____ .
(d) The probability of each outcome remains _____ for each trial.

3. If an experiment has n outcomes that are assigned the payoffs $m_1, m_2, \ldots, m_n$, occurring with probabilities $p_1, p_2, \ldots, p_n$, then $E = m_1 p_1 + m_2 p_2 + \cdots + m_n p_n$ is the _____ _____ .

4. A random variable on a sample space S is a rule that assigns _____ _____ to each element in S.

5. If X is a random variable assuming the values $x_1, x_2, \ldots x_n$, then $E(X) = x_1 p(x_1) + x_2 p(x_2) + \cdots + x_n p(x_n)$ is the _____ _____ of X.

REVIEW EXERCISES   Answers to odd-numbered problems begin on page AN-29.

In Problems 1–14 on page 408 use the tree diagram below to find the indicated probability:

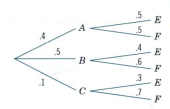

1. $P(E|A)$　　　2. $P(F|A)$　　　3. $P(E|B)$　　　4. $P(F|B)$

5. $P(E|C)$　　　6. $P(F|C)$　　　7. $P(E)$　　　8. $P(F)$

9. $P(A|E)$　　　10. $P(A|F)$　　　11. $P(B|E)$　　　12. $P(B|F)$

13. $P(C|E)$　　　14. $P(C|F)$

15. The table below indicates a survey conducted by a deodorant producer:

	Like the Deodorant	Did Not Like the Deodorant	No Opinion
Group I	180	60	20
Group II	110	85	12
Group III	55	65	7

Let the events E, F, G, H, and K be defined as follows:

　E:　Customer likes the deodorant

　F:　Customer does not like the deodorant

　G:　Customer is from group I

　H:　Customer is from group II

　K:　Customer is from group III

Find

(a) $P(E|G)$　　　(b) $P(G|E)$　　　(c) $P(H|E)$

(d) $P(K|E)$　　　(e) $P(F|G)$　　　(f) $P(G|F)$

(g) $P(H|F)$　　　(h) $P(K|F)$

16. **Quality Control**   Three machines in a factory, A_1, A_2, A_3, produce 55%, 30%, and 15% of total production, respectively. The percentage of defective output of these machines is 1%, 2%, and 3%, respectively. An item is chosen at random and it is defective. What is the probability that it came from machine A_1? From A_2? From A_3?

17. **Test for Cancer**   A lung cancer test has been found to have the following reliability. The test detects 85% of the people who have cancer and does not detect 15% of these people. Among the noncancerous group it detects 92% of the people not having cancer, whereas 8% of this group are detected erroneously as having lung cancer. Statistics show that about 1.8% of the population has cancer. Suppose an individual is given the test for lung cancer and it detects the disease. What is the probability that the person actually has cancer?

18. What is the expected number of girls in families having exactly 3 children?

19. **Advertising**   Management believes that 1 out of 5 people watching a television advertisement about their new product will purchase the product. Five people who watched the advertisement are picked at random. What is

the probability that none of these people will purchase the product? That exactly 3 will purchase the product?

20. **Baseball**   Suppose that the probability of a player hitting a home run is $\frac{1}{20}$. In 5 tries what is the probability that the player hits at least 1 home run?

21. **Guessing on a True–False Exam**   In a 12-item true–false examination, a student guesses on each question.

(a) What is the probability that the student will obtain all correct answers?

(b) If 7 correct answers constitute a passing grade, what is the probability that the student will pass?

(c) What are the odds in favor of passing?

22. **Guessing on a True–False Exam**   In a 20-item true–false examination, a student guesses on each question.

(a) What is the probability that the student will obtain all correct answers?

(b) If 12 correct answers constitute a passing grade, what is the probability that the student will pass?

(c) What are the odds in favor of passing?

23. **Throwing Dice**   Find the probability of throwing a sum of 11 at least 3 times in 5 throws of a pair of fair dice.

24. **Evaluating a Game**   In a certain game a player has the probability $\frac{1}{7}$ of winning a prize worth $90 and the probability $\frac{1}{3}$ of winning another prize worth $50. What is the expected cost of the game for the player?

25. **Evaluating a Game**   Frank pays $0.70 to play a certain game. He draws 2 balls (together) from a bag containing 2 red balls and 4 green balls. He receives $1 for each red ball that he draws. Has he paid too much? By how much?

26. **Playing the Lottery**   In a lottery 1000 tickets are sold at $0.25 each. There are 3 cash prizes: $100, $50, and $30. Alice buys 8 tickets.

(a) What would have been a fair price for a ticket?

(b) How much extra did Alice pay?

27. **Evaluating a Game**   The figure at the top of page 409 shows a spinning game for which a person pays $0.30 to purchase an opportunity to spin the dial. The numbers in the figure indicate the amount of payoff and its corresponding probability. Find the expected value of this game. Is the game fair?

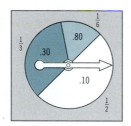

28. Evaluating a Game Consider the 3 boxes in the figure. The game is played in 2 stages. The first stage is to choose a ball from box A. If the result is a ball marked I, then we go to box I, and select a ball from there. If the ball is marked II, then we select a ball from box II. The number drawn on the second stage is the gain. Find the expected value of this game.

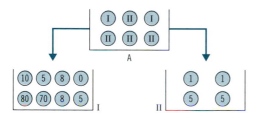

29. What is the expected number of heads that will turn up if a biased coin, $P(H) = \frac{1}{4}$, is tossed 200 times?

30. European Roulette A European roulette wheel has only 37 compartments, 18 red, 18 black, and 1 green. A player will be paid $2 (including his $1 bet) if he picks correctly the color of the compartment in which the ball finally rests. Otherwise, he loses $1. Is the game fair to the player?

31. Blood Testing A group of 1000 people is subjected to a blood test that can be administered in 2 ways: (1) each person can be tested separately (in this case 1000 tests are required) or (2) the blood samples of 30 people can be pooled and analyzed together. If we use the second way and the test is negative, then 1 test suffices for 30 people. If the test is positive, each of the 30 people can then be tested separately, and, in all, $30 + 1$ tests are required for the 30 people. Assume the probability p that the test is positive is the same for all people and that the people to be tested are independent.

(a) What is the probability that the test for a pooled sample of 30 people will be positive?
(b) What is the expected number of tests necessary under plan 2?

Mathematical Questions from Professional Exams

1. Actuary Exam—Part II What is the probability that 10 independent tosses of an unbiased coin result in no fewer than 1 head and no more than 9 heads?

(a) $(\frac{1}{2})^9$ (b) $1 - 11(\frac{1}{2})^9$ (c) $1 - 11(\frac{1}{2})^{10}$
(d) $1 - (\frac{1}{2})^9$ (e) $1 - (\frac{1}{2})^{10}$

2. CPA Exam The Stat Company wants more information on the demand for its products. The following data are relevant:

Units Demanded	Probability of Unit Demand	Total Cost of Units Demanded
0	.10	$0
1	.15	1.00
2	.20	2.00
3	.40	3.00
4	.10	4.00
5	.05	5.00

What is the total expected value or payoff with perfect information?

(a) $2.40   (b) $7.40   (c) $9.00   (d) $9.15

3. CPA Exam Your client wants your advice on which of 2 alternatives he should choose. One alternative is to sell an investment now for $10,000. Another alternative is to hold the investment 3 days, after which he can sell it for a certain selling price based on the following probabilities:

Selling Price	Probability
$5,000	.4
$8,000	.2
$12,000	.3
$30,000	.1

Using probability theory, which of the following is the most reasonable statement?

(a) Hold the investment 3 days because the expected value of holding exceeds the current selling price.
(b) Hold the investment 3 days because of the chance of getting $30,000 for it.
(c) Sell the investment now because the current selling price exceeds the expected value of holding.
(d) Sell the investment now because there is a 60% chance that the selling price will fall in 3 days.

4. CPA Exam The Polly Company wishes to determine the amount of safety stock that it should maintain for product D that will result in the lowest cost.

The following information is available:

Stockout cost	$80 per occurrence
Carrying cost of safety stock	$2 per unit
Number of purchase orders	5 per year

The available options open to Polly are as follows:

Units of Safety Stock	10	20	30	40	50	55
Probability	50%	40%	30%	20%	10%	5%

The number of units of safety stock that will result in the lowest cost is

(a) 20   (b) 40   (c) 50   (d) 55

5. CPA Exam The ARC Radio Company is trying to decide whether to introduce as a new product a wrist "radiowatch" designed for shortwave reception of exact time as broadcast by the National Bureau of Standards. The "radiowatch" would be priced at $60, which is exactly twice the variable cost per unit to manufacture and sell it. The incremental fixed costs necessitated by introducing this new product would amount to $240,000 per year. Subjective estimates of the probable demand for the product are shown in the following probability distribution:

Annual Demand	6,000	8,000	10,000	12,000	14,000	16,000
Probability	.2	.2	.2	.2	.1	.1

The expected value of demand for the new product is

(a) 11,000 units   (b) 10,200 units   (c) 9000 units
(d) 10,600 units   (e) 9800 units

6. CPA Exam In planning its budget for the coming year, King Company prepared the following payoff probability distribution describing the relative likelihood of monthly sales volume levels and related contribution margins for product A:

Monthly Sales Volume	Contribution Margin	Probability
4,000	$ 80,000	.20
6,000	120,000	.25
8,000	160,000	.30
10,000	200,000	.15
12,000	240,000	.10

What is the expected value of the monthly contribution margin for product A?

(a) $140,000   (b) $148,000
(c) $160,000   (d) $180,000

7. CPA Exam A decision tree has been formulated for the possible outcomes of introducing a new product line. Branches related to alternative 1 reflect the possible payoffs from introducing the product without an advertising campaign. The branches for alternative 2 reflect the possible payoffs with an advertising campaign costing $40,000.

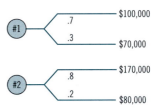

The expected values of alternatives 1 and 2, respectively, are

(a) #1: $(.7 \times \$100,000) + (.3 \times \$70,000)$
    #2: $(.8 \times \$170,000) + (.2 \times \$80,000)$
(b) #1: $(.7 \times \$100,000) + (.3 \times \$70,000)$
    #2: $(.8 \times \$130,000) + (.2 \times \$40,000)$
(c) #1: $(.7 \times \$100,000) + (.3 \times \$70,000)$
    #2: $(.8 \times \$170,000) + (.2 \times \$80,000) - \$40,000$
(d) #1: $(.7 \times \$100,000) + (.3 \times \$70,000) - \$40,000$
    #2: $(.8 \times \$170,000) + (.2 \times \$80,000) - \$40,000$

8. CPA Exam A battery manufacturer warrants its automobile batteries to perform satisfactorily for as long as the owner keeps the car. Auto industry data show that only 20% of car buyers retain their cars for 3 years or more. Historical data suggest the following:

Number of Years Owned	Probability of Battery Failure	Battery Exchange Costs	Percentage of Failed Batteries Returned
Less than 3 years	.4	$50	75%
3 years or more	.6	$20	50%

If 50,000 batteries were sold this year, what is the estimated warranty cost?

(a) $375,000   (b) $435,000
(b) $500,000   (d) $660,000

Chapter 9

Statistics

9.1 Bar Graphs; Pie Charts

9.2 Organization of Data

9.3 Measures of Central Tendency

9.4 Measures of Dispersion

9.5 The Normal Distribution

Chapter Review

Statistics is the science of collecting, organizing, analyzing, and interpreting numerical facts. By making observations, statisticians collect **data** in the form of measurements or counts. A measurable characteristic is called a **variable.** If a variable can assume any real value between certain limits, it is called a **continuous variable.** It is called a **discrete variable** if it can assume only a finite set of values or as many values as there are whole numbers. Examples of continuous variables are weight, height, length, time, etc. Examples of discrete variables are the number of votes a candidate gets, the number of cars sold, etc. We shall discuss only discrete data in this chapter.

The **organization of data** involves the presentation of the collected measurements or counts in a form suitable for determining logical conclusions. Usually, tables or graphs are used to represent the collected data. The **analysis of data** is the process of extracting, from given measurements or counts, related and relevant information from which a brief numerical description can be formulated. In this process we use concepts known as the *mean, median, range, variance,* and *standard deviation.* By **interpretation of data** we mean the art of drawing conclusions from the analysis of the data. This involves the formulation of predictions concerning a large collection of objects based on the information available from a small collection of similar objects.

In collecting data concerning varied characteristics, it is often impossible or impractical to observe an entire group. Instead of examining an entire group, called the **population,** a small segment, called the **sample,** is chosen. It would be difficult, for example, to question all cigarette smokers in order to study the effects of smoking. Therefore, appropriate samples of smokers are usually selected for questioning.

411

The method of selecting the sample is extremely important if we want the results to be reliable. All members of the population under investigation should have an equal probability of being selected; otherwise, a **biased sample** could result. For example, if we want to study the relationship between smoking cigarettes and lung cancer, we cannot choose a sample of smokers who all live in the same location. The individuals chosen might have dozens of characteristics peculiar to their area, which would give a false impression with regard to all smokers.

Samples collected in such a way that each population item is equally likely to be chosen are called **random samples.** Of course, there are many random samples that can be chosen from a population. By combining the results of more than one random sample of a population, it is possible to obtain a more accurate representation of the population.

If a sample is representative of a population, important conclusions about the population can often be inferred from analysis of the sample. The phase of statistics dealing with conditions under which such inference is valid is called **inductive statistics** or **statistical inference.** Since such inference cannot be absolutely certain, the language of probability is often used in stating conclusions. Thus when a meteorologist makes a forecast, weather data collected over a large region are studied and, based on this study, the weather forecast is given in terms of chances. A typical forecast might be "There is a 20% chance of rain tomorrow."

To summarize, in statistics we are interested in four principles: collecting data or information, organizing it, analyzing it, and interpreting it.

9.1 Bar Graphs; Pie Charts*

There are two popular methods for graphically displaying data: bar graphs and pie charts. A bar graph will show the number or the percent of data that are in each category, while a pie chart will show only the percent of data in each category.

We begin with an example showing how to construct a bar graph from data.

EXAMPLE 1   Constructing a Bar Graph

For the fiscal year 1993 (October 1992–September 1993), the federal government spent a total of $1444 billion. The breakdown of expenditures (in billions of dollars) is given in Table 1 at the top of page 413. Construct a bar graph of the data.

---

* Based on material from *College Algebra: Graphing and Data Analysis,* Sullivan and Sullivan, Prentice-Hall, 1998. Used here with the permission of the author and the publisher.

TABLE 1

Social Security, Medicare, and other retirement	$500
National defense, veterans, and foreign affairs	$344
Net interest (interest on the public debt)	$199
Physical, human, and community development	$119
Social programs	$254
Law enforcement and general government	$28

Source: Department of the Treasury

SOLUTION A horizontal axis is used to indicate the category of spending and a vertical axis is used to represent the amount spent in each category. For each category of spending we draw rectangles of equal width whose height represents the amount spent in the category. The rectangles will not touch each other. See Figure 1.

Figure 1

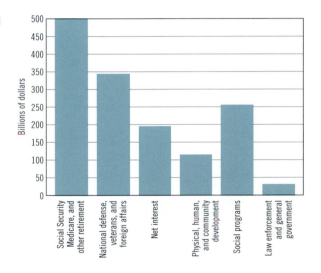

Now we show how to construct a pie chart using the data given in Table 1.

EXAMPLE 2 Constructing a Pie Chart

Use the data given in Table 1 to construct a pie chart.

SOLUTION To construct a pie chart a circle is divided into sectors, one sector for each category of data. The size of each sector is proportional to the total amount spent. Since Social Security, Medicare, and other retirement is $500 billion and total spending is $1444 billion, the percent of data in this category is $500/1444 \approx 0.35 = 35\%$. Therefore, Social Security, Medicare, and other retirement will make up 35% of the pie chart. Since a circle has 360°, the degree measure of the sector for this category of spending is 0.35(360°) = 126°. Following this procedure for the remaining categories of spending, we obtain Table 2.

Table 2

Category	Spending (in billions)	Percent of Total Spending*	Degree Measure of Sector*
Social Security, Medicare, and other retirement	$500	0.35 = 35%	126
National defense, veterans, and foreign affairs	$344	0.24 = 24%	86
Net interest (interest on the public debt)	$199	0.14 = 14%	50
Physical, human, and community development	$119	0.08 = 8%	29
Social programs	$254	0.18 = 18%	65
Law enforcement and general government	$ 28	0.02 = 2%	7

* The data in column three does not add up to 100% due to rounding. Similarly, the data in column four does not add up to 360° due to rounding.

To construct a pie chart by hand, we use a protractor to approximate the angles for each sector. To construct a pie chart with a computer, select a spreadsheet program that has the capability of drawing pie charts. See Figure 2.

Figure 2

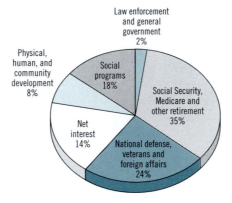

One reason for graphing data by drawing bar graphs or pie charts is to quickly determine certain information about the data.

EXAMPLE 3   Analyzing a Bar Graph

The bar graph in Figure 3 represents the sources of revenue for the United States federal government in its fiscal year 1993. Answer the questions below using the bar graph.

Figure 3

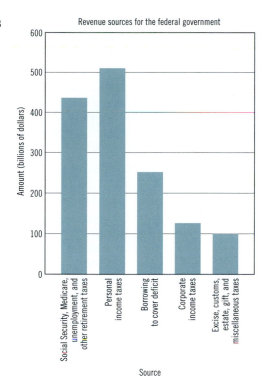

Revenue sources for the federal government

(a) What is the largest source of revenue for the federal govenment?
(b) What is the smallest source of revenue for the federal government?
(c) How much revenue does the government collect from all sources, excluding borrowing?
(d) What is the total revenue of the U.S. federal government?

SOLUTION

(a) The largest source of revenue for the federal government is personal income taxes. The government collects about $510 billion from this source.
(b) The smallest source of revenue for the federal government is excise, customs, estate, gift, and miscellaneous taxes. The government collects about $100 billion from this source.
(c) The revenue collected, excluding borrowing, totals approximately $430 + $510 + $120 + $100 = $1160 billion.
(d) The remaining amount represents borrowing of approximately $250 billion. The total revenue is therefore $1160 + $250 = $1410 billion.

EXERCISE 9.1   Answers to odd-numbered problems begin on page AN-30.

In Problems 1–6 list some possible ways to choose random samples for each study.

1. A study to determine opinion about a certain television program.

2. A study to detect defective radio resistors.

3. A study of the opinions of people toward Medicare.

4. A study to determine opinions about an election of a U.S. president.

5. A study of the number of savings accounts per family in the United States.

6. A national study of the monthly budget for a family of four.

7. The following is an example of a biased sample: In a study of political party preferences, poorly dressed interviewers obtained a significantly greater proportion of answers favoring Democratic party candidates in their samples than did their well-dressed and wealthier-looking counterparts. Give two more examples of biased samples.

8. In a study of the number of savings accounts per family, a sample of accounts totaling less than $10,000 was taken and, from the owners of these accounts, information about the total number of accounts owned by all family members was obtained. Criticize this sample.

9. It is customary for newsreporters to sample the opinions of a few people to find out how the population at large feels about the events of the day. A reporter questions people on a downtown street corner. Is there anything wrong with such an approach?

10. In 1936 the *Literary Digest* conducted a poll to predict the presidential election. Based on its poll it predicted the election of Landon over Roosevelt. In the actual election, Roosevelt won. The sample was taken by drawing the mailing list from telephone directories and lists of car owners. What was wrong with the sample?

11. **On-Time Performance**   The bar graph below represents the overall percentage of reported flight operations arriving on time for ten different airlines.

   (a) Which airline has the highest percentage of on-time flights?
   (b) Which airline has the lowest percentage of on-time flights?
   (c) What percentage of United Airlines' flights are on time?

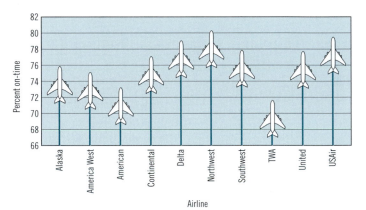

Source: United States Department of Transportation.

12. **Income Required for a Loan**   The bar graph below shows the minimum annual income required for a $100,000 loan using interest rates available on 12/1/96. Taxes and insurance are assumed to be $230 monthly.

   (a) What minimum annual income is needed to qualify for a 5/1 year ARM (Adjustable Rate Mortgage)?
   (b) Which loan type requires the most annual income? What minimum annual income is required for this loan type?

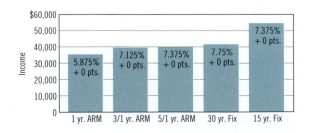

13. **Consumer Price Index**   The Consumer Price Index (CPI) is an index that measures inflation. It is calculated by obtaining the prices of a market basket of goods each month. The market basket, along with the percentages of each product, is given in the following pie chart:

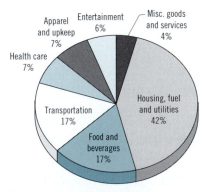

Source: Bureau of Labor Statistics.

(a) What is the largest component of the CPI?

(b) What is the smallest component of the CPI?

(c) Senior citizens spend about 14% of their income on health care. Why do you think they feel the CPI weight for health care is too low?

14. **Asset Allocation**   According to financial planners, an individual's investment mix should change over a person's lifetime. The longer an individual's time horizon, the more the individual should invest in stocks. A financial planner suggested that Jim's retirement portfolio be diversified according to the mix provided in the pie chart below, since Jim has 40 years to retirement.

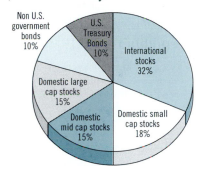

(a) How much should Jim invest in stocks?

(b) How much should Jim invest in bonds?

(c) How much should Jim invest in domestic (U.S.) stocks?

(d) The return on bonds over long periods of time is less than that of stocks. Explain why you think the financial planner recommended what she did for bonds.

15. **Household Income**   The data below represent the median income of households (in dollars) by region of the country for 1995.

Region	Median Income
Northwest	36,111
Midwest	35,839
South	30,942
West	35,979

Source: U.S. Bureau of the Census, March 1996 Current Population Survey.

(a) Draw a bar graph of the data.

(b) Draw a pie chart of the data.

(c) Which chart seems to summarize the data better?

(d) Which region has the highest median income?

(e) Which region has the lowest median income?

16. **Household Income**   The data at the top of the column to the right represent the median income (in dollars) of families by type of household for 1995.

(a) Draw a bar graph of the data.

(b) Draw a pie chart of the data.

Family Household	Median Income
Married-couple families	47,129
Female householder, no husband present	21,348
Male householder, no wife present	33,534

Source: U.S. Bureau of the Census, March 1996 Current Population Survey.

(c) Which chart seems to summarize the data better?

(d) Which household type has the highest median income?

(e) Which household type has the lowest median income?

17. **Busing Revenue**   The data below represent the total operating revenues (in thousands of dollars) of regional class I motor carriers of passengers for the first quarter of 1996.

Motor Carrier	Total Revenue
Academy Lines, Inc.	5,203
Bonanza Bus Lines	3,599
Connecticut Limousine	4,769
Hudson Transit	5,763
Kerrville Bus Company	4,269
New Jersey Transit	48,524
Peter Pan Bus Lines	7,914
Texas, New Mexico & Oklahoma Coaches	4,983

Source: Bureau of Transportation Statistics.

(a) Draw a bar graph of the data.

(b) Draw a pie chart of the data.

(c) Which chart seems to summarize the data better?

(d) Which motor carrier has the highest total revenue?

(e) Which motor carrier has the lowest total revenue?

18. **Busing Revenue**   The data below represent the total operating revenues (in thousands of dollars) of regional class I motor carriers of passengers for the first quarter of 1995.

Motor Carrier	Total Revenue
Academy Lines, Inc.	4,605
Bonanza Bus Lines	3,437
Connecticut Limousine	4,846
Hudson Transit	5,793
Kerrville Bus Company	4,271
New Jersey Transit	48,113
Peter Pan Bus Lines	8,061
Texas, New Mexico & Oklahoma Coaches	4,708

Source: Bureau of Transportation Statistics.

(a) Draw a bar graph of the data.
(b) Draw a pie chart of the data.
(c) Which chart seems to summarize the data better?
(d) Which motor carrier has the highest total revenue?
(e) Which motor carrier has the lowest total revenue?

19. Causes of Death   The data below represent the causes of death for 15–24 year olds in 1995.

Cause of Death	Number
Accidents and adverse effects	13,532
Homicide and legal intervention	6,827
Suicide	4,789
Malignant neoplasms	1,599
Diseases of heart	964
Human immunodefeficiency virus infection	643
Congenital anomalies	425
Chronic obstructive pulmonary diseases	220
Pneumonia and influenza	193
Cerebrovascular diseases	166
All other causes	4,211

Source: National Center for Health Statistics, 1996.

(a) Draw a bar graph of the data.
(b) Draw a pie chart of the data.
(c) Which chart seems to summarize the data better?
(d) What was the leading cause of death for 15–24 year olds in 1995?

20. Licensed Drivers   The data below represent the number of licensed drivers in the Great Lake States in 1994.

State	Number of Licensed Drivers
Illinois	7,502,201
Indiana	3,806,329
Michigan	6,601,924
Minnesota	2,705,701
Ohio	7,142,173
Wisconsin	3,554,003

Source: Each state's authorities.

(a) Draw a bar graph of the data.
(b) Draw a pie chart of the data.
(c) Which chart seems to summarize the data better?
(d) Which state has the most licensed drivers?
(e) Which state has the fewest licensed drivers?

9.2 ORGANIZATION OF DATA

Often studies result in data represented by a large collection of numbers. If the data are to be interpreted, they must be organized. In this section we discuss the organization of data. One way to organize data is by using a frequency table and line chart.

Frequency Tables; Line Charts

We begin with an example in which the weights of 71 children have been recorded.

EXAMPLE 1   Listing Data in a Table; Forming a Frequency Table

Table 3 lists the weights of a random sample of 71 children selected from a group of 10,000. List the data from smallest to highest in a Table and form a frequency table.

Table 3   Weights of 71 Students, in Pounds

69	71	71	55	52	55	58	58	58	62	67	94
82	94	95	89	89	104	93	93	58	62	67	62
94	85	92	75	75	79	75	82	94	105	115	104
105	109	94	92	89	85	85	89	95	92	105	71
72	72	79	79	85	72	79	119	89	72	72	69
79	79	69	93	85	93	79	85	85	69	79	

SOLUTION Certain information available from the sample becomes more evident once the data are ordered according to some scheme. If the 71 measurements are written from smallest to highest, we obtain Table 4.

Table 4

52	55	55	58	58	58	58	62	62	62	67	67
69	69	69	69	71	71	71	72	72	72	72	72
75	75	75	79	79	79	79	79	79	79	79	82
82	85	85	85	85	85	85	85	89	89	89	89
89	92	92	92	93	93	93	93	94	94	94	94
94	95	95	104	104	105	105	105	109	115	119	

Once Table 4 has been constructed, it becomes easy to present the data in a **frequency table.** This is done as follows: Tally marks are used to record the occurrence of weights. Then the **frequency** f with which each weight occurs is listed. See Table 5.

Table 5

Score	Tally	Frequency, f	Score	Tally	Frequency, f
52	/	1	85	卌 //	7
55	//	2	89	卌	5
58	////	4	92	///	3
62	///	3	93	////	4
67	//	2	94	卌	5
69	////	4	95	//	2
71	///	3	104	//	2
72	卌	5	105	///	3
75	///	3	109	/	1
79	卌 ///	8	115	/	1
82	//	2	119	/	1

A graphical representation of the data in Table 5 is provided by using a **line chart,** which is obtained in the following way: The vertical axis (*y*-axis) denotes the frequency f and the horizontal axis (*x*-axis) denotes the weight data. Then points are plotted according to the information in Table 5 and a vertical line is drawn from each point to the horizontal axis. See Figure 4.

Figure 4

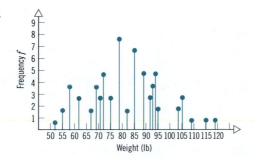

Now Work Problems

1(a)–(b)

When data are collected and few repeated entries are obtained, the data are usually easier to organize by grouping them and constructing a histogram.

Grouping Data; Histograms

Table 6 lists the monthly electric bills of 71 residential customers, beginning with the smallest bill.

Table 6 Monthly Electric Bills

52.30	55.61	55.71	58.01	58.41	58.51	58.91	62.33	62.50	62.71
67.13	67.23	69.51	69.67	69.80	69.82	71.34	71.65	71.83	72.15
72.22	72.41	72.59	72.67	75.11	75.71	75.82	79.03	79.06	79.09
79.15	79.28	79.32	79.51	79.62	82.32	82.61	85.09	85.13	85.25
85.31	85.41	85.51	85.58	89.21	89.32	89.49	89.61	89.78	92.41
92.63	92.89	93.05	93.19	93.28	93.91	94.17	94.28	94.31	94.52
94.71	95.32	95.51	104.31	104.71	105.21	105.37	105.71	109.34	115.71
119.38									

The first step in grouping data is to calculate the *range*.

The **range** of a set of numbers is the difference between the largest and the smallest number in the set. Thus,

$$\text{Range} = (\text{Largest value}) - (\text{Smallest value})$$

For the data in Table 6 the range is

$$\text{Range} = 119.38 - 52.30 = 67.08$$

To group this data, we divide the range into intervals of equal size, called **class intervals.** Table 7 shows the data using 14 class intervals, each of size 5; Table 8 shows the same data using 7 class intervals, each of size 10. Then the frequency, the number of bills that fell into each class interval, is tallied.

Table 7

	Class Interval	Tally	Frequency
1	50– 54.99	/	1
2	55– 59.99	𝑇𝐻𝐿 /	6
3	60– 64.99	///	3
4	65– 69.99	𝑇𝐻𝐿 /	6
5	70– 74.99	𝑇𝐻𝐿 ///	8
6	75– 79.99	𝑇𝐻𝐿 𝑇𝐻𝐿 /	11
7	80– 84.99	//	2
8	85– 89.99	𝑇𝐻𝐿 𝑇𝐻𝐿 //	12
9	90– 94.99	𝑇𝐻𝐿 𝑇𝐻𝐿 //	12
10	95– 99.99	//	2
11	100–104.99	//	2
12	105–109.99	////	4
13	110–114.99		0
14	115–119.99	//	2

Table 8

	Class Interval	Tally	Frequency
1	50– 59.99	𝑇𝐻𝐿 //	7
2	60– 69.99	𝑇𝐻𝐿 ////	9
3	70– 79.99	𝑇𝐻𝐿 𝑇𝐻𝐿 𝑇𝐻𝐿 ////	19
4	80– 89.99	𝑇𝐻𝐿 𝑇𝐻𝐿 ////	14
5	90– 99.99	𝑇𝐻𝐿 𝑇𝐻𝐿 ////	14
6	100–109.99	𝑇𝐻𝐿 /	6
7	110–119.99	//	2

The class intervals shown in Tables 7 and 8 each begin at 50 and end at 119.99, so as to include all the data from Table 6. The first number in a class interval is called the **lower class limit;** the second number is called the **upper class limit.** We choose these limits so that each item in Table 6 can be assigned to one and only one class interval. The **midpoint** of a class interval is defined as

$$\text{Midpoint} = \frac{\text{Upper class limit} + \text{Lower class limit}}{2}$$

The **class width** is the difference between consecutive lower class limits.

When the data are represented in the form of Table 7 (or Table 8), they are said to be **grouped data.** Notice that once raw data are converted to grouped data, it is impossible to retrieve or recover the original data. The best we can do is to choose the midpoint of each class interval as a representative for each class. In Table 7, for example, the actual scores of 105.21, 105.37, 105.71, and 109.34 are viewed as being represented by the midpoint of the class interval from 105 to 109.99, namely, (105 + 109.99)/2 = 107.50.

Next we present the grouped data of Table 8 in a graph, called a **histogram.**

EXAMPLE 2   Building a Histogram

Build a histogram for the grouped data of Table 8.

SOLUTION   To build a histogram for the data in Table 8, we construct a set of adjoining rectangles having as base the size of the class interval and as height the frequency of occurrence of data in that particular interval. The center of the base is the midpoint of each class interval. Figure 5 shows the histogram for the data in Table 8.

Figure 5

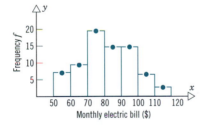

If we connect all the midpoints of the tops of the rectangles in Figure 5, we obtain a line graph called a **frequency polygon.** (In order not to leave the graph hanging, we will connect it to the horizontal axis on each side.) See Figure 6.

Figure 6

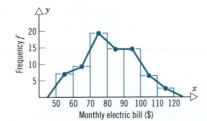

Now Work Problems

3(a)–(g)

Sometimes it is useful to learn how many cases fall below (or above) a certain value. For the data of Table 8, we do this as follows: Start at the lowest class interval (50–59.99) and note how many scores are in this interval. The number is 7. So we write 7 in the column labeled *cf* (cumulative frequency) of Table 9 in the row for 50–59.99. Next, we list how many scores fall in the next class interval (60–69.99), 9, and place the cumulative total $7 + 9 = 16$ in the *cf* column. The process is continued. The bottom entry of the last column should equal the total number of scores in the sample. The numbers in the column *cf* are called the **cumulative (less than) frequencies.**

Table 9

Class Interval	Tally	*f*	*cf*
50– 59.99	𝈫𝈫 //	7	7
60– 69.99	𝈫𝈫 ////	9	16
70– 79.99	𝈫𝈫 𝈫𝈫 𝈫𝈫 ////	19	35
80– 89.99	𝈫𝈫 𝈫𝈫 ////	14	49
90– 99.99	𝈫𝈫 𝈫𝈫 ////	14	63
100–109.99	𝈫𝈫 /	6	69
110–119.99	//	2	71

The graph in which the horizontal axis represents class intervals and the vertical axis represents cumulative frequencies is called the **cumulative (less than) frequency distribution.** See Figure 7 for the cumulative frequency distribution for the data from Table 9. Notice that the points plotted are connected by lines to aid in visualizing the graph.

Figure 7

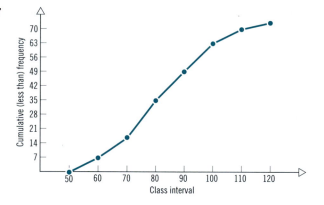

EXERCISE 9.2   Answers to odd-numbered problems begin on page AN-31.

1. The following scores were made on a 60-item test:

25	30	34	37	41	42	46	49	53
26	31	34	37	41	42	46	50	53
28	31	35	37	41	43	47	51	54
29	32	36	38	41	44	48	52	54
30	33	36	39	41	44	48	52	55
30	33	37	40	42	45	48	52	

(a) Set up a frequency table for the above data. What is the range?
(b) Draw a line chart for the data.
(c) Draw a histogram for the data using a class interval of size 2.
(d) Draw the frequency polygon for this histogram.
(e) Find the cumulative (less than) frequencies.
(f) Draw the cumulative (less than) frequency distribution.

2. For Table 5 in the text:

   (a) Draw a histogram.

   (b) Draw the frequency polygon.

   (c) Find the cumulative (less than) frequencies.

   (d) Draw the cumulative (less than) frequency distribution.

3. Licensed Drivers in Florida The histogram below represents the number of licensed drivers between the ages of 20 and 84 in the state of Florida.

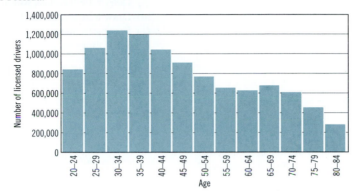

   (a) Determine the number of class intervals.

   (b) What is the lower class limit of the first class interval? What is the upper class limit of the first class interval?

   (c) Determine the class width.

   (d) How many licensed drivers are 70 to 84 years old?

   (e) Which class interval has the most licensed drivers?

   (f) Which class interval has the fewest licensed drivers?

   (g) Draw a frequency polygon for the given data.

4. IQ Scores The histogram below represents the IQ scores of students enrolled in College Algebra at a local university.

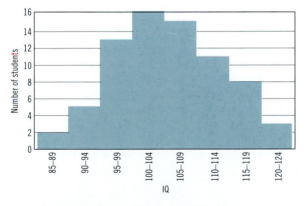

   (a) Determine the number of class intervals.

   (b) What is the lower class limit of the first class interval? What is the upper class limit of the first class interval?

   (c) Determine the class width.

   (d) How many students have an IQ between 100 and 104?

   (e) How many students have an IQ above 110?

   (f) How many students are enrolled in College Algebra?

   (g) Draw a frequency polygon for the given data.

5. Licensed Drivers in Tennessee The frequency table below provides the number of licensed drivers between the ages of 20 and 84 in the state of Tennessee in 1994.

Age	Number of Licensed Drivers
20–24	345,941
25–29	374,629
30–34	428,748
35–39	439,137
40–44	414,344
45–49	372,814
50–54	292,460
55–59	233,615
60–64	204,235
65–69	181,977
70–74	150,347
75–79	100,068
80–84	50,190

Source: FHWA.

   (a) Determine the number of class intervals.

   (b) What is the lower class limit of the first class interval? What is the upper class limit of the first class interval?

   (c) Determine the class width.

   (d) Draw a histogram of the data.

   (e) Draw a frequency polygon of the data.

   (f) Which age group has the most licensed drivers?

   (g) Which age group has the fewest licensed drivers?

6. Licensed Drivers in Hawaii The frequency table at the top of page 424 provides the number of licensed drivers between the ages of 20 and 84 in the state of Hawaii in 1994.

Age	Number of Licensed Drivers
20–24	65,951
25–29	78,119
30–34	91,976
35–39	92,557
40–44	87,430
45–49	75,978
50–54	55,199
55–59	39,678
60–64	35,650
65–69	33,885
70–74	26,125
75–79	14,990
80–84	6,952

Source: FHWA.

(a) Determine the number of class intervals.
(b) What is the lower class limit of the first class interval? What is the upper class limit of the first class interval?
(c) Determine the class width.
(d) Draw a histogram of the data.
(e) Draw a frequency polygon of the data.
(f) Which age group has the most licensed drivers?
(g) Which age group has the fewest licensed drivers?

7. Undergraduate Tuition   The data below represent the cost of undergraduate tuition at four-year colleges for

Tuition (Dollars)	Number of 4-Year Colleges
0–999	10
1000–1999	7
2000–2999	45
3000–3999	66
4000–4999	84
5000–5999	84
6000–6999	97
7000–7999	118
8000–8999	138
9000–9999	110
10,000–10,999	104
11,000–11,999	82
12,000–12,999	61
13,000–13,999	34
14,000–14,999	29

Source: The College Board, New York, NY, Annual Survey of Colleges 1992 and 1993.

1992–1993 having tuition amounts ranging from $0 through $14,999.

(a) Determine the number of class intervals.
(b) What is the lower class limit of the first class interval? What is the upper class limit of the first class interval?
(c) Determine the class width.
(d) Draw a histogram of the data.
(e) Draw a frequency polygon of the data.
(f) What range of tuition occurs most frequently?

8. Undergraduate Tuition   The data below represent the cost of undergraduate tuition at four-year colleges for 1993–1994 having tuition amounts ranging from $0 through $14,999.

Tuition (Dollars)	Number of 4-Year Colleges
0–999	8
1000–1999	5
2000–2999	28
3000–3999	48
4000–4999	76
5000–5999	65
6000–6999	81
7000–7999	96
8000–8999	112
9000–9999	118
10,000–10,999	106
11,000–11,999	90
12,000–12,999	70
13,000–13,999	59
14,000–14,999	23

Source: The College Board, New York, NY, Annual Survey of Colleges 1993 and 1994.

(a) Determine the number of class intervals.
(b) What is the lower class limit of the first class interval? What is the upper limit of the first class interval?
(c) Determine the class width.
(d) Draw a histogram of the data.
(e) Draw a frequency polygon of the data.
(f) What range of tuition occurs most frequently?

Technology Exercises

Most graphing calculators are able to draw histograms. Check your user's manual to determine the appropriate command. Use a graphing calculator to solve Problems 1–6.

1. The table at the top of the left column on page 425 gives the birth rate per 1000 people in 20 states. (*Source:* U.S. Census Bureau, Statistical Abstract of the United States,

1994). Make a frequency distribution for the data using 8 classes of equal width, and draw the frequency histogram.

State	AK	AL	AR	CA	CT	DC
Birth Rate	20.5	15.4	15.0	20.1	14.8	19.7

State	HI	IL	IN	KY	MA	ME
Birth Rate	17.6	16.8	15.3	14.6	14.7	13.6

State	MI	NH	OH	PA	RI	TN
Birth Rate	16.0	14.8	15.2	14.1	14.7	15.0

State	VT	WI
Birth Rate	14.0	14.5

2. The table below gives the heart disease death rate per 100,000 people in 20 states in 1991. (*Source:* U.S. Census Bureau, Statistical Abstract of the United States, 1994.) Make a frequency distribution for the data using 4 classes of equal width, and draw the frequency histogram.

State	AK	AL	AR	CA	CT
Heart Disease Rate	83	322	346	222	291

State	DC	HI	IL	IN	KY
Heart Disease Rate	312	180	309	299	322

State	MA	ME	MI	NH	OH
Heart Disease Rate	285	300	295	246	320

State	PA	RI	TN	VT	WI
Heart Disease Rate	362	323	313	259	290

3. The table below gives the cancer death rate per 100,000 people in 20 states in 1991. (*Source:* U.S. Census Bureau, Statistical Abstract of the United States, 1994.) Make a frequency distribution for the data using 5 classes of equal width, and draw the frequency histogram.

State	AK	AL	AR	CA	CT
Cancer Rate	88	216	236	165	214

State	DC	HI	IL	IN	KY
Cancer Rate	259	146	212	214	230

State	MA	ME	MI	NH	OH
Cancer Rate	230	238	206	203	222

State	PA	RI	TN	VT	WI
Cancer Rate	251	236	213	197	209

4. The table below gives the energy expenditures per person during 1991 in 20 states. (*Source:* U.S. Census Bureau, Statistical Abstract of the United States, 1994.) Make a frequency distribution for the data using 5 classes of equal width, and draw the frequency histogram.

State	AK	AL	AR	CA
Energy Expenditures	3249	2029	1975	1562

State	CT	DC	HI	IL
Energy Expenditures	1885	1899	1793	1863

State	IN	KY	MA	ME
Energy Expenditures	2125	1936	1767	2057

State	MI	NH	OH	PA
Energy Expenditures	1786	1727	1928	1863

State	RI	TN	VT	WI
Energy Expenditures	1747	1872	1930	1645

5. The following table gives the average hourly earnings of production workers in manufacturing during 1993 in 20 states. (*Source:* U.S. Census Bureau, Statistical Abstract of the United States, 1994.) Make a frequency distribution for the data using 6 classes of equal width, and draw the frequency histogram.

State	AK	AL	AR	CA
Hourly Earnings	11.14	10.36	9.36	12.37

State	CT	DC	HI	IL
Hourly Earnings	13.01	13.18	11.98	12.04

State	IN	KY	MA	ME
Hourly Earnings	13.17	11.48	12.36	11.40

State	MI	NH	OH	PA
Hourly Earnings	15.35	11.61	14.05	12.09

State	RI	TN	VT	WI
Hourly Earnings	10.22	10.33	11.81	12.17

6. The table on the right gives the number of hazardous waste sites in 20 states. (*Source:* U.S. Census Bureau, *Statistical Abstract of the United States, 1994.*) Make a frequency distribution for the data using 5 classes of equal width, and draw the frequency histogram.

State	AK	AL	AR	CA
Hazardous Waste Sites	8	14	12	95

State	CT	DC	HI	IL
Hazardous Waste Sites	15	0	3	37

State	IN	KY	MA	ME
Hazardous Waste Sites	33	20	31	10

State	MI	NH	OH	PA
Hazardous Waste Sites	76	17	36	99

State	RI	TN	VT	WI
Hazardous Waste Sites	12	15	8	40

9.3 MEASURES OF CENTRAL TENDENCY

The idea of taking an *average* is familiar to practically everyone. Often we hear people talk about average salary, average height, average grade, and so on. The idea of averages is so commonly used it should not surprise you to learn that several kinds of averages have been introduced in statistics.

Averages are called *measures of central tendency* because they estimate the "center" of the data collected. The three most common measures of central tendency are the *arithmetic mean, median,* and *mode*.

Mean

The **arithmetic mean,** or **mean,** of a set of real numbers $x_1, x_2, \ldots, x_n$ is denoted by $\overline{X}$ and is defined as

$$\overline{X} = \frac{x_1 + x_2 + \cdots + x_n}{n} \tag{1}$$

where n is the number of items being averaged.

EXAMPLE 1   Computing the Mean

The grades of a student on eight 100-point examinations were 70, 65, 69, 85, 94, 62, 79, and 100. Find the mean.

SOLUTION   In this example $n = 8$. The mean of this set of grades is

$$\overline{X} = \frac{70 + 65 + 69 + 85 + 94 + 62 + 79 + 100}{8} = 78$$

An interesting fact about the mean is that the sum of deviations of each item from the mean is zero. In Example 1 the deviation of each score from the mean $\overline{X} = 78$ is $(100 - 78)$, $(94 - 78)$, $(85 - 78)$, $(79 - 78)$, $(70 - 78)$, $(69 - 78)$, $(65 - 78)$, and $(62 - 78)$. Table 10 lists each score, the mean, and the deviation from the mean. If we add the deviations from the mean, we obtain a sum of zero.

Table 10

Score	Mean	Deviation from Mean
62	78	−16
65	78	−13
69	78	−9
70	78	−8
79	78	1
85	78	7
94	78	16
100	78	22
		Sum of Deviations: 0

For any set of data the following result is true:

The sum of the deviations from the mean is zero.

As a matter of fact, we could have defined the mean as that real number for which the sum of the deviations is zero.

Another interesting fact about the mean is given below.

If Y is any guessed or assumed mean (which may be any real number) and if d_j denotes the deviation of each item of the data from the assumed mean ($d_j = x_j - Y$), then the actual mean is

$$\overline{X} = Y + \frac{d_1 + d_2 + \cdots + d_n}{n} \tag{2}$$

Look again at Example 1. We know that the actual mean is 78. Suppose we had guessed the mean to be 52. Then, using Formula (2), we obtain

$$\overline{X} = 52 + \frac{(100 - 52) + (94 - 52) + (85 - 52) + (79 - 52) + (70 - 52) + (69 - 52) + (65 - 52) + (62 - 52)}{8}$$

$$= 52 + 26 = 78$$

which agrees with the mean computed in Example 1.

A method for computing the mean for grouped data given in a frequency table is illustrated by the following example.

EXAMPLE 2   Finding the Mean of Grouped Data

Find the mean for the grouped data given in Table 8 (repeated in the first two columns of Table 11 below).

SOLUTION

1. Take the midpoint (m_i) of each of the class intervals as a reference point and enter the result in column 3 of Table 11. For example, the midpoint of the class interval 80–89.99 is 85.
2. Next, multiply the entry in column 3 by the frequency f_i for that class interval and enter the product in column 4, which is labeled $f_i m_i$.
3. Add the entries in column 4; sum of $f_i m_i = 5775$.

Table 11

Class Interval	f_i	m_i	$f_i m_i$
50– 59.99	7	55	385
60– 69.99	9	65	585
70– 79.99	19	75	1425
80– 89.99	14	85	1190
90– 99.99	14	95	1330
100–109.99	6	105	630
110–119.99	2	115	230
$n = 71$			$5775 = $ Sum of $f_i m_i$

The mean $\overline{X}$ is then computed by dividing the sum of $f_i m_i$ by the number n of entries. That is,

$$\overline{X} = \frac{5775}{71} = 81.34$$

■

For grouped data, the mean $\overline{X}$ is

$$\overline{X} = \frac{\Sigma f_i m_i}{n} \tag{3}$$

where

Σ Means add the entries $f_i m_i$
$f_i = $ Number of entries in the ith class interval
$m_i = $ Midpoint of ith class interval
$n = $ Number of items

When data are grouped, the original data are lost due to grouping. As a result, the number obtained by using (3) is only an approximation to the actual mean. The reason for this is that using (3) amounts to computing the weighted average midpoint of a class interval (weighted by the frequency of scores in that interval) and therefore cannot be a computation for $\overline{X}$ exactly.

Median

The **median** of a set of real numbers arranged in order of magnitude is the middle value if the number of items is odd, and it is the mean of two middle values if the number of items is even.

EXAMPLE 3   Finding the Median of a Set of Data

(a) The set of data 2, 2, 3, 4, 5, 7, 7, 7, 11 has median 5.
(b) The set of data 2, 2, 3, 3, 4, 5, 7, 7, 7, 11 has median 4.5 since

$$\frac{4 + 5}{2} = 4.5$$

Finding the median for grouped data requires more work. As with the mean, the median for grouped data only approximates the actual median that would have been obtained prior to grouping the data. We use Example 4 below to show the steps.

EXAMPLE 4   Finding the Median for Grouped Data

Find the median for the grouped data in Table 12.

Table 12

	Class Interval	Tally	Frequency
1	50– 59.99	𝑇𝐻𝐿 //	7
2	60– 69.99	𝑇𝐻𝐿 ////	9
3	70– 79.99	𝑇𝐻𝐿 𝑇𝐻𝐿 𝑇𝐻𝐿 ////	19
4	80– 89.99	𝑇𝐻𝐿 𝑇𝐻𝐿 ////	14
5	90– 99.99	𝑇𝐻𝐿 𝑇𝐻𝐿 ////	14
6	100–109.99	𝑇𝐻𝐿 /	6
7	110–119.99	//	2

SOLUTION

Step 1 Find the interval containing the median.

The median is that number that has 50% of the data items below it and 50% of the data items above it. Since 50% of 71 (the number of data items) is 35.5, we start counting the tallies, beginning with the interval 50–59.99, until we come as close as we can to 35.5. This brings us through the interval 70–79.99, since through this interval there are 35 data items. Thus the median will lie in the interval 80–89.99.

Step 2 In the interval containing the median, count the number p of items remaining to reach the median.

Since we have accounted for 35 data items, there is left $35.5 - 35 = 0.5$ items. Thus, $p = 0.5$.

Step 3 If q is the frequency for the interval containing the median and i is its size, then the interpolation factor is

$$\textbf{interpolation factor} = \frac{p}{q} \cdot i$$

For the grouped data in Table 12, we have $q = 14$ and $i = 10$. Thus

$$\text{interpolation factor} = \frac{0.5}{14} \cdot (10) = 0.36$$

Step 4 The median M is

$$\textbf{M} = \begin{bmatrix} \textbf{lower limit of interval} \\ \textbf{containing the median} \end{bmatrix} + \textbf{[interpolation factor]}$$

For the grouped data in Table 12, the median M is

$$\text{M} = 80 + 0.36 = 80.36$$

To summarize,

The median M for grouped data is given by

$$M = \begin{bmatrix} \textbf{lower limit of interval} \\ \textbf{containing the median} \end{bmatrix} + \begin{bmatrix} \dfrac{p}{q} \cdot i \end{bmatrix} \tag{3}$$

where

p = Number required to reach the median from the lower limit of the interval containing the median

q = Number of data entries in this interval

i = Size of this interval

As with the mean, the median is an approximation to the actual median since it is obtained from grouped data. For Example 4, if we go back to the original data listed in Table 6, we obtain the true median $M = 82.32$.

The median of a set of data or grouped data is sometimes called the **fiftieth percentile** and is denoted by C_{50} to indicate that 50% of the data are below it and 50% are above it. Similarly, we can define C_{25}, or the first quartile, and C_{75}, or the third quartile.

Mode

The **mode** of a set of real numbers is the value that occurs with the greatest frequency exceeding a frequency of 1.

The mode does not necessarily exist, and if it does, it is not always unique.

EXAMPLE 5 **Set of Data with No Mode**

The set of data 2, 3, 4, 5, 7, 15 has no mode.

 **Now Work Problem 1** ∎

EXAMPLE 6 **A Bimodal Set of Data**

The set of data 2, 2, 2, 3, 3, 7, 7, 7, 11, 15 has two modes, 2 and 7, and is called **bimodal.** ∎

When data have been listed in a frequency table, the mode is defined as the midpoint of the interval consisting of the largest number of cases. For example, the mode for the data in Table 8, page 420, is 75 (the midpoint of the interval 70–79.99).

EXAMPLE 7 **Finding the Mode**

For the data listed in Table 5, page 419, the mode is 79 (8 is the highest frequency). ∎

Of the three measures of central tendency considered so far, the mean is the most important, the most reliable, and the one most frequently used. The reason for this is that it is easy to understand, easy to compute, and uses all the data in the collection. If two samples are chosen from the same population, the two *means* corresponding to the two samples will not generally differ by as much as the two *medians* of these samples.

EXERCISE 9.3   Answers to odd-numbered problems begin on page AN-36.

In Problems 1–8 compute the mean, median, and mode of the given set of data.

1. 21, 25, 43, 36

2. 16, 18, 24, 30

3. 55, 55, 80, 92, 70

4. 90, 80, 82, 82

5. 65, 82, 82, 95, 70

6. 62, 71, 83, 90, 75

7. 48, 65, 80, 92, 80

8. 95, 90, 91, 82

9. If an investor purchased 50 shares of IBM stock at $85 per share, 90 shares at $105 per share, 120 shares at $110 per share, and another 75 shares at $130 per share, what is the average cost per share?

10. If a farmer sells 120 bushels of corn at $4 per bushel, 80 bushels at $4.10 per bushel, 150 bushels at $3.90 per bushel, and 120 bushels at $4.20 per bushel, what is the average income per bushel?

11. The annual salaries of five faculty members in the mathematics department at a large university are $34,000, $35,000, $36,000, $36,500, and $65,000. Compute the mean and median. Which measure describes the situation more realistically? If you were among the four lower-paid members, which measure would you use to describe the situation? What if you were the one making $65,000?

12. According to an article in the *Wall Street Journal,* for companies having fewer than 5000 employees the average sales per employee were as follows:

Size of Company (Number of Employees)	Sales per Employee (Thousands of Dollars)
1–4	112
5–19	128
20–99	127
100–499	118
500–4999	120

Estimate the mean sales per employee for all firms having fewer than 5000 employees.

13. The distribution of the monthly earnings of 1155 secretaries in May 1994 in the Chicago metropolitan area is summarized in the table to the right. Find the mean salary and the median salary.

Monthly Earnings, $	Number of Secretaries
950–1199.99	25
1200–1449.99	55
1450–1699.99	325
1700–1949.99	410
1950–2199.99	215
2200–2449.99	75
2450–2699.99	50

14. Use the data in the table to estimate the mean net worth of U.S. families in each year. Use a class midpoint of $1 million for the class 500,000 or more.

Net Worth ($)	Percentage of Families		
	1985	1990	1995
0–4,999	33	24	11
5,000–9,999	5	7	5
10,000–24,999	12	10	12
25,000–49,999	16	19	22
50,000–99,999	17	19	22
100,000–249,999	12	14	17
250,000–499,999	3	4	6
500,000 or more	2	3	5

15. Licensed Drivers in Tennessee   Refer to the data provided in Problem 5, Exercise 9.2 and find the average age of a licensed driver in Tennessee.

16. Licensed Drivers in Hawaii   Refer to the data provided in Problem 6, Exercise 9.2 and find the average age of a licensed driver in Hawaii.

17. Undergraduate Tuition 1992–1993   Refer to the data provided in Problem 7, Exercise 9.2 and find the average tuition at a 4-year college in 1992–1993.

18. Undergraduate Tuition 1993–1994   Refer to the data provided in Problem 8, Exercise 9.2 and find the average tuition at a 4-year college in 1993–1994.

In Problems 19–22 use each graph to determine the mean and median.

19.

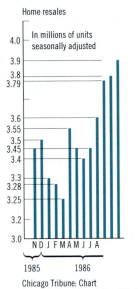

Home resales

In millions of units seasonally adjusted

N D J F M A M J J A

1985   1986

Chicago Tribune: Chart
Source: National Association of Realtors

20.

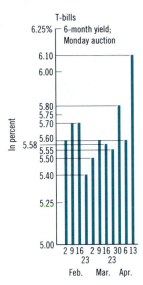

T-bills

6.25% ⌐ 6-month yield;
Monday auction

In percent

2 9 16 2 9 16 30 6 13
23   23
Feb.   Mar. Apr.

21.

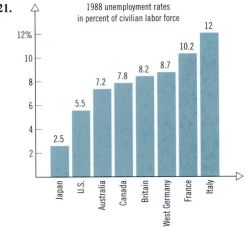

1988 unemployment rates
in percent of civilian labor force

Japan 2.5, U.S. 5.5, Australia 7.2, Canada 7.8, Britain 8.2, West Germany 8.7, France 10.2, Italy 12

Source: The Boston Company.

22. Direct investment in the United States.

Top countries investing in billions of dollars for 1988
(includes ownership of at least 10% of a company)

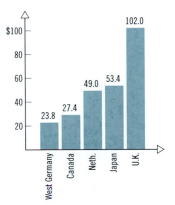

West Germany 23.8, Canada 27.4, Neth. 49.0, Japan 53.4, U.K. 102.0

Source: U.S. Department of Commerce.

9.4 MEASURES OF DISPERSION

EXAMPLE 1   Comparing Mean and Median for a Set of Data

Find the mean and median for each of the following sets of data:

$$S_1: \quad 4, 6, 8, 10, 12, 14, 16$$

$$S_2: \quad 4, 7, 9, 10, 11, 13, 16$$

SOLUTION   For S_1, the mean $\overline{X}_1$ and median M_1 are

$$\overline{X}_1 = \frac{4 + 6 + 8 + 10 + 12 + 14 + 16}{7} = \frac{70}{7} = 10 \qquad M_1 = 10$$

For S_2, the mean $\overline{X}_2$ and median M_2 are

$$\overline{X}_2 = \frac{4 + 7 + 9 + 10 + 11 + 13 + 16}{7} = \frac{70}{7} = 10 \qquad M_2 = 10$$

■

Notice that each set of scores has the same mean and the same median. Now look at Figure 8. Do you see that the data in S_1 seem to be more spread out from the mean 10 than those in S_2?

Figure 8

We seek a way to measure the extent to which scores are spread out. Such measures are called *measures of dispersion.*

Range

The simplest measure of dispersion is the **range,** which we have already defined as the difference between the largest value and the smallest value. For each of the sets S_1 and S_2 in Example 1, the range is $16 - 4 = 12$. We conclude that the range is a poor measure of dispersion since it depends on only two data items and tells us nothing about how the rest of the data are spread out.

Variance

Another measure of dispersion is the deviation from the mean. Recall that this measure is characterized by the fact that if the deviations from the mean of each data item are all added, the result is zero. We need a measure that will give us an idea of how much deviation is involved without having these deviations sum to zero.

By squaring each deviation from the mean, adding them, and dividing by the number of data items, we obtain an average squared deviation, called the **variance** of the set of data. The formula for the variance, which is denoted by σ^2*, is

$$\sigma^2 = \frac{(x_1 - \overline{X})^2 + (x_2 - \overline{X})^2 + \cdots + (x_n - \overline{X})^2}{n} = \frac{\Sigma(x_i - \overline{X})^2}{n}$$

where $\overline{X}$ is the mean of the data $x_1, x_2, \ldots, x_n$ and n is the number of data items.

* The lowercase Greek letter sigma.

EXAMPLE 2   Finding the Variance for a Set of Data

Calculate the variance for sets S_1 and S_2 of Example 1.

SOLUTION   For $S_1, \overline{X} = 10$ so that

$$\sigma^2 = \frac{(4 - 10)^2 + (6 - 10)^2 + (8 - 10)^2 + (10 - 10)^2 + (12 - 10)^2 + (14 - 10)^2 + (16 - 10)^2}{7}$$

$= 16$

For $S_2, \overline{X} = 10$ so that

$$\sigma^2 = \frac{(4 - 10)^2 + (7 - 10)^2 + (9 - 10)^2 + (10 - 10)^2 + (11 - 10)^2 + (13 - 10)^2 + (16 - 10)^2}{7}$$

$= 13.14$

Since the variance for set S_1 is larger than the variance for set S_2, we conclude that the data in set S_1 are more widely dispersed than the data in set S_2, confirming what we saw in Figure 8.

■

Standard Deviation

In computing the variance, we square the deviations from the mean. This means, for example, that if our data represent dollars, then the variance has the units "dollars squared." To remedy this, we use the square root of the variance, called the *standard deviation*.

Standard Deviation

The **standard deviation** of a set of n data items, $x_1, x_2, \ldots, x_n$, whose mean is $\overline{X}$, is defined as

$$\sigma = \sqrt{\frac{(x_1 - \overline{X})^2 + (x_2 - \overline{X})^2 + \cdots + (x_n - \overline{X})^2}{n}} = \sqrt{\frac{\Sigma(x_i - \overline{X})^2}{n}}$$

For the data in Example 1 the standard deviation for S_1 is

$$\sigma = \sqrt{\frac{36 + 16 + 4 + 0 + 4 + 16 + 36}{7}} = \sqrt{\frac{112}{7}} = \sqrt{16} = 4$$

and the standard deviation for S_2 is

$$\sigma = \sqrt{\frac{36 + 9 + 1 + 0 + 1 + 9 + 36}{7}} = \sqrt{\frac{92}{7}} = \sqrt{13.14} = 3.625$$

Again, the fact that the standard deviation of the set S_2 is less than the standard deviation of the set S_1 means that the data of S_2 are more clustered around the mean than those of S_1.

EXAMPLE 3   Finding the Standard Deviation

Find the standard deviation for the data

$$100, 90, 90, 85, 80, 75, 75, 75, 70, 70, 65, 65, 60, 40, 40, 40$$

SOLUTION   The mean is

$$\overline{X} = \frac{100 + 2 \cdot 90 + 85 + 80 + 3 \cdot 75 + 2 \cdot 70 + 2 \cdot 65 + 60 + 3 \cdot 40}{16} = 70$$

The deviations from the mean and their squares are given in Table 13. The standard deviation is

$$\sigma = \sqrt{\frac{4950}{16}} = \frac{70.4}{4} = 17.6$$

 Now Work Problem 3

EXAMPLE 4   Finding the Standard Deviation

Find the standard deviation for the data

$$80, 80, 80, 80, 75, 75, 75, 75, 70, 70, 65, 65, 60, 60, 55, 55$$

SOLUTION   Here the mean is $\overline{X} = 70$ for the 16 scores. Table 14 gives the deviations from the mean and their squares. The standard deviation is

$$\sigma = \sqrt{\frac{1200}{16}} = \sqrt{75} = 8.7$$

Table 13

Scores, x	Deviation from the Mean, $x - \overline{X}$	Deviation Squared, $(x - \overline{X})^2$
40	−30	900
40	−30	900
40	−30	900
60	−10	100
65	−5	25
65	−5	25
70	0	0
70	0	0
75	5	25
75	5	25
75	5	25
80	10	100
85	15	225
90	20	400
90	20	400
100	30	900
Mean = 70 $n = 16$	Sum = 0	Sum = 4950

Table 14

Scores, x	Deviation from the Mean, $x - \overline{X}$	Deviation Squared, $(x - \overline{X})^2$
55	−15	225
55	−15	225
60	−10	100
60	−10	100
65	−5	25
65	−5	25
70	0	0
70	0	0
75	5	25
75	5	25
75	5	25
75	5	25
80	10	100
80	10	100
80	10	100
80	10	100
Mean = 70 $n = 16$	Sum = 0	Sum = 1200

These two examples show that although the samples have the same mean, 70, and the same sample size, 16, the data in Example 3 deviate further from the mean than do the data in Example 4.

> In general, a relatively small standard deviation indicates that the measures tend to cluster close to the mean, and a relatively high standard deviation shows that the measures are widely scattered from the mean.

Standard Deviation for Grouped Data

> To find the standard deviation for grouped data, we use the formula
>
> $$\sigma = \sqrt{\frac{(m_1 - \overline{X})^2 \cdot f_1 + (m_2 - \overline{X})^2 \cdot f_2 + \cdots + (m_k - \overline{X})^2 \cdot f_k}{n}}$$
>
> $$= \sqrt{\frac{\Sigma (m_i - X)^2 \cdot f_i}{n}}$$
>
> where $m_1, m_2, \ldots, m_k$ are the class midpoints; $f_1, f_2, \ldots, f_k$ are the respective frequencies; n is the number of data items, that is, $n = f_1 + f_2 + \cdots + f_k$; and $\overline{X}$ is the mean.

EXAMPLE 5   Finding Standard Deviation for Grouped Data

Find the standard deviation for the grouped data given in Table 15.

Table 15

	Class Interval	Tally	Frequency
1	50– 59.99	7HK //	7
2	60– 69.99	7HK ////	9
3	70– 79.99	7HK 7HK 7HK ////	19
4	80– 89.99	7HK 7HK ////	14
5	90– 99.99	7HK 7HK ////	14
6	100–109.99	7HK /	6
7	110–119.99	//	2

SOLUTION   We have already found (see Example 2, p. 428) that the mean for the grouped data is

$$\overline{X} = 81.34$$

The class midpoints are 55, 65, 75, 85, 95, 105, and 115. The deviations of the mean from the class midpoints, their squares, and the products of the squares by the respective frequencies are listed in Table 16. The standard deviation is

$$\sigma = \sqrt{\frac{16{,}447.99}{71}} = \sqrt{231.66} = 15.22$$

Table 16

Class Midpoint	f_i	$m_i - \overline{X}$	$(m_i - \overline{X})^2$	$(m_i - \overline{X})^2 \cdot f_i$
115	2	33.7	1,135.69	2271.38
105	6	23.7	561.69	3370.14
95	14	13.7	187.69	2627.66
85	14	3.7	13.69	191.66
75	19	−6.3	39.69	754.11
65	9	−16.3	265.69	2391.21
55	7	−26.3	691.69	4841.83
Sum	71			16,447.99

A little computation shows that the sum of the deviations of the approximate mean from the class midpoints is not exactly zero. This is due to the fact that we are using an approximation to the mean. Remember, we cannot compute the exact mean for grouped data.

Chebychev's Theorem

Suppose we are observing an experiment with numerical outcomes and that the experiment has mean $\overline{X}$ and standard deviation σ. We wish to estimate the probability that a randomly chosen outcome lies within k units of the mean.

Chebychev's Theorem*

For any distribution of numbers with mean $\overline{X}$ and standard deviation σ, the probability that a randomly chosen outcome lies between $\overline{X} - k$ and $\overline{X} + k$ is at least

$$1 - \frac{\sigma^2}{k^2}.$$

* Named after the nineteenth-century Russian mathematician P. L. Chebychev.

EXAMPLE 6   Using Chebychev's Theorem

Suppose that an experiment with numerical outcomes has mean 4 and standard deviation 1. Use Chebychev's theorem to estimate the probability that an outcome lies between 2 and 6.

SOLUTION   Here, $\overline{X} = 4$, $\sigma = 1$. Since we wish to estimate the probability that an outcome lies between 2 and 6, the value of k is $k = 6 - \overline{X} = 6 - 4 = 2$ (or $k = \overline{X} - 2 = 4 - 2 = 2$). Then by Chebychev's theorem, the desired probability is at least

$$1 - \frac{\sigma^2}{k^2} = 1 - \frac{1}{2^2} = 1 - \frac{1}{4} = .75$$

That is, we expect at least 75% of the outcomes of this experiment to lie between 2 and 6.

 Now Work Problem 19

EXAMPLE 7   Using Chebychev's Theorem

An office supply company sells boxes containing 100 paper clips. Because of the packaging procedure, not every box contains exactly 100 clips. From previous data it is known that the average number of clips in a box is indeed 100 and the standard deviation is 2.8. If the company ships 10,000 boxes, estimate the number of boxes having between 94 and 106 clips, inclusive.

SOLUTION   Our experiment involves counting the number of clips in the box. For this experiment we have $\overline{X} = 100$ and $\sigma = 2.8$. Therefore, by Chebychev's theorem the fraction of boxes having between $100 - 6$ and $100 + 6$ clips ($k = 6$) should be at least

$$1 - \frac{(2.8)^2}{6^2} = 1 - .22 = .78$$

That is, we expect at least 78% of 10,000 boxes, or about 7800 boxes to have between 94 and 106 clips.

The importance of Chebychev's theorem stems from the fact that it applies to *any* data—only the mean and standard deviation must be known. However, the estimate is a crude one. Other results (such as the *normal distribution* given later) produce more accurate estimates about the probability of falling within k units of the mean.

EXERCISE 9.4   Answers to odd-numbered problems begin on page AN-36.

1. Use histograms (a) and (b) to determine by inspection which distribution has the larger variance.

2. Use histograms (b) and (c) to determine by inspection which distribution has the larger variance.

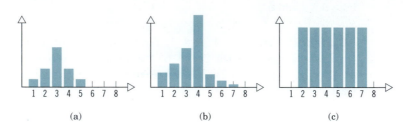

(a)   (b)   (c)

In Problems 3–8 compute the standard deviation for each set of data.

3. 4, 5, 9, 9, 10, 14, 25

4. 6, 8, 10, 10, 11, 12, 18

5. 62, 58, 70, 70

6. 55, 65, 80, 80, 90

7. 85, 75, 62, 78, 100

8. 92, 82, 75, 75, 82

In Problems 9 and 10 calculate the mean and the standard deviation.

9.

Class	Frequency
10–16	1
17–23	3
24–30	10
31–37	12
38–44	5
45–51	2

10.

Class	Frequency
0–3	2
4–7	5
8–11	8
12–15	6
16–19	3

11. The lifetimes of six light bulbs are 968, 893, 769, 845, 922, and 815 hours. Calculate the mean lifetime and the standard deviation.

12. A group of 25 applicants for admission to Midwestern University made the following scores on the quantitative part of an aptitude test:

591	570	425	472	555
490	415	479	517	570
606	614	542	607	441
502	506	603	488	460
550	551	420	590	482

Find the mean and standard deviation of these scores.

13. **Fishing**   The number of salmon caught in each of two rivers over the past 15 years is as follows:

River I		River II	
Number Caught	Years	Number Caught	Years
500–1499	4	750–1249	2
1500–2499	8	1350–1799	3
2500–3499	2	1800–2249	4
3500–4499	1	2250–2699	4
		2700–3149	2

Which river should be preferred for fishing?

14. **Charge Accounts**   A department store takes a sample of its customer charge accounts and finds the following:

Outstanding Balance	Number of Accounts
0–49	15
50–99	41
100–149	80
150–199	60
200–249	8

Find the mean and the standard deviation of the outstanding balances.

15. **Licensed Drivers in Tennessee**   Refer to the data provided in Problem 5, Exercise 9.2 and find the standard deviation.

16. Licensed Drivers in Hawaii   Refer to the data provided in Problem 6, Exercise 9.2 and find the standard deviation.

17. Undergraduate Tuition 1992–1993   Refer to the data provided in Problem 7, Exercise 9.2 and find the standard deviation.

18. Undergraduate Tuition 1993–1994   Refer to the data provided in Problem 8, Exercise 9.2 and find the standard deviation.

19. Suppose that an experiment with numerical outcomes has mean 25 and standard deviation 3. Use Chebychev's theorem to tell what percent of outcomes lie

(a) Between 19 and 31.
(b) Between 20 and 30.

(c) Between 16 and 34.
(d) Less than 19 or more than 31.
(e) Less than 16 or more than 34.

20. A watch company determines that the number of defective watches in each box averages 6 with standard deviation 2. Suppose that 1000 boxes are produced. Estimate the number of boxes having between 0 and 12 defective watches.

21. Sales   The average sale at a department store is $51.25, with a standard deviation of $8.50. Find the smallest interval such that by Chebychev's theorem at least 90% of the store's sales fall within it.

9.5 THE NORMAL DISTRIBUTION

Frequency polygons or frequency distributions can assume almost any shape or form, depending on the data. However, the data obtained from many experiments often follow a common pattern. For example, heights of adults, weights of adults, test scores, and coin tossing all lead to data that have the same kind of frequency distribution. This distribution is referred to as the **normal distribution** or the **Gaussian distribution.** Because it occurs so often in practical situations, it is generally regarded as the most important distribution, and much statistical theory is based on it. The graph of the normal distribution, called the **normal curve,** is the bell-shaped curve shown in Figure 9.

Some properties of the normal distribution are listed below.

Figure 9

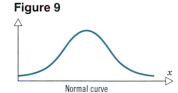

Normal curve

Figure 10

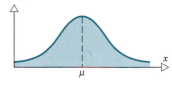

1. Normal curves are bell-shaped and are symmetric with respect to a vertical line at the mean μ. See Figure 10.
2. The mean, median, and mode of a normal distribution are equal.
3. Regardless of the shape, the area enclosed by the normal curve and the x-axis is always equal to 1 square unit. The shaded region in Figure 10 has an area of 1 square unit.

Figure 11

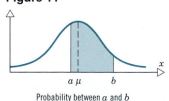

Probability between a and b
= area of the shaded region.

4. The probability that an outcome of a normally distributed experiment is between a and b equals the area under the associated normal curve from $x = a$ to $x = b$. See the shaded region in Figure 11.
5. The standard deviation of a normal distribution plays a major role in describing the area under the normal curve. As shown in Figure 12, on page 442, the standard deviation is related to the area under the normal curve as follows:

(a) About 68.27% of the total area under the curve is within 1 standard deviation of the mean (from $\mu - \sigma$ to $\mu + \sigma$).
(b) About 95.45% of the total area under the curve is within 2 standard deviations of the mean (from $\mu - 2\sigma$ to $\mu + 2\sigma$).
(c) About 99.73% of the total area under the curve is within 3 standard deviations of the mean (from $\mu - 3\sigma$ to $\mu + 3\sigma$).

It is also worth noting that, in theory, the normal curve will never touch the x-axis but will extend to infinity in either direction.

 **Now Work Problem 1**

Figure 12

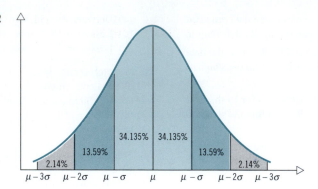

34.135% 34.135%

13.59%   13.59%

2.14%   2.14%

$\mu - 3\sigma$ $\mu - 2\sigma$ $\mu - \sigma$ μ $\mu - \sigma$ $\mu - 2\sigma$ $\mu - 3\sigma$

EXAMPLE 1   Using the Normal Curve to Analyze IQ Score

At Jefferson High School the average IQ score of the 1200 students is 100, with a standard deviation of 15. The IQ scores have a normal distribution.

(a) How many students have an IQ between 85 and 115?
(b) How many students have an IQ between 70 and 130?
(c) How many students have an IQ between 55 and 145?
(d) How many students have an IQ under 55 or over 145?
(e) How many students have an IQ over 145?

SOLUTION   See Figure 13.

Figure 13

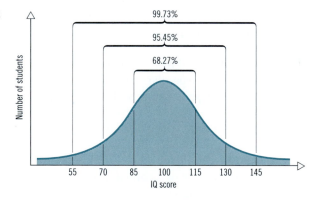

(a) The IQ scores have a normal distribution and the mean is 100. Since the standard deviation σ is 15, then 1σ either side of the mean is from 85 to 115. By Property 5(a) we know that 68.27% of 1200, or

$$(0.6827)(1200) = 819 \text{ students}$$

have IQs between 85 and 115.

(b) The scores from 70 to 130 extend 2σ ($= 30$) either side of the mean. By Property 5(b) we know that 95.45% of 1200, or

$$(0.9545)(1200) = 1145 \text{ students}$$

have IQs between 70 and 130.

(c) The scores from 55 to 145 extend 3σ ($=45$) either side of the mean. By Property 5(c) we know that 99.73% of 1200, or

$$(0.9973)(1200) = 1197 \text{ students}$$

have IQs between 55 and 145.

(d) There are three students ($1200 - 1197$) who have scores that are not between 55 and 145.

(e) One or two students have IQs above 145.

◼

A normal distribution is completely determined by the mean μ and the standard deviation σ. Hence, normal distributions of data with different means or different standard deviations give rise to different shapes of the normal curve.

Figure 14 indicates how the normal curve changes when the standard deviation changes. Each normal curve has the same mean 0.

As the standard deviation increases, the normal curve spreads out [Figure 14(c)], indicating a greater likelihood for the outcomes to be spread out. A compressed curve [Figure 14(a), (b)] indicates that the outcomes are more likely to be close to the mean.

Figure 14

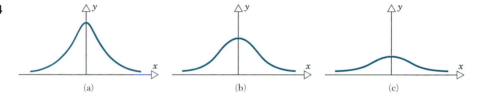

(a)   (b)   (c)

Standard Normal Curve

It would be a hopeless task to attempt to set up separate tables for areas under a normal curve for every conceivable value of μ and σ. Fortunately, we are able to transform all the observations to one table—the table corresponding to the so-called **standard normal curve,** which is the normal curve for which $\mu = 0$ and $\sigma = 1$. This is accomplished by introducing new data, called *Z-scores*.

Z-Score

The **Z-score** of a data item x is defined as

$$Z = \frac{\text{Difference between } x \text{ and } \mu}{\text{Standard deviation}} = \frac{x - \mu}{\sigma} \tag{1}$$

where

$x =$ Original data

$\mu =$ Mean of the original data

$\sigma =$ Standard deviation of the original data

The new data obtained using (1) will always have a *zero mean* and a *standard deviation* of one. Such data are said to be expressed in **standard units** or **standard scores.**

By expressing data in terms of standard units, it becomes possible to make a comparison of distributions.

EXAMPLE 2   Finding *Z*-Scores

On a test, 80 is the mean and 7 is the standard deviation. What is the *Z*-score of a score of

(a) 88?   (b) 62?

Interpret your results. Graph the normal curve.

SOLUTION

(a) Here, 88 is the original score. Using (1) with $x = 88$, $\mu = 80$, $\sigma = 7$, we get

$$Z = \frac{x - \mu}{\sigma} = \frac{88 - 80}{7} = \frac{8}{7} = 1.1429$$

(b) Here, 62 is the original score. Using (1) with $x = 62$, $\mu = 80$, and $\sigma = 7$, we get

$$Z = \frac{62 - 80}{7} = \frac{-18}{7} = -2.5714$$

The *Z*-score of 1.1429 tells us that the original score of 88 is 1.1429 standard deviations *above* the mean. The *Z*-score of -2.5714 tells us that the original score of 62 is 2.5714 standard deviations *below* the mean. A negative *Z*-score always means that the score is below the mean. See Figure 15 for a graph of the normal curve.

Figure 15

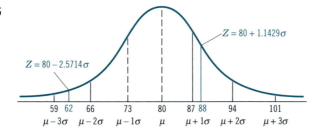

$$Z = 80 + 1.1429\sigma$$
$$Z = 80 - 2.5714\sigma$$

| 59 | 62 | 66 | 73 | 80 | 87 | 88 | 94 | 101 |
| $\mu - 3\sigma$ | | $\mu - 2\sigma$ | $\mu - 1\sigma$ | μ | $\mu + 1\sigma$ | | $\mu + 2\sigma$ | $\mu + 3\sigma$ |

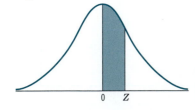

Now Work Problem 5

Figure 16

0   Z

The curve in Figure 16 with mean $\mu = 0$ and standard deviation $\sigma = 1$ is the standard normal curve. For this curve the areas between $Z = -1$ and 1, $Z = -2$ and 2, $Z = -3$ and 3 are equal, respectively, to 68.27%, 95.45%, and 99.73% of the total area under the curve, which is 1. To find the areas cut off between other points, we proceed as in the following example.

EXAMPLE 3   Using the Standard Normal Curve

(a) Find the area, that is, find the proportion of cases, included between 0 and 0.6 on a standard normal curve. Refer to the shaded area in Figure 17(a).

(b) Find the area, that is, find the proportion of cases, included between 0.6 and 1.86 on a standard normal curve. Refer to the shaded area in Figure 17(b).

Figure 17

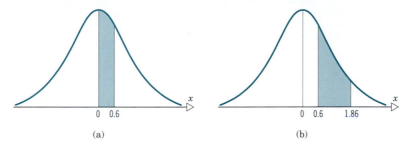

(a)   (b)

SOLUTION   We use the **standard normal curve table,** provided in Table III in the back of the book.

(a) To find the area between 0 and 0.6, we find $Z = 0.6$ in the table. Corresponding to $Z = 0.6$ is the value 0.2257, which is the area between the mean 0 and 0.6. In other words, 22.57% of the cases will lie between 0 and 0.6.

(b) We begin by checking the table to find the area of the curve cut off between the mean and a point equivalent to a standard score of 0.6 from the mean. This value is 0.2257, as we found in Part (a). Next, we continue down the table in the left-hand column until we come to a standard score of 1.8. By looking across the row to the column below 0.06, we find that 0.4686 of the area is included between the mean and 1.86. Then the area of the curve between these two points is the difference between the two areas, $0.4686 - 0.2257$, which is 0.2429. We can then state that approximately 24.29% of the cases fall between 0.6 and 1.86, or that *the probability of a score falling between these two points is about .2429.* ∎

 **Now Work Problem 7(a)**

In the next example we take two points that are on different sides of the mean.

EXAMPLE 4   Using the Standard Normal Curve

We want to determine the area of the standard normal curve that falls between a standard score of -0.39 and one of 1.86.

SOLUTION   See Figure 18.

Figure 18

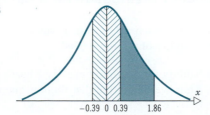

There are no values for negative standard scores in Table III. Because of the symmetry of normal curves, standard scores equal in absolute value, give equal areas when taken from the mean. From Table III we find that a standard score for 0.39 cuts off an area of 0.1517 between it and the mean. A standard score of 1.86 includes 0.4686 of the area of the curve between it and the mean. The area included between both points is then equal to the sum of these two areas, 0.1517 + 0.4686, which is 0.6203. Thus approximately 62.03% of the area is between −0.39 and 1.86. In other words, the probability of a score falling between these two points is about .6203.

EXAMPLE 5   Comparing Exam Scores

A student receives a grade of 82 on a final examination in biology for which the mean is 73 and the standard deviation is 9. In his final examination in sociology, for which the mean grade is 81 and the standard deviation is 15, he receives an 89. In which examination is his relative standing higher?

SOLUTION   In their present form these distributions are not comparable since they have different means and, more important, different standard deviations. In order to compare the data, we transform the data to standard scores. For the biology test data the Z-score for the student's examination score of 82 is

$$Z = \frac{82 - 73}{9} = \frac{9}{9} = 1$$

For the sociology test data, the Z-score for the student's examination score of 89 is

$$Z = \frac{89 - 81}{15} = \frac{8}{15} = 0.533$$

This means the student's score in the biology exam is 1 standard unit above the mean, while his score in the sociology exam is 0.533 standard unit above the mean. Hence, his **relative standing** is higher in biology.

Now Work Problem 19

The Normal Curve as an Approximation to the Binomial Distribution

We start with an example.

EXAMPLE 6   Finding the Frequency Distribution for a Binomial Probability

Consider an experiment in which a fair coin is tossed 10 times. Find the frequency distribution for the probability of tossing a head.

SOLUTION   The probability for obtaining exactly k heads is given by a binomial distribution $b(10, k; \frac{1}{2})$. Thus we obtain the distribution given in Table 17. If we graph this frequency distribution, we obtain the line chart shown in Figure 19. When we connect the tops of the lines of the line chart, we obtain a *normal curve,* as shown.

Table 17

No. of Heads	Probability $b(10, k; \frac{1}{2})$
0	.0010
1	.0098
2	.0439
3	.1172
4	.2051
5	.2461
6	.2051
7	.1172
8	.0439
9	.0098
10	.0010

Figure 19

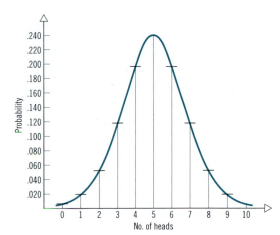

This particular distribution for $n = 10$ and $p = \frac{1}{2}$ is not a result of the choice of n or p. As a matter of fact, the line chart for any binomial probability $b(n, k; p)$ will give an approximation to a normal curve. You should verify this for the cases in which $n = 15$, $p = .3$, and $n = 8$, $p = \frac{3}{4}$.

Probabilities associated with binomial experiments are readily obtainable from the formula $b(n, k; p)$ when n is small. If n is large, we can compute the binomial probabilities by an approximating procedure using a normal curve. It turns out that the normal distribution provides a very good approximation to the binomial distribution when n is large or p is close to $\frac{1}{2}$.

The mean μ for the binomial distribution is given by $\mu = np$ (see Expected Value for Bernoulli Trials, page 393). Moreover, it can be shown that the standard deviation is $\sigma = \sqrt{npq}$.

EXAMPLE 7   Quality Control

A company manufactures 60,000 pencils each day. Quality control studies have shown that, on the average, 4% of the pencils are defective. A random sample of 500 pencils is selected from each day's production and tested. What is the probability that in the sample there are

(a) At least 12 and no more than 24 defective pencils?
(b) 32 or more defective pencils?

SOLUTION

(a) Since $n = 500$ is very large, it is appropriate to use a normal curve approximation for the binomial distribution. Thus with $n = 500$ and $p = .04$,

$$\mu = np = 500(.04) = 20 \qquad \sigma = \sqrt{npq} = \sqrt{500(.04)(.96)} = 4.38$$

To find the approximate probability that the number of defective pencils in a sample is at least 12 and no more than 24, we find the area under a normal curve from $x = 12$ to $x = 24$. See Figure 20.

Figure 20

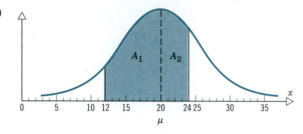

Areas A_1 and A_2 are found by converting to Z-scores and using the standard normal curve table, Table III.

$$x = 12: \quad Z_1 = \frac{x - \mu}{\sigma} = \frac{12 - 20}{4.38} = -1.83 \qquad A_1 = .4664$$

$$x = 24: \quad Z_2 = \frac{x - \mu}{\sigma} = \frac{24 - 20}{4.38} = .91 \qquad A_2 = .3186$$

$$\text{Total area} = A_1 + A_2 = .4664 + .3186 = .785$$

Thus the approximate probability of the number of defective pencils in the sample being at least 12 and no more than 24 is .785.

(b) We want to find the area A_2 indicated in Figure 21. We know that the area to the right of the mean is .5, and if we subtract the area A_1 from .5, we will obtain A_2. Therefore, we find the area A_1:

$$Z = \frac{x - \mu}{\sigma} = \frac{32 - 20}{4.38} = 2.74 \qquad A_1 = .4969$$

Then

$$A_2 = .5 - A_1 = .5 - .4969 = .0031$$

Thus the approximate probability of finding 32 or more defective pencils in the sample is .0031.

Figure 21

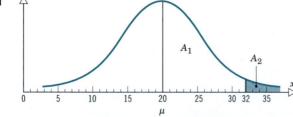

EXERCISE 9.5 Answers to odd-numbered problems begin on page AN-37.

In Problems 1–4 determine μ and σ by inspection for each normal curve.

1.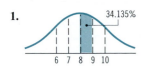

2.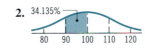

3.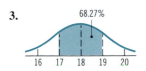

4.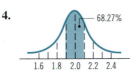

7. Given the following Z-scores on a standard normal distribution, find the area from the mean to each score.

   (a) 0.89   (b) 1.10   (c) 2.50
   (d) 3.00   (e) −0.75   (f) −2.31
   (g) 0.80   (h) 3.03

8. An instructor assigns grades in an examination according to the following procedure:

   A   if score exceeds $\mu + 1.6\sigma$

   B   if score is between $\mu + 0.6\sigma$ and $\mu + 1.6\sigma$

   C   if score is between $\mu - 0.3\sigma$ and $\mu + 0.6\sigma$

   D   if score is between $\mu - 1.4\sigma$ and $\mu - 0.3\sigma$

   F   if score is below $\mu - 1.4\sigma$

What percent of the class receives each grade, assuming that the scores are normally distributed?

5. Given a normal distribution with a mean of 13.1 and a standard deviation of 9.3, find the Z-score equivalent of the following scores in this distribution:

$$7, 9, 13, 15, 29, 37, 41$$

6. Given a normal distribution with a mean of 15.2 and a standard deviation of 5.1, find the Z-score equivalent of the following scores in this distribution:

$$8, 9, 15, 16, 22, 23, 25$$

In Problems 9–12 use Table III to find the area of each shaded region under the standard normal curve.

9.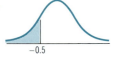

10.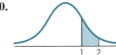

11.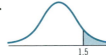

12.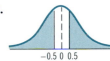

13. Women's Heights The average height of 2000 women in a random sample is 64 inches. The standard deviation is 2 inches. The heights have a normal distribution.

   (a) How many women in the sample are between 62 and 66 inches tall?

   (b) How many women in the sample are between 60 and 68 inches tall?

   (c) How many women in the sample are between 58 and 70 inches tall?

14. Weight of Corn Flakes in a Box Corn flakes come in a box that says it holds a mean weight of 16 ounces of cereal. The standard deviation is 0.1 ounce. Suppose that the manufacturer packages 600,000 boxes with weights that have a normal distribution.

   (a) How many boxes weigh between 15.9 and 16.1 ounces?

   (b) How many boxes weigh between 15.8 and 16.2 ounces?

   (c) How many boxes weigh between 15.7 and 16.3 ounces?

   (d) How many boxes weigh under 15.7 or over 16.3 ounces?

   (e) How many boxes weigh under 15.7 ounces?

15. Student Weights The weight of 100 college students closely follows a normal distribution with a mean of 130 pounds and a standard deviation of 5.2 pounds.

   (a) How many of these students would you expect to weigh at least 142 pounds?

   (b) What range of weights would you expect to include the middle 70% of the students in this group?

16. Life Expectancy of Clothing If the average life of a certain make of clothing is 40 months with a standard deviation of 7 months, what percentage of these clothes can be expected to last from 28 months to 42 months? Assume that clothing lifetime follows a normal distribution.

17. Life Expectancy of Shoes Records show that the average life expectancy of a pair of shoes is 2.2 years with a standard deviation of 1.7 years. A manufacturer guarantees that shoes lasting less than a year are replaced free. For every 1000 pairs sold, how many pairs should the manufacturer expect to replace free? Assume a normal distribution.

18. **Movie Theater Attendance** The attendance over a weekly period of time at a movie theater is normally distributed with a mean of 10,000 and a standard deviation of 1000 persons. Find

(a) The number in the lowest 70% of the attendance figures.

(b) The percent of attendance figures that falls between 8500 and 11,000 persons.

(c) The percent of attendance figures that differs from the mean by 1500 persons or more.

19. **Comparing Test Scores** Colleen, Mary, and Kathleen are vying for a position as editor. Colleen, who is tested with group I, gets a score of 76 on her test; Mary, who is tested with group II, gets a score of 89; and Kathleen, who is tested with group III, gets a score of 21. If the

average score for group I is 82, for group II is 93, and for group III is 24, and if the standard deviation for each group is 7, 2, and 9, repectively, which person has the highest relative standing?

20. In Mathematics 135 the average final grade is 75.0 and the standard deviation is 10.0. The professor's grade distribution shows that 15 students with grades from 68.0 to 82.0 received Cs. Assuming the grades follow a normal distribution, how many students are in Mathematics 135?

21. Draw the line chart and frequency curve for the probability of a head in an experiment in which a biased coin is tossed 15 times and the probability that a head occurs is .3. [*Hint:* Find $b(15, k; .30)$ for $k = 0, 1, \ldots, 15.$]

22. Follow the same directions as in Problem 21 for an experiment in which a biased coin is tossed 8 times and the probability that a head appears is $\frac{3}{4}$.

In Problems 23–28 suppose a binomial experiment consists of 750 trials and the probability of success for each trial is .4. Then

$$\mu = np = 300 \quad and \quad \sigma = \sqrt{npq} = \sqrt{(750)(.4)(.6)} = 13$$

Approximate the probability of obtaining the number of successes indicated by using a normal curve.

23. 285–315

24. 280–320

25. 300 or more

26. 300 or less

27. 325 or more

28. 275 or less

29. **Lifetime Batting Averages** A baseball player has a lifetime batting average of .250. If, in a season, this player comes to bat 300 times, what is the probability that at least 80 and no more than 90 hits occur? What is the probability that 85 or more hits occur?

30. **Hitting a Target** A skeet shooter has a long-established probability of hitting a target of .75. If, in a particular session, 200 attempts are made, what is the probability

that at least 135 and no more than 160 are successful? What is the probability that 160 or more are successful?

31. **Quality Control** A company manufactures 100,000 packages of jelly beans each week. On average, 1% of the packages do not seal properly. A random sample of 500 packages is selected at the end of the week. What is the probability that in this sample at least 10 are not properly sealed?

Technology Exercises

1. Graph the standard normal curve on a graphing calculator or a computer. For what value of x does the function assume its maximum? The equation is given by

$$y = \frac{1}{\sqrt{2\pi}} e^{-(1/2)x^2}$$

2. Graph the normal curve with $\mu = 10$ and $\sigma = 2$ on a graphing calculator or a computer. For what value of x does the function assume its maximum? The equation is given by

$$y = \frac{1}{2\sqrt{2\pi}} e^{-(1/8)(x-10)^2}$$

CHAPTER REVIEW

IMPORTANT TERMS AND CONCEPTS

bar graph 412
pie chart 413
frequency table 419
range 420
class interval 420
histogram 421
frequency polygon 421

mean 426, 428
median 429, 430
mode 431
standard deviation 435
standard deviation for
   grouped data 437

Chebychev's theorem 438
normal distribution 441
standard normal curve 443
Z-score 443

IMPORTANT FORMULAS

Mean for Ungrouped Scores $\overline{X} = \dfrac{x_1 + x_2 + \cdots + x_n}{n}$

Mean for Grouped Scores $\overline{X} = \dfrac{\Sigma f_i m_i}{n}$

Standard Deviation for Ungrouped Scores

$$\sigma = \sqrt{\dfrac{(x_1 - \overline{X})^2 + (x_2 - \overline{X})^2 + \cdots + (x_n - \overline{X})^2}{n}} = \sqrt{\dfrac{\Sigma(x_i - \overline{X})^2}{n}}$$

Standard Deviation for Grouped Scores

$$\sigma = \sqrt{\dfrac{(m_1 - \overline{X})^2 \cdot f_1 + (m_2 - \overline{X})^2 \cdot f_2 + \cdots + (m_k - \overline{X})^2 \cdot f_k}{n}} = \sqrt{\dfrac{\Sigma(m_i - \overline{X})^2 \cdot f_i}{n}}$$

Z-Score $Z = \dfrac{x - \mu}{\sigma}$

TRUE–FALSE ITEMS Answers are on page AN-37.

T_____ F_____ **1.** The range of a set of numbers is the difference between the standard deviation and the mean.

T_____ F_____ **2.** Two sets of scores can have the same mean and median, yet be different.

T_____ F_____ **3.** A relatively small standard deviation indicates that measures are widely scattered from the mean.

T_____ F_____ **4.** The sum of the deviations from the mean is zero for ungrouped data.

T_____ F_____ **5.** For the normal distribution, approximately 68.27% of the total area under the curve is within 2 standard deviations of the mean.

FILL IN THE BLANKS Answers are on page AN-37.

1. The three most common measures of central tendency are

  (a) _____ (b) _____

  (c) _____ .

2. The square root of the variance is called _____ .

3. The graph of the normal distribution has a _____ shape.

4. The formula $\dfrac{x - \mu}{\sigma}$ is called the _____ of x.

5. The formula $1 - \dfrac{\sigma^2}{k^2}$ measures the probability that a randomly chosen variable lies between _____ and _____ .

REVIEW EXERCISES   Answers to odd-numbered problems begin on page AN-38.

1. The following data show how office workers in Chicago get to work:

Means of Transportation	Percentage
Ride alone	64
Car pool	5
Ride bus	30
Other	1

Construct a bar graph and a pie chart, and compare them to see which one seems more informative to you.

2. To study their attitudes toward a new product, 1000 people were interviewed. Their response is given in the following table:

Attitude	No. of Responses
Do not like	420
Like	360
Like very much	220

Construct a bar graph and a pie chart, and compare them to see which one seems more informative to you.

3. Test Scores   The following scores were made on a math exam:

80	99	82	21	100	55	80	26	78	52
12	73	20	44	72	63	19	85	33	66
78	42	87	90	30	10	48	75	83	77
63	85	69	80	14	87	66	52	17	60
74	70	73	95	89	14	92	8	100	72

(a) Set up a frequency table for the above data. What is the range?
(b) Draw a line chart for the data.
(c) Draw a histogram for the data using a class interval of size 5 beginning with 4.5.
(d) Draw the frequency polygon for the histogram.
(e) Find the cumulative (less than) frequencies and draw the cumulative (less than) frequency distribution.

4. Tax Rates   The following table gives the percentage of marginal tax rates for married couples filing jointly for taxable income for 1998. (*Source:* U.S. Internal Revenue Service).

Income	Marginal Rates in Percent
0–42,350	15
42,350–102,300	28
102,300–155,950	31
155,950–278,450	36
Over 278,450	39.6

Graph the data using a bar graph.

5. Find the mean, median, and mode for each of the following sets of measurements.

(a) 12, 10, 8, 2, 0, 4, 10, 5, 4, 4, 8, 0
(b) 195, 5, 2, 2, 2, 2, 1, 0
(c) 2, 5, 5, 7, 7, 7, 9, 9, 11

6. In which of the sets of data in Problem 5 is the mean a poor measure of central tendency? Why?

7. Give an example of two sets of scores for which the means are the same and the standard deviations are different.

8. Give one advantage of the mean over the median. Give an example.

9. In seven different rounds of golf, Joe scores 74, 72, 76, 81, 77, 76, and 73. What is the standard deviation of his scores?

10. A normal distribution has a mean of 25 and a standard deviation of 5.

(a) What proportion of the scores fall between 20 and 30?
(b) What proportion of the scores will lie above 35?

11. A set of 600 scores is normally distributed. How many scores would you expect to find:

   (a) Between $\pm 1\sigma$ of the mean?
   (b) Between 1σ and 3σ above the mean?
   (c) Between $\pm \frac{2}{3}\sigma$ of the mean?

12. **Average Life of a Dog** The average life expectancy of a dog is 14 years, with a standard deviation of about 1.25 years. Assuming that the life spans of dogs are normally distributed, approximately how many dogs will die before reaching the age of 10 years, 4 months?

13. Use Table III to calculate the area under the normal curve between

   (a) $Z = -1.35$ and $Z = -2.75$
   (b) $Z = 1.2$ and $Z = 1.75$

14. **Comparing Test Scores** Bob got an 89 on the final exam in mathematics and a 79 on the sociology exam. In the mathematics class the average grade was 79 with a standard deviation of 5, and in the sociology class the average grade was 72 with a standard deviation of 3.5. Assuming that the grades in both subjects were normally distributed, in which class did Bob rank higher?

15. Suppose it is known that the number of items produced in a factory has a mean of 40. If the variance of a week's production is known to equal 25, then what can be said about the probability that this week's production will be between 30 and 50?

16. From past experience a teacher knows that the test scores of students taking an examination have a mean of 75 and a variance of 25. What can be said about the probability that a student will score between 65 and 85?

Mathematical Questions from Professional Exams

1. **Actuary Exam—Part II** Under the hypothesis that a pair of dice are fair, the probability is approximately .95 that the number of 7s appearing in 180 throws of the dice will lie within $30 \pm K$. What is the value of K?

   (a) 2   (b) 4   (c) 6   (d) 8   (e) 10

2. **Actuary Exam—Part II** If X is normally distributed with mean μ and variance μ^2 and if $P(-4 < X < 8) = .9974$, then $\mu =$

   (a) 1   (b) 2   (c) 4   (d) 6   (e) 8

3. **Actuary Exam—Part II** A manufacturer makes golf balls whose weights average 1.62 ounces, with a standard deviation of 0.05 ounce. What is the probability that the weight of a group of 100 balls will lie in the interval 162 ± 0.5 ounces?

   (a) .18   (b) .34   (c) .68   (d) .84   (e) .96

Chapter 10

Markov Chains; Games

10.1 Markov Chains and Transition Matrices

10.2 Regular Markov Chains

10.3 Absorbing Markov Chains

10.4 Two-Person Games

10.5 Mixed Strategies

10.6 Optimal Strategy in Two-Person Zero-Sum Games with 2 × 2 Matrices

Chapter Review

This chapter is divided into two parts: Markov chains (10.1–10.3) and games (10.4–10.6). These parts are completely independent of one another. Thus one may be covered and the other skipped without any difficulty. Both topics rely on probability and matrices.

A Markov chain is a probability model that consists of repeated trials of an experiment in which some change or transition is occurring, such as consumers changing their preferred choice of a soft drink. Information about how this change takes place is then used to predict the future market share.

Games provide a mathematical way to express the strategies available to people or corporations. Ways to arrive at a "best" strategy are discussed.

10.1 MARKOV CHAINS AND TRANSITION MATRICES

In Chapter 8 we introduced Bernoulli trials. In discussing Bernoulli trials, we made the assumption that the outcome of each trial is *independent* of the outcome of any previous trial.

Here we discuss another type of probability model, called a *Markov chain,* where there is some connection between one trial and the next. Markov chains have been shown to have applications in many areas, among them business, psychology, sociology, and biology.

Loosely speaking, a Markov chain or process is one in which what happens next is governed by what happened immediately before. At any stage a Markov experiment is in one of a finite number of states, with the next stage of the experiment consisting of movement to a possibly different state. The probability of moving to a certain state depends only on the state previously occupied and does not vary with time. Here are some situations that may be thought of as Markov chains:

1. There are yearly population shifts between a city and its surrounding suburbs. At any time, a person is either in the city or in the suburbs. So there are two states here with movement between them. As we track the population over time, we can ask: What percentage of the population is, say, in the city?
2. Several detergents compete in the market. Each year there is some shift of customer loyalty from one brand to another. At any point in time the state of the experiment would correspond to the brand of detergent he or she uses.
3. A psychology experiment consists of placing a mouse in a maze composed of rooms. The mouse moves from room to room, and we think of this movement as moving from state to state.

Figure 1

Consider a maze consisting of four connecting rooms as shown in Figure 1. The rooms are numbered 1, 2, 3, 4 for convenience, and each room contains pulsating lights of a different color. The experiment consists of releasing a mouse in a particular room and observing its behavior.

We assume an observation is made whenever a movement occurs or after a fixed time interval, whichever comes first. Since the movement of the mouse is random in nature, we will use probabilistic terms to describe it.

We will refer to the rooms as states. For example, the probability p_{12} that the mouse moves from state 1 to state 2 might be $p_{12} = \frac{1}{2}$, while the probability of moving from state 1 to state 3 might be $p_{13} = \frac{1}{4}$. We are using p_{ij} to represent the probability of moving from state i to state j in one observation interval. So p_{11} would represent the probability that the mouse remains in room 1 for one observation interval; p_{12} represents the probability the mouse moves from room 1 to room 2 in one observation interval; and so on. Note that $p_{14} = 0$ since there is no direct passage between rooms 1 and 4.

A convenient way of writing the probabilities p_{ij} is to display them in a matrix. The resulting matrix

$$P = \begin{bmatrix} p_{11} & p_{12} & p_{13} & p_{14} \\ p_{21} & p_{22} & p_{23} & p_{24} \\ p_{31} & p_{32} & p_{33} & p_{34} \\ p_{41} & p_{42} & p_{43} & p_{44} \end{bmatrix}$$

is called the **transition matrix** of the experiment.

In our example there are four states and P is a 4×4 matrix. Were we to assign values to the remaining probabilities, a possible choice for the transition matrix P

might be

$$P = \begin{array}{c} \\ 1 \\ 2 \\ 3 \\ 4 \end{array} \begin{array}{cccc} 1 & 2 & 3 & 4 \\ \left[\begin{array}{cccc} \frac{1}{4} & \frac{1}{2} & \frac{1}{4} & 0 \\ \frac{1}{6} & \frac{2}{3} & 0 & \frac{1}{6} \\ \frac{1}{3} & 0 & \frac{1}{3} & \frac{1}{3} \\ 0 & \frac{1}{4} & \frac{1}{2} & \frac{1}{4} \end{array} \right] \end{array}$$

where the rows and columns are indexed by the four states (rooms).

We can display the entries in P using a tree diagram. See Figure 2. Thus the entries in P are conditional probabilities representing the probabilities of where the mouse will go next given that we know where it is now. This essential idea of movement from state to state with attached probabilities forms the basis of a *Markov chain.*

Figure 2

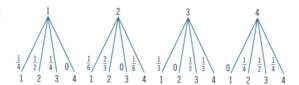

Markov Chain

A **Markov chain** is a sequence of experiments, each of which results in one of a finite number of states that we label 1, 2, . . . , m.

If p_{ij} is the probability of moving from state i to state j, then the **transition matrix** $\mathbf{P} = [p_{ij}]$ of a Markov chain is the $m \times m$ matrix

$$P = \begin{bmatrix} p_{11} & p_{12} & \cdots & p_{1m} \\ \cdot & \cdot & & \cdot \\ \cdot & \cdot & & \cdot \\ \cdot & \cdot & & \cdot \\ p_{m1} & p_{m2} & \cdots & p_{mm} \end{bmatrix}$$

Notice that the transition matrix P is a square matrix with entries that are always between 0 and 1, inclusive, since they represent probabilities. Also, the sum of the entries in every row is 1 since, as in the mouse and maze example, once the mouse is in a given room, it either stays there or moves to one of the other rooms.

Now Work Problem 1

Computing State Distributions

The transition matrix contains the information necessary to predict what happens next, given that we know what happened before. It remains to specify what the state of affairs was at the start of the experiment. For example, what room was the mouse placed in at the start? We use a row vector for this purpose.

Initial Probability Distribution

In a Markov chain with m states, the **initial probability distribution** is a $1 \times m$ row vector $v^{(0)}$ whose ith entry is the probability the experiment was in state i at the start.

For example, if the mouse was equally likely to be placed in any one of the rooms at the start, then we would have $v^{(0)} = [\frac{1}{4} \quad \frac{1}{4} \quad \frac{1}{4} \quad \frac{1}{4}]$. Whereas, if it was decided to always place the mouse initially in room 1, then we would have $v^{(0)} = [1 \quad 0 \quad 0 \quad 0]$.

Probability Vector

A **probability vector** is a vector whose entries are nonnegative and sum to 1.

We conclude that the initial probability distribution $v^{(0)}$ of a Markov chain is a probability row vector.

EXAMPLE 1   Finding a Transition Matrix

Look again at the maze in Figure 1. Suppose we assign the following transition probabilities:

$$\text{From room 1 to} \quad \left\{ \begin{matrix} 1 & 2 & 3 & 4 \\ \frac{1}{3} & \frac{1}{3} & \frac{1}{3} & 0 \end{matrix} \right\}$$

Here, the mouse starts in room 1, and $\frac{1}{3}$ of the time it remains there during the time interval of observation, $\frac{1}{3}$ of the time it enters room 2, and $\frac{1}{3}$ of the time it enters room 3. Since it cannot go to room 4 directly from room 1, the probability assignment is 0. Similarly, the transition probabilities in moving from room 2, room 3, and room 4 may be given as follows:

$$\text{From room 2 to} \quad \left\{ \begin{matrix} 1 & 2 & 3 & 4 \\ \frac{1}{3} & \frac{1}{3} & 0 & \frac{1}{3} \end{matrix} \right\}$$

$$\text{From room 3 to} \quad \left\{ \begin{matrix} 1 & 2 & 3 & 4 \\ \frac{1}{3} & 0 & \frac{1}{3} & \frac{1}{3} \end{matrix} \right\}$$

$$\text{From room 4 to} \quad \left\{ \begin{matrix} 1 & 2 & 3 & 4 \\ 0 & \frac{1}{3} & \frac{1}{3} & \frac{1}{3} \end{matrix} \right\}$$

(a) Find the transition matrix P.
(b) If the initial placement of the mouse is in room 4, find the initial probability distribution.
(c) What are the probabilities of being in each room after two observations?

SOLUTION

(a) The transition matrix P is

$$P = [p_{ij}] = \begin{matrix} & \begin{matrix} 1 & 2 & 3 & 4 \end{matrix} \\ \begin{matrix} 1 \\ 2 \\ 3 \\ 4 \end{matrix} & \begin{bmatrix} \frac{1}{3} & \frac{1}{3} & \frac{1}{3} & 0 \\ \frac{1}{3} & \frac{1}{3} & 0 & \frac{1}{3} \\ \frac{1}{3} & 0 & \frac{1}{3} & \frac{1}{3} \\ 0 & \frac{1}{3} & \frac{1}{3} & \frac{1}{3} \end{bmatrix} \end{matrix}$$

(b) Next, since the initial placement of the mouse is in room 4, the initial probability distribution, denoted by $v^{(0)}$, is

$$v^{(0)} = [p_1^{(0)} \quad p_2^{(0)} \quad p_3^{(0)} \quad p_4^{(0)}] = [0 \quad 0 \quad 0 \quad 1]$$

(c) To answer (c), we use a tree diagram. See Figure 3.

Figure 3

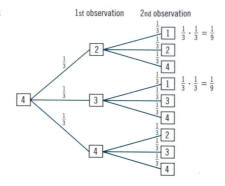

The numbers in each square refer to the room occupied. From this tree diagram we deduce, for example, that the mouse will be in state 1 after two observations with probability $\frac{1}{9} + \frac{1}{9} = \frac{2}{9}$. That is,

$$\text{Probability of moving from 4 to 1 in two stages} = \frac{2}{9}$$

Similarly,

Probability of moving from 4 to 2 in two stages $= \frac{1}{3} \cdot \frac{1}{3} + \frac{1}{3} \cdot \frac{1}{3} = \frac{2}{9}$

Probability of moving from 4 to 3 in two stages $= \frac{1}{3} \cdot \frac{1}{3} + \frac{1}{3} \cdot \frac{1}{3} = \frac{2}{9}$

Probability of moving from 4 to 4 in two stages $= \frac{1}{3} \cdot \frac{1}{3} + \frac{1}{3} \cdot \frac{1}{3} + \frac{1}{3} \cdot \frac{1}{3} = \frac{1}{3}$ (1)

 **Now Work Problem 3**

■

We can record the results of Example 1(c) by writing $v^{(2)} = [\frac{2}{9} \quad \frac{2}{9} \quad \frac{2}{9} \quad \frac{1}{3}]$. Using this notation, the row vector $v^{(k)}$ would be used to record the probabilities of being in the various states after k trials. The ith entry of $v^{(k)}$ is the probability of being in state i at stage k. For example, $v^{(4)} = [\frac{1}{8} \quad \frac{1}{8} \quad 0 \quad \frac{3}{4}]$ indicates that the probability of being in state 3 at stage 4 of the experiment is 0.

Instead of following the procedure of Example 1 to find $v^{(2)}$, it is possible to compute $v^{(k)}$ directly from the transition matrix P and the initial probability distribution $v^{(0)}$.

> **Probability Distribution after k Stages**
>
> In a Markov chain the probability distribution $v^{(k)}$ after k stages is
>
> $$v^{(k)} = v^{(k-1)}P \qquad (2)$$
>
> where P is the transition matrix.

Formula (2) shows that

$$v^{(1)} = v^{(0)} P$$
$$v^{(2)} = v^{(1)} P$$
$$v^{(3)} = v^{(2)} P$$

and so on. Thus the succeeding distribution can always be derived from the previous one by multiplying by the transition matrix.

EXAMPLE 2   **Using Formula (2)**

Using the information given in Example 1, use (2) to find the probability distribution after two observations.

SOLUTION   The initial probability distribution is $v^{(0)} = [0\ 0\ 0\ 1]$, so by (2),

$$v^{(1)} = v^{(0)}P = [0\ 0\ 0\ 1] \begin{bmatrix} \frac{1}{3} & \frac{1}{3} & \frac{1}{3} & 0 \\ \frac{1}{3} & \frac{1}{3} & 0 & \frac{1}{3} \\ \frac{1}{3} & 0 & \frac{1}{3} & \frac{1}{3} \\ 0 & \frac{1}{3} & \frac{1}{3} & \frac{1}{3} \end{bmatrix} = [0\ \frac{1}{3}\ \frac{1}{3}\ \frac{1}{3}]$$

Using (2) again gives

$$v^{(2)} = v^{(1)}P = [0\ \frac{1}{3}\ \frac{1}{3}\ \frac{1}{3}] \begin{bmatrix} \frac{1}{3} & \frac{1}{3} & \frac{1}{3} & 0 \\ \frac{1}{3} & \frac{1}{3} & 0 & \frac{1}{3} \\ \frac{1}{3} & 0 & \frac{1}{3} & \frac{1}{3} \\ 0 & \frac{1}{3} & \frac{1}{3} & \frac{1}{3} \end{bmatrix} = [\frac{2}{9}\ \frac{2}{9}\ \frac{2}{9}\ \frac{1}{3}]$$

This way of obtaining $v^{(2)}$ agrees with the results found in Example 1(c). ■

 **Now Work Problem 5**

A few computations using (2) lead to another result.

$$v^{(1)} = v^{(0)}P$$
$$v^{(2)} = v^{(1)}P = [v^{(0)}P]P = v^{(0)}P^2$$
$$v^{(3)} = v^{(2)}P = [v^{(0)}P^2]P = v^{(0)}P^3$$
$$v^{(4)} = v^{(3)}P = [v^{(0)}P^3]P = v^{(0)}P^4$$

In each line above we have substituted the result of the preceding line. These calculations can be continued to obtain the following result.

Probability Distribution after k Stages

In a Markov chain the probability distribution $v^{(k)}$ after k stages is

$$v^{(k)} = v^{(0)}P^k \tag{3}$$

where P^k is the kth power of the transition matrix.

Thus the conclusion of Example 2 could also have been arrived at by squaring the transition matrix and computing

$$v^{(2)} = v^{(0)}P^2 = [0\ 0\ 0\ 1] \begin{bmatrix} \frac{1}{3} & \frac{2}{9} & \frac{2}{9} & \frac{2}{9} \\ \frac{2}{9} & \frac{1}{3} & \frac{2}{9} & \frac{2}{9} \\ \frac{2}{9} & \frac{2}{9} & \frac{1}{3} & \frac{2}{9} \\ \frac{2}{9} & \frac{2}{9} & \frac{2}{9} & \frac{1}{3} \end{bmatrix} = [\frac{2}{9}\ \frac{2}{9}\ \frac{2}{9}\ \frac{1}{3}]$$

EXAMPLE 3 Population Movement

Suppose that the city of Oaklawn is experiencing a movement of its population to the suburbs. At present, 85% of the total population lives in the city and 15% lives in the suburbs. But each year 7% of the city people move to the suburbs, while only 1% of the suburb people move back to the city. Assuming that the total population (city and suburbs together) remains constant, what percent of the total will remain in the city after 5 years?

SOLUTION This problem can be expressed as a sequence of experiments in which each experiment measures the proportion of people in the city and the proportion of people in the suburbs.

In the $(n + 1)$st year these proportions will depend for their value only on the proportions in the nth year and not on the proportions found in earlier years. Thus we have an experiment that can be represented as a Markov chain.

The initial probability distribution for this Markov chain is

$$\begin{array}{cc} \text{City} & \text{Suburbs} \\ v^{(0)} = [.85 & .15] \end{array}$$

That is, initially, 85% of the people reside in the city and 15% in the suburbs.

The transition matrix P is

$$P = \begin{array}{c} \text{City} \\ \text{Suburbs} \end{array} \begin{array}{cc} \text{City} & \text{Suburbs} \\ \begin{bmatrix} .93 & .07 \\ .01 & .99 \end{bmatrix} \end{array}$$

That is, each year 7% of the city people move to the suburbs (so that 93% remain in the city) and 1% of the suburb people move to the city (so that 99% remain in the suburbs).

To find the probability distribution after 5 years, we need to compute $v^{(5)}$. We'll use Formula (2) five times.

$$v^{(1)} = v^{(0)}P = [.85 \quad .15] \begin{bmatrix} .93 & .07 \\ .01 & .99 \end{bmatrix} = [.792 \quad .208]$$

$$v^{(2)} = v^{(1)}P = [.792 \quad .208] \begin{bmatrix} .93 & .07 \\ .01 & .99 \end{bmatrix} = [.73864 \quad .26136]$$

$$v^{(3)} = v^{(2)}P = [.73864 \quad .26136] \begin{bmatrix} .93 & .07 \\ .01 & .99 \end{bmatrix} = [.68955 \quad .31045]$$

$$v^{(4)} = v^{(3)}P = [.68955 \quad .31045] \begin{bmatrix} .93 & .07 \\ .01 & .99 \end{bmatrix} = [.64439 \quad .35561]$$

$$v^{(5)} = v^{(4)}P = [.64439 \quad .35561] \begin{bmatrix} .93 & .07 \\ .01 & .99 \end{bmatrix} = [.60284 \quad .39716]$$

Thus, after 5 years, 60.28% of the residents live in the city and 39.72% live in the suburbs.

Now Work Problem 9

Example 3 leads us to inquire whether the situation in Oaklawn ever stabilizes. That is, after a certain number of years is an equilibrium reached? Also, does the equilibrium, if attained, depend on what the initial distribution of the population was, or is it independent of the initial state? We deal with these questions in the next section.

EXERCISE 10.1   Answers to odd-numbered problems begin on page AN-40.

---

1. Explain why the matrix below cannot be the transition matrix for a Markov chain.

$$\begin{bmatrix} 0 & 1 & 0 \\ \frac{1}{3} & \frac{1}{3} & \frac{1}{3} \\ \frac{1}{2} & -\frac{1}{2} & 0 \end{bmatrix}$$

2. Explain why the matrix below cannot be the transition matrix for a Markov chain.

$$\begin{bmatrix} 1 & \frac{1}{2} & \frac{1}{3} & \frac{1}{4} \\ 0 & 1 & 0 & 0 \\ 0 & \frac{1}{2} & \frac{1}{2} & 0 \\ 1 & 0 & 0 & 0 \end{bmatrix}$$

3. Consider a Markov chain with transition matrix

$$\begin{array}{cc} & \text{State 1} \quad \text{State 2} \end{array}$$
$$\begin{array}{c} \text{State 1} \\ \text{State 2} \end{array} \begin{bmatrix} \frac{1}{3} & \frac{2}{3} \\ \frac{1}{4} & \frac{3}{4} \end{bmatrix}$$

(a) What does the entry $\frac{1}{4}$ in this matrix represent?
(b) Assuming that the system is initially in state 1, find the probability distribution one observation later. What is it two observations later?
(c) Assuming that the system is initially in state 2, find the probability distribution one observation later. What is it two observations later?

4. Consider a Markov chain with transition matrix

$$\begin{array}{cc} & \text{State 1} \quad \text{State 2} \end{array}$$
$$\begin{array}{c} \text{State 1} \\ \text{State 2} \end{array} \begin{bmatrix} .3 & .7 \\ .4 & .6 \end{bmatrix}$$

(a) What does the entry .7 in the matrix represent?
(b) Assuming that the system is initially in state 1, find the probability distribution two observations later.
(c) Assuming that the system is initially in state 2, find the probability distribution two observations later.

5. Consider the transition matrix of Problem 4. If the initial probability distribution is [.25   .75], what is the probability distribution after two observations?

6. Consider a Markov chain with transition matrix

$$P = \begin{bmatrix} .7 & .2 & .1 \\ .6 & .2 & .2 \\ .4 & .1 & .5 \end{bmatrix}$$

If the initial distribution is [.25   .25   .5], what is the probability distribution in the next observation?

7. Find the values of a, b, and c that will make the following matrix a transition matrix for a Markov chain:

$$\begin{bmatrix} .2 & a & .4 \\ b & .6 & .3 \\ 0 & c & 0 \end{bmatrix}$$

8. In the maze of Figure 1 (page 455) if the initial probability distribution is $v^{(0)} = [\frac{1}{2} \ 0 \ \frac{1}{2} \ 0]$, find the probability distribution after two observations.

9. In Example 3 if the initial probability distribution for Oaklawn is $v^{(0)} = [.7 \ .3]$, what is the population distribution after 5 years?

10. **Mass Transit**   A new rapid transit system has just been installed. It is anticipated that each week 90% of the commuters who used the rapid transit will continue to do so. Of those who traveled by car, 5% will begin to use the rapid transit instead.

(a) Explain why the above is a Markov chain.
(b) Set up the 2 × 2 matrix P with columns and rows labeled R (rapid transit) and C (car) to display these transitions.
(c) Compute P^2 and P^3.

11. **Voting Patterns**   The voting pattern for a certain group of cities is such that 60% of the Democratic (D) mayors were succeeded by Democrats and 40% by Republicans (R). Also, 30% of the Republican mayors were succeeded by Democrats and 70% by Republicans.

(a) Explain why the above is a Markov chain.
(b) Set up the 2 × 2 matrix P with columns and rows labeled D and R to display these transitions.
(c) Compute P^2 and P^3.

12. **Maze Experiment**   Consider the maze with nine rooms shown in the figure on page 462 at the top of the left column. The system consists of the maze and a mouse. We assume that the following learning pattern exists: If the mouse is in room 1, 2, 3, 4, or 5, it moves with equal probability to any room that the maze permits; if it is in room 8, it moves directly to room 9; if it is in room 6, 7, or 9, it remains in that room.

1	2	3
4	5	6
7	8	9

(a) Explain why the above experiment is a Markov chain.
(b) Construct the transition matrix P.

13. **Market Penetration**   A company is promoting a certain product, say brand X wine. The result of this is that 75% of the people drinking brand X wine over any given period of 1 month continue to drink it the next month; of those people drinking other brands of wine in the period of 1 month, 35% change over to the promoted wine the next month. We would like to know what fraction of wine drinkers will drink brand X after 2 months if 50% drink brand X wine now.

14. A professor either walks or drives to a university. He never drives 2 days in a row, but if he walks 1 day, he is just as likely to walk the next day as to drive his car. Show that this forms a Markov chain and give the transition matrix.

15. **Insurance**   Suppose that, during the year 1998, 45% of the drivers in a certain metropolitan area had Travelers automobile insurance, 30% had General American insurance, and 25% were insured by some other companies. Suppose also that a year later: (*i*) of those who had been insured by Travelers in 1998, 92% continued to be insured by Travelers, but 8% had switched their insurance to General American; (*ii*) of those who had been insured by Gen-

eral American in 1998, 90% continued to be insured by General American, but 4% had switched to Travelers and 6% had switched to some other companies; (*iii*) of those who had been insured by some other companies in 1998, 82% continued but 10% had switched to Travelers, and 8% had switched to General American. Using these data, answer the following questions:

(a) What percentage of drivers in the metropolitan area were insured by Travelers and General American in 1999?
(b) If these trends continued for one more year, what percentage of the drivers were insured by Travelers and General American in 2000?

16. If A is a transition matrix, what about A^2? A^3? What do you conjecture about A^n?

17. Let

$$A = \begin{bmatrix} a_{11} & a_{12} \\ a_{21} & a_{22} \end{bmatrix}$$

be a transition matrix and

$$u = [u_1 \quad u_2]$$

be a probability row vector. Prove that uA is a probability vector.

18. Let

$$P = \begin{bmatrix} p_{11} & p_{12} \\ p_{21} & p_{22} \end{bmatrix}$$

be a transition matrix. Prove that $v^{(k)} = v^{(k-1)}P$.

Technology Exercises

Most graphing calculators are able to store and multiply matrices. Check your user's manual to see how to enter a matrix into the memory of your calculator.

Use a graphing calculator to solve Problems 1–4.

1. Consider the transition matrix

$$P = \begin{bmatrix} .5 & .2 & .1 & .2 \\ .3 & .3 & .2 & .2 \\ .1 & .5 & .1 & .3 \\ .25 & .25 & .25 & .25 \end{bmatrix}$$

If the initial probability distribution is $v = [.4 \quad .3 \quad .1 \quad .2]$, what is the probability distribution after ten observations?

2. Consider the transition matrix

$$P = \begin{bmatrix} .7 & .15 & .05 & .1 \\ .15 & .7 & .05 & .1 \\ .1 & .05 & .7 & .15 \\ .15 & .05 & .1 & .7 \end{bmatrix}$$

If the initial probability distribution is $v = [.15 \quad .15 \quad .55 \quad .15]$, what is the probability distribution after eight observations?

3. Consider the maze with six rooms shown in the figure. The system consists of the maze and a mouse. We assume the following learning pattern exists: If the mouse is in room 1, 2, 3, or 4, it moves with equal probability to any room the maze permits; if it is in room 5, it moves directly to room 6; if it is in room 6, it remains in that room.

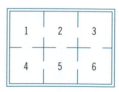

1	2	3
4	5	6

(a) Construct the transition matrix P.

(b) If the mouse is initially in room 2, what is the initial probability distribution?

(c) What is the probability distribution for the mouse after 10 stages?

(d) What is the most likely room for the mouse to be in after 10 stages?

4. Consider the maze with five rooms shown in the figure. The system consists of the maze and a mouse. We assume the following learning pattern exists: If the mouse is in room 1, 2, or 4, it moves with equal probability to any room the maze permits; if it is in room 3, it moves directly to room 5; if it is in room 5, it remains in that room.

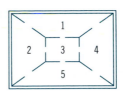

(a) Construct the transition matrix P.

(b) If the mouse is initially in room 2, what is the initial probability distribution?

(c) What is the probability distribution for the mouse after 10 stages?

(d) What is the most likely room for the mouse to be in after 10 stages?

10.2 REGULAR MARKOV CHAINS

One important question about Markov chains is: What happens in the long run? Does the distribution of the states tend to stabilize over time? In this section we investigate the conditions under which a Markov chain produces an *equilibrium*, or *steady-state*, situation. We also give a procedure for finding this equilibrium distribution, when it exists.

Powers of the Transition Matrix

We begin by examining *powers* of the transition matrix P. Recall that the (i, j)th entry p_{ij} of P is the probability of moving from state i to state j in any one step. Are there corresponding interpretations for the entries in any *power* $P^2, P^3, P^4, \ldots$ of the transition matrix?

To motivate the answer, we use the following example.

EXAMPLE 1   Consumer Loyalty

A company is promoting a certain product, say, brand X wine. The result of this is that 75% of the people drinking brand X wine over any given period of 1 month continue to drink it the next month; of those people drinking other brands of wine in the period of 1 month, 35% change over to the promoted wine the next month.

The transition matrix P of this experiment is

$$
P = \begin{array}{c} \\ \text{Brand X } E_1 \\ \text{Other brands } E_2 \end{array}
\begin{array}{c} \overset{\displaystyle \text{Brand X}}{E_1} \quad \overset{\displaystyle \text{Other brands}}{E_2} \\ \begin{bmatrix} .75 & .25 \\ .35 & .65 \end{bmatrix} \end{array}
$$

(a) Find the probability of passing from E_1 to either E_1 or E_2 in two stages.

(b) Find the probability of passing from E_2 to either E_1 or E_2 in two stages.

SOLUTION A tree diagram depicting two stages of this experiment is given in Figure 4.

Figure 4

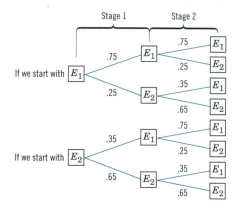

(a) The probability $p_{11}^{(2)}$ of proceeding from E_1 to E_1 in two stages is

$$p_{11}^{(2)} = (.75)(.75) + (.25)(.35) = .65$$

The probability $p_{12}^{(2)}$ from E_1 to E_2 in two stages is

$$p_{12}^{(2)} = (.75)(.25) + (.25)(.65) = .35$$

(b) The probability $p_{21}^{(2)}$ from E_2 to E_1 in two stages is

$$p_{21}^{(2)} = (.35)(.75) + (.65)(.35) = .49$$

The probability $p_{22}^{(2)}$ from E_2 to E_2 in two stages is

$$p_{22}^{(2)} = (.35)(.25) + (.65)(.65) = .51$$

If we square the transition matrix P, we obtain

$$
P^2 = \begin{bmatrix} .75 & .25 \\ .35 & .65 \end{bmatrix} \begin{bmatrix} .75 & .25 \\ .35 & .65 \end{bmatrix}
$$

$$
= \begin{bmatrix} (.75)(.75) + (.25)(.35) & (.75)(.25) + (.25)(.65) \\ (.35)(.75) + (.65)(.35) & (.35)(.25) + (.65)(.65) \end{bmatrix}
$$

$$
= \begin{bmatrix} .65 & .35 \\ .49 & .51 \end{bmatrix}
$$

and we notice that

$$
P^2 = \begin{bmatrix} p_{11}^{(2)} & p_{12}^{(2)} \\ p_{21}^{(2)} & p_{22}^{(2)} \end{bmatrix}
$$

Thus for Example 1 the *square* of the transition matrix gives the probabilities for moving from one state to another state in *two* stages.

This is true in general, and higher powers of the transition matrix carry similar interpretations.

> **Probability of Passing from State i to State j in n Stages**
>
> If P is the transition matrix of a Markov chain, then the (i, j)th entry of P^n (nth power of P) gives the probability of passing from state i to state j in n stages.

Regular Chains

We now turn to the question about the long-term behavior of a Markov chain.

EXAMPLE 2   **Examining Long-Term Behavior**

Use the transition matrix P given in Example 1 and compute P^2 through P^{12}.*

SOLUTION   Some of the powers of the transition matrix P are given below.

$$P = \begin{bmatrix} .7500 & .2500 \\ .3500 & .6500 \end{bmatrix} \qquad P^7 = \begin{bmatrix} .5840 & .4160 \\ .5824 & .4176 \end{bmatrix}$$

$$P^2 = \begin{bmatrix} .6500 & .3500 \\ .4900 & .5100 \end{bmatrix} \qquad P^8 = \begin{bmatrix} .5836 & .4164 \\ .5830 & .4170 \end{bmatrix}$$

$$P^3 = \begin{bmatrix} .6100 & .3900 \\ .5460 & .4540 \end{bmatrix} \qquad P^9 = \begin{bmatrix} .5834 & .4166 \\ .5832 & .4168 \end{bmatrix}$$

$$P^4 = \begin{bmatrix} .5940 & .4060 \\ .5684 & .4316 \end{bmatrix} \qquad P^{10} = \begin{bmatrix} .5834 & .4166 \\ .5833 & .4167 \end{bmatrix}$$

$$P^5 = \begin{bmatrix} .5876 & .4124 \\ .5774 & .4226 \end{bmatrix} \qquad P^{11} = \begin{bmatrix} .5834 & .4166 \\ .5833 & .4167 \end{bmatrix}$$

$$P^6 = \begin{bmatrix} .5850 & .4150 \\ .5809 & .4191 \end{bmatrix} \qquad P^{12} = \begin{bmatrix} .5833 & .4167 \\ .5833 & .4167 \end{bmatrix}$$

We notice the interesting fact that the powers P^n are converging and stabilizing around a fixed matrix. Also, the rows of that matrix are equal. Finally, if we let

$$\mathbf{t} = [.5833 \quad .4167]$$

and compute $\mathbf{t}P$, we find

$$\mathbf{t}P = [.5833 \quad .4167] \begin{bmatrix} .75 & .25 \\ .35 & .65 \end{bmatrix} = [.5833 \quad .4167]$$

That is, $\mathbf{t}P = \mathbf{t}$. So, the row vector $\mathbf{t}$ is *fixed* by P.

Will this always happen? And what interpretation can be placed on the vector $\mathbf{t}$? Before giving the answers, we need a definition.

---

*A graphing calculator such as the Casio 7700fx or the TI-83 or TI-86 makes this easy.

> **Regular Markov Chain**
>
> A Markov chain is said to be **regular** if, for some power of its transition matrix P, all of the entries are positive.

EXAMPLE 3   Identifying Regular Markov Chains

The transition matrix

$$P = \begin{bmatrix} \frac{1}{2} & \frac{1}{2} \\ 1 & 0 \end{bmatrix}$$

is regular since the square of P, namely,

$$P^2 = \begin{bmatrix} \frac{1}{2} & \frac{1}{2} \\ 1 & 0 \end{bmatrix}^2 = \begin{bmatrix} \frac{1}{2} & \frac{1}{2} \\ 1 & 0 \end{bmatrix}\begin{bmatrix} \frac{1}{2} & \frac{1}{2} \\ 1 & 0 \end{bmatrix} = \begin{bmatrix} \frac{3}{4} & \frac{1}{4} \\ \frac{1}{2} & \frac{1}{2} \end{bmatrix}$$

has all positive entries.

■

EXAMPLE 4   Identifying Markov Chains That Are Not Regular

The matrix

$$P = \begin{bmatrix} 1 & 0 \\ \frac{3}{4} & \frac{1}{4} \end{bmatrix}$$

is not regular since every power of P will always have $p_{12} = 0$.

The matrix

$$\begin{bmatrix} 0 & 1 \\ 1 & 0 \end{bmatrix}$$

is another example of a transition matrix that is not regular.

■

Fixed Probability Vector and Equilibrium

We now turn to the question about the long-term behavior of a Markov chain. Notice that the results stated below hold only if the Markov chain is regular.

> **Long-Term Behavior of a Regular Markov Chain**
>
> Let P be the transition matrix of a regular Markov chain.
>
> (a) The matrices P^n approach a fixed matrix T as n gets large. We write $P^n \rightarrow T$ (as n gets large).
>
> (b) The rows of T are all identical and equal to a probability row vector $\mathbf{t}$.
>
> (c) $\mathbf{t}$ is the unique probability vector that satisfies $\mathbf{t}P = \mathbf{t}$.
>
> (d) If $v^{(0)}$ is any initial distribution, then $v^{(n)} \rightarrow \mathbf{t}$ as n gets large.

Because of (c), **t** is called the **fixed probability vector** of the transition matrix P. Because of (d), **t** represents an equilibrium state since the state distributions $v^{(n)}$ approach **t** as time goes on, regardless of the initial distribution $v^{(0)}$. For example, the computations in Example 2 show that in the long run 58.33% of wine drinkers will be drinking brand X wine.

An important fact to note is that it is not necessary to compute higher and higher powers P^n of the transition matrix to find T and, in the process, **t**. Result (c) allows us to find **t** by solving a system of equations.

EXAMPLE 5   Finding the Probability Vector of a Markov Chain

Find the probability vector **t** of the Markov chain whose transition matrix P is

$$P = \begin{bmatrix} .93 & .07 \\ .01 & .99 \end{bmatrix}$$

(This is the transition matrix of Example 3, page 460, which represents population movement.)

SOLUTION   First, we observe that P is regular since $P = P^1$ has all positive entries.

Because $\mathbf{t} = [t_1 \quad t_2]$ is a probability vector, we must have $t_1 + t_2 = 1$. Because of (c), we must have $\mathbf{t}P = \mathbf{t}$. Thus

$$[t_1 \quad t_2] \begin{bmatrix} .93 & .07 \\ .01 & .99 \end{bmatrix} = [t_1 \quad t_2] \qquad \text{and} \qquad t_1 + t_2 = 1$$

$$[.93t_1 + .01t_2 \quad .07t_1 + .99t_2] = [t_1 \quad t_2] \qquad \text{and} \qquad t_1 + t_2 = 1$$

$$.93t_1 + .01t_2 = t_1 \qquad .07t_1 + .99t_2 = t_2 \qquad\qquad t_1 + t_2 = 1$$

$$-0.7t_1 + .01t_2 = 0 \qquad .07t_1 - .01t_2 = 0 \qquad\qquad t_1 + t_2 = 1$$

The two equations on the left are equivalent so, in fact, we have a system of two equations containing two variables, namely,

$$.07t_1 - .01t_2 = 0 \qquad \text{and} \qquad t_1 + t_2 = 1$$

or, equivalently,

$$\begin{cases} 7t_1 - t_2 = 0 \\ t_1 + t_2 = 1 \end{cases}$$

Adding, we find that $8t_1 = 1$ so

$$t_1 = \tfrac{1}{8} \qquad \text{and} \qquad t_2 = \tfrac{7}{8}$$

Thus the fixed probability vector of P is $\mathbf{t} = [\tfrac{1}{8} \quad \tfrac{7}{8}]$. ◼

In terms of Example 3, page 460, we interpret $\mathbf{t} = [\tfrac{1}{8} \quad \tfrac{7}{8}]$ as follows: in the long run, $\tfrac{1}{8}$ or 12.5% of the population will live in the city, while $\tfrac{7}{8}$ or 87.5% will live in the suburbs.

Now Work Problem 1

EXAMPLE 6   Finding the Fixed Probability Vector

Find the fixed probability vector **t** of the transition matrix

$$P = \begin{bmatrix} \frac{1}{2} & 0 & \frac{1}{2} \\ \frac{1}{2} & \frac{1}{2} & 0 \\ \frac{1}{3} & \frac{1}{3} & \frac{1}{3} \end{bmatrix}$$

SOLUTION   First, P is regular, since P^2 has all positive entries (you should verify this). Let $\mathbf{t} = [t_1 \quad t_2 \quad t_3]$ be the fixed vector. Then

$$\mathbf{t}P = \mathbf{t} \qquad \text{and} \qquad t_1 + t_2 + t_3 = 1$$

$$[t_1 \quad t_2 \quad t_3] \begin{bmatrix} \frac{1}{2} & 0 & \frac{1}{2} \\ \frac{1}{2} & \frac{1}{2} & 0 \\ \frac{1}{3} & \frac{1}{3} & \frac{1}{3} \end{bmatrix} = [t_1 \quad t_2 \quad t_3] \qquad \text{and} \qquad t_1 + t_2 + t_3 = 1$$

$$\tfrac{1}{2}t_1 + \tfrac{1}{2}t_2 + \tfrac{1}{3}t_3 = t_1 \qquad \tfrac{1}{2}t_2 + \tfrac{1}{3}t_3 = t_2 \qquad \tfrac{1}{2}t_1 + \tfrac{1}{3}t_3 = t_3 \qquad t_1 + t_2 + t_3 = 1$$

Upon simplifying,

$$-\tfrac{1}{2}t_1 + \tfrac{1}{2}t_2 + \tfrac{1}{3}t_3 = 0 \qquad -\tfrac{1}{2}t_2 + \tfrac{1}{3}t_3 = 0 \qquad \tfrac{1}{2}t_1 - \tfrac{2}{3}t_3 = 0 \qquad t_1 + t_2 + t_3 = 1$$

Removing fractions, we obtain the following system of equations:

$$\begin{cases} t_1 + t_2 + t_3 = 1 & (1) \\ 3t_1 - 3t_2 - 2t_3 = 0 & (2) \\ \quad\;\; -3t_2 + 2t_3 = 0 & (3) \\ 3t_1 \qquad\;\; - 4t_3 = 0 & (4) \end{cases}$$

Note that the sum of Equations (4) and (3) yields Equation (2), so that we really have a system of three equations containing three variables. We'll solve by substitution. From (3), $t_2 = \tfrac{2}{3}t_3$ and from (4), $t_1 = \tfrac{4}{3}t_3$. Substituting into (1), we obtain

$$\frac{4}{3}t_3 + \frac{2}{3}t_3 + t_3 = 1 \qquad t_1 + t_2 + t_3 = 1; t_1 = \frac{4}{3}t_3; t_2 = \frac{2}{3}t_3$$

$$3t_3 = 1$$

$$t_3 = \frac{1}{3}$$

Thus $t_1 = \tfrac{4}{3}t_3 = \tfrac{4}{9}$, $t_2 = \tfrac{2}{3}t_3 = \tfrac{2}{9}$, and $t_3 = \tfrac{1}{3}$. The fixed probability vector **t** of the transition matrix P is $\mathbf{t} = [\tfrac{4}{9} \quad \tfrac{2}{9} \quad \tfrac{1}{3}]$.

Now Work Problem 5

So, one feature of a regular Markov chain is that the long-run distribution **t** can be found simply by solving a system of equations. A more remarkable feature is that the same equilibrium **t** is reached no matter what the initial distribution is. We sketch the reason for this in the 2×2 case.

Suppose $v^{(0)} = [a \quad b]$ is *any* initial distribution of a regular Markov chain with transition matrix P. Let $\mathbf{t} = [t_1 \quad t_2]$ be the fixed vector. Then, by (a), we have

$$P^n \longrightarrow T = \begin{bmatrix} t_1 & t_2 \\ t_1 & t_2 \end{bmatrix}$$

so that

$$v^{(0)}P^n \longrightarrow v^{(0)}T$$

But $v^{(n)} = v^{(0)}P^n$. Thus

$$v^{(n)} \longrightarrow v^{(0)}T$$

Now

$$v^{(0)}T = [a \quad b] \begin{bmatrix} t_1 & t_2 \\ t_1 & t_2 \end{bmatrix} = [(a+b)t_1 \quad (a+b)t_2]$$

Since $v^{(0)}$ is a probability vector, it follows that $a + b = 1$.
  Thus, $v^{(0)}T = [t_1 \quad t_2]$ and

$$v^{(n)} \longrightarrow v^{(0)}T = [t_1 \quad t_2] = \mathbf{t}$$

That is, the same $\mathbf{t}$ is reached regardless of the entries used in $v^{(0)}$.

EXAMPLE 7 Consumer Loyalty

Consider a certain community with three grocery stores. Within this community (we assume that the population is fixed) there always exists a shift of customers from one grocery store to another. A study was made on January 1, and it was found that $\frac{1}{4}$ of the population shopped at store I, $\frac{1}{3}$ at store II, and $\frac{5}{12}$ at store III. Each month store I retains 90% of its customers and loses 10% of them to store II. Store II retains 5% of its customers and loses 85% of them to store I and 10% of them to store III. Store III retains 40% of its customers and loses 50% of them to store I and 10% to store II. The transition matrix P is

$$
\begin{array}{c}
\phantom{P = \text{II}} \quad \begin{array}{ccc} \text{I} & \text{II} & \text{III} \end{array} \\
P = \begin{array}{c} \text{I} \\ \text{II} \\ \text{III} \end{array} \begin{bmatrix} .90 & .10 & 0 \\ .85 & .05 & .10 \\ .50 & .10 & .40 \end{bmatrix}
\end{array}
$$

We would like to answer the following questions:

(a) What proportion of customers will each store retain by February 1?
(b) By March 1?
(c) Assuming the same pattern continues, what will be the long-run distribution of customers among the three stores?

SOLUTION

(a) To answer the first question, we note that the initial probability distribution is
  $v^{(0)} = [\frac{1}{4} \quad \frac{1}{3} \quad \frac{5}{12}]$. By February 1, the probability distribution is

$$v^{(1)} = v^{(0)}P = [\tfrac{1}{4} \quad \tfrac{1}{3} \quad \tfrac{5}{12}] \begin{bmatrix} .90 & .10 & .00 \\ .85 & .05 & .10 \\ .50 & .10 & .40 \end{bmatrix} = [.7167 \quad .0833 \quad .2000]$$

(b) To find the probability distribution after 2 months (March 1), we compute $v^{(2)}$:

$$v^{(2)} = v^{(0)}P^2 = \begin{bmatrix} \frac{1}{4} & \frac{1}{3} & \frac{5}{12} \end{bmatrix} \begin{bmatrix} .895 & .095 & .010 \\ .8575 & .0975 & .045 \\ .735 & .095 & .170 \end{bmatrix} = \begin{bmatrix} .8158 & .0958 & .0883 \end{bmatrix}$$

Since P^2 has all positive entries, we conclude P is regular.

(c) To find the long-run distribution, we determine the fixed probability vector $\mathbf{t}$ of the regular transition matrix P. Let $\mathbf{t} = \begin{bmatrix} t_1 & t_2 & t_3 \end{bmatrix}$. Then $\mathbf{t}P = \mathbf{t}$ and $t_1 + t_2 + t_3 = 1$, so that

$$\begin{bmatrix} t_1 & t_2 & t_3 \end{bmatrix} \begin{bmatrix} .90 & .10 & .00 \\ .85 & .05 & .10 \\ .50 & .10 & .40 \end{bmatrix} = \begin{bmatrix} t_1 & t_2 & t_3 \end{bmatrix} \qquad \text{and} \qquad t_1 + t_2 + t_3 = 1$$

The solution is found to be

$$\begin{bmatrix} t_1 & t_2 & t_3 \end{bmatrix} = \begin{bmatrix} .8889 & .0952 & .0159 \end{bmatrix}$$

Thus in the long run store I will have about 88.9% of all customers, store II will have 9.5%, and store III will have 1.6%.

 **Now Work Problem 9**

EXAMPLE 8 Spread of Rumor

A U.S. senator has decided whether to vote yes or no on an important bill pending in Congress and conveys this decision to an aide. The aide then passes this news on to another individual, who passes it on to a friend, and so on, each time to a new individual. Assume p is the probability that any one person passes on the information opposite to the way he or she heard it. Then $1 - p$ is the probability a person passes on the information the same way he or she heard it. With what probability will the nth person receive the information as a yes vote?

SOLUTION This can be viewed as a Markov chain model. Although it is not intuitively obvious, we shall find that the answer is essentially independent of p, provided $p \neq 0$ or 1.

To obtain the transition matrix, we observe that two states are possible: a yes is heard or a no is heard. The transition from a yes to a no or from a no to a yes occurs with probability p. The transition from yes to yes or from no to no occurs with probability $1 - p$. Thus the transition matrix P is

$$\begin{array}{cc} & \begin{array}{cc} \text{Yes} & \text{No} \end{array} \\ \begin{array}{c} \text{Yes} \\ \text{No} \end{array} & \begin{bmatrix} 1 - p & p \\ p & 1 - p \end{bmatrix} \end{array}$$

The probability that the nth person will receive the information in one state or the other is given by successive powers of the matrix P, that is, by the matrices P^n. In fact, the answer in this case rapidly approaches $t_1 = \frac{1}{2}, t_2 = \frac{1}{2}$, after any considerable number of people are involved. This can easily be shown by verifying that the fixed probability vector is $\begin{bmatrix} \frac{1}{2} & \frac{1}{2} \end{bmatrix}$. Hence, no matter what the senator's initial decision is, eventually (after enough information exchange), half the people hear that the senator is going to vote yes and half hear that the senator is going to vote no.

An interesting interpretation of this result applies to successive roll calls in Congress on the same issue. We let p represent the probability that a member changes his or her mind on an issue from one roll call to the next. Then the above example shows that if enough roll calls are forced, the Congress will eventually be near an even split on the issue. This could perhaps serve as a model for explaining a minority party's use of the parliamentary device of delaying actions.

EXERCISE 10.2   Answers to odd-numbered problems begin on page AN-40.

In Problems 1–6 determine which of the given matrices are regular. For those that are, find the fixed probability vector.

1. $\begin{bmatrix} \frac{1}{2} & \frac{1}{2} \\ 1 & 0 \end{bmatrix}$

2. $\begin{bmatrix} \frac{1}{2} & \frac{1}{2} \\ 0 & 1 \end{bmatrix}$

3. $\begin{bmatrix} 0 & 1 \\ \frac{1}{4} & \frac{3}{4} \end{bmatrix}$

4. $\begin{bmatrix} \frac{1}{3} & \frac{2}{3} \\ 1 & 0 \end{bmatrix}$

5. $\begin{bmatrix} 1 & 0 & 0 \\ \frac{1}{4} & \frac{1}{2} & \frac{1}{4} \\ 0 & 1 & 0 \end{bmatrix}$

6. $\begin{bmatrix} \frac{1}{4} & \frac{3}{4} & 0 \\ \frac{1}{2} & 0 & \frac{1}{2} \\ 0 & 1 & 0 \end{bmatrix}$

7. Show that the transition matrix P of Example 8 has the fixed probability vector $\begin{bmatrix} \frac{1}{2} & \frac{1}{2} \end{bmatrix}$.

8. Verify the result we obtained in Example 7, part (c).

9. **Consumer Loyalty**   A grocer stocks her store with three types of detergents, A, B, C. When brand A is sold out, the probability is .7 that she stocks up with brand A and .15 each with brands B and C. When she sells out brand B, the probability is .8 that she will stock up again with brand B and .1 each with brands A and C. Finally, when she sells out brand C, the probability is .6 that she will stock up with brand C and .2 each with brands A and B. Find the transition matrix. In the long run what is the stock of detergents?

10. **Consumer Loyalty**   A housekeeper buys three kinds of cereal: A, B, C. She never buys the same cereal in successive weeks. If she buys cereal A, then the next week she buys cereal B. However, if she buys either B or C, then the next week she is three times as likely to buy A as the other brand. Find the transition matrix. In the long run how often does she buy each of the three brands?

11. **Voting Loyalty**   In England, of the sons of members of the Conservative party, 70% vote Conservative and the rest vote Labour. Of the sons of Labourites, 50% vote Labour, 40% vote Conservative, and 10% vote Socialist. Of the sons of Socialists, 40% vote Socialist, 40% vote Labour, and 20% vote Conservative. What is the probability that the grandson of a Labourite will vote Socialist? What is the membership distribution in the long run?

12. If $\begin{bmatrix} \frac{1}{3} & 0 & \frac{1}{3} & \frac{1}{3} \end{bmatrix}$ is the fixed probability vector of a matrix P, can P be regular?

13. **Family Traits**   The probabilities that a blond mother will have a blond, brunette, or redheaded daughter are .6,

.2, and .2, respectively. The probabilities that a brunette mother will have a blond, brunette, or redheaded daughter are .1, .7, and .2, respectively. And the probabilities that a redheaded mother will have a blond, brunette, or redheaded daughter are .4, .2, and .4, respectively. What is the probability that a blond woman is the grandmother of a brunette? If the population of women is now 50% brunettes, 30% blonds, and the rest redheads, what will the distribution be:

(a) After two generations?   (b) In the long run?

14. **Educational Trends**   Use the data given in the table and assume that the indicated trends continue in order to answer the questions below.

		Maximum Education Children Achieve		
		College	H.S.	E.S.
Highest Educational Level of Parents	**College**	80%	18%	2%
	H.S.	40%	50%	10%
	E.S.	20%	60%	20%

(a) What is the transition matrix?
(b) What is the probability that a grandchild of a college graduate is a college graduate?
(c) What is the probability that the grandchild of a high school graduate finishes only elementary school?
(d) If at present 30%, 40%, and 30% of the population are college, high school, and elementary school graduates, respectively, what will be the distribution of the grandchildren of the present population?
(e) What will the long-run distribution be?

Technology Exercises

Use a graphing calculator to compute powers of the following transition matrices.
Stop when there is no change. Use this procedure to find the fixed probability
vector **t** *of each transition matrix.*

1.
$$\begin{bmatrix} .93 & .07 \\ .01 & .99 \end{bmatrix}$$

2.
$$\begin{bmatrix} .90 & .10 \\ .10 & .90 \end{bmatrix}$$

3.
$$\begin{bmatrix} \frac{1}{2} & 0 & \frac{1}{2} \\ \frac{1}{2} & \frac{1}{2} & 0 \\ \frac{1}{3} & \frac{1}{3} & \frac{1}{3} \end{bmatrix}$$

4.
$$\begin{bmatrix} \frac{1}{4} & \frac{1}{4} & \frac{1}{2} \\ \frac{1}{2} & \frac{1}{2} & 0 \\ \frac{1}{4} & \frac{1}{2} & \frac{1}{4} \end{bmatrix}$$

5.
$$\begin{bmatrix} .1 & 0 & .3 & .6 \\ .5 & .4 & .1 & 0 \\ .2 & .3 & .2 & .3 \\ .3 & .2 & .2 & .3 \end{bmatrix}$$

6.
$$\begin{bmatrix} .5 & .4 & .1 & 0 \\ .3 & .3 & .3 & .1 \\ .2 & .2 & .2 & .4 \\ .1 & .1 & .4 & .4 \end{bmatrix}$$

10.3 ABSORBING MARKOV CHAINS

We have already seen examples of transition matrices of Markov chains in which there are states that are impossible to leave. Such states are called *absorbing*. For example, in the transition matrix below, state 2 is absorbing since, once it is reached, it is impossible to leave.

$$\begin{array}{cc} & \text{State 1} \quad \text{State 2} \end{array}$$
$$\begin{array}{c} \text{State 1} \\ \text{State 2} \end{array} \begin{bmatrix} \frac{1}{4} & \frac{3}{4} \\ 0 & 1 \end{bmatrix}$$

Absorbing State; Absorbing Markov Chain

In a Markov chain if p_{ij} denotes the probability of going from state E_i to state E_j, then E_i is called an **absorbing state** if $p_{ii} = 1$. A Markov chain is said to be an **absorbing Markov chain** if and only if it contains at least one absorbing state and it is possible to go from *any* nonabsorbing state to an absorbing state in one or more stages.

Thus an absorbing state will capture the process and will not allow any state to pass from it.

 **Now Work Problem 1**

EXAMPLE 1   Identifying Absorbing Markov Chains

Consider two Markov chains with P_1 and P_2 as their transition matrices:

$$P_1 = \begin{array}{c} E_1 \\ E_2 \\ E_3 \\ E_4 \end{array} \begin{array}{cccc} E_1 & E_2 & E_3 & E_4 \\ \begin{bmatrix} .4 & .2 & .4 & 0 \\ 0 & 1 & 0 & 0 \\ .1 & 0 & .5 & .4 \\ .1 & 0 & .3 & .6 \end{bmatrix} \end{array} \qquad P_2 = \begin{array}{c} E_1 \\ E_2 \\ E_3 \end{array} \begin{array}{ccc} E_1 & E_2 & E_3 \\ \begin{bmatrix} 1 & 0 & 0 \\ 0 & \frac{1}{4} & \frac{3}{4} \\ 0 & \frac{1}{3} & \frac{2}{3} \end{bmatrix} \end{array}$$

Determine whether either or both are absorbing chains.

SOLUTION   The chain having P_1 as its transition matrix is absorbing since the second state is an absorbing state and it is possible to pass from each of the other states to the second. Specifically, it is possible to pass from the first state directly to the second state and from either the third or the fourth state to the first state and then to the second. On the other hand, the matrix P_2 is an example of a nonabsorbing matrix since it is impossible to go from the nonabsorbing state E_2 to the absorbing state E_1.

When working with an absorbing Markov chain, it is convenient to rearrange the states so that the absorbing states come first and the nonabsorbing states follow.

Once this rearrangement is made, the transition matrix can be subdivided into four submatrices:

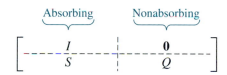

Here, I is an identity matrix, $\mathbf{0}$ denotes a matrix having all 0 entries, and the matrices S and Q are the two submatrices corresponding to the absorbing and nonabsorbing states.

EXAMPLE 2   Subdividing a Transition Matrix

Rearrange the transition matrix given below and find the submatrices S and Q:

$$
\begin{array}{c}
 \\
E_1 \\
E_2 \\
E_3 \\
E_4 \\
E_5
\end{array}
\begin{array}{ccccc}
E_1 & E_2 & E_3 & E_4 & E_5 \\
\left[\begin{array}{ccccc}
0 & .5 & 0 & 0 & .5 \\
0 & 0 & .9 & 0 & .1 \\
0 & 0 & 0 & .7 & .3 \\
0 & 0 & 0 & 1 & 0 \\
0 & 0 & 0 & 0 & 1
\end{array}\right]
\end{array}
$$

SOLUTION   First, we move the absorbing states E_4 and E_5 so they appear first.

$$
\begin{array}{c}
 \\
E_4 \\
E_5 \\
E_1 \\
E_2 \\
E_3
\end{array}
\begin{array}{ccccc}
E_4 & E_5 & E_1 & E_2 & E_3 \\
\left[\begin{array}{ccccc}
1 & 0 & 0 & 0 & 0 \\
0 & 1 & 0 & 0 & 0 \\
0 & .5 & 0 & .5 & 0 \\
0 & .1 & 0 & 0 & .9 \\
.7 & .3 & 0 & 0 & 0
\end{array}\right]
\end{array}
$$

Then the subdivided matrix is

$$
\begin{bmatrix}
1 & 0 & 0 & 0 & 0 \\
0 & 1 & 0 & 0 & 0 \\
\hline
0 & .5 & 0 & .5 & 0 \\
0 & .1 & 0 & 0 & .9 \\
.7 & .3 & 0 & 0 & 0
\end{bmatrix}
\qquad
\begin{bmatrix}
I & 0 \\
\hline
S & Q
\end{bmatrix}
$$

so that

$$
S = \begin{bmatrix} 0 & .5 \\ 0 & .1 \\ .7 & .3 \end{bmatrix}
\qquad
Q = \begin{bmatrix} 0 & .5 & 0 \\ 0 & 0 & .9 \\ 0 & 0 & 0 \end{bmatrix}
$$

■

Gambler's Ruin Problem

Consider the following game involving two players, sometimes called the *gambler's ruin problem.* Player I has $3 and player II has $2. They flip a fair coin; if it is a head, player I pays player II $1, and if it is a tail, player II pays player I $1. The total amount of money in the game is, of course, $5. We would like to know how long the game will last; that is, how long it will take for one of the players to go broke or win all the money. (This game can easily be generalized by assuming that player I has M dollars and player II has N dollars.)

In this experiment how much money a player has after any given flip of the coin depends only on how much he had after the previous flip and will not depend (directly) on how much he had in the preceding stages of the game. This experiment can thus be represented by a Markov chain.

For the *gambler's ruin problem* the game does not have to involve flipping a coin. That is, the probability that player I wins does not have to equal the probability that player II wins. Also, questions can be raised as to what happens if the stakes are doubled, how long the game can be expected to last, and so on.

The coin being flipped is fair so that a probability of $\frac{1}{2}$ is assigned to each outcome. The states are the amounts of money each player has at each stage of the game. Note that each player can increase or decrease the amount of money he has by only $1 at a time.

The transition matrix P is then

$$
P = \begin{array}{c}
\begin{array}{ccccccc}
 & 0 & 1 & 2 & 3 & 4 & 5
\end{array} \\
\begin{array}{c}
0 \\ 1 \\ 2 \\ 3 \\ 4 \\ 5
\end{array}
\begin{bmatrix}
1 & 0 & 0 & 0 & 0 & 0 \\
\frac{1}{2} & 0 & \frac{1}{2} & 0 & 0 & 0 \\
0 & \frac{1}{2} & 0 & \frac{1}{2} & 0 & 0 \\
0 & 0 & \frac{1}{2} & 0 & \frac{1}{2} & 0 \\
0 & 0 & 0 & \frac{1}{2} & 0 & \frac{1}{2} \\
0 & 0 & 0 & 0 & 0 & 1
\end{bmatrix}
\end{array}
$$

Here, $p_{10} = \frac{1}{2}$ is the probability of starting with $1 and losing it. Also, p_{00} is the probability of having $0 given that a player has started with $0. This is a sure event since, once a player is in state 0, he stays there forever (he is broke). Similarly, p_{55}

represents the probability of having \$5, given that a player started with \$5, which is again a sure event (the player has won all the money).

With regard to this problem, the following questions are of interest:

(a) Given that one gambler is in a nonabsorbing state, what is the expected number of times he will hold between \$1 and \$4 inclusive before the termination of the game? That is, on the average, how many times will the process be in nonabsorbing states?
(b) What is the expected length of the process (game)?
(c) What is the probability that an absorbing state is reached (that is, that one gambler will eventually be wiped out)?

To answer these questions, let's look at the transition matrix P. We rearrange this matrix so that the two absorbing states will appear in the first two rows:

$$
P = \begin{array}{c} \\ 0 \\ 5 \\ 1 \\ 2 \\ 3 \\ 4 \end{array}
\begin{array}{c} \begin{array}{cccccc} 0 & 5 & 1 & 2 & 3 & 4 \end{array} \\
\left[\begin{array}{cc:cccc}
1 & 0 & 0 & 0 & 0 & 0 \\
0 & 1 & 0 & 0 & 0 & 0 \\
\hdashline
\frac{1}{2} & 0 & 0 & \frac{1}{2} & 0 & 0 \\
0 & 0 & \frac{1}{2} & 0 & \frac{1}{2} & 0 \\
0 & 0 & 0 & \frac{1}{2} & 0 & \frac{1}{2} \\
0 & \frac{1}{2} & 0 & 0 & \frac{1}{2} & 0
\end{array}\right]
\end{array}
$$

We write P in the form

$$
P = \left[\begin{array}{c:c} I_2 & \mathbf{0} \\ \hdashline S & Q \end{array}\right]
$$

where

$$
I_2 = \begin{bmatrix} 1 & 0 \\ 0 & 1 \end{bmatrix} \qquad \mathbf{0} = \begin{bmatrix} 0 & 0 & 0 & 0 \\ 0 & 0 & 0 & 0 \end{bmatrix}
$$

$$
S = \begin{bmatrix} \frac{1}{2} & 0 \\ 0 & 0 \\ 0 & 0 \\ 0 & \frac{1}{2} \end{bmatrix} \qquad Q = \begin{bmatrix} 0 & \frac{1}{2} & 0 & 0 \\ \frac{1}{2} & 0 & \frac{1}{2} & 0 \\ 0 & \frac{1}{2} & 0 & \frac{1}{2} \\ 0 & 0 & \frac{1}{2} & 0 \end{bmatrix}
$$

As we indicated earlier, we can do this to any transition matrix of an absorbing Markov chain. If r of the states are absorbing, the transition matrix P can be written as

$$
P = \left[\begin{array}{c:c} I_r & \mathbf{0} \\ \hdashline S & Q \end{array}\right]
$$

where I_r is the $r \times r$ identity matrix, $\mathbf{0}$ is the zero matrix of dimension $r \times s$, S is of dimension $s \times r$, and Q is of dimension $s \times s$.

In order to answer the questions raised about gambler's ruin problem, we need the following result:

Fundamental Matrix of an Absorbing Markov Chain

For an absorbing Markov chain that has a transition matrix P of the form

$$P = \left[\begin{array}{c|c} I_r & 0 \\ \hline S & Q \end{array}\right]$$

where S is of dimension $s \times r$ and Q is of dimension $s \times s$, define the matrix T, called the **fundamental matrix** of the Markov chain, to be

$$T = [I_s - Q]^{-1} \tag{1}$$

The entries of T give the expected number of times the process is in each non-absorbing state, provided the process began in a nonabsorbing state.

In the gambler's ruin problem the fundamental matrix T is

$$T = \left[\begin{bmatrix} 1 & 0 & 0 & 0 \\ 0 & 1 & 0 & 0 \\ 0 & 0 & 1 & 0 \\ 0 & 0 & 0 & 1 \end{bmatrix} - \begin{bmatrix} 0 & \frac{1}{2} & 0 & 0 \\ \frac{1}{2} & 0 & \frac{1}{2} & 0 \\ 0 & \frac{1}{2} & 0 & \frac{1}{2} \\ 0 & 0 & \frac{1}{2} & 0 \end{bmatrix}\right]^{-1} = \begin{array}{c} 1 \\ 2 \\ 3 \\ 4 \end{array}\begin{array}{cccc} \!\!\!1 & \!\!2 & \!\!3 & \!\!4 \\ \begin{bmatrix} 1.6 & 1.2 & .8 & .4 \\ 1.2 & 2.4 & 1.6 & .8 \\ .8 & 1.6 & 2.4 & 1.2 \\ .4 & .8 & 1.2 & 1.6 \end{bmatrix} \end{array} \tag{2}$$

This provides the answers to question (a): the entry .8 in row 3, column 1, indicates that .8 is the expected number of times the player will have \$1 if he started with \$3. In the fundamental matrix T the column headings indicate present money, while the row headings indicate money started with.

Expected Number of Steps before Absorption

The expected number of steps before absorption for each nonabsorbing state is found by adding the entries in the corresponding row of the fundamental matrix T.

So to answer question (b), we again look at the matrix T in (2). The expected number of games before absorption (when one of the players wins or loses all the money) can be found by adding the entries in each row of T. Thus if a player starts with \$3, the expected number of games before absorption is

$$.8 + 1.6 + 2.4 + 1.2 = 6.0$$

That is, on the average, we expect this game to end on the seventh play.

If a player starts with \$1, the expected number of games before absorption is

$$1.6 + 1.2 + .8 + .4 = 4.0$$

Our last result deals with the probability of being absorbed.

> **Probability of Being Absorbed**
>
> The (i, j)th entry in the matrix product $T \cdot S$ gives the probability that, starting in nonabsorbing state i, we reach the absorbing state j.

Thus to answer question (c), we find the product of the matrices T and S:

$$
T \cdot S =
\begin{bmatrix}
1.6 & 1.2 & .8 & .4 \\
1.2 & 2.4 & 1.6 & .8 \\
.8 & 1.6 & 2.4 & 1.2 \\
.4 & .8 & 1.2 & 1.6
\end{bmatrix}
\begin{bmatrix}
\frac{1}{2} & 0 \\
0 & 0 \\
0 & 0 \\
0 & \frac{1}{2}
\end{bmatrix}
=
\begin{matrix}
& \begin{matrix} 0 & \;\; 5 \end{matrix} \\
\begin{matrix} 1 \\ 2 \\ 3 \\ 4 \end{matrix} &
\begin{bmatrix}
.8 & .2 \\
.6 & .4 \\
.4 & .6 \\
.2 & .8
\end{bmatrix}
\end{matrix}
$$

The entry in row 3, column 2, indicates the probability is .6 that the player starting with $3 will win all the money. The entry in row 2, column 1, indicates the probability is .6 that a player starting with $2 will lose all his money.

The above techniques are applicable in general to any transition matrix P of an absorbing Markov chain.

In the gambler's ruin problem we assumed player I started with $3 and player II with $2. Furthermore, we assumed the probability of player I winning $1 was .5. Table 1 gives probabilities for ruin and expected length for other kinds of betting situations for which the bet is 1 unit.

Table 1

Probability that Player I Wins	Amount of Units Player I Starts with	Amount of Units Player II Starts with	Probability that Player I Goes Broke	Expected Length of Game	Expected Gain of Player I
.50	9	1	.1	9	0
.50	90	10	.1	900	0
.50	900	100	.1	90,000	0
.50	8000	2000	.2	16,000,000	0
.45	9	1	.210	11	−1.1
.45	90	10	.866	765.6	−76.6
.45	99	1	.182	171.8	−17.2
.40	90	10	.983	441.3	−88.3
.40	99	1	.333	161.7	−32.3

Suppose, for example, that player I starts with $90 and player II with $10, with player I having a probability of .45 of winning (the game being unfavorable to player I). If at each trial, the stake is $1, Table 1 shows that the probability is .866 that player I is ruined. If the same game is played with player I having $9 to start and player II having $1 to start, the probability that player I is ruined drops to .210.

Now Work Problem 11

Model: The Rise and Fall of Stock Prices*

This model involves using Markov chains to analyze the movement of stock prices. The stocks (shares) were those whose daily progress was recorded by the London *Financial Times*. At any point in time a given stock was classified as being in one of three states: increase (*I*), decrease (*D*), or no change (*N*), depending on whether its price had risen, fallen, or remained the same compared to the previous trading day. The intent of the model was to study the long-range movement of stocks through these states. Data reporting the past history of stock prices over a period of 1097 trading days were used to statistically fit a transition matrix to the model. The transition matrix used was

$$P = \begin{array}{c} \\ I \\ D \\ N \end{array} \begin{array}{c} \begin{array}{ccc} I & D & N \end{array} \\ \left[\begin{array}{ccc} .586 & .074 & .340 \\ .070 & .638 & .292 \\ .079 & .064 & .857 \end{array} \right] \end{array}$$

Thus a share that increased one day would have probability .586 of also increasing the next day and a probability of .340 of registering no change the next day. Note that the tendency to remain in the same state is strongest for the no-change state since p_{33} is the largest of the diagonal elements.

The fixed probability vector **t** for the transition matrix can be computed to be

$$\mathbf{t} = [.157 \quad .154 \quad .689]$$

Thus in the long run a stock will have increased its price 15.7% of the time, decreased its price 15.4% of the time, and remained unchanged 68.9% of the time.

How long will a stock increase before its price begins to fall? Or, in general, how long will a stock be in a given state before it moves to another one? These questions are highly reminiscent of those encountered while studying absorbing chains, yet here the transition matrix *P* is not absorbing. Thus the techniques used earlier do not apply.

Suppose we wanted to know on the average how long a stock would spend in states *I* or *N* before arriving in state *D*. We could then assume state *D* is an absorbing one and replace the current probability p_{22} with 1 and make all other elements in the second row of *P* zero. The new matrix *P'* we have created is

$$P' = \begin{array}{c} \\ I \\ D \\ N \end{array} \begin{array}{c} \begin{array}{ccc} I & D & N \end{array} \\ \left[\begin{array}{ccc} .586 & .074 & .340 \\ 0 & 1 & 0 \\ .079 & .064 & .857 \end{array} \right] \end{array}$$

which is the matrix of an absorbing chain.

Subdividing this matrix, we obtain

$$\begin{array}{c} \\ D \\ I \\ N \end{array} \begin{array}{c} \begin{array}{ccc} D & I & N \end{array} \\ \left[\begin{array}{c|cc} 1 & 0 & 0 \\ \hline .074 & .586 & .340 \\ .064 & .079 & .857 \end{array} \right] \end{array}$$

* Adapted from Myles M. Dryden, "Share Price Movements: A Markovian Approach," *Journal of Finance, 24* (March 1969), pp. 49–60.

with

$$Q = \begin{array}{c} \\ I \\ N \end{array} \begin{array}{cc} I & N \\ \left[\begin{array}{cc} .586 & .340 \\ .079 & .857 \end{array} \right] \end{array}$$

The fundamental matrix of this absorbing Markov chain is

$$(I - Q)^{-1} = \left[\begin{array}{cc} .414 & -.340 \\ -.079 & .143 \end{array} \right]^{-1} = \begin{array}{c} \\ I \\ N \end{array} \begin{array}{cc} I & N \\ \left[\begin{array}{cc} 4.42 & 10.51 \\ 2.44 & 12.80 \end{array} \right] \end{array}$$

The entries of $(I - Q)^{-1}$ are interpreted as follows. The (i, j)th entry gives the average time spent in state j having started in state i before reaching the absorbing state D for the first time.

So, a stock currently increasing would spend an average of 4.4 days increasing and 10.5 days not changing before declining. Hence, a total of 14.9 days on the average would elapse before an increasing stock first began to decrease. Likewise, if a stock is currently exhibiting no change, the second row of the above matrix shows that an average of $2.44 + 12.80 = 15.24$ days would go by before it began to decrease.

Suppose a stock is increasing and then either levels off or declines. How long will it take before it starts rising again? In general, given we are in state i and leave, what is the average time that elapses before we return again to state i? The answer, which we state without proof, is given in the box below.

Average Time between Visits

The average amount of time elapsed between visits to state i (called **mean recurrence time**) is given by the reciprocal of the ith component of the fixed probability vector **t**.

Here

$$\mathbf{t} = [.157 \quad .154 \quad .689]$$

So we can then compute

State	Mean Recurrence Time (Days)
I	6.37
D	6.49
N	1.45

In this model, then, days on which a share's price fails to change follow each other fairly closely.

EXERCISE 10.3   Answers to odd-numbered problems begin on page AN-40.

In Problems 1–6 state which of the given matrices represent absorbing Markov chains.

1. $\begin{bmatrix} 0 & 1 \\ \frac{1}{4} & \frac{3}{4} \end{bmatrix}$

2. $\begin{bmatrix} 1 & 0 \\ \frac{1}{3} & \frac{2}{3} \end{bmatrix}$

3. $\begin{bmatrix} 1 & 0 & 0 \\ \frac{1}{8} & \frac{5}{8} & \frac{2}{8} \\ 0 & 0 & 1 \end{bmatrix}$

4. $\begin{bmatrix} 0 & 0 & 1 \\ 1 & 0 & 0 \\ 0 & 1 & 0 \end{bmatrix}$

5. $\begin{bmatrix} 0 & 1 & 0 & 0 \\ 1 & 0 & 0 & 0 \\ 0 & 0 & 1 & 0 \\ \frac{1}{4} & 0 & \frac{3}{4} & 0 \end{bmatrix}$

6. $\begin{bmatrix} \frac{1}{3} & \frac{1}{3} & 0 & \frac{1}{3} \\ 0 & \frac{1}{4} & \frac{1}{4} & \frac{1}{2} \\ 0 & 0 & 1 & 0 \\ 0 & \frac{1}{2} & 0 & \frac{1}{2} \end{bmatrix}$

7. Find the fundamental matrix T of the absorbing Markov chain in Problem 3. Also, find S and $T \cdot S$.

8. Follow the directions of Problem 7 for the matrix in Problem 6.

9. Suppose that for a certain absorbing Markov chain the fundamental matrix T is found to be

$$\begin{array}{cc} & \begin{array}{ccc} \$1 & \$2 & \$3 \end{array} \\ \begin{array}{c} \$1 \\ \$2 \\ \$3 \end{array} & \begin{bmatrix} 1.5 & 1.0 & 0.5 \\ 1.0 & 2.0 & 1.0 \\ 0.5 & 1.0 & 1.5 \end{bmatrix} \end{array}$$

(a) What is the expected number of times a person will have $3, given that she started with $1? With $2?
(b) If a player starts with $3, how many games can she expect to play before absorption?

10. Use the matrix T from Problem 9 to answer the following questions:

(a) What is the expected number of times a person will have $2 given that she started with $1? With $2?
(b) If a player starts with $1, how many games can she expect to play before the game ends?

11. **Gambler's Ruin Problem**   A person repeatedly bets $1 each day. If he wins, he wins $1 (he receives his bet of $1 plus winnings of $1). He stops playing when he goes broke, or when he accumulates $3. His probability of winning is .4 and of losing is .6. What is the probability of eventually accumulating $3 if he starts with $1? With $2?

12. The following data were obtained from the admissions office of a 2-year junior college. Of the first-year class (F), 75% became sophomores (S) the next year and 25% dropped out (D). Of those who were sophomores during a particular year, 90% graduated (G) by the following year and 10% dropped out.

(a) Set up a Markov chain with states D, F, G, and S that describes the situation.
(b) How many states are absorbing?
(c) Determine the matrix T.

(d) Determine the probability that an entering first-year student will eventually graduate.

13. **Gambler's Ruin Problem**   Colleen wants to purchase a $4000 used car, but has only $1000 available. Not wishing to finance the purchase, she makes a series of wagers in which the winnings equal whatever is bet. The probability of winning is .4 and the probability of losing is .6. In a daring strategy Colleen decides to bet all her money or at least enough to obtain $4000 until she loses everything or has $4000. That is, if she has $1000, she bets $1000; if she has $2000, she bets $2000; if she has $3000, she bets $1000.

(a) What is the expected number of wagers placed before the game ends?
(b) What is the probability that Colleen is wiped out?
(c) What is the probability that Colleen wins the amount needed to purchase the car?

14. Answer the questions in Problem 13 if the probability of winning is .5.

15. Answer the questions in Problem 13 if the probability of winning is .6.

16. **Military Strategy**   Three armored cars, A, B, and C, are engaged in a three-way battle. Armored car A has probability $\frac{1}{3}$ of destroying its target, B has probability $\frac{1}{2}$ of destroying its target, and C has probability $\frac{1}{6}$ of destroying its target. The armored cars fire at the same time and each fires at the strongest opponent not yet destroyed. Using as states the surviving cars at any round, set up a Markov chain and answer the following questions:

(a) How many states are in this chain?
(b) How many are absorbing?
(c) Find the expected number of rounds fired.
(d) Find the probability that A survives.

17. **Stock Price Model**   In the model discussed on pages 478–479 a stock currently showing no change would, on the average, stay how long in that state?

18. Stock Price Model   Refer to the model discussed on pages 478–479. By making state *I* an absorbing state, compute the average number of days elapsed before a stock currently decreasing would start to increase again.

Technology Exercises

Use a graphing calculator to solve Problems 1 and 2.

1. Suppose that for a certain absorbing Markov chain the transition matrix *P* is given by

$$P = \begin{bmatrix} 1 & 0 & 0 & 0 & 0 \\ 0 & 1 & 0 & 0 & 0 \\ 0 & 0 & .05 & .95 & 0 \\ 0 & .2 & 0 & .5 & .3 \\ .75 & .25 & 0 & 0 & 0 \end{bmatrix}$$

Determine the matrices I_r, S, and Q. Use a graphing calculator to find the fundamental matrix T. Determine the product $T \cdot S$.

2. Consider the maze with five rooms shown in the figure. The system consists of the maze and a mouse. We assume the following learning pattern exists: If the mouse is in room 2, 3, or 4, it moves with equal probability to any room the maze permits; if it is in room 1, it gets caught in the trap and remains there; if it is in room 5, it remains in that room with the cheese forever.

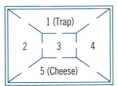

(a) Construct the transition matrix *P*.
(b) How many rooms represent an absorbing state for this Markov chain?
(c) If a mouse begins in room 3, what is the probability that it will find the cheese in 5 steps? Use a graphing calculator to determine the powers of matrix *P*.
(d) Determine the matrices I_r, S, and Q. Use a graphing calculator to find the fundamental matrix T. Determine the product $T \cdot S$.
(e) How many steps, on the average, will the mouse make in the maze before it ends up in the trap?

10.4 TWO-PERSON GAMES

Game theory, as a branch of mathematics, is concerned with the analysis of human behavior in conflicts of interest. In other words, game theory gives mathematical expression to the strategies of opposing players and offers techniques for choosing the best possible strategy. In most parlor games it is relatively easy to define winning and losing and, on this basis, to quantify the best strategy for each player. However, game theory is not merely a tool for the gambler so that he or she can take advantage of the odds; nor is it merely a method for winning games like tic-tac-toe or matching pennies.

For example, when union and management sit down at the bargaining table to discuss contracts, each has definite strategies. Each will bluff, persuade, and try to discover the other's strategy, while at the same time trying to prevent the discovery of his or her own. If enough information is known, results from the theory of games can help determine what is the best possible strategy for each player. Another application of game theory can be made to politics. If two people are vying for the same political office, each has various campaign strategies. If it is possible to determine the impact of alternate strategies on the voters, the theory of games can help to find the best strategy (usually the one that gains the most votes, while losing the least votes). Thus game theory can be used in certain situations of conflict to indicate how people should behave to achieve certain goals. Of course, game theory does not tell us how people actually behave. Game theory is the study, then, of strategy in conflict situations.

> **Two-Person Game**
>
> Any conflict or competition between two people is called a **two-person game.**

Let's consider some examples of two-person games.

EXAMPLE 1   Example of a Two-Person Game

In a game similar to matching pennies, player I picks heads or tails and player II attempts to guess the choice. Player I will pay player II \$3 if both choose heads; player I will pay player II \$2 if both choose tails. If player II guesses incorrectly, he will pay player I \$5.

We use Example 1 to illustrate some terminology. First, since two players are involved, this is a two-person game. Next, notice that no matter what outcome occurs (*HH, HT, TH, TT*), whatever is lost (or gained) by player I is gained (or lost) by player II. Such games are called **two-person zero-sum games.**

If we denote the gains of player I by positive entries and his losses by negative entries, we can display this game in a 2×2 matrix as

$$
\begin{array}{c c}
 & \begin{array}{cc} H & T \end{array} \\
\begin{array}{c} H \\ T \end{array} &
\begin{bmatrix} -3 & 5 \\ 5 & -2 \end{bmatrix}
\end{array}
$$

This matrix is called the **game matrix** or **payoff matrix,** and each entry a_{ij} of the matrix is termed a **payoff.**

Conversely, any $m \times n$ matrix $A = [a_{ij}]$ can be regarded as the game matrix for a two-person zero-sum game in which player I chooses any one of the m rows of A and simultaneously player II chooses any one of the n columns of A. The entry in the row and column chosen is the payoff.

We will assume that the game is played repeatedly, and that the problem facing each player is what choice he should make so that he gains the most benefit. Thus, player I wishes to maximize his winnings and player II wishes to minimize his losses. By a strategy of player I for a given matrix game A, we mean the decision by player I to select rows of A in some manner.

Now Work Problem 1

EXAMPLE 2   Finding a Best Strategy

Consider a two-person zero-sum game given by the matrix

$$
\begin{bmatrix} 3 & 6 \\ -2 & -3 \end{bmatrix}
$$

in which the entries denote the winnings of player I. The game consists of player I choosing a row and player II simultaneously choosing a column, with the intersection of row and column giving the payoff for this play in the game. For example, if player I chooses row 1 and player II chooses column 2, then player I wins \$6; if player I chooses row 2 and player II chooses column 1, then player I loses \$2.

It is immediately evident from the matrix that this particular game is biased in favor of player I, who will always choose row 1 since she cannot lose by doing so. Similarly, player II, recognizing that player I will choose row 1, will always choose column 1, since her losses are then minimized.

Thus the *best strategy* for player I is row 1 and the *best strategy* for player II is column 1. When both players employ their best strategy, the result is that player I wins $3. This amount is called the *value* of the game. Notice that the payoff $3 is the minimum of the entries in its row and is the maximum of the entries in its column.

Strictly Determined Game

A game defined by a matrix is said to be **strictly determined** if and only if there is an entry of the matrix that is the smallest element in its row and is also the largest element in its column. This entry is then called a **saddle point** and is the **value** of the game.

If a game has a positive value, the game favors player I. If a game has a negative value, the game favors player II. Any game with a value of 0 is termed a **fair game.**

If a game matrix has a saddle point, it can be shown that the row containing the saddle point is the best strategy for player I and the column containing the saddle point is the best strategy for player II. This is why such games are called *strictly determined games.* Such games are also called **games of pure strategy.**

Of course, a matrix may have more than one saddle point, in which case each player has more than one best strategy available. However, the value of the game is always the same no matter how many saddle points the matrix may have.

EXAMPLE 3   Identifying a Strictly Determined Game

Determine whether the game defined by the matrix below is strictly determined.

$$\begin{bmatrix} 3 & 0 & -2 & -1 \\ 2 & -3 & 0 & -1 \\ 4 & 2 & 1 & 0 \end{bmatrix}$$

SOLUTION First, we look at each row and find the smallest entry in each row:

Row 1: -2 Row 2: -3 Row 3: 0

Next, we check to see if any of the above elements are also the largest in their column. The element -2 in row 1 is not the largest entry in column 3; the element -3 in row 2 is not the largest entry in column 2; however, the element 0 in row 3 is the largest entry in column 4. Thus, this game is strictly determined. Its value is 0, so the game is fair.

The game of Example 3 is represented by a 3×4 matrix. This means that player I has three strategies open to her, while player II can choose from four strategies.

Now Work Problem 5

EXAMPLE 4   Selecting the Best Business Site

Two competitive franchising firms, Alpha Products and Omega Industries, are each planning to add an outlet in a certain city. It is possible for the site to be located either in the center of the city or in a large suburb of the city. If both firms decide to build in the center of the city, Alpha Products will show an annual profit of $1000 more than the profit of Omega Industries. If both firms decide to locate their outlets in the suburb, then it is determined that Alpha Products' profit will be $2000 less than the profit of Omega Industries. If Alpha locates in the suburb and Omega in the city, then Alpha will show a profit of $4000 more than Omega. Finally, if Alpha locates in the city and Omega in the suburb, then Alpha will have a profit of $3000 less than Omega's. Is there a best site for each firm to locate its outlet? In this case, by *best site* we mean the one that produces the most relative profit.

SOLUTION   If we assign rows as Alpha strategies and columns as Omega strategies and if we use positive entries to denote the gain of Alpha over Omega and negative entries for the gain of Omega over Alpha, then the game matrix for this situation is

$$
\begin{array}{c}
 & \text{Omega} \\
\text{Alpha} \quad
\begin{array}{cc}
 & \text{City} \quad \text{Suburb}
\end{array} \\
\begin{array}{c}
\text{City} \\
\text{Suburb}
\end{array}
\begin{bmatrix}
1 & -3 \\
4 & -2
\end{bmatrix}
\end{array}
$$

where the entries are in thousands of dollars.

This game is strictly determined and the saddle point is -2, which is the value of the game. Thus, if both firms locate in the suburb, this results in the best competition. This is so since Omega will always choose to locate in the suburb, guaranteeing a larger profit than Alpha's. This being the case, Alpha, in order to minimize this larger profit of Omega, must always choose the suburb. Of course, the game is not fair since it is favorable to Omega.

■

EXERCISE 10.4   Answers to odd-numbered problems begin on page AN-41.

In Problems 1–4 write the game matrix that corresponds to each two-person conflict situation.

1. Katy and Colleen simultaneously each show one or two fingers. If they show the same number of fingers, Katy pays Colleen one dime. If they show a different number of fingers, Colleen pays Katy one dime.

2. Katy and Colleen simultaneously each show one or two fingers. If the total number of fingers shown is even, Katy pays Colleen that number of dimes. If the total number of fingers shown is odd, Colleen pays Katy that number of dimes.

3. Katy and Colleen, simultaneously and independently, each write down one of the numbers 1, 4, or 7. If the sum of the numbers is even, Katy pays Colleen that number of dimes. If the sum of the numbers is odd, Colleen pays Katy that number of dimes.

4. Katy and Colleen, simultaneously and independently, each write down one of the numbers 3, 6, or 8. If the sum of the numbers is even, Katy pays Colleen that number of dimes. If the sum of the numbers is odd, Colleen pays Katy that number of dimes.

*In Problems 5–14 determine which of the two-person zero-sum games are strictly
determined. For those that are, find the value of the game. All entries are the winnings of player I, who plays rows.*

5. $\begin{bmatrix} -2 & 4 \\ -3 & 6 \end{bmatrix}$

6. $\begin{bmatrix} 8 & 0 \\ 0 & -1 \end{bmatrix}$

7. $\begin{bmatrix} 4 & 3 \\ 3 & 2 \end{bmatrix}$

8. $\begin{bmatrix} -6 & -2 \\ 0 & 0 \end{bmatrix}$

9. $\begin{bmatrix} 2 & 0 & -1 \\ 3 & 6 & 0 \\ 1 & 3 & 7 \end{bmatrix}$

10. $\begin{bmatrix} 2 & 3 & -2 \\ -2 & 0 & 4 \\ 0 & -3 & -2 \end{bmatrix}$

11. $\begin{bmatrix} 1 & 0 & 3 \\ -1 & 2 & 1 \\ 2 & 2 & 3 \end{bmatrix}$

12. $\begin{bmatrix} 1 & -3 & -2 \\ 2 & 5 & 4 \\ 2 & 3 & 2 \end{bmatrix}$

13. $\begin{bmatrix} 6 & 4 & -2 & 0 \\ -1 & 7 & 5 & 2 \\ 1 & 0 & 4 & 4 \end{bmatrix}$

14. $\begin{bmatrix} 8 & 6 & 4 & 0 \\ -1 & 6 & 5 & -2 \\ 0 & 1 & 3 & 3 \end{bmatrix}$

15. For what values of a is the matrix below strictly determined?

$$\begin{bmatrix} a & 8 & 3 \\ 0 & a & -9 \\ -5 & 5 & a \end{bmatrix}$$

16. Show that the matrix below is strictly determined for any
choice of a, b, or c.

$$\begin{bmatrix} a & a \\ b & c \end{bmatrix}$$

17. Find necessary and sufficient conditions for the matrix
below to be strictly determined.

$$\begin{bmatrix} a & 0 \\ 0 & b \end{bmatrix}$$

10.5 MIXED STRATEGIES

EXAMPLE 1   Determining if a Game Is Strictly Determined

Consider a two-person zero-sum game given by the matrix

$$\begin{bmatrix} 6 & 0 \\ -2 & 3 \end{bmatrix}$$

in which the entries denote the winnings of player I. Is this game strictly determined?
If so, find its value.

SOLUTION   The smallest entry in each row is

Row 1: 0   Row 2: -2

The entry 0 in row 1 is not the largest element in its column; similarly, the entry -2
in row 2 is not the largest element in its column. Thus, this game is not strictly determined.

Remember that a two-person game is not usually played just once. With this in mind,
player I in Example 1 might decide always to play row 1 since she may win $6 at best

and win $0 at worst. Does this mean she should always employ this strategy? If she does, player II would catch on and begin to choose column 2 since this strategy limits his losses to $0. However, after a while, player I would probably start choosing row 2 to obtain a payoff of $3. Thus in a nonstrictly determined game it would be advisable for the players to *mix* their strategies rather than to use the same one all the time. That is, a random selection is desirable. Indeed, to make certain that the other player does not discover the pattern of moves, it may be best not to have any pattern at all. For instance, player I may elect to play row 1 in 40% of the plays (that is, with probability .4), while player II elects to play column 2 in 80% of the plays (that is, with probability .8). This idea of mixing strategies is important and useful in game theory. Games in which each player's strategies are **mixed** are termed **mixed-strategy games.**

Suppose we know the probability for each player to choose a certain strategy. What meaning can be given to the term *payoff of a game* if mixed strategies are used? Since the payoff has been defined for a pair of pure strategies and in a mixed-strategy situation we do not know what strategy is being used, it is not possible to define a payoff for a single game. However, in the long run we do know how often each strategy is being used, and we can use this information to compute the *expected payoff* of the game.

In Example 1 if player I chooses row 1 in 50% of the plays and row 2 in 50% of the plays, and if player II chooses column 1 in 30% of the plays and column 2 in 70% of the plays, the expected payoff of the game can be computed. For example, the strategy of row 1, column 1, is chosen $(.5)(.3) = .15$ of the time. This strategy has a payoff of $6, so that the expected payoff will be $($6)(.15) = 0.90. Table 2 summarizes the entire process. Thus, the expected payoff E of this game, when the given strategies are employed, is $1.65, which makes the game favorable to player I.

Table 2

Strategy	Payoff	Probability	Expected Payoff
Row 1—Column 1	6	.15	$0.90
Row 2—Column 1	-2	.15	-0.30
Row 1—Column 2	0	.35	0.00
Row 2—Column 2	3	.35	1.05
Totals		1.00	$1.65

 **Now Work Problem 1**

If we look very carefully at the above derivation, we get a clue as to how the expected payoff of a game that is not strictly determined should be defined.

Let's consider a game defined by the 2×2 matrix

$$A = \begin{bmatrix} a_{11} & a_{12} \\ a_{21} & a_{22} \end{bmatrix}$$

Let the strategy for player I, who plays rows, be denoted by the row vector

$$P = [p_1 \quad p_2]$$

and the strategy for player II, who plays columns, be denoted by the column vector

$$Q = \begin{bmatrix} q_1 \\ q_2 \end{bmatrix}$$

The probability that player I wins the amount a_{11} is p_1q_1. Similarly, the probabilities that she wins the amounts a_{12}, a_{21}, and a_{22} are p_1q_2, p_2q_1, and p_2q_2, respectively. If we denote by $E(P, Q)$ the expectation of player I, that is, the expected value of the amount she wins when she uses strategy P and player II uses strategy Q, then

$$E(P, Q) = p_1a_{11}q_1 + p_1a_{12}q_2 + p_2a_{21}q_1 + p_2a_{22}q_2$$

Using matrix notation, the above can be expressed as

$$E(P, Q) = PAQ$$

In general, if A is an $m \times n$ game matrix, we are led to the following definition:

Expected Payoff

   The **expected payoff** E of a two-person zero-sum game, defined by the matrix A, in which the row vector P and column vector Q define the respective strategy probabilities of player I and player II is

$$E = PAQ$$

EXAMPLE 2   Finding the Expected Payoff

Find the expected payoff of the game matrix

$$A = \begin{bmatrix} 3 & -1 \\ -2 & 1 \\ 1 & 0 \end{bmatrix}$$

if player I and player II decide on the strategies

$$P = \begin{bmatrix} \frac{1}{3} & \frac{1}{3} & \frac{1}{3} \end{bmatrix} \qquad Q = \begin{bmatrix} \frac{1}{3} \\ \frac{2}{3} \end{bmatrix}$$

SOLUTION   The expected payoff E of this game is

$$E = PAQ = \begin{bmatrix} \frac{1}{3} & \frac{1}{3} & \frac{1}{3} \end{bmatrix} \begin{bmatrix} 3 & -1 \\ -2 & 1 \\ 1 & 0 \end{bmatrix} \begin{bmatrix} \frac{1}{3} \\ \frac{2}{3} \end{bmatrix}$$

$$= \begin{bmatrix} \frac{2}{3} & 0 \end{bmatrix} \begin{bmatrix} \frac{1}{3} \\ \frac{2}{3} \end{bmatrix} = \frac{2}{9}$$

Thus the game is biased in favor of player I and has an expected payoff of $\frac{2}{9}$.

Now Work Problem 7

   Most games are not strictly determined. That is, most games do not give rise to best pure strategies for each player. Examples of games that are not strictly determined are matching pennies (see Example 1, Section 10.4), bridge, poker, and so on. In the next section, we discuss a technique for finding optimal strategies for games that are not strictly determined.

EXERCISE 10.5　Answers to odd-numbered problems begin on page AN-41.

1. For the game of Example 1 find the expected payoff E if player I chooses row 1 in 30% of the plays and player II chooses column 1 in 40% of the plays.

2. For the game of Example 2 find the expected payoff E if player I chooses row 1 with probability .3 and row 2 with probability .4, while player II chooses column 1 half the time.

In Problems 3–6 find the expected payoff of the game matrix $\begin{bmatrix} 4 & 0 \\ 2 & 3 \end{bmatrix}$ *for the given strategies.*

3. $P = \begin{bmatrix} \frac{1}{2} & \frac{1}{2} \end{bmatrix}$　$Q = \begin{bmatrix} \frac{1}{2} \\ \frac{1}{2} \end{bmatrix}$

4. $P = \begin{bmatrix} \frac{1}{2} & \frac{1}{2} \end{bmatrix}$　$Q = \begin{bmatrix} \frac{3}{4} \\ \frac{1}{4} \end{bmatrix}$

5. $P = \begin{bmatrix} \frac{1}{4} & \frac{3}{4} \end{bmatrix}$　$Q = \begin{bmatrix} \frac{1}{2} \\ \frac{1}{2} \end{bmatrix}$

6. $P = \begin{bmatrix} 0 & 1 \end{bmatrix}$　$Q = \begin{bmatrix} 0 \\ 1 \end{bmatrix}$

In Problems 7–10 find the expected payoff of each game matrix.

7. $\begin{bmatrix} 4 & 0 \\ -3 & 6 \end{bmatrix}$;　$P = \begin{bmatrix} \frac{2}{3} & \frac{1}{3} \end{bmatrix}$　$Q = \begin{bmatrix} \frac{1}{3} \\ \frac{2}{3} \end{bmatrix}$

8. $\begin{bmatrix} 1 & -1 \\ -2 & 3 \end{bmatrix}$;　$P = \begin{bmatrix} \frac{1}{4} & \frac{3}{4} \end{bmatrix}$　$Q = \begin{bmatrix} \frac{1}{3} \\ \frac{2}{3} \end{bmatrix}$

9. $\begin{bmatrix} 1 & 0 & 0 \\ 0 & 1 & 0 \\ 0 & 0 & 1 \end{bmatrix}$;　$P = \begin{bmatrix} \frac{1}{3} & \frac{1}{3} & \frac{1}{3} \end{bmatrix}$　$Q = \begin{bmatrix} \frac{1}{3} \\ \frac{1}{3} \\ \frac{1}{3} \end{bmatrix}$

10. $\begin{bmatrix} 4 & -1 & 0 \\ 2 & 3 & 1 \end{bmatrix}$;　$P = \begin{bmatrix} \frac{1}{3} & \frac{2}{3} \end{bmatrix}$　$Q = \begin{bmatrix} \frac{2}{3} \\ \frac{1}{6} \\ \frac{1}{6} \end{bmatrix}$

11. Show that in a 2×2 game matrix $\begin{bmatrix} a_{11} & a_{12} \\ a_{21} & a_{22} \end{bmatrix}$ the only games that are not strictly determined are those for which either

(a) $a_{11} > a_{12}$　$a_{11} > a_{21}$　$a_{21} < a_{22}$　$a_{12} < a_{22}$

or

(b) $a_{11} < a_{12}$　$a_{11} < a_{21}$　$a_{21} > a_{22}$　$a_{12} > a_{22}$

10.6 OPTIMAL STRATEGY IN TWO-PERSON ZERO-SUM GAMES WITH 2 × 2 MATRICES

We have already seen that the best strategy for two-person zero-sum games that are strictly determined is found in the row and column containing the saddle point. Suppose the game is not strictly determined.

In 1927 John von Neumann, along with E. Borel, initiated research in the theory of games and proved that, even in nonstrictly determined games, there is a single course of action that represents the best strategy. In practice this means that in order to avoid always using a single strategy, a player in a game may instead choose plays randomly according to a fixed probability. This has the effect of making it impossible for the opponent to know what the player will do, since even the player will not know until the final moment. That is, by selecting a strategy randomly according to the laws of probability, the actual strategy chosen at any one time cannot be known even to the one choosing it.

For example, in the Italian game of *Morra* each player shows one, two, or three fingers and simultaneously calls out his guess as to what the sum of his and his opponent's fingers is. It can be shown that if he guesses four fingers each time and varies his own moves so that every 12 times he shows one finger 5 times, two fingers 4 times, and three fingers 3 times, he will, at worst, break even (in the long run).

EXAMPLE 1   Determining Optimal Strategies

Consider the nonstrictly determined game

$$A = \begin{bmatrix} 1 & -1 \\ -2 & 3 \end{bmatrix}$$

in which player I plays rows and player II plays columns. Determine the optimal strategy for each player.

SOLUTION   If player I chooses row 1 with probability p, then she must choose row 2 with probability $1 - p$. If player II chooses column 1, player I then expects to earn

$$E_I = p + (-2)(1 - p) = 3p - 2 \tag{1}$$

Similarly, if player II chooses column 2, player I expects to earn

$$E_I = (-1)p + 3(1 - p) = -4p + 3 \tag{2}$$

We graph these two equations using E_I as the vertical axis and p as the horizontal axis. See Figure 5.

Figure 5

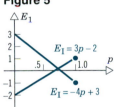

Player I wants to maximize her expected earnings so she should maximize the minimum expected gain. This occurs when the two lines intersect since for any other choice of p, one or the other of the two expected earnings is less. Thus solving Equations (1) and (2) simultaneously, we obtain

$$3p - 2 = -4p + 3$$
$$7p = 5$$
$$p = \frac{5}{7}$$

The optimal strategy for player I is therefore

$$P = \begin{bmatrix} \frac{5}{7} & \frac{2}{7} \end{bmatrix}$$

Similarly, suppose player II chooses column 1 with probability q (and therefore column 2 with probability $1 - q$). If player I chooses row 1, player II's expected earnings are

$$E_{II} = (-1)q + (1)(1 - q) = 1 - 2q$$

If player I chooses row 2, player II's expected earnings are

$$E_{II} = (2)q + (-3)(1 - q) = 5q - 3$$

The optimal strategy for player II occurs when

$$1 - 2q = 5q - 3$$
$$-7q = -4$$
$$q = \frac{4}{7}$$

Figure 6

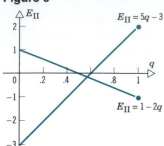

See Figure 6.

The optimal strategy for player II is thus

$$Q = \begin{bmatrix} \frac{4}{7} \\ \frac{3}{7} \end{bmatrix}$$

The expected payoff E corresponding to these optimal strategies is

$$E = PAQ = \begin{bmatrix} \frac{5}{7} & \frac{2}{7} \end{bmatrix} \begin{bmatrix} 1 & -1 \\ -2 & 3 \end{bmatrix} \begin{bmatrix} \frac{4}{7} \\ \frac{3}{7} \end{bmatrix} = \frac{1}{7}$$

Optimal Strategies in a Game

Now consider a two-person zero-sum game given by the 2×2 matrix

$$A = \begin{bmatrix} a_{11} & a_{12} \\ a_{21} & a_{22} \end{bmatrix}$$

in which player I chooses row strategies and player II chooses column strategies.

Using the method illustrated above, it can be shown that the optimal strategy for player I is given by

$$P = \begin{bmatrix} p_1 & p_2 \end{bmatrix}$$

where

Optimal Strategy for Player I

$$p_1 = \frac{a_{22} - a_{21}}{a_{11} + a_{22} - a_{12} - a_{21}} \qquad p_2 = \frac{a_{11} - a_{12}}{a_{11} + a_{22} - a_{12} - a_{21}} \tag{3}$$

with

$$a_{11} + a_{22} - a_{12} - a_{21} \neq 0$$

Notice that $p_1 + p_2 = 1$, as expected.

Similarly, the optimal strategy for player II is given by

$$Q = \begin{bmatrix} q_1 \\ q_2 \end{bmatrix}$$

where

Optimal Strategy for Player II

$$q_1 = \frac{a_{22} - a_{12}}{a_{11} + a_{22} - a_{12} - a_{21}} \qquad q_2 = \frac{a_{11} - a_{21}}{a_{11} + a_{22} - a_{12} - a_{21}} \tag{4}$$

with

$$a_{11} + a_{22} - a_{12} - a_{21} \neq 0$$

Notice that $q_1 + q_2 = 1$, as expected.

The expected payoff E of the game corresponding to these optimal strategies is

---

Expected Payoff of the Game

$$E = PAQ = \frac{a_{11} \cdot a_{22} - a_{12} \cdot a_{21}}{a_{11} + a_{22} - a_{12} - a_{21}}$$

EXAMPLE 2   Finding Optimal Strategies

For the game matrix

$$\begin{bmatrix} 1 & -1 \\ -2 & 3 \end{bmatrix}$$

determine the optimal strategies for player I and player II, and find the expected payoff of the game.

SOLUTION   Using Formula (3), we have

$$p_1 = \frac{3 - (-2)}{1 + 3 - (-1) - (-2)} = \frac{5}{7}$$

$$p_2 = \frac{1 - (-1)}{1 + 3 - (-1) - (-2)} = \frac{2}{7}$$

Thus player I's optimal strategy is to select row 1 with probability $\frac{5}{7}$ and row 2 with probability $\frac{2}{7}$. Also, by (4), player II's optimal strategy is

$$q_1 = \frac{3 - (-1)}{1 + 3 - (-1) - (-2)} = \frac{4}{7}$$

$$q_2 = \frac{1 - (-2)}{1 + 3 - (-1) - (-2)} = \frac{3}{7}$$

Player II's optimal strategy is to select column 1 with probability $\frac{4}{7}$ and column 2 with probability $\frac{3}{7}$. The expected payoff of the game is

$$E = \frac{1 \cdot 3 - (-1)(-2)}{1 + 3 - (-1) - (-2)} = \frac{1}{7}$$

Thus in the long run the game is favorable to player I. ∎

The results obtained in Example 2 are in agreement with those obtained earlier using the graphical technique.

Now Work Problem 1

EXAMPLE 3 Finding Optimal Strategies

Find the optimal strategy for each player, and determine the expected payoff of the game given by the matrix

$$\begin{bmatrix} 6 & 0 \\ -2 & 3 \end{bmatrix}$$

SOLUTION Using the graphical technique or Formulas (3) and (4), we find player I's optimal strategy to be

$$p_1 = \tfrac{5}{11} \qquad p_2 = \tfrac{6}{11}$$

Player II's optimal strategy is

$$q_1 = \tfrac{3}{11} \qquad q_2 = \tfrac{8}{11}$$

The expected payoff of the game is

$$E = \tfrac{18}{11} = 1.64$$

Thus the game favors player I, whose optimal strategy is $[\tfrac{5}{11} \quad \tfrac{6}{11}]$.

■

Applications

EXAMPLE 4 Election Strategy

In a presidential campaign, there are two candidates, a Democrat (D) and a Republican (R), and two types of issues, domestic issues and foreign issues. The units assigned to each candidate's strategy are given in the table. We assume that positive entries indicate a strength for the Democrat, while negative entries indicate a weakness. We also assume that a strength of one candidate equals a weakness of the other so that the game is zero-sum. What is the best strategy for each candidate? What is the expected payoff of the game?

		Republican	
		Domestic	Foreign
Democrat	Domestic	4	-2
	Foreign	-1	3

SOLUTION Notice first that this game is not strictly determined. If D chooses to always play foreign issues, then R will counter with domestic issues, in which case D would also talk about domestic issues, in which case, etc., etc. There is no *single* strategy either can use. We compute that the optimal strategy for the Democrat is

$$p_1 = \frac{3 - (-1)}{4 + 3 - (-2) - (-1)} = \frac{4}{10} = .4 \qquad p_2 = \frac{4 - (-2)}{10} = .6$$

The optimal strategy for the Republican is

$$q_1 = \frac{3 - (-2)}{10} = .5 \qquad q_2 = \frac{4 - (-1)}{10} = .5$$

Thus the best strategy for the Democrat is to spend 40% of her time on domestic issues and 60% on foreign issues, while the Republican should divide his time evenly between the two issues.

The expected payoff of the game is

$$E = \frac{3 \cdot 4 - (-1)(-2)}{10} = \frac{10}{10} = 1.0$$

Thus no matter what the Republican does, the Democrat gains at least 1.0 unit by employing her best strategy.

 **Now Work Problem 7**

EXAMPLE 5   War Game

In a naval battle, attacking bomber planes are trying to sink ships in a fleet protected by an aircraft carrier with fighter planes. The bombers can attack either high or low, with a low attack giving more accurate results. Similarly, the aircraft carrier can send its fighters at high altitudes or low altitudes to search for the bombers. If the bombers avoid the fighters, credit the bombers with 8 points; if the bombers and fighters meet, credit the bombers with -2 points. Also, credit the bombers with 3 additional points for flying low (since this results in more accurate bombing). Find optimal strategies for the bombers and the fighters. What is the expected payoff of the game?

SOLUTION   First, we set up the game matrix. Designate the bombers as playing rows and the fighters as playing columns. Also, each entry of the matrix will denote winnings of the bombers. Then the game matrix is

$$\begin{array}{c} & \text{Fighters} \\ & \begin{array}{cc} \text{Low} & \text{High} \end{array} \\ \text{Bombers} \begin{array}{c} \text{Low} \\ \text{High} \end{array} & \begin{bmatrix} 1 & 11 \\ 8 & -2 \end{bmatrix} \end{array}$$

The reason for a 1 in row 1, column 1, is that -2 points are credited for the planes meeting, but 3 additional points are credited to the bombers for a low flight.

Next, using Formula (3) and (4), the optimal strategies for the bombers $[p_1 \quad p_2]$ and for the fighters $\begin{bmatrix} q_1 \\ q_2 \end{bmatrix}$ are

$$p_1 = \frac{-10}{-20} = \frac{1}{2} \qquad p_2 = \frac{-10}{-20} = \frac{1}{2}$$

$$q_1 = \frac{-13}{-20} = \frac{13}{20} \qquad q_2 = \frac{-7}{-20} = \frac{7}{20}$$

The expected payoff of the game is

$$E = \frac{-2 - 88}{-20} = \frac{-90}{-20} = 4.5$$

Thus the game is favorable to the bombers if both players employ their optimal strategies.

The bombers can decide whether to fly high or low by flipping a fair coin and flying high whenever heads appear. The fighters can decide whether to fly high or low by using a bowl with 13 black balls and 7 white balls. Each day, a ball should be selected at random and then replaced. If the ball is black, they will go low; if it is white, they will go high.

∎

EXERCISE 10.6   Answers to odd-numbered problems begin on page AN-41.

In Problems 1–6 find the optimal strategy for each player and determine the expected payoff of each 2 × 2 game by using graphical techniques. Check your answers by using Formulas (3) and (4).

1. $\begin{bmatrix} 1 & 2 \\ 4 & 1 \end{bmatrix}$

2. $\begin{bmatrix} 2 & 4 \\ 3 & -2 \end{bmatrix}$

3. $\begin{bmatrix} -3 & 2 \\ 1 & 0 \end{bmatrix}$

4. $\begin{bmatrix} 3 & -2 \\ -1 & 2 \end{bmatrix}$

5. $\begin{bmatrix} 2 & -1 \\ -1 & 4 \end{bmatrix}$

6. $\begin{bmatrix} 5 & 4 \\ -3 & 7 \end{bmatrix}$

7. In Example 4 suppose the candidates are assigned the following weights for each issue:

		Republican	
		Domestic	**Foreign**
Democrat	**Domestic**	4	−1
	Foreign	0	3

What is each candidate's best strategy? What is the expected payoff of the game and whom does it favor?

8. **War Game**   For the situation described in Example 5, credit the bomber with 4 points for avoiding the fighters and with −6 points for meeting the fighters. Also, grant the bombers 2 additional points for flying low. What are the optimal strategies and the expected payoff of the game? Give instructions to the fighters and bombers as to how they should decide whether to fly high or low.

9. **Strategies for Spies**   A spy can leave an airport through two exits, one a relatively deserted exit and the other an exit heavily used by the public. His opponent, having been notified of the spy's presence in the airport, must guess which exit he will use. If the spy and opponent meet at the deserted exit, the spy will be killed; if the two meet at the heavily used exit, the spy will be arrested. Assign a payoff of 30 points to the spy if he avoids his opponent by using the deserted exit and of 10 points to the spy if he avoids his opponent by using the busy exit. Assign a payoff of −100 points to the spy if he is killed and −2 points if he is arrested. What are the optimal strategies and the expected payoff of the game?

10. Prove Formulas (3) and (4).

11. In the game matrix

$$\begin{bmatrix} a_{11} & a_{12} \\ a_{21} & a_{22} \end{bmatrix}$$

what can be said if $a_{11} + a_{22} - a_{12} - a_{21} = 0$?

CHAPTER REVIEW

IMPORTANT TERMS AND CONCEPTS

Markov chain 456
transition matrix 456
initial probability distribution 456
probability vector 457
probability distribution after k
  stages 458, 459
probability of passing from state
  i to state j in n stages 465

regular Markov chain 466
fixed probability vector 467
absorbing state 472
absorbing Markov chain 472
gambler's ruin problem 474
fundamental matrix of an
  absorbing Markov chain 476
expected number of steps
  before absorption 476

probability of being
  absorbed 477
two-person game 481
game matrix 482
strictly determined game 483
saddle point 483
mixed-strategy games 486
expected payoff 487, 491
optimal strategies 490

TRUE–FALSE ITEMS   Answers are on page AN-41.

T_____ F_____ **1.** The matrix $\begin{bmatrix} 1 & 0 \\ -1 & 1 \end{bmatrix}$ is a transition matrix for a Markov chain.

T_____ F_____ **2.** A Markov chain with the transition matrix $\begin{bmatrix} \frac{1}{2} & \frac{1}{2} \\ 0 & 1 \end{bmatrix}$ is regular.

T_____ F_____ **3.** A Markov chain with the transition matrix $\begin{bmatrix} 0 & 1 \\ \frac{1}{3} & \frac{2}{3} \end{bmatrix}$ is absorbing.

T_____ F_____ **4.** If P is the transition matrix of a Markov chain, then P^2 gives the probability of moving from one state to another state in two stages.

T_____ F_____ **5.** The value of a strictly determined game is unique.

T_____ F_____ **6.** In a two-person zero-sum game whatever is gained (lost) by player I is lost (gained) by player II.

T_____ F_____ **7.** In mixed-strategy games the expected payoff of the game depends on the strategy each player uses.

FILL IN THE BLANKS   Answers are on page AN-41.

1. In a Markov chain with m states the initial probability distribution is a row vector of dimension _____ .

2. A probability vector is a vector whose entries are _____ and sum up to _____ .

3. In a Markov chain with transition matrix P the probability distribution $v^{(k)}$ after k observations satisfies _____ .

4. A Markov chain is said to be regular if, for some power of its transition matrix P, all entries are _____ .

5. Each entry of a game matrix is called a _____ .

REVIEW EXERCISES   Answers to odd-numbered problems begin on page AN-41.

1. Find the fixed probability vector of

(a) $\begin{bmatrix} \frac{1}{4} & \frac{3}{4} \\ \frac{1}{2} & \frac{1}{2} \end{bmatrix}$ (b) $\begin{bmatrix} \frac{1}{3} & \frac{2}{3} \\ \frac{2}{3} & \frac{1}{3} \end{bmatrix}$ (c) $\begin{bmatrix} .7 & .1 & .2 \\ .6 & .1 & .3 \\ .4 & .2 & .4 \end{bmatrix}$

2. Define and explain in words the meaning of a *regular* transition matrix. Give an example of such a matrix and of a matrix that is not regular.

3. Customer Loyalty   Three beer distributors, A, B, and C, each presently holds $\frac{1}{3}$ of the beer market. Each wants to increase its share of the market, and to do so, each introduces a new brand. During the next year, it is learned

(a) that A keeps 50% of its business and loses 20% to B and 30% to C.
(b) that B keeps 40% of its business and loses 40% to A and 20% to C.
(c) that C keeps 25% of its business and loses 50% to A and 25% to B.

Assuming this trend continues, after 2 years what share of the market does each distributor have? In the long run what is each distributor's share?

4. If the current share of the market for each distributor A, B, and C in Problem 3 is A: 25%, B: 25%, C: 50%, and the market trend is the same, answer the same questions.

5. Sales Strategy   A representative of a book publishing company has to cover three universities, U_1, U_2, and U_3. She never sells at the same university in successive months. If she sells at university U_1, then the next month she sells at U_2. However, if she sells at either U_2 or U_3, then the next month she is three times as likely to sell at U_1 as at the other university. In the long run how often does she sell at each of the universities?

6. Family Traits   Assume that the probability of a fat father having a fat son is .7 and that of a skinny father having a skinny son is .4. What is the probability of a fat father being the great-grandfather of a fat great-grandson? In the long run what will be the distribution? Does it depend on the initial physical state of the fathers?

7. Gambler's Ruin Problem   Suppose a man has $2, which he is going to bet $1 at a time until he either loses all his money or he wins $5. Assume he wins with a probability of .45 and he loses with a probability of .55. Construct the transition matrix for this game.

(a) On the average, how many times will the process be in nonabsorbing states?

(b) What is the expected length of the game?

(c) What is the probability the man loses all his money or wins $5?

Hint: The fundamental matrix T of the transition matrix P is

$$T = \begin{bmatrix} 1.584283 & 1.062332 & .635282 & .285877 \\ 1.298406 & 2.360738 & 1.411737 & .635282 \\ 0.949001 & 1.725456 & 2.360738 & 1.062332 \\ .52195 & .949001 & 1.298406 & 1.584282 \end{bmatrix}$$

8. Determine which of the following two-person zero-sum games are strictly determined. For those that are, find the value of the game.

(a) $\begin{bmatrix} 5 & 3 \\ 2 & 4 \end{bmatrix}$ (b) $\begin{bmatrix} 29 & 15 \\ 79 & 3 \end{bmatrix}$ (c) $\begin{bmatrix} 50 & 75 \\ 30 & 15 \end{bmatrix}$

(d) $\begin{bmatrix} 7 & 14 \\ 9 & 13 \end{bmatrix}$ (e) $\begin{bmatrix} 0 & 2 & 4 \\ 4 & 6 & 10 \\ 16 & 14 & 12 \end{bmatrix}$

9. Find the expected payoff of the game below for the given strategies:

$$\begin{bmatrix} -1 & 1 \\ 1 & -1 \end{bmatrix}$$

(a) $P = [\frac{1}{3} \quad \frac{2}{3}]$ $Q = \begin{bmatrix} 1 \\ 0 \end{bmatrix}$

(b) $P = [0 \quad 1]$ $Q = \begin{bmatrix} \frac{1}{2} \\ \frac{1}{2} \end{bmatrix}$

(c) $P = [\frac{1}{2} \quad \frac{1}{2}]$ $Q = \begin{bmatrix} \frac{1}{2} \\ \frac{1}{2} \end{bmatrix}$

10. **Investment Strategy**   An investor has a choice of two investments, A and B. The percentage gain of each investment over the next year depends on whether the economy is "up" or "down." This investment information is displayed in the following payoff matrix.

	Economy up	Economy down
Invest in A	-5	20
Invest in B	18	0

(a) Find the best investment allocation. That is, find the row player's optimal strategy in the corresponding matrix game.

(b) Find the percentage gain the investor is assured of when using this optimal strategy. That is, find the expected payoff of the corresponding matrix game.

11. **Investment Strategy**   There are two possible investments, A and B, and two possible states of the economy,

"inflation" and "recession." The estimated percentage increases in the value of the investments over the coming year for each possible state of the economy are shown in the following payoff matrix.

	Inflation	Recession
Invest in A	10	5
Invest in B	-5	20

(a) Find the investor's optimal strategy $P = [p_1 \quad p_2]$.

(b) Find the expected payoff of this game.

(c) What return can the investor expect if the optimal strategy is followed?

12. **Real Estate Development**   A real estate developer has bought a large tract of land in Cook County. He is considering using some of the land for apartments, some for a shopping center, and some for houses. It is not certain whether the Cook County government will build a highway near his property or not. His financial advisor provides an estimate for the percentage profit to be made in each case and these percentages are given in the following table:

Builder	Government	
	Highway	No Highway
Apartments	25%	5%
Shopping center	20%	15%
Houses	10%	20%

What percentage of the land should he use for each of the three categories, assuming that the government of Cook County is an active opponent?

13. **Betting Strategy**   The Pistons are going to play the Bulls in a basketball game. If you place your bet in Detroit, you can get 3 to 2 odds for a bet on the Bulls; and if you place your bet in Chicago, you can get 1 to 1 odds for a bet on the Pistons. This information is displayed in the following payoff matrix in which the entries represent the payoffs in dollars for each $1 bet.

	Bulls win	Pistons win
Bet on Bulls	1.5	-1
Bet on Pistons	-1	1

Think of this situation as a game in which you are the row player and find your optimal strategy $P = [p_1 \quad p_2]$ and the expected payoff of the game.

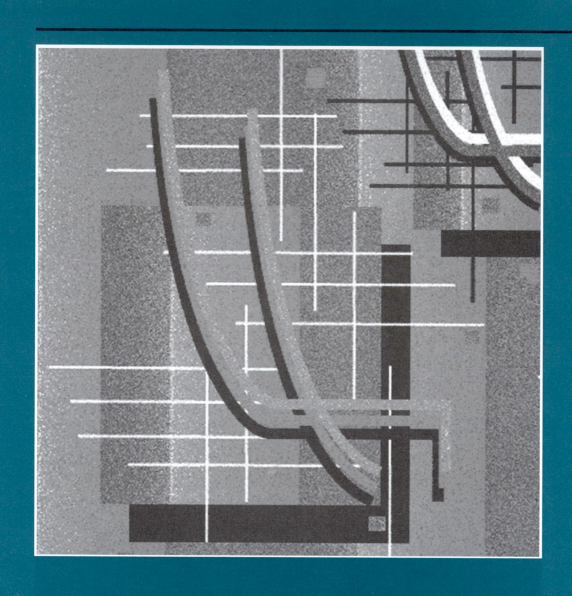

PART THREE

Discrete Mathematics

Chapter 11 Logic and Logic Circuits
Chapter 12 Relations, Functions, and Induction
Chapter 13 Graphs and Trees

Logic and Logic Circuits

11.1 Propositions

11.2 Truth Tables

11.3 Implications; The Biconditional Connective; Tautologies

11.4 Arguments

11.5 Logic Circuits

Chapter Review

In this chapter we survey many of the fundamental concepts found in the area of mathematics called *logic.* There are several reasons for studying logic. Two of the more important ones are (1) to gain proficiency in correct mathematical reasoning and (2) to apply the tool of logic to situations such as the design of *logic circuits* used in computers and other electronic devices.

In mathematics the words "not" and "or" and the phrases "if . . . , then . . . ," "if and only if," and so on are used extensively. A knowledge of the exact meaning of these words is necessary before we can make precise the laws of inference and deduction that are constantly used in mathematics. The study of logic will enable you to gain a basic understanding of what constitutes a mathematical argument. This will eliminate common errors made in mathematical, as well as nonmathematical, arguments.

We hope this chapter will give you some indication of the usefulness of logic in uncovering ambiguities and nonsequiturs. Furthermore, we hope this chapter offers evidence in favor of Church's remark that "the value of logic . . . is not that it supports a particular system, but that the process of logical organization of any system (empiricist or otherwise) serves to test its internal consistency, to verify its logical adequacy to its declared purpose, and to isolate and clarify the assumptions on which it rests."*

*Church, *Introduction to Mathematical Logic,* Vol. 1, Princeton University Press, Princeton, NJ, 1956, p. 55.

In the last two sections of this chapter we apply the concepts of logic to a problem in the application of conflicting rules and to the design of electronic circuits.

11.1 PROPOSITIONS

Propositions

English and other natural languages are composed of various words and phrases with distinct functions that have a bearing on the meaning of the sentences in which they occur.

English sentences may be classified as *declarative, interrogative, exclamatory,* or *imperative.* In the study of logic, we assume that we are able to recognize a declarative sentence (or *statement*) and form an opinion as to whether it is true or false.

Proposition

    A **proposition** is a declarative sentence that can be meaningfully classified as either true or false.

EXAMPLE 1   Determining Whether a Sentence Is a Proposition

The price of an IBM personal computer was $1800 on June 16, 1995.
    This is a proposition, although few of us can say whether it is true or false.

EXAMPLE 2   Determining Whether a Sentence Is a Proposition

The earth is round.
    This sentence records a possible fact about reality and is a proposition. Some people would classify this proposition as true and others as false, depending on what the word *round* means to them. (Does it mean simply "curved," or "perfectly spherical"?) Only in an ideal situation can we unequivocally state to which truth category a proposition belongs.

EXAMPLE 3   Determining Whether a Sentence Is a Proposition

What is the exchange rate from United States dollars to German marks?
    This is not a proposition—it is an interrogative sentence.

EXAMPLE 4 **Determining Whether a Sentence Is a Proposition**

The prices of most stocks on the New York Stock Exchange rose during the period 1929–1931.

This is a proposition. It happens to be false. ∎

In an ideal situation, a proposition could be easily and decisively classified as true or false. However, as Examples 1 and 2 illustrate, it is often difficult to classify propositions in this way, because of unclear meanings, ambiguous situations, differences of opinion, etc. In the mathematical treatment of logic, we avoid these difficulties by assuming "for the sake of argument" either the truth or the falsity of certain propositions to draw conclusions about other propositions, using *symbolic logic.*

Compound Propositions

Consider the proposition "Jones is handsome and Smith is selfish." This sentence is obtained by joining the two propositions "Jones is handsome," "Smith is selfish" by the word "and."

Compound Proposition; Connectives

A **compound proposition** is a proposition formed by connecting two or more propositions or by negating a single proposition. The words and phrases (or symbols) used to form compound propositions are called **connectives.**

Some of the connectives used in English are *or; either . . . or; and; but; if . . . , then; not.*

Conjunction

Let p and q denote propositions. The compound proposition p *and* q is called the **conjunction** *of* p *and* q and is denoted symbolically by

$$p \wedge q$$

We define $p \wedge q$ to be true when both p and q are true and to be false otherwise.

EXAMPLE 5 **Forming a Compound Proposition**

Consider the two statements

p: Washington, D.C., is the capital of the United States.

q: Hawaii is the fiftieth state of the United States.

The conjunction of p and q is

$p \wedge q$:   Washington, D.C., is the capital of the United States and
Hawaii is the fiftieth state of the United States.

Since both p and q are true statements, we conclude that the compound statement $p \wedge q$ is true.

■

EXAMPLE 6   Forming a Compound Proposition

Consider the two statements

p: Washington, D.C., is the capital of the United States.

q: Vermont is the largest of the fifty states.

The compound proposition $p \wedge q$ is

$p \wedge q$:   Washington, D.C., is the capital of the United States and
Vermont is the largest of the fifty states.

Since the statement q is false, the compound statement $p \wedge q$ is false—even though p is true.

■

> **Inclusive Disjunction**
>
> Let p and q be any propositions. The compound proposition **p or q** is called the **inclusive disjunction of p and q,** read as "p or q or both," is denoted symbolically by
>
> $$p \vee q$$
>
> We define $p \vee q$ to be true if *at least one* of the propositions p, q is true. That is, $p \vee q$ is true if both p and q are true, if p is true and q is false, or if p is false and q is true. It is false only if both p and q are false.

EXAMPLE 7   Forming a Compound Proposition

Consider the propositions

p: XYZ Company is the largest producer of nails in the world.

q: Mines, Ltd., has three uranium mines in Nevada.

The compound proposition $p \vee q$ is

$p \vee q$:   XYZ Company is the largest producer of nails in the world
or Mines, Ltd., has three uranium mines in Nevada.

■

Exclusive Disjunction

The English word "or" can be used in two different ways—as an inclusive ("and/or") or exclusive ("either/or") disjunction. The correct meaning is usually inferred from the context in which the word is used. However, when it is important to be precise (as it often is in mathematics, business, science, etc.) we must carefully distinguish between the two meanings of "or."

Exclusive Disjunction

Let p and q be any proposition. The **exclusive disjunction of p and q,** read as "either p or q but not both," is denoted symbolically by

$$p \vee q$$

We define $p \vee q$ to be true if exactly one of the propositions p, q is true. That is, $p \vee q$ is true $\overline{\text{if}}$ p is true and q is false, or if p is false and q is true. It is false if both p and q are false or if both p and q are true.

EXAMPLE 8   Forming a Compound Proposition

Consider the propositions

  p: XYZ Company earned $3.20 per share in 1980.

  q: XYZ Company paid a dividend of $1.20 per share in 1980.

The inclusive disjunction of p and q is

  $p \vee q$: XYZ Company earned $3.20 per share in 1980 or XYZ Company paid a dividend of $1.20 per share in 1980, or both.

The exclusive disjunction of p and q is

  $p \vee q$: XYZ Company earned $3.20 per share in 1980 or XYZ Company paid a dividend of $1.20 per share in 1980, but *not* both.

EXAMPLE 9   Analyzing a Compound Proposition Using the Connective "or"

Consider the compound proposition

  p: This weekend I will meet Caryl or Mary.

The use of the connective "or" is not clear here. If the "or" means inclusive disjunction, then I will be meeting at least one and possibly both persons. If the "or" means exclusive disjunction, then I will be meeting only one person.

Negation; Quantifiers

> **Negation**
>
> If p is any proposition, the **negation of p,** denoted by
>
> $$\sim p$$
>
> and read as *"not p,"* is a proposition that is false when p is true and true when p is false.

The negation of p is sometimes called the *denial* of p. The symbol $\sim$ is called the **negation operator.**

The definition of $\sim p$ assumes that p and $\sim p$ cannot both be true. In classical logic this assumption is known as the **law of contradiction.**

EXAMPLE 10   Negating a Proposition

Consider the proposition

$\quad\quad\quad\quad p$:   One share of XYZ stock is worth less than \$85.

The negation of p is

$\quad\quad\quad\quad \sim p$:   One share of XYZ stock is worth at least \$85.

The sentence "A share of XYZ stock is worth more than \$85" is *not* a correct statement of the negation of p.

 **Now Work Problem 11**

■

A **quantifier** is a word or phrase telling how many (Latin *quantus*). English quantifiers include "all," "none," "some," and "not all." The quantifiers "all," "every," and "each" are interchangeable. The quantifiers "some," "there exist(s)," and "at least one" are also interchangeable.

EXAMPLE 11   Using Different Quantifiers to Give the Same Meaning

The following propositions all have the same meaning:

$\quad\quad\quad\quad\quad\quad p$:   All people are intelligent.

$\quad\quad\quad\quad\quad\quad q$:   Every person is intelligent.

$\quad\quad\quad\quad\quad\quad r$:   Each person is intelligent.

$\quad\quad\quad\quad\quad\quad s$:   Any person is intelligent.

■

EXAMPLE 12 **Studying the Negation of a Proposition**

The negation of the proposition

p: All students are intelligent.

is

$\sim p$:   Some students are not intelligent.
$\sim p$:   There exists a student who is not intelligent.
$\sim p$:   At least one student is not intelligent.

The negation of

q: No student is intelligent.

is

$\sim q$:   At least one student is intelligent.

Note that "No student is intelligent" is *not* the negation of p; "All students are intelligent" is *not* the negation of q.

EXERCISE 11.1   Answers to odd-numbered problems begin on page AN-42.

In Problems 1–8 determine which are propositions.

1. The cost of shell egg futures was up on June 18, 1990.

2. The gross national product exceeded one billion dollars in 1935.

3. What a portfolio!

4. Why did you buy XYZ Company stock?

5. The earnings of XYZ Company doubled last year.

6. Where is the new mine of Mines, Ltd.?

7. Jones is guilty of murder in the first degree.

8. What a hit!

In Problems 9–16 negate each proposition.

9. A fox is an animal.

10. The outlook for bonds is not good.

11. I am buying stocks and bonds.

12. Rob is selling his apartment building and his business.

13. No one wants to buy my house.

14. Everyone has at least one television set.

15. Some people have no car.

16. Jones is permitted not to see that all votes are not counted.

In Problems 17–24 let p denote "John is an economics major" and let q denote "John is a sociology minor." State each proposition as a simple sentence.

17. $p \vee q$

18. $p \underline{\vee} q$

19. $p \wedge q$

20. $\sim p$

21. $\sim p \vee \sim q$

22. $\sim(\sim q)$

23. $\sim p \vee q$

24. $\sim p \wedge q$

11.2 TRUTH TABLES

Truth Values and Truth Tables

The **truth value** of a proposition is either **true** (denoted by T) or **false** (denoted by F). A **truth table** is a table that shows the truth value of a compound proposition for all possible cases.

For example, consider the conjunction of any two propositions p and q. Recall that $p \wedge q$ is false if either p is false or q is false, or if both p and q are false. There are four possible cases.

1. p is true and q is true.   **3.** p is false and q is true.
2. p is true and q is false.   **4.** p is false and q is false.

These four cases are listed in the first two columns of Table 1, which is the truth table for $p \wedge q$. For convenience, the cases for p and q will always be listed in this order.

The truth values of $\sim p$ are given in Table 2.

Using the previous definitions of inclusive disjunction and exclusive disjunction, we obtain truth tables for $p \vee q$ and $p \underline{\vee} q$. See Table 3.

Table 1

	p	q	$p \wedge q$
Case 1	T	T	T
Case 2	T	F	F
Case 3	F	T	F
Case 4	F	F	F

Table 2

p	$\sim p$
T	F
F	T

Table 3

p	q	$p \vee q$	$p \underline{\vee} q$
T	T	T	F
T	F	T	T
F	T	T	T
F	F	F	F

Besides using the connectives $\wedge$, $\vee$, $\underline{\vee}$, $\sim$ one at a time to form compound propositions, we can use them together to form more complex statements.

For example, Table 4 is the truth table for $(p \vee q) \underline{\vee} (\sim p)$. The parentheses are used to indicate that $\vee$ and $\sim$ are applied before $\underline{\vee}$.

Table 4

p	q	$p \vee q$	$\sim p$	$(p \vee q) \underline{\vee} (\sim p)$
T	T	T	F	T
T	F	T	F	T
F	T	T	T	F
F	F	F	T	T

Observe that the first columns of the table are for the component propositions p, q, . . . and that there are enough rows in the table to allow for all possible combinations of T and F. (For two components p and q, 4 rows are necessary; for three components p, q, and r, 8 rows would be necessary, and so on; for n statements, 2^n rows are needed.)

Table 4 has five columns, and each column corresponds to a stage in constructing the compound proposition, beginning with the simple components p, q, . . . At each stage one or more components constructed in earlier stages are combined, until we obtain the final result in the last column.

We outline the steps to obtain the truth table for the given compound proposition:

p	q	$p \vee q$	$\sim p$	$(p \vee q) \vee (\sim p)$
T	T			
T	F			
F	T			
F	F			

Each stage of the proposition is written on the top row, to the right of its intermediate stages; there is a column underneath each component or connective. Truth values are then entered in the truth table, one step at a time:

Stage 1

p	q	$p \vee q$	$\sim p$	$(p \vee q) \vee (\sim p)$
T	T	T		
T	F	T		
F	T	T		
F	F	F		

Stage 2

p	q	$p \vee q$	$\sim p$	$(p \vee q) \vee (\sim p)$
T	T	T	F	
T	F	T	F	
F	T	T	T	
F	F	F	T	

Stage 3

p	q	$p \vee q$	$\sim p$	$(p \vee q) \vee (\sim p)$
T	T	T	F	T
T	F	T	F	T
F	T	T	T	F
F	F	F	T	T

The truth table of the compound proposition then consists of the original columns under p and q and the fifth column entered into the table in the last stage.

EXAMPLE 1   Constructing Truth Tables

Construct the truth table for $(p \vee \sim q) \wedge p$. The component parts of this proposition are p, $\sim q$, and $p \vee \sim q$.

SOLUTION

p	q	$\sim q$	$p \vee \sim q$	$(p \vee \sim q) \wedge p$
T	T	F	T	T
T	F	T	T	T
F	T	F	F	F
F	F	T	T	F

The parentheses are used to indicate that $\vee$ is applied before $\wedge$ and the $\sim$ applies only to q. Notice that the truth table is the same as for p alone.

EXAMPLE 2   Constructing Truth Tables

Construct the truth table for $\sim(p \vee q) \vee (\sim p \wedge \sim q)$.

SOLUTION

p	q	$\sim p$	$\sim q$	$p \vee q$	$\sim(p \vee q)$	$(\sim p \wedge \sim q)$	$\sim(p \vee q) \vee (\sim p \wedge \sim q)$
T	T	F	F	T	F	F	F
T	F	F	T	T	F	F	F
F	T	T	F	T	F	F	F
F	F	T	T	F	T	T	T

The parentheses indicate that the first $\sim$ symbol negates $p \vee q$ [not p alone and not $(p \vee q) \vee (\sim p \wedge \sim q)$]. Notice that the statements $\sim(p \vee q)$, $\sim p \wedge \sim q$, and $\sim(p \vee q) \vee (\sim p \wedge \sim q)$ all have the same truth table.

Now Work Problem 11

The next example is of a truth table involving three components, p, q, and r.

EXAMPLE 3   Constructing Truth Tables

Construct the truth table for $p \wedge (q \vee r)$.

SOLUTION

p	q	r	$q \vee r$	$p \wedge (q \vee r)$
T	T	T	T	T
T	T	F	T	T
T	F	T	T	T
T	F	F	F	F
F	T	T	T	F
F	T	F	T	F
F	F	T	T	F
F	F	F	F	F

Logical Equivalence; The Laws of Logic

Very often, two propositions stated in different ways have the same meaning. For example, in law, "Jones agreed and is obligated to paint Smith's house" has the same meaning as "It was agreed and contracted by Jones that Jones would paint the house belonging to Smith and Jones is therefore required to paint the aforesaid house."

> **Logically Equivalent**
>
> If two propositions a and b have the same truth values in every possible case, the propositions are called **logically equivalent.** This relationship is denoted by $a \equiv b$.

EXAMPLE 4   Using Truth Tables to Show That Two Statements Are Logically Equivalent

Show that $\sim(p \wedge q)$ is logically equivalent to $\sim p \vee \sim q$.

SOLUTION   Construct the truth table as shown below.

1	2	3	4	5	6	7
p	q	$p \wedge q$	$\sim(p \wedge q)$	$\sim p$	$\sim q$	$\sim p \vee \sim q$
T	T	T	F	F	F	F
T	F	F	T	F	T	T
F	T	F	T	T	F	T
F	F	F	T	T	T	T

Since the entries in columns 4 and 7 of the truth table are the same, the two propositions are logically equivalent.

∎

EXAMPLE 5   Using Truth Tables to Show That Two Statements Are Logically Equivalent

Show that $\sim p \wedge \sim q$ is logically equivalent to $\sim(p \vee q)$.

SOLUTION   Construct the truth table as shown. The entries under $\sim p \wedge \sim q$ and $\sim(p \vee q)$ are the same, so the two propositions are logically equivalent.

p	q	$p \vee q$	$\sim p$	$\sim q$	$\sim p \wedge \sim q$	$\sim(p \vee q)$
T	T	T	F	F	F	F
T	F	T	F	T	F	F
F	T	T	T	F	F	F
F	F	F	T	T	T	T

📖 **Now Work Problem 23**

∎

The following laws are especially useful in the study of logic circuits. They can be proved using truth tables.

Idempotent Laws

For any proposition p,

$$p \wedge p \equiv p \qquad p \vee p \equiv p$$

Here, $p \wedge p \equiv p$ since $p \wedge p$ is true when p is true and false when p is false. A similar argument shows that $p \vee p \equiv p$.

Commutative Laws

For any two propositions p and q,

$$p \wedge q \equiv q \wedge p \qquad p \vee q \equiv q \vee p$$

The commutative laws state that, if two or more propositions are combined using the connective *and*, changing the order in which the components are connected does not change the meaning of the compound proposition. The same is true of the connective *or*.

For example, if we consider the two statements

p: Mrs. Jones is attractive.

q: Mr. Jones is intelligent.

we see that the compound propositions

$p \wedge q$:   Mrs. Jones is attractive and Mr. Jones is intelligent.

$q \wedge p$:   Mr. Jones is intelligent and Mrs. Jones is attractive.

have the same meaning.

Associative Laws

For any three propositions p, q, r,

$$(p \wedge q) \wedge r \equiv p \wedge (q \wedge r) \qquad (p \vee q) \vee r \equiv p \vee (q \vee r)$$

Because of the associative laws, it is possible to omit the parentheses when using the same connective more than once. For instance, we can write

$$p \wedge q \wedge r \quad \text{for} \quad (p \wedge q) \wedge r$$

and

$$p \wedge q \wedge r \wedge s \quad \text{for} \quad [(p \wedge q) \wedge r] \wedge s$$

the same comments hold for $\vee$. Note, however, that the parentheses cannot be omitted when using $\wedge$ and $\vee$ together; see Problems 34 and 35.

Distributive Laws

For any three propositions p, q, r,

$$p \vee (q \wedge r) \equiv (p \vee q) \wedge (p \vee r)$$
$$p \wedge (r \vee q) \equiv (p \wedge r) \vee (p \wedge q)$$

Here, $p \vee (q \wedge r) \equiv (p \vee q) \wedge (p \vee r)$ means that *p or (q and r)* has the same meaning as *(p or q) and (p or r)*. Also, $p \wedge (r \vee q) \equiv (p \wedge r) \vee (p \wedge q)$ means that the compound proposition *p and (r or q)* has the same meaning as *(p and r) or (p and q)*.

EXAMPLE 6   **Using the Distributive Law to Show That Two Statements Are Logically Equivalent**

Consider the three propositions

p: Betsy will do her homework.

q: Betsy will wash her car.

r: Betsy will read a book.

The first distributive law, namely,

$$p \vee (q \wedge r) \equiv (p \vee q) \wedge (p \vee r)$$

says that these two statements are logically equivalent:

1. Betsy will do her homework, or she will wash her car and read a book.
2. Betsy will do her homework or wash her car, and Betsy will do her homework or read a book.

■

De Morgan's* Laws

For any two propositions p and q,

$$\sim(p \vee q) \equiv \sim p \wedge \sim q \qquad \sim(p \wedge q) \equiv \sim p \vee \sim q$$

That is, the compound proposition p *or* q is false only when p and q are both false. Similarly, p *and* q is false when either p or q (or both) is false. De Morgan's laws are proved using the truth tables in Examples 4 and 5.

EXAMPLE 7   Using De Morgan's Laws to Negate Statements

Negate the compound statements:

(a) The first child is a girl and the second child is a boy.
(b) Tonight I will study or I will go bowling.

SOLUTION

(a) Let p and q represent the components:

$$p: \quad \text{The first child is a girl.}$$
$$q: \quad \text{The second child is a boy.}$$

To negate the statement $p \wedge q$ is to find $\sim(p \wedge q)$. By De Morgan's law,

$$\sim(p \wedge q) \equiv \sim p \vee \sim q$$

*Augustus De Morgan (1806–1871), British mathematician and logician, was born in Madura, India, the son of a British army officer. He was graduated from Trinity College in Cambridge, England in 1827, but was denied a teaching position there for refusing to subscribe to religious tests. He was, however, appointed to a mathematics professorship at the newly opened University of London. He is best known for his work *Formal Logic,* which appeared in 1847. He also wrote papers on the foundations of algebra, philosophy of mathematical methods, and probability, as well as several successful elementary textbooks.

The negation can be stated as

The first child is not a girl or the second child is not a boy.

(b) We represent p and q as

p: I will study.

q: I will go bowling.

We want to negate the statement $p \vee q$. By De Morgan's law,

$$\sim (p \vee q) \equiv \sim p \wedge \sim q$$

The negation can be stated as

Tonight I will not study and I will not go bowling.

Absorption Laws

For any two propositions p and q,

$$p \vee (p \wedge q) \equiv p \qquad p \wedge (p \vee q) \equiv p$$

Here, $p \vee (p \wedge q) \equiv p$ means that p *or* (p *and* q) is logically equivalent to p. Similarly, $p \wedge (p \vee q) \equiv p$ means that p *and* (p *or* q) is logically equivalent to p.

EXAMPLE 8   Using Truth Tables to Prove the Distributive Law

Use truth tables to prove the distributive law: $p \vee (q \wedge r) \equiv (p \vee q) \wedge (p \vee r)$.

SOLUTION   Since the entries in the last two columns of the truth table below are the same, the two propositions are logically equivalent.

p	q	r	$q \wedge r$	$p \vee q$	$p \vee r$	$p \vee (q \wedge r)$	$(p \vee q) \wedge (p \vee r)$
T	T	T	T	T	T	T	T
T	T	F	F	T	T	T	T
T	F	T	F	T	T	T	T
T	F	F	F	T	T	T	T
F	T	T	T	T	T	T	T
F	T	F	F	T	F	F	F
F	F	T	F	F	T	F	F
F	F	F	F	F	F	F	F

EXAMPLE 9   Using Truth Tables to Prove the Idempotent Law

Use truth tables to prove the idempotent law: $p \wedge p \equiv p$.

SOLUTION   The truth table is

p	$p \wedge p$
T	T
F	F

EXERCISE 11.2   Answers to odd-numbered problems begin on page AN-42.

In Problems 1–16 construct a truth table for each compound proposition.

1. $p \vee \sim q$
2. $\sim p \vee \sim q$
3. $\sim p \wedge \sim q$
4. $\sim p \wedge q$
5. $\sim (\sim p \wedge q)$
6. $(p \vee \sim q) \wedge \sim p$
7. $\sim (\sim p \vee \sim q)$
8. $(p \vee \sim q) \wedge (q \wedge \sim p)$
9. $(p \vee \sim q) \wedge p$
10. $p \wedge (q \vee \sim q)$
11. $(p \vee q) \wedge (p \wedge \sim q)$
12. $(p \wedge \sim q) \vee (q \wedge \sim p)$
13. $(p \wedge q) \vee (\sim p \wedge \sim q)$
14. $(p \overline{\wedge} q) \vee (p \wedge r)$
15. $(p \wedge \sim q) \underline{\vee} r$
16. $(\sim p \vee q) \wedge \sim r$

In Problems 17–22 construct a truth table for each law. (Note: Some of these are done as examples.)

17. Idempotent laws
18. Commutative laws
19. Associative laws
20. Distributive laws
21. Absorption laws
22. De Morgan's laws

In Problems 23–26 show that the given propositions are logically equivalent.

23. $p \wedge (\sim q \vee q)$ and p
24. $p \vee (q \wedge \sim q)$ and p
25. $\sim (\sim p)$ and p
26. $p \wedge q$ and $q \wedge p$

In Problems 27–30 construct a truth table for each proposition.

27. $p \wedge (q \wedge \sim p)$
28. $(p \wedge q) \vee p$
29. $[(p \wedge q) \vee (\sim p \wedge \sim q)] \wedge p$
30. $(\sim p \wedge \sim q \wedge r) \vee (p \wedge q \wedge r)$

In Problems 31–33 use the propositions

p: Smith is an ex-convict. q: Smith is rehabilitated.

to give examples, using English sentences, of each law.

31. Idempotent laws
32. Commutative laws
33. De Morgan's laws

34. Use the distributive and commutative laws to show that

$$(p \vee q) \wedge r \equiv (p \wedge r) \vee (q \wedge r)$$

35. Use the distributive and commutative laws to show that

$$(p \wedge q) \vee r \equiv (p \vee r) \wedge (q \vee r)$$

36. The statement

The actor is intelligent or handsome and talented.

could be interpreted to mean either

   a: The actor is intelligent, or he is handsome and talented.

   or

   b: The actor is intelligent or handsome, and he is talented.

Describe a case in which *a* is true and *b* is false.

37. The statement

Michael will sell his car and buy a bicycle or rent a truck.

could be interpreted to mean either

   a: Michael will sell his car, and he will buy a bicycle or rent a truck.

   or

   b: Michael will sell his car and buy a bicycle, or he will rent a truck.

Find a case in which *b* is true and *a* is false.

In Problems 38–40 use De Morgan's laws to negate each proposition.

38. Mike can hit the ball well and he can pitch strikes.

39. Katy is a good volleyball player and is not conceited.

40. The baby is crying or talking all the time.

11.3 IMPLICATIONS; THE BICONDITIONAL CONNECTIVE; TAUTOLOGIES

The Conditional Connective

Consider the following compound proposition: "If I get an A in math, then I will continue to study." The above sentence states a condition under which I will continue to study. Another example of such a proposition is: "If we get an offer of $100,000, we will sell the house."

Such propositions occur frequently in mathematics, and an understanding of them is important.

Implication; Conditional Connective

   If *p* and *q* are any two propositions, then we call the proposition

If p, then q

an **implication** or **conditional statement** and the connective *if . . . , then* the **conditional connective.**

   We denote the conditional connective symbolically by $\Rightarrow$ and the implication *If p, then q* by $p \Rightarrow q$. In the implication $p \Rightarrow q$, *p* is called the **hypothesis** and *q* is called the **conclusion.** The implication $p \Rightarrow q$ can also be read as follows:

1. *p* implies *q*.
2. *p* is sufficient for *q*.
3. *p* only if *q*.

4. *q* is necessary for *p*.
5. *q*, if *p*.

To arrive at a truth table for implication, we consider the following situation. Suppose we make the statement

If XYZ common stock reaches $90 per share, it will be sold.

Table 5

p	q	$p \Rightarrow q$
T	T	T
T	F	F
F	T	T
F	F	T

When is the statement true and when is it false? Clearly, if XYZ stock reaches $90 per share and it is not sold, the implication is false. It is also clear that if XYZ stock reaches $90 per share and it is sold, the implication is true. In other words, if the hypothesis and conclusion are both true, the implication is true; if the hypothesis is true and the conclusion false, the implication is false.

But what happens when the hypothesis is false? If XYZ stock does not reach $90 a share the stock might be sold or it might not be sold. In either case we would not say that a person making the above statement was a liar. That is, we would not accuse the person of making a false statement. For this reason we say that an implication with a false hypothesis is not false and, therefore, is true.) The truth table for implication is given in Table 5.

As in the previous section we express a compound proposition symbolically by replacing each component statement and connective by an appropriate symbol.

For example, denoting "I study" and "I will pass" by a and b, respectively, the proposition "If I study, then I will pass" is written as $a \Rightarrow b$. This proposition can also be read as "A sufficient condition for passing is to study."

To understand "implication" better, look at an implication as a conditional promise. If the promise is broken, the implication is false; otherwise, it is true. For this reason the only circumstance under which the implication $p \Rightarrow q$ is false is when p is true and q is false, or, *it is the case that p but not q.*

EXAMPLE 1 **Studying the Meaning of an Implication**

Consider the implication

If you are a thief, then you will go to jail.

If you are a thief and you do go to jail, the implication is true. If you are a thief and you do not go to jail, the promise is broken; thus, the implication is false. If you are not a thief, we have no way of telling what would happen if you were a thief. Since the premise is not tested, it is not broken; thus, the implication is true. ◾

The word *then* in an implication merely serves to separate the conclusion from the hypothesis—it can be, and often is, omitted.

The implication $p \Rightarrow q$ can be expressed in the symbols defined previously. In fact, Table 6 shows that $p \Rightarrow q$ is logically equivalent to the compound proposition $\sim p \vee q$.

Table 6

p	q	$\sim p$	$\sim p \vee q$	$p \Rightarrow q$
T	T	F	T	T
T	F	F	F	F
F	T	T	T	T
F	F	T	T	T

Converse, Contrapositive, and Inverse

Suppose we start with the implication $p \Rightarrow q$ and then interchange the roles of p and q, obtaining the implication, "If q, then p."

Table 7

p	q	$p \Rightarrow q$	$q \Rightarrow p$
T	T	T	T
T	F	F	T
F	T	T	F
F	F	T	T

> **Converse**
>
> The implication *If q, then p* is called the **converse** of the implication *If p, then q.* That is, $q \Rightarrow p$ is the converse of $p \Rightarrow q$.

The truth tables for the implication $p \Rightarrow q$ and its converse $q \Rightarrow p$ are compared in Table 7.

Notice that $p \Rightarrow q$ and $q \Rightarrow p$ are not equivalent. As an illustration, consider the following example.

EXAMPLE 2   **Forming the Converse of an Implication**

Consider the statements

p: You are a thief.

q: You will go to jail.

The implication $p \Rightarrow q$ states that

If you are a thief, you will go to jail.

The converse of this implication, namely, $q \Rightarrow p$, states that

If you go to jail, you are a thief.

To say that thieves go to jail is not the same as saying that everyone who goes to jail is a thief.

■

This example illustrates that the truth of an implication does not imply the truth of its converse. Many common fallacies in thinking arise from confusing an implication with its converse.

> **Contrapositive**
>
> The implication *If not q, then not p*, written as $\sim q \Rightarrow \sim p$, is called the **contrapositive** of the implication $p \Rightarrow q$.

EXAMPLE 3 **Studying the Contrapositive of an Implication**

Consider the statements

$$p: \quad \text{You are a thief.}$$

$$q: \quad \text{You will go to jail.}$$

The implication $p \Rightarrow q$ states "If you are a thief, then you will go to jail." The contrapositive, namely $\sim q \Rightarrow \sim p$, is "If you do not go to jail, then you are not a thief." ∎

> **Inverse**
>
> The implication *If not p, then not q,* written as $\sim p \Rightarrow \sim q$, is called the **inverse** of the implication *If p, then q.*

EXAMPLE 4 **Studying the Inverse of an Implication**

Consider the statements

$$p: \quad \text{You are a thief.}$$

$$q: \quad \text{You will go to jail.}$$

The implication $p \Rightarrow q$ states "If you are a thief, then you will go to jail." The inverse of $p \Rightarrow q$, namely $\sim p \Rightarrow \sim q$, is "If you are not a thief, then you will not go to jail." ∎

EXAMPLE 5 **Constructing Truth Tables**

The truth tables for $p \Rightarrow q$, $q \Rightarrow p$, $\sim p \Rightarrow \sim q$, and $\sim q \Rightarrow \sim p$ are given in Table 8.

Table 8

State-ments		Implication	Converse			Inverse	Contrapositive
p	q	$p \Rightarrow q$	$q \Rightarrow p$	$\sim p$	$\sim q$	$\sim p \Rightarrow \sim q$	$\sim q \Rightarrow \sim p$
T	T	T	T	F	F	T	T
T	F	F	T	F	T	T	F
F	T	T	F	T	F	F	T
F	F	T	T	T	T	T	T

Notice that the entries under Implication and Contrapositive are the same; also, the entries under Converse and Inverse are the same. We conclude that

$$p \Rightarrow q \equiv \, \sim q \Rightarrow \, \sim p \qquad q \Rightarrow p \equiv \, \sim p \Rightarrow \, \sim q$$

Thus, we have shown that an implication and its contrapositive are logically equivalent. Also, the converse and inverse of an implication are logically equivalent.

The Biconditional Connective

The compound statement "p if and only if q" is another way of stating the conjunction of two implications: "if p, then q and if q, then p." For example, consider

$$p: \quad \text{Jill is happy.}$$
$$q: \quad \text{Jack is attentive.}$$

If we say "Jill is happy if, and only if, Jack is attentive," we mean that if Jill is happy, then Jack is attentive, and if Jack is attentive, then Jill is happy. In symbols, we could write this as

$$(p \Rightarrow q) \wedge (q \Rightarrow p)$$

Biconditional Connective

The connective *if and only if* is called the **biconditional connective** and is denoted by the symbol $\Leftrightarrow$. The compound proposition

$$p \Leftrightarrow q \qquad (p \text{ if and only if } q)$$

is equivalent to

$$(p \Rightarrow q) \wedge (q \Rightarrow p)$$

The statement $p \Leftrightarrow q$ also may be read as "p is necessary and sufficient for q," or as "p implies q and q implies p." (The abbreviation "iff" for "if and only if" is also sometimes used.)

We can restate the definition of the biconditional connective by its truth table (Table 9).

Table 9

p	q	$p \Rightarrow q$	$q \Rightarrow p$	$(p \Rightarrow q) \wedge (q \Rightarrow p)$	$p \Leftrightarrow q$
T	T	T	T	T	T
T	F	F	T	F	F
F	T	T	F	F	F
F	F	T	T	T	T

EXAMPLE 6   Using the Biconditional Connective

Let p and q denote the statements

$$p: \quad \text{Joe is happy.}$$

$$q: \quad \text{Joe is not studying.}$$

Then "A *necessary condition* for Joe to be happy is that Joe is not studying" means "If Joe is happy, he is not studying." Moreover, "A *sufficient condition* for Joe to be happy is that Joe is not studying" means "If Joe is not studying, he is happy."

Thus, if both implications ($p \Rightarrow q$, $q \Rightarrow p$) are true statements, then a necessary and sufficient condition for Joe to be happy is that Joe is not studying.

In other words, Joe is happy if and only if he is not studying.   ■

Tautologies

In Section 11.1 we saw how compound propositions can be obtained from simple propositions by using connectives. By using symbols of grouping (parentheses and brackets), we can form more complicated propositions. We denote by $P(p, q, \ldots)$ a compound proposition, where $p, q, \ldots$ are the components of the compound proposition. Some examples of such propositions are

$$\sim(p \wedge q) \qquad p \vee \sim q \qquad (p \wedge \sim q) \qquad (\sim p \wedge q) \sim p \vee \sim(\sim q \wedge \sim r)$$

Ordinarily, when we write a compound proposition, we cannot be certain of the truth of that proposition unless we know the truth or falsity of the component propositions. A compound proposition that is true regardless of the truth values of its components is called a tautology.

Tautology

A **tautology** is a compound proposition $P(p, q, \ldots)$ that is true in every possible case.

Examples of tautologies are

$$p \vee \sim p \qquad p \Rightarrow (p \vee q)$$

Tables 10 and 11 show the truth tables of these tautologies.

Table 10

p	$\sim p$	$p \vee \sim p$
T	F	T
F	T	T

Table 11

p	q	$p \vee q$	$p \Rightarrow (p \vee q)$
T	T	T	T
T	F	T	T
F	T	T	T
F	F	F	T

Many other examples of tautologies are provided by the logical equivalences in Section 11.2. For instance, to say that

$$\sim(p \vee q) \equiv \sim p \wedge \sim q$$

is to say that the biconditional proposition

$$\sim(p \vee q) \Leftrightarrow \sim p \wedge \sim q$$

is a tautology.

Substitution

Let $P(p, q, \ldots)$ be a compound proposition with p as one of its components. If we replace p with another proposition h having the same truth values, then we obtain a new compound proposition $P(h, q, \ldots)$ that has the same truth value as $P(p, q, \ldots)$. This leads to the following principle:

The Law of Substitution

Suppose p and h are propositions and $h \equiv p$. If h is substituted for p in the compound proposition $P(p, q, \ldots)$, a logically equivalent proposition is obtained:

$$P(p, q, \ldots) \equiv P(h, q, \ldots)$$

The law of substitution is often used to obtain new versions of existing tautologies. It is particularly useful in mathematics for constructing tautologies of a certain kind, known as **valid arguments** or **proofs.** This is the topic of Section 11.4.

EXERCISE 11.3   Answers to odd-numbered problems begin on page AN-44.

In Problems 1–10 write the converse, contrapositive, and inverse of each statement.

1. $\sim p \Rightarrow q$

2. $\sim p \Rightarrow \sim q$

3. $\sim q \Rightarrow \sim p$

4. $p \Rightarrow \sim q$

5. If it is raining, the grass is wet.

6. It is raining if it is cloudy.

7. If it is not raining, it is not cloudy.

8. If it is not cloudy, then it is not raining.

9. Rain is sufficient for it to be cloudy.

10. Rain is necessary for it to be cloudy.

11. Give a verbal sentence that describes

   (a) $p \Rightarrow q$ (b) $q \Rightarrow p$ (c) $\sim p \Rightarrow q$

using the components

   p: Jack studies psychology.
   q: Mary studies sociology.

12. Show that

$$p \Rightarrow q \equiv \sim q \Rightarrow \sim p$$

using the fact that $p \Rightarrow q \equiv \sim p \vee q$.

13. Show that

$$p \Rightarrow (q \vee r) \equiv (p \wedge \sim q) \Rightarrow r$$

(a) using a truth table; (b) using De Morgan's laws and the fact that $p \Rightarrow q \equiv \sim p \vee q$.

14. Show that

$$(p \wedge q) \Rightarrow r \equiv (p \wedge \sim r) \Rightarrow \sim q$$

using the same two methods as in Problem 13.

In Problems 15–24 construct a truth table for each statement.

15. $\sim p \lor (p \land q)$ **16.** $\sim p \land (p \lor q)$ **17.** $p \lor (\sim p \land q)$ **18.** $(p \lor q) \land \sim q$ **19.** $\sim p \Rightarrow q$

20. $(p \lor q) \Rightarrow p$ **21.** $\sim p \lor p$ **22.** $p \land \sim p$ **23.** $p \land (p \Rightarrow q)$ **24.** $p \lor (p \Rightarrow q)$

In Problems 25–28 prove that the following have only truth value T.

25. $p \land (q \land r) \Leftrightarrow (p \land q) \land r$

26. $p \lor (q \lor r) \Leftrightarrow (p \lor q) \lor r$

27. $p \land (p \lor q) \Leftrightarrow p$

28. $p \lor (p \land q) \Leftrightarrow p$

In Problems 29–34 let p be "The examination is hard" and q be "The grades are low." Write each symbolically.

29. If the examination is hard, the grades are low.

30. The examination is not hard nor are the grades low.

31. The grades are not low, and the examination is not hard.

32. The examination is not hard, and the grades are low.

33. The grades are low only if the examination is hard.

34. The grades are low if the examination is hard.

11.4 ARGUMENTS

Valid Arguments

In an argument or a proof we are not concerned with the *truth* of the conclusion, but rather with whether the conclusion does or does not follow from the premises. If the conclusion follows from the premises, we say that our reasoning is *valid;* if it does not, we say that our reasoning is *invalid.* For example, from the two

> **Premises:**   (1) All college students are intelligent.
>
>   (2) All freshmen are college students.

follows the

> **Conclusion:**   All freshmen are intelligent.

Now the last statement certainly is not regarded generally as true, but the reasoning leading to it is valid. *If both of the premises are true, the conclusion is also true.*

Argument

An **argument** or **proof** consists of a set of propositions $p_1, p_2, \ldots, p_n$, called the **premises** or **hypotheses,** and a proposition q, called the **conclusion.** An argument is **valid** if and only if the conclusion is true whenever the premises are all true. An argument that is *not* valid is called a **fallacy** or an **invalid** argument.

Direct and Indirect Proof

We will limit our discussion to two types of argument: direct proof and indirect proof.

In a **direct proof** we go through a chain of propositions, beginning with the hypotheses and leading to the desired conclusion.

EXAMPLE 1 **Using a Direct Proof to Prove a Statement**

Suppose it is true that

> Either John obeyed the law or John was punished, but not both.

and

> John was not punished.

Prove that

> John obeyed the law.

SOLUTION   Let p and q denote the propositions

p: John obeyed the law.

q: John was punished.

We can write the premises as

$$p \underline{\vee} q \quad \text{and} \quad \sim q$$

Since $\sim q$ is true, then, by the law of contradiction (see page 506), q is false. Since $p \underline{\vee} q$ is true, either p is true or q is true. Thus, p must be true, so John obeyed the law. ■

More examples of direct proof are given later, but before we look at them, let's discuss two laws of logic that are useful in direct proofs.

Law of Detachment

If the implication $p \Rightarrow q$ is true and p is true, then q must be true.

See Table 5, Section 11.3, for an illustration of this law.

Law of Syllogism

Let p, q, r be three propositions. If

$$p \Rightarrow q \quad \text{and} \quad q \Rightarrow r$$

are both true, then $p \Rightarrow r$ is true.

Table 12 illustrates this law.

Table 12

p	q	r	$p \Rightarrow q$	$q \Rightarrow r$	$p \Rightarrow r$	$(p \Rightarrow q) \wedge (q \Rightarrow r)$	$(p \Rightarrow q) \wedge (q \Rightarrow r) \Rightarrow (p \Rightarrow r)$
T	T	T	T	T	T	T	T
T	T	F	T	F	F	F	T
T	F	T	F	T	T	F	T
T	F	F	F	T	F	F	T
F	T	T	T	T	T	T	T
F	T	F	T	F	T	F	T
F	F	T	T	T	T	T	T
F	F	F	T	T	T	T	T

EXAMPLE 2   Using a Direct Proof to Prove a Statement

Suppose it is true that

> It is snowing.

and

> If it is warm, then it is not snowing.

and

> If it is not warm, then I cannot go swimming.

Prove that

> I cannot go swimming.

SOLUTION   Let p, q, r represent the statements

p: It is snowing.

q: It is warm.

r: I can go swimming.

Our premises are the propositions

$$p \qquad q \Rightarrow {\sim}p \qquad {\sim}q \Rightarrow {\sim}r$$

We want to prove that

$$\sim r$$

is true. Since $q \Rightarrow {\sim}p$ is true, its contrapositive

$$p \Rightarrow {\sim}q$$

is also true. Using the law of syllogism, since $p \Rightarrow \sim q$ and $\sim q \Rightarrow \sim r$, we see that

$$p \Rightarrow \sim r$$

But we know p is true. By the law of detachment $\sim r$ is true. That is, I cannot go swimming. ∎

EXAMPLE 3 **Using a Direct Proof to Prove a Statement**

Suppose it is true that

> If Dan comes, so will Bill.

and

> If Sandy will not come, then Bill will not come.

Show that

> If Dan comes, then Sandy will come.

SOLUTION Let p, q, r denote the propositions

p: Dan comes.
q: Bill will come.
r: Sandy will come.

The premises are

$$p \Rightarrow q \qquad \text{and} \qquad \sim r \Rightarrow \sim q$$

If we assume $\sim r \Rightarrow \sim q$ is true, then the contrapositive $q \Rightarrow r$ is true also. Thus,

$$p \Rightarrow q \qquad \text{and} \qquad q \Rightarrow r$$

By the law of syllogism it is true that

$$p \Rightarrow r$$

That is, if Dan comes, then Sandy will come. ∎

EXAMPLE 4 **Using a Direct Proof to Prove a Statement**

Suppose it is true that

> If I enjoy studying, then I will study.

and

> I will do my homework or I will not study.

and

> I will not do my homework.

Show that

I do not enjoy studying.

SOLUTION   Let p, q, r denote the three propositions

p: I enjoy studying.

q: I will study.

r: I will do my homework.

Then we know that

$$\sim r \qquad r \vee \sim q \qquad p \Rightarrow q$$

are true. We want to prove that $\sim p$ is true. Since $\sim r$ is true, then r is false. Also, either r or $\sim q$ is true. Thus, $\sim q$ is true. Since $p \Rightarrow q$, we have

$$\sim q \Rightarrow \sim p$$

is true. Hence, $\sim p$ must be true.

■

In the examples of direct proof that we have just seen, one proceeds by the laws of logic from the premises to the conclusion.

In any valid proof one shows that an implication $p \Rightarrow q$ is a tautology. To do this by the method of **indirect proof,** we tentatively suppose that q is false (equivalently, that $\sim q$ is true) and show that we are thereby led to a **contradiction**—that is, a logically impossible situation. This contradiction can be resolved only by abandoning the supposition that the conclusion is false. Therefore, it must be true.

An indirect proof is also called a **proof by contradiction.** The concept is illustrated in the next two examples.

EXAMPLE 5   **Using Indirect Proof to Prove a Statement**

Prove the result of Example 1 using an indirect proof.

SOLUTION   To prove that the conclusion p is true, we suppose instead that p is false, hoping to be able to show that this leads to a contradiction.

Thus, we assume that

$$\sim p \qquad \sim q \qquad p \vee q$$

all are true. Either p is true or q is true. But both p and q are false. This is impossible. thus, we have reached a contradiction, which means that p must be true.

■

Now Work Problem 1

EXAMPLE 6   **Using Indirect Proof to Prove a Statement**

Suppose that

If I am lazy, I do not study.

and

I study or I enjoy myself.

and

I do not enjoy myself.

Prove that

I am not lazy.

SOLUTION Let p, q, r be the statements

p: I am lazy.

q: I study.

r: I enjoy myself.

Assume that

$$\sim r \qquad p \Rightarrow \sim q \qquad q \vee r$$

are true.

We want to show that $\sim p$, the conclusion, is true. In an indirect proof we assume that $\sim p$ is false — that is, that p is true. If p and $p \Rightarrow \sim q$ are true, then $\sim q$ is true (by the law of detachment). That is, q is false. One of the premises is $\sim r$; thus, r is also false. But $q \vee r$ is true, which is impossible if q and r are both false. This contradiction tells us that we have incorrectly assumed that $\sim p$ is false. Thus, $\sim p$ is true.

■

Model: Life Insurance*

The following problem was first studied by Edmund C. Berkeley in 1936 and was solved by the use of principles of logic. His employer, the Prudential Life Insurance Company, had the procedure that when any policyholder requested a change in the schedule of premium payments, one of two rules was involved as company policy. The question was raised as to whether these two rules were logically equivalent. That is, did both rules give rise to the same payment arrangements, or did the use of one rule over the other give different payment arrangements?

Berkeley was of the opinion that the two rules, in some instances, gave different directions to the policyholder. An example of a typical clause found in one of the rules was this: If a policyholder was making premium payments several times a year with one of the payments falling due on the policy anniversary, and if he requested the schedule be changed to one annual payment on the anniversary date, and if his payments were in full up to a date not the anniversary date, and if he made this request more than 2 months after the issue date, and if his request also came within 2 months after the policy anniversary date, then a certain action was to be taken!

Berkeley replaced this complicated part of one of the rules by the compound proposition

$$p \wedge q \wedge r \wedge s \wedge t \Rightarrow A$$

*John E. Pfeiffer, "Symbolic Logic," *Scientific American* (December 1960).

where p, q, r, s, t are the five statements and A is the action called for. By doing the same with all parts of both rules and by using the laws of logic, Berkeley was able to demonstrate an inconsistency of application of one rule over the other. In fact, there turned out to be four situations in which contradictory actions occurred.

As a result of Berkeley's effort, Prudential replaced the two cumbersome rules by a simple one.

Since this incident, similar situations involving a maze of if's, and's, but's, and implications, particularly in the areas of legal contracts between corporations, have been checked for accuracy and consistency (to eliminate loopholes, etc.) by the use of logic.

EXERCISE 11.4   Answers to odd-numbered problems begin on page AN-45.

Prove the statements in Problems 1–4 first by using a direct proof and then by using an indirect proof.

1. When it rains, John does not go to school. John is going to school. Show that it is not raining.

2. If I do not go to work, I will go fishing. I will not go fishing. Show that I will go to work.

3. If Smith is elected president, Kuntz will be elected secretary. If Kuntz is elected secretary, then Brown will not be elected treasurer. Smith is elected president. Show that Brown is not elected treasurer.

4. Either Laura is a good girl or Rob is a good boy. If Danny cries, then Laura is not a good girl. Rob is not a good boy. Does Danny cry?

In Problems 5–8 determine whether the arguments are valid.

5. Hypotheses: When students study, they receive good grades.
These students do not study.
Conclusion: These students do not receive good grades.

6. Hypotheses: If Danny is affluent, he is either a snob or a hypocrite, but not both.
Danny is a snob and is not a hypocrite.
Conclusion: Danny is not affluent.

7. Hypotheses: If Tami studies, she will not fail this course.

If she does not play with her dolls too often, she will study.
Tami failed the course.
Conclusion: She played with her dolls too often.

8. Hypotheses: If John invests his money wisely, he will be rich.
If John is rich, he will buy an expensive car.
John bought an expensive car.
Conclusion: John invested his money wisely.

11.5 LOGIC CIRCUITS

Circuits and Gates

A **logic circuit** is a type of electrical circuit widely used in computers and other electronic devices (such as calculators, digital watches, and compact disc players). The simplest logic circuits, called **gates,** have the following properties.

1. Current flows to the circuit through one or two connectors called **input lines,** and from the circuit through a connector called the **output line.**

2. The current in any of the input or output lines may have either of two possible voltage levels. The higher level is denoted by 1 and the lower level by 0. (Level 1 is sometimes referred to as **On** or **True**; Level 0 may be referred to as **Off** or **False**.)
3. The voltage level of the output line depends on the voltage level(s) of the input line(s), according to rules of logic similar to the principles discussed in Section 11.2.

Figure 1

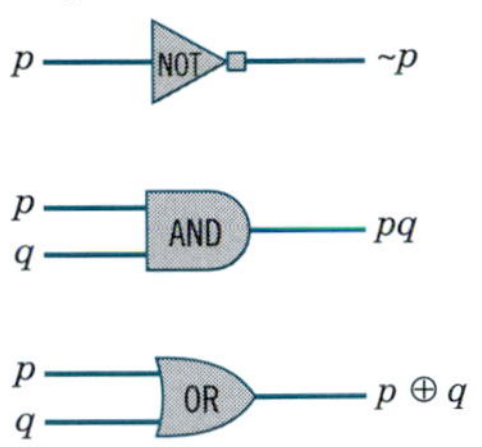

The three most basic types of gates are the **inverter** (or **NOT gate**), the **AND gate,** and the **OR gate.**

The standard symbols for inverters, AND gates, and OR gates appear in Figure 1. In each case, the input lines appear at the left of the diagram and the output line at the right. The label on each line denotes the voltage level or *truth value* of that line; the symbols $\sim p$, pq (or $p \wedge q$), and $p \oplus q$ (or $p \vee q$) are defined by the output tables in Tables 13, 14, and 15. Note that $\sim p = 1 - p$ and

$$p \oplus q = \begin{cases} p & \text{if} & p \geq q \\ q & \text{if} & q > p \end{cases} = \max \{p, q\}$$

Table 13

p	$\sim p$
1	0
0	1

$(\sim p = 1 - p)$

Table 14

p	q	pq
1	1	1
1	0	0
0	1	0
0	0	0

Table 15

p	q	$p \oplus q$
1	1	1
1	0	1
0	1	1
0	0	0

Computer designers and electrical engineers frequently use one of the symbols $\tilde{p}$, $\bar{p}$, or p' in place of $\sim p$.

More complex logic circuits are constructed by connecting two or more gates.

EXAMPLE 1   Constructing a Logic Circuit

Construct a logic circuit whose inputs are p, q, r and whose output is $(pq) \oplus \sim r$. When is this output equal to 1?

Figure 2

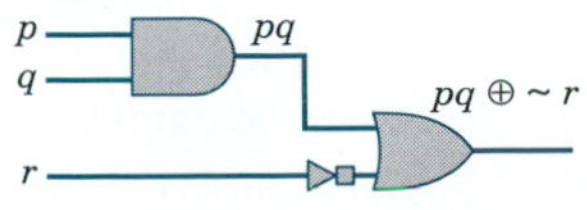

SOLUTION   Apparently, we need to connect an AND gate, an inverter, and an OR gate. Figure 2 shows how.

(Just as we write $ab + c$ for $(ab) + c$ in ordinary algebra, we can write $pq \oplus r$ for $(pq) \oplus r$ and $pq \oplus \sim r$ for $(pq) \oplus (\sim r)$ in circuit algebra. Note that $p(q \oplus r) \neq pq \oplus r$.)

Table 16 is the output table for $pq \oplus \sim r$. This table confirms what we would expect: the output is 1 when $p = q = 1$ (regardless of whether r is 0 or 1) and when $r = 0$ (regardless of the values of p and q).

Table 16

p	q	r	pq	$\sim r$	$pg \oplus \sim r$
1	1	1	1	0	1
1	1	0	1	1	1
1	0	1	0	0	0
1	0	0	0	1	1
0	1	1	0	0	0
0	1	0	0	1	1
0	0	1	0	0	0
0	0	0	0	1	1

 **Now Work Problem 1**

Redesigning Circuits

Since the output tables for circuits have the same form as the truth tables for propositions (with 1 and 0 corresponding to T and F, pq corresponding to $p \wedge q$, and $p \oplus q$ corresponding to $p \vee q$), the principles of logic formulated in Section 11.2 apply. We can translate these principles as follows:

1. Idempotent laws

$$pp = p \qquad\qquad p \oplus p = p$$

2. Associative laws

$$(pq)r = p(qr) \qquad\qquad (p \oplus q) \oplus r = p \oplus (q \oplus r)$$

3. Commutative laws

$$pq = qp \qquad\qquad p \oplus q = q \oplus p$$

4. Distributive laws

$$p \oplus qr = (p \oplus q)(p \oplus r) \qquad p(q \oplus r) = pq \oplus pr$$

5. De Morgan's laws

$$\sim(p \oplus q) = (\sim p)(\sim q) \qquad \sim(pq) = \sim p \oplus \sim q$$

6. Absorption laws

$$p \oplus pq = p \qquad\qquad p(p \oplus q) = p$$

These principles are often useful in simplifying the design of circuits. If two circuits perform the same function and one has fewer gates and connectors than the other, the simpler circuit is preferred because it is less expensive to manufacture or purchase, and also because it is less likely to fail due to overheating or physical stress. Very often it will also be more efficient in terms of power consumption and speed of operation.

EXAMPLE 2   Redesigning a Logic Circuit

Figure 3

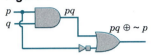

Figure 4

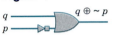

Redesign the circuit of Example 1 so that $r = p$; simplify the circuit.

SOLUTION   We can modify the circuit as in Figure 3. The solid dot is used in circuit diagrams to show where a connector branches.
   Using the distributive laws, we can simplify:

$$pq \oplus {\sim}p = (p \oplus {\sim}p)(q \oplus {\sim}p) = 1 \cdot (q \oplus {\sim}p) = q \oplus {\sim}p$$

(Recall that ${\sim}p = 1$ if $p = 0$, and vice versa.) The circuit can therefore be replaced with the one in Figure 4. ∎

EXAMPLE 3   Redesigning a Logic Circuit

Find the output of the circuit in Figure 5 and simplify its design.

Figure 5

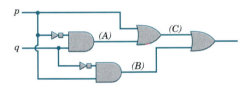

SOLUTION   The output is $B \oplus C$, where

$$A = ({\sim}p)q, \qquad B = p({\sim}q), \qquad C = p \oplus A$$

Thus,

$$C = p \oplus ({\sim}p)q = (p \oplus {\sim}p)(p \oplus q) = 1 \cdot (p \oplus q) = p \oplus q$$

and the output is

$$B \oplus C = p({\sim}q) \oplus (p \oplus q)$$

This expression can be simplified to

$$p({\sim}q) \oplus (p \oplus q) = [p({\sim}q) \oplus p] \oplus q = p \oplus q$$

Thus, the circuit can be replaced by a single OR gate. ∎

NAND, NOR, and XOR

The standard symbols for **NAND, NOR,** and **XOR gates,** and their outputs, are shown in Figure 6. (The names are abbreviations for NOT-AND, NOT-OR, and Exclusive-OR, respectively.) Table 17 is the output table for XOR.

Figure 6

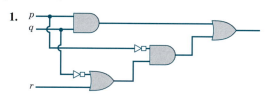

Table 17

p	q	$p \vee q$
1	1	0
1	0	1
0	1	1
0	0	0

EXERCISE 11.5   Answers to odd-numbered problems begin on page AN-45.

In Problems 1–4 determine when the output of each circuit is 1; use a truth table if necessary.

1.

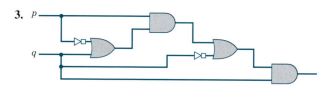

2.

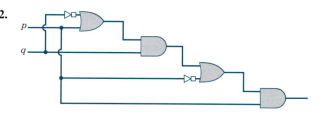

3.

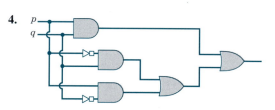

4.

In Problems 5–8 construct a circuit corresponding to each expression.

5. $(\sim p \oplus \sim q)(p \oplus q)$

6. $(p \oplus \sim q)(\sim p)$

7. $\sim (p \oplus q)(\sim p)$

8. $(\sim p \oplus q)[(\sim p)(\sim q)]$

9. Design simpler circuits having the same outputs as the circuits in Problems 1, 3, 5, and 7.

10. Design simpler circuits having the same outputs as the circuits in Problems 2, 4, 6, and 8.

11. Design a logic circuit that can be turned on or off from either of two switches. (That is, if the inputs are p and q, the output can be changed from 1 to 0 or from 0 to 1 by changing the level of either p or q, but not both.)

12. Design a circuit, consisting of at least two gates, whose output is always 1.

13. Design a circuit, consisting of at least two gates, whose output is always 0.

14. Using an OR gate, an AND gate, and an inverter, design a circuit that will work as an XOR gate. [*Hint:* What combination of $\cdot$, $\oplus$, and $\sim$ has the same truth table as $\vee$?]

15. Design a circuit that will work as an OR gate, using

(a) Two inverters and a NAND gate

(b) Three NAND gates

[*Hint:* (a) Use De Morgan's laws. (b) How can a NAND gate be substituted for an inverter?]

16. Design a circuit that will work as an AND gate using

    (a) Two inverters and a NOR gate

    (b) Three NOR gates

17. Show that a circuit whose output is

$$pq \oplus pr \oplus q(\sim r)$$

can be replaced by one whose output is

$$pr \oplus q(\sim r)$$

[*Hint:* $pq = pq(r \oplus \sim r) = pqr \oplus pq(\sim r)$]

18. Show that

    (a) $(p \oplus q)(p \oplus r)(q \oplus \sim r) = (p \oplus r)(q \oplus \sim r)$

    (b) $(p \oplus r)(q \oplus \sim r) = p(\sim r) \oplus qr$

CHAPTER REVIEW

IMPORTANT TERMS AND CONCEPTS

proposition 502
connective 503
conjunction ($\wedge$) 503
inclusive disjunction ($\vee$) 504
exclusive disjunction ($\underline{\vee}$) 505
negation ($\sim$) 506
law of contradiction 506
quantifier 506
truth value 508
truth table 508
logically equivalent ($\equiv$) 511
idempotent laws 512
commutative laws 512
associative laws 512
distributive laws 512

De Morgan's laws 514
absorption laws 515
implication ($\Rightarrow$) 517
conditional connective ($\Rightarrow$) 517
hypothesis (premise) 517
conclusion 517
sufficient condition 517
necessary condition 517
converse 519
contrapositive 519
inverse 520
biconditional connective ($\Leftrightarrow$) 521
tautology 522
law of substitution 523
argument 524

valid, invalid 524
fallacy 524
direct proof 525
law of detachment 525
law of syllogism 525
indirect proof 528
proof by contradiction 528
logic circuit 530
gate 530
inverter 531
AND gate 531
OR gate 531
NAND gate 534
NOR gate 534
XOR gate 534

IMPORTANT FORMULAS

$$p \wedge q \equiv q \wedge p$$
$$p \vee q \equiv q \vee p$$
$$(p \wedge q) \wedge r \equiv p \wedge (q \wedge r)$$
$$(p \vee q) \vee r \equiv p \vee (q \vee r)$$

$$p \vee (q \wedge r) \equiv (p \vee q) \wedge (p \vee r)$$
$$p \wedge (r \vee q) \equiv (p \wedge r) \vee (p \wedge q)$$
$$\sim(p \wedge q) \equiv \sim p \vee \sim q$$
$$\sim(p \vee q) \equiv \sim p \wedge \sim q$$

TRUE–FALSE ITEMS   Answers are on page AN-47.

T_____F_____ **1.** The negation of the statement "Some salesmen are intelligent" is "All salesmen are intelligent."

T_____F_____ **2.** The statement $\sim(p \wedge q)$ is logically equivalent to $\sim p \vee \sim q$.

T_____F_____ **3.** The statement $\sim(p \vee q)$ is logically equivalent to $\sim p \wedge \sim q$.

T_____F_____ **4.** The statement $\sim p \Rightarrow q$ is logically equivalent to $q \Rightarrow p$.

T_____F_____ **5.** The statement $p \Leftrightarrow q$ is logically equivalent to $(p \Rightarrow q) \wedge (\sim q \Rightarrow \sim p)$.

FILL IN THE BLANKS   Answers are on page AN-47.

1. The compound proposition *p or q* is called the inclusive disjunction of *p* and *q* and is denoted by _____ .

2. The negation of *p* is denoted by _____ .

3. If two propositions have the same truth values in every possible case, the propositions are said to be _____ .

4. The two parts of an argument are the _____ and the _____ .

5. The output of a logic circuit is either _____ or _____ .

REVIEW EXERCISES   Answers to odd-numbered problems begin on page AN-47.

In Problems 1–4 circle each correct answer; some questions may have more than one correct answer.

1. Which of the following negate the statement below?

    p: All people are rich.

    (a) Some people are rich.
    (b) Some people are poor.
    (c) Some people are not rich.
    (d) No person is rich.

2. Which of the following negate the statement below?

    p: It is either hot or humid.

    (a) It is neither hot nor humid.
    (b) Either it is not hot or it is not humid.
    (c) It is not hot and it is not humid.
    (d) It is hot, but not humid.

3. Which of the following statements are logically equivalent to the statement below?

    $$(\sim p \vee q) \wedge r$$

    (a) $(p \Rightarrow q) \wedge r$
    (b) $\sim p \vee (q \wedge r)$
    (c) $(\sim p \Rightarrow q) \wedge r$
    (d) $(\sim p \vee r) \wedge (q \wedge r)$

4. Which of the following statements are logically equivalent to the statement below?

    $$p \wedge \sim q$$

    (a) $\sim q \wedge p$
    (b) $p \vee q$
    (c) $\sim p \vee q$
    (d) $q \Rightarrow \sim p$

In Problems 5–8 negate each proposition.

5. Some people are rich.

6. All people are rich.

7. Danny is not tall and Mary is short.

8. Neither Mike nor Katy is big.

In Problems 9–12 construct a truth table for each compound proposition.

9. $(p \wedge q) \vee \sim p$

10. $(p \vee q) \wedge p$

11. $\sim p \vee (p \vee \sim q)$

12. $\sim p \Rightarrow (p \vee q)$

In Problems 13–15 let p stand for "I will pass the course" and let q stand for "I will do homework regularly." Put each statement into symbolic form.

13. I will pass the course if I do homework regularly.

14. Passing this course is a sufficient condition for me to do homework regularly.

15. I will pass this course if and only if I do homework regularly.

16. Write the converse, contrapositive, and inverse of the statement

    If it is not sunny, it is cold.

17. Prove the following by the use of direct proof: If I do not

paint the house, I will go bowling. I will not go bowling. Show that I will paint the house.

18. Using reasons from logic with the following premises, give a valid argument to answer the question, "Is Laura a good girl?"

    (a) Rob is a bad boy or Danny is crying.
    (b) If Laura is a good girl, then Rob is not a bad boy.
    (c) Danny is not crying.

19. Determine whether the following are logically equivalent:

$$\sim p \vee q \qquad p \Rightarrow q$$

20. Determine whether the two statements below are logically equivalent:

$$(p \Rightarrow q) \wedge (\sim q \vee p) \qquad p \Leftrightarrow q$$

21. Show that the output of an XOR gate is equal to

$$(p \oplus q)[\sim(pq)]$$

Use this fact to construct an XOR gate by connecting an OR gate, an AND gate, and a NAND gate.

22. Construct an XOR gate by connecting two NOR gates and an AND gate.

Chapter 12

Relations, Functions, and Induction

12.1 Relations

12.2 Functions

12.3 Sequences

12.4 Mathematical Induction

12.5 Recurrence Relations

Chapter Review

In this chapter we introduce the concepts of a relation and a function, which are fundamental in mathematics and applications. These concepts form the basis for the study of graphs and trees as introduced in Chapter 13.

12.1 RELATIONS

The word relation is a common term used in mathematics to indicate a relationship between two objects. Relationships occur everywhere. Two people belonging to the same family may be related as father-and-son, mother-and-son, brother-and-sister, husband-and-wife, and so forth. In mathematics two numbers may be related by being equal, or by one being greater than the other. Two sets may be related as one being a subset of the other, etc. These are only a few examples of relations. Later in this section, and in the exercises, we present more examples. First we start with a formal definition.

Definition of Relation

> **Relation**
>
> Let A and B be two nonempty sets. A **relation** R from A to B is a set R of ordered pairs (a, b) where $a \in A$ and $b \in B$. For every such ordered pair we write aRb, read "a is related to b by R." If the relation R is from the set A to itself, we say R is a **relation on A.***

EXAMPLE 1   **Studying Relations Between Two Sets**

Let $A = \{1, 2, 7\}$ and $B = \{2, 5\}$. List the elements of each relation R defined below.

(a) $a \in A$ is related to $b \in B$, that is, aRb if, and only if, $a < b$.
(b) $a \in A$ is related to $b \in B$, that is, aRb if, and only if, a and b are both odd numbers.
(c) $a \in A$ is related to $b \in B$, that is, aRb if, and only if, $(a + b)$ is an even number.

SOLUTION

(a) Since $1 \in A$ is less than $2 \in B$, then $1R2$. Similarly $1R5$ and $2R5$. Therefore,

$$R = \{(1, 2), (1, 5), (2, 5)\}$$

Note that $7 \in A$ is not related to $5 \in B$ since 7 is not less than 5.
(b) Since $1 \in A$ and $5 \in B$ are both odd, then $1R5$. Similarly, $7R5$. Therefore,

$$R = \{(1, 5), (7, 5)\}$$

Note that $2 \in A$ is not related to $5 \in B$ since 2 is not odd.
(c) Here, $1 \in A$ is related to $5 \in B$ since $1 + 5 = 6$ is even. Therefore, $1R5$. Similarly, $2R2$ and $7R5$. Therefore,

$$R = \{(1, 5), (2, 2), (7, 5)\}$$

Note that $1 + 2 = 3$ is not even; thus, $1 \in A$ is not related to $2 \in B$. ∎

EXAMPLE 2   **Studying Relations Between Two Sets**

Let A be the set of all integers. We define a relation R on A as follows: aRb if, and only if, a and b have the same remainder when divided by 2.† Is $5R3$? Is $-1R1$? Is $7R4$?

SOLUTION   Since both $5 \div 2$ and $3 \div 2$ give 1 as a remainder, then $5R3$. Since $-1 = -1 \cdot 2 + 1$ and $1 = 0 \cdot 2 + 1$, we conclude that both $-1 \div 2$ and $1 \div 2$ give 1 as a remainder; therefore, $-1R1$. Since $7 \div 2$ gives 1 as a remainder and $4 \div 2$ gives 0 as a remainder, 7 is not related to 4. ∎

 **Now Work Problem 1**

---

*R is sometimes called a *binary relation R* from A to B since the elements of the set R are ordered *pairs*.

†This relation is called the "congruence modulo 2" relation and may be denoted by $a \equiv b$ (mod 2).

Applications to Computer Science

In computer science the binary digits 0 and 1 are called **bits.** An ordered collection of consecutive bits is called a **binary word.** Since computers store information in the form of binary words, they are important in the field of computer science. Of special importance are binary words of length 7, called **bytes,*** because a byte is the smallest addressable memory area in the computer, that is, the smallest unit that can be located in memory.

The sets S and S^* defined below are referred to in the exercises and in later sections of this chapter.

Let $S = \{0, 1\}$. We use 0 and 1 to form binary words. Some binary words are 0, 00, 01, 11, 10, 000, 001, 010, 100, 110, 1101, 10011, and 11000011.

Now let S^* denote the set of all binary words. S^*, clearly, contains an infinite number of elements and hence makes its listing impractical. Next, for every word a in S^* let

$$\text{Length } (a) = \text{The number of bits in } a$$

For example, if $a = 101011$, then length $(a) = 6$. With this in mind we can define a relation R on S^* by considering the number of bits in each element in S^* as follows: For a and b in S^*, aRb if, and only if, length $(a) =$ length (b). Thus, for instance, (000) R (111) since length (000) = length (111) = 3; and (1001) R (0110) since both 1001 and 0110 have length equal to 4.

EXAMPLE 3   Studying the Length of Binary Words

Let the set S^* and the relation R be the ones defined above. Let 01 be in S^*. List all elements in S^* that are related to 01 under R.

SOLUTION   01 is of length 2. The binary words in S^* related to 01 under R are also of length 2. The list of all such words is 00, 01, 10, 11. ∎

Properties of Relations

A relation R on a set A often satisfies certain properties. Three of these properties, reflexivity, symmetry, and transitivity, are defined as follows:

1. R is **reflexive** if, for each element $a \in A$, we have aRa.
2. R is **symmetric** if, for $a, b \in A$, whenever aRb, we have bRa.
3. R is **transitive** if, for $a, b, c \in A$, whenever aRb and bRc, we have aRc.

EXAMPLE 4   Studying Properties of a Relation

Let R be a relation defined on the set $A = \{a, b, c\}$. For each definition of R given below determine whether R is reflexive, symmetric, or transitive.

---

* On some computers, such as the IBM computers, a byte is 8 bits long.

(a) $R = \{(a, a), (a, b), (b, b), (c, c), (b, c)\}$.

(b) $R = \{(c, a), (a, c), (c, c), (c, b), (b, c)\}$.

(c) $R = \{(a, b), (b, a), (a, c), (c, a), (b, c), (c, b), (a, a), (b, b), (c, c)\}$.

SOLUTION

(a) Since (a, a), (b, b), and $(c, c) \in R$, then R is reflexive. Because $(a, b) \in R$, but $(b, a) \notin R$, then R is not symmetric. And because (a, b) and $(b, c) \in R$, but $(a, c) \notin R$, then R is not transitive.

(b) This relation is not reflexive because $(a, a) \notin R$. It is not transitive because (a, c) and (c, a) are in R, but (a, a) is not. However, it is symmetric. (Why?)

(c) R is reflexive, symmetric, and transitive. (Why?)

 **Now Work Problem 23**

EXAMPLE 5   **Studying Properties of a Relation**

Let $A = \{1, 2, 3\}$ and let R be a relation defined on A. Suppose further that the ordered pairs $(1, 1)$, $(1, 2)$, and $(2, 3)$ belong to R. What ordered pairs, besides the ones given above, must belong to R in order to make R

(a) Reflexive?   (b) Symmetric?   (c) Transitive?

SOLUTION

(a) R will be reflexive if every element in A is related to itself. The ordered pairs $(2, 2)$ and $(3, 3)$ must also belong to R to make it reflexive.

(b) R will be symmetric if, whenever (a, b) is in R, so is (b, a). Obviously, $(1, 1)$ satisfies this condition. Now since $(1, 2) \in R$, then, for R to be symmetric, $(2, 1)$ must also belong to R. Similarly, since $(2, 3) \in R$, then, for R to be symmetric, $(3, 2)$ must also belong to R.

(c) R will be transitive if, whenever (a, b) and (b, c) are in R, so is (a, c). Now $(1, 2)$ and $(2, 3) \in R$ implies that, for R to be transitive, $(1, 3)$ must also belong to R.

EXERCISE 12.1   Answers to odd-numbered problems begin on page AN-47.

1. Let $A = \{x, y, z\}$, $B = \{1, 3, 5\}$. A relation R from A to B is defined as

$$R = \{(x, 5), (y, 1), (z, 3)\}$$

Answer true or false:

$xR5$, $yR3$, $zR3$, $zR1$, $yR1$,
$xR1$, $xR3$, $yR5$, $zR5$

2. Let $A = \{1, 2, 3\}$, $B = \{a, b, c, d\}$. A relation R from A to B is defined as

$$R = \{(1, a), (1, c), (2, b), (3, a), (3, c)\}$$

Answer true or false:

$1Ra$, $1Rb$, $1Rc$, $2Ra$, $2Rb$,
$2Rc$, $3Ra$, $3Rb$, $3Rc$

3. Let $A = \{2, 3, 5\}$, $B = \{2, 4, 6, 10\}$. A relation R from A to B is given as follows:

$2R2$, $2R4$, $2R6$, $2R10$, $3R6$, $5R10$

Write R as a set of ordered pairs.

4. Let $C = \{\text{Apple, TRS, IBM, Atari}\}$, $M = \{1000, 360, 370, 800, \text{II}\}$. Let R, the relation from C to M, be given as follows:

(Apple) R (II), (TRS) R (1000), (IBM) R (370), (Atari) R (800)

Write R as a set of ordered pairs.

5. Let $D = \{1, 2, 3, 4, 5\}$ and $C = \{1, 4, 9, 16, 25, 36, 49, 64, 81\}$. The relation R from D to C is defined as follows:

$$xRy \text{ means } y = x^2$$

Replace the "?" by the appropriate value:

$1R?, \quad 2R?, \quad 3R?, \quad 4R?, \quad 5R?,$
$?R4, \quad ?R25$

6. Let $F = \{-2, -10, -1, 0, 1, 2\}$. The relation R on F is defined as follows:

$$aRb \text{ means } a \geq b$$

Answer true or false:

$-2R - 2, \quad -2R - 1, \quad -1R - 2, \quad -1R0,$
$0R0, \quad 1R2, \quad 2R1$

7. The **Cartesian product** of two nonempty sets A and B, denoted by $A \times B$, is the set consisting of all ordered pairs (a, b) with $a \in A$ and $b \in B$. That is,

$$A \times B = \{(a, b) \mid a \in A, b \in B\}$$

Let $A = \{1, 2, 5\}$ and $B = \{2, 4, 7\}$.
   (a) Find the Cartesian product $A \times B$.
   (b) Define the relation "less than," call it R, from the set A to the set B. Write R as a set of ordered pairs.
   (c) What can you say about the set R and $A \times B$?

8. Repeat Parts (b) and (c) of Problem 7, but now let R be the relation "greater than or equal to."

9. Let $S = \{a, b\}$ and $S^2 = $ set of all words on S of length 2.
   (a) List all elements of S^2.
   (b) The relation L on S^2 is defined this way: vLw means that the first letter in v is the same as the first letter in w where v and w are in S^2. Write L as a set of ordered pairs.

10. Repeat Problem 9(b) for L defined as follows: vLw means the first letter of v is not the same as the first letter of w.

In Problems 11–18 write the set of ordered pairs in the relation R.

11. Let $A = \{1, 2, 3\}$, $B = \{2, 4, 6, 8\}$. Define R to be the relation from the set A to the set B where aRb means a is a factor of b.

12. Let $A = \{1, 2, 3, 4, 5, 6\}$. Let R be a relation on A where aRb means $a \equiv b \pmod 2$(see footnote page 539).

13. Let $A = \{1, 2, 3, 4, 5, 6, 7, 8, 9, 10\}$; R is a relation on A where aRb means a and b are either both even or both odd.

14. Let $A = \{1, 2, 3, 4, 5\}$; R is a relation on A where aRb means ab is even.

15. Let $A = \{0, 1, 2, 3\}$ and R on A be defined as follows: aRb means $a \leq b$.

16. Let $A = \{1, 2, 3, 4, 5, 9\}$ and R on A be defined as follows: aRb means $b = a + 2$.

17. Let $D = \{0, 1, 2, 3, 4, 5, 9, 15, 16\}$ and R on D be defined as follows: xRy means $x = \sqrt{y}$.

18. Let $I = \{-4, -2, -1, 0, 1, 2, 4\}$ and R on I be defined as: xRy means $x = y^2$.

For a relation R from the set A to the set B, the inverse relation *of R, denoted R^{-1}, is a relation from B to A defined as follows:*

$$bR^{-1}a \quad \text{means} \quad aRb$$

For Problems 19–22 find R^{-1}.

19. $R = \{(1, 1), (1, 2), (2, 5), (3, 7)\}$.

20. $R = \{(a, a), (a, b), (b, a), (b, c)\}$.

21. $R = \{(*, 42), (H, 72), (/, 47), (x, 88)\}$.

22. R is the relation "$\leq$" on integers.

In Problems 23–30 determine whether the given relation R on the set A is reflexive, symmetric, or transitive.

23. Let R on a set A of real numbers be defined as aRb means $a = b$.

24. Let R on the set A of real numbers be defined as aRb means $a < b$.

25. Let R be the relation of congruence modulo 2 on the set A of integers (see Example 2).

26. Let A be the set of integers and R a relation on A be defined as aRb means a divides b (or a is a factor of b).

27. Let A be the set of all lines in the plane and R be defined by l_1Rl_2 means l_1 is a line parallel to l_2.

28. R is defined as follows: if an arrow connects a point to another, then the two points are related. (See the figure.) That is,

$$R = \{(a, a), (a, b), (b, b), (b, a)\}$$

29. R is defined as follows: if an arrow connects a point to another, then the two points are related. (See the figure.)

That is,

$$R = \{(A, A), (A, B), (B, A), (B, B), (B, C), (C, B),$$
$$(C, C), (A, C)\}$$

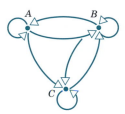

30. Let $A = \{1, 2, 3\}$ and $R = \{(2, 2)\}$.

31. Consider the set $A = \{t, u, v, w\}$ and the relation R on A defined as $R = \varnothing$. (That is, R is the empty relation. In other words, no relation exists between any pair of elements of A.) Show that R is symmetric and transitive but not reflexive.

12.2 FUNCTIONS

In this section we discuss one of the most important concepts in mathematics and its applications: the concept of a function. As we will see, a function is a special type of a relation.

Definition of a Function

> **Function**
>
> Let A and B be two nonempty sets. A **function** f from A into B is a relation from A to B that relates every element in A to only one element in B. The set A is called the **domain** of the function. For each element a in A, the corresponding element b in B, related to a by f, is called the **image** of a. The set of all images of the elements of the domain is called the **range** of the function.

Since there may be elements in B that are the image of no a in A, it follows that the range of a function is often a proper subset of B.

Functions are often denoted by letters such as f, F, g, and G. If f is a function from A into B, then for each element a in A the corresponding image in the set B is denoted by the symbol $f(a)$ and is read "f of a." (Note that $f(a)$ does not mean f times a.) We can think of a function f as associating elements in A to elements in B and visualize f by means of a simple diagram as we do in Figure 1.

Figure 1

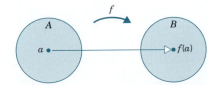

EXAMPLE 1 **Determining the Values of a Given Function**

Let $A = \{1, 3, 5\}$ and $B = \{2, 4\}$. Define the function f from A into B, as shown in Figure 2. Find $f(1), f(3), f(5)$, and the range of f.

Figure 2

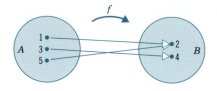

SOLUTION Following the arrows, we have

$$f(1) = 2 \qquad f(3) = 4 \qquad f(5) = 2 \qquad \text{Range of } f = \{2, 4\}$$

■

EXAMPLE 2 **Determining the Values of a Given Function**

Let $A = \{1, 2, 3\}$ and $B = \{a, b, c, d, e\}$. Define the function f from A into B, as shown in Figure 3. Find $f(1), f(2), f(3)$, and the range of f.

Figure 3

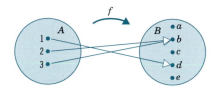

SOLUTION Following the arrows in Figure 3, we find

$$f(1) = d \qquad f(2) = b \qquad f(3) = b \qquad \text{Range of } f = \{b, d\}$$

Because b and d are the only elements in B that are assigned to elements in A, we observe that the range of f is a proper subset of B.

✍ **Now Work Problem 1**

■

EXAMPLE 3 **Determining When a Relation Is a Function**

The diagrams in Figure 4 show a correspondence by which elements of $A = \{1, 2, 3\}$ are associated to elements in $B = \{w, x, y, z\}$. Which of these diagrams define a function from A into B?

Figure 4

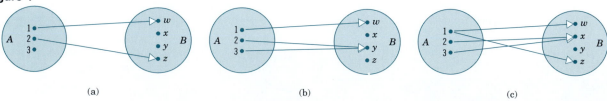

(a)   (b)   (c)

SOLUTION In Figure 4(a) not all elements in A are associated with elements in B. For example, the element 3 in A has no image in B. This is indicated by the absence of an arrow going from 3 in A to an element in B. Thus Figure 4(a) does not define a function from A into B. In Figure 4(b) every element in A is associated with a unique element in B. Thus Figure 4(b) defines a function. In Figure 4(c) the uniqueness of images is violated. For example, the element $1 \in A$ has two images. This is shown by the two arrows emanating from $1 \in A$ to $w \in B$ and to $z \in B$. Thus, Figure 4(c) does not define a function.

■

When the sets A and B are infinite, a complete arrow diagram of a function f cannot be drawn. In such cases functions are sometimes defined by a formula, and the images of f can then be found from the formula defining f.

EXAMPLE 4 **Determining Values Of a Function From a Formula**

Let f be a function from the set of real numbers to itself defined by the formula

$$f(x) = 2x - 1$$

Find $f(0), f(0.5), f(-1), f(1), f(-10), f(100)$.

SOLUTION $f(0)$ is obtained from substituting 0 for x in $2x - 1$. That is,

$$f(0) = 2(0) - 1 = -1$$

Similarly,

$$f(0.5) = 2(0.5) - 1 = 0$$
$$f(-1) = 2(-1) - 1 = -3$$
$$f(1) = 2(1) - 1 = 1$$
$$f(-10) = 2(-10) - 1 = -21$$
$$f(100) = 2(100) - 1 = 199$$

 **Now Work Problem 3**

■

EXAMPLE 5 **Hamming Distance Function**

In a branch of computer science, called *coding theory,* a function called the *Hamming distance function* is of importance. Refer to Section 8.2. This function gives a measure of the difference between two binary words that have the same length. We define the Hamming distance function H as follows: Let (u, v) be a pair of binary words of the same length. We compare u and v position by position and define

$H(u, v) =$ the number of positions in which u and v have different bits

For example,

$$H(0101, 1010) = 4$$

because 0101 and 1010 differ in all four positions. On the other hand,

$$H(1110, 1010) = 1$$

because 1110 and 1010 differ in only one position, namely, the second position.

∎

Special Types of Functions

In Example 1 we notice that the two elements 1 and 5 in the domain A are assigned the same element 2 in the range B. And in Example 2 we notice that the range of f is not the entire set B. This shows that functions can be of different types. In the remainder of this section we introduce functions that have special properties. These functions are important in mathematics and its applications.

One-to-One Function; Injective

A function f from A into B is **one-to-one,** or **injective,** if for all elements a, b in A such that $a \neq b$ we have $f(a) \neq f(b)$. That is, if no two elements of A are assigned to the same element in B, or, equivalently, if each element of the range corresponds to exactly one element of the domain, then f is one-to-one.

EXAMPLE 6   Checking Whether a Function Is One-to-One

Let $A = \{1, 2, 3\}$, $B = \{a, b, c, d\}$, and let $f(1) = a$, $f(2) = d$, $f(3) = c$. Is f one-to-one?

SOLUTION   Yes, because the different elements 1, 2, 3, in A are assigned to the different elements a, d, c, respectively, in B.

∎

EXAMPLE 7   Checking Whether a Function Is One-to-One

Figure 5 defines a function f from the set $A = \{a, b\}$ into the set $B = \{x, y\}$. Is f one-to-one?

Figure 5

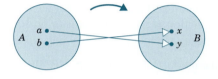

SOLUTION   Yes, because the two arrows show that a is assigned to y and that b, which is different from a, is assigned to x, which is different from y.

∎

EXAMPLE 8   Checking Whether a Function Is One-to-One

Let $f(x) = x^2$, x any real number. Show that f is not one-to-one.

SOLUTION   We need only show that two distinct numbers are assigned to the same number under f. Choose the distinct values $x_1 = 1$ and $x_2 = -1$. Now

$$f(x_1) = f(1) = (1)^2 = 1$$
$$f(x_2) = f(-1) = (-1)^2 = 1$$

Thus, we see that $f(1) = f(-1) = 1$. Therefore, f is not one-to-one.

> **Onto Function; Surjective**
>
> A function f from A into B is **onto**, or **surjective**, if every element of B is the image of some element in A, that is, if $B = $ range of f.

EXAMPLE 9   Checking Whether a Function Is Onto

The diagrams in Figure 6 define the functions f and g from the set $A = \{1, 2, 3, 4\}$ into the set $B = \{a, b, c, d\}$. Which of these functions is onto?

Figure 6

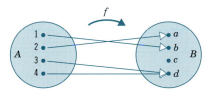   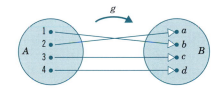

SOLUTION   Under f, $c \in B$ is not the image of any element in A. Thus, f is not onto. Under g, every element in B is an image. Therefore, g is onto.

EXAMPLE 10   Checking Whether a Function Is Onto

Let $f(x) = x^2$, x any real number. Show that f is not onto.

SOLUTION   Since we cannot find a real number whose square is negative, then the set of negative real numbers is not in the range. Thus f is not onto.

> **Bijective Function**
>
> If a function f from A into B is both one-to-one and onto we say that f is **bijective**.

If a function f is bijective, then there will be another function g that will "undo" what f did. In a way, g will be the opposite of f as the following definition shows:

Inverse Function

Let f be a function from A into B. The function g from B to A is called the **inverse** of f if $g(b) = a$, whenever $f(a) = b$. The function g is usually denoted by f^{-1}, read "f inverse."

Note that if f is one-to-one and onto, then the inverse g of f exists. Also, the range of f is the domain of g, and the range of g is the domain of f.

EXAMPLE 11   Finding the Inverse of a Function

Let $A = \{1, 2, 3\}$, $B = \{a, b, c\}$. Define f from A into B as

$$f(1) = a, \qquad f(2) = b, \qquad f(3) = c$$

Since f is both one-to-one and onto, it has an inverse. The inverse f^{-1} from B into A is

$$f^{-1}(a) = 1, \qquad f^{-1}(b) = 2, \qquad f^{-1}(c) = 3$$

 Now Work Problem 25

EXERCISE 12.2   Answers to odd-numbered problems begin on page AN-48.

1. Let $A = \{a, b, c\}$ and $B = \{1, 2, 3, 4\}$. Let f be a function from A into B, as shown in the figure below. Find $f(a)$, $f(b)$, $f(c)$. Also find the range of f.

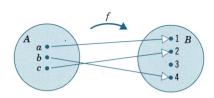

2. Let $A = \{0, 1, 2\}$ and $B = \{1, 10, 100\}$. Let f be a function from A into B, as shown in the figure below. Find $f(0)$, $f(1)$, $f(2)$. Also find the range of f.

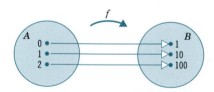

3. Let f be a function from the set of real numbers into itself defined by

$$f(x) = 5 - x$$

Find $f(0), f(5), f(-5), f(10), f(-10)$.

4. Let f be a function defined by

$$f(x) = \frac{-1}{x - 1}$$

where x is any real number and $x \neq 1$. Find $f(-2), f(-1)$, $f(0), f(2)$.

In Problems 5–8, f assigns elements of A = {x, y, z} into elements of B = {1, 2, 3, 4}.
State whether f defines a function. If it doesn't, explain why.

5. $f(x) = 3, f(x) = 1, f(z) = 4.$

6. $f(x) = 1, f(y) = 2, f(z) = 3.$

7. $f(x) = 4, f(y) = 2, f(z) = 4.$

8. $f(y) = 2, f(y) = 3, f(z) = 1.$

In Problems 9–12, A = {0, 1, 2} and B = {1, 2, 3, 4}. Use the correspondence
pictures in the accompanying figures to determine whether f is a function.

9.

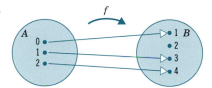

10.

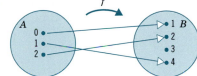

11.

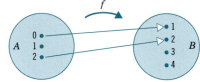

12.

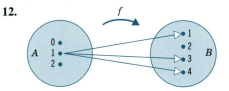

13. Let l be a relation from the set of all binary words into the set $M = \{0, 1, 2, 3, \ldots\}$, defined by

$$l(w) = \text{Length of } w$$

(see Section 12.1). Is l a function?

14. Let f be a relation that sends binary words into the set $\{0, 1\}$ defined by

$$f(w) = \begin{cases} 1, & \text{if } w \text{ contains an odd number of 0s} \\ 0, & \text{if } w \text{ contains an odd number of 1s} \end{cases}$$

Is f a function?

15. Let f be a relation that sends binary words into the set $M = \{0, 1, 2, 3, \ldots\}$ defined by

$$f(w) = \text{Number of 0s in } w$$

Is f a function?

16. Let H be the Hamming distance function. (See Example 5.) Find $H(1000, 0001), H(1011, 1011), H(10, 01), H(101, 010), H(1010101, 1010111).$

In Problems 17–20, f is a function from A into B pictured in the accompanying fig-
ures. State whether f is one-to-one, onto, or bijective.

17.

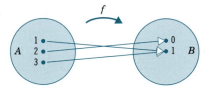

18.

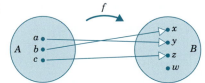

19.

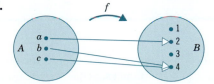

20.

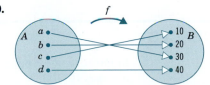

21. Show that the Hamming distance function on the set of all pairs of binary words of length n is not one-to-one.

22. Show that the Hamming distance function on the set of all pairs of binary words of length n is not onto. [*Hint:* Can you find binary words u and v of length n such that $H(u, v) \geq n$?]

23. Let f be a function from the set of integers into the set of integers given by the formula $f(n) = 2n$. Is f onto?

24. Let f be the function from the set A into the set B given in the figure below. (a) Find $f(1), f(2), f(3)$. (b) Show that f is bijective.

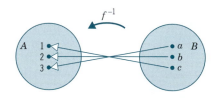

25. Refer to the figure of Problem 24. Find $f^{-1}(a)$, $f^{-1}(b)$, $f^{-1}(c)$.

26. Let f be a bijective function defined by

$$f(1) = 1, \quad f(2) = 4, \quad f(3) = 9, \quad f(4) = 16$$

Find

$$f^{-1}(1), \quad f^{-1}(4), \quad f^{-1}(9), \quad f^{-1}(16)$$

Let f be a function from A into B and let g be a function from B into C. The composition of f and g, denoted by $g \circ f$, read "g of f," results in a new function from A into C and is given by

$$(g \circ f)(a) = g(f(a))$$

where a is in A. That is, the function $g \circ f$ associates a in A to f(a) in B first and then associates the element f(a) in B to the element g(f(a)) in C. See the figure below.

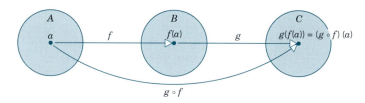

27. Use the figure below to find

$$(g \circ f)(1), \quad (g \circ f)(2), \quad (g \circ f)(3)$$

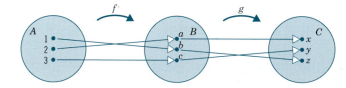

28. If f is a bijective function from A into B, then f has an inverse f^{-1} from B into A. This means that $f^{-1} \circ f$ is a function from A into A. Such a function has the property

$$(f^{-1} \circ f)(a) = a \quad \text{and} \quad (f \circ f^{-1})(a) = a$$

We call this the **identity function.** Let $f(x) = x + 1$ be a function from the set of real numbers to the set of real numbers. If $f^{-1}(x) = x - 1$, show that $(f^{-1} \circ f)(x) = x$ and $(f \circ f^{-1})(x) = x$.

12.3 SEQUENCES

In this section we present a special type of function—one whose domain is the set of numbers $\{0, 1, 2, 3, \ldots\}$. These functions are called **sequences.** Sequences are useful in both mathematics and computer science.

> **Sequence**
>
> Let S be a set. A **sequence** is a function s from the set $\{0, 1, 2, 3, \ldots\}$ into S. If S is the set of real numbers, then s is called a **sequence of reals.**

For every number n in the set $\{0, 1, 2, 3, \ldots\}$, $s(n)$ will be the value assigned to n by the function s. We call $s(n)$ the nth term of the sequence and denote it by s_n [instead of $s(n)$]. We will also use the symbol (s_n) to denote the sequence itself.

EXAMPLE 1   Listing Terms of a Sequence

Let the sequence (s_n) be given by

$$s_n = n^2$$

List the first four terms of (s_n).

SOLUTION

$$s_0 = 0^2 = 0$$
$$s_1 = 1^2 = 1$$
$$s_2 = 2^2 = 4$$
$$s_3 = 3^2 = 9$$

EXAMPLE 2   Listing Terms of a Sequence

Let $s_n = (-1)^n$. Write out the first four terms of (s_n).

SOLUTION

$$s_0 = (-1)^0 = 1$$
$$s_1 = (-1)^1 = -1$$
$$s_2 = (-1)^2 = 1$$
$$s_3 = (-1)^3 = -1$$

Note that the range of s is $\{-1, 1\}$.

 **Now Work Problem 1**

EXAMPLE 3   Listing Terms of a Sequence

Let (s_n) be a sequence given by $s_n = a$. List s_0, s_1, s_2, s_3, etc.

SOLUTION

$$s_0 = a, \; s_1 = a, \; s_2 = a, \; s_3 = a, \text{ etc.}$$

Such a sequence is called a **constant sequence.**

EXAMPLE 4   Listing Terms of a Sequence

For $(s_n) = \left(\dfrac{1}{n}\right)$, $n \neq 0$, write out the first four terms of s_n.

SOLUTION

$$s_1 = \tfrac{1}{1} = 1, \qquad s_2 = \tfrac{1}{2}, \qquad s_3 = \tfrac{1}{3}, \qquad s_4 = \tfrac{1}{4}$$

Sometimes values of a sequence may be matrices as the following example shows.

EXAMPLE 5   Listing Terms of a Sequence

Let the matrix-values sequence (M_n) be given by

$$M_n = \begin{bmatrix} 1+n & -n \\ 0 & (-1)^n \end{bmatrix}$$

Write out the terms M_0, M_1, and M_2.

SOLUTION

$$M_0 = \begin{bmatrix} 1+0 & 0 \\ 0 & (-1)^0 \end{bmatrix} = \begin{bmatrix} 1 & 0 \\ 0 & 1 \end{bmatrix}$$

$$M_1 = \begin{bmatrix} 1+1 & -1 \\ 0 & (-1)^1 \end{bmatrix} = \begin{bmatrix} 2 & -1 \\ 0 & -1 \end{bmatrix}$$

$$M_2 = \begin{bmatrix} 1+2 & -2 \\ 0 & (-1)^2 \end{bmatrix} = \begin{bmatrix} 3 & -2 \\ 0 & 1 \end{bmatrix}$$

 Now Work Problem 9

EXERCISE 12.3   Answers to odd-numbered problems begin on page AN-48.

1. Suppose a sequence (s_n) is defined by $s_n = (-1)^n/n$, $n \neq 0$. Write out $s_1, s_2, s_3, s_4, s_{100}$.

2. Answer the question in Problem 1 for $s_n = (-1)^{n+1}/n$.

3. Consider a sequence (b_n) given by $b_n = n/(n+1)$
  (a) List $b_0, b_1, b_2, b_3, b_4, b_5$.
  (b) Find $b_{n+1} - b_n$, for $n = 0, 1, 2$.

4. Answer the question in Problem 3 for $b_n = -n/(n+1)$.

5. Find the first 8 terms of the sequence $a_k = 2^k$.

6. Answer the question in Problem 5 for $a_k = 2^{-k}$.

7. List the first 7 terms of the sequence $s_n = n!$.

8. Let $M_n = \begin{bmatrix} 1 & n^2 \\ n & 1 \end{bmatrix}$. Find M_0, M_1, M_2, and M_3.

9. Let $M_n = \begin{bmatrix} 1 & 1-n & 0 \\ 1-n & n & n-1 \\ 0 & n-1 & n+1 \end{bmatrix}$. Find M_0, M_1, M_2, and M_3.

10. Let $M_n = \begin{bmatrix} 0 & 1 \\ 1 & 0 \end{bmatrix}^n$ where $\begin{bmatrix} 0 & 1 \\ 1 & 0 \end{bmatrix}^n$ stands for $\begin{bmatrix} 0 & 1 \\ 1 & 0 \end{bmatrix}$ multiplied by itself n times. Find M_1, M_2, M_3, and M_{10}.

In Problems 11–16, write a formula for the nth term.

11. $1, -1, 1, -1, \ldots$

12. $2, 0, 2, 0, 2, 0, \ldots$

13. $1, 3, 5, 7, 9, \ldots$

14. $0, 5, 25, 125, \ldots$

15. $1, \frac{1}{2}, \frac{1}{3}, \frac{1}{4}, \frac{1}{5}, \ldots$

16. $1, \frac{1}{3}, \frac{1}{9}, \frac{1}{27}, \ldots$

17. Towers of Hanoi* The number of moves required in solving the puzzle of the Towers of Hanoi for n disks is given by

$$(s_n) = 2^n - 1$$

Find $s_0, s_1, s_2, s_3, s_4, s_5, s_6, s_7,$ and s_8.

18. Suppose we have a set of straight lines where no two of the lines are parallel and no three of the lines go through the same point. The number of distinct regions that such a set of n lines divides the plane into is given by the nth term of the sequence

$$(s_n) = \binom{n+1}{2} + 1, n \geq 1$$

Find $s_1, s_2, s_3,$ and s_4.

19. The Fibonacci sequence† is given by

$$(s_n) = \frac{1}{\sqrt{5}}\left[\left(\frac{1+\sqrt{5}}{2}\right)^{n+1} - \left(\frac{1-\sqrt{5}}{2}\right)^{n+1}\right]$$

Find $s_0, s_1, s_2, s_3, s_4, s_5, s_6, s_7,$ and s_8.

12.4 MATHEMATICAL INDUCTION

Mathematical induction is used to establish that mathematical statements involving positive integers are true for *all* positive integers. For example, we can use mathematical induction to prove that the following statement:

$$1 + 2 + 3 + \cdots + n = \frac{n(n+1)}{2} \tag{1}$$

holds for all positive integers n.

Before describing the method of mathematical induction, let us try to realize its power. To do this, we use Equation (1) and substitute the various possible values of $n = 1, 2, 3, \ldots$ to construct the following table:

Value of n	Left-Hand Side of Equation (1)	Right-Hand Side of Equation (1)	Formula in Equation (1)
1	1	$\frac{1(1+1)}{2} = 1$	Holds
2	$1 + 2 = 3$	$\frac{2(2+1)}{2} = 3$	Holds
3	$1 + 2 + 3 = 6$	$\frac{3(3+1)}{2} = 6$	Holds
4	$1 + 2 + 3 + 4 = 10$	$\frac{4(4+1)}{2} = 10$	Holds
etc.			

* See Example 3, Section 12.5.
† See Example 2, Section 12.5.

We make two observations about this table:

1. It is impossible to substitute all possible values of n because there is an infinite number of them (as many as the number of positive integers). This means that our table can never be complete, and so Equation (1) can never be proven by substituting all the various values of n.
2. Although the pattern of the rightmost column suggests that $n(n + 1)/2$ is valid, we can never be sure that Equation (1) does not fail for some untried value of n.

By mathematical induction, however, we will prove that Equation (1) holds for all positive integers n. We will show this in Example 2.

EXAMPLE 1   Finding Values of a Statement

Let $S(n)$ be the statement

$$1 + 2 + 3 + \cdots + n = \frac{n(n + 1)}{2}$$

(a) Write $S(1)$. Is $S(1)$ true?
(b) Write $S(2)$. Is $S(2)$ true?
(c) Write $S(k)$. (k is a positive integer.)
(d) Write $S(k + 1)$.

SOLUTION   It is clear that $S(n)$ is the statement about the sum of the first n positive integers.

(a) $S(1)$ is the statement about the sum of the first positive integer. By Formula (1), $S(1)$ is the statement

$$1 = \frac{1(1 + 1)}{2}$$

which is true.

(b) $S(2)$ is the statement about the sum of the first two positive integers. By Formula (1), $S(2)$ is the statement

$$1 + 2 = \frac{2(2 + 1)}{2}$$

which is also true.

(c) $S(k)$ is the statement about the sum of the first k positive integers. By Formula (1), $S(k)$ is the statement

$$1 + 2 + 3 + \cdots + k = \frac{k(k + 1)}{2}$$

(d) $S(k + 1)$ is the statement about the sum of the first $(k + 1)$ positive integers. This is obtained by substituting $k + 1$ for n in Equation (1), to get

$$1 + 2 + 3 + \cdots + k + (k + 1) = \frac{(k + 1)[(k + 1) + 1]}{2} \tag{2}$$

■

We are now ready to state the principle of mathematical induction.

Principle of Mathematical Induction

Let $S(n)$ be a statement that involves positive integers $n = 1, 2, 3, \ldots$. If we can show that the following two conditions are satisfied:

Condition I The statement $S(1)$ is true.

Condition II Let k be any positive integer. If assuming $S(k)$ is true implies that $S(k + 1)$ is also true,

then $S(n)$ must be true for *all* positive integers.

Notice that from Condition I, $S(1)$ is true. By Condition II, $S(2)$ is true. Again using Condition II, $S(3)$ is true, and so on, for all natural numbers.

EXAMPLE 2 **Using Mathematical Induction To Prove Formula (1)**

Show that

$$1 + 2 + 3 + \cdots + n = \frac{n(n + 1)}{2} \tag{3}$$

is true for all positive integers n.

SOLUTION   We need to show first that $S(1)$ is true. Because $1 = 1(1 + 1)/2$, Condition I is satisfied. To show that Condition II is true, we assume that Formula (3) holds for some positive integer k. That is, we assume $S(k)$ is true for some positive integer k and, based on this assumption, we must show that $S(k + 1)$ is true. We look at the sum of the first $k + 1$ positive integers:

$$1 + 2 + 3 + \cdots + k + (k + 1) = [1 + 2 + 3 + \cdots + k] + (k + 1)$$

$$= \underset{\uparrow}{\frac{k(k + 1)}{2}} + (k + 1)$$

By our assumption that $S(k)$ is true

$$= \frac{k(k + 1) + 2(k + 1)}{2}$$

$$= \underset{\uparrow}{\frac{(k + 1)[k + 2]}{2}}$$

Take $(k + 1)$ as a common factor.

$$= \frac{(k + 1)[(k + 1) + 1]}{2}$$

which is $S(k + 1)$. [See Formula (2).] Thus, Condition II is also satisfied. And by the principle of mathematical induction, Equation (3) is true for all positive integers n.

Now Work Problem 7

EXAMPLE 3   Using Mathematical Induction

Let $S(n)$ be the following statement describing the sum of the first n positive odd integers:

$$1 + 3 + 5 + \cdots + (2n - 1) = n^2 \qquad (4)$$

(a) Write $S(1)$ and show it is true.   (b) Write $S(2)$ and show it is true.
(c) Write $S(k)$.   (d) Write $S(k + 1)$.
(e) Show, using mathematical induction, that $S(n)$ is true for all positive integers n.

SOLUTION

(a) $S(1)$ describes the sum of the first positive odd integer; Equation (4) gives $1 = 1^2$. This is true; therefore, $S(1)$ is true.
(b) $S(2)$ describes the sum of the first two positive odd integers; Equation (4) gives $1 + 3 = 2^2$. Because both sides are equal to 4, $S(2)$ is true.
(c) $S(k)$ describes the sum of the first k positive odd integers. Equation (4) gives $1 + 3 + 5 + \cdots + (2k - 1) = k^2$.
(d) $S(k + 1)$ describes the sum of the first $k + 1$ positive odd integers. Equation (4) gives

$$1 + 3 + 5 + \cdots + (2k - 1) + (2k + 1) = (k + 1)^2$$

(e) Since, from (a) above, $S(1)$ is true, Condition I is satisfied. To show that Condition II is satisfied, we assume that $S(k)$ is true [see (c) above] and we show that this implies $S(k + 1)$ [see (d) above] is true. So we consider the sum of the first $(k + 1)$ positive odd integers:

$$1 + 3 + 5 + \cdots + (2k - 1) + (2k + 1)$$
$$= \underbrace{[1 + 3 + 5 + \cdots + (2k - 1)]}_{\substack{= \; k^2 \text{ since we} \\ \text{assumed } S(k) \text{ holds}}} + (2k + 1)$$
$$= k^2 + 2k + 1$$
$$= (k + 1)^2$$

Thus, $S(k + 1)$ holds and so Condition II is satisfied, and Formula (4) is true for all positive integers.

EXAMPLE 4   Using Mathematical Induction to Prove an Inequality

Prove by mathematical induction that $5^n > n^2$ for all positive integers n.

SOLUTION We need to show that our statement is true for $n = 1$. Because $5^1 > 1^2$, Condition I is satisfied. To show that Condition II is satisfied, we assume that $5^k > k^2$ for some positive integer k and, based on this assumption, we must show that $5^{k+1} > (k + 1)^2$.

$$5^{k+1} = 5(5^k)$$
$$> 5k^2$$
$\uparrow$
By our assumption that $5^k > k^2$
$$= k^2 + 2k^2 + 2k^2$$
$\uparrow$
We want to compare $5k^2$ to $(k + 1)^2 = k^2 + 2k + 1$.
$$> k^2 + 2k + 1$$
$\uparrow$
Since k is an integer ≥ 1, $2k^2 \geq 2k$ and $2k^2 > 1$.
$$= (k + 1)^2$$

Thus, $5^{k+1} > (k + 1)^2$, and Condition II is also satisfied. By the principle of mathematical induction, we have also proved $5^n > n^2$ for all positive integers n.

■

EXERCISE 12.4 Answers to odd-numbered problems begin on page AN-48.

1. Let $S(n)$ be the statement

$$n < 2^n \qquad n \text{ is a positive integer}$$

(a) Write $S(1)$. Is $S(1)$ true?
(b) Write $S(2)$. Is $S(2)$ true?
(c) Write $S(k)$.
(d) Write $S(k + 1)$.

2. Let $S(n)$ be the statement

$$1 \cdot 2 + 2 \cdot 3 + 3 \cdot 4 + \cdots + n \cdot (n + 1) =$$
$$\frac{n(n + 1)(n + 2)}{3}$$

(a) Write $S(1)$. Is $S(1)$ true?
(b) Write $S(2)$. Is $S(2)$ true?
(c) Write $S(k)$.
(d) Write $S(k + 1)$.

3. Let $S(n)$ be the statement

$$1^2 + 2^2 + 3^2 + \cdots + n^2 = \frac{n(n + 1)(2n + 1)}{6}$$

(a) Write $S(1)$. Is $S(1)$ true?
(b) Write $S(5)$. Is $S(5)$ true?
(c) Write $S(k)$.
(d) Write $S(k + 1)$.

4. Let $S(n)$ be the statement

$$\frac{1}{1 \cdot 2} + \frac{1}{2 \cdot 3} + \frac{1}{3 \cdot 4} + \cdots + \frac{1}{n(n + 1)} = \frac{n}{n + 1}$$

(a) Write $S(1)$. Is $S(1)$ true?
(b) Write $S(3)$. Is $S(3)$ true?
(c) Write $S(k)$.
(d) Write $S(k + 1)$.

5. Let $S(n)$ be the statement

$$1 + 2 + 2^2 + \cdots + 2^n = 2^{n+1} - 1$$

(a) Write $S(1)$. Is $S(1)$ true?
(b) Write $S(5)$. Is $S(5)$ true?
(c) Write $S(k)$.
(d) Write $S(k + 1)$.

6. Let $S(n)$ be the statement

$$-1 + 1 - 1 + 1 - \cdots + (-1)^n = \frac{(-1)^n - 1}{2}$$

(a) Write $S(1)$. Is $S(1)$ true?
(b) Write $S(2)$. Is $S(2)$ true?
(c) Write $S(100)$. Is $S(100)$ true?
(d) Write $S(k)$.
(e) Write $S(k + 1)$.

In Problems 7–28 prove by mathematical induction that the statement is true for all positive integers n.

7. $1 \cdot 2 + 2 \cdot 3 + 3 \cdot 4 + \cdots + n \cdot (n + 1) = \dfrac{n(n + 1)(n + 2)}{3}$

8. $1^2 + 2^2 + 3^2 + \cdots + n^2 = \dfrac{n(n + 1)(2n + 1)}{6}$

9. $\dfrac{1}{1 \cdot 2} + \dfrac{1}{2 \cdot 3} + \dfrac{1}{3 \cdot 4} + \cdots + \dfrac{1}{n \cdot (n + 1)} = \dfrac{n}{n + 1}$

10. $2 + 4 + 6 + \cdots + 2n = n \cdot (n + 1)$

11. $2 + 5 + 8 + \cdots + (3n - 1) = \dfrac{n(3n + 1)}{2}$

12. $n < 2^n$

13. $2 + 4 + 6 + \cdots + 2n = n(n + 1)$

14. $1 + 5 + 9 + \cdots + (4n + 1) = (n + 1)(2n + 1)$

15. $-1 + 2^2 - 3^2 + 4^2 - 5^2 + \cdots + (-1)^n n^2 = (-1)^n \dfrac{n(n + 1)}{2}$

16. $\dfrac{1}{1 \cdot 3} + \dfrac{1}{3 \cdot 5} + \dfrac{1}{5 \cdot 7} + \cdots + \dfrac{1}{(2n - 1) \cdot (2n + 1)} = \dfrac{n}{2n + 1}$

17. $\dfrac{1}{1 \cdot 4} + \dfrac{1}{4 \cdot 7} + \dfrac{1}{7 \cdot 10} + \cdots + \dfrac{1}{(3n - 2) \cdot (3n + 1)} = \dfrac{n}{3n + 1}$

18. $1^3 + 2^3 + 3^3 + \cdots + n^3 = \left(\dfrac{n(n + 1)}{2} \right)^2$

19. $\dfrac{1}{1 \cdot 2 \cdot 3} + \dfrac{1}{2 \cdot 3 \cdot 4} + \dfrac{1}{3 \cdot 4 \cdot 5} + \cdots + \dfrac{1}{n \cdot (n + 1) \cdot (n + 2)} = \dfrac{n(n + 3)}{4(n + 1)(n + 2)}$

20. $1 \cdot 2 + 2 \cdot 2^2 + 3 \cdot 2^3 + \cdots + n \cdot 2^n = (n - 1)2^{n+1} + 2$

21. $1 + 3 + 3^2 + \cdots + 3^n = \frac{1}{2}(3^{n+1} - 1)$

22. $3 + 10 + 17 + \cdots + (7n - 4) = \frac{1}{2}n(7n - 1)$

23. $1 + 2n \le 3^n$

24. $\underbrace{\sqrt{2 + \sqrt{2 + \sqrt{2 + \ldots + \sqrt{2}}}}}_{n \text{ roots}} < 2$

25. $\dfrac{1}{2} \cdot \dfrac{3}{4} \cdot \dfrac{5}{6} \cdots \dfrac{2n - 1}{2n} \le \dfrac{1}{\sqrt{3n + 1}}$

26. 3 is a factor of $4^n - 1$

27. 3 is a factor of $n^3 + 2n$

28. 57 is a factor of $7^{n+2} + 8^{2n+1}$

12.5 RECURRENCE RELATIONS

Thus far we have defined a sequence by giving a general formula for its *n*th term or by writing a few of its terms. Often this approach is not possible. An alternative is to write the sequence by finding a relationship among its terms. Such a relationship is called a **recurrence relation.** Sequences defined this way normally arise when each term of the sequence depends upon the previous terms of the sequence. Defining a sequence recursively requires both giving the recurrence relation and specifying the first few terms of the sequence. Specifying the first few terms of the sequence is called setting **initial conditions.** An example will help clarify these concepts.

EXAMPLE 1 Computing Terms of a Sequence

Compute the first six terms of the sequence defined by the following recurrence relation and initial conditions:

(a) Recurrence relation: $s_n = 2s_{n-1} + 1$ for all integers $n \ge 1$
Initial condition: $s_0 = 1$

(b) Recurrence relation: $s_n = s_{n-1} + s_{n-2}$ for all integers $n \ge 2$
Initial conditions: $s_0 = 1$, $s_1 = 2$

(c) Recurrence relation: $s_n = s_{n-1}^2 + ns_{n-2} + s_{n-3}$ for all integers $n \geq 3$
Initial conditions: $s_0 = 1$, $s_1 = 2$, $s_2 = -1$

(d) Recurrence relation: $s_n = \begin{cases} \dfrac{s_{n-1}}{2} & \text{if } s_{n-1} \text{ is even} \\ 3s_{n-1} + 1 & \text{if } s_{n-1} \text{ is odd} \end{cases}$

Initial condition: $s_0 = 126$

SOLUTION

(a) Here the initial condition is given as $s_0 = 1$. Therefore, we need to compute $s_1, s_2,$ $s_3, s_4,$ and s_5. Substituting $n = 1$ in the recurrence relation gives

$$s_1 = 2s_{1-1} + 1 = 2s_0 + 1 = 2(1) + 1 = 3$$
$$\uparrow$$
$$s_0 = 1$$

Similarly,

$$s_2 = 2s_{2-1} + 1 = 2s_1 + 1 = 2(3) + 1 = 7$$
$$\uparrow$$
$$s_1 = 3$$

$$s_3 = 2s_{3-1} + 1 = 2s_2 + 1 = 2(7) + 1 = 15$$
$$\uparrow$$
$$s_2 = 7$$

$$s_4 = 2s_{4-1} + 1 = 2s_3 + 1 = 2(15) + 1 = 31$$
$$\uparrow$$
$$s_3 = 15$$

and

$$s_5 = 2s_{5-1} + 1 = 2s_4 + 1 = 2(31) + 1 = 63$$
$$\uparrow$$
$$s_4 = 31$$

Thus, the first six terms of this sequence are 1, 3, 7, 15, 31, 63.

(b) Here, the initial conditions are given as $s_0 = 1$ and $s_1 = 2$; therefore, we need to compute $s_2, s_3, s_4,$ and s_5. Substituting $n = 2$ in the recurrence relation gives

$$s_2 = s_{2-1} + s_{2-2} = s_1 + s_0 = 2 + 1 = 3$$
$$\uparrow$$
$$s_0 = 1 \text{ and } s_1 = 2$$

Similarly,

$$s_3 = s_{3-1} + s_{3-2} = s_2 + s_1 = 3 + 2 = 5$$
$$\uparrow$$
$$s_1 = 2 \text{ and } s_2 = 3$$

$$s_4 = s_{4-1} + s_{4-2} = s_3 + s_2 = 5 + 3 = 8$$
$$\uparrow$$
$$s_2 = 3 \text{ and } s_3 = 5$$

and

$$s_5 = s_{5-1} + s_{5-2} = s_4 + s_3 = 8 + 5 = 13$$
$$\uparrow$$
$$s_3 = 5 \text{ and } s_4 = 8$$

Thus, the first six terms of this sequence are 1, 2, 3, 5, 8, 13.

(c) Here, the initial conditions are given as $s_0 = 1$, $s_1 = 2$, and $s_2 = -1$; therefore, we need to compute s_3, s_4, and s_5. Substituting $n = 3$ in the recurrence relation gives

$$s_3 = s_{3-1}^2 + 3s_{3-2} + s_{3-3}$$
$$= s_2^2 + 3s_1 + s_0 = (-1)^2 + 3(2) + 1 = 1 + 6 + 1 = 8$$
$$\uparrow$$
$$s_0 = 1, s_1 = 2, \text{ and } s_2 = -1$$

Similarly,

$$s_4 = s_{4-1}^2 + 4s_{4-2} + s_{4-3} = s_3^2 + 4s_2 + s_1 = 8^2 + 4(-1) + 2$$
$$\uparrow$$
$$s_1 = 2, s_2 = -1, \text{ and } s_3 = 8$$
$$= 64 - 4 + 2 = 62,$$

and

$$s_5 = s_{5-1}^2 + 5s_{5-2} + s_{5-3} = s_4^2 + 5s_3 + s_2 = (62)^2 + 5(8) - 1$$
$$\uparrow$$
$$s_2 = -1, s_3 = 8, \text{ and } s_4 = 62$$
$$= 3844 + 40 - 1 = 3883$$

Thus, the first six terms of this sequence are $1, 2, -1, 8, 62, 3883$.

(d) Here, the initial condition is $s_0 = 126$; therefore, we need to compute s_1, s_2, s_3, s_4, s_5.

Since s_0 is even,

$$s_1 = \frac{s_0}{2} = \frac{126}{2} = 63$$

Since s_1 is odd,

$$s_2 = 3s_1 + 1 = 3(63) + 1 = 190$$

Similarly,

$$s_3 = \frac{s_2}{2} = \frac{190}{2} = 95$$

$$s_4 = 3s_3 + 1 = 3(95) + 1 = 286$$

$$s_5 = \frac{s_4}{2} = \frac{286}{2} = 143$$

Thus, the first six terms of this sequence are $126, 63, 190, 95, 286, 143$.

Now Work Problem 3

These examples lead us to the following definition.

Recurrence Relation

A **recurrence relation** for the sequence $s_0, s_1, s_2, s_3, \ldots$ is a formula that relates each term s_n to previous terms of the sequence. The **initial condition** for a recurrence relation is a set of values of the first few terms of the sequence.

The next two examples are two well-known sequences that are defined recursively.

EXAMPLE 2   Finding How Fast Rabbits Multiply

In 1202 A.D., a European mathematician named Leonardo Fibonacci posed the following problem:

Assume we start with a single pair of newborn rabbits and that each pair does not reproduce during its first month of life, but after its first month, it produces one new pair (a male and a female) at the end of every month. If we further assume that no rabbits die, how many pairs of rabbits will there be at the end of n months?

SOLUTION   Let r_n denote the number of rabbit pairs at the end of the nth month. Since we started with one pair, then we have $r_0 = 1$. And, since this pair is not fertile during its first month, then at the end of the first month we will still have (this) one pair, that is, $r_1 = 1$. At the end of the second month this pair will produce another pair and so there will be two pairs, that is, $r_2 = r_0 + r_1 = 1 + 1 = 2$. At the end of the third month the one-month-old pair does not produce but the old pair does, the one we started with; thus there will be one newborn pair in addition to the previous two pairs, that is,

$$r_3 = r_2 + r_1 = 2 + 1 = 3$$

At the end of the fourth month there will be two pairs that are reproductive (as many as there was at the end of the second month, which is 2), which will produce two pairs, and the one-month-old pair that cannot produce. Thus,

$$r_4 = r_3 + r_2 = 3 + 2 = 5$$

Continuing in this manner, we find that at the end of the nth month there will be r_{n-2} new pairs born plus r_{n-1} pairs (those include the one-month-olds). That is,

Recurrence relation: $r_n = r_{n-1} + r_{n-2}$
Initial condition: $r_0 = 1, \qquad r_1 = 1$

 **Now Work Problem 15**

Fibonacci numbers occur in many applications. For example, the number of binary words that do not contain the pattern 00, follows a Fibonacci sequence.

EXAMPLE 3   The Towers of Hanoi

This is a puzzle about three poles and n disks of increasing sizes called the Towers of Hanoi puzzle. Initially, all of the disks (all have holes at their centers) are placed on the first pole, as shown in Figure 7(a). We want to transfer all the disks from the first pole to the third, so that they appear as in Figure 7(b). If we can move only one disk at a time and if we are not allowed to place a larger disk on top of a smaller one, what is the minimum number of moves required to transfer a tower of n disks from the first pole to the third pole?

Figure 7

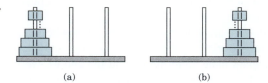

(a)   (b)

SOLUTION   We begin by letting m_n be the minimum number of moves required to move the n disks from the first pole to the third. Obviously, $m_0 = 0$. $m_1 = 1$ because we can move one disk from the first pole to the third in one move. Notice that this is the minimum number of moves required to transfer one disk from the first pole to the third. To find m_2, we proceed as follows: First, we move the smaller disk from the first pole to the second, leaving the larger one on the first pole [Figures 8(a)–8(b)]. Second, we move the larger disk from the first pole to the third [Figures 8(b)–(c)]. Third, and last, we move the smaller disk from the second pole to the third [Figures 8(c)–(d)].

Figure 8

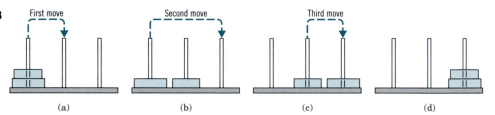

(a)   (b)   (c)   (d)

Thus, a minimum of three moves is needed to transfer two disks from the first pole to the third. This is the minimum number of moves because if we follow a different set of moves to transfer the two disks from the first pole to the third without violating the rules of the puzzle, we end up with more than three moves (Try it!). Thus, m_2 requires the following: m_1 moves from the first pole to the second (moves for transferring the smaller disk from the first pole to the second), one move of the larger disk from the first pole to the third, and another m_1 moves from the second pole to the third (moves for transferring the smaller disk from the second pole to the third). Thus,

$$m_2 = 2m_1 + 1 = 2 \cdot 1 + 1 = 3$$

For a tower of three disks the minimum number of moves m_3 required to transfer the three disks from the first pole to the third is computed as follows: m_2 moves are required to transfer the top two disks from the first pole to the second [Figures 9(a), 9(b)], one move of the large disk from the first pole to the third [Figure 9(c)], and another m_2 moves of the smaller two disks from the second pole to the third [Figure 9(d)]. Thus,

$$m_3 = 2m_2 + 1 = 2 \cdot 3 + 1 = 7$$

Figure 9

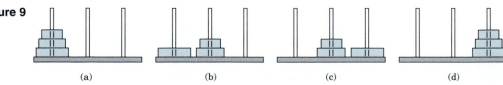

(a)   (b)   (c)   (d)

Continuing our reasoning* in this way, it can be shown that

Recurrence relation: $m_n = 2m_{n-1} + 1$ for all integers $n \geq 2$

Initial condition: $m_0 = 0,$ $m_1 = 1$

More examples on recursively defined sequences are found in the exercises.

EXERCISE 12.5   Answers to odd-numbered problems begin on page AN-51.

In Problems 1–14 a recurrence relation and an initial condition are given that define a sequence (s_n). Find the first six terms of (s_n).

1. $s_n = 2s_{n-1}$ for all integers $n \geq 1$

$s_0 = 1$

2. $s_n = s_{n-1} + n$ for all integers $n \geq 1$

$s_0 = 1$

3. $s_n = ns_{n-1} - 1$ for all integers $n \geq 1$

$s_0 = 1$

4. $s_n = s_{n-1} + 2^n$ for all integers $n \geq 1$

$s_0 = -1$

5. $s_n = (n + 1)s_{n-1}$ for all integers $n \geq 1$

$s_0 = 2$

6. $s_n = s_{n-1} + n^2$ for all integers $n \geq 1$

$s_0 = 1$

7. $s_n = s_{n-2} + s_{n-1} + \binom{n}{2}$ for all integers $n \geq 2$

$s_0 = 1, s_1 = 1$

8. $s_n = 2s_{n-2} + s_{n-1}^2$ for all integers $n \geq 2$

$s_0 = 1, s_1 = -1$

9. $s_n = s_{n-1}s_{n-2}$ for all integers $n \geq 2$

$s_0 = 1, s_1 = 2$

10. $s_n = (s_{n-1} + s_{n-2})^{1/2}$ for all integers $n \geq 2$

$s_0 = 3, s_1 = 6$

11. $s_n = s_{n-1} + s_{n-2} + s_{n-3}$ for all integers $n \geq 3$

$s_0 = 1, s_1 = 2, s_2 = 2$

12. $s_n = s_{n-1} + s_{n-2} - 2s_{n-3}$ for all integers $n \geq 3$

$s_0 = 2, s_1 = 1, s_2 = 1$

13. $s_n = ns_{n-1} + s_{n-2} - s_{n-3}$ for all integers $n \geq 3$

$s_0 = 1, s_1 = -1, s_2 = -1$

14. $s_n = (-1)^n s_{n-1} + ns_{n-2} + n^2 s_{n-3}$ for all integers $n \geq 3$

$s_0 = 1, s_1 = 1, s_2 = 1$

15. Use the recurrence relation and initial condition for the Fibonacci number sequence $r_0, r_1, r_2, r_3, \ldots$ defined in Example 2 to compute $r_{10}, r_{11}, r_{12}, r_{13},$ and r_{14}.

16. Use the recurrence relation and initial condition for the Towers of Hanoi sequence $m_0, m_1, m_2, m_3, \ldots$ defined in Example 3 to compute $m_6, m_7, m_8,$ and m_9.

17. (a) Write all binary words of lengths 0, 1, 2, 3, and 4 that do not contain the bit pattern 00.
(b) For all integers $n \geq 2$, let $s_n = $ the number of binary words of length n that do not contain the pattern 00. Find a recurrence relation relating s_n to s_{n-1} and s_{n-2}.

18. Repeat Problem 17 for the binary words that do not contain the bit pattern 11.

19. Suppose $1000 is invested in an account paying 10% interest compounded annually. Assume that no withdrawals are made, and let A_n be the amount in the account after n years.

(a) Compute $A_0, A_1, A_2, A_3,$ and A_4.
(b) Find a recurrence relation relating A_n to A_{n-1} for all integers $n \geq 1$.

20. Repeat Problem 19 for $2000 with annually compounded interest of 5.5%.

---

* A proof of this requires mathematical induction.

21. Consider a set of n lines in a plane no two of which are parallel and no three of which intersect at the same point. Let p_n denote the number of distinct regions formed by these lines. See the figure.

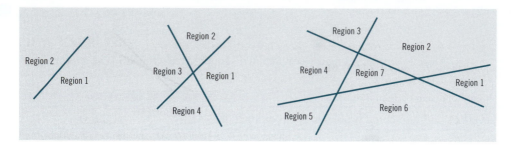

(a) Find p_1, p_2, p_3, and p_4.

(b) Find a recurrence relation relating p_n to p_{n-1} for all integers $n \geq 2$.

CHAPTER REVIEW

IMPORTANT TERMS AND CONCEPTS

relation 539	function 543	inverse function 548
bit 540	domain 543	composition of functions
binary word 540	image 543	550
byte 540	range 543	identity function 550
reflexive 540	one-to-one 546	sequence 551
symmetric 540	injective 546	constant sequence 551
transitive 540	onto 547	mathematical induction 553
Cartesian product 542	surjective 547	recurrence relation 558
inverse relation 542	bijective 547	initial condition 558

TRUE–FALSE ITEMS   Answers are on page AN-51.

T_____ F_____ **1.** For all relations R from A to B and all $a \in A$, $b \in B$ if aRb, then bRa.

T_____ F_____ **2.** The range and the domain of all functions are the same.

T_____ F_____ **3.** A bijective function is both one-to-one and onto.

T_____ F_____ **4.** One way to define a sequence is to give a formula for its nth term.

T_____ F_____ **5.** Condition II of the principle of mathematical induction states that if the statement $S(k)$ is true for some positive integer k, then so is $S(k + 1)$.

T_____ F_____ **6.** Every recurrence relation must have some initial condition.

FILL IN THE BLANKS   Answers are on page AN-51.

1. A relation R from A to B is the set R of _____ _____ where $a \in A$ $b \in B$.

2. A _____ from the set X into the set Y is a relation

that associates with each element of X exactly one element of Y.

3. A one-to-one and onto function is called _____ .

4. A sequence is a function whose domain is the set of _____ _____ .

5. _____ _____ is used to prove that a statement $S(n)$, which involves natural numbers, is true for all n.

6. A formula that relates each term of a sequence of previous terms is called a _____ _____ . Specifying the first few terms is giving the _____ _____ .

REVIEW EXERCISES   Answers to odd-numbered problems begin on page AN-51.

1. Let $A = \{1, 2, 3, 8, 9, 27\}$ and R be a relation on A defined by aRb if, and only if, a is the cube of b. Write R as a set of ordered pairs.

2. State the inverse relation R^{-1} of the relation R of Problem 1. Write R^{-1} as a set of ordered pairs.

3. Which of the following relations defines a function from $A = \{0, 1, 2\}$ to $B = \{0, 1, 2, 3\}$?

(a) $f(0) = 0, f(1) = 3, f(2) = 1$
(b) $g(0) = 1, g(1) = 1, g(2) = 1$
(c) $h(0) = 1, h(1) = 2$
(d) $F(0) = 3, F(1) = 2, F(0) = 1, F(2) = 3$

4. Suppose $S(n)$ is the statement

$$1 + 3 + 3^2 + 3^3 + \cdots + 3^{n-1} = \tfrac{1}{2}(3^n - 1).$$

Find $S(1), S(2), S(3)$. Use mathematical induction to prove that $S(n)$ is true for all n.

5. Find the first 10 terms of each sequence

(a) $a_k = (-1)^k + k$ (b) $a_k = (-1)^k \cdot k$

6. Find the first 5 terms of the sequence defined by the following recurrence relation and initial conditions:

$$s_n = (n - 2)s_{n-2} + s_{n-1} \qquad n \geq 2$$

$$s_0 = 1, \qquad s_1 = 0$$

7. Find the first 7 terms of the sequence defined by the following recurrence relation and initial conditions:

$$s_n = s_{n-3} \cdot (s_{n-2} + s_{n-3}) \qquad n \geq 3$$

$$s_0 = 2, \qquad s_1 = 2, \qquad s_2 = 1$$

8. Suppose that you start with a single pair of newborn rabbits and that each pair reproduces a new pair (a male and a female) every two months except the first two months of its life. If no rabbits die, how many rabbits will there be at the end of the

(a) Fourth month?   (b) Sixth month?
(c) Tenth month?

Chapter 13

Graphs and Trees

13.1 Graphs

13.2 Paths and Connectedness

13.3 Eulerian and Hamiltonian Circuits

13.4 Trees

13.5 Directed Graphs

13.6 Matrix Applications to Graphs and Directed Graphs

Chapter Review

Many situations that occur in computer science, operations research, the physical sciences, and economics are of a combinatorial nature and can be analyzed by using techniques found in an area of mathematics called *graph theory*. In this chapter we present a brief introduction and explain some of the more elementary results of graph theory and their applications.

13.1 GRAPHS

Before giving a formal definition of a graph, we point out that the term *graph* has two quite different meanings. One definition of a graph is the one we studied in Chapters 1 and 11 and used when we graphed straight lines and linear inequalities.

Definition of Graph

In this chapter a **graph** will mean a set of points (called **vertices**) with possibly one or more curves or lines (called **edges**) connecting a point to itself or a pair of points. For a given pair of vertices, our concern is not what the edge looks like, but rather whether the two vertices have an edge joining them or not. For example, the two diagrams in Figure 1 represent the same graph: each has 4 vertices v_1, v_2, v_3, and v_4 and 8 edges

$e_1, e_2, e_3, e_4, e_5, e_6, e_7,$ and e_8. Note that the edges e_5 and e_6 intersect in the graph in Figure 1(a), but their intersection is not a vertex. Edges e_2 and e_7 are two different edges connecting the same pair of vertices (v_2 and v_3). Such edges are called **parallel edges.** (Note that this is a departure from the usual definition of parallel lines.) Edge e_8 is called a **loop** because it connects a vertex, v_3, to itself.

Figure 1

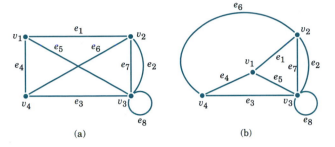

(a)   (b)

Graph

A **graph** G consists of the following two finite sets: (1) a nonempty set V of vertices and (2) a set E of edges, where each edge either connects two vertices or connects a vertex to itself.

A vertex that is not connected to any other vertex or to itself is called an **isolated vertex.**

EXAMPLE 1   **Analyzing a Graph**

For the graph given in Figure 2

(a) Write the vertex set   (b) Write the edge set.
(c) Find all isolated vertices.   (d) Find all loops.
(e) Find all parallel edges.

Figure 2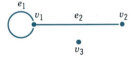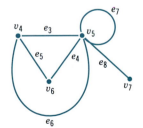

SOLUTION

(a) The vertex set is

$$V = \{v_1, v_2, v_3, v_4, v_5, v_6, v_7\}$$

(b) The edge set is

$$E = \{e_1, e_2, e_3, e_4, e_5, e_6, e_7, e_8\}$$

(c) v_3 is the only vertex that is isolated.
(d) e_1 and e_7 are the only loops.
(e) e_3 and e_6 are parallel edges since they both connect v_4 and v_5.

Simple Graph

A graph with no parallel edges and no loops is called a **simple graph.**

EXAMPLE 2   Studying Simple Graphs

Draw all simple graphs with 3 vertices.

SOLUTION   In Figure 3 we list all possible graphs having 3 vertices that are free of parallel edges and loops.

Figure 3

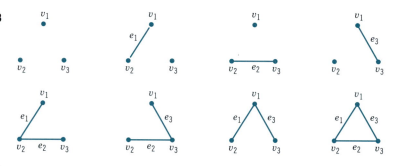

Now Work Problem 3

EXAMPLE 3   The Königsberg Bridge Problem

Figure 4

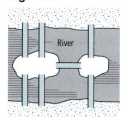

This is one of the oldest problems involving graphs. The town of Königsberg (in eastern Europe) is crossed by the Pregel river with two islands in the river. The islands are connected to each other by one bridge. The larger island is connected to each bank by two bridges. The smaller island is connected to each bank by one bridge. Thus, there is a total of seven bridges. See Figure 4. The townspeople wondered whether one can begin at one of the banks or islands, walk across each bridge exactly once, and return to the starting point. This problem is equivalent to the following: Let each mass of land be represented by a vertex and each bridge by an edge to obtain a graph as in Figure 5 where A and C are the river banks, B is the larger island, D is the smaller island, and the 7 edges represent the 7 bridges. Can one start at any one of the vertices A, B, C, or D, traverse each edge in Figure 5 exactly once, and return to the starting vertex? A

Figure 5

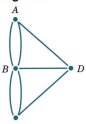

Swiss mathematician named Leonhard Euler* (1707–1783) studied this problem and in 1736 was able to answer it. We give his answer in Section 13.3.

Computer, telephone, and transportation route networks (to name a few) can be represented by graphs. Inspecting or analyzing their graphs usually determines connecting edges for optimization purposes. Figure 6 shows a typical communication network.

Figure 6

EXAMPLE 4   Pascal Data Types

Relationships can be depicted by graphs. For example, the hierarchy of data types in the Pascal language is given in the graph in Figure 7.

Figure 7

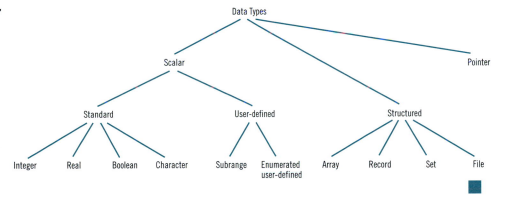

In the remainder of this section we discuss some definitions and results associated with a graph.

Formula for Degrees and Edges

> **Degree**
>
> Let G be a graph and v a vertex of G. The **degree** of v, denoted deg(v), is the number of edges emanating from v. An edge that is a loop is counted twice.

Note that a vertex v is isolated if and only if deg(v) = 0.

*Pronounced "Oiler."

EXAMPLE 5   Determining the Degree of a Vertex of a Graph

Figure 8

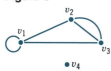

Find the degree of each vertex in the graph shown in Figure 8.

SOLUTION   There are two edges connecting the vertex v_1 to the vertices v_2 and v_3, respectively, and there is a loop at the vertex v_1; thus, $\deg(v_1) = 4$. Similarly, $\deg(v_2) = 3$; $\deg(v_3) = 3$; and $\deg(v_4) = 0$, since v_4 is isolated.

It turns out that there is a relation between the number of edges in a graph and the sum of the degrees of the vertices in the graph. If we add all the degrees, every edge will be counted twice. Thus, we have the following theorem:

> Let G be a graph with vertices $v_1, v_2, \ldots, v_n$. Then the sum of the degrees of all the vertices of G equals twice the number of edges in G. That is,
>
> $$\deg(v_1) + \deg(v_2) + \cdots + \deg(v_n) = 2 \cdot (\text{The number of edges in } G) \quad (1)$$

Note that this theorem says that the sum of the degrees of the vertices is even, and hence the number of odd degree vertices must be even.

EXAMPLE 6   Studying Properties of Graphs

Can we draw a graph G with 3 vertices v_1, v_2, and v_3 where

(a) $\deg(v_1) = 1,$ $\deg(v_2) = 2,$ and $\deg(v_3) = 2$?
(b) $\deg(v_1) = 2,$ $\deg(v_2) = 1,$ and $\deg(v_3) = 1$?
(c) $\deg(v_1) = 0,$ $\deg(v_2) = 0,$ and $\deg(v_3) = 4$?

Justify your answers, and if yes, draw a graph G with the desired characteristics.

SOLUTION

(a) No, since $\deg(v_1) + \deg(v_2) + \deg(v_3) = 1 + 2 + 2 = 5$, which is an odd number. By the above theorem, such a graph does not exist.
(b) Yes, $\deg(v_1) + \deg(v_2) + \deg(v_3) = 4$, which is an even number. Then two possibilities for G are given in Figure 9. In each case

$$\deg(v_1) = 2, \qquad \deg(v_2) = 1, \qquad \text{and } \deg(v_3) = 1$$

Figure 9

Other graphs are possible.

(c) Yes, $\deg(v_1) + \deg(v_2) + \deg(v_3) = 4$, which is an even number. Then one possibility for G is the graph shown in Figure 10. Here $\deg(v_1) = 0$, $\deg(v_2) = 0$, and $\deg(v_3) = 4$ (because of the two loops at v_3). Other graphs are possible.

Figure 10

v_2 v_3 v_1

 **Now Work Problem 9**

EXERCISE 13.1   Answers to odd-numbered problems begin on page AN-51.

In Problems 1 and 2 write the vertices and edges of each graph and find all loops, all isolated vertices, and all parallel edges.

1.

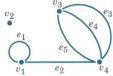

2.

3. Draw all simple graphs that have two vertices.

4. Draw all simple graphs that have four vertices and six edges.

5. Find the degree of each vertex of the graphs in Problems 1 and 2.

6. Can we draw a graph with vertices v_1, v_2, v_3, v_4, v_5, v_6, v_7, and v_8 of degrees 2, 2, 3, 4, 5, 5, 6, and 8, respectively? Justify your answer.

7. Let G be a graph with vertices v_1, v_2, v_3, v_4, v_5, and v_6 of degrees 1, 2, 3, 3, 4, and 5, respectively. How many edges does G have? Justify your answer.

8. Can we draw a graph with vertices v_1, v_2, v_3, and v_4 of degrees

   (a) 1, 2, 3, and 3, respectively?

   (b) 1, 1, 1, and 4, respectively?

   (c) 1, 2, 3, and 4, respectively?

Justify your answers.

9. Can we draw a simple graph with vertices v_1, v_2, v_3, and v_4 of degrees 1, 2, 3, and 4, respectively? Justify your answer.

10. Can we draw a simple graph with vertices v_1, v_2, v_3, v_4, and v_5 of degrees

   (a) 2, 3, 3, 3, and 5, respectively?

   (b) 1, 1, 1, 2, and 3, respectively?

Justify your answers.

11. Give an example of a simple graph (a) having no vertices of even degree, (b) having no vertices of odd degree.

In Problems 12 and 13 use the following definition:

A **complete graph of order n,** denoted by K_n, is a graph that has n vertices and every vertex is connected to every other vertex by exactly one edge.

12. Draw K_1, K_2, K_3, K_4, and K_5.

13. How many edges does K_6 have?

In Problems 14 and 15 use the following definition:

A graph G is called a **subgraph** of a graph H if every vertex in G is also a vertex in H and every edge in G is also an edge in H.

14. Find five subgraphs of

15. Find three subgraphs that contain all vertices of

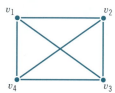

In Problems 16–18 use the following definition:

Let G be a simple graph. The **complement** of G is the simple graph $\overline{G}$ with the same vertices as G, but contains only those edges that are missing in G.

16. Find the complement of

17. Find the complement of

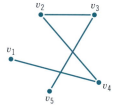

18. Find the complement of

13.2 PATHS AND CONNECTEDNESS

Figure 11

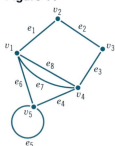

In this section we introduce some definitions and give some results regarding a special type of graph, the **connected graph.** Intuitively speaking, a graph is connected if it is possible to go from any vertex to any other vertex by traversing the edges of the graph. Many problems in graph theory involve the question of whether or not a graph is connected.

Going from a vertex to another in a graph amounts to following a sequence of neighboring edges. For example, in the graph in Figure 11 we can go from the vertex v_1 to the vertex v_3 by traversing the edges e_1 and e_2. This we write as an alternating sequence of adjacent vertices and edges: $v_1 e_1 v_2 e_2 v_3$. Or we can take a longer route as follows: $v_1 e_7 v_4 e_4 v_5 e_5 v_5 e_6 v_1 e_7 v_4 e_3 v_3$. Obviously, there are several other routes that can also be followed to go from v_1 to v_3.

> ### Path
>
> Let u and v be two vertices in a graph G. A **path** from u to v is an alternating sequence of adjacent vertices and edges of G. This sequence begins at u and ends in v.

If u and v are the same vertex (that is, $u = v$), then the path with no edges is **trivial** and is denoted by u or v.

A **simple path** from u to v is a path with no repeated vertices.

A **circuit** (or **cycle**) is a path that begins and ends at the same vertex and has no repeated edges.

A **simple circuit** is a path with no repeated edges and no repeated vertices other than the first and the last.

EXAMPLE 1   Determining Properties of Graphs

Consult the graph in Figure 12 to determine which of the following sequences are paths, simple paths, circuits, and simple circuits.

(a) $v_1 e_1 v_2 e_6 v_4 e_3 v_3 e_2 v_2$ (b) $v_1 e_1 v_2 e_2 v_3 e_3 v_4 e_4 v_5$

(c) $v_1 e_8 v_4 e_3 v_3 e_7 v_1 e_8 v_4$ (d) $v_5 e_5 v_1 e_8 v_4 e_3 v_3 e_2 v_2 e_6 v_4 e_4 v_5$

(e) $v_2 e_2 v_3 e_3 v_4 e_4 v_5 e_5 v_1 e_1 v_2$

Figure 12

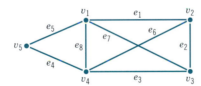

SOLUTION

(a) This is a path from v_1 to v_2. Vertex v_2 repeats; therefore, it is not a simple path. The first and the last vertices are different; therefore, it is not a circuit.

(b) This is a path from v_1 to v_5. No vertex repeats; therefore, this is a simple path. (Note that a simple path is never a circuit.)

(c) Here, edge e_8 repeats. Therefore, this is just a path from v_1 to v_4.

(d) This path starts and ends at v_5. It has no repeated edges. Vertex v_4 repeats; therefore, it is a circuit.

(e) This path starts and ends at v_2. It has no repeated edges and no repeated vertices (other than v_2); therefore, it is a simple circuit.

✍ **Now Work Problem 3**

We are now ready to define a connected graph.

> **Connected Graph**
>
> Let G be a graph. Then G is said to be a **connected graph** if for every pair of distinct vertices u and v ($u \neq v$) in G, there is a path between u and v.

EXAMPLE 2   Determining When Graphs Are Connected

Let G_1, G_2, and G_3 be the graphs given in Figure 13. Determine which of these graphs is connected.

Figure 13

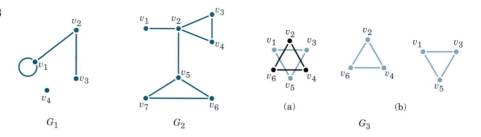

G_1 G_2 (a)   (b) G_3

SOLUTION   In G_1 there is no path from the vertex v_4 to any other vertex. Thus, G_1 is not connected. In G_2 there is a path between every distinct pair of vertices. Thus, G_2 is connected. For G_3, (a) and (b) in Figure 13 are the same graph. (b) shows that G_3 consists of two separate parts. That is, there is no path between, say, v_1 and v_2. Thus, G_3 is not connected. You should not be misled by the graph of G_3 given in (a), which seems to show that G_3 is connected. (Recall that two edges may intersect but not necessarily at a vertex.) ∎

We end this section by stating without proof two results that may help in determining the connectedness of a given graph.

> **Theorem I**
>
> Let G be a connected graph with n vertices. Then G must have at least $n - 1$ edges.

> **Theorem II**
>
> Let G be a simple graph with n vertices (that is, G has no loops and no parallel edges). If G has more than $\dbinom{n-1}{2}$ edges, then it must be connected.

EXERCISE 13.2   Answers to odd-numbered problems begin on page AN-52.

In Problems 1–4 a graph is given. Identify the specified paths as not simple; simple; circuits; or simple circuits.

1. (a) $v_3 e_3 v_4 e_4 v_5 e_6 v_2 e_7 v_4$
    (b) $v_1 e_5 v_5 e_6 v_2 e_1 v_1 e_5 v_5$
    (c) $v_1 e_1 v_2 e_6 v_5 e_4 v_4 e_3 v_3$
    (d) $v_3 e_2 v_2 e_1 v_1 e_5 v_5 e_6 v_2 e_7 v_4 e_3 v_3$
    (e) $v_5 e_6 v_2 e_2 v_3 e_3 v_4 e_4 v_5$

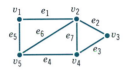

2. (a) $v_3 e_4 v_3$
    (b) $v_1 e_1 v_2 e_2 v_3$
    (c) $v_2 e_1 v_1 e_3 v_3 e_4 v_3 e_2 v_2$

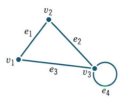

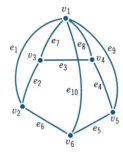

3. (a) $v_1 e_1 v_2 e_2 v_1$
    (b) $v_1 e_1 v_2 e_5 v_3 e_4 v_1$
    (c) $v_1 e_3 v_1$
    (d) $v_1 e_1 v_2 e_6 v_2 e_2 v_1 e_3 v_1$

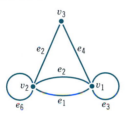

4. (a) $v_1 e_1 v_2 e_2 v_3 e_3 v_4 e_8 v_1$
    (b) $v_1 e_{10} v_6 e_6 v_2 e_2 v_3 e_7 v_1 e_1 v_2 e_6 v_6$

5. For the graph below find

    (a) Four different simple paths.
    (b) Four different circuits that are not simple.
    (c) Four different simple circuits.

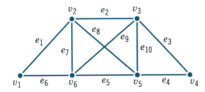

6. Find six simple paths from v_1 to v_4 in the graph.

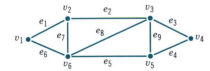

7. For the graph in Problem 6 find a circuit that is not simple, contains all vertices, and begins at v_1.

8. For the graph in Problem 6 how many simple circuits that contain all vertices and begin at v_1 are there?

9. Draw a simple circuit consisting of

    (a) Only one edge
    (b) Only two edges

10. Draw a graph that is connected, but removing one edge makes it disconnected.

11. Let G be a connected graph and let e be an edge. If removing e from G results in a disconnected graph, then e is called a *bridge*. Find all bridges for each of the following graphs.

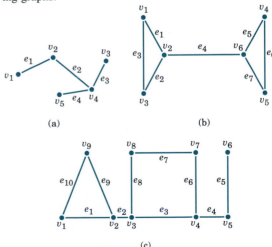

12. Can there be a connected graph G with $n - 1$ edges and n vertices where the deletion of one edge disconnects G? [*Hint:* Use Theorem I.]

(Refer to Problem 14 of Section 13.1.) A subgraph G of H is a connected component of H if and only if

1. *G is connected.*
2. *H does not have a connected subgraph that contains G as its subgraph.*

13. Find a connected component of the following graph.

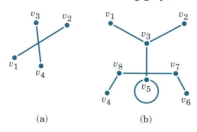

14. Repeat Problem 13 for the following graphs.

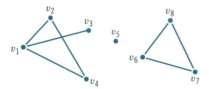

(a)   (b)

15. Suppose *H* is a simple graph with

    (a) 6 vertices and 11 edges. Is *H* disconnected? Why?
    (b) 6 vertices and 10 edges. Is *H* disconnected? Why?

16. Use induction to prove Theorem I.

17. **Shortest Path**   The vertices in the graph below represent cities, and each edge represents a road linking two cities. For each edge the length of the road is indicated in the picture. The shortest path from one city to another is the path with smallest sum of the lengths of its edges. Find the shortest path from Firsttown to Sixthtown.

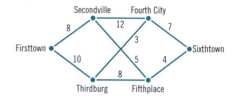

18. **Shortest Path**   Repeat Problem 17 for the following graph.

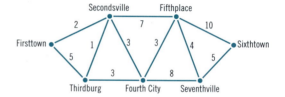

13.3 EULERIAN AND HAMILTONIAN CIRCUITS

In this section we present results that will help us solve the Königsberg bridge problem and problems similar to it. As noted in Section 13.1, Euler studied this problem and in 1736 published its solution. We begin with the following definition of an **Eulerian circuit.**

Eulerian Circuit

> **Eulerian Circuit**
>
>   Let *G* be a connected graph. A circuit in *G* that contains every edge in *G* is called an **Eulerian circuit.**

Thus, an Eulerian circuit is a path that begins and ends at the same vertex, traverses every vertex at least once, and every edge exactly once.

EXAMPLE 1   Identifying Eulerian Circuits

Consider the graphs G_1 to G_6 given in Figure 14.

G_1 contains no Eulerian circuit because there is no edge connecting vertex v_4 to the rest of the graph; that is, because G_1 is not connected. G_2 contains no Eulerian circuit because any circuit will use e_1 twice. The circuit $v_1 e_1 v_2 e_2 v_1$ in G_3 is Eulerian. In G_4 the circuit $v_1 e_1 v_2 e_2 v_3 e_3 v_1$ is Eulerian. G_5 contains no Eulerian circuit. (Try it!) $v_1 e_1 v_2 e_2 v_3 e_3 v_4 e_4 v_2 e_5 v_5 e_6 v_1$ is an Eulerian circuit contained in G_6.

Figure 14

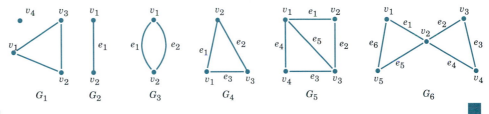

G_1 G_2 G_3 G_4 G_5 G_6

 Now Work Problem 1

We make the following observations regarding Example 1:

1. G_1 suggests that a disconnected graph cannot contain an Eulerian circuit.
2. G_2 and G_5, both of which contain no Eulerian circuit, have vertices of odd degree. Both vertices in G_2 are of degree 1. Vertices v_1 and v_3 in G_5 are of degree 3.
3. G_3, G_4, and G_6, which contain an Eulerian circuit, have all vertices of even degree. In G_3, v_1 and v_2 are of degree 2. In G_4, v_1, v_2, and v_3 are all of degree 2. In G_6, v_1, v_3, v_4, and v_5 are all of degree 2, and v_2 is of degree 4.

These observations are not accidental. Indeed, there is a criterion for determining when a graph has an Eulerian circuit. We state this criterion without proving it. Later we will present an algorithm for finding an Eulerian circuit for those graphs that possess one.

Criterion for an Eulerian Circuit

Theorem III

Let G be a graph. G contains an Eulerian circuit if, and only if,

1. G is connected.
2. Every vertex in G is of even degree.

EXAMPLE 2   Deciding When a Graph Contains an Eulerian Circuit

Show that the graphs in Figure 15 contain no Eulerian circuit.

Figure 15

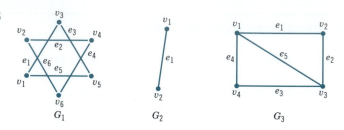

G_1 G_2 G_3

SOLUTION G_1 is not connected. Therefore, by Theorem III it cannot contain an Eulerian circuit. G_2 is connected, but vertices v_1 and v_2 are of degree 1, which is odd. Therefore, by Theorem III, G_2 cannot contain an Eulerian circuit.

G_3 is also connected but vertices v_1 and v_3 are each of degree 3, which is odd. Therefore, by Theorem III, G_3 cannot contain an Eulerian circuit.   ∎

Now Work Problem 3

We are ready, at this point, to give the solution to the Königsberg bridge problem outlined in Section 13.1, Example 3.

EXAMPLE 3   Revisiting the Königsberg Bridge Problem

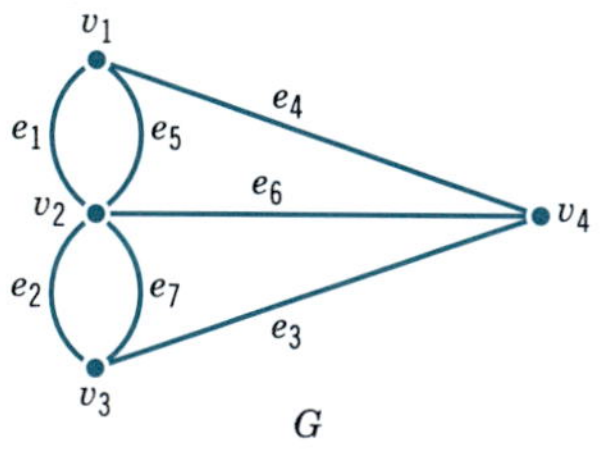

In Section 13.1, Example 3, it was pointed out that the town of Königsberg can be represented by the graph G in Figure 16, where each vertex represents a mass of land and each edge represents a bridge. The problem can then be stated as follows: Is there a route that starts at a vertex, traverses every edge exactly once and returns to that vertex? In other words, does graph G contain an Eulerian circuit?

Figure 16

SOLUTION   In the graph G in Figure 16, every vertex is of odd degree. Vertices v_1, v_4, and v_3 are of degree 3, and vertex v_2 is of degree 5. Therefore, by Theorem III, G does not contain an Eulerian circuit. That is, it is not possible to start at one of the banks or islands, cross every bridge exactly once, and return to the starting place.   ∎

Our discussion continues with an algorithm that gives a way to construct Eulerian circuits when they exist.

Fleury's Algorithm

If G is a graph that contains an Eulerian circuit, then this circuit can be obtained as follows.

1. Select any vertex in G as an initial vertex. From this initial vertex traverse edges in G, removing all edges that are traversed and all isolated vertices that are generated in the process.
2. If edge e is a bridge connecting vertices u and v (see Problem 11, Section 13.2), then e can be selected only if the removal of e causes either u or v or both to become isolated.

We illustrate this procedure by applying it to a graph that contains an Eulerian circuit.

EXAMPLE 4   Finding an Eulerian Circuit of a Graph

Figure 17

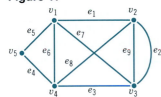

Use Fleury's algorithm to find an Eulerian circuit of the graph G given in Figure 17.

SOLUTION G is connected and every vertex in G is of even degree. Therefore, by Theorem III, G contains an Eulerian circuit. To obtain this circuit, we start at v_1 (we could start at any other vertex) and traverse edge e_1, e_7, e_6, or e_5. We choose edge e_1 and, thus, remove it. Figure 18(a) shows edge e_1 removed. (We denote removed edges by dashed lines.) At v_2 there are three untraversed edges (e_2, e_8, and e_9; these are the solid lines at v_2 in Figure 18(a)). We can choose to traverse any one of them. We choose e_8 and thus remove it. Figure 18(b) shows e_8 removed. Next, we traverse e_3 and remove

Figure 18

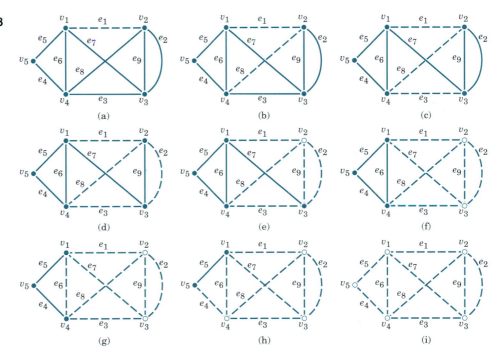

it. Figure 18(c) shows e_3 removed. Note that we could equally choose e_4 or e_6. At this point we are at v_3 with edges e_2, e_9, and e_7 untraversed. Inspecting the remaining graph [Figure 18(c)], that is, the graph without the dashed (removed) edges, we find that e_7 is a bridge, which means that its removal disconnects the remaining graph. So we do not choose e_7 (unless we have no other choice). This means that we can choose only either edge e_9 or e_2. We select e_2, traverse, and remove it [Figure 18(d)]. We are back at v_2. The only untraversed edge at this vertex is e_9. We traverse e_9 and remove it. This leaves v_2 isolated from the rest of the graph. In Figure 18(e) we draw v_2 as a blank circle to denote that it is a removed vertex. We proceed with our edge traversal in the following order: e_7, e_6, e_4, and e_5 as diagrammed in Figure 18(f) to 18(i). Thus the circuit that we obtain is

$$v_1 e_1 v_2 e_8 v_4 e_3 v_3 e_2 v_2 e_9 v_3 e_7 v_1 e_6 v_4 e_4 v_5 e_5 v_1$$

Remember that this is not the only Eulerian circuit in G.

 **Now Work Problem 5**

Hamiltonian Circuits

A related problem to that of finding an Eulerian circuit, in which all edges of a graph appear exactly once, is that of finding a circuit in which all vertices of a graph appear exactly once. Such a circuit is called a **Hamiltonian circuit** honoring the Irish mathematician Sir William R. Hamilton (1805–1864), who introduced this problem in 1859 (see Problem 10). Note that a Hamiltonian circuit includes every vertex of a graph exactly once (except that the first and last are the same) and may or may not include every edge.

EXAMPLE 5   Finding a Hamiltonian Circuit

Figure 19

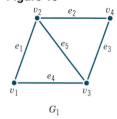

G_1

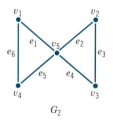

G_2

Which of the graphs given in Figure 19 has a Hamiltonian circuit? Give the circuits for the graphs that contain them.

SOLUTION   In G_1 we can start at v_1 and traverse edge e_1 to v_2, e_2 to v_4, e_3 to v_3, and then e_4 to v_1. Thus, G_1 has a Hamiltonian circuit given by $v_1 e_1 v_2 e_2 v_4 e_3 v_3 e_4 v_1$. Notice how all vertices appear in this circuit but not all edges. Edge e_5 is not used in this circuit.

For G_2 we argue as follows: If we start at v_1, v_2, v_3, or v_4 and visit every vertex, vertex v_5 will be visited twice. And if we start at v_5 itself, then in going from the vertices v_1 or v_4 to v_3 or v_2, respectively, or vice-versa, v_5 will be passed again before completing the circuit. To complete the circuit, we must come back to v_5, thus traversing it three times. All this means that G_2 has no Hamiltonian circuit. ∎

While there is a criterion for determining whether or not a graph contains an Eulerian circuit (Theorem III), a similar criterion does not exist for Hamiltonian circuits. There is, however, a procedure that can be used to determine that some graphs contain Hamiltonian circuits. We state this procedure without proof.

Theorem IV

Let G be a simple connected graph with n vertices where $n \geq 3$. If the sum of the degrees of each pair of its nonadjacent vertices is greater than or equal to n, then G has a Hamiltonian circuit.

The converse of Theorem IV is not true. That is, if a simple graph G with $n \geq 3$ vertices has a Hamiltonian circuit, then the sum of the degrees of each pair of its nonadjacent vertices *does not have to be* greater than or equal to n. This is demonstrated in the following example.

EXAMPLE 6   Demonstrating that the Converse of Theorem IV Is Not True

Figure 20

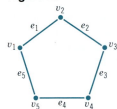

For the graph in Figure 20, $n = 5$ and the circuit $v_1e_1v_2e_2v_3e_3v_4e_4v_5e_5v_1$ is Hamiltonian. Yet the sum of the degrees of any two nonadjacent vertices is 4, which is not greater than or equal to 5. For instance, $\deg(v_1) + \deg(v_3) = 4$, $\deg(v_1) + \deg(v_4) = 4$, and so on.

■

Thus, Theorem IV is a sufficient, but not a necessary, condition for a simple graph to possess a Hamiltonian circuit.

EXAMPLE 7   Determining Hamiltonian Circuits

In graph G_1 in Figure 19 the sum of the degrees of each pair of nonadjacent vertices is 4, $[\deg(v_1) + \deg(v_4)]$, which is equal to the number of vertices. Thus, G_1 has a Hamiltonian circuit (see Example 5).

■

EXERCISE 13.3   Answers to odd-numbered problems begin on page AN-52.

1. Determine which of the following graphs contain an Eulerian circuit. If it does, then find an Eulerian circuit for the graph.

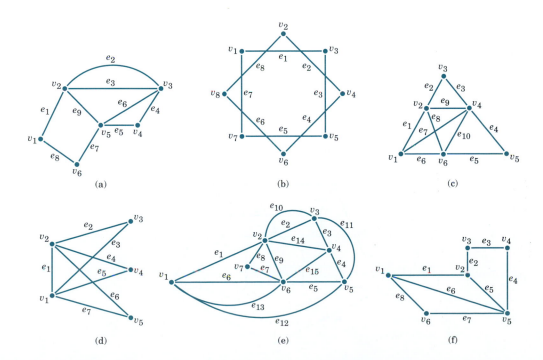

2. Does the complete graph K_4 (see Problem 12 in Exercise 13.1) contain an Eulerian circuit?

3. Does the complete graph K_5 contain an Eulerian circuit?

4. Does the complete graph K_n contain an Eulerian circuit? Explain.

5. A city consists of two land masses, situated on both banks of a river and three islands connected to each other and to the banks as shown in the figure. Is there a way to start at any point and make a round trip through all land masses, crossing each bridge exactly once? If so, how can this be done?

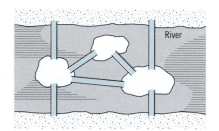

6. Give an example of a graph for which every vertex has an even degree but which does not contain an Eulerian circuit.

7. The following is a floor plan of a certain house. (*A* denotes the outside of the house.) Is it possible to start outside the house, or in any one of the rooms, take a tour through the house, passing through each doorway exactly once, and return to the starting place? [*Hint:* Represent the floor plan of the house with a graph where the outside and each room is represented by a vertex and each doorway by an edge.]

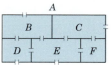

8. Repeat Problem 7 for the following floor plan.

9. Find a Hamiltonian circuit for each of the following graphs.

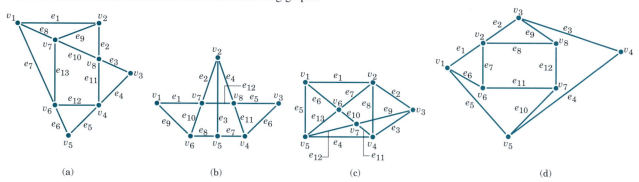

(a)	(b)	(c)	(d)

10. **Hamilton's Puzzle**   In 1859 Sir William Hamilton presented the following problem: Suppose each vertex in the graph represents a big international city and that each edge represents a transportation route. Can a salesperson make a round trip where he or she visits every city exactly once?

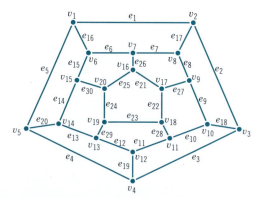

11. Show that the following graphs do not have a Hamiltonian circuit.

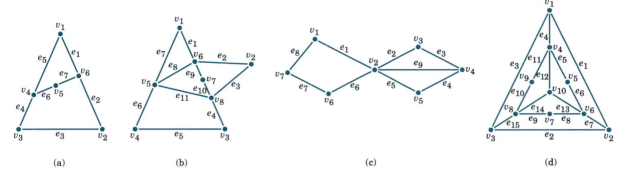

(a)   (b)   (c)   (d)

12. Determine whether or not the graph contains a Hamiltonian circuit. Find such a circuit if the graph has one.

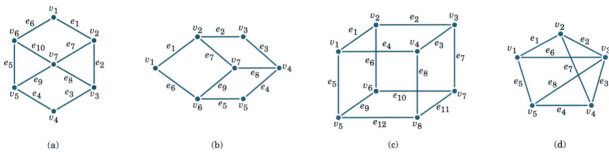

(a)   (b)   (c)   (d)

13. Give an example of a graph that contains both Eulerian and Hamiltonian circuits.

14. Give an example of a graph that contains an Eulerian circuit but does not contain a Hamiltonian circuit.

15. Give an example of a graph that contains a Hamiltonian circuit but does not contain an Eulerian circuit.

16. Which of the following complete graphs has a Hamiltonian circuit? (Refer to Problem 12, Exercise 13.1.)

$$K_2, K_3, K_n \quad \text{where } n > 3$$

17. The Chinese Postman Problem   Suppose a mail carrier traverses all of the roads in a city, starting and ending at the same vertex. We seek a circuit of minimum total length that covers all of the roads (edges). If every vertex is even, this minimal circuit is an Eulerian circuit, otherwise we must repeat some of the edges. This is called the Chinese Postman Problem in honor of the Chinese mathematician Guan Meigu, who proposed it. Find the closed circuit of minimal length for the city below:

18. The Chinese Postman Problem   Repeat Problem 17 for the following graph.

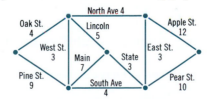

13.4 TREES

Another class of graphs, called **trees,** are graphs that have no circuits (thus, they cannot have parallel edges or loops). Trees arise frequently in computer science. In this section we present a brief exposure to the topic of trees.

Tree

Let T be a graph. T is called a **tree** if, and only if,

(a) T is connected.
(b) T contains no circuits (except trivial ones).

T is called a **trivial tree** if it consists of a single vertex.

If after relabeling all the edges and vertices of two trees, they correspond, we say the trees are **identical.**

EXAMPLE 1   Examples of Trees

Draw all distinct trees that have

(a) One vertex.
(b) Two vertices.
(c) Three vertices.
(d) Four vertices.

SOLUTION

(a) A tree consisting of one vertex is the trivial tree T_1 shown in Figure 21(a).
(b) Figure 21(b) shows the tree T_2 that has two vertices.
(c) Figure 21(c) shows the tree T_3 with three vertices.
(d) Here, the tree T_4 may look like either one of the trees drawn in Figure 21(d).

Figure 21

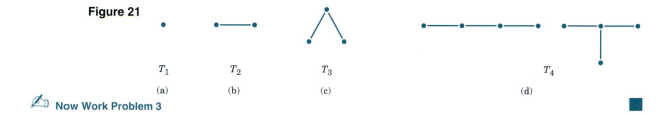

T_1 T_2 T_3 T_4

(a)   (b)   (c)   (d)

Now Work Problem 3

EXAMPLE 2   Determining When a Graph Does Not Represent a Tree

State why each graph in Figure 22 does not represent a tree.

Figure 22

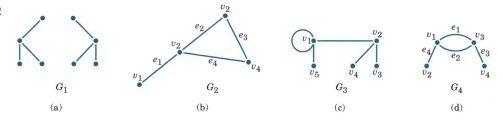

G_1 (a) G_2 (b) G_3 (c) G_4 (d)

SOLUTION   In Figure 22 the graph G_1 in (a) is not connected, so it is not a tree. The graph G_2 in (b) has the nontrivial circuit $v_2e_2v_3e_3v_4e_4v_2$. Therefore, it is not a tree. The graph G_3 in (c) contains a loop at v_1. Thus, it is not a tree. The graph G_4 in (d) has two edges e_1 and e_2 that are parallel. Hence, G_4 is not a tree. ■

The next theorem, which we present without a proof, is a useful way to characterize trees. In essence, it says that a tree is a connected graph with the fewest number of edges, meaning that if one edge is removed, then the graph becomes disconnected.

> **Theorem V**
>
> Let G be a connected graph that has n vertices. G is a tree if, and only if, it has exactly $n - 1$ edges.

Rooted Tree

A special type of tree, called a **rooted tree,** is useful in computer science. We next give the definition of this tree and some terms related to it. Let T be a tree and let u, v, and w be vertices in T.

1. If v is distinguished from the other vertices in T, then T is called a **rooted** tree and v is called a **root.**
2. If u is adjacent to v (v need not necessarily be a root) but farther from the root than v, then u is called the **child** of v. If u and w are the only children of v with u located to the left of v and w to the right of v, then u and w are called, respectively, the **left** and the **right child** of v.
3. If T is rooted and if every vertex in T has left and right children, either a left or a right child, or no children, then T is called a **binary tree.**
4. If the vertex u has no children, then u is called a **leaf** (or a **terminal** vertex). If u, which is not a root, has either one or two children, then u is called an **internal vertex.**
5. The **descendants** of the vertex u is the set consisting of all the children of u together with the descendants of those children.

EXAMPLE 3   Studying Properties of a Tree

Figure 23

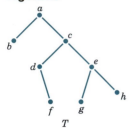

T

Consider the rooted tree T in Figure 23.

(a) What is the root of T?
(b) Is T a binary tree? If so, find the left and the right children of every vertex.
(c) Find the leaves and the internal vertices of T.
(d) Find the descendants of the vertices a and c.

SOLUTION

(a) Vertex a is distinguished as the only vertex located at the top of the tree. Therefore, a is the root.
(b) Yes, every vertex has two children, one child, or no children. The following table indicates the children of each vertex.

Vertex	Left Child	Right Child
a	b	c
b	None	None
c	d	e
d	None	f
e	g	h
f	None	None
g	None	None
h	None	None

(c) The leaves are those vertices that have no children. These are b, f, g, and h. The internal vertices are those that have either one child or two children. These are c, d, and e.
(d) The descendants of a are b, c, d, e, f, g, h. The descendants of c are d, e, f, g, h.

Let T be a binary tree, and let u and v be two vertices in T. The **subtree rooted at u** is the tree consisting of the root u, all its descendants, and all the edges connecting them. If u is the left child of v, then the subtree rooted at u is called the **left subtree of v rooted at u,** and if u is the right child of v, then the subtree rooted at u is called the **right subtree of v rooted at u.** If u is a leaf, then the subtree rooted at u is called a **trivial subtree.**

EXAMPLE 4   Studying a Binary Tree

Consider the binary tree in Figure 24.

Figure 24

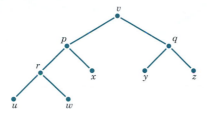

(a) Find the left and right subtrees of v.
(b) Find the subtree rooted at r.

Figure 25

SOLUTION

(a) The left subtree of v is given in Figure 25.
    The right subtree of v is given in Figure 26.
(b) The subtree rooted at r is given in Figure 27.

Figure 26

Figure 27

Applications in Computer Science

ALGEBRAIC EXPRESSIONS

Binary trees are used to represent algebraic expressions. The vertices of the tree are labeled with the numbers, variables, or operations that make up the expression. The leaves of the tree can be labeled only with numbers or variables. Operations such as addition, subtraction, multiplication, division, or exponentiation can be assigned only to internal vertices, or to a root. The operation at each vertex operates on its left and right subtrees from left to right. The next two examples illustrate such uses of binary trees.

EXAMPLE 5   Using Binary Trees to Represent Algebraic Expressions

Use a binary tree to represent the expression

(a) $x * y$ ("$*$" means multiplication)
(b) $(x + y)/z$ ("$/$" means division)
(c) $[(x - y) ** 2]/(x + y)$ ("$**$" means exponentiation)

SOLUTION

Figure 28

(a) In this expression the first term is x, the second term is y, and the operation is $*$. Therefore, our tree, shown in Figure 28, must be rooted at $*$ and must have two subtrees, one for each term.
(b) Here, we first add x and y and then divide by z. This means our tree must have $/$ as its root and two subtrees: a left subtree rooted at $+$ that adds x to y and a right subtree rooted at z. The tree is shown in Figure 29.
(c) To get the first term, which is $(x - y) ** 2$, we first subtract y from x, then square. To get the second term, which is $x + y$, we add x to y. Finally, we divide these two terms. This means our tree must be rooted at $/$ and must have two subtrees as shown in Figure 30.

Figure 29

Figure 30

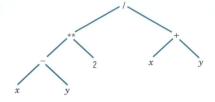

✍ Now Work Problem 29

■

THE BINARY SEARCH TREE

Suppose we have a set of numbers. We will call them *keys*. We are interested in two of the many operations that can be performed on this set:

1. Ordering (or sorting) the set.
2. Searching the ordered set to locate a certain key and, in the event of not finding the key in the set, adding it at the right position so that the ordering of the set is maintained.

In the next example we describe a method that uses trees to perform the operations outlined above. This method involves storing the list in a tree where each element is stored at a vertex.

EXAMPLE 6   Using a Binary Search Tree to Perform Operations

(a) Use a binary tree to store the elements of the following list of numbers in an increasing order: 7, 10, 21, 3, 24, 23.
(b) Use the tree constructed in (a) to search for the number 19 in the list. If the number is not found in the list, update the list by adding the number to it.

SOLUTION

(a) We begin by selecting any number from the list to be the root of our binary tree. Say we select 10 to be this root. We draw the left and the right children of 10 as shown in Figure 31(a).

Figure 31

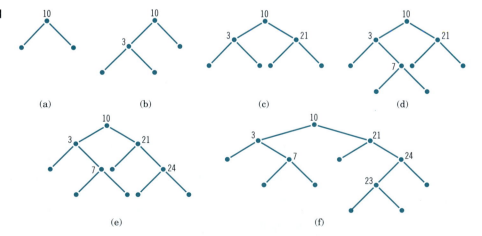

Then we pick another number from the list, say, 3. Since 3 is less than 10, we label the left child of 10 with 3 and then draw the left and the right children of 3, as Figure 31(b) shows.

Next, we pick another number from the list, say, 21. To position 21 on the tree, we start at the root 10 and compare it to 21. Since 21 is greater than 10, we move down to the right child of 10. This child is unlabeled, so we label it with 21 and then draw the left and the right children of this vertex. See Figure 31(c).

We continue in this fashion. We choose our next element from the list, say, 7. We again start at the root 10 and compare it to 7. Since 7 is less than 10, we move down to the left child of 10, which is the vertex labeled 3. We compare 7 to 3. Since 7 is greater than 3, then we move further down to the right child of 3. This child is unlabeled, so we label it with 7 and draw the left and the right children of 7. See Figure 31(d).

Say our next choice from the list is the number 24. Because 24 is greater than 10, we move down to the right child 21 of 10. Because 24 is greater than 21, we move further down to the right child of 21. This child is unlabeled. Therefore, we label it 24, and then draw the left and the right children of 24. See Figure 31(e).

Finally, the last number left in the list is 23. We start at the root 10. 23 is greater than 10, so we move down to the right child of 10, which is labeled 21. Since 23 is greater than 21, then we move further down to the right child of 21, which is labeled 24. Because 23 is less than 24, we move even further down to the left child of 24, which is unlabeled. So, we label it with 23 and then draw the right and the left children of 23. See Figure 31(f). Figure 31(f) is called a **binary search tree.**

(b) To search for the number 19 in the order list, we begin at the root 10. Because 19 is greater than 10, we move down to the right child of 10, which is labeled 21. Because 19 is less than 21, we move down to the left child of 21, which is unlabeled. See Figure 31(f). Since we have reached an unlabeled vertex, we conclude that the number 19 is not found in the list, and we label this vertex 19 and draw the left and the right children of 19. See Figure 32. This adds the key 19 to the list while maintaining the increasing order of the numbers in the list.

Figure 32

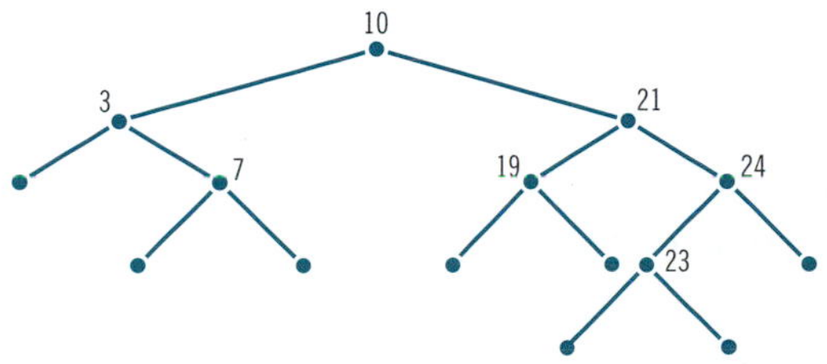

EXERCISE 13.4   Answers to odd-numbered problems begin on page AN-52.

1. Draw three distinct trees that have five vertices.

2. Draw three distinct binary trees that have five vertices.

3. Draw three distinct binary trees that have seven vertices.

4. Draw three distinct binary trees that have eight vertices.

In Problems 5–12 either draw a graph with the given properties or explain why such a graph cannot exist.

5. A tree that has 7 vertices and 7 edges.

6. A connected graph that has 7 vertices and 7 edges.

7. A tree that has 6 vertices with the sum of the degrees of the vertices being 10.

8. A tree that has (a) 6 vertices and 8 edges, (b) 6 edges and 8 vertices.

9. A tree with all vertices of degree 2.

10. A tree that has 10 vertices of degree 3, 3, 3, 3, 1, 1, 1, 1, 1, 1.

11. A connected graph that has 5 edges and 6 vertices and that is not a tree (Theorem V!)

12. A tree that has 7 vertices.

13. Does a tree with 120 vertices and 119 edges exist? Explain.

14. Does a connected graph with 3 vertices and 1 edge exist? Explain.

15. Draw three distinct rooted trees that have 4 vertices.

16. Draw four distinct rooted trees that have 5 vertices.

In Problems 17 and 18 find all leaves (or terminal vertices) and internal vertices for the given graph.

17.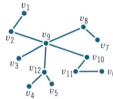

18.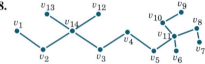

19. In a tree can a leaf (or terminal vertex) be of degree 2? Explain.

20. In a tree can an internal vertex be of degree 1? Explain.

21. Give an example of a nontrivial tree that has no internal vertices.

In Problems 22 and 23, given a tree T, find

(a) All descendants of the vertices x and y.
(b) The left and the right subtrees of the vertices x and y.
(c) The subtree rooted at x.

22.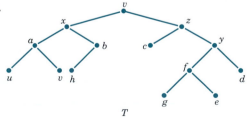

23.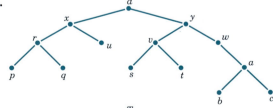

24. Give an example of a binary tree that has neither a left nor a right subtree.

In Problems 25–28 a binary tree is given. Find the algebraic expression represented by the tree.

25.

26.

27.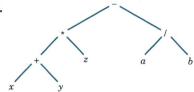

28.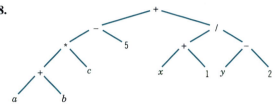

In Problems 29 and 30 an algebraic expression is given. Use a binary tree to represent the expression.

29. $[(a + b)/c] + d$

30. $\{1/[(a + b) ** 2]\} - [(a + 1) * c] + b$

In Problems 31 and 32 a list is given. For this list:

(a) *Construct a binary search tree that stores the set.*
(b) *Use the tree constructed in (a) to search for the given key. If the key is not found in the set, then add it to the set at the right location.*

31. Bob, Mary, Peter, Frank, Michael, Sue, Angela. (The order must be alphabetic.) Key is Paul.

32. 25, 99, 112, 56, 72, 65. (The order must be increasing.) Key is 132.

> *Use the definition below in Problems 33 and 34.*
>
> Let G be a connected graph. If a subgraph T of G is a tree that contains all vertices of G, then T is called a **spanning tree** for G.

33. (a) Is the tree (i) a spanning tree for the graph (ii)?

(i) (ii)

(b) Is the tree (i) a spanning tree for the graph (ii)?

(i) (ii)

34. Find two spanning trees for the graph below.

35. Minimum Spanning Tree   A spanning tree of minimum total length is called a **minimum spanning tree.** The graph below represents a computer network at Your State University. The numbers represent the distance between two computers. The computers must be interconnected by a cable. If there is no line between two computers they are incompatible, and can't communicate directly. Find the minimum length of cable required.

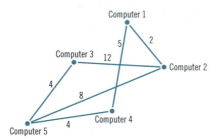

36. Minimum Phone Line Cost   Mayor Ilove Math of the city Graphville wants to connect five subdivisions with phone lines to ensure that all the subdivisions can communicate with each other. The distances between the subdivisions are given in the graph below. What is the minimum length of telephone cable required?

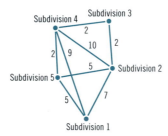

13.5 DIRECTED GRAPHS

So far our discussion in this chapter has been about different types of graphs that are undirected, meaning none of the edges have a direction; thus, crossing an edge is possible in both directions, back and forth. In this section we discuss a new type of graph called a **directed graph.** In a directed graph every edge is assigned a certain direction, so that crossing an edge is possible only in the direction assigned. Directed graphs are used in many situations. For example, one-way streets are represented by directed graphs where the vertices are the intersections and the edges are the streets. Flowcharts can be viewed as directed graphs where the vertices represent the instructions, and the edges represent the flow of control. In Section 13.6 we discuss applications of directed graphs to business and the social sciences using matrices.

Definition of a Directed Graph

> **Directed Graph**
>
> Let D be a graph. If every edge in D has a direction, then D is called a **directed graph** or **digraph,** and its edges are called **arcs.** The vertex where an arc starts is called the **initial point,** and the vertex where the arc ends is called the **terminal point.** When the directions of the edges in D are disregarded, the resulting graph obtained is called the **underlying graph** of D.

EXAMPLE 1   Studying Digraphs

Figure 33

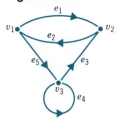

Consider the digraph in Figure 33.

(a) Give the initial and terminal points of each arc.
(b) Draw the underlying graph.

SOLUTION

(a) The following table gives all the arcs with their initial and terminal points.

Arc	Initial Point	Terminal Point
e_1	v_1	v_2
e_2	v_2	v_1
e_3	v_3	v_2
e_4	v_3	v_3
e_5	v_1	v_3

Figure 34

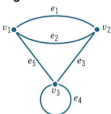

(b) Figure 34 shows the underlying graph.

✍ **Now Work Problem 1(a)**

■

EXAMPLE 2   Studying Digraphs

Draw the digraph with the following specifications:

Arc	Initial Point	Terminal Point
e_1	v_1	v_2
e_2	v_3	v_2
e_3	v_3	v_4
e_4	v_4	v_1
e_5	v_2	v_4
e_6	v_4	v_2
e_7	v_3	v_3

Figure 35

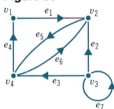

SOLUTION   Figure 35 shows the graph.

■

The directed edges in a digraph may be described in terms of the number of edges that begin and end at a vertex.

> **Indegree; Outdegree**
>
> Let v be a vertex in the digraph D. The **indegree of v,** denoted by indeg(v), is the number of arcs in D whose terminal point is v. The **outdegree of v,** denoted by outdeg(v), is the number of arcs in D whose initial point is v.

EXAMPLE 3   Studying Digraph

In the digraph in Figure 36, find the indegree and the outdegree of each vertex.

Figure 36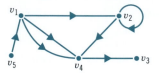

SOLUTION   The answer is given in the following table:

Vertex	Indegree	Outdegree
v_1	1	3
v_2	2	2
v_3	1	0
v_4	3	1
v_5	0	1

 **Now Work Problem 5**

Note that the loop at v_2 is counted as an arc coming out of and going into v_2.

Definition of a Directed Path

> **Directed Path**
>
> A **directed path** in a digraph D is a sequence of vertices and edges such that the terminal point of one arc is the initial point of the next. If, in D, a directed path exists from the vertex u to the vertex v, then v is said to be **reachable** from u.

EXAMPLE 4   Studying Digraphs

Consider the digraph in Figure 37.

Figure 37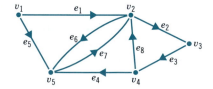

(a) Find three distinct directed paths from v_1 to v_5.
(b) Is v_1 reachable from v_4? Is v_1 reachable from v_5?

SOLUTION

(a) One directed path from v_1 to v_5 is $v_1 e_1 v_2 e_6 v_5$. A second directed path is $v_1 e_1 v_2 e_2 v_3 e_3 v_4 e_4 v_5$. A third directed path is $v_1 e_1 v_2 e_2 v_3 e_3 v_4 e_8 v_2 e_6 v_5$.

(b) Since the indegree of v_1 is 0, then v_1 is not reachable from any other vertex. ∎

 **Now Work Problem 9**

Digraphs where every vertex is reachable from any other vertex are important. We discuss them next.

Strongly Connected; Weakly Connected

Let D be a digraph. If every vertex in D is reachable from any other vertex in D, then D is called **strongly connected.** If the underlying graph of D is connected, then D is said to be **weakly connected.**

EXAMPLE 5   Classifying Digraphs

Which of the two digraphs in Figure 38 is strongly connected and which one is weakly connected?

Figure 38

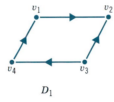

D_1

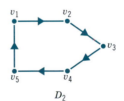

D_2

SOLUTION In D_1, v_3 is not reachable from v_1 or v_2, and the underlying graph is connected. Therefore, D_1 is weakly connected. In D_2 every vertex is reachable from the others. Therefore, D_2 is strongly connected. ∎

Orientable Graphs

Suppose a graph represents a network of streets in a certain town where the vertices represent the intersections of the streets and the edges represent the streets connecting these intersections. An important question is whether or not the graph can be directed so that the streets can be made into one-way streets and still be able to get from any intersection to any other. That is, we want to determine whether or not the resulting digraph is strongly connected. We refer to this process of directing the edges of a graph as *orientation*.

> **Orientation**
>
>   Let G be a graph. If every edge in G can be given a direction such that the result is a strongly connected digraph, then G is said to be **orientable.** An edge of a connected graph is called **a bridge** if its removal results in a disconnected graph.

The following theorem, stated without proof, is a criterion that characterizes an orientable graph.

> **Theorem VI**
>
>   A graph G is orientable if, and only if, it is connected and has no bridges.

Next, we present a procedure that produces an orientation for an orientable graph G. We remind you that a spanning tree of a graph G is a subgraph of G, which is a tree that contains all the vertices of G. (See Problem 33, Section 13.4.)

Our procedure consists of the following steps.

Step 1 This step is referred to as a **depth-first search.** It involves the generation of a spanning tree T of the graph G as follows: We start at a vertex and search through the edges and vertices of G for a path. Then we extend this path as long as possible to a spanning tree. Because such a tree must include all vertices of G, if, during our search, a vertex is reached beyond which no new vertices can be reached, we back up along the path obtained, as far as necessary, to a vertex where branching to a new vertex is possible. We continue this process until a spanning tree of G is formed. During the search every vertex and edge encountered is numbered. Such numbering is called a **depth-first numbering** of G. Example 6 below explains further the process described in this step.

Step 2 We assign directions to the edges of T formed in Step 1 in the following way: Each edge of the tree T is directed from lower to higher depth-first number vertices.

Step 3 We direct all other vertices (not included in the tree T) from higher to lower depth-first number vertices.

We clarify this procedure by considering the following example.

EXAMPLE 6   **Determining When a Graph Is Orientable**

Figure 39

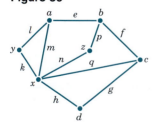

Determine whether or not the graph drawn in Figure 39 is orientable. If it is, then use the above procedure to orient the edges so that the resulting graph is strongly connected.

SOLUTION   Since the graph is connected and has no bridges, then it is orientable. To orient it, we proceed as follows:

Step 1 Our aim is to generate a spanning tree. We begin by selecting any vertex in the graph; we pick vertex a and label it v_1. Then we traverse edge e to vertex b, labeling them e_1 and v_2, respectively. We continue in this fashion, avoiding all edges that will produce a circuit, to obtain the path $v_1e_1v_2e_2v_3e_3v_4e_4v_5$. See Figure 40(a). In this figure we denote the untraversed edges by dashed lines and the unvisited vertices by a blank circle. In this path we stop at vertex d (labeled v_5) because we cannot go any further to reach a new vertex. Notice that if we continue the path from v_5 to the already visited vertex v_3, we would form the circuit whose vertices are v_3, v_4, and v_5 and thus no longer have a tree. Remember that trees are circuit-free connected graphs. Now, because no new vertices can be reached from v_5, we back up to vertex v_4, from which we can branch to either vertex y or vertex z. We choose to traverse edge k to vertex y and label them e_5 and v_6, respectively. See Figure 40(b).

Figure 40

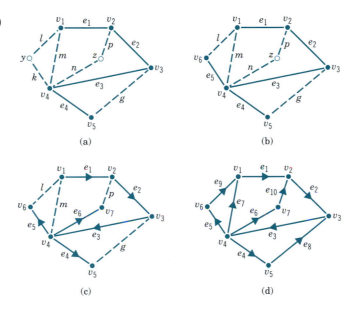

(a) (b) (c) (d)

At v_6 no new vertices can be branched. So here again we back up to v_4, from which we reach vertex z. We traverse edge n to this vertex and label them e_6 and v_7, respectively. See Figure 40(c).

This completes the generation of our spanning tree because the tree with vertices v_1, v_2, v_3, v_4, v_5, v_6, and v_7 and edges e_1, e_2, e_3, e_4, e_5, and e_6, which we just formed, is a subgraph of the given graph, is connected, and contains all vertices of the given graph. Thus, both our depth-first search and depth-first numbering are complete.

Step 2 We direct the edges of the generated spanning tree from lower- to higher-numbered vertices as shown in Figure 40(c). Note that no direction is assigned to the dashed lines at this point.

Step 3 We finally direct the edges that are not part of our spanning tree (formed in Step 1); that is, we direct the dashed lines in Figure 40(c) from higher- to lower-numbered vertices. See Figure 40(d).

The digraph thus produced is strongly connected because every vertex is reachable from any other vertex.

EXERCISE 13.5   Answers to odd-numbered problems begin on page AN-53.

In Problems 1 and 2 a digraph is given.

(a) Find the initial and terminal points of each arc.
(b) Find the indegrees and the outdegrees of all vertices.

1.

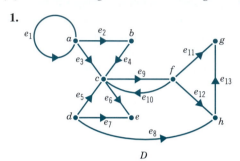

2.

In Problems 3 and 4 draw the digraph with the given specifications.

3.

Arc	Initial Point	Terminal Point
e_1	v_1	v_2
e_2	v_2	v_2
e_3	v_2	v_3
e_4	v_1	v_3
e_5	v_1	v_4
e_6	v_4	v_1

4.

Arc	Initial Point	Terminal Point
e_1	v_2	v_1
e_2	v_3	v_2
e_3	v_3	v_4
e_4	v_4	v_1
e_5	v_1	v_3
e_6	v_3	v_1
e_7	v_4	v_2

In Problems 5 and 6 draw the digraph with the given specifications. (Note that the sum of the indegrees, or outdegrees, of all vertices is equal to the number of the arcs in the digraph.)

5.

Vertex	Indegree	Outdegree
v_1	0	1
v_2	2	2
v_3	2	0
v_4	2	1
v_5	1	3
v_6	1	1

6.

Vertex	Indegree	Outdegree
v_1	1	2
v_2	2	0
v_3	2	2
v_4	0	1

7. Draw a digraph with three vertices where each vertex has indegree 2.

8. Draw a digraph with three vertices where each vertex has outdegree 2.

In Problems 9 and 10 a digraph D is given.

(a) Is v_5 reachable from v_1?   (b) Is v_1 reachable from v_5?

Justify your answers.

9.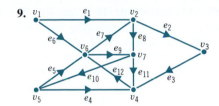

10.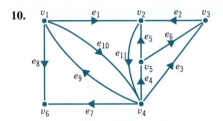

11. Find three distinct directed paths from v_1 to v_6 in the digraph in Problem 9.

12. Is the digraph in Problem 9 strongly connected? Justify your answer.

13. Is the digraph in Problem 10 strongly connected? Justify your answer.

15. Let $A = \{2, 3, 4, 9, 36\}$ and let R define the relation on A as follows: xRy if, and only if, x divides y. Draw a digraph D that represents R where the vertices of D are the elements of A and an arc from u to v in D means uRv.

16. Repeat Problem 15 for $A = \{1, 2, 5, 8, 9\}$ and R defined on A as follows: aRb if and only if a is less than b.

14. A digraph D is called a **tournament** if it is loop-free and has exactly one arc between any two vertices. Thus, the vertices may represent contestants competing in a match, and an arc from vertex v to u may mean u won over v. Given the following tournament:
(a) What contestant won the greatest number of games?
(b) Find a path that does not start with the contestant in (a) and that contains all contestants.

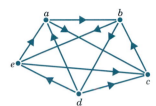

In Problems 17–20 determine whether or not the given graph is orientable, and if it is, then use the procedure given in this section to orient the edges so that the resulting graph is strongly connected.

17.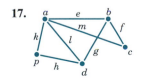

18.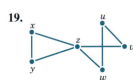

19. **20.**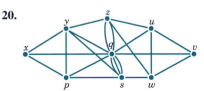

21. Network Flow A directed graph is said to describe a **network flow** if each directed edge is associated with its maximum capacity. The graph below represents a network of oil pipelines distributing oil from Houston, Texas to other parts of the country. For each pipeline the maximum capacity (in millions of gallons of oil per hour) is indicated. What is the maximum amount of oil we can transfer from Houston to Los Angeles?

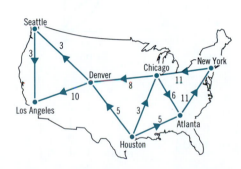

22. Repeat the Problem 21 for the following graph.

23. Maximum Number of Flights   Alpha Airlines wants to arrange the maximum number of connecting flights between New York and Los Angeles. Connecting flights must stop in Atlanta and then stop in Dallas or Denver. Because of limited landing space, Alpha is limited to making the number of flights as shown in graph below. What is the maximum number of connecting flights daily from New York to Los Angeles?

24. Gary–Chicago–Milwaukee Corridor Capacity   The capacity of each of the highways connecting Gary, Indiana to Milwaukee, Wisconsin is shown on the directed graph below in tens of thousands of cars per day. What is the maximum number of cars that can be sent through this corridor?

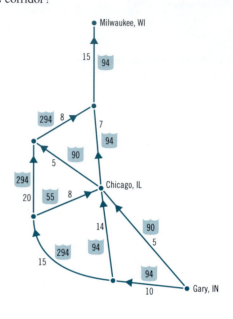

13.6 MATRIX APPLICATIONS TO GRAPHS AND DIRECTED GRAPHS

In order to better visualize the information given in a graph and to be able to draw conclusions about the situation depicted by a graph, we use a matrix. This will allow us to take full advantage of the computer for tedious calculations.

To determine the matrix representation of a graph, we use a square matrix in which the entry in row i column j is the number of edges joining the ith vertex with the jth vertex. A 0 denotes that the vertices are not joined by an edge. For example, the graphs in Figure 41 have the following matrix representations:

Figure 41

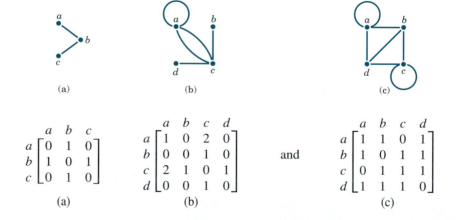

EXAMPLE 1   Representing a Graph by a Matrix

Figure 42

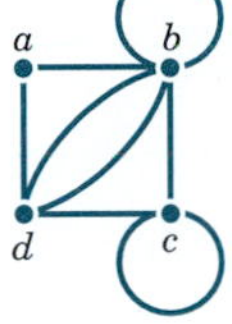

Determine the matrix representation for the graph in Figure 42.

SOLUTION   We note that vertices b and c have a loop, which means that diagonal entries corresponding to vertices b and c are equal to 1. The vertices b and d are connected with parallel edges so their corresponding entry will be equal to 2. The matrix representation is

$$\begin{array}{c} \\ a \\ b \\ c \\ d \end{array} \begin{array}{cccc} a & b & c & d \\ \left[\begin{array}{cccc} 0 & 1 & 0 & 1 \\ 1 & 1 & 1 & 2 \\ 0 & 1 & 1 & 1 \\ 1 & 2 & 1 & 0 \end{array}\right] \end{array}$$

Conversely, if we are given a matrix representation we can determine the graph corresponding to the matrix.

EXAMPLE 2   Determining the Graph from a Matrix

Determine the graph corresponding to the matrix representation

$$\begin{array}{c} \\ a \\ b \\ c \\ d \end{array} \begin{array}{cccc} a & b & c & d \\ \left[\begin{array}{cccc} 1 & 0 & 0 & 2 \\ 0 & 1 & 1 & 0 \\ 0 & 1 & 0 & 0 \\ 2 & 0 & 0 & 0 \end{array}\right] \end{array}$$

Figure 43

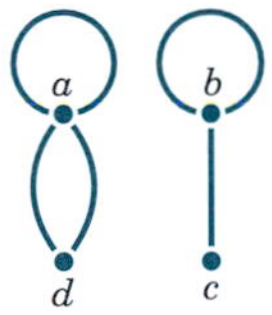

SOLUTION   We start with the four vertices a, b, c, and d. The diagonal entries indicate that vertices a and b have loops. The only vertices that are connected are vertices b and c with a single edge, and vertices a and d with 2 edges. Figure 43 shows a graph with these properties.

Perfect Communication Model

In a perfect communication model, we assume the existence of a group of cities a, b, c, . . . and a mode of communication between them such as highways or long distance telephone lines. In such a model, two cities may or may not communicate with each other, and there may be more than one line connecting two cities. The following example illustrates this.

EXAMPLE 3   Representing Telephone Lines by a Matrix

Connecting four cities, there exist direct telephone lines as depicted in the graph in Figure 44. The number of lines directly connecting a with b is one; directly connecting

Figure 44

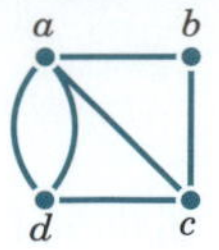

a with c is one; directly connecting a and d is two; and one line each connects b with c and d with c. We can represent the graph of direct connections by the matrix

$$\begin{array}{c} \\ a \\ b \\ c \\ d \end{array} \begin{array}{c} \begin{array}{cccc} a & b & c & d \end{array} \\ \begin{bmatrix} 0 & 1 & 1 & 2 \\ 1 & 0 & 1 & 0 \\ 1 & 1 & 0 & 1 \\ 2 & 0 & 1 & 0 \end{bmatrix} \end{array}$$

in which a zero is used to denote the number of connections of a city with itself.

If $A = [a_{ij}]$ is the matrix representation of a perfect communication model, then $a_{ij} = a_{ji}$. Such matrices are called **symmetric.**

Matrix Representation of a Simple Graph

When we consider simple graphs, that is, graphs without parallel edges and without loops, we notice that the matrix representation of such a graph is always a matrix of 0's and 1's. Such a matrix is called an **incidence matrix.**

Business Communication Model

Figure 45

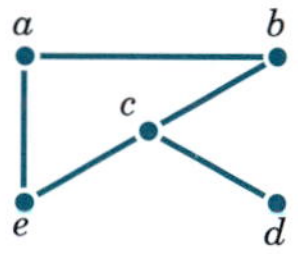

The business communication model in Figure 45 illustrates a situation in which a, b, c, d and e are officials of a business structure and the line between them indicates a communication relation between them. For example, a line might mean that memos are exchanged. The matrix representation for the graph in Figure 45 is

$$\begin{array}{c} \\ a \\ b \\ c \\ d \\ e \end{array} \begin{array}{c} \begin{array}{ccccc} a & b & c & d & e \end{array} \\ \begin{bmatrix} 0 & 1 & 0 & 0 & 1 \\ 1 & 0 & 1 & 0 & 0 \\ 0 & 1 & 0 & 1 & 1 \\ 0 & 0 & 1 & 0 & 0 \\ 1 & 0 & 1 & 0 & 0 \end{bmatrix} \end{array}$$

> It is characteristic of the matrix representing business communication models (simple graphs) that they are symmetric incidence matrices.

Matrix Representation of a Digraph

To determine the matrix representative of a digraph, we use a square matrix in which the entry in row i column j is the number of directed edges starting at the ith vertex and ending at the jth vertex. A 0 denotes that the vertices are not joined by an edge.

Figure 46

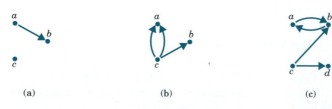

(a)   (b)   (c)

For example, the graphs in Figure 46 have the matrix representations

$$
\begin{array}{c}
\begin{array}{ccc} a & b & c \end{array} \\
\begin{array}{c} a \\ b \\ c \end{array}
\left[\begin{array}{ccc}
0 & 1 & 0 \\
0 & 0 & 0 \\
0 & 0 & 0
\end{array}\right],
\end{array}
\qquad
\begin{array}{c}
\begin{array}{ccc} a & b & c \end{array} \\
\begin{array}{c} a \\ b \\ c \end{array}
\left[\begin{array}{ccc}
0 & 0 & 0 \\
0 & 0 & 0 \\
2 & 1 & 0
\end{array}\right]
\end{array}
\qquad \text{and} \qquad
\begin{array}{c}
\begin{array}{cccc} a & b & c & d \end{array} \\
\begin{array}{c} a \\ b \\ c \\ d \end{array}
\left[\begin{array}{cccc}
0 & 1 & 0 & 0 \\
1 & 0 & 0 & 0 \\
0 & 1 & 0 & 1 \\
0 & 0 & 0 & 0
\end{array}\right]
\end{array}
$$

(a)   (b)   (c)

Dominance

Figure 47

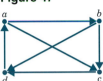

Here we study organizations of people in which we assume that for every pair of people one of them either dominates (has influence over) or is dominated by (is influenced by) the other. If we use a vertex to represent a person and an edge to represent dominance or influence so that $a \rightarrow b$ means that a has influence over b, then organizations with this property can be illustrated with a directed graph, called a **dominance digraph.**

Such a situation occurs in round-robin tournaments in which each team must play every other team once and no ties are allowed. This means that for every pair of teams, one wins and the other loses. For example, in Figure 47 we have a group of four teams a, b, c, and d, in which

team a has won over b and c

team b has won over c and d

team c has won over d

team d has won over a

where $a \rightarrow b$ indicates a has won over b. The matrix representation for this digraph is

$$
\begin{array}{c}
\begin{array}{cccc} a & b & c & d \end{array} \\
\begin{array}{c} a \\ b \\ c \\ d \end{array}
\left[\begin{array}{cccc}
0 & 1 & 1 & 0 \\
0 & 0 & 1 & 1 \\
0 & 0 & 0 & 1 \\
1 & 0 & 0 & 0
\end{array}\right]
\end{array}
$$

Other situations that give rise to a dominance digraph occur in groups in which investigation of every pair of people reveals the dominant or influential one. For example, in a group of four people a, b, c, and d, we might write all possible pairings and determine which in each pair is the dominant one. The following table illustrates this.

Pair	ab	ac	ad	bc	bd	cd
Dominant One	a	a	d	b	b	c

A dominance digraph can be used to represent this situation if by $a \rightarrow b$ we mean a dominates b in the pairing ab. See Figure 47. The matrix representation for this digraph is

$$
\begin{array}{c}
\begin{array}{cccc} a & b & c & d \end{array} \\
\begin{array}{c} a \\ b \\ c \\ d \end{array}
\left[\begin{array}{cccc}
0 & 1 & 1 & 0 \\
0 & 0 & 1 & 1 \\
0 & 0 & 0 & 1 \\
1 & 0 & 0 & 0
\end{array}\right]
\end{array}
$$

Every situation that gives rise to a dominance digraph has the property that for every pair of vertices either $a \rightarrow b$ or $b \rightarrow a$ but not both. This means the matrix representation of a dominance digraph has entries that are either 0's or 1's, that is, it is an incidence matrix. Moreover, in a dominance digraph, if a dominates b, then b cannot dominate a. Thus, if $A = [a_{ij}]$ is the incidence matrix for a dominance digraph, then $a_{ij} = 0$ implies $a_{ji} = 1$ and $a_{ij} = 1$ implies $a_{ji} = 0$, $i \neq j$. Such an incidence matrix is said to be **asymmetric.**

> The matrix representation of a dominance digraph is always an asymmetric incidence matrix.

EXAMPLE 4   Studying Dominance in a Gang

Figure 48

A gang consisting of five members a, b, c, d, and e is interviewed by a sociologist and the resulting dominance digraph is shown in Figure 48. Determine the incidence matrix for this digraph.

SOLUTION   We notice from the digraph that member a dominates b, c, and e; b dominates d and e; c dominates b, d, and e; d dominates a and e; and e dominates no one.

The incidence matrix for this situation is

$$A = \begin{array}{c} \\ a \\ b \\ c \\ d \\ e \end{array} \begin{array}{c} \begin{array}{ccccc} a & b & c & d & e \end{array} \\ \left[\begin{array}{ccccc} 0 & 1 & 1 & 0 & 1 \\ 0 & 0 & 0 & 1 & 1 \\ 0 & 1 & 0 & 1 & 1 \\ 1 & 0 & 0 & 0 & 1 \\ 0 & 0 & 0 & 0 & 0 \end{array} \right] \end{array}$$

From the sum of the entries of the matrix in Example 4 in a row, we obtain the number of one-stage dominances of a person. For example, the sum of the entries in the first row is 3, so a dominates three people in one stage; c also dominates three people in one stage. Any person who dominates everyone in one stage is called a **consensus leader** of the group.

If there is no person that dominates everyone in one stage, as in our example, we look to find the person who dominates the most people in one stage, and call this individual a **leader.** In our example, there are two people, a and c, who dominate the most people in one stage. Which one should be designated a leader?

To answer this, we introduce the notion of **two-stage dominance.** Notice that a dominates b and b dominates d. Here a does not dominate d. However, since $a \rightarrow b$ and $b \rightarrow d$, we say that a has two-stage dominance or two-stage influence over d. In this case we write $a \rightarrow b \rightarrow d$ to indicate that a dominates d in two stages. In the case of a tie for the most number of one-stage dominances, we agree that a person who dominates the most people in one or two stages will be designated as a **leader** of the group.

It is important to realize that there is no transitive property for dominance. Thus if a dominates b and b dominates c, it may or may not follow that a dominates c.

But how can the number of two-stage dominances of one individual be determined? As we shall see, the square matrix A^2 of the incidence matrix A plays a major role in answering this question. For the incidence matrix of Example 4,

$$A^2 = \begin{array}{c} \\ a \\ b \\ c \\ d \\ e \end{array} \begin{array}{c} \begin{matrix} a & b & c & d & e \end{matrix} \\ \begin{bmatrix} 0 & 1 & 0 & 2 & 2 \\ 1 & 0 & 0 & 0 & 1 \\ 1 & 0 & 0 & 1 & 2 \\ 0 & 1 & 1 & 0 & 1 \\ 0 & 0 & 0 & 0 & 0 \end{bmatrix} \end{array}$$

How should the matrix A^2 be interpreted?

From the incidence matrix A, we see that a dominates c and c dominates b so that a has two-stage dominance over b as the matrix A^2 indicates. The entry "2" in row a column d indicates that a has two-stage dominance over d in two ways, namely $a \rightarrow b \rightarrow d$ and $a \rightarrow c \rightarrow d$. Thus, *the square of an incidence matrix gives the number of two-stage dominances for each entry.*

If we add the matrices A and A^2, the total number of ways that people can be dominated by a in either one or two stages is the sum of the entries in row a, while the sum of the entries in the column a gives the total number of ways people can dominate a in one or two stages.

$$A + A^2 = \begin{array}{c} \\ a \\ b \\ c \\ d \\ e \end{array} \begin{array}{c} \begin{matrix} a & b & c & d & e \end{matrix} \\ \begin{bmatrix} 0 & 2 & 1 & 2 & 3 \\ 1 & 0 & 0 & 1 & 2 \\ 1 & 1 & 0 & 2 & 3 \\ 1 & 1 & 1 & 0 & 2 \\ 0 & 0 & 0 & 0 & 0 \end{bmatrix} \end{array}$$

Thus we see that a can dominate in 8 ways in one or two stages; b can dominate in 4 ways in one or two stages; c can dominate in 7 ways in one or two stages; and d can dominate in 5 ways in one or two stages. Person e dominates no one in one or two stages. Also a can be dominated in 3 ways in one or two stages; b can be dominated in 4 ways; c in 2 ways; d in 5 ways; and e in 10 ways, each in one or two stages.

In order to determine who the leader of the gang is (recall that a and c each dominated four people in one stage), we might call a the leader since he dominates the most people in one or two stages.

If we cube the matrix A, obtaining the matrix A^3, namely

$$A^3 = \begin{bmatrix} 2 & 0 & 0 & 1 & 3 \\ 0 & 1 & 1 & 0 & 1 \\ 1 & 1 & 1 & 0 & 2 \\ 0 & 1 & 0 & 2 & 2 \\ 0 & 0 & 0 & 0 & 0 \end{bmatrix}$$

we have the number of three-stage dominances in the group. The matrix A^4 gives the number of four-stage dominances, and so on. In general, we state the following result:

Theorem VII

Let A be an incidence matrix of dimension $n \times n$, and let A^k be the kth power of A. Then the entry in row i and column j, a_{ij}, is the number of k-stage dominances of person i over person j.

For example, the matrix A^3 above indicates that person a has three-stage dominance over person e in exactly 3 ways, namely $a \rightarrow b \rightarrow d \rightarrow e$, $a \rightarrow c \rightarrow d \rightarrow e$, and $a \rightarrow c \rightarrow b \rightarrow e$. Notice that $a \rightarrow b \rightarrow e \rightarrow e$ is not a three-stage dominance.

Theorem VIII

Let A be an incidence matrix of dimension $n \times n$. The sum of the entries in the ith row of the matrix

$$A + A^2 + \cdots + A^k$$

gives the total number of ways the ith person can dominate in one, two, . . . , up to k stages.

EXAMPLE 5   Determining the Winner in a Tournament

Figure 49

In a group of five teams a, b, c, d, e, suppose we know that $a \rightarrow b$ (a wins over b), $a \rightarrow c$, $b \rightarrow c$, $b \rightarrow d$, $b \rightarrow e$, $c \rightarrow d$, $d \rightarrow a$, $e \rightarrow a$, $e \rightarrow c$, and $e \rightarrow d$. Draw a digraph for this situation. Who should be declared winner of this tournament?

SOLUTION   The digraph is given in Figure 49. The incidence matrix is

$$
A = \begin{array}{c} \\ a \\ b \\ c \\ d \\ e \end{array}
\begin{array}{c} \begin{array}{ccccc} a & b & c & d & e \end{array} \\
\left[\begin{array}{ccccc}
0 & 1 & 1 & 0 & 0 \\
0 & 0 & 1 & 1 & 1 \\
0 & 0 & 0 & 1 & 0 \\
1 & 0 & 0 & 0 & 0 \\
1 & 0 & 1 & 1 & 0
\end{array} \right] \end{array}
$$

The number of games each team won (dominates) is the sum of entries of their row. Thus, team a won 2 games, b won 3, c won 1, d won 1, and e won 3. Since there is a tie for first between teams b and e, there is no clear winner.

To see which team should be declared the winner, we might look at the matrix A^2 to determine the number of two-stage wins for each team. Now

$$
A^2 = \begin{array}{c} \\ a \\ b \\ c \\ d \\ e \end{array}
\begin{array}{c} \begin{array}{ccccc} a & b & c & d & e \end{array} \\
\left[\begin{array}{ccccc}
0 & 0 & 1 & 2 & 1 \\
2 & 0 & 1 & 2 & 0 \\
1 & 0 & 0 & 0 & 0 \\
0 & 1 & 1 & 0 & 0 \\
1 & 1 & 1 & 1 & 0
\end{array} \right] \end{array}
$$

Thus, team b has a total of 5 two-stage wins, while team e has a total of 4 two-stage wins. Also team a has a total of 4 two-stage wins while team d has only 2. Thus, we might award the top prize to team b, followed by team e, team a, team d, and lastly team c.

From Theorem VII, we know that a nonzero entry in row i, column j of the kth power of the incidence matrix A implies that we can go from person i to person j using at most k stages (steps). If we eliminate all repeated vertices from this connection, we then have a path from i to j.

Referring to Theorem VIII and applying the result to simple graphs, we can state the following result, which serves as a test for determining whether a graph is connected or disconnected.

Theorem IX

Let G be a simple graph with n vertices and incidence matrix A. Then G is connected if and only if every entry of the matrix

$$A + A^2 + \cdots + A^{n-1}$$

is nonzero.

EXAMPLE 6   Studying Internal Organization Structure

Figure 50

Figure 50 depicts the internal communication graph as it exists in an organization headed by five officials, which we will denote by a, b, c, d, and e. Find the incidence matrix of this graph, and determine whether the graph is connected.

SOLUTION   The incidence matrix of the given graph is:

$$
A = \begin{array}{c c}
 & \begin{array}{c c c c c} a & b & c & d & e \end{array} \\
\begin{array}{c} a \\ b \\ c \\ d \\ e \end{array} &
\left[\begin{array}{c c c c c}
0 & 1 & 1 & 1 & 0 \\
1 & 0 & 1 & 0 & 1 \\
1 & 1 & 0 & 1 & 0 \\
1 & 0 & 1 & 0 & 0 \\
0 & 1 & 0 & 0 & 0
\end{array} \right]
\end{array}
$$

To determine whether this graph is connected, we must find

$$A + A^2 + A^3 + A^4$$

Using a graphing calculator, we find

$$
A + A^2 + A^3 + A^4 = \begin{array}{c c}
 & \begin{array}{c c c c c} a & b & c & d & e \end{array} \\
\begin{array}{c} a \\ b \\ c \\ d \\ e \end{array} &
\left[\begin{array}{c c c c c}
23 & 18 & 23 & 16 & 8 \\
18 & 20 & 18 & 16 & 6 \\
23 & 18 & 23 & 16 & 8 \\
16 & 16 & 16 & 14 & 4 \\
8 & 6 & 8 & 4 & 4
\end{array} \right]
\end{array}
$$

Since no zero entry appears, the graph in Figure 50 is connected.

Theorem IX can be restated in the language of directed graphs.

> **Theorem X**
>
> Let G be a digraph with n vertices and incidence matrix A. Then G is strongly connected if and only if every entry of the matrix
>
> $$A + A^2 + \cdots + A^{n-1}$$
>
> is nonzero.

EXAMPLE 7   Determining When a Digraph Is Strongly Connected

Figure 51

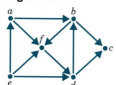

Determine whether the digraph given in Figure 51 is strongly connected by using Theorem X. Are there any three-step paths from e to c? If so, how many?

SOLUTION   The incidence matrix for the digraph is given by

$$
A = \begin{array}{c} \\ a \\ b \\ c \\ d \\ e \\ f \end{array}
\begin{array}{c} \begin{array}{cccccc} a & b & c & d & e & f \end{array} \\
\left[\begin{array}{cccccc}
0 & 1 & 0 & 0 & 0 & 1 \\
0 & 0 & 1 & 0 & 0 & 1 \\
0 & 0 & 0 & 0 & 0 & 0 \\
0 & 1 & 1 & 0 & 0 & 0 \\
1 & 0 & 0 & 1 & 0 & 1 \\
0 & 0 & 0 & 1 & 0 & 0
\end{array}\right] \end{array}
$$

To check for connectivity, we find the matrix $A + A^2 + A^3 + A^4 + A^5$:

$$
A + A^2 + A^3 + A^4 + A^5 = \left[\begin{array}{cccccc}
0 & 3 & 5 & 3 & 0 & 4 \\
0 & 1 & 3 & 2 & 0 & 2 \\
0 & 0 & 0 & 0 & 0 & 0 \\
0 & 2 & 4 & 1 & 0 & 2 \\
1 & 6 & 9 & 6 & 0 & 6 \\
0 & 2 & 3 & 2 & 0 & 1
\end{array}\right]
$$

Therefore, according to Theorem X, this digraph is not strongly connected. In fact, a closer inspection of the incidence matrix A reveals that row c contains only zeros, so we can never leave vertex c. At the same time we can never enter vertex e (the column corresponding to e contains only zeros).

Calculation of A^3 reveals that there are 3 three-step paths from e to c.

$$
A^3 = \begin{array}{c} \\ a \\ b \\ c \\ d \\ e \\ f \end{array}
\begin{array}{c} \begin{array}{cccccc} a & b & c & d & e & f \end{array} \\
\left[\begin{array}{cccccc}
0 & 1 & 1 & 1 & 0 & 0 \\
0 & 1 & 1 & 0 & 0 & 0 \\
0 & 0 & 0 & 0 & 0 & 0 \\
0 & 0 & 0 & 1 & 0 & 0 \\
0 & 1 & 3 & 1 & 0 & 2 \\
0 & 0 & 1 & 0 & 0 & 1
\end{array}\right] \end{array}
$$

EXERCISE 13.6 Answers to odd-numbered problems begin on page AN-54.

1. For the following graphs, write the corresponding matrices.

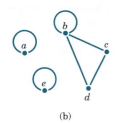
(a)

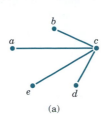
(b)

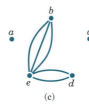
(c)

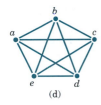
(d)

2. For the following matrices, find the corresponding graphs.

(a)
$$\begin{array}{c} \\ a \\ b \\ c \\ d \end{array} \begin{array}{cccc} a & b & c & d \\ \left[\begin{array}{cccc} 1 & 2 & 0 & 1 \\ 2 & 0 & 1 & 0 \\ 0 & 1 & 0 & 3 \\ 1 & 0 & 3 & 0 \end{array}\right] \end{array}$$

(b)
$$\begin{array}{c} \\ a \\ b \\ c \\ d \end{array} \begin{array}{cccc} a & b & c & d \\ \left[\begin{array}{cccc} 0 & 1 & 0 & 0 \\ 1 & 0 & 0 & 0 \\ 0 & 0 & 0 & 0 \\ 0 & 0 & 0 & 1 \end{array}\right] \end{array}$$

(c)
$$\begin{array}{c} \\ a \\ b \\ c \end{array} \begin{array}{ccc} a & b & c \\ \left[\begin{array}{ccc} 2 & 0 & 0 \\ 0 & 1 & 1 \\ 0 & 1 & 1 \end{array}\right] \end{array}$$

(d)
$$\begin{array}{c} \\ a \\ b \\ c \\ d \\ e \end{array} \begin{array}{ccccc} a & b & c & d & e \\ \left[\begin{array}{ccccc} 1 & 1 & 1 & 0 & 1 \\ 1 & 0 & 2 & 0 & 0 \\ 1 & 2 & 0 & 1 & 0 \\ 0 & 0 & 1 & 0 & 0 \\ 1 & 0 & 0 & 0 & 1 \end{array}\right] \end{array}$$

3. For the following digraphs, write the corresponding matrices.

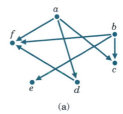
(a)

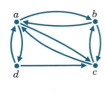
(b)

(c)

(d)

4. For the following matrices, find the corresponding digraphs.

(a)
$$\begin{array}{c} \\ a \\ b \\ c \\ d \end{array} \begin{array}{cccc} a & b & c & d \\ \left[\begin{array}{cccc} 0 & 1 & 1 & 1 \\ 0 & 0 & 1 & 1 \\ 0 & 0 & 0 & 1 \\ 0 & 0 & 0 & 0 \end{array}\right] \end{array}$$

(b)
$$\begin{array}{c} \\ a \\ b \\ c \\ d \end{array} \begin{array}{cccc} a & b & c & d \\ \left[\begin{array}{cccc} 0 & 1 & 0 & 0 \\ 1 & 0 & 1 & 0 \\ 0 & 0 & 0 & 0 \\ 1 & 0 & 1 & 0 \end{array}\right] \end{array}$$

(c)
$$\begin{array}{c} \\ a \\ b \\ c \end{array} \begin{array}{ccc} a & b & c \\ \left[\begin{array}{ccc} 0 & 1 & 0 \\ 1 & 0 & 1 \\ 0 & 1 & 0 \end{array}\right] \end{array}$$

(d)
$$\begin{array}{c} \\ a \\ b \\ c \\ d \\ e \end{array} \begin{array}{ccccc} a & b & c & d & e \\ \left[\begin{array}{ccccc} 0 & 1 & 1 & 0 & 1 \\ 0 & 0 & 0 & 1 & 1 \\ 0 & 1 & 0 & 1 & 1 \\ 1 & 0 & 0 & 0 & 1 \\ 0 & 0 & 0 & 0 & 0 \end{array}\right] \end{array}$$

5. Four people, a, b, c, d, live in an apartment building. If a hears a piece of gossip, he will pass it on to b and d; b passes gossip to c; c passes gossip on to a; d never gossips. Write a directed graph for this situation. What is the matrix representation?

6. Write a communication matrix for the group: Caryl, Tami, and Laura, if Laura communicates with Caryl and Tami and Caryl communicates with Tami. Find the total number of ways Laura can communicate with Caryl through no other person, or one other person. In how many ways can Caryl get one-stage feedback information?

7. For the dominance matrix
$$A = \begin{array}{c} \\ a \\ b \\ c \end{array} \begin{array}{ccc} a & b & c \\ \left[\begin{array}{ccc} 0 & 1 & 1 \\ 0 & 0 & 1 \\ 0 & 0 & 0 \end{array}\right] \end{array}$$

find the two-stage dominances and interpret your answer. Interpret $A + A^2$.

8. For the dominance matrix
$$A = \begin{array}{c} \\ a \\ b \\ c \\ d \end{array} \begin{array}{cccc} a & b & c & d \\ \left[\begin{array}{cccc} 0 & 1 & 0 & 0 \\ 0 & 0 & 1 & 1 \\ 1 & 0 & 0 & 0 \\ 1 & 0 & 1 & 0 \end{array}\right] \end{array}$$

find the two-stage dominances. What total do a and c dominate in one or two stages? What total are dominated by d in two stages?

9. In a basketball tournament composed of seven teams, the final standings showed that

Team	Won	Lost
a	4	2
b	4	2
c	4	2
d	3	3
e	3	3
f	2	4
g	1	5

It is known that

a beat d, e, f, g

b beat a, c, d, f

c beat a, e, f, g

d beat c, f, g

e beat b, d, g

f beat e, g

g beat b

Who should be declared the winner?

10. In trying to determine the flavor of ice cream Mr. Polansky likes best, a survey team decides to use paired comparisons. The questionnaire establishes that Mr. Polansky would choose Vanilla over Neopolitan, Pecan, Fudge Ripple, and Cherry. Chocolate was preferred over Vanilla, Strawberry, Neopolitan, and Fudge Ripple. Strawberry was chosen over Vanilla, Pecan, Fudge Ripple, and Cherry. Pecan won out over Chocolate, Neopolitan, and Cherry. Fudge Ripple was preferred to Pecan and Cherry; Neopolitan was preferred over Fudge Ripple, Cherry, and Strawberry. Finally, Cherry was chosen over Chocolate. What flavor does Mr. Polansky like best?

11. (a) Find the matrix representation, A, of the graph below.

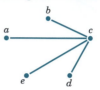

   (b) Find the square of the matrix A. What does it represent?

   (c) Find the number of paths of length 2 from c to c.

12. (a) Find the matrix representation, A, of the graph below.

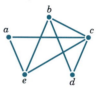

   (b) Find the cube of the matrix A. What does it represent?

   (c) Find the number of paths of length 3 from a to b.

CHAPTER REVIEW

IMPORTANT TERMS AND CONCEPTS

vertex 566
edge 566
graph 566, 567
parallel edges 567
loop 567
isolated vertex 567
simple graph 568
degree 569
complete graph 571
subgraph 572
complement 572
path 573
trivial 573
simple path 573
circuit 573
simple circuit 573
connected graph 574

Eulerian circuit 576
Fleury's algorithm 578
Hamiltonian circuit 580
tree 584
trivial tree 584
identical trees 584
rooted tree 585
child 585
binary tree 585
leaf 585
internal vertex 585
descendants 585
subtree 586
binary search tree 589
spanning tree 591
directed graph 592
digraph 592

arc 592
initial point 592
terminal point 592
underlying graph 592
indegree 593
outdegree 593
directed path 594
reachable 594
strongly connected 595
weakly connected 595
orientation 596
bridge 596
depth-first search 596
depth-first numbering 596
tournament 600
perfect communication model 601
dominance graph 603

IMPORTANT FORMULA

$\deg(v_1) + \deg(v_2) + \cdots + \deg(v_n) = 2 \cdot$ (The number of edges in G)

TRUE–FALSE ITEMS   Answers are on page AN-55.

T_____ F_____ **1.** The sum of the degrees of the vertices of a graph may be any number.

T_____ F_____ **2.** If a graph G has n vertices and $n - 1$ edges, then G is necessarily connected.

T_____ F_____ **3.** If the degree of every vertex in a graph G is even, then G must contain an Eulerian circuit.

T_____ F_____ **4.** A Hamiltonian circuit need not include every edge.

T_____ F_____ **5.** No tree can contain a circuit.

T_____ F_____ **6.** In a strongly connected digraph D any vertex in D is reachable from any other vertex in D.

FILL IN THE BLANKS   Answers are on page AN-55.

1. A graph consists of a set V of _____ and a set E of _____ .

2. G is a simple graph if G has no _____ and no _____ edges.

3. A circuit that contains every edge is said to be _____ .

4. Trees are graphs that are _____ -free.

5. In a digraph every edge has a _____ .

6. A graph G is orientable if its edges can be _____ in such a way that the resulting digraph is _____ _____ .

REVIEW EXERCISES   Answers to odd-numbered problems begin on page AN-55.

1. Can we draw a simple graph with vertices v_1, v_2, v_3, and v_4 of degrees 2, 2, 3, and 3, respectively? Justify your answer.

2. Draw the complete graph K_8. (See Exercise 13.1, Problem 12.)

3. Consult the figure.

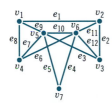

Find

(a) A simple path between v_1 and v_3 containing v_7
(b) A circuit starting at v_1 and containing v_7 and v_3
(c) A simple circuit starting at v_1 and containing v_7

4. How many simple circuits, that contain all vertices, are there in K_4?

5. The figure shows the outline of a city built on both banks of the river and on the three islands connected to the banks and to each other by the bridges shown. Can a tourist make a round trip through all land masses, crossing each bridge once? Justify your answer.

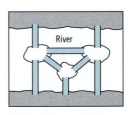

6. Does the graph in the figure contain a Hamiltonian circuit? Justify your answer.

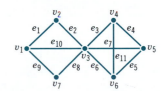

7. Use Fleury's algorithm to find an Eulerian circuit for the graph shown in the figure.

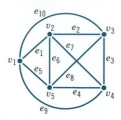

8. (a) Use a binary tree to store alphabetically the elements of the following list: one, two, three, four, five, six.
   (b) Use the tree in (a) to search for the word "ten." If this word is not in the list, add it to the list at the right position without losing the alphabetical order.

9. Use a binary tree to represent the following algebraic expression

$$[(a - b)/d] + (d - 1)/a$$

10. Draw a rooted tree that has 3 internal vertices and 2 leaves.

11. Use the graph in the figure to find two spanning trees.

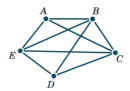

12. Draw a digraph with 5 vertices where each vertex has outdegree 2.

13. Is the digraph in the figure strongly connected? Why or why not?

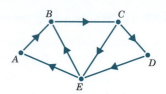

14. Determine whether or not the graph shown in the figure is orientable. If it is, use the procedure of Section 13.5 to orient the edges in such a way that the resulting graph is strongly connected.

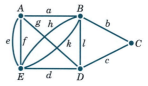

Appendix A

Review Topics from Algebra and Geometry*

A.1 Basic Algebra

A.2 Exponents and Logarithms

A.3 Geometric Sequences

A.1 BASIC ALGEBRA

Classification of Numbers

It is helpful to classify the various kinds of numbers. The **counting numbers,** or **natural numbers,** are the numbers 1, 2, 3, 4, ... (The three dots, called an **ellipsis,** indicate that the pattern continues indefinitely.) As their name implies, these numbers are often used to count things. For example, there are 26 letters in our alphabet; there are 100 cents in a dollar. The **whole numbers** are the numbers 0, 1, 2, 3, . . . , that is, the counting numbers together with 0.

> **Integers**
> The **integers** are the numbers . . . , $-3, -2, -1, 0, 1, 2, 3, \ldots$

These numbers prove useful in many situations. For example, if your checking account has $10 in it and you write a check for $15, you can represent the current balance as $-\$5$.

Notice that the counting numbers are included among the whole numbers. Each time we expand a number system, such as from the whole numbers to the integers, we do so in order to be able to handle new, and usually more complicated, problems. Thus the integers allow us to solve problems requiring both positive and negative counting numbers, such as profit/loss, height above/below sea level, temperature above/below 0°F, and so on.

*Based on material from *College Algebra,* 5th ed., by Michael Sullivan. Used here with the permission of the author and Prentice-Hall, Inc.

But integers alone are not sufficient for *all* problems. For example, they do not answer the questions, "What part of a dollar is 38 cents?" or "What part of a pound is 5 ounces?" To answer such questions, we enlarge our number system to include *rational numbers.* For example, $\frac{38}{100}$ answers the question, "What part of a dollar is 38 cents?" and $\frac{5}{16}$ answers the question, "What part of a pound is 5 ounces?"

Rational Number

A **rational number** is a number that can be expressed as a quotient a/b of two integers. The integer a is called the **numerator,** and the integer b, which cannot be 0, is called the **denominator.**

Examples of rational numbers are $\frac{3}{4}, \frac{5}{2}, \frac{0}{4}, -\frac{2}{3}$, and $\frac{100}{3}$. Since $a/1 = a$ for any integer a, it follows that the rational numbers contain the integers as a special case.

Real Numbers

Rational numbers may be represented as **decimals.** For example, the rational numbers $\frac{3}{4}, \frac{5}{2}, -\frac{2}{3}$, and $\frac{7}{66}$ may be represented as decimals by merely carrying out the indicated division:

$$\frac{3}{4} = 0.75 \qquad \frac{5}{2} = 2.5 \qquad -\frac{2}{3} = -0.666\ldots \qquad \frac{7}{66} = 0.1060606\ldots$$

Notice that the decimal representations of $\frac{3}{4}$ and $\frac{5}{2}$ terminate, or end. The decimal representations of $-\frac{2}{3}$ and $\frac{7}{66}$ do not terminate, but they do exhibit a pattern of repetition. For $-\frac{2}{3}$ the 6 repeats indefinitely; for $\frac{7}{66}$ the block 06 repeats indefinitely. It can be shown that every rational number may be represented by a decimal that either terminates or is nonterminating with a repeating block of digits, and vice versa.

On the other hand, there are decimals that do not fit into either of these categories. Such decimals represent **irrational numbers.** For example, the decimal 0.12345678910111213141.4. . . , in which we write the positive integers successively one after the other, will neither repeat (think about it) nor terminate, and so, represents an irrational number. Thus every irrational number may be represented by a decimal that neither repeats nor terminates.

Irrational numbers occur naturally. For example, consider the isosceles right triangle whose legs are each of length 1; see Figure 1. The number that equals the length of the hypotenuse is the positive number whose square is 2. It can be shown that this number, symbolized by $\sqrt{2}$, is an irrational number.

Also, the number that equals the ratio of the circumference C to the diameter d of any circle, denoted by the symbol π (the Greek letter pi), is an irrational number. See Figure 2.

The irrational numbers $\sqrt{2}$ and π have decimal representations that begin as follows:

$$\sqrt{2} = 1.414213\ldots \qquad \pi = 3.14159\ldots$$

In practice irrational numbers are generally represented by approximations. For example, using the symbol $\approx$ (read as "approximately equal to"), we can write

$$\sqrt{2} \approx 1.4142 \qquad \pi \approx 3.1416$$

Other examples of irrational numbers are $\sqrt{3}, \sqrt{5}, \sqrt{7}, \sqrt[3]{2}, \sqrt[3]{3}$, and so on.

Figure 1

Figure 2 $\pi = C/d$

> **Real Numbers**
>
> Together, the rational numbers and irrational numbers form the **real numbers.**

Thus every decimal may be represented by a real number (either rational or irrational) and, conversely, every real number may be represented by a decimal. It is this feature of real numbers that gives them their practicality. In the physical world many changing quantities such as the length of a heated rod, the velocity of a falling object, and so on, are assumed to pass through every possible magnitude from the initial one to the final one as they change. Real numbers in the form of decimals provide a convenient way to measure such quantities as they change. Figure 3 shows the relationship of various types of numbers.

Figure 3

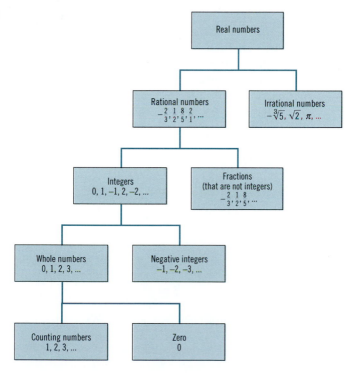

Properties of Real Numbers

As an aid to your review of real numbers, we list below several important properties and notations. The letters a, b, c, d represent real numbers; any exceptions will be noted as they occur.

1. *Commutative Properties*
   (a) $a + b = b + a$ (b) $a \cdot b = b \cdot a$
2. *Associative Properties*
   (a) $a + b + c = (a + b) + c = a + (b + c)$
   (b) $a \cdot b \cdot c = (a \cdot b) \cdot c = a \cdot (b \cdot c)$

3. *Distributive Property*
$$a \cdot (b + c) = (a \cdot b) + (a \cdot c)$$

4. *Arithmetic of Ratios*

   (a) $\dfrac{a}{b} = a \cdot \dfrac{1}{b}$ $b \neq 0$

   (b) $\dfrac{a}{b} + \dfrac{c}{d} = \dfrac{(a \cdot d) + (b \cdot c)}{b \cdot d}$ $b \neq 0, \quad d \neq 0$

   (c) $\dfrac{a}{b} \cdot \dfrac{c}{d} = \dfrac{a \cdot c}{b \cdot d}$ $b \neq 0, \quad d \neq 0$

   (d) $\dfrac{a}{b} \div \dfrac{c}{d} = \dfrac{a}{b} \cdot \dfrac{d}{c} = \dfrac{a \cdot d}{b \cdot c}$ $b \neq 0, \quad c \neq 0, \quad d \neq 0$

5. *Rules for Division*

   $$0 \div a = 0 \quad \text{or} \quad \frac{0}{a} = 0 \quad \text{for any real number } a \text{ different from } 0$$

   $$a \div a = 1 \quad \text{or} \quad \frac{a}{a} = 1 \quad \text{for any real number } a \text{ different from } 0$$

   Note: Division by zero is not allowed. One reason is to avoid the following difficulty: $2/0 = x$ means to find x such that $0 \cdot x = 2$. But $0 \cdot x = 0$ for all x, so there is no number x such that $2/0 = x$.

6. *Cancellation Properties*

   (a) If $a \cdot c = b \cdot c$ and c is not 0, then $a = b$.

   (b) If b and c are not 0, then $\dfrac{a \cdot c}{b \cdot c} = \dfrac{a}{b}$.

7. *Product Law*
   If $a \cdot b = 0$, then either $a = 0$ or $b = 0$.

8. *Rules of Signs*

   (a) $a \cdot (-b) = -(a \cdot b)$ (b) $(-a) \cdot b = -(a \cdot b)$

   (c) $(-a) \cdot (-b) = a \cdot b$ (d) $-(-a) = a$

9. *Agreement: Notations*

   (a) Given $a \cdot b + c$ or $c + a \cdot b$, we agree to multiply $a \cdot b$ first, and then add c.

   (b) A *mixed number* $3\frac{5}{8}$ means $3 + \frac{5}{8} = 3.625$; 3 *times* $\frac{5}{8}$ is written as $3(\frac{5}{8})$ or $(3)(\frac{5}{8})$ or $3 \cdot \frac{5}{8} = \frac{15}{8} = 1\frac{7}{8}$.

10. *Exponents*
    For any positive integer n and any real number x we define

    $$x^1 = x, \quad x^2 = x \cdot x, \quad \ldots, \quad x^n = \underbrace{x \cdot x \cdot \ldots \cdot x}_{n \text{ factors}}$$

    For any real number $x \neq 0$ we define

    $$x^0 = 1, \quad x^{-1} = \frac{1}{x}, \quad x^{-2} = \frac{1}{x^2}, \quad \ldots, \quad x^{-n} = \frac{1}{x^n}$$

    A review of exponents and logarithms may be found in Section A.2.

The Real Number Line

It can be shown that there is a one-to-one correspondence between real numbers and points on a line. That is, every real number corresponds to a point on the line and, conversely, each point on the line has a unique real number associated with it. We establish this correspondence of real numbers with points on a line in the following manner.

We start with a line that is, for convenience, drawn horizontally. We pick a point on the line and label it O, for **origin.** Then we pick another point some fixed distance to the right of O and label it U, for unit, as shown in Figure 4.

Figure 4   Horizontal line

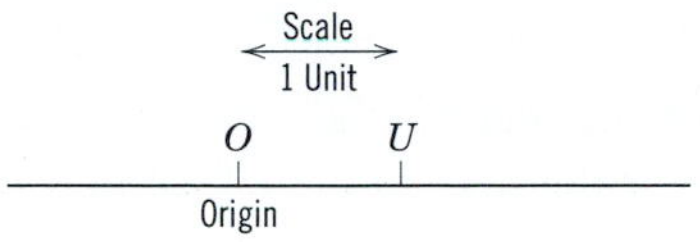

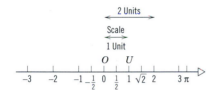

The fixed distance, which may be 1 inch, 1 centimeter, 1 light-year, or any unit distance, determines the **scale.** We associate the real number 0 with the origin O and the number 1 with the point U. Refer now to Figure 5. The point to the right of U that is twice as far from O as U is associated with the number 2. The point to the right of U that is three times as far from O as U is associated with the number 3. The point midway between O and U is assigned the number 0.5, or $\frac{1}{2}$. Corresponding points to the left of the origin O are assigned the numbers $-\frac{1}{2}$, -1, -2, -3, and so on, depending on how far they are from O. Notice in Figure 5 that we placed an arrowhead on the right end of the line to indicate the direction in which the assigned numbers increase. Figure 5 also shows the points associated with the irrational numbers $\sqrt{2}$ and π.

Figure 5   The real number line

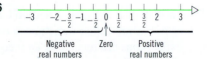

Coordinate; Real Number Line

The real number associated with a point P is called the **coordinate** of P, and the line whose points have been assigned coordinates is called the **real number line.**

The real number line divides the real numbers into three classes, as shown in Figure 6.

Figure 6

1. The **negative real numbers** are the coordinates of points to the left of the origin O.
2. The real number **zero** is the coordinate of the origin O.
3. The **positive real numbers** are the coordinates of points to the right of the origin O.

Figure 7

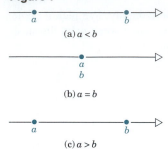

(a) $a < b$

(b) $a = b$

(c) $a > b$

Negative and positive numbers have the following multiplication properties:

1. The product of two positive numbers is a positive number.
2. The product of two negative numbers is a positive number.
3. The product of a positive number and a negative number is a negative number.

Inequality Symbols

An important property of the real number line follows from the fact that, given two numbers (points) a and b, either a is to the left of b, a equals b, or a is to the right of b. See Figure 7.

If a is to the left of b, we say "a is less than b" and write $a < b$. If a is to the right of b, we say "a is greater than b" and write $a > b$. Of course, if a equals b, we write $a = b$. If a is either less than or equal to b, we write $a \le b$. Similarly, $a \ge b$ means a is either greater than or equal to b. Collectively, the symbols $<, >, \le, \ge$ are called **inequality symbols.** A formal definition of *less than* and *greater than* follows.

> Let a and b be two real numbers. We say that *a is less than b,* written $a < b$, or, equivalently, that *b is greater than a,* written $b > a$, if there is a positive number p so that $a + p = b$.

Note that $a < b$ and $b > a$ mean the same thing. Thus it does not matter whether we write $2 < 3$ or $3 > 2$.

EXAMPLE 1 **Using Inequality Symbols**

(a) $3 < 7$ (b) $-5 > -16$ (c) $-6 < 0$
(d) $-8 < -4$ (e) $4 > -1$ (f) $8 > 0$

In Example 1(a) we conclude that $3 < 7$ either because 3 is to the left of 7 on the real number line or because $3 + 4 = 7$.

Similarly, we conclude in Example 1(b) that $-5 > -16$ either because -5 lies to the right of -16 on the real number line or because $-16 + 11 = -5$.

Look again at Example 1. You may find it useful to observe that the inequality symbol always points in the direction of the smaller number.

✍ **Now Work Problem 3**

Statements of the form $a < b$ or $b > a$ are called **strict inequalities,** while statements of the form $a \le b$ or $b \ge a$ are called **nonstrict inequalities.** An **inequality** is a statement in which two expressions are related by an inequality symbol. The expressions are referred to as the sides of the inequality.

Based on the discussion thus far, we conclude that

> $a > 0$ is equivalent to a is positive
>
> $a < 0$ is equivalent to a is negative

Thus we sometimes read $a > 0$ by saying that "a is positive." If $a \geq 0$, then either $a > 0$ or $a = 0$, and we may read this as "a is nonnegative."

Suppose a and b are two real numbers and $a < b$. We shall use the notation $a < x < b$ to mean that x is a number *between* a and b. Thus the expression $a < x < b$ is equivalent to the two inequalities $a < x$ and $x < b$. Similarly, the expression $a \leq x \leq b$ is equivalent to the two inequalities $a \leq x$ and $x \leq b$. The remaining two possibilities, $a \leq x < b$ and $a < x \leq b$, are defined analogously.

Although it is acceptable to write $3 \geq x \geq 2$, it is preferable to reverse the inequality symbols and write instead $2 \leq x \leq 3$, so that, as you read from left to right, the values go from smaller to larger.

A statement such as $2 \leq x \leq 1$ is false because there is no number x for which $2 \leq x$ and $x \leq 1$. Finally, we never mix inequality symbols, as in $2 \leq x \geq 3$.

EXAMPLE 2   Graphing Inequalities

(a) In the inequality $x > 4$, x is any number greater than 4. In Figure 8 we use an open circle at 4 to indicate that the number 4 is not part of the graph.

(b) In the inequality $4 < x \leq 6$, x is any number between 4 and 6, including 6 but excluding 4. In Figure 9 we use a solid dot at 6 to indicate that 6 is part of the graph.

Figure 8

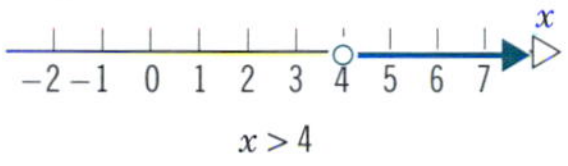

$x > 4$

Figure 9

$4 < x \leq 6$

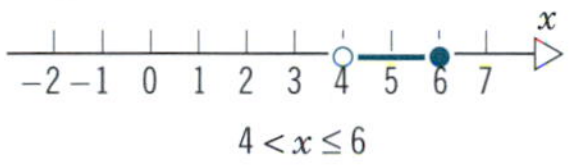

Now Work Problem 7

Inequalities have the following properties:

1. *Addition Property*

If $a \leq b$, then $a + c \leq b + c$ for any choice of c.

That is, the addition of a number to each side of an inequality will not affect the sense or direction of the inequality.

2. *Multiplication Properties*

(a) If $a \leq b$ and $c > 0$, then $a \cdot c \leq b \cdot c$.

(b) If $a \leq b$ and $c < 0$, then $a \cdot c \geq b \cdot c$.

When multiplying each side of an inequality by a number, the sense or direction of the inequality remains the same if we multiply by a positive number; it is reversed if we multiply by a negative number.

EXAMPLE 3   Using Properties of Inequalities

(a) Since $2 < 3$, then $2 + 5 < 3 + 5$, or $7 < 8$.

(b) Since $2 < 3$ and $6 > 0$, then $2 \cdot 6 < 3 \cdot 6$, or $12 < 18$.

(c) Since $2 < 3$ and $-4 < 0$, then $2 \cdot (-4) > 3 \cdot (-4)$, or $-8 > -12$.

Equations and Inequalities

An **equation** is a statement involving one or more variables and an "equals" sign (=). To **solve** an equation means to find all possible numbers that the variables can assume to make the statement true. The set of all such numbers is called the **solution.** Two equations with the same solution are called **equivalent equations.**

EXAMPLE 4   Solving an Equation

Solve the equation: $2x + 3(x + 2) = 1$

SOLUTION

$$2x + 3(x + 2) = 1$$
$$2x + (3x + 6) = 1 \qquad \text{Remove parentheses using the Distributive Property.}$$
$$(2x + 3x) + 6 = 1 \qquad \text{Use the Associative Property.}$$
$$5x + 6 = 1 \qquad \text{Combine } 2x + 3x = (2 + 3)x = 5x.$$
$$5x = -5 \qquad \text{Subtract 6 from each side.}$$
$$x = -1 \qquad \text{Divide each side by 5.}$$

Check: $2(-1) + 3(-1 + 2) = -2 + 3(1) = -2 + 3 = 1$ ■

An **inequality** is a statement involving one or more variables and one of the inequality symbols ($<$, $\leq$, $>$, $\geq$). To **solve** an inequality means to find all possible numbers that the variables can assume to make the statement true. The set of all such numbers is called the **solution.** Two inequalities with the same solution are called **equivalent inequalities.**

To find the solution of an inequality, we apply the properties for inequalities.

EXAMPLE 5   Solving an Inequality

Solve the inequality: $x + 2 \leq 3x - 5$. Graph the solution.

SOLUTION

$$x + 2 \leq 3x - 5$$
$$-2x \leq -7 \qquad \text{Subtract 2 and then } 3x \text{ from each side.}$$
$$x \geq \tfrac{7}{2} \qquad \text{Multiply by } -\tfrac{1}{2} \text{ and remember to reverse the inequality because we multiplied by a negative number.}$$

Figure 10

The solution is all real numbers to the right of $\tfrac{7}{2}$, including $\tfrac{7}{2}$. See Figure 10 for the graph of the solution.   ■

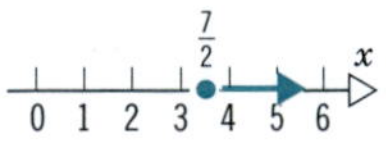

EXERCISE A.1   Answers to odd-numbered problems begin on page AN-56.

In Problems 1–6 write each statement as an inequality.

1. x is positive

2. z is negative

3. x is less than 4

4. y is greater than -6

5. x is less than or equal to 2

6. x is greater than or equal to 3

In Problems 7–16 graph the numbers x, if any, on the real number line.

7. $x \geq -1$

8. $x < 1$

9. $x \geq 4$ and $x < 6$

10. $x > 3$ and $x \leq 7$

11. $x \leq 0$ or $x < 6$

12. $x > 0$ or $x \geq 5$

13. $x \leq -2$ and $x > 1$

14. $x \geq 4$ and $x < -2$

15. $x \leq -2$ or $x > 1$

16. $x \geq 4$ or $x < -2$

In Problems 17–24 solve each equation.

17. $x + 5 = 7$

18. $x + 6 = 2$

19. $6 - x = 0$

20. $6 + x = 0$

21. $3(2 - x) = 9$

22. $5(x + 1) = 10$

23. $4x + 3 = 2x - 5$

24. $5x - 8 = 2x + 1$

In Problems 25–32 solve each inequality. Graph the solutions.

25. $x + 5 \leq 2$

26. $4 - x \geq 3$

27. $3x + 5 \geq 2$

28. $14x - 21x + 16 \leq 3x - 2$

29. $-3x + 5 \leq 2$

30. $8 - 2x \leq 5x - 6$

31. $6x - 3 \geq 8x + 5$

32. $-3x \leq 2x + 5$

A.2 EXPONENTS AND LOGARITHMS

Exponents

Integer exponents provide a convenient notation for repeated multiplication of a real number.

If a is a real number and n is a positive integer, then the **symbol a^n** represents the product of n factors of a. That is,

$$a^n = \underbrace{a \cdot a \cdot \ldots \cdot a}_{n \text{ factors}}$$

where it is understood that $a^1 = a$.

In the expression a^n, read "a raised to the power n," a is called the **base** and n is called the **exponent** or **power.**

If n is a positive integer and $a \neq 0$, then we define

$$a^{-n} = \frac{1}{a^n} \qquad a \neq 0$$

Also, we define

$$a^0 = 1 \qquad a \neq 0$$

EXAMPLE 1   Evaluating Exponents

(a) $2^3 = 2 \cdot 2 \cdot 2 = 8$ (b) $3^{-2} = \dfrac{1}{3^2} = \dfrac{1}{9}$ (c) $5^0 = 1$

 Now Work Problem 1

The **principal nth root of a real number a,** written $\sqrt[n]{a}$, is defined as follows:

(a) If $a > 0$ and n is even, then $\sqrt[n]{a}$ is the positive number x for which $x^n = a$, $x > 0$.
(b) If $a < 0$ and n is even, then $\sqrt[n]{a}$ does not exist.
(c) If n is odd, then $\sqrt[n]{a}$ is the number x for which $x^n = a$.
(d) $\sqrt[n]{0} = 0$ for any n.

If $n = 2$, we write $\sqrt{a}$ in place of $\sqrt[2]{a}$.

EXAMPLE 2   Evaluating Principal nth Roots

(a) $\sqrt[3]{8} = 2$ because $2^3 = 8$
(b) $\sqrt{64} = 8$ because $8^2 = 64$ and $8 > 0$
(c) $\sqrt[3]{-27} = -3$ because $(-3)^3 = -27$

 Now Work Problem 7

We use principal roots to define exponents that are rational. If a is a real number and $n \geq 2$ is an integer, then

$$a^{1/n} = \sqrt[n]{a}$$

provided $\sqrt[n]{a}$ exists.

If a is a real number and m, n are integers containing no common factors with $n \geq 2$, then

$$a^{m/n} = \sqrt[n]{a^m} = (\sqrt[n]{a})^m$$

provided $\sqrt[n]{a}$ exists.

EXAMPLE 3   Evaluating Rational Exponents

(a) $4^{3/2} = (\sqrt{4})^3 = 2^3 = 8$ (b) $(-8)^{4/3} = (\sqrt[3]{-8})^4 = (-2)^4 = 16$
(c) $(32)^{-2/5} = (\sqrt[5]{32})^{-2} = 2^{-2} = \frac{1}{4}$

 Now Work Problem 13

The Laws of Exponents establish some rules for working with exponents.

> **Laws of Exponents**
>
> If a and b are positive real numbers and r and s are rational numbers, then
>
> $$a^r \cdot a^s = a^{r+s} \qquad (a^r)^s = a^{rs} \qquad (ab)^r = a^r b^r \qquad a^{-r} = \frac{1}{a^r}, \quad a \neq 0$$

But what is the meaning of a^x, where the base a is a positive real number and the exponent x is an irrational number? Although a rigorous definition requires methods discussed in calculus, the basis for the definition is easy to follow: Select a rational number r that is formed by truncating (removing) all but a finite number of digits from the irrational number x. Then it is reasonable to expect that

$$a^x \approx a^r$$

For example, take the irrational number $\pi \approx 3.14159.$ . . . An approximation to a^π is

$$a^\pi \approx a^{3.14}$$

where the digits after the hundredths position have been removed from the value for π. A better approximation would be

$$a^\pi \approx a^{3.14159}$$

where the digits after the hundred-thousandths position have been removed. Continuing in this way, we can obtain approximations to a^π to any desired degree of accuracy.

Most scientific calculators have an $\boxed{x^y}$ key (or a $\boxed{y^x}$ key) for working with exponents. To use this key, first enter the base x, then press the $\boxed{x^y}$ key, enter y, and press the $\boxed{=}$ key.

EXAMPLE 4 **Evaluating Exponents with a Calculator**

Using a calculator with an $\boxed{x^y}$ key, evaluate

(a) $2^{1.4}$ (b) $2^{1.41}$ (c) $2.^{1.414}$ (d) $2^{1.4142}$ (e) $2^{\sqrt{2}}$

SOLUTION

(a) $2^{1.4} \approx 2.6390158$ (b) $2^{1.41} \approx 2.6573716$
(c) $2^{1.414} \approx 2.6647497$ (d) $2.^{1.4142} \approx 2.6651191$
(e) $2^{\sqrt{2}} \approx 2.6651441$

✎ **Now Work Problem 19**

It can be shown that the Laws of Exponents hold for real exponents.

Logarithms

We have given meaning to a number a raised to the power x. Suppose $N = a^x$, $a > 0$, $a \neq 1$, x real. The definition of a logarithm is based on this exponential relationship.

The **logarithm to the base a of N** is symbolized by $\log_a N$ and is defined by

$$\log_a N = x \quad \text{if and only if} \quad N = a^x$$

EXAMPLE 5 **Applying the Definition of a Logarithm**

(a) $\log_2 8 = x$ means $2^x = 8 = 2^3$, so $x = 3$
(b) $\log_3 9 = x$ means $3^x = 9 = 3^2$, so $x = 2$

(c) $\log_{10} N = 3$ means $N = 10^3 = 1000$
(d) $\log_4 N = 2$ means $N = 4^2 = 16$
(e) $\log_a 16 = 2$ means $16 = a^2$, so $a = 4$
(f) $\log_a 3 = -1$ means $3 = a^{-1}$, so $a = \frac{1}{3}$

 **Now Work Problems**

25, 29, and 33

Thus logarithms are another name for exponents.

EXAMPLE 6   **Changing Exponents to Logarithms**

(a) If $100 = (1.1)^n$, then $n = \log_{1.1} 100$
(b) If $350 = (1.25)^n$, then $n = \log_{1.25} 350$
(c) If $1000 = (1.005)^n$, then $n = \log_{1.005} 1000$

To evaluate the expressions shown in Example 6, we use a scientific calculator. Such calculators have a key labeled $\boxed{\log}$ to evaluate logarithms to the base 10. To use the calculator, we need to know the following formula involving changing the base.

> **Change-of-Base Formula**
>
> $$\log_a M = \frac{\log_{10} M}{\log_{10} a}$$

To use your calculator to evaluate $\log_{10} M$, enter $\boxed{M}$ and press $\boxed{\log}$.

Proof of Change-of-Base Formula   Let

$$x = \log_a M$$

Then, by the definition of logarithm

$$a^x = M$$

Take the $\log_{10}$ of both sides

$$\log_{10} a^x = \log_{10} M$$

or

$$x \cdot \log_{10} a = \log_{10} M$$

so

$$x = \frac{\log_{10} M}{\log_{10} a}$$

Since $x = \log_a M$, we have

$$\log_a M = \frac{\log_{10} M}{\log_{10} a}$$ ■

EXAMPLE 7   Using the Change-of-Base Formula

(a) $\log_{1.1} 100 = \dfrac{\log_{10} 100}{\log_{10} 1.1} = \dfrac{2}{0.0414} = 48.3177$

(b) $\log_{1.25} 350 = \dfrac{\log_{10} 350}{\log_{10} 1.25} = \dfrac{2.5441}{0.0969} = 26.2519$

(c) $\log_{1.005} 1000 = \dfrac{\log_{10} 1000}{\log_{10} 1.005} = \dfrac{3}{0.002166} = 1385.0021$

■

Comparing Examples 6 and 7, we find that:

(a) $100 = (1.1)^n$ means $n = \log_{1.1} 100 = 48.3177$
(b) $350 = (1.25)^n$ means $n = \log_{1.25} 350 = 26.2519$
(c) $1000 = (1.005)^n$ means $n = \log_{1.005} 1000 = 1385.0021$

 **Now Work Problem 37**

EXERCISE A.2   Answers to odd-numbered problems begin on page AN-56.

In Problems 1–18 evaluate each expression.

1. 4^3

2. 8^2

3. 2^{-3}

4. 4^{-2}

5. 8^0

6. $(-3)^0$

7. $\sqrt{16}$

8. $\sqrt{9}$

9. $\sqrt[3]{27}$

10. $\sqrt[3]{-1}$

11. $\sqrt[4]{16}$

12. $\sqrt{0}$

13. $8^{2/3}$

14. $9^{3/2}$

15. $16^{-3/2}$

16. $8^{-2/3}$

17. $(-8)^{-2/3}$

18. $(-27)^{-4/3}$

In Problems 19–24 approximate each number using a calculator.

19. (a) $3^{2.2}$ (b) $3^{2.23}$ (c) $3^{2.236}$ (d) $3^{\sqrt{5}}$

20. (a) $5^{1.7}$ (b) $5^{1.73}$ (c) $5^{1.732}$ (d) $5^{\sqrt{3}}$

21. (a) $2^{3.14}$ (b) $2^{3.141}$ (c) $2^{3.1415}$ (d) 2^{π}

22. (a) $2^{2.7}$ (b) $2^{2.71}$ (c) $2^{2.718}$ (d) 2^e

23. (a) $3.1^{2.7}$ (b) $3.14^{2.71}$ (c) $3.141^{2.718}$ (d) π^e

24. (a) $2.7^{3.1}$ (b) $2.71^{3.14}$ (c) $2.718^{3.141}$ (d) e^{π}

In Problems 25–28 evaluate each expression.

25. $\log_3 27$

26. $\log_2 16$

27. $\log_2 \frac{1}{2}$

28. $\log_3 \frac{1}{9}$

In Problems 29–32 find N.

29. $\log_2 N = 3$

30. $\log_2 N = -2$

31. $\log_3 N = -1$

32. $\log_8 N = \frac{1}{3}$

In Problems 33–36 find a.

33. $\log_a 8 = 3$ **34.** $\log_a 4 = -1$ **35.** $\log_a 9 = 2$ **36.** $\log_a 16 = -2$

In Problems 37–46 use a calculator to evaluate each logarithm.

37. $\log_{1.1} 200$ **38.** $\log_{1.05} 200$ **39.** $\log_{1.005} 1000$ **40.** $\log_{1.001} 10$

41. $\log_{1.002} 20$ **42.** $\log_{1.005} 30$ **43.** $\log_{1.0005} 500$ **44.** $\log_{1.0001} 1000$

45. $\log_{1.003} 500$ **46.** $\log_{1.006} 500$

A.3 GEOMETRIC SEQUENCES

A **sequence** is a rule that assigns a real number to each positive integer.

A sequence is often represented by listing its values in order. For example, the sequence whose rule is to assign to each positive integer its reciprocal may be represented as

$$s_1 = 1, \qquad s_2 = \frac{1}{2}, \qquad s_3 = \frac{1}{3}, \qquad s_4 = \frac{1}{4}, \ldots, \qquad s_n = \frac{1}{n}, \ldots$$

or merely as the list

$$1, \quad \frac{1}{2}, \quad \frac{1}{3}, \quad \frac{1}{4}, \ldots, \quad \frac{1}{n}, \ldots$$

The list never ends, as the dots indicate. The real numbers in this ordered list are called the **terms** of the sequence. For the sequence above we have a rule for the nth term, namely, $s = 1/n$, so it is easy to find any term of the sequence. In this case we represent the sequence by placing braces around the formula for the nth term, writing $\{s_n\} = \{1/n\}$.

EXAMPLE 1   Writing the Terms of a Sequence

Write the first six terms of the sequence below.

$$\{a_n\} = \left\{\frac{n-1}{n}\right\}$$

SOLUTION

$$a_1 = 0$$
$$a_2 = \tfrac{1}{2}$$
$$a_3 = \tfrac{2}{3}$$
$$a_4 = \tfrac{3}{4}$$
$$a_5 = \tfrac{4}{5}$$
$$a_6 = \tfrac{5}{6}$$

EXAMPLE 2   Writing the Terms of a Sequence

Write the first six terms of the sequence below.

$$\{b_n\} = \left\{(-1)^{n-1}\left(\frac{2}{n}\right)\right\}$$

SOLUTION

$$b_1 = 2$$
$$b_2 = -1$$
$$b_3 = \tfrac{2}{3}$$
$$b_4 = -\tfrac{1}{2}$$
$$b_5 = \tfrac{2}{5}$$
$$b_6 = -\tfrac{1}{3}$$

 **Now Work Problem 5**

A second way of defining a sequence is to assign a value to the first term (or the first few terms) and specify the nth term by a formula or equation that involves one or more of the terms preceding it. Sequences defined this way are said to be defined **recursively,** and the rule or formula is called a **recursive formula.**

EXAMPLE 3   Writing the Terms of a Recursively Defined Sequence

Write the first five terms of the recursively defined sequence given below.

$$s_1 = 1, \qquad s_n = 4s_{n-1}$$

SOLUTION   The first term is given as $s_1 = 1$. To get the second term, we use $n = 2$ in the formula to get $s_2 = 4s_1 = 4 \cdot 1 = 4$. To get the third term, we use $n = 3$ in the formula to get $s_3 = 4s_2 = 4 \cdot 4 = 16$. To get the new term requires that we know the value of the preceding term. The first five terms are

$$s_1 = 1$$
$$s_2 = 4 \cdot 1 = 4$$
$$s_3 = 4 \cdot 4 = 16$$
$$s_4 = 4 \cdot 16 = 64$$
$$s_5 = 4 \cdot 64 = 256$$

EXAMPLE 4   Writing the Terms of a Recursively Defined Sequence

Write the first five terms of the recursively defined sequence given below.

$$u_1 = 1, \qquad u_2 = 1, \qquad u_{n+2} = u_n + u_{n+1}$$

SOLUTION We are given the first two terms. To get the third term requires that we know each of the previous two terms. Thus

$$u_1 = 1$$
$$u_2 = 1$$
$$u_3 = u_1 + u_2 = 2$$
$$u_4 = u_2 + u_3 = 1 + 2 = 3$$
$$u_5 = u_3 + u_4 = 2 + 3 = 5$$

The sequence defined in Example 4 is called a **Fibonacci sequence,** and the terms of this sequence are called **Fibonacci numbers.** These numbers appear in a wide variety of applications.

EXAMPLE 5 **Writing the Terms of a Recursively Defined Sequence**

Write the first five terms of the recursively defined sequence given below.

$$f_1 = 1, \qquad f_{n+1} = (n + 1)f_n$$

SOLUTION Here,

$$f_1 = 1$$
$$f_2 = 2f_1 = 2 \cdot 1 = 2$$
$$f_3 = 3f_2 = 3 \cdot 2 = 6$$
$$f_4 = 4f_3 = 4 \cdot 6 = 24$$
$$f_5 = 5f_4 = 5 \cdot 24 = 120$$

 **Now Work Problems**

13 and 21

Geometric Sequences

When the ratio of successive terms of a sequence is always the same nonzero number, the sequence is called **geometric.** Thus a **geometric sequence*** may be defined recursively as $a_1 = a$, $a_{n+1}/a_n = r$, or as:

Geometric Sequence

$$a_1 = a, \qquad a_{n+1} = ra_n$$

where $a = a_1$ and $r \neq 0$ are real numbers. The number a is the **first term,** and the nonzero number r is called the **common ratio.**

---

* Sometimes called a **geometric progression.**

Thus the terms of a geometric sequence with first term a and common ratio r follow the pattern

$$a, \quad ar, \quad ar^2, \quad ar^3, \quad \ldots$$

EXAMPLE 6   Identifying a Geometric Sequence

The sequence

$$2, \quad 6, \quad 18, \quad 54, \quad 162, \ldots$$

is geometric since the ratio of successive terms is 3. The first term is 2, and the common ratio is 3.

∎

EXAMPLE 7   Finding the First Term and Common Ratio of a Geometric Sequence

Show that the sequence below is geometric. Find the first term and the common ratio.

$$\{t_n\} = \{4^n\}$$

SOLUTION   The first term is $t_1 = 4^1 = 4$. The $(n + 1)$st and nth terms are

$$t_{n+1} = 4^{n+1} \qquad \text{and} \qquad t_n = 4^n$$

Their ratio is

$$\frac{t_{n+1}}{t_n} = \frac{4^{n+1}}{4^n} = 4$$

Because the ratio of successive terms is a nonzero number, the sequence $\{t_n\}$ is a geometric sequence with common ratio 4.

∎

EXAMPLE 8   Finding the First Term and Common Ratio of a Geometric Sequence

Show that the sequence below is geometric. Find the first term and the common ratio.

$$\{s_n\} = \{2^{-n}\}$$

SOLUTION   The first term is $s_1 = 2^{-1} = \frac{1}{2}$. The $(n + 1)$st and nth terms of the sequence $\{s_n\}$ are

$$s_{n+1} = 2^{-(n+1)} \qquad \text{and} \qquad s_n = 2^{-n}$$

Their ratio is

$$\frac{s_{n+1}}{s_n} = \frac{2^{-(n+1)}}{2^{-n}} = 2^{-n-1+n} = 2^{-1} = \frac{1}{2}$$

Because the ratio of successive terms is a nonzero number, the sequence $\{s_n\}$ is geometric with common ratio $\frac{1}{2}$.

∎

Adding the First n Terms of a Geometric Sequence

The next result gives us a formula for finding the sum of the first n terms of a geometric sequence.

Sum of the First n Terms of a Geometric Sequence

Let $\{a_n\}$ be a geometric sequence with first term a and common ratio r. The sum S_n of the first n terms of $\{a_n\}$ is

$$S_n = a\left(\frac{1 - r^n}{1 - r}\right) \qquad r \neq 0, 1 \tag{1}$$

Formula (1) may be derived as follows:

$$S_n = a + ar + \cdots + ar^{n-1} \tag{2}$$

Multiply each side by r to obtain

$$rS_n = ar + ar^2 + \cdots + ar^n \tag{3}$$

Now subtract (3) from (2). The result is

$$S_n - rS_n = a - ar^n$$
$$(1 - r)S_n = a(1 - r^n)$$

Since $r \neq 1$, we can solve for S_n:

$$S_n = a\left(\frac{1 - r^n}{1 - r}\right)$$

EXAMPLE 9   **Adding the Terms of a Geometric Sequence**

Find the sum S_n of the first n terms of the sequence $\{(\frac{1}{2})^n\}$; that is, find

$$\frac{1}{2} + \frac{1}{4} + \frac{1}{8} + \cdots + \left(\frac{1}{2}\right)^n$$

SOLUTION The sequence $\{(\frac{1}{2})^n\}$ is a geometric sequence with $a = \frac{1}{2}$ and $r = \frac{1}{2}$. The sum S_n we seek is the sum of the first n terms of the sequence, so we use Formula (1) to get

$$
\begin{aligned}
S_n &= \frac{1}{2} + \frac{1}{4} + \frac{1}{8} + \cdots + \left(\frac{1}{2}\right)^n \\
&= \frac{1}{2}\left[\frac{1 - (\frac{1}{2})^n}{1 - \frac{1}{2}}\right] \\
&= \frac{1}{2}\left[\frac{1 - (\frac{1}{2})^n}{\frac{1}{2}}\right] \\
&= 1 - \left(\frac{1}{2}\right)^n
\end{aligned}
$$

■

EXERCISE A.3   Answers to odd-numbered problems begin on page AN-56.

In Problems 1–12 write the first five terms of each sequence.

1. $\{n\}$

2. $\{n^2 + 1\}$

3. $\left\{\dfrac{n}{n+1}\right\}$

4. $\left\{\dfrac{2n+3}{2n-1}\right\}$

5. $\{(-1)^{n+1}n^2\}$

6. $\left\{(-1)^{n-1}\left(\dfrac{n}{2n-1}\right)\right\}$

7. $\left\{\dfrac{2^n}{3^n+1}\right\}$

8. $\left\{\left(\dfrac{4}{3}\right)^n\right\}$

9. $\left\{\dfrac{(-1)^n}{(n+1)(n+2)}\right\}$

10. $\left\{\dfrac{3^n}{n}\right\}$

11. $\left\{\dfrac{n}{e^n}\right\}$

12. $\left\{\dfrac{n^2}{2^n}\right\}$

In Problems 13–26 a sequence is defined recursively. Write the first five terms.

13. $a_1 = 1;\quad a_{n+1} = 2 + a_n$

14. $a_1 = 3;\quad a_{n+1} = 5 - a_n$

15. $a_1 = -2;\quad a_{n+1} = n + a_n$

16. $a_1 = 1;\quad a_{n+1} = n - a_n$

17. $a_1 = 5;\quad a_{n+1} = 2a_n$

18. $a_1 = 2;\quad a_{n+1} = -a_n$

19. $a_1 = 3;\quad a_{n+1} = \dfrac{a_n}{n}$

20. $a_1 = -2;\quad a_{n+1} = n + 3a_n$

21. $a_1 = 1;\quad a_2 = 2;\quad a_{n+2} = a_n a_{n+1}$

22. $a_1 = -1;\quad a_2 = 1;\quad a_{n+2} = a_{n+1} + na_n$

23. $a_1 = A;\quad a_{n+1} = a_n + d$

24. $a_1 = A;\quad a_{n+1} = ra_n,\quad r \neq 0$

25. $a_1 = \sqrt{2};\quad a_{n+1} = \sqrt{2 + a_n}$

26. $a_1 = \sqrt{2};\quad a_{n+1} = \sqrt{a_n/2}$

In Problems 27–36 a geometric sequence is given. Find the common ratio and write out the first four terms. Also, find the sum of the first n terms.

27. $\{2^n\}$

28. $\{(-4)^n\}$

29. $\left\{-3\left(\dfrac{1}{2}\right)^n\right\}$

30. $\left\{\left(\dfrac{5}{2}\right)^n\right\}$

31. $\left\{\dfrac{2^{n-1}}{4}\right\}$

32. $\left\{\dfrac{3^n}{9}\right\}$

33. $\{2^{n/3}\}$

34. $\{3^{2n}\}$

35. $\left\{\dfrac{3^{n-1}}{2^n}\right\}$

36. $\left\{\dfrac{2^n}{3^{n-1}}\right\}$

Using LINDO to Solve Linear Programming Problems

A number of available software packages can be used to solve linear programming problems. Three of the more popular ones are these:

GAMS—General Algebraic Modeling System.   This is a general-purpose optimization system specifically designed for modeling and solving large, complex linear, nonlinear, and mixed-integer programming problems. It is available for PC, 386/486, workstation, or mainframe.

GINO—General INteractive Optimizer.   This system can solve linear and nonlinear constrained optimization problems on a PC, Mac, 386, or mainframe.

LINDO—Linear INteractive and Discrete Optimizer.   This is the most popular interactive linear, integer, and quadratic programming system on the market. It is available for PC and Mac.

Because LINDO is the most popular software package, we have chosen to include a brief section illustrating how it can be used to solve some linear programming problems. LINDO was designed to allow users to do simple problems in an easy, cost-efficient manner. At the other extreme, LINDO has been used to solve real industrial programs involving more than 10,000 rows and several thousand variables. In solving such problems, LINDO uses a *revised* simplex method—one that exploits the special character of large linear programming problems. As a result, the intermediate tableaus obtained using LINDO may differ from those obtained manually.

EXAMPLE 1   Using LINDO

Use LINDO to maximize

$$P = 3x_1 + 4x_2$$

subject to

$$2x_1 + 4x_2 \leq 120$$
$$2x_1 + 2x_2 \leq 80$$
$$x_1 \geq 0 \qquad x_2 \geq 0$$

LINDO SOLUTION

The objective function

```
MAX      3 X1 + 4 X2
```

The constraints: numbered 2 and 3

```
SUBJECT TO
        2)    2 X1 + 4 X2 <=    120
        3)    2 X1 + 2 X2 <=     80
END
```

Instruction to begin

```
: tabl
```

The initial tableau SLK 2 is the slack variable associated with constraint number 2. ROW 1 ART is the objective row. The column for *P* is not written since pivoting does not affect it. The column on the far right is RHS. The bottom row ART is added by LINDO to represent the pivot objective of minimizing the sum of the infeasibilities.

```
THE TABLEAU
        ROW (BASIS)         X1        X2     SLK   2     SLK   3
         1 ART           -3.000    -4.000      .000        .000       .000
         2 SLK    2       2.000     4.000     1.000        .000    120.000
         3 SLK    3       2.000     2.000      .000       1.000     80.000
ART      3 ART           -3.000    -4.000      .000        .000       .000
```

Instruction to pivot The pivot element is in row 3, column X1. X1 will become a basic variable.

```
: piv
        X1 ENTERS AT VALUE    40.000   IN   ROW   3   OBJ. VALUE=   120.00
```

After pivoting, SLK 2 and X1 are basic variables. *P* = 120 at this stage; SLK 2 = 40; X1 = 40; SLK 3 = 0; X2 = 0.

```
: tabl
THE TABLEAU
        ROW (BASIS)         X1        X2     SLK   2     SLK   3
         1 ART             .000    -1.000      .000       1.500    120.000
         2 SLK    2         .000     2.000     1.000      -1.000     40.000
         3       x1        1.000     1.000      .000        .500     40.000
```

The pivot element is in row 2, column X2. After pivoting, X2 and X1 are basic variables.

```
: piv
        X2 ENTERS AT VALUE    20.000   IN   ROW   2   OBJ. VALUE=   140.00
: tabl

THE TABLEAU
        ROW (BASIS)         X1        X2     SLK   2     SLK   3
         1 ART             .000      .000      .500       1.000    140.000
         2       X2         .000     1.000      .500       -.500     20.000
         3       X1        1.000      .000     -.500       1.000     20.000
: piv
```

This is a final tableau. The solution is *P* = 140, X1 = 20, X2 = 20.

```
LP OPTIMUM FOUND AT STEP      2

        OBJECTIVE FUNCTION VALUE

        1)      140.00000

VARIABLE         VALUE          REDUCED COST
    X1        20.000000            .000000
    X2        20.000000            .000000
```

The value of the slack variables are SLK 2 = 0, SLK 3 = 0.

```
ROW      SLACK OR SURPLUS     DUAL PRICES
    2)          .000000           .500000
    3)          .000000          1.000000
```

Two pivots were used.

```
NO. ITERATIONS=        2
```

Compare the steps in Example 1, using LINDO, with the steps given in Example 1, Section 4.2.

EXAMPLE 2   Using LINDO

Use LINDO to maximize

$$P = 20x_1 + 15x_2$$

subject to

$$x_1 + x_2 \geq 7$$
$$9x_1 + 5x_2 \leq 45$$
$$2x_1 + x_2 \geq 8$$
$$x_1 \geq 0 \qquad x_2 \geq 0$$

SOLUTION

```
MAX       20 X1 + 15 X2
SUBJECT TO
       2)    X1 + X2 >=    7
       3)    9 X1 + 5 X2 <=    45
       4)    2 X1 + X2 >=    8
END

: tabl

THE TABLEAU
     ROW  (BASIS)        X1        X2  SLK    2  SLK    3  SLK    4
       1 ART          -20.000   -15.000    .000      .000      .000      .000
       2 SLK    2      -1.000    -1.000   1.000      .000      .000    -7.000
       3 SLK    3       9.000     5.000    .000     1.000      .000    45.000
       4 SLK    4      -2.000    -1.000    .000      .000     1.000    -8.000
ART    4 ART           -3.000    -2.000   1.000      .000     1.000   -15.000

: piv
     X1 ENTERS AT VALUE    5.0000     IN ROW    3 OBJ. VALUE=   100.00

: tabl

THE TABLEAU
     ROW  (BASIS)        X1        X2  SLK    2  SLK    3  SLK    4
       1 ART             .000    -3.889    .000     2.222      .000   100.000
       2 SLK    2        .000     -.444   1.000     .111      .000    -2.000
       3       X1       1.000      .556    .000     .111      .000     5.000
       4 SLK    4        .000      .111    .000     .222     1.000     2.000
ART    4 ART             .000     -.444   1.000     .111      .000    -2.000

: piv
     X2 ENTERS AT VALUE    4.5000     IN ROW    2 OBJ. VALUE=   117.50

: tabl

THE TABLEAU
     ROW  (BASIS)        X1        X2  SLK    2  SLK    3  SLK    4
       1 ART             .000      .000  -8.750    1.250      .000   117.500
       2       X2        .000     1.000  -2.250    -.250      .000     4.500
       3       X1       1.000      .000   1.250     .250      .000     2.500
       4 SLK    4        .000      .000    .250     .250     1.000     1.500
ART    4 ART             .000      .000  -8.750    1.250

: piv
SLK    2 ENTERS AT VALUE    2.0000     IN ROW    3 OBJ. VALUE=   117.50
```

```
: tabl
THE TABLEAU
     ROW  (BASIS)          X1         X2  SLK   2  SLK   3    SLK   4
       1 ART            7.000       .000      .000   3.000      .000   135.000
       2      X2        1.800      1.000      .000    .200      .000     9.000
       3 SLK  2          .800       .000     1.000    .200      .000     2.000
       4 SLK  4         -.200       .000      .000    .200     1.000     1.000

: piv
 LP OPTIMUM FOUND AT STEP        3

          OBJECTIVE FUNCTION VALUE

       1)       135.00000

    VARIABLE           VALUE           REDUCED COST
       X1            .000000           7.000000
       X2           9.000000            .000000

     ROW    SLACK OR SURPLUS        DUAL PRICES
      2)         2.000000            .000000
      3)          .000000           3.000000
      4)         1.000000            .000000

 NO. ITERATIONS=       3
```

Compare the final tableau found using LINDO with the final tableau of Example 1, Section 4.4, namely,

$$
\begin{array}{c|cccccc|c}
\text{BV} & P & x_1 & x_2 & s_1 & s_2 & s_3 & \text{RHS} \\
\hline
s_1 & 0 & \frac{4}{5} & 0 & 1 & \frac{1}{5} & 0 & 2 \\
x_2 & 0 & \frac{9}{5} & 1 & 0 & \frac{1}{5} & 0 & 9 \\
s_3 & 0 & -\frac{1}{5} & 0 & 0 & \frac{1}{5} & 1 & 1 \\
\hline
P & 1 & 7 & 0 & 0 & 3 & 0 & 135
\end{array}
$$

The solution, in both cases, is $P = 135$, $x_2 = 9$, $x_1 = 0$, $s_1 = $ SLK $2 = 2$, $s_2 = $ SLK $3 = 0$, $s_3 = $ SLK $4 = 1$. Notice also that the intermediate tableaus are different. This is because LINDO uses a *revised* simplex method that requires additional computations not easily performed manually.

EXAMPLE 3   Using LINDO

Use LINDO to minimize

$$z = 5x_1 + 6x_2$$

subject to

$$
\begin{aligned}
x_1 + x_2 &\le 10 \\
x_1 + 2x_2 &\ge 12 \\
2x_1 + x_2 &\ge 12 \\
x_1 &\ge 3 \\
x_1 \ge 0 \qquad x_2 &\ge 0
\end{aligned}
$$

SOLUTION

```
MIN      5 X1 + 6 X2
SUBJECT TO
        2)    X1 + X2 <=    10
        3)    X1 + 2 X2 >=    12
        4)    2 X1 + X2 >=    12
        5)    X1 >=    3
END
: tabl
THE TABLEAU
     ROW  (BASIS)        X1        X2   SLK   2   SLK   3   SLK   4   SLK   5
       1 ART          5.000     6.000     .000      .000      .000      .000      .000
       2 SLK    2     1.000     1.000    1.000      .000      .000      .000    10.000
       3 SLK    3    -1.000    -2.000     .000     1.000      .000      .000   -12.000
       4 SLK    4    -2.000    -1.000     .000      .000     1.000      .000   -12.000
       5 SLK    5    -1.000      .000     .000      .000      .000     1.000    -3.000
 ART   5 ART         -4.000    -3.000     .000     1.000     1.000     1.000   -27.000
: piv
     X1 ENTERS AT VALUE    6.0000     IN ROW    4 OBJ. VALUE= -30.000
: tabl
THE TABLEAU
     ROW  (BASIS)        X1        X2   SLK   2   SLK   3   SLK   4   SLK   5
       1 ART           .000     3.500     .000      .000     2.500      .000   -30.000
       2 SLK    2      .000      .500    1.000      .000      .500      .000     4.000
       3 SLK    3      .000    -1.500     .000     1.000     -.500      .000    -6.000
       4        X1    1.000      .500     .000      .000     -.500      .000     6.000
       5 SLK    5      .000      .500     .000      .000     -.500     1.000     3.000
 ART   5 ART           .000    -1.500     .000     1.000     -.500      .000    -6.000
  : piv
     X2 ENTERS AT VALUE    4.0000     IN ROW    3 OBJ. VALUE= -44.000
: tabl
THE TABLEAU
     ROW  (BASIS)        X1        X2   SLK   2   SLK   3   SLK   4   SLK   5
       1 ART           .000      .000     .000     2.333     1.333      .000   -44.000
       2 SLK    2      .000      .000    1.000      .333      .333      .000     2.000
       3        X2      .000     1.000     .000     -.667      .333      .000     4.000
       4        X1    1.000      .000     .000      .333     -.667      .000     4.000
       5 SLK    5      .000      .000     .000      .333     -.667     1.000     1.000
 ART   5 ART           .000      .000     .000     2.333     1.333
: piv
LP OPTIMUM FOUND AT STEP         2
OBJECTIVE FUNCTION VALUE
        1)      44.000000

  VARIABLE         VALUE          REDUCED COST
        X1       4.000000            .000000
        X2       4.000000            .000000

     ROW    SLACK OR SURPLUS        DUAL PRICES
       2)       2.000000             .000000
       3)        .000000           -2.333333
       4)        .000000           -1.333333
       5)       1.000000             .000000

NO. ITERATIONS=         2
```

Compare this result, using LINDO, to the result obtained in Example 2, Section 4.4. Note again that the same solution is found: $z = 44$, $x_1 = 4$, $x_2 = 4$, $s_1 = $ SLK 2 = 2, $s_2 = $ SLK 3 = 0, $s_3 = $ SLK 4 = 0, $s_4 = $ SLK 5 = 1. Also, note again the different intermediate tableaus.

EXERCISE B.1   Answers to odd-numbered problems begin on page AN-56.

In Problems 1–24 use LINDO (or any other software package) to solve each linear programming problem.

1. Maximize
$$P = 3x_1 + 2x_2 + x_3$$

subject to
$$3x_1 + x_2 + x_3 \leq 30$$
$$5x_1 + 2x_2 + x_3 \leq 24$$
$$x_1 + x_2 + 4x_3 \leq 20$$
$$x_1 \geq 0 \quad x_2 \geq 0 \quad x_3 \geq 0$$

2. Maximize
$$P = x_1 + 4x_2 + 3x_3 + x_4$$

subject to
$$2x_1 + x_2 \leq 10$$
$$3x_1 + x_2 + x_3 + 2x_4 \leq 18$$
$$x_1 + x_2 + x_3 + x_4 \leq 14$$
$$x_1 \geq 0 \quad x_2 \geq 0 \quad x_3 \geq 0 \quad x_4 \geq 0$$

3. Maximize
$$P = 3x_1 + x_2 + x_3$$

subject to
$$x_1 + x_2 + x_3 \leq 6$$
$$2x_1 + 3x_2 + 4x_3 \leq 10$$
$$x_1 \geq 0 \quad x_2 \geq 0 \quad x_3 \geq 0$$

4. Maximize
$$P = 3x_1 + x_2 + x_3$$

subject to
$$x_1 + x_2 + x_3 \leq 8$$
$$2x_1 + x_2 + 4x_3 \geq 6$$
$$x_1 \geq 0 \quad x_2 \geq 0$$

5. Maximize
$$P = 2x_1 + x_2 + 3x_3$$

subject to
$$x_1 + x_2 - x_3 \leq 10$$
$$x_2 + x_3 \leq 4$$
$$x_1 \geq 0 \quad x_2 \geq 0 \quad x_3 \geq 0$$

6. Maximize
$$P = 2x_1 + 2x_2 + 3x_3$$

subject to
$$x_1 - x_2 + x_3 \leq 6$$
$$x_1 \leq 4$$
$$x_1 \geq 0 \quad x_2 \geq 0 \quad x_3 \geq 0$$

7. Maximize
$$P = x_1 + x_2 + x_3$$

subject to
$$x_1 + x_2 + x_3 \leq 6$$
$$4x_1 + x_2 \geq 12$$
$$x_1 \geq 0 \quad x_2 \geq 0 \quad x_3 \geq 0$$

8. Maximize
$$P = 2x_1 + x_2 + 3x_3$$

subject to
$$-x_1 + x_2 + x_3 \geq -6$$
$$2x_1 - 3x_2 \geq -12$$
$$x_1 \geq 0 \quad x_2 \geq 0 \quad x_3 \geq 0$$

9. Maximize
$$P = 2x_1 + x_2 + 3x_3$$

subject to
$$5x_1 + 2x_2 + x_3 \leq 20$$
$$6x_1 + x_2 + 4x_3 \leq 24$$
$$x_1 + x_2 + 4x_3 \leq 16$$
$$x_1 \geq 0 \quad x_2 \geq 0 \quad x_3 \geq 0$$

10. Maximize
$$P = 3x_1 + 2x_2 + x_3$$

subject to
$$3x_1 + 2x_2 - x_3 \leq 10$$
$$x_1 - x_2 + 3x_3 \leq 12$$
$$2x_1 + x_2 + x_3 \leq 6$$
$$x_1 \geq 0 \quad x_2 \geq 0 \quad x_3 \geq 0$$

11. Maximize

$$P = 2x_1 + 3x_2 + x_3$$

subject to

$$x_1 + x_2 + x_3 \le 50$$
$$3x_1 + 2x_2 + x_3 \le 10$$
$$x_1 \ge 0 \qquad x_2 \ge 0 \qquad x_3 \ge 0$$

12. Maximize

$$P = 4x_1 + 4x_2 + 2x_3$$

subject to

$$3x_1 + x_2 + x_3 \le 10$$
$$x_1 + x_2 + 3x_3 \le 5$$
$$x_1 \ge 0 \qquad x_2 \ge 0 \qquad x_3 \ge 0$$

13. Maximize

$$P = 2x_1 + x_2 + x_3$$

subject to

$$-2x_1 + x_2 - 2x_3 \le 4$$
$$x_1 - 2x_2 + x_3 \le 2$$
$$x_1 \ge 0 \qquad x_2 \ge 0 \qquad x_3 \ge 0$$

14. Maximize

$$P = 4x_1 + 2x_2 + 5x_3$$

subject to

$$x_1 + 3x_2 + 2x_3 \le 30$$
$$2x_1 + x_2 + 3x_3 \le 12$$
$$x_1 \ge 0 \qquad x_2 \ge 0 \qquad x_3 \ge 0$$

15. Maximize

$$P = 2x_1 + x_2 + 3x_3$$

subject to

$$x_1 + 2x_2 + x_3 \le 25$$
$$3x_1 + 2x_2 + 3x_3 \le 30$$
$$x_1 \ge 0 \qquad x_2 \ge 0 \qquad x_3 \ge 0$$

16. Maximize

$$P = 6x_1 + 3x_2 + 2x_3$$

subject to

$$2x_1 + 2x_2 + 3x_3 \le 30$$
$$2x_1 + 2x_2 + x_3 \le 12$$
$$x_1 \ge 0 \qquad x_2 \ge 0 \qquad x_3 \ge 0$$

17. Maximize

$$P = 2x_1 + 4x_2 + x_3 + x_4$$

subject to

$$2x_1 + x_2 + 2x_3 + 3x_4 \le 12$$
$$2x_2 + x_3 + 2x_4 \le 20$$
$$2x_1 + x_2 + 4x_3 \le 16$$
$$x_1 \ge 0 \qquad x_2 \ge 0 \qquad x_3 \ge 0 \qquad x_4 \ge 0$$

18. Maximize

$$P = 2x_1 + 4x_2 + x_3$$

subject to

$$-x_1 + 2x_2 + 3x_3 \le 6$$
$$-x_1 + 4x_2 + 5x_3 \le 5$$
$$-x_1 + 5x_2 + 7x_3 \le 7$$
$$x_1 \ge 0 \qquad x_2 \ge 0 \qquad x_3 \ge 0$$

19. Maximize

$$P = 2x_1 + x_2 + x_3$$

subject to

$$x_1 + 2x_2 + 4x_3 \le 20$$
$$2x_1 + 4x_2 + 4x_3 \le 60$$
$$3x_1 + 4x_2 + x_3 \le 90$$
$$x_1 \ge 0 \qquad x_2 \ge 0 \qquad x_3 \ge 0$$

20. Maximize

$$P = x_1 + 2x_2 + 4x_3$$

subject to

$$8x_1 + 5x_2 - 4x_3 \le 30$$
$$-2x_1 + 6x_2 + x_3 \le 5$$
$$-2x_1 + 2x_2 + x_3 \le 15$$
$$x_1 \ge 0 \qquad x_2 \ge 0 \qquad x_3 \ge 0$$

21. Maximize

$$P = x_1 + 2x_2 + 4x_3 - x_4$$

subject to

$$5x_1 + 4x_3 + 6x_4 \le 20$$
$$4x_1 + 2x_2 + 2x_3 + 8x_4 \le 40$$
$$x_1 \ge 0 \qquad x_2 \ge 0 \qquad x_3 \ge 0 \qquad x_4 \ge 0$$

22. Maximize

$$P = x_1 + 2x_2 - x_3 + 3x_4$$

subject to

$$2x_1 + 4x_2 + 5x_3 + 6x_4 \le 24$$
$$4x_1 + 4x_2 + 2x_3 + 2x_4 \le 4$$
$$x_1 \ge 0 \qquad x_2 \ge 0 \qquad x_3 \ge 0 \qquad x_4 \ge 0$$

23. Minimize

$$z = x_1 + x_2 + x_3 + x_4 + x_5 + x_6 + x_7$$

subject to

$$4x_1 + 2x_2 + x_3 \quad\quad + 2x_5 + x_6 \quad\quad \geq 75$$
$$x_2 + 2x_3 + 3x_4 \quad\quad + x_6 \quad\quad \geq 110$$
$$x_5 + x_6 + 2x_7 \geq 50$$
$$x_1 \geq 0 \quad x_2 \geq 0 \quad x_3 \geq 0 \quad x_4 \geq 0$$
$$x_5 \geq 0 \quad x_6 \geq 0 \quad x_7 \geq 0$$

24. Minimize

$$z = x_1 + x_2 + x_3 + x_4 + x_5 + x_6 + x_7$$

subject to

$$4x_1 + x_2 + 2x_3 \quad\quad + 2x_5 \quad\quad + x_7 \geq 75$$
$$2x_2 + x_3 + 3x_4 \quad\quad\quad + x_7 \geq 180$$
$$x_5 + 2x_6 + x_7 \geq 50$$
$$x_1 \geq 0 \quad x_2 \geq 0 \quad x_3 \geq 0 \quad x_4 \geq 0$$
$$x_5 \geq 0 \quad x_6 \geq 0 \quad x_7 \geq 0$$

Tables

Table I Amount of an Annuity

Table II Present Value of an Annuity

Table III Standard Normal Curve Table

Table I Amount of an Annuity
(a) Annual Compounding

No. of Periods n	8% per annum $\dfrac{(1 + 0.08)^n - 1}{0.08}$	$\left[\dfrac{(1 + 0.08)^n - 1}{0.08}\right]^{-1}$	10% per annum $\dfrac{(1 + 0.10)^n - 1}{0.10}$	$\left[\dfrac{(1 + 0.10)^n - 1}{0.10}\right]^{-1}$	12% per annum $\dfrac{(1 + 0.12)^n - 1}{0.12}$	$\left[\dfrac{(1 + 0.12)^n - 1}{0.12}\right]^{-1}$
1	1.00000000	1.00000000	1.00000000	1.00000000	1.00000000	1.00000000
2	2.08000000	0.48076923	2.10000000	0.47619048	2.12000000	0.47169811
3	3.24640000	0.30803351	3.31000000	0.30211480	3.37440000	0.29634898
4	4.50611200	0.22192080	4.64100000	0.21547080	4.77932800	0.20923444
5	5.86660096	0.17045645	6.10510000	0.16379748	6.35284736	0.15740973
6	7.33592904	0.13631539	7.71561000	0.12960738	8.11518904	0.12322572
7	8.92280336	0.11207240	9.48717100	0.10540550	10.0890117	0.09911774
8	10.6366276	0.09401476	11.4358881	0.08744402	12.2996931	0.08130284
9	12.4875578	0.08007971	13.5794769	0.07364054	14.7756563	0.06767889
10	14.4865625	0.06902949	15.9374246	0.06274539	17.5487351	0.05698416
11	16.6454875	0.06007634	18.5311671	0.05396314	20.6545833	0.04841540
12	18.9771265	0.05269502	21.3842838	0.04676332	24.1331333	0.04143681
13	21.4952966	0.04652181	24.5227121	0.04077852	28.0291093	0.03567720
14	24.2149203	0.04129685	27.9749834	0.03574622	32.3926024	0.03087125
15	27.1521139	0.03682954	31.7724817	0.03147378	37.2797147	0.02682424
16	30.3242830	0.03297687	35.9497299	0.02781662	42.7532804	0.02339002
17	33.7502257	0.02962943	40.5447029	0.02466413	48.8836741	0.02045673
18	37.4502437	0.02670210	45.5991731	0.02193022	55.7497150	0.01793731
19	41.4462632	0.02412763	51.1590904	0.01954687	63.4396808	0.01576300
20	45.7619643	0.02185221	57.2749995	0.01745962	72.0524424	0.01387878
21	50.4229214	0.01983225	64.0024994	0.01562439	81.6987355	0.01224009
22	55.4567552	0.01803207	71.4027494	0.01400506	92.5025838	0.01081051
23·	60.8932956	0.01642217	79.5430243	0.01257181	104.602894	0.00955996
24	66.7647592	0.01497796	88.4973268	0.01129978	118.155241	0.00846344
25	73.1059400	0.01367878	98.3470594	0.01016807	133.333870	0.00749997
26	79.9544151	0.01250713	109.181765	0.00915904	150.333934	0.00665186
27	87.3507684	0.01144810	121.099942	0.00825764	169.374007	0.00590409
28	95.3388298	0.01048891	134.209936	0.00745101	190.698887	0.00524387

Table I *(Continued)*

(a) Annual Compounding

No. of Periods n	8% per annum		10% per annum		12% per annum	
	$\dfrac{(1 + 0.08)^n - 1}{0.08}$	$\left[\dfrac{(1 + 0.08)^n - 1}{0.08}\right]^{-1}$	$\dfrac{(1 + 0.10)^n - 1}{0.10}$	$\left[\dfrac{(1 + 0.10)^n - 1}{0.10}\right]^{-1}$	$\dfrac{(1 + 0.12)^n - 1}{0.12}$	$\left[\dfrac{(1 + 0.12)^n - 1}{0.12}\right]^{-1}$
29	103.965936	0.00961854	148.630930	0.00672807	214.582754	0.00466021
30	113.283211	0.00882743	164.494023	0.00607925	241.332684	0.00414366
31	123.345868	0.00810728	181.943425	0.00549621	271.292606	0.00368606
32	134.213537	0.00745081	201.137767	0.00497172	304.847719	0.00328033
33	145.950620	0.00685163	222.251442	0.00449941	342.429446	0.00292031
34	158.626670	0.00630411	245.476699	0.00407371	384.520979	0.00260064
35	172.316804	0.00580326	271.024368	0.00368971	431.663496	0.00231662
36	187.102148	0.00534467	299.126805	0.00334306	484.463116	0.00206414
37	203.070320	0.00492440	330.039486	0.00302994	543.598690	0.00183959
38	220.315945	0.00453894	364.043434	0.00274692	609.830533	0.00163980
39	238.941221	0.00418513	401.447778	0.00249098	684.010197	0.00146197
40	259.056519	0.00386016	442.592556	0.00225941	767.091420	0.00130363

(b) Monthly Compounding

No. of Periods n	8% per annum		10% per annum		12% per annum	
	$\dfrac{(1 + \frac{0.08}{12})^n - 1}{\frac{0.08}{12}}$	$\left[\dfrac{(1 + \frac{0.08}{12})^n - 1}{\frac{0.08}{12}}\right]^{-1}$	$\dfrac{(1 + \frac{0.10}{12})^n - 1}{\frac{0.10}{12}}$	$\left[\dfrac{(1 + \frac{0.10}{12})^n - 1}{\frac{0.10}{12}}\right]^{-1}$	$\dfrac{(1 + \frac{0.12}{12})^n - 1}{\frac{0.12}{12}}$	$\left[\dfrac{(1 + \frac{0.12}{12})^n - 1}{\frac{0.12}{12}}\right]^{-1}$
12	12.4499260	0.08032176	12.5655681	0.07958255	12.6825030	0.07884879
24	25.9331897	0.03856062	26.4469154	0.03781159	26.9734649	0.03707347
36	40.5355577	0.02466970	41.7818211	0.02393385	43.0768784	0.02321431
48	56.3499150	0.01774626	58.7224919	0.01702925	61.2226078	0.01633384
60	73.4768562	0.01360973	77.4370723	0.01291371	81.6696699	0.01224445
72	92.0253250	0.01086657	98.1113137	0.01019250	104.709931	0.00955019
84	112.113307	0.00891955	120.950418	0.00826785	130.672274	0.00765273
96	133.868583	0.00747001	146.181076	0.00684083	159.927293	0.00625284
108	157.429535	0.00635205	174.053713	0.00574535	192.892579	0.00518423
120	182.946035	0.00546609	204.844979	0.00488174	230.038689	0.00434709
132	210.580392	0.00474878	238.860493	0.00418654	271.895856	0.00367788
144	240.508386	0.00415786	276.437876	0.00361745	319.061559	0.00313419
156	272.920390	0.00366407	317.950103	0.00314515	372.209054	0.00268666
168	308.022573	0.00324652	363.809201	0.00274869	432.096982	0.00231430
180	346.038221	0.00288985	414.470347	0.00241272	499.580198	0.00200168
192	387.209149	0.00258258	470.436376	0.00212569	575.621974	0.00173725
204	431.797243	0.00231590	532.262781	0.00187877	661.307751	0.00151216
216	480.086127	0.00208296	600.563217	0.00166510	757.860630	0.00131950
228	532.382965	0.00187835	676.015602	0.00147926	866.658830	0.00115386
240	589.020414	0.00169773	759.368837	0.00131688	989.255365	0.00101086
252	650.358744	0.00153761	851.450246	0.00117447	1127.40021	0.00088700
264	716.788125	0.00139511	953.173781	0.00104913	1283.06528	0.00077938
276	788.731112	0.00126786	1065.54910	0.00093848	1458.47257	0.00068565
288	866.645331	0.00115387	1189.69158	0.00084055	1656.12591	0.00060382
300	951.026392	0.00105150	1326.83341	0.00075367	1878.84663	0.00053224
312	1042.41104	0.00095931	1478.33577	0.00067644	2129.81391	0.00046952
324	1141.38057	0.00087613	1645.70241	0.00060764	2412.61013	0.00041449
336	1248.56452	0.00080092	1830.59453	0.00054627	2731.27198	0.00036613
348	1364.64468	0.00073279	2034.84726	0.00049144	3090.34813	0.00032359
360	1490.35944	0.00067098	2260.48793	0.00044238	3494.96413	0.00028613

Table II   Present Value of an Annuity

(a) Annual Compounding

No. of Periods n	8% per annum $\dfrac{1-(1+0.08)^{-n}}{0.08}$	$\left[\dfrac{1-(1+0.08)^{-n}}{0.08}\right]^{-1}$	10% per annum $\dfrac{1-(1+0.10)^{-n}}{0.10}$	$\left[\dfrac{1-(1+0.10)^{-n}}{0.10}\right]^{-1}$	12% per annum $\dfrac{1-(1+0.12)^{-n}}{0.12}$	$\left[\dfrac{1-(1+0.12)^{-n}}{0.12}\right]^{-1}$
1	0.92592593	1.08000000	0.90909091	1.10000000	0.89285714	1.12000000
2	1.78326475	0.56076923	1.73553719	0.57619048	1.69005102	0.59169811
3	2.57709699	0.38803351	2.48685199	0.40211480	2.40183127	0.41634898
4	3.31212684	0.30192080	3.16986545	0.31547080	3.03734935	0.32923444
5	3.99271004	0.25045645	3.79078677	0.26379748	3.60477620	0.27740973
6	4.62287966	0.21631539	4.35526070	0.22960738	4.11140732	0.24322572
7	5.20637006	0.19207240	4.86841882	0.20540550	4.56375654	0.21911774
8	5.74663894	0.17401476	5.33492620	0.18744402	4.96763977	0.20130284
9	6.24688791	0.16007971	5.75902382	0.17364054	5.32824979	0.18767889
10	6.71008140	0.14902949	6.14456711	0.16274539	5.65022303	0.17698416
11	7.13896426	0.14007634	6.49506101	0.15396314	5.93769913	0.16841540
12	7.53607802	0.13269502	6.81369182	0.14676332	6.19437423	0.16143681
13	7.90377594	0.12652181	7.10335620	0.14077852	6.42354842	0.15567720
14	8.24423698	0.12129685	7.36668746	0.13574622	6.62816823	0.15087125
15	8.55947869	0.11682954	7.60607951	0.13147378	6.81086449	0.14682424
16	8.85136916	0.11297687	7.82370864	0.12781662	6.97398615	0.14339002
17	9.12163811	0.10962943	8.02155331	0.12466413	7.11963049	0.14045673
18	9.37188714	0.10670210	8.20141210	0.12193022	7.24967008	0.13793731
19	9.60359920	0.10412763	8.36492009	0.11954687	7.36577686	0.13576300
20	9.81814741	0.10185221	8.51356372	0.11745962	7.46944362	0.13387878
21	10.0168032	0.09983225	8.64869429	0.11562439	7.56200324	0.13224009
22	10.2007437	0.09803207	8.77154026	0.11400506	7.64464575	0.13081051
23	10.3710589	0.09642217	8.88321842	0.11257181	7.71843370	0.12955997
24	10.5287583	0.09497796	8.98474402	0.11129978	7.78431581	0.12846344
25	10.6747762	0.09367878	9.07704002	0.11016807	7.84313911	0.12749997
26	10.8099780	0.09250713	9.16094547	0.10915904	7.89565992	0.12665186
27	10.9351648	0.09144810	9.23722316	0.10825764	7.94255350	0.12590409
28	11.0510785	0.09048891	9.30656651	0.10745101	7.98442277	0.12524387
29	11.1584060	0.08961854	9.36960591	0.10672807	8.02180604	0.12466021
30	11.2577833	0.08882743	9.42691447	0.10607925	8.05518397	0.12414366
31	11.3497994	0.08810728	9.47901315	0.10549621	8.08498569	0.12368606
32	11.4349994	0.08745081	9.52637559	0.10497172	8.11159436	0.12328033
33	11.5138884	0.08685163	9.56943236	0.10449941	8.13535211	0.12292031
34	11.5869337	0.08630411	9.60857487	0.10407371	8.15656438	0.12260064
35	11.6545682	0.08580326	9.64415897	0.10368971	8.17550391	0.12231662
36	11.7171928	0.08534467	9.67650816	0.10334306	8.19241421	0.12206414
37	11.7751785	0.08492440	9.70591651	0.10302994	8.20751269	0.12183959
38	11.8288690	0.08453894	9.73265137	0.10274692	8.22099347	0.12163980
39	11.8785824	0.08418513	9.75695579	0.10249098	8.23302988	0.12146197
40	11.9246133	0.08386016	9.77905072	0.10225941	8.24377668	0.12130363

(b) Monthly Compounding

No. of Periods n	8% per annum $\dfrac{1-(1+\frac{0.08}{12})^{-n}}{\frac{0.08}{12}}$	$\left[\dfrac{1-(1+\frac{0.08}{12})^{-n}}{\frac{0.08}{12}}\right]^{-1}$	10% per annum $\dfrac{1-(1+\frac{0.10}{12})^{-n}}{\frac{0.10}{12}}$	$\left[\dfrac{1-(1+\frac{0.10}{12})^{-n}}{\frac{0.10}{12}}\right]^{-1}$	12% per annum $\dfrac{1-(1+\frac{0.12}{12})^{-n}}{\frac{0.12}{12}}$	$\left[\dfrac{1-(1+\frac{0.12}{12})^{-n}}{\frac{0.12}{12}}\right]^{-1}$
12	11.4957818	0.08698843	11.3745084	0.08791589	11.2550775	0.08884879
24	22.1105436	0.04522729	21.6708548	0.04614493	21.2433873	0.04707347
36	31.9118055	0.03133637	30.9912356	0.03226719	30.1075050	0.03321431
48	40.9619129	0.02441292	39.4281601	0.02536258	37.9739595	0.02633384
60	49.3184333	0.02027639	47.0653691	0.02124704	44.9550384	0.02224445
72	57.0345221	0.01753324	53.9786655	0.01852584	51.1503915	0.01955019
84	64.1592611	0.01558621	60.2366674	0.01660118	56.6484528	0.01765273

Table II *(Continued)*
(b) Monthly Compounding

No. of Periods n	8% per annum		10% per annum		12% per annum	
	$\dfrac{1 - (1 + \frac{0.08}{12})^{-n}}{\frac{0.08}{12}}$	$\left[\dfrac{1 - (1 + \frac{0.08}{12})^{-n}}{\frac{0.08}{12}}\right]^{-1}$	$\dfrac{1 - (1 + \frac{0.10}{12})^{-n}}{\frac{0.10}{12}}$	$\left[\dfrac{1 - (1 + \frac{0.10}{12})^{-n}}{\frac{0.10}{12}}\right]^{-1}$	$\dfrac{1 - (1 + \frac{0.12}{12})^{-n}}{\frac{0.12}{12}}$	$\left[\dfrac{1 - (1 + \frac{0.12}{12})^{-n}}{\frac{0.12}{12}}\right]^{-1}$
96	70.7379704	0.01413668	65.9014885	0.01517416	61.5277030	0.01625284
108	76.8124971	0.01301871	71.0293549	0.01407869	65.8577898	0.01518423
120	82.4214808	0.01213276	75.6711634	0.01321507	69.7005220	0.01434709
132	87.6006002	0.01141545	79.8729861	0.01251988	73.1107518	0.01367788
144	92.3827995	0.01082453	83.6765283	0.01195078	76.1371575	0.01313419
156	96.7984979	0.01033074	87.1195419	0.01147848	78.8229389	0.01268666
168	100.875784	0.00991318	90.2362006	0.01108203	81.2064335	0.01231430
180	104.640592	0.00955652	93.0574389	0.01074605	83.3216640	0.01200168
192	108.116871	0.00924925	95.6112588	0.01045902	85.1988236	0.01173725
204	111.326733	0.00898257	97.9230083	0.01021210	86.8647075	0.01151216
216	114.290596	0.00874963	100.015633	0.00999843	88.3430948	0.01131950
228	117.027313	0.00854501	101.909902	0.00981259	89.6550886	0.01115386
240	119.554292	0.00836400	103.624619	0.00965022	90.8194164	0.01101086
252	121.887607	0.00820428	105.176801	0.00950780	91.8526982	0.01088700
264	124.042099	0.00806178	106.581856	0.00938246	92.7696833	0.01077938
276	126.031475	0.00793453	107.853730	0.00927182	93.5834610	0.01068565
288	127.868388	0.00782054	109.005405	0.00917389	94.0356475	0.01060382
300	129.564523	0.00771816	110.047230	0.00908701	94.9465513	0.01053224
312	131.130668	0.00762598	110.990629	0.00900977	95.5153208	0.01046952
324	132.576786	0.00754280	111.844605	0.00894098	96.0200749	0.01041449
336	133.912076	0.00746759	112.617635	0.00887960	96.4680186	0.01036613
348	135.145031	0.00739946	113.317392	0.00882477	96.8655458	0.01032359
360	136.283494	0.00733765	113.950820	0.00877572	97.2183311	0.01028613

Table III   Standard Normal Curve Table
$Z = Z\text{-Score}$

An entry in the table is the area under the curve between $Z = 0$ and a positive value of Z. Areas for negative values of A are obtained by symmetry.

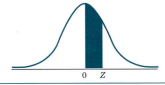

Z	0.00	0.01	0.02	0.03	0.04	0.05	0.06	0.07	0.08	0.09
0.0	0.0000	0.0040	0.0080	0.0120	0.0160	0.0199	0.0239	0.0279	0.0319	0.0359
0.1	0.0398	0.0438	0.0478	0.0517	0.0557	0.0596	0.0636	0.0675	0.0714	0.0753
0.2	0.0793	0.0832	0.0871	0.0910	0.0948	0.0987	0.1026	0.1064	0.1103	0.1141
0.3	0.1179	0.1217	0.1255	0.1293	0.1331	0.1368	0.1406	0.1433	0.1480	0.1517
0.4	0.1554	0.1591	0.1628	0.1664	0.1700	0.1736	0.1772	0.1808	0.1844	0.1879
0.5	0.1915	0.1950	0.1985	0.2019	0.2054	0.2088	0.2123	0.2157	0.2190	0.2224
0.6	0.2257	0.2291	0.2324	0.2357	0.2389	0.2422	0.2454	0.2486	0.2517	0.2549
0.7	0.2580	0.2611	0.2642	0.2673	0.2703	0.2734	0.2764	0.2794	0.2823	0.2852
0.8	0.2881	0.2910	0.2939	0.2967	0.2995	0.3023	0.3051	0.3078	0.3106	0.3133
0.9	0.3159	0.3186	0.3212	0.3238	0.3264	0.3289	0.3315	0.3340	0.3365	0.3389
1.0	0.3413	0.3438	0.3461	0.3485	0.3508	0.3531	0.3554	0.3577	0.3599	0.3621
1.1	0.3642	0.3665	0.3686	0.3708	0.3729	0.3749	0.3770	0.3790	0.3810	0.3830
1.2	0.3849	0.3869	0.3888	0.3907	0.3925	0.3944	0.3962	0.3980	0.3997	0.4015
1.3	0.4032	0.4049	0.4066	0.4082	0.4099	0.4115	0.4131	0.4147	0.4162	0.4177
1.4	0.4192	0.4207	0.4222	0.4236	0.4251	0.4265	0.4279	0.4292	0.4306	0.4319

Table III *(Continued)*
Z = Z-Score

An entry in the table is the area under the curve between $Z = 0$ and a positive
value of Z. Areas for negative values of A are obtained by symmetry.

Z	0.00	0.01	0.02	0.03	0.04	0.05	0.06	0.07	0.08	0.09
1.5	0.4332	0.4345	0.4357	0.4370	0.4382	0.4394	0.4406	0.4418	0.4429	0.4441
1.6	0.4452	0.4463	0.4474	0.4484	0.4495	0.4505	0.4515	0.4525	0.4535	0.4545
1.7	0.4554	0.4564	0.4573	0.4582	0.4591	0.4599	0.4608	0.4616	0.4625	0.4633
1.8	0.4641	0.4649	0.4656	0.4664	0.4671	0.4678	0.4686	0.4693	0.4699	0.4706
1.9	0.4713	0.4719	0.4726	0.4732	0.4738	0.4744	0.4750	0.4756	0.4761	0.4767
2.0	0.4772	0.4778	0.4783	0.4788	0.4793	0.4798	0.4803	0.4808	0.4812	0.4817
2.1	0.4821	0.4826	0.4830	0.4834	0.4838	0.4842	0.4846	0.4850	0.4854	0.4857
2.2	0.4861	0.4864	0.4868	0.4871	0.4875	0.4878	0.4881	0.4884	0.4887	0.4890
2.3	0.4893	0.4896	0.4898	0.4901	0.4904	0.4906	0.4909	0.4911	0.4913	0.4916
2.4	0.4918	0.4920	0.4922	0.4925	0.4927	0.4929	0.4931	0.4932	0.4934	0.4936
2.5	0.4938	0.4940	0.4941	0.4943	0.4945	0.4946	0.4948	0.4949	0.4951	0.4952
2.6	0.4953	0.4955	0.4956	0.4957	0.4959	0.4960	0.4961	0.4962	0.4963	0.4964
2.7	0.4965	0.4966	0.4967	0.4968	0.4969	0.4970	0.4971	0.4972	0.4973	0.4974
2.8	0.4974	0.4975	0.4976	0.4977	0.4977	0.4978	0.4979	0.4979	0.4980	0.4981
2.9	0.4981	0.4982	0.4982	0.4983	0.4984	0.4984	0.4985	0.4985	0.4986	0.4986
3.0	0.4987	0.4987	0.4987	0.4988	0.4988	0.4989	0.4989	0.4989	0.4990	0.4990

Answers To Odd-Numbered Problems

CHAPTER 1

Exercise 1.1 (page 15)

1. $A = (4, 2)$, $B = (6, 2)$, $C = (5, 3)$, $D = (-2, 1)$,
$E = (-2, -3)$, $F = (3, -2)$, $G = (6, -2)$, $H = (5, 0)$

3.

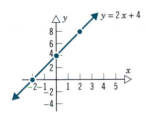

The set of points of the form $(2, y)$, where y is a real number, is a vertical line passing through 2 on the x-axis.

5.

x	0	-2	2	-2	4	-4
y	4	0	8	0	12	-4

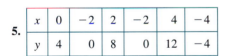

7.

x	0	3	2	-2	4	-4
y	-6	0	-2	-10	2	-14

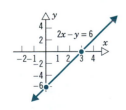

9. $\frac{1}{2}$; We interpret the slope to mean that for every 2-unit change in x, y will change by 1 unit.

11. -1; We interpret the slope to mean that for every 1-unit change in x, y will change by -1 unit.

13. Slope $= 3$

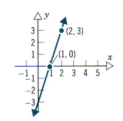

15. Slope $= \dfrac{-1}{2}$

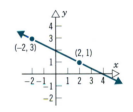

17. Slope $= 0$

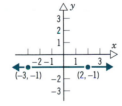

19. Undefined slope (vertical line)

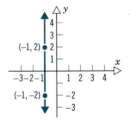

21. Slope $= \dfrac{\sqrt{3} - 3}{1 - \sqrt{2}} \approx 3.0611$

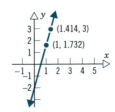

23.

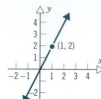

25.

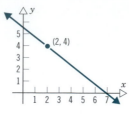

59. Slope $= -1$; y-intercept $= (0, 1)$

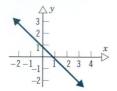

27.

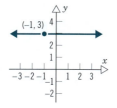

29.

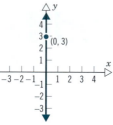

61. Slope is undefined; no y-intercept

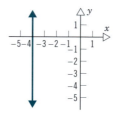

31. $x - 2y = 0$ **33.** $x + y = 2$ **35.** $2x - y = -9$

37. $2x + 3y = -1$ **39.** $x - 2y = -5$

41. $2x + y = 3$ **43.** $3x - y = -12$

45. $4x - 5y = 0$ **47.** $x - 2y = 2$

49. $x = 1$ **51.** $y = 4$

53. Slope $= 2$; y-intercept $= (0, 3)$

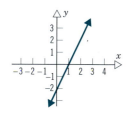

63. Slope $= 0$; y-intercept $= (0, 5)$

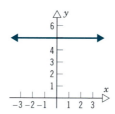

55. Slope $= 2$; y-intercept $= (0, -2)$

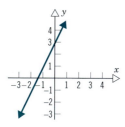

65. Slope $= 1$; y-intercept $= (0, 0)$

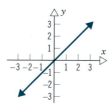

57. Slope $= \frac{2}{3}$; y-intercept $= (0, -2)$

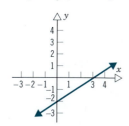

67. Slope $= \frac{3}{2}$; y-intercept $= (0, 0)$ **69.** $y = 0$

71. $y = -3$

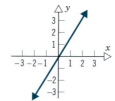

73. $9C - 5F = -160$ (where C represents degrees Celsius and F represents degrees Fahrenheit). When $F = 70$, $C = 21\frac{1}{9} \approx 21.111$.

75. $P = 0.50x - 100$ dollars

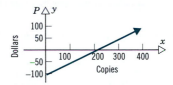

77. $C = 0.10494x + 9.36$, $0 \le x \le 400$. When $x = 100$, $C = \$19.85$, when $x = 300$, $C = \$40.84$. The slope of the line means that for every 1-unit change in kilowatt-hour, the cost of electricity increases by 10.494¢.

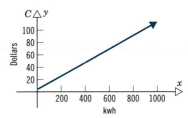

79. $C = 0.2x + 280$ dollars   **81.** $w = 5h - 190$

83. (a) $w = -4.75d + 347.5$ (where d represents the day of the month and w represents million gallons of water)
(b) 281 million gallons
(c) The slope means that each day the reservoir loses 4.75 million gallons of water.

Technology Exercises (page 18)

1. x-intercept $= (16.67, 0)$, y-intercept $= (0, 25)$

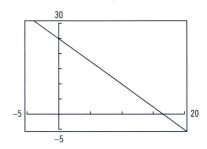

3. x-intercept $= (.25, 0)$, y-intercept $= (0, -530)$

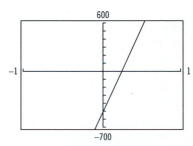

5. x-intercept $= (2.83, 0)$, y-intercept $= (0, 2.56)$

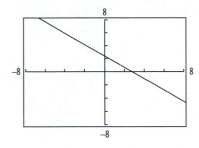

7. x-intercept $= (.78, 0)$, y-intercept $= (0, -1.41)$

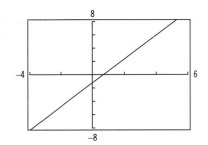

9.

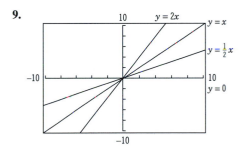

11. Lines of the form $y = 2x + b$ are parallel.

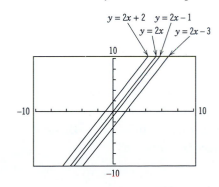

13. b   **15.** d   **17.** $y = x + 2$ or $x - y = -2$

19. $y = -\frac{1}{3}x + 1$ or $x + 3y = 3$

21. $y = -\frac{2}{3}x$ or $2x + 3y = 0$

Exercise 1.2 *(page 23)*

1. Parallel   **3.** Intersecting   **5.** Coincident

7. Parallel   **9.** Intersecting   **11.** Intersecting

13. (3, 2)   **15.** (3, 1)

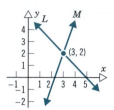

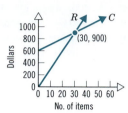

17. (1, 0)   **19.** (2, 1)

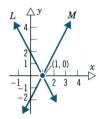

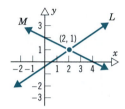

21. $(-1, 1)$ **23.** $(4, -2)$

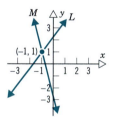

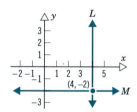

25. $m_1 = \frac{1}{3}$, $m_2 = -3$, $m_1 m_2 = -1$

27. $m_1 = -\frac{1}{2}$, $m_2 = 2$, $m_1 m_2 = -1$

29. $m_1 = -\frac{1}{4}$, $m_2 = 4$, $m_1 m_2 = -1$

31. $y = 2x - 3$ or $2x - y = 3$

33. $y = -\frac{1}{2}x + \frac{9}{2}$ or $x + 2y = 9$

35. $y = 4x + 6$ or $4x - y = -6$

37. $y = 2x$ or $2x - y = 0$ **39.** $x = 4$

41. $y = -\frac{1}{2}x - \frac{5}{2}$ or $x + 2y = -5$

43. $y = -\frac{1}{2}x + \frac{19}{30}$ or $15x + 30y = 19$ **45.** $t = 4$

47. $y = \frac{5}{19}x - \frac{85}{19}$ or $5x - 19y = 85$

Exercise 1.3 *(page 31)*

1. (a) \$80,000 (b) \$95,000 (c) \$105,000 (d) \$120,000

3. $C = 500\,t + 8000$ (where $t = 0$ corresponds to the year 1995 and C represents the average cost)

In 2000, $t = 5$ and $C = \$10,500$. The slope means that each year the average cost of a compact car increases by \$500.

5. (a) SAT $= -7t + 592$ (where $t = 0$ corresponds to the year 1994)

(b) 550

7. $x = 30$ **9.** $x = 500$

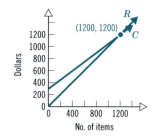

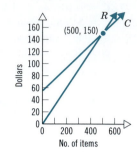

11. 1200 items   **13.** 200 newspapers

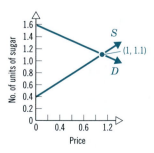

15. 20 caramels and 30 creams; increase the number of caramels to obtain a profit

17. \$50,000 in AA bonds and \$100,000 in S&L Certificates

19. 1510 adults (and 1090 children)

21. 30 cc of 15% solution and 70 cc of 5% solution

23. $p = \$1$ **25.** $p = \$10$

27. Market price is 1; supply at the market price is 1.1; the point of intersection occurs at the market price, i.e., the price at which the amount of sugar that sellers will supply matches the amount of sugar that consumers will purchase.

29. $D = 23 - 4p$

Chapter Review
True–False Items *(page 34)*

1. F   **2.** T   **3.** T   **4.** F   **5.** F   **6.** T

7. F   **8.** T   **9.** F   **10.** F

Fill in the Blanks *(page 34)*

1. x-coordinate, y-coordinate   **2.** undefined, zero

3. negative   **4.** parallel   **5.** coincident

6. perpendicular   **7.** intersecting

Review Exercises *(page 34)*

1.
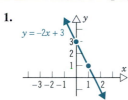

3.
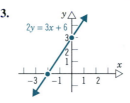

5. $x + 2y = 5$ **7.** $3x + 2y = 0$

9. $y = -3x + 5$ or $3x + y = 5$

11. $y = 4$ **13.** $y = -\frac{5}{2}x + 5$ or $5x + 2y = 10$

15. $y = -\frac{4}{3}x - 4$ or $4x + 3y = -12$

17. $y = -\frac{2}{3}x - \frac{1}{3}$ or $2x + 3y = -1$

19. $y = \frac{3}{2}x + \frac{21}{2}$ or $3x - 2y = -21$

21. Slope $= -4\frac{1}{2}$; y-intercept: $(0, 9)$

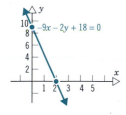

23. Slope $= -2$; y-intercept: $(0, 4\frac{1}{2})$

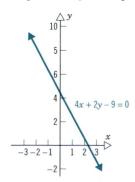

25. Parallel   **27.** Intersecting   **29.** Coincident

31. $(5, 1)$ **33.** $(1, 3)$

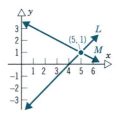 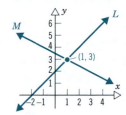

35. $(-2, 1)$

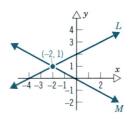

37. \$78,571.43 in B-rated bonds and \$11,428.57 in the bank

39. (a) & (b)

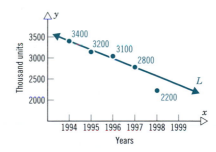

(c) $y - 3400 = -200(x - 1994)$ or $y = -200x + 402,200$, using the points $(1994, 3400)$ and $(1997, 2800)$

(d) 2400 thousand units (Answer varies with choice of L.)

(e) The slope means that each year the number of units sold decreases by approximately 200 thousand.

Mathematical Questions from Professional Exams *(page 36)*

1. b   **2.** a   **3.** d   **4.** b

5. c   **6.** c   **7.** b   **8.** b

CHAPTER 2

Exercise 2.1 *(page 49)*

1. $4 + 1 = 5; 10 - 2 = 8$: solution

3. $6 - 2 = 4; 2 - \frac{3}{2} = \frac{1}{2}$: solution

5. $3 - 3 + 4 = 4; 1 + 1 - 2 = 0; -2 - 6 = -8$: solution

7. $x = 6, y = 3$ **9.** $x = 3, y = 2$

11. $x = 8, y = -4$ **13.** $x = 4, y = -2$

15. Inconsistent   **17.** $x = \frac{1}{2}, y = \frac{3}{4}$

19. x is any real number, $y = 2 - \frac{1}{2}x$ *(Alternatively:* $x = 4 - 2y, y$ is any real number)

21. $x = \frac{1}{10}, y = \frac{2}{5}$ **23.** $x = \frac{3}{2}, y = 1$

25. x is any real number, $y = \frac{5}{3} - \frac{2}{3}x$ *(Alternatively:* $x = \frac{5}{2} - \frac{3}{2}y, y$ is any real number)

27. $x = \frac{4}{3}, y = \frac{1}{5}$ **29.** $x = 8, y = 2, z = 0$

31. $x = 2, y = -1, z = 1$ **33.** Inconsistent

35. x is any real number, $y = \frac{4}{5}x - \frac{7}{5}, z = \frac{1}{5}x + \frac{2}{5}$ *(Alternatively:* $x = \frac{5}{4}y + \frac{7}{4}, z = \frac{1}{4}y + \frac{3}{4}, y$ is any real number or $x = 5z - 2, y = 4z - 3, z$ is any real number)

37. Inconsistent   **39.** $x = 1, y = 3, z = -2$

41. $x = -3, y = \frac{1}{2}, z = 1$

43. 61 and 20 (or $\frac{181}{5}$ and $\frac{224}{5}$)

45. Length = 30 ft, width = 15 ft

47. Cheeseburger: \$1.55; shake: \$0.85

49. \$31,250 in AA bonds; \$18,750 in S & L Certificates

51. 8 nickels   **53.** 60 cc of 10% acid; 40 cc of 30% acid

55. 3260 adults

57. \$37,500 in the first (10%) investment; \$12,500 in the second (12%) investment

59. 100 acres of corn; 900 acres of soybeans

61. 40 of the \$20 sets; 160 of the \$15 sets

Technology Exercises *(page 51)*

1. $x = -2.54, y = 28.81$

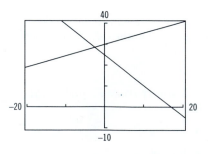

3. $x = 3.07, y = -.22$

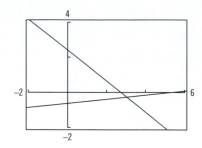

Exercise 2.2 *(page 62)*

1. $\begin{bmatrix} 2 & -3 & | & 5 \\ 1 & -1 & | & 3 \end{bmatrix}$ **3.** $\begin{bmatrix} 2 & 1 & | & -6 \\ 3 & 1 & | & -1 \end{bmatrix}$

5. $\begin{bmatrix} 2 & -1 & -1 & | & 0 \\ 1 & -1 & 1 & | & 1 \\ 3 & -1 & 0 & | & 2 \end{bmatrix}$

7. $\begin{bmatrix} 2 & -3 & 1 & | & 7 \\ 1 & 1 & -1 & | & 1 \\ 2 & 2 & -3 & | & -4 \end{bmatrix}$

9. $\begin{bmatrix} 4 & -1 & 2 & -1 & | & 4 \\ 1 & 1 & 0 & 0 & | & -6 \\ 0 & 2 & -1 & 1 & | & 5 \end{bmatrix}$

11. $\begin{bmatrix} 1 & -1 & 1 & -1 & | & 0 \\ 2 & 3 & -1 & 4 & | & 5 \end{bmatrix}$

13. (a) $\begin{bmatrix} 1 & -3 & -5 & | & -2 \\ 0 & 1 & 6 & | & 9 \\ -3 & 5 & 4 & | & 6 \end{bmatrix}$

(b) $\begin{bmatrix} 1 & -3 & -5 & | & -2 \\ 0 & 1 & 6 & | & 9 \\ 0 & -4 & -11 & | & 0 \end{bmatrix}$

(c) $\begin{bmatrix} 1 & -3 & -5 & | & -2 \\ 0 & 1 & 6 & | & 9 \\ 0 & 0 & 13 & | & 36 \end{bmatrix}$

15. (a) $\begin{bmatrix} 1 & -3 & 4 & | & 3 \\ 0 & 1 & -2 & | & 0 \\ -3 & 3 & 4 & | & 6 \end{bmatrix}$

(b) $\begin{bmatrix} 1 & -3 & 4 & | & 3 \\ 0 & 1 & -2 & | & 0 \\ 0 & -6 & 16 & | & 15 \end{bmatrix}$

(c) $\begin{bmatrix} 1 & -3 & 4 & | & 3 \\ 0 & 1 & -2 & | & 0 \\ 0 & 0 & 4 & | & 15 \end{bmatrix}$

17. (a) $\begin{bmatrix} 1 & -3 & 2 & | & -6 \\ 0 & 1 & -1 & | & 8 \\ -3 & -6 & 4 & | & 6 \end{bmatrix}$

(b) $\begin{bmatrix} 1 & -3 & 2 & | & -6 \\ 0 & 1 & -1 & | & 8 \\ 0 & -15 & 10 & | & -12 \end{bmatrix}$

(c) $\begin{bmatrix} 1 & -3 & 2 & | & -6 \\ 0 & 1 & -1 & | & 8 \\ 0 & 0 & -5 & | & 108 \end{bmatrix}$

19. (a) $\begin{bmatrix} 1 & -3 & 1 & | & -2 \\ 0 & 1 & 4 & | & 2 \\ -3 & 1 & 4 & | & 6 \end{bmatrix}$

(b) $\begin{bmatrix} 1 & -3 & 1 & | & -2 \\ 0 & 1 & 4 & | & 2 \\ 0 & -8 & 7 & | & 0 \end{bmatrix}$

(c) $\begin{bmatrix} 1 & -3 & 1 & | & -2 \\ 0 & 1 & 4 & | & 2 \\ 0 & 0 & 39 & | & 16 \end{bmatrix}$

21. $\begin{cases} x + 2y = 5 \\ \quad\quad y = -1 \end{cases}$
Consistent $x = 7, y = -1$

23. $\begin{cases} x + 2y + 3z = 1 \\ \quad\quad y + 4z = 2 \\ 0x + 0y + 0z = 3 \end{cases}$
Inconsistent

25. $\begin{cases} x \quad\quad + 2z = -1 \\ \quad y - 4z = -2 \\ 0x + 0y + 0z = 0 \end{cases}$
Consistent
$x = 2z - 1, y = 4z - 2, z$ is any real number

27. $\begin{cases} x_1 + 2x_2 - x_3 + x_4 = 1 \\ \quad\quad x_2 + 4x_3 + x_4 = 2 \\ \quad\quad\quad\quad x_3 + 2x_4 = 3 \end{cases}$
Consistent
$x_1 = 24 - 17x_4, x_2 = -10 + 7x_4, x_3 = 3 - 2x_4, x_4$ is any real number

29. $\begin{cases} x_1 + 2x_2 \quad\quad + 4x_4 = 2 \\ \quad\quad x_2 + x_3 + 3x_4 = 3 \\ 0x_1 + 0x_2 + 0x_3 + 0x_4 = 0 \end{cases}$
Consistent
$x_1 = -4 + 2x_3 + 2x_4, x_2 = 3 - x_3 - 3x_4, x_3$ and x_4 are any real numbers

31. $\begin{cases} x_1 - 2x_2 \quad\quad + x_4 = -2 \\ \quad\quad x_2 - 3x_3 + 2x_4 = 2 \\ \quad\quad\quad\quad x_3 - x_4 = 0 \\ 0x_1 + 0x_2 + 0x_3 + 0x_4 = 0 \end{cases}$
Consistent
$x_1 = 2 + x_4, x_2 = 2 + x_4, x_3 = x_4, x_4$ is any real number

33. $x = 2, y = 4$ **35.** $x = 2, y = 1$

37. $x = 2, y = 1$ **39.** $x = 2, y = -3$

41. $x = \frac{1}{2}, y = \frac{1}{3}$ **43.** $x = \frac{3}{2}, y = \frac{2}{3}$ **45.** $x = 2, y = 3$

47. $x = \frac{2}{3}, y = \frac{1}{3}$ **49.** $x = 1, y = 4, z = 0$

51. $x = -1, y = 1, z = 2$ **53.** $x = 2, y = -1, z = 1$

55. $x = \frac{1}{2}, y = \frac{1}{4}, z = \frac{3}{4}$

57. $x = \frac{1}{3}, y = \frac{2}{3}, z = 1$ **59.** No solution

61. Infinitely many solutions

63. No solution   **65.** Unique solution

67. Unique solution   **69.** No solution

71. 22.5 lb of cashews

73. 10 two-student work stations and 6 three-student work stations

75. $5.56

77. 40 orchestra seats, 260 main seats, and 200 balcony seats

79. $3000 invested at 6%, $1500 invested at 8%, and $2000 invested at 9%

81. 20 cases of orange juice, 12 cases of tomato juice, 6 cases of pineapple juice

83. 2 packages of first type, 10 packages of second type, 4 packages of third type

85. 5 units of food I, 5 units of food II, $\frac{5}{3}$ units of food III

87. 4 assorted cartons, 8 mixed cartons, 5 single cartons

89. 3 large cans, 2 mammoth cans, 4 giant cans

Technology Exercises (page 65)

1. $\begin{bmatrix} -2 & 0 & -8 & | & 0 \\ -2 & 1 & -4 & | & 1 \\ 3 & -1 & 0 & | & 1 \end{bmatrix}$

3. $\begin{bmatrix} 1 & 0 & 4 & | & 0 \\ 1 & 1 & 8 & | & 1 \\ 3 & -1 & 0 & | & 1 \end{bmatrix}$

5. REF: $\begin{bmatrix} 1 & -1 & 0.5 & | & 1 \\ 0 & 1 & 3 & | & 0 \\ 0 & 0 & 1 & | & 0.22 \end{bmatrix}$ or $\begin{bmatrix} 1 & -1 & \frac{1}{2} & | & 1 \\ 0 & 1 & 3 & | & 0 \\ 0 & 0 & 1 & | & \frac{2}{9} \end{bmatrix}$;

RREF: $\begin{bmatrix} 1 & 0 & 0 & | & 0.22 \\ 0 & 1 & 0 & | & -0.67 \\ 0 & 0 & 1 & | & 0.22 \end{bmatrix}$ or $\begin{bmatrix} 1 & 0 & 0 & | & \frac{2}{9} \\ 0 & 1 & 0 & | & -\frac{2}{3} \\ 0 & 0 & 1 & | & \frac{2}{9} \end{bmatrix}$

$x = .22, y = -.67, z = .22$ $x = \frac{2}{9}, y = -\frac{2}{3}, z = \frac{2}{9}$

7. REF: $\begin{bmatrix} 1 & 1 & 1 & | & 4 \\ 0 & 1 & 1 & | & 2 \\ 0 & 0 & 1 & | & 3 \end{bmatrix}$; RREF $\begin{bmatrix} 1 & 0 & 0 & | & 2 \\ 0 & 1 & 0 & | & -1 \\ 0 & 0 & 1 & | & 3 \end{bmatrix}$

$x = 2, y = -1, z = 3$

9. REF: $\begin{bmatrix} 1 & 1 & 1 & 1 & | & 20 \\ 0 & 1 & 1 & 1 & | & 0 \\ 0 & 0 & 1 & 1 & | & 13 \\ 0 & 0 & 0 & 1 & | & -4 \end{bmatrix}$;

RREF $\begin{bmatrix} 1 & 0 & 0 & 0 & | & 20 \\ 0 & 1 & 0 & 0 & | & -13 \\ 0 & 0 & 1 & 0 & | & 17 \\ 0 & 0 & 0 & 1 & | & -4 \end{bmatrix}$

$x_1 = 20, x_2 = -13,$
$x_3 = 17, x_4 = -4$

Exercise 2.3 (page 76)

1. Not in reduced row-echelon form

3. Not in reduced row-echelon form

5. Not in reduced row-echelon form

7. In reduced row-echelon form

9. In reduced row-echelon form

11. In reduced row-echelon form

13. Infinitely many solutions

15. One solution   **17.** Infinitely many solutions

19. Infinitely many solutions

21. Infinitely many solutions   **23.** One solution

25. Infinitely many solutions

27. Infinitely many solutions

29. $\begin{bmatrix} 1 & 0 & | & 2 \\ 0 & 1 & | & 1 \end{bmatrix}$; $x = 2, y = 1$

31. $\begin{bmatrix} 1 & 0 & | & 0 \\ 0 & 1 & | & 0 \\ 0 & 0 & | & 1 \end{bmatrix}$; inconsistent

33. $\begin{bmatrix} 1 & -2 & | & 4 \\ 0 & 0 & | & 0 \\ 0 & 0 & | & 0 \end{bmatrix}$; infinitely many solutions: $x = 4 + 2y$, y is any real number

35. $\begin{bmatrix} 1 & 0 & 0 & | & -3 \\ 0 & 1 & 3 & | & 5 \\ 0 & 0 & 0 & | & 0 \end{bmatrix}$; infinitely many solutions: $x = -3, y = 5 - 3z, z$ is any real number

37. $\begin{bmatrix} 1 & 0 & 0 & | & 3 \\ 0 & 1 & 0 & | & 2 \\ 0 & 0 & 1 & | & -4 \end{bmatrix}$; $x = 3, y = 2, z = -4$

39. $\begin{bmatrix} 1 & 0 & 0 & 0 & | & -17 \\ 0 & 1 & 0 & 0 & | & 24 \\ 0 & 0 & 1 & 0 & | & 33 \\ 0 & 0 & 0 & 1 & | & 14 \end{bmatrix}$; $x_1 = -17, x_2 = 24,$ $x_3 = 33, x_4 = 14$

41. $\begin{bmatrix} 1 & 0 & 0 & -\frac{1}{3} & | & \frac{4}{3} \\ 0 & 1 & 0 & -\frac{11}{15} & | & \frac{14}{15} \\ 0 & 0 & 1 & \frac{4}{15} & | & -\frac{16}{15} \end{bmatrix}$;

infinitely many solutions; we can solve for x_1, x_2, and x_3 in terms of x_4: $x_1 = \frac{4}{3} + \frac{1}{3} x_4, x_2 = \frac{14}{15} + \frac{11}{15} x_4,$ $x_3 = -\frac{16}{15} - \frac{4}{15} x_4$

43. $\begin{bmatrix} 1 & -1 & 1 & | & 0 \\ 0 & 0 & 0 & | & 1 \end{bmatrix}$; inconsistent

45. $\begin{bmatrix} 1 & 0 & \frac{7}{12} & | & 0 \\ 0 & 1 & -\frac{1}{4} & | & 0 \\ 0 & 0 & 0 & | & 1 \end{bmatrix}$; inconsistent

47. $\begin{bmatrix} 1 & 0 & 0 & 0 & | & 1 \\ 0 & 1 & 0 & 0 & | & 2 \\ 0 & 0 & 1 & 0 & | & 0 \\ 0 & 0 & 0 & 1 & | & 1 \end{bmatrix}$; $x_1 = 1, x_2 = 2, x_3 = 0, x_4 = 1$

49. $\begin{bmatrix} 1 & 0 & 0 & | & 0 \\ 0 & 1 & 1 & | & 0 \\ 0 & 0 & 0 & | & 1 \end{bmatrix}$; inconsistent

51. $\begin{bmatrix} 1 & 0 & 0 & | & 0 \\ 0 & 1 & -1 & | & -6 \\ 0 & 0 & 0 & | & 0 \end{bmatrix}$; infinitely many solutions: $x = 0, y = -6 + z, z$ is any real number

53.

No. of Liters 10% Solution	No. of Liters 30% Solution	No. of Liters 50% Solution
55	15	30
50	25	25
45	35	20
40	45	15
35	55	10
30	65	5

55. The information is not sufficient to determine the prices; possibilities include:

Hamburger	Fries	Cola
$2.13	$0.89	$0.62
$2.10	$0.90	$0.65
$2.07	$0.91	$0.68
$2.04	$0.92	$0.71
$2.01	$0.93	$0.74
$1.98	$0.94	$0.77
$1.95	$0.95	$0.80
$1.92	$0.96	$0.83
$1.89	$0.97	$0.86
$1.86	$0.98	$0.89

57.

Treasury Bills	Corporate Bonds	Junk Bonds
$12,500	$12,500	$0
$13,500	$10,500	$1,000
$14,500	$8,500	$2,000
$15,500	$6,500	$3,000
$16,500	$4,500	$4,000
$17,500	$2,500	$5,000
$18,500	$500	$6,000

(b)

Treasury Bills	Corporate Bonds	Junk Bonds
$500	$11,500	$13,000
$1,500	$9,500	$14,000
$2,500	$7,500	$15,000
$3,500	$5,500	$16,000
$4,500	$3,500	$17,000
$5,500	$1,500	$18,000

59. (a) If they invest all $25,000 in Treasury bills, their return will be $1750 per year, greater than the $1500 they require.

(c) As the required return increases, the amount they can invest in Treasury bills diminishes, and the amount they must invest in junk bonds increases.

Technology Exercises (page 78)

1. REF: $\begin{bmatrix} 1 & 2 & 1 & 3 & | & 4 \\ 0 & 1 & .33 & 1.67 & | & .67 \\ 0 & 0 & 1 & 2 & | & 8 \\ 0 & 0 & 0 & 1 & | & 3 \end{bmatrix}$ or $\begin{bmatrix} 1 & 2 & 1 & 3 & | & 4 \\ 0 & 1 & \frac{1}{3} & \frac{5}{3} & | & \frac{2}{3} \\ 0 & 0 & 1 & 2 & | & 8 \\ 0 & 0 & 0 & 1 & | & 3 \end{bmatrix}$; RREF: $\begin{bmatrix} 1 & 0 & 0 & 0 & | & 3 \\ 0 & 1 & 0 & 0 & | & -5 \\ 0 & 0 & 1 & 0 & | & 2 \\ 0 & 0 & 0 & 1 & | & 3 \end{bmatrix}$; $x_1 = 3, x_2 = -5, x_3 = 2, x_4 = 3$

3. REF: $\begin{bmatrix} 1 & 1 & -1 & -1 & | & 6 \\ 0 & 1 & -3 & -1 & | & 12 \\ 0 & 0 & 1 & 1 & | & -10 \\ 0 & 0 & 0 & 1 & | & -19 \end{bmatrix}$; RREF: $\begin{bmatrix} 1 & 0 & 0 & 0 & | & -24 \\ 0 & 1 & 0 & 0 & | & 20 \\ 0 & 0 & 1 & 0 & | & 9 \\ 0 & 0 & 0 & 1 & | & -19 \end{bmatrix}$; $x_1 = -24, x_2 = 20, x_3 = 9, x_4 = -19$

5. REF: $\begin{bmatrix} 1 & -2 & -5 & 0 & | & -25 \\ 0 & 1 & 1 & 0 & | & 15 \\ 0 & 0 & 1 & 4 & | & -15 \\ 0 & 0 & 0 & 1 & | & -10 \end{bmatrix}$; RREF: $\begin{bmatrix} 1 & 0 & 0 & 0 & | & 80 \\ 0 & 1 & 0 & 0 & | & -10 \\ 0 & 0 & 1 & 0 & | & 25 \\ 0 & 0 & 0 & 1 & | & -10 \end{bmatrix}$; $x_1 = 80, x_2 = -10, x_3 = 25, x_4 = -10$

Exercise 2.4 (page 87)

1. 2×2 **3.** 2×3 **5.** 3×2 **7.** 3×2

9. 2×1 **11.** 1×1

13. False. Two equal matrices must have the same dimensions.

15. True   **17.** True   **19.** True   **21.** True

23. False. The sum of two 2×1 matrices is a 2×1 matrix.

25. $\begin{bmatrix} 1 & 1 \\ 6 & 7 \end{bmatrix}$ **27.** $\begin{bmatrix} 6 & 18 & 0 \\ 12 & -6 & 3 \end{bmatrix}$

29. $\begin{bmatrix} 2 & 4 & -8 \\ 1 & -4 & -1 \end{bmatrix}$ **31.** $\begin{bmatrix} 13a & 54 \\ -2b & -7 \\ -2c & -6 \end{bmatrix}$

33. $\begin{bmatrix} 3 & -5 & 4 \\ 5 & 3 & 3 \end{bmatrix}$ **35.** $\begin{bmatrix} 13 & -6 & -7 \\ -6 & 1 & -7 \end{bmatrix}$

37. $\begin{bmatrix} 9 & -5 & -6 \\ 1 & 1 & -3 \end{bmatrix}$ **39.** $\begin{bmatrix} -2 & -17 & 32 \\ 28 & 14 & 23 \end{bmatrix}$

41. $\begin{bmatrix} 5 & -2 & 3 \\ -12 & 1 & -5 \end{bmatrix}$ **43.** $\begin{bmatrix} 23 & -7 & -18 \\ -17 & -1 & -17 \end{bmatrix}$

45. $A + B = \begin{bmatrix} 2+1 & -3+(-2) & 4+0 \\ 0+5 & 2+1 & 1+2 \end{bmatrix}$
$= \begin{bmatrix} 3 & -5 & 4 \\ 5 & 3 & 3 \end{bmatrix}$;

$B + A = \begin{bmatrix} 1+2 & -2+(-3) & 0+4 \\ 5+0 & 1+2 & 2+1 \end{bmatrix}$
$= \begin{bmatrix} 3 & -5 & 4 \\ 5 & 3 & 3 \end{bmatrix}$; so

$A + B = \begin{bmatrix} 3 & -5 & 4 \\ 5 & 3 & 3 \end{bmatrix} = B + A$

47. $A + (-A) = \begin{bmatrix} 2 & -3 & 4 \\ 0 & 2 & 1 \end{bmatrix} + \begin{bmatrix} -2 & 3 & -4 \\ 0 & -2 & -1 \end{bmatrix}$
$= \begin{bmatrix} 2+(-2) & -3+3 & 4+(-4) \\ 0+0 & 2+(-2) & 1+(-1) \end{bmatrix}$
$= \begin{bmatrix} 0 & 0 & 0 \\ 0 & 0 & 0 \end{bmatrix} = \mathbf{0}$

49. $2B + 3B = \begin{bmatrix} 2 & -4 & 0 \\ 10 & 2 & 4 \end{bmatrix} + \begin{bmatrix} 3 & -6 & 0 \\ 15 & 3 & 6 \end{bmatrix}$

$= \begin{bmatrix} 5 & -10 & 0 \\ 25 & 5 & 10 \end{bmatrix} = 5B$

51. $x = -4, z = 4$ **53.** $x = 5, y = 1$

55. $x = 4, y = -6, z = 6$

57.

	$\frac{1}{2}''$	$1''$	$2''$
Steel	25	45	35
Aluminum	13	20	23

or

	Steel	Aluminum
$\frac{1}{2}''$	25	13
$1''$	45	20
$2''$	35	23

59.

	<$25,000	≥$25,000
Dem.	351	203
Rep.	271	215
Ind.	73	55

or

	Dem.	Rep.	Ind.
<$25,000	351	271	73
≥$25,000	203	215	55

61.

	LAS	ENG	EDUC
Male	250	225	80
Female	250	75	120

Technology Exercises (page 89)

1. $\begin{bmatrix} -2 & 1 & 7 & 5 \\ 4 & 6 & 7 & 5 \\ -3.5 & 8 & -4 & 13 \\ 12 & -1 & 7 & 6 \end{bmatrix}$

3. $\begin{bmatrix} 19 & -11 & -14 & -15 \\ -12 & -13 & -21 & -17 \\ 15.5 & -24 & 19 & -39 \\ -29 & 10 & -14 & -11 \end{bmatrix}$

5. $\begin{bmatrix} 37 & -17 & 30 & 15 \\ 4 & 9 & 13 & 1 \\ 20.5 & 16 & -3 & 31 \\ 29 & 18 & 22 & 43 \end{bmatrix}$

Exercise 2.5 (page 96)

1. $[14]$ **3.** $[4]$ **5.** $[18 \quad -8]$ **7.** $\begin{bmatrix} 4 & 2 \\ 2 & 8 \end{bmatrix}$

9. $[4 \quad 6]$ **11.** $\begin{bmatrix} 4 & -14 \\ 12 & -32 \end{bmatrix}$ **13.** $\begin{bmatrix} 4 & 2 \\ 2 & 8 \\ 9 & 8 \end{bmatrix}$

15. $\begin{bmatrix} 9 & 1 \\ 5 & 4 \\ 11 & 7 \end{bmatrix}$ **17.** BA is defined; dimension is 3×4.

19. AB is not defined.   **21.** $(BA)C$ is not defined.

23. $BA + A$ is defined; dimension is 3×4.

25. $DC + B$ is defined; dimension is 3×3.

27. $\begin{bmatrix} -1 & 10 & -1 \\ -4 & 16 & -8 \end{bmatrix}$ **29.** $\begin{bmatrix} 11 & 5 \\ 13 & -9 \end{bmatrix}$

31. $\begin{bmatrix} 6 & 10 \\ 8 & 2 \\ -4 & 5 \end{bmatrix}$ **33.** $\begin{bmatrix} 3 & -1 \\ 4 & 2 \end{bmatrix}$

35. $\begin{bmatrix} 8 & 4 & 22 \\ 4 & 32 & 16 \end{bmatrix}$ **37.** $\begin{bmatrix} -14 & 7 \\ -20 & -6 \end{bmatrix}$

39. $\begin{bmatrix} 10 & 30 & 37 \\ 15 & 16 & 50 \\ -6 & 20 & -8 \end{bmatrix}$

41. $D(CB) =$

$\begin{bmatrix} 1 & 0 & 4 \\ 0 & 1 & 2 \\ 0 & -1 & 1 \end{bmatrix}\left(\begin{bmatrix} 3 & 1 \\ 4 & -1 \\ 0 & 2 \end{bmatrix}\begin{bmatrix} 1 & 2 & 3 \\ -1 & 4 & -2 \end{bmatrix}\right)$

$= \begin{bmatrix} 1 & 0 & 4 \\ 0 & 1 & 2 \\ 0 & -1 & 1 \end{bmatrix}\begin{bmatrix} 2 & 10 & 7 \\ 5 & 4 & 14 \\ -2 & 8 & -4 \end{bmatrix}$

$= \begin{bmatrix} -6 & 42 & -9 \\ 1 & 20 & 6 \\ -7 & 4 & -18 \end{bmatrix};$

$(DC)B =$

$\left(\begin{bmatrix} 1 & 0 & 4 \\ 0 & 1 & 2 \\ 0 & -1 & 1 \end{bmatrix}\begin{bmatrix} 3 & 1 \\ 4 & -1 \\ 0 & 2 \end{bmatrix}\right)\begin{bmatrix} 1 & 2 & 3 \\ -1 & 4 & -2 \end{bmatrix}$

$= \begin{bmatrix} 3 & 9 \\ 4 & 3 \\ -4 & 3 \end{bmatrix}\begin{bmatrix} 1 & 2 & 3 \\ -1 & 4 & -2 \end{bmatrix}$

$= \begin{bmatrix} -6 & 42 & -9 \\ 1 & 20 & 6 \\ -7 & 4 & -18 \end{bmatrix};$ so $D(CB) = (DC)B$

43. $AB = \begin{bmatrix} 5 & -2 \\ 6 & 4 \end{bmatrix};$ $BA = \begin{bmatrix} 7 & -3 \\ 6 & 2 \end{bmatrix}$

45. $A = \begin{bmatrix} 2 & 1 \\ -\frac{1}{2} & -\frac{1}{2} \end{bmatrix}$

47. $\begin{bmatrix} 1 & 2 & 5 \\ 2 & 4 & 10 \\ -1 & -2 & -5 \end{bmatrix}\begin{bmatrix} 1 & 2 & 5 \\ 2 & 4 & 10 \\ -1 & -2 & -5 \end{bmatrix} = \begin{bmatrix} 0 & 0 & 0 \\ 0 & 0 & 0 \\ 0 & 0 & 0 \end{bmatrix}$

49. The possibilities are: $a = 0, b = 0$; $a = -1, b = 0$; $a = -\frac{1}{2}, b = \frac{1}{2}$; $a = -\frac{1}{2}, b = -\frac{1}{2}$.

51. $[\frac{1}{3} \quad \frac{2}{3}]$

53. (a) PQ represents the matrix of raw materials needed to fill the order: $PQ = [138 \quad 189 \quad 97 \quad 399]$

(b) QC represents the matrix of costs to produce each product:

$QC = \begin{bmatrix} 311 \\ 653 \\ 614 \end{bmatrix}$

(c) PQC represents the total cost to produce the order: $PQC = \$13,083$

Technology Exercises (page 98)

1. $\begin{bmatrix} .5 & 16 & -30 & 25 \\ 21 & 14 & 28 & 64 \\ 19.5 & 8 & -23 & 33 \\ -9.5 & 45 & -9 & 83 \end{bmatrix}$

3. $\begin{bmatrix} 31.5 & 251 & -31.5 & 143 \\ 861 & 350 & 791 & 420 \\ 369.5 & 115 & 206.5 & 215 \\ 412.5 & 882 & 451.5 & 491 \end{bmatrix}$

5. $\begin{bmatrix} 66 & 74 & 94 & 38 \\ 71 & 13 & 106 & 28 \\ 165 & 124.5 & 79 & 52 \\ 158 & -3 & 152 & 46 \end{bmatrix}$

7. $\begin{bmatrix} -5 & 23 & -102 & 44 \\ -108 & -56 & -70 & 122 \\ 108 & -152 & -67 & 36 \\ -346 & 279 & -249 & 187 \end{bmatrix}$

9. (a) $A^2 = \begin{bmatrix} -2.95 & -.5 & -1.21 & 1.6 & -.49 \\ -2.8 & 2.56 & 2.54 & 1.4 & 1.44 \\ 5.2 & 8.6 & 1.11 & -2.6 & -.26 \\ -1.5 & -1 & .3 & 1 & .3 \\ 3.6 & -1.8 & -.63 & -1.8 & -.18 \end{bmatrix}$;

$A^{10} = \begin{bmatrix} 433.0971 & -1583.5617 & -369.8169 & -216.5481 & -141.6383 \\ 1207.6949 & 5998.4376 & 2563.2274 & -603.8474 & 1045.7047 \\ -1423.0228 & 8065.0704 & 4271.7984 & 711.5114 & 2023.3641 \\ 479.4764 & -246.5381 & -231.5431 & -239.7372 & -140.5573 \\ -899.3405 & -3064.3130 & -1627.5015 & 449.6703 & -698.6086 \end{bmatrix}$;

(Rounded to four decimal places)

$A^{15} = \begin{bmatrix} -2247.8448 & -118449.3176 & -69122.3644 & 1123.9224 & -32134.3035 \\ -11094.8095 & 542433.5133 & 249928.4018 & 5547.4047 & 110820.2098 \\ 100366.7830 & 831934.5635 & 386083.7906 & -50183.3915 & 165597.1358 \\ -18782.2970 & -38432.0936 & -18654.0856 & 9391.1485 & -7395.9599 \\ -1633.6655 & -306291.0222 & -130154.5685 & 816.8327 & -56015.4536 \end{bmatrix}$

(Rounded to four decimal places)

(b) $A^2 = \begin{bmatrix} 1.16 & 1.42 & -.408 & .29 \\ 1.25 & -1.26 & .506 & .91 \\ 13.9 & -5.29 & .632 & 7.8 \\ -2.74 & -.53 & .6 & -.788 \end{bmatrix}$;

$A^{10} = \begin{bmatrix} -31.1428 & 14.0580 & 1.0561 & -16.1157 \\ 28.4909 & -.3398 & -3.2689 & 14.6394 \\ -50.8954 & 149.5593 & -30.8932 & -34.3845 \\ 52.3686 & -47.5375 & 5.6205 & 29.5703 \end{bmatrix}$;

(Rounded to four decimal places)

$A^{15} = \begin{bmatrix} 111.7124 & -169.5588 & 31.0626 & 68.5252 \\ -155.7116 & 145.0638 & -15.6641 & -86.5465 \\ -621.4964 & -218.2318 & 157.5564 & -283.4745 \\ -1.3722 & 252.7427 & -69.4323 & -20.2157 \end{bmatrix}$

(Rounded to four decimal places)

(c) $A^2 = \begin{bmatrix} 1 & 0 & 1 \\ 1 & 2 & 1 \\ 2 & 2 & 2 \end{bmatrix}$; $A^{10} = \begin{bmatrix} 171 & 170 & 171 \\ 341 & 342 & 341 \\ 512 & 512 & 512 \end{bmatrix}$;

$A^{15} = \begin{bmatrix} 5461 & 5462 & 5461 \\ 10923 & 10922 & 10923 \\ 16384 & 16384 & 16384 \end{bmatrix}$

Exercise 2.6 (*page 105*)

1. $\begin{bmatrix} 1 & 2 \\ 2 & 3 \end{bmatrix}\begin{bmatrix} -3 & 2 \\ 2 & -1 \end{bmatrix} = \begin{bmatrix} 1 & 0 \\ 0 & 1 \end{bmatrix} = I_2$

3. $\begin{bmatrix} -1 & -2 \\ 3 & 4 \end{bmatrix}\begin{bmatrix} 2 & 1 \\ -\frac{3}{2} & -\frac{1}{2} \end{bmatrix} = \begin{bmatrix} 1 & 0 \\ 0 & 1 \end{bmatrix} = I_2$

5. $\begin{bmatrix} 1 & 2 & 3 \\ 2 & 3 & 4 \\ 1 & 2 & 1 \end{bmatrix}\begin{bmatrix} -\frac{5}{2} & 2 & -\frac{1}{2} \\ 1 & -1 & 1 \\ \frac{1}{2} & 0 & -\frac{1}{2} \end{bmatrix} = \begin{bmatrix} 1 & 0 & 0 \\ 0 & 1 & 0 \\ 0 & 0 & 1 \end{bmatrix} = I_3$

7. $\begin{bmatrix} 5 & -7 \\ -2 & 3 \end{bmatrix}$ **9.** $\begin{bmatrix} 4 & -1 \\ 3 & -1 \end{bmatrix}$

11. $\begin{bmatrix} 1.5 & -0.5 \\ -2 & 1 \end{bmatrix}$ **13.** $\begin{bmatrix} 0 & 0 & 1 \\ 0 & 1 & 0 \\ 1 & 0 & 0 \end{bmatrix}$

15. $\begin{bmatrix} \frac{4}{9} & \frac{1}{9} & \frac{1}{9} \\ \frac{4}{3} & -\frac{2}{3} & \frac{1}{3} \\ \frac{7}{9} & -\frac{5}{9} & \frac{4}{9} \end{bmatrix}$ **17.** $\begin{bmatrix} 1 & -1 & 2 \\ -1 & 2 & -3 \\ -1 & 1 & -1 \end{bmatrix}$

19. $\begin{bmatrix} 2 & -1 & -1 & -2 \\ -1 & 1 & 1 & 2 \\ -2 & 1 & 2 & 3 \\ -1 & 1 & 1 & 1 \end{bmatrix}$

21. $\begin{bmatrix} 4 & 6 & | & 1 & 0 \\ 2 & 3 & | & 0 & 1 \end{bmatrix} \rightarrow \begin{bmatrix} 4 & 6 & | & 1 & 0 \\ \boxed{0 & 0} & | & -\frac{1}{2} & 0 \end{bmatrix}$

23. $\begin{bmatrix} -8 & 4 & | & 1 & 0 \\ -4 & 2 & | & 0 & 1 \end{bmatrix} \rightarrow \begin{bmatrix} -8 & 4 & | & 1 & 0 \\ \boxed{0 & 0} & | & -\frac{1}{2} & 0 \end{bmatrix}$

25. $\begin{bmatrix} 1 & 1 & 1 & | & 1 & 0 & 0 \\ 3 & -4 & 2 & | & 0 & 1 & 0 \\ \boxed{0 & 0 & 0} & | & 0 & 0 & 1 \end{bmatrix}$

27. $\begin{bmatrix} 2 & -1 \\ -1 & 1 \end{bmatrix}$ **29.** $\begin{bmatrix} \frac{1}{3} & \frac{1}{6} \\ 0 & \frac{1}{4} \end{bmatrix}$ **31.** No inverse

33. $\begin{bmatrix} \frac{7}{2} & -1 & -\frac{3}{2} \\ 1 & -1 & -1 \\ \frac{1}{2} & 1 & \frac{1}{2} \end{bmatrix}$

35. $A^{-1} = \begin{bmatrix} \frac{1}{5} & \frac{2}{5} \\ \frac{2}{5} & -\frac{1}{5} \end{bmatrix}$; $B^{-1} = \begin{bmatrix} -\frac{1}{5} & \frac{3}{5} \\ \frac{2}{5} & -\frac{1}{5} \end{bmatrix}$

$A^{-1} - B^{-1} = \begin{bmatrix} \frac{2}{5} & -\frac{1}{5} \\ 0 & 0 \end{bmatrix}$

37. $A = \begin{bmatrix} 1 & 3 & 2 \\ 2 & 7 & 3 \\ 1 & 0 & 6 \end{bmatrix}$; $B = \begin{bmatrix} 2 \\ 1 \\ 3 \end{bmatrix}$;

$A^{-1}B = \begin{bmatrix} 42 & -18 & -5 \\ -9 & 4 & 1 \\ -7 & 3 & 1 \end{bmatrix}\begin{bmatrix} 2 \\ 1 \\ 3 \end{bmatrix}$

39. $x = 36$, $y = -14$ **41.** $x = 2$, $y = 1$

43. $x = 88$, $y = -36$ **45.** $x = \frac{14}{9}$, $y = \frac{26}{3}$, $z = \frac{65}{9}$

47. $x = \frac{20}{3}$, $y = 24$, $z = \frac{56}{3}$ **49.** $x = -\frac{14}{9}$, $y = \frac{10}{3}$, $z = \frac{16}{9}$

51. $\begin{bmatrix} a & b \\ c & d \end{bmatrix}\begin{bmatrix} d & -b \\ -c & a \end{bmatrix} = \begin{bmatrix} ad - bc & 0 \\ 0 & -bc + ad \end{bmatrix} =$

$\begin{bmatrix} \Delta & 0 \\ 0 & \Delta \end{bmatrix}$; thus, $\begin{bmatrix} a & b \\ c & d \end{bmatrix}\begin{bmatrix} \dfrac{d}{\Delta} & \dfrac{-b}{\Delta} \\ \dfrac{-c}{\Delta} & \dfrac{a}{\Delta} \end{bmatrix} = \begin{bmatrix} 1 & 0 \\ 0 & 1 \end{bmatrix}$

53. $\begin{bmatrix} 1 & 5 \\ 2 & 0 \end{bmatrix}^{-1} = \frac{1}{-10}\begin{bmatrix} 0 & -5 \\ -2 & 1 \end{bmatrix} = \begin{bmatrix} 0 & \frac{1}{2} \\ \frac{1}{5} & -\frac{1}{10} \end{bmatrix}$

55. $\begin{bmatrix} 1 & 2 \\ 8 & 15 \end{bmatrix}^{-1} = -1\begin{bmatrix} 15 & -2 \\ -8 & 1 \end{bmatrix} = \begin{bmatrix} -15 & 2 \\ 8 & -1 \end{bmatrix}$

Technology Exercises (*page 107*)

1. $\begin{bmatrix} .0054 & .0509 & -.0066 \\ .0104 & -.0186 & .0095 \\ -.0193 & .0116 & .0344 \end{bmatrix}$
(rounded to four decimal places)

3. $\begin{bmatrix} .0249 & -.0360 & -.0057 & .0059 \\ -.0171 & .0521 & .0292 & -.0305 \\ .0206 & .0081 & -.0421 & .0005 \\ -.0175 & .0570 & .0657 & .0619 \end{bmatrix}$
(rounded to four decimal places)

5. $x = 4.5666$, $y = -6.4436$, $z = -24.0747$

7. $x = -1.1874$, $y = 2.4568$, $z = 8.2650$

9. $\begin{bmatrix} .25 & -.0625 & -.28125 & .3125 & .09375 \\ -1.5 & .125 & 3.0625 & -2.625 & .3125 \\ 1.75 & -.1875 & -2.84375 & 2.9375 & -.71875 \\ -.5 & .375 & .6875 & -.875 & .4375 \\ -1.25 & .3125 & 2.40625 & -2.5625 & .53125 \end{bmatrix}$;

$\begin{bmatrix} 1 & 0 & 0 & 0 & 0 \\ 0 & 1 & 0 & 0 & 0 \\ 0 & 0 & 1 & 0 & 0 \\ 0 & 0 & 0 & 1 & 0 \\ 0 & 0 & 0 & 0 & 1 \end{bmatrix}$

Exercise 2.7: Application 1 (*page 114*)

1. A's wages = C's wages = \$30,000; B's wages = $\frac{3}{4}(C$'s wages) = \$22,500

3. A's wages = $\frac{2}{5}(C$'s wages) = \$12,000; B's wages = $\frac{11}{15}(C$'s wages) = \$22,000; C's wages = \$30,000

5. $\begin{bmatrix} 203.282 \\ 166.977 \\ 137.847 \end{bmatrix}$

7. Farmer's wages = $\frac{4}{5}$(Rancher's wages) = \$20,000; Builder's wages = $\frac{18}{25}$(Rancher's wages) = \$18,000; Tailor's wages = $\frac{12}{25}$(Rancher's wages) = \$12,000; Rancher's wages = \$25,000

9. $X = \begin{bmatrix} 160 \\ 75.38 \end{bmatrix}$

Technology Exercises *(page 114)*

1. $\begin{bmatrix} 7.58065 \\ 9.27419 \\ 16.00806 \\ 4.51613 \end{bmatrix}$ **3.** $\begin{bmatrix} 7.70008 \\ 10.11655 \\ 3.64413 \\ 1.04118 \end{bmatrix}$ **5.** $\begin{bmatrix} 495.858 \\ 463.861 \\ 679.199 \\ 512.930 \\ 468.156 \end{bmatrix}$

Exercise 2.7: Application 2 *(page 118)*

1. (a) LIFE IS HARD   (b) ET TU BRUTE

3. A CORRECT ANSWER

5. (a) (I) 41 23 70 45 41 23 62 41 64 36 19 11 59 39 7 4
94 60
(II) 13 69 −21 20 98 −35 1 123 18 8 44 −13 1 32
0 1 141 24
(b) (I) 85 50 71 43 90 54 99 61 43 24 59 32 45 24 69
41 67 43
(II) 20 140 −27 15 153 −12 15 138 −16 5 62 −5
18 132 −21 13 139 −7
(c) (I) 64 36 49 31 75 47 65 37 72 43 75 47 57 35 77
46 95 57 24 13 39 22
(II) 20 93 −35 13 143 −7 19 141 −23 14 146 −9
9 120 −2 15 159 −11 9 89 −6 5 171 16

Exercise 2.7: Application 3 *(page 121)*

1.

Dept.	Total Costs	Direct Costs	Indirect Costs	
			S_1	S_2
S_1	$3109.09	$2000	$345.45	$763.64
S_2	$2290.91	$1000	$1036.36	$254.55
P_1	$3354.54	$2500	$345.45	$509.09
P_2	$2790.91	$1500	$1036.36	$254.55
P_3	$3854.54	$3000	$345.45	$509.09
Totals			$3109.07	$2290.92

Service charges allocated to P_1, P_2, & P_3 = $2999.99 ≈
$3000 = Direct costs of S_1 & S_2.

Exercise 2.7: Application 4 *(page 125)*

1. $\begin{bmatrix} 4 & 3 \\ 1 & 1 \\ 2 & 0 \end{bmatrix}$ **3.** $\begin{bmatrix} 1 & 0 & 1 \\ 11 & 12 & 4 \end{bmatrix}$ **5.** $[8 \quad 6 \quad 3]$

7. (a) $y = \frac{54}{35}x + \frac{27}{5}$ (b) 17.743 thousand units

9. $y = \frac{334}{223}x + \frac{8058}{223}$

11. (a) *Not* symmetric (b) Symmetric (c) *Not* symmetric
Yes; a symmetric matrix *must* be square.

Chapter Review
True−False Items *(page 127)*

1. T **2.** F **3.** F **4.** T **5.** F **6.** F **7.** F

Fill in the Blanks *(page 127)*

1. 3×2 **2.** one; infinitely many **3.** rows; columns
4. inverse **5.** 3×3 **6.** 5×5

Review Exercises *(page 127)*

1. $\begin{bmatrix} -1 & 3 & 16 \\ 3 & 15 & 8 \\ 5 & 10 & 29 \end{bmatrix}$ **3.** $\begin{bmatrix} -3 & 9 & 48 \\ 9 & 45 & 24 \\ 15 & 30 & 87 \end{bmatrix}$

5. $\begin{bmatrix} -1 & 15 & 66 \\ 13 & 59 & 34 \\ 21 & 42 & 103 \end{bmatrix}$ **7.** $\begin{bmatrix} -20 & 0 & 70 \\ 10 & 80 & 30 \\ 20 & 40 & 210 \end{bmatrix}$

9. $\begin{bmatrix} -\frac{7}{2} & -\frac{3}{2} & \frac{25}{2} \\ 3 & \frac{9}{2} & \frac{11}{2} \\ -\frac{37}{2} & -10 & 19 \end{bmatrix}$ **11.** $\begin{bmatrix} 19 & 36 & 38 \\ 26 & 77 & 73 \\ 73 & 160 & 206 \end{bmatrix}$

13. $\begin{bmatrix} 16 & 32 & 27 \\ 16 & 10 & 19 \\ -104 & -80 & -113 \end{bmatrix}$

15. $\begin{bmatrix} 304 & 465 & 416 \\ 584 & 922 & 786 \\ 1648 & 2315 & 2160 \end{bmatrix}$ **17.** $\begin{bmatrix} \frac{1}{3} & 0 \\ \frac{2}{3} & 1 \end{bmatrix}$

19. $\begin{bmatrix} 1 & -3 & 2 \\ -3 & 3 & -1 \\ 2 & -1 & 0 \end{bmatrix}$ **21.** $\begin{bmatrix} \frac{1}{16} & \frac{1}{32} & \frac{1}{4} \\ \frac{3}{16} & \frac{3}{32} & -\frac{1}{4} \\ -\frac{3}{16} & \frac{13}{32} & \frac{1}{4} \end{bmatrix}$

23. $\left[\begin{array}{ccc|ccc} 1 & 2 & -3 & 1 & 0 & 0 \\ 0 & -2 & 14 & -4 & 1 & 0 \\ 0 & 0 & 0 & 3 & 0 & 1 \end{array}\right]$ The matrix has no inverse.

25. $x = \frac{14}{9}, y = \frac{26}{9}$ **27.** $x = -103, y = 32, z = 9$

29. $x = 29, y = -10, z = -1$

31. $\left[\begin{array}{ccc|c} 1 & 1 & -1 & 2 \\ 0 & 1 & -1 & 1 \\ 0 & 0 & 0 & \frac{1}{2} \end{array}\right]$ There is no solution.

33. $x = 9, y = -\frac{56}{3}, z = -\frac{37}{3}$

35. $x = 29, y = 8, z = -24$

37. $x = \frac{10}{7} + \frac{3}{7}z, y = -\frac{9}{7} + \frac{5}{7}z$; three sample solutions:
$x = 1, y = -2, z = -1; x = 4, y = 3, z = 6; x = 7,$
$y = 8, z = 13$

39. $x = 1 - \frac{3}{5}z, y = 2 + \frac{4}{5}z$; three sample solutions: $x = 4,$
$y = -2, z = -5; x = 1, y = 2, z = 0; x = -2, y = 6,$
$z = 5$

41. $\begin{bmatrix} 1 & 0 & | & 4 \\ 0 & 1 & | & 2 \\ \boxed{0 & 0} & | & \frac{7}{2} \end{bmatrix}$ There is no solution.

43. $y = -z, x = w$

45. 20 caramels, 30 creams; increase the number of caramels to obtain a profit.

47.

Pounds of Almonds	Pounds of Cashews	Pounds of Peanuts
5	60	35
20	40	40
35	20	45
50	0	50

49. (a)

Treasury Bills	Corporate Bonds	Junk Bonds
$36,000	$3,000	$1,000
$36,500	$2,000	$1,500
$37,000	$1,000	$2,000
$37,500	$0	$2,500

(b)

Treasury Bills	Corporate Bonds	Junk Bonds
$10,000	$30,000	0
$11,000	$28,000	$1,000
$12,000	$26,000	$2,000
$13,000	$24,000	$3,000
$14,000	$22,000	$4,000
$25,000	$0	$15,000

(c)

Treasury Bills	Corporate Bonds	Junk Bonds
$0	$25,000	$15,000
$1,000	$23,000	$16,000
$2,000	$21,000	$17,000
$3,000	$19,000	$18,000
$10,000	$5,000	$25,000
$12,500	$0	$27,500

Mathematical Questions from Professional Exams (page 129)

1. b   **2.** b   **3.** d   **4.** c

CHAPTER 3

Exercise 3.1 (page 139)

1. $x \geq 0$

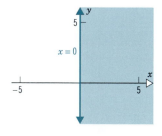

3. $x \leq 4$

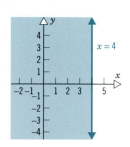

9. $5x + y \leq 10$

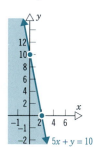

11. $x + 5y \leq 5$

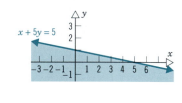

5. $y \geq 1$

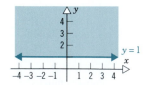

7. $2x + 3y \leq 6$

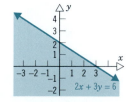

13. Only $P_1 = (3, 8)$ is part of the system's graph.

15. Only $P_3 = (5, 1)$ is part of the system's graph.

17. None of the points are part of the system's graph.

19. *b*   **21.** *b*   **23.** *b*   **25.** *c*

27. Bounded; corner points:
(0, 0), (0, 2), (2, 0)

29. Bounded; corner points:
(0, 2), (3, 0), (2, 0)

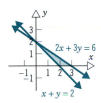

31. Bounded; corner points:
(0, 2), (0, 8), (2, 6), (5, 0), (2, 0)

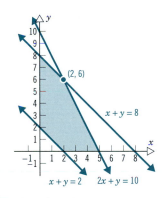

33. Bounded; corner points:
$(0, 2), (0, 4), (\frac{24}{7}, \frac{12}{7}), (4, 0), (2, 0)$

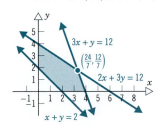

35. Bounded; corner points:
$(0, \frac{1}{2}), (0, 5), (10, 0), (1, 0)$

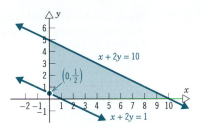

37. Bounded; corner points:
$(0, 1), (0, 3), (\frac{4}{5}, \frac{3}{5})$

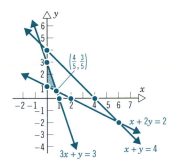

39. (a) $\begin{cases} 4x + 8y \le 960 \\ 12x + 8y \le 1440 \\ x \ge 0, y \ge 0 \end{cases}$ or $\begin{cases} x + 2y \le 240 \\ 3x + 2y \le 360 \\ x \ge 0, y \ge 0 \end{cases}$

(b) Corner points: (0, 0), (0, 120), (60, 90), (120, 0)

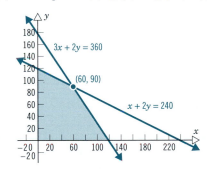

41. (a) $\begin{cases} 3x + 2y \le 80 \\ 4x + 3y \le 120 \\ x \ge 0, y \ge 0 \end{cases}$

(b) Corner points: (0, 0), (0, 40), $(\frac{80}{3}, 0)$

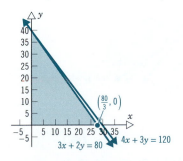

43. (a) $\begin{cases} x \geq 15 \\ y \leq 10 \\ x + y \leq 25 \\ y \geq 0 \end{cases}$ Where x thousand dollars are invested in Treasury bills and y thousand dollars are invested in corporate bonds.

(b) Corner points: $(15, 0)$, $(15, 10)$, $(25, 0)$

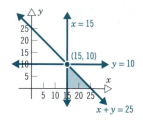

(c) $(15, 0)$: invest \$15,000 in Treasury bills and do not invest the remaining \$10,000.
$(15, 10)$: invest \$15,000 in Treasury bills and \$10,000 in corporate bonds.
$(25, 0)$: invest all \$25,000 in Treasury bills.

45. (a) $\begin{cases} x + 2y \geq 5 \\ 5x + y \geq 16 \\ x \geq 0, y \geq 0 \end{cases}$ Where $x =$ no. of units of first grain and $y =$ no. of units of second grain.

(b) Corner points: $(0, 16)$, $(3, 1)$, $(5, 0)$

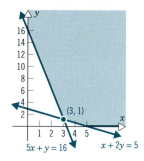

47. (a) $\begin{cases} 5x + 4y \geq 85 \\ 3x + 3y \geq 70 \\ 2x + 3y \geq 50 \\ x \geq 0, y \geq 0 \end{cases}$ Where $x =$ no. of ounces of food A and $y =$ no. of ounces of food B.

(b) Corner points: $(0, \frac{70}{3})$, $(20, \frac{10}{3})$, $(25, 0)$

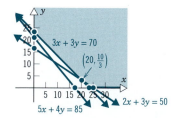

Technology Exercises *(page 142)*

1. Corner points: $(-1, 2)$, $(2, -1)$, $(2, 2)$

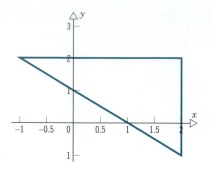

3. Corner points: $(1, 0)$, $(1, 1.33)$, $(3, 0)$

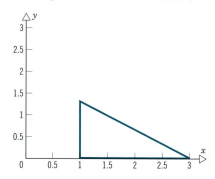

5. Corner points: $(0, -1)$, $(0, 5)$, $(3, 2)$

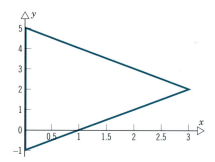

Exercise 3.2 *(page 150)*

1. The maximum is 38 at $(7, 8)$; the minimum is 10 at $(2, 2)$.

3. The maximum is 15 at $(7, 8)$; the minimum is 4 at $(2, 2)$.

5. The maximum is 55 at $(7, 8)$; the minimum is 14 at any point on the line segment joining $(2, 2)$ and $(8, 1)$.

7. The maximum is 53 at $(7, 8)$; the minimum is 14 at $(2, 2)$.

9. The maximum is 81 at $(8, 1)$; the minimum is 22 at $(2, 2)$.

11. $(13, 0)$, $(3, 0)$, $(0, 4)$ [*Note:* The set of feasible points is unbounded above.]

13. (0, 10), (5, 10), (15, 0), (0, 0)

15. (0, 8), (10, 8), (10, 0), (3, 0), (0, 4)

17. The maximum is 14 at (0, 2).

19. The maximum is 15 at (3, 0).

21. The maximum is 56 at (0, 8).

23. The maximum is 58 at (6, 4).

25. The minimum is 0 at (0, 0).

27. The minimum is 4 at (2, 0).

29. The minimum is 4 at (2, 0).

31. The minimum is $\frac{3}{2}$ at $(0, \frac{1}{2})$.

33. The maximum is 10 at any point on the line segment joining (0, 10) and (10, 0). The minimum is $\frac{20}{3}$ at $(\frac{10}{3}, \frac{10}{3})$.

35. The maximum is 50 at (10, 0); the minimum is 20 at (0, 10).

37. The maximum is 40 at (0, 10); the minimum is $\frac{70}{3}$ at $(\frac{10}{3}, \frac{10}{3})$.

39. The maximum is 100 at (10, 0); the minimum is 10 at (0, 10).

41. The maximum is 192 at (4, 4); the minimum is 54 at (3, 0).

43. The maximum is 58 at (4, 5); the minimum is 12 at (0, 2).

45. The maximum is 240 at (3, 10).

47. The maximum is 216 at (2, 10).

49. Make 90 packages of the low-grade mixture and 105 packages of the high-grade mixture (for a maximum profit of $69).

Exercise 3.3 *(page 157)*

1. Plant x acres of soybeans, where $24 \leq x \leq 30$, and $y = 60 - 2x$ acres of corn (for a maximum profit of $9000).

3. Manufacture 15 units of the first product and 25 units of the second product (for a maximum profit of $2100).

5. Invest $7500 in the first security and $22,500 in the second security (for a maximum income of $2550).

7. Employ the first repairman for 2 weeks and the second repairman for 3 weeks (for a minimum cost of $1160).

9. Produce 700 units of item A and 400 units of item B (for a maximum legal profit of $1450).

11. Add 5 oz of supplement I and 1 oz of supplement II to each 100 oz of feed (for a minimum cost of $0.19 for the supplements used per 100 oz of feed).

13. The price per roll of high-grade carpet should be between $520 and $620.

Technology Exercises *(page 159)*

1. Corner points: (3, 0), (1, 2), (4, 5), (6, 3) Minimum is $z = 6$ at $x = 1$, $y = 2$. Maximum is $z = 24.75$ at $x = 6$, $y = 3$.

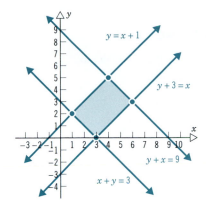

3. Corner points: (.55, 5.27), (4, 7), (5.33, 4.33), (1.5, .5) Minimum is $z = 5.875$ at $x = 1.5$, $y = .5$. Maximum is $z = 24.07$ at $x = 5.33$, $y = 4.33$.

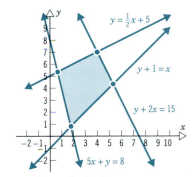

5. Corner points: (.55, 5.27), (4, 7), (6.67, 1.67), (5, 0), (1.6, 0) Minimum is $z = 5.6$ at $x = 1.6$, $y = 0$. Maximum is $z = 25.43$ at $x = 6.67$, $y = 1.67$.

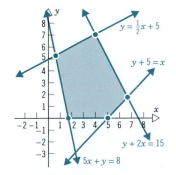

Chapter Review
True–False Items (page 160)

1. T   **2.** F   **3.** T   **4.** T   **5.** T   **6.** F

Fill in the Blanks (page 160)

1. half-plane   **2.** objective   **3.** feasible
4. bounded   **5.** corner point

Review Exercises (page 160)

1.

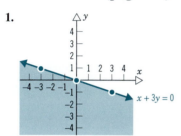

3.

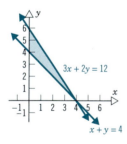

5. Only $P_2 = (2, -6)$ is part of the system's graph.

7. a

9. Corner points: (0, 4), (0, 6), (4, 0); bounded graph

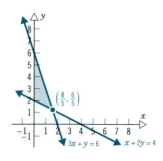

11. Corner points: (0, 2), (0, 6), $(\frac{8}{5}, \frac{6}{5})$; bounded graph

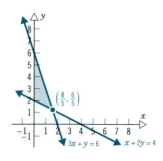

13. Corner points: (0, 3), (0, 4), (2, 3), (4, 0), (2, 0); bounded
graph

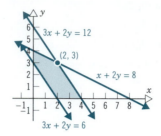

15–21. Corner poins: (0, 10),
(0, 20), $(\frac{40}{3}, \frac{40}{3})$, (20, 0),
(10, 0)

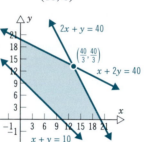

15. Maximum $z = \frac{80}{3}$ at
$(x, y) = (\frac{40}{3}, \frac{40}{3})$

17. Minimum $z = 20$ at
$(x, y) = (0, 10)$

19. Maximum $z = 40$ at
any point (x, y) on the
line segment joining
$(\frac{40}{3}, \frac{40}{3})$ and (20, 0)

21. Minimum $z = 20$ at
$(x, y) = (10, 0)$

23. Maximum $z = 235$ at $(x, y) = (5, 8)$; Minimum $z = 60$
at any point (x, y) on the line segment joining (0, 3) and
(4, 0)

25. Maximum $z = 155$ at $(x, y) = (5, 4)$; Minimum $z = 0$ at
$(x, y) = (0, 0)$

27. Maximum $z = 42$ at $(x, y) = (9, 8)$

29. Maximum $z = 24$ at $(x, y) = (8, 8)$

31. Minimum $z = 8$ at $(x, y) = (0, 4)$

33. Katy should buy $7\frac{1}{2}$ lb of food A and $11\frac{1}{4}$ lb of food B
each month (for a minimum bill of $18.75).

35. Make 8 downhill skis and 24 cross-county skis for a maximum profit of $1760.

37. The maximum weekly profit is $7200.

39. The maximum profit is $160 (by producing 20 dozen cans
of food A and 10 dozen cans of food B).

Mathematical Questions from Professional
Exams (page 162)

1. b   **2.** a   **3.** c   **4.** c   **5.** d   **6.** c   **7.** c
8. b   **9.** b   **10.** a   **11.** b   **12.** e   **13.** c
14. b

CHAPTER 4

Exercise 4.1 (page 175)

1. Standard form   **3.** Not in standard form

5. Not in standard form   **7.** Not in standard form

9. Standard form

11. Cannot be modified to be in standard form

13. Cannot be modified to be in standard form

15. Maximize $P = 2x_1 + x_2 + 3x_3$ subject to the constraints
$x_1 - x_2 - x_3 \leq 6$, $-2x_1 + 3x_2 \leq 12$, $x_3 \leq 2$, $x_1 \geq 0$,
$x_2 \geq 0$, and $x_3 \geq 0$.

17.

BV	P	x_1	x_2	x_3	s_1	s_2	s_3	RHS
s_1	0	5	2	1	1	0	0	20
s_2	0	6	1	4	0	1	0	24
s_3	0	1	1	4	0	0	1	16
P	1	−2	−1	−3	0	0	0	0

19.

BV	P	x_1	x_2	s_1	s_2	s_3	RHS
s_1	0	2.2	−1.8	1	0	0	5
s_2	0	0.8	1.2	0	1	0	2.5
s_3	0	1	1	0	0	1	0.1
P	1	−3	−5	0	0	0	0

21.

BV	P	x_1	x_2	x_3	s_1	s_2	RHS
s_1	0	1	1	1	1	0	50
s_2	0	3	2	1	0	1	10
P	1	−2	−3	−1	0	0	0

23.

BV	P	x_1	x_2	x_3	s_1	s_2	s_3	RHS
s_1	0	3	1	4	1	0	0	5
s_2	0	1	1	0	0	1	0	5
s_3	0	2	−1	1	0	0	1	6
P	1	−3	−4	−2	0	0	0	0

25. Modified version:

Maximize $P = x_1 + 2x_2 + 5x_3$

Subject to the constraints

$x_1 - 2x_2 - 3x_3 \leq 10$
$-3x_1 - x_2 + x_3 \leq 12$
$x_1 \geq 0 \quad x_2 \geq 0 \quad x_3 \geq 0$

Initial tableau:

BV	P	x_1	x_2	x_3	s_1	s_2	RHS
s_1	0	1	−2	−3	1	0	10
s_2	0	−3	−1	1	0	1	12
P	1	−1	−2	−5	0	0	0

27. Modified version:

Maximize $P = 2x_1 + 3x_2 + x_3 + 6x_4$

Subject to the constraints

$$-x_1 + x_2 + 2x_3 + x_4 \le 10$$
$$-x_1 + x_2 - x_3 + x_4 \le 8$$
$$x_1 + x_2 + x_3 + x_4 \le 9$$
$$x_1 \ge 0 \quad x_2 \ge 0 \quad x_3 \ge 0 \quad x_4 \ge 0$$

Initial tableau:

BV	P	x_1	x_2	x_3	x_4	s_1	s_2	s_3	RHS
s_1	0	-1	1	2	1	1	0	0	10
s_2	0	-1	1	-1	1	0	1	0	8
s_3	0	1	1	1	1	0	0	1	9
P	1	-2	-3	-1	-6	0	0	0	0

29. New tableau:

BV	P	x_1	x_2	s_1	s_2	RHS
x_2	0	$\frac{1}{2}$	1	$\frac{1}{2}$	0	150
s_2	0	2	0	-1	1	180
P	1	0	0	1	0	300

New system: $x_2 = -\frac{1}{2}x_1 - \frac{1}{2}s_1 + 150$
$\qquad\qquad s_2 = -2x_1 + s_1 + 180$
$\qquad\qquad P = -s_1 + 300$

Current values: $P = 300,\ x_2 = 150,\ s_2 = 180$

31. New tableau:

BV	P	x_1	x_2	x_3	s_1	s_2	s_3	RHS
s_1	0	-2	0	0	1	0	-1	6
s_2	0	$\frac{5}{4}$	$-\frac{3}{2}$	0	0	1	$-\frac{1}{4}$	$\frac{55}{2}$
x_3	0	$\frac{3}{4}$	$\frac{1}{2}$	1	0	0	$\frac{1}{4}$	$\frac{9}{2}$
P	1	$\frac{5}{4}$	$-\frac{1}{2}$	0	0	0	$\frac{3}{4}$	$\frac{27}{2}$

New system: $s_1 = 2x_1 + s_3 + 6$
$\qquad\qquad s_2 = -\frac{5}{4}x_1 + \frac{3}{2}x_2 + \frac{1}{4}s_3 + \frac{55}{2}$
$\qquad\qquad x_3 = -\frac{3}{4}x_1 - \frac{1}{2}x_2 - \frac{1}{4}s_3 + \frac{9}{2}$
$\qquad\qquad P = -\frac{5}{4}x_1 + \frac{1}{2}x_2 - \frac{3}{4}s_3 + \frac{27}{2}$

Current values: $P = \frac{27}{2},\ s_1 = 6,\ s_2 = \frac{55}{2},\ x_3 = \frac{9}{2}$

33. New tableau:

BV	P	x_1	x_2	x_3	x_4	s_1	s_2	s_3	s_4	RHS
s_1	0	-3	0	1	0	1	0	0	0	20
x_4	0	2	0	0	1	0	1	0	0	24
s_3	0	0	-3	1	0	0	0	1	0	28
s_4	0	-2	-3	0	0	0	-1	0	1	0
P	1	7	-2	-3	0	0	4	0	0	96

New system: $s_1 = 3x_1 - x_3 + 20$
$\qquad\qquad x_4 = -2x_1 - s_2 + 24$
$\qquad\qquad s_3 = 3x_2 - x_3 + 28$
$\qquad\qquad s_4 = 2x_1 + 2x_3 + s_2$
$\qquad\qquad P = -7x_1 + 2x_2 + 3x_3 - 4s_2 + 96$

Current values: $P = 96,\ s_1 = 20,\ x_4 = 24,\ s_3 = 28,$
$\qquad\qquad s_4 = 0$

Technology Exercises (page 179)

1.

BV	P	x_1	x_2	s_1	s_2	RHS
s_1	0	1.67	0	1	$-.67$	100
s_2	0	.33	1	0	.33	50
P	1	-1	0	0	1	150

3.

BV	P	x_1	x_2	s_1	s_2	RHS
s_1	0	.25	1	.25	0	25
s_2	0	.75	0	-1.25	1	10
P	1	-1.25	0	.75	0	75

Exercise 4.2 (page 192)

1. (b); the pivot element is 1 in row 1, column 2

3. (a); the solution is $P = \frac{256}{7}$, $x_1 = \frac{32}{7}$, $x_2 = 0$ **5.** (c)

7. (b); the pivot element is 1 in row 3, column 4

9. The maximum is $P = \frac{204}{7} = 29\frac{1}{7}$ when $x_1 = \frac{24}{7}$, $x_2 = \frac{12}{7}$.

11. The maximum is $P = 8$ when $x_1 = \frac{2}{3}$, $x_2 = \frac{2}{3}$.

13. The maximum is $P = 6$ when $x_1 = 2$, $x_2 = 0$.

15. There is no maximum for P; the feasible region is unbounded.

17. The maximum is $P = 30$ when $x_1 = 0$, $x_2 = 0$, $x_3 = 10$.

19. The maximum is $P = 42$ when $x_1 = 1$, $x_2 = 10$, $x_3 = 0$, $x_4 = 0$.

21. The maximum is $P = 40$ when $x_1 = 20$, $x_2 = 0$, $x_3 = 0$.

23. The maximum is $P = 50$ when $x_1 = 0$, $x_2 = 15$, $x_3 = 5$, $x_4 = 0$.

25. The maximum profit is $1500 when the manufacturer makes 400 of Jean I, 0 of Jean II, and 50 of Jean III.

27. The maximum profit is $190 from the sale of 0 of product A, 40 of product B, and 75 of product C.

29. The maximum revenue is $275,000 when 200,000 gal of regular, 0 gal of premium, and 25,000 gal of super premium are mixed.

31. The maximum return is $7500 when she invests $45,000 in stocks, $15,000 in corporate bonds, and $30,000 in municipal bonds.

33. The maximum profit is $14,400 when 180 acres of crop A, 20 acres of crop B, and 0 acres of crop C are planted.

35. The maximum revenue is $2800 for 50 cans of can I, no cans of can II, and 70 cans of can III.

37. The maximum profit is $12,000 from 1200 television cabinets and no stereo or radio cabinets.

39. The maximum profit is $30,000 when no TVs are shipped from Chicago, 375 TVs are shipped from New York, and no TVs are shipped from Denver.

Technology Exercises (page 196)

1. Maximum is $P = 167.86$ at $x_1 = 14.29$, $x_2 = 32.14$.

3. Maximum is $P = 8$ at $x_1 = .67$, $x_2 = .67$.

Exercise 4.3 (page 203)

1. Standard form   **3.** Not in standard form

5. Not in standard form

7. Maximize $P = 2y_1 + 6y_2$ subject to $y_1 + 2y_2 \le 2$, $y_1 + 3y_2 \le 3$, $y_1 \ge 0$, $y_2 \ge 0$.

9. Maximize $P = 5y_1 + 4y_2$ subject to $y_1 + 2y_2 \le 3$, $y_1 + y_2 \le 1$, $y_1 \le 1$, $y_1 \ge 0$, $y_2 \ge 0$.

11. Maximize $P = 60y_1 + 90y_2$ subject to $y_1 + 3y_2 \le 3$, $y_1 + 2y_2 \le 4$, $y_1 + y_2 \le 1$, $2y_1 + 2y_2 \le 2$, $y_1 \ge 0$, $y_2 \ge 0$.

13. $C = 6$, $x_1 = 0$, $x_2 = 2$ **15.** $C = 12$, $x_1 = 0$, $x_2 = 4$

17. $C = \frac{21}{5}$, $x_1 = \frac{8}{5}$, $x_2 = 0$, $x_3 = \frac{13}{5}$

19. $C = 5$, $x_1 = 1$, $x_2 = 1$, $x_3 = 0$, $x_4 = 0$ ($C = 5$ also when $x_1 = 1$, $x_2 = 0$, $x_3 = 0$, $x_4 = 1$)

21. The minimum cost is $0.22 when 2 P pills and 4 Q pills are taken as supplements.

23. The minimum cost is $290 when A = 20 units, B = 30 units, and C = 150 units.

25. The minimum cost is $65.20 when 4 lunch #1, 3 lunch #2, and 2 lunch #3 are ordered.

Technology Exercises (page 205)

1. Minimium is $C = 167.86$ at $x_1 = 14.29$, $x_2 = 32.14$.

3. Minimum is $C = 8$ at $x_1 = .67$, $x_2 = .67$.

Exercise 4.4 (page 218)

1. The maximum is $P = 44$ when $x_1 = 4$, $x_2 = 8$.

3. The maximum is $P = 27$ when $x_1 = 9$, $x_2 = x_3 = 0$.

5. The minimum is $z = \frac{20}{3}$ when $x_1 = x_2 = 0$, $x_3 = \frac{20}{3}$.

7. The maximum is $P = 7$ when $x_1 = 1$, $x_2 = 2$.

9. The minimum total shipping charge is $150,000; ship 100 engines from M1 to A1, 300 engines from M1 to A2, 400 engines from M2 to A1, and no engines from M2 to A2.

11. The minimum preparation cost is $7.50, using $\frac{5}{8}$ unit of food I, $\frac{25}{4}$ units of food II, and no units of food III.

13. The minimum cost is $965 when 55 sets are shipped from W_1 to R_1 and 75 sets are shipped from W_2 to R_2.

15. The minimum cost is $120 when 2 units of A and 14 units of B are used.

Chapter Review:
True−False Items (page 220)

1. T   **2.** F   **3.** T   **4.** F   **5.** T   **6.** T

Fill in the Blanks (page 220)

1. slack variables   **2.** column   **3.** $\geq$
4. Von Neumann duality

Review Exercises (page 221)

1. The maximum is $P = 22,500$ when $x_1 = 0$, $x_2 = 100$, $x_3 = 50$.

3. The maximum is $P = 352$ when $x_1 = 0$, $x_2 = \frac{6}{5}$, $x_3 = \frac{28}{5}$.

5. The minimum is $C = 7$ when $x_1 = 3$, $x_2 = 1$.

7. The minimum is $C = 350$ when $x_1 = 0$, $x_2 = 50$, $x_3 = 50$. ($C = 350$ also when $x_1 = 25$, $x_2 = 0$, and $x_3 = 75$).

9. The maximum is $P = 12,250$ when $x_1 = 0$, $x_2 = 5$, $x_3 = 25$.

11. Make $83\frac{1}{3}$ lb of hamburger patties and 500 lb of picnic patties (for a maximum of $583\frac{1}{3}$ lb of meat used).

13. The maximum profit is approx. $5714.29 for cultivation of 0 acres of corn, 0 acres of wheat, and $\frac{1000}{7} \approx 143$ acres of soybeans.

Mathematical Questions from Professional Exams (page 222)

1. c   **2.** d   **3.** c   **4.** a   **5.** b   **6.** a   **7.** c

8. d   **9.** d   **10.** a   **11.** d

CHAPTER 5

Exercise 5.1 (page 229)

1. 60%   **3.** 110%   **5.** 6%   **7.** 0.25%   **9.** 0.25
11. 1   **13.** .065   **15.** 0.734   **17.** 150   **19.** 18
21. 105   **23.** 5%   **25.** 160%   **27.** 250
29. $333\frac{1}{3}$ **31.** $10   **33.** $45   **35.** $150
37. 10%   **39.** $33\frac{1}{3}$%   **41.** $13\frac{1}{3}$%   **43.** $1140
45. $1680   **47.** $1263.16   **49.** $2380.95
51. $471.70   **53.** $3\frac{1}{4}$ yr

55. The discounted loan at 9% has less interest for 6 months.

57. You should choose the simple interest loan at 6.3%.

59. She should choose the simple interest loan at 12.3%.

61. $978,750   **63.** $2,884,275

Exercise 5.2 (page 237)

1. $1270.24   **3.** $647.51   **5.** $854.36   **7.** $95.14
9. $456.97

11. (a) Amount = $1295.03; Interest earned = $295.03
    (b) Amount = $1302.26; Interest earned = $302.26
    (c) Amount = $1306.05; Interest earned = $306.05
    (d) Amount = $1308.65; Interest earned = $308.65

13. (a) $1266.77   (b) $1425.76   (c) $1604.71

15. (a) $3947.05   (b) $3115.83

17. (a) 8.16%   (b) 12.68%

19. 25.99%

21. Approx. $11\frac{1}{2}$ yr

23. The 10% loan compounded monthly has less interest.

25. $1759.11   **27.** 5.35%   **29.** 6.82%

31. $6\frac{1}{4}$% compounded annually

33. 9% compounded monthly   **35.** $109,400

37. $656.07   **39.** $29,137.83   **41.** $42,640.10

43. 19,918   **45.** Yes, the stock returns 7.26% quarterly.

47. $18,508.09   **49.** $10,810.76   **51.** 9.07%

53. Approx. $15\frac{1}{4}$ yr   **55.** $940.90   **57.** $858.73

59. Approx. $22\frac{3}{4}$ yr   **61.** Approx. 11.2 yrs

Exercise 5.3 (page 247)

1. $1593.74   **3.** $5073   **5.** $7867.22   **7.** $6977
9. $113,201.03   **11.** $147.05   **13.** $1868.68
15. $4121.33   **17.** $200.46   **19.** $2088.11
21. $62,822.56   **23.** $9126.56   **25.** $524.04
27. $22,192.08   **29.** $205,367.98   **31.** $1655.75
33. (a) $180,611.12   (b) $2395.33
35. Approx. 34 yr

Exercise 5.4 (page 255)

1. $15,495.62   **3.** $856.60   **5.** $85,135.64
7. $229,100; $25,239.51   **9.** $470.73   **11.** $530.76

13. $2008.18   **15.** $25,906.15

17.

Loan	Monthly Payment
8%, 20 yr	~ $794.62
9%, 25 yr	~ $797.24

The 9% loan for 25 yr has the larger monthly payment. Clearly, the interest is greater on the 9% loan for 25 yr, since the monthly payment is larger, and there are more monthly payments.

Loan	Equity after 10 Years
8%, 20 yr	~$54,506.24
9%, 25 yr	~$41,397.39

After 10 yr, the equity from the 8%, 20 yr, loan is greater.

19. $55.82   **21.** (a) $1207.64 (b) $36.24

23. (a) $15,200 (b) $60,800 (c) $489.21 (d) $115,315.60 (e) 199 mo = 16 yr 7 mo (f) $56,452.79

25. $332.79

27. Monthly payment: $474.01; Total interest paid: $4752.48

29. For a 30-yr mortgage: Monthly payment: $966.37; Total interest paid: $235,893.20
For a 15-yr mortgage: Monthly payment: $1189.89; Total interest paid: $102,180.20

31. Approx. 4 yr, 8 mo

Exercise 5.5 *(page 259)*

1. Leasing is preferable.

3. Machine A is preferable. (Machine A's annual cost is $125.56 *less* than its labor savings, but machine B costs $36.86 *more* each year than its labor savings.)

5. $1086.46

Chapter Review
True–False Items *(page 260)*

1. T   **2.** T   **3.** F   **4.** F

Fill in the Blanks *(page 260)*

1. proceeds   **2.** present value   **3.** annuity
4. amortized

Review Exercises *(page 261)*

1. $I = $36, A = 436 **3.** $125.12

5. The interest from loan (a) is $1080, while the interest from loan (b) is $1044.55. The 10% loan compounded monthly costs Mike less.

7. $71.36   **9.** $404.81

11. (a) $545.22   (b) $103,566   (c) $23,501.79

13. Monthly payment: $1049; Equity after 10 yr: $21,575.51

15. $119,431.77   **17.** $108,003.59   **19.** $10,078.44

21. $37.98   **23.** $141.22   **25.** 9.38%

27. $1156.60   **29.** $2087.09   **31.** $330.74

Mathematical Questions from Professional Exams *(page 262)*

1. b   **2.** c   **3.** b   **4.** b   **5.** d   **6.** a   **7.** c

---

CHAPTER 6

Exercise 6.1 *(page 275)*

1. True   **3.** False   **5.** False   **7.** True   **9.** True

11. {2, 3}   **13.** {1, 2, 3, 4, 5}   **15.** ∅

17. {a, b, d, e, f, q}

19. (a) {0, 1, 2, 3, 5, 7, 8}   (b) {5}   (c) {5}
(d) {0, 1, 2, 3, 4, 6, 7, 8, 9}   (e) {4, 6, 9}
(f) {0, 1, 5, 7}   (g) ∅   (h) {5}

21. (a) {b, c, d, e, f, g}   (b) {c}

(c) {a, h, i, j, k, l, m, . . . , x, y, z}
(d) {a, b, d, e, f, g, h, i, . . . , x, y, z}

23. (a) $\bar{A} \cap B$ (b) $(\bar{A} \cap \bar{B}) \cup C$

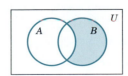

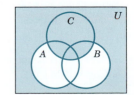

(c) and (d) $A \cap (A \cup B) = A \cup (A \cap B)$

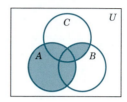

(e) and (f) $(A \cup B) \cap (A \cup C) = A \cup (B \cap C)$

(g) $(A \cap B)$ $(A \cap \bar{B})$

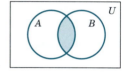

$A = (A \cap B) \cup (A \cap \bar{B})$

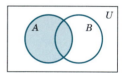

(h) $(A \cap B)$ $(\bar{A} \cap B)$

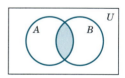   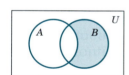

$B = (A \cap B) \cup (\bar{A} \cap B)$

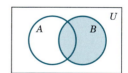

25. $\{x \mid x$ is both a customer of IBM and a member of the board of directors of IBM$\}$

27. $\{x \mid x$ is either a customer of IBM or a stockholder of IBM$\}$

29. $\bar{A} \cap D = \{x \mid x$ is not a customer of IBM and is a stockholder of IBM$\}$

31. $M \cap S = \{$All male college students who smoke$\}$

33. $\bar{M} \cup \bar{F} = \{$All college students who are female or not freshmen$\}$

35. $F \cap S \cap M = \{$All male freshmen college students who smoke$\}$

37. $\varnothing, \{a\}, \{b\}, \{c\}, \{a, b\}, \{a, c\}, \{b, c\}, \{a, b, c\}$

Exercise 6.2 *(page 280)*

1. 6   **3.** 3   **5.** 6   **7.** 5   **9.** 2   **11.** 10

13. 452   **15.** 24   **17.** 34   **19.** 15   **21.** 54

23. 3   **25.** (a) 536   (b) 317   (c) 134

27. (a) 259   (b) 455   (c) 227   (d) 76   (e) 118   (f) 93   (g) 912

29. (a) 40   (b) 35   (c) 40   (d) 205   (e) 155

31. There are eight possible blood types:

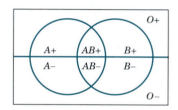

33. 46

35. $\varnothing, \{a\}, \{b\}, \{c\}, \{d\}, \{a, b\}, \{a, c\}, \{a, d\}, \{b, c\}, \{b, d\}, \{c, d\}, \{a, b, c\}, \{a, b, d\}, \{a, c, d\}, \{b, c, d\}, \{a, b, c, d\}$ There are 16 $(=2^4)$ subsets of $\{a, b, c, d\}$.

Exercise 6.3 *(page 285)*

1. 8   **3.** 24   **5.** 864   **7.** 36   **9.** 720

11. 960

13. 6 people: 720 ways; 8 people: 40,320 ways

15. No repeated letters: 360 code words; repeated letters: 1296 code words

17. 5040   **19.** $4^{10} \cdot 2^{15} = 2^{35} = 34,359,738,368$

21. (a) 6,760,000   (b) 3,407,040   (c) 3,276,000

23. 16   **25.** 60   **27.** $2^8 = 256$

29. $50^3 = 125,000$ **31.** 8

Exercise 6.4 *(page 291)*

1. 60   **3.** 120   **5.** 90   **7.** 9   **9.** 56   **11.** 42

13. 40,320   **15.** 1   **17.** 56   **19.** 1

21. (a) $6! = 720$ (b) $5! = 120$ (c) $4! = 24$

23. $P(10, 4) = 5040$ **25.** $P(9, 5) = 15{,}120$

27. $P(12, 8) = 19{,}958{,}400$

29. $P(1500, 3) = 3{,}368{,}253{,}000$

31. $P(15, 4) = 32{,}760$ **33.** $5! = 120$

Exercise 6.5 *(page 297)*

1. 15   **3.** 21   **5.** 5   **7.** 28   **9.** 56

11. 2380   **13.** 1140   **15.** 15   **17.** 60

19. 26,046,720   **21.** 10   **23.** 1,192,052,400

25. 75,287,520   **27.** 1,217,566,350   **29.** 60

31. 10,626

33. $P(50, 15) = \frac{50!}{35!} \approx 2.9 \times 10^{24}$

35. 5040

Exercise 6.6 *(page 304)*

1. (a) 1024   (b) 210   (c) 56   (d) 968

3. (a) 63   (b) 35   (c) 1   **5.** 70   **7.** 1260

9. 4,989,600   **11.** 27,720

13. (a) 280   (b) 280   (c) 640

15. $\dfrac{30!}{(6!)^5} \approx 1.37 \times 10^{18}$ **17.** 93   **19.** 826

21. Even number of 1s: 128; Odd number of 1s: 128

Exercise 6.7 *(page 311)*

1. $x^5 + 5x^4y + 10x^3y^2 + 10x^2y^3 + 5xy^4 + y^5$

3. $x^3 + 9x^2y + 27xy^2 + 27y^3$

5. $16x^4 - 32x^3y + 24x^2y^2 - 8xy^3 + y^4$

7. 10   **9.** 405   **11.** 32   **13.** 1023   **15.** 512

17. $\binom{10}{7} = \binom{9}{7} + \binom{9}{6}$ and $\binom{9}{7} = \binom{8}{7} + \binom{8}{6}$,
so $\binom{10}{7} = \binom{8}{7} + \binom{8}{6} + \binom{9}{6}$
Also, $\binom{8}{7} = \binom{7}{7} + \binom{7}{6}$ and $\binom{7}{7} = 1 = \binom{6}{6}$,
so $\binom{10}{7} = \binom{6}{6} + \binom{7}{6} + \binom{8}{6} + \binom{9}{6}$

19. $\binom{12}{6}$

21. $k \cdot \binom{n}{k} = k \cdot \dfrac{n!}{k!(n-k)!} = k \cdot \dfrac{n!}{k \cdot (k-1)!(n-k)!}$

$= \dfrac{n \cdot (n-1)!}{(k-1)!(n-k)!}$

$= n \cdot \dfrac{(n-1)!}{(k-1)!((n-1)-(k-1))!}$

$= n \cdot \binom{n-1}{k-1}$

Alternatively, suppose we wish to select a team of k people from a class of n students, and designate one of the team members as "team leader." We could do this by selecting the team [in $\binom{n}{k}$ ways] and then designate the leader (k ways), so there are $k \cdot \binom{n}{k}$ possible results, *Or*, we could first choose the leader (n choices) and then choose the remaining $k-1$ team members [$\binom{n-1}{k-1}$ ways], so there are $n \cdot \binom{n-1}{k-1}$ possible results. Since these two methods count the same set of results, the two answers must be the same, i.e., $k \cdot \binom{n}{k} = n \cdot \binom{n-1}{k-1}$.

Chapter Review
True – False Items *(page 312)*

1. T   **2.** T   **3.** F   **4.** F   **5.** T   **6.** F   **7.** F

Fill in the Blanks *(page 313)*

1. disjoint   **2.** permutation   **3.** combination

4. Pascal's   **5.** binomial coefficients

6. Binomial Theorem   **7.** $\binom{5}{2} 2^2 = 40$

Review Exercises *(page 313)*

1. None of these.   **3.** None of these.

5. None of these.   **7.** $\subset, \subseteq$ **9.** $\subseteq, =$

11. None of these.   **13.** $\subset, \subseteq$ **15.** None of these.

17. (a) $\{3, 6, 8, 9\}$ (b) $\{6\}$ (c) $\{2, 3, 6, 7\}$

19. 3   **21.** (a) 45   (b) 33   (c) 50   **23.** 120

25. 10   **27.** 6   **29.** 72

31. The maximum number of words is 12; 6 words are possible.

33. 218,400   **35.** (a) 525   (b) 1715

37. $(4! \cdot 5! \cdot 6!) \cdot 3! = 12{,}441{,}600$ **39.** 20,790

41. 240   **43.** 30   **45.** 24

47. (a) 4845   (b) 5700   (c) 7805   **49.** 924

51. $x^4 + 8x^3 + 24x^2 + 32x + 16$ **53.** 560

CHAPTER 7

Exercise 7.1 (page 325)

1. (a) $\{H, T\}$ (b) $\{0, 1, 2\}$ (c) $\{M, D\}$

3. $S = \{HH, HT, TH, TT\}$

5. $S = \{HHH, HHT, HTH, HTT, THH, THT, TTH, TTT\}$

7. $S = \{HH1, HH2, HH3, HH4, HH5, HH6, HT1, HT2,$ $HT3, HT4, HT5, HT6, TH1, TH2, TH3, TH4, TH5,$ $TH6, TT1, TT2, TT3, TT4, TT5, TT6\}$

9. $S = \{RA, RB, RC, GA, GB, GC\}$

11. $S = \{AA, AB, AC, BA, BB, BC, CA, CB, CC\}$

13. $S = \{AA1, AA2, AA3, AA4, AB1, AB2, AB3, AB4, BA1,$ $BA2, BA3, BA4, BB1, BB2, BB3, BB4, AC1, AC2,$ $AC3, AC4, CA1, CA2, CA3, CA4, BC1, BC2, BC3,$ $BC4, CB1, CB2, CB3, CB4, CC1, CC2, CC3, CC4\}$

15. $S = \{RA1, RA2, RA3, RA4, RB1, RB2, RB3, RB4, RC1,$ $RC2, RC3, RC4, GA1, GA2, GA3, GA4, GB1, GB2,$ $GB3, GB4, GC1, GC2, GC3, GC4\}$

17. $2^4 = 16$ **19.** $6^3 = 216$ **21.** $\dfrac{52 \cdot 51}{2} = 1326$

23. 676   **25.** 1, 2, 3, and 6   **27.** 2

29. $P(H) = \frac{3}{4}, P(T) = \frac{1}{4}$

31. $P(1) = P(3) = P(5) = \frac{2}{9}, P(2) = P(4) = P(6) = \frac{1}{9}$

33–37. The sample space is $S = \{(x, y) | x = 1, \ldots, 6,$ $y = 1, \ldots, 6\}$, for which the probability of each event is $\frac{1}{36}$.

33. $\frac{2}{36} = \frac{1}{18}$ **35.** $\frac{4}{36} = \frac{1}{9}$ **37.** $\frac{6}{36} = \frac{1}{6}$

39–45. The sample space is $S = \{(x, y) | x = 1, \ldots, 6,$ $y = H$ or $T\}$, for which the probability of each event is $\frac{1}{12}$.

39. $\frac{1}{2}$ **41.** $\frac{1}{6}$ **43.** $\frac{5}{6}$ **45.** $\frac{1}{3}$ **47.** $\frac{1}{4}$

49.

Outcome	HH	HT	TH	TT
Probability	$\frac{1}{4}$	$\frac{1}{4}$	$\frac{1}{4}$	$\frac{1}{4}$

51.

Outcome	1H	1T	2H	2T	3H	3T	4H	4T	5H	5T	6H	6T
Probability	$\frac{1}{12}$	$\frac{1}{12}$	$\frac{1}{12}$	$\frac{1}{12}$	$\frac{1}{12}$	$\frac{1}{12}$	$\frac{1}{12}$	$\frac{1}{12}$	$\frac{1}{12}$	$\frac{1}{12}$	$\frac{1}{12}$	$\frac{1}{12}$

53.

Outcome	HHHH	HHHT	HHTH	HHTT	HTHH	HTHT	HTTH	HTTT
Probability	$\frac{1}{16}$	$\frac{1}{16}$	$\frac{1}{16}$	$\frac{1}{16}$	$\frac{1}{16}$	$\frac{1}{16}$	$\frac{1}{16}$	$\frac{1}{16}$

Outcome	THHH	THHT	THTH	THTT	TTHH	TTHT	TTTH	TTTT
Probability	$\frac{1}{16}$	$\frac{1}{16}$	$\frac{1}{16}$	$\frac{1}{16}$	$\frac{1}{16}$	$\frac{1}{16}$	$\frac{1}{16}$	$\frac{1}{16}$

55. $\{HTTT, TTTT\}$ **57.** $\{HHHT, HHTH, HHTT, HTHH, HTHT, HTTH, THHH, THHT, THTH, TTHH\}$

59.

Outcome	RRR	RRL	RLR	LRR	RLL	LRL	LLR	LLL
Probability	$\frac{1}{12}$	$\frac{1}{6}$	$\frac{1}{12}$	$\frac{1}{12}$	$\frac{1}{6}$	$\frac{1}{6}$	$\frac{1}{12}$	$\frac{1}{6}$

(a) $\frac{1}{3}$ (b) $\frac{1}{6}$ (c) $\frac{1}{2}$ (d) $\frac{1}{2}$

61. $P(C_1) = \frac{1}{4}, P(C_2) = \frac{1}{2}, P(C_3) = \frac{1}{4}$

Exercise 7.2 (page 335)

1. .75   **3.** .65   **5.** .5   **7.** $\frac{1}{52}$ **9.** $\frac{13}{52} = \frac{1}{4}$

11. $\frac{12}{52} = \frac{3}{13}$ **13.** $\frac{20}{52} = \frac{5}{13}$ **15.** $\frac{48}{52} = \frac{12}{13}$

17. Yes, the events "Sum is 2" and "Sum is 12" are mutually exclusive. $P(\text{"Sum is 2" or "Sum is 12"}) = \frac{2}{36} = \frac{1}{18}$

19. $\frac{3}{23}$ **21.** $\frac{7}{23}$ **23.** $\frac{8}{23}$ **25.** $\frac{11}{23}$ **27.** $\frac{6}{36} = \frac{1}{6}$

29. .30   **31.** .2   **33.** (a) .7 (b) .3 (c) .2 (d) .3

35. (a) .68 (b) .58 (c) .32

37. (a) .57 (b) .95 (c) .83 (d) .38 (e) .29 (f) .05 (g) .78 (h) .71

39. $\frac{65}{150} = \frac{13}{30} \approx .433$ **41.** $\frac{3}{4}$ **43.** $\frac{5}{12}$ **45.** $\frac{1}{2}$

47. The odds are 3 to 2 for E; 2 to 3 against E.

49. The odds are 3 to 1 for F; 1 to 3 against F.

51. 1 to 5; 1 to 17; 2 to 7   **53.** 23 to 27

55. $P(A \text{ or } B \text{ wins}) = \frac{1}{3} + \frac{2}{5} = \frac{11}{15}$; the odds that either A or B wins are 11 to 4.

57. $P(E \cup F) = P(E) + P(\overline{E} \cap F) = P(E) + P(F) - P(E \cap F)$

59.
$$P(E) = \frac{a}{b}(1 - P(E)) = \frac{a}{b} - \frac{a}{b}P(E)$$

$$P(E) + \frac{a}{b}P(E) = \frac{a}{b}$$

$$\left(1 + \frac{a}{b}\right)P(E) = \frac{a}{b}$$

$$\frac{b + a}{b}P(E) = \frac{a}{b}$$

$$P(E) = \frac{a}{b} \cdot \frac{b}{a + b}$$

$$P(E) = \frac{a}{a + b}$$

Technology Exercises (page 338)

1. $P(H) = .50,\ P(T) = .50$

3. $P(H) = .75,\ P(T) = .25$

5. $P(R) = \frac{1}{3},\ P(Y) = \frac{2}{15},\ P(W) = \frac{8}{15}$

Exercise 7.3 (page 344)

1. .940   **3.** .664   **5.** (a) .3125 (b) .03125

7. (a) .00463 (b) .126

9. $P(\text{All 5 are defective}) = \frac{6 \cdot 5 \cdot 4 \cdot 3 \cdot 2}{50 \cdot 49 \cdot 48 \cdot 47 \cdot 46} \approx .00000283$;
$P(\text{At least 2 are defective}) = \frac{218,246}{2,118,760} \approx .103$

11. $1 - \frac{1320}{1728} = \frac{17}{72} \approx .236$

13. $1 - \frac{970,200}{1,000,000} = \frac{149}{5000} = .0298$

15. $1 - \frac{365 \cdot 364 \cdot 363 \cdot \ldots \cdot 266}{365^{100}} \approx .999999692751$

17. $\frac{1}{2}$ **19.** $\frac{1}{5}$ **21.** $\frac{11}{13} \approx .846$ **23.** .0388

25. $\frac{342,132,219}{63,501,355,960} \approx .00539$ **27.** $\frac{105}{512} \approx .205$

Exercise 7.4 (page 352)

1. .5   **3.** $\frac{3}{7}$ **5.** .3   **7.** .5

9. $P(E|F) = .25;\ P(F|E) = .5$ **11.** .5   **13.** $\frac{4}{13}$

15. (a) $\frac{1}{2}$ (b) $\frac{2}{3}$ **17.** .69   **19.** .9   **21.** .2

23. .1   **25.** .2   **27.** $\frac{2}{3} \approx .67$ **29.** $\frac{1}{2}$

31. $P(HHHH) = \frac{1}{16}$; yes: $P(?H??) = \frac{1}{8}$ **33.** $\frac{25}{204}$ **35.** $\frac{1}{5}$

37. (a) $\frac{1}{26}$ (b) $\frac{1}{2}$ (c) $\frac{1}{13}$ **39.** $\frac{3}{8}$ or 37.5%   **41.** .40

43. .24   **45.** .10   **47.** .08   **49.** $\frac{5}{12}$ **51.** $\frac{1}{3}$

53. (a) $\frac{5}{11}$ (b) $\frac{6}{23}$ (c) $\frac{5}{11}$ (d) $\frac{5}{12}$ (e) $\frac{14}{25}$ (f) $\frac{1}{4}$

55. (a) .7175 (b) .3300 (c) .0714 (d) .1939 (e) .2362
(f) .3574 (g) .3796 (h) .3137

57. $\frac{2}{3}$, or approx. 66.7%   **59.** .5   **61.** 71.5%

63. $P(E|E) = \dfrac{P(E \cap E)}{P(E)} = \dfrac{P(E)}{P(E)} = 1$ when $P(E) \neq 0$

65. $P(E|S) = \dfrac{P(E \cap S)}{P(S)} = \dfrac{P(E)}{1} = P(E)$

Exercise 7.5 (page 360)

1. .24   **3.** .125   **5.** No

7. (a) .2 (b) .4 (c) .08 (d) .52   **9.** $\frac{4}{147}$

11. $P(E|F) = .5$; no   **13.** No   **15.** Yes

17. (a) $\frac{1}{4}$ (b) $\frac{1}{4}$ (c) $\frac{1}{16}$ **19.** No

21. $P(A) = \frac{1}{4} + \frac{1}{4} = \frac{1}{2}$; $P(B) = \frac{1}{4} + \frac{1}{4} = \frac{1}{2}$;
$P(C) = \frac{1}{4} + \frac{1}{4} = \frac{1}{2}$
$P(A \cap B) = \frac{1}{4} = P(A) \cdot P(B)$, so A and B are
independent.
$P(A \cap C) = \frac{1}{4} = P(A) \cdot P(C)$, so A and C are
independent.
$P(B \cap C) = \frac{1}{4} = P(B) \cdot P(C)$, so B and C are
independent.

23. (a) $\frac{9}{16}$ (b) $\frac{1}{16}$ (c) $\frac{3}{8}$

25. (a) $\frac{27}{64}$ (b) $\frac{9}{64}$

27. (a) .6561 (b) .0486 (c) .9963

29. (a) $\frac{9}{25}$ (b) $\frac{12}{25}$

31. (a) $\frac{8}{65} \approx .1231$ (b) $\frac{1}{103} \approx .0097$ (c) No (d) No
(e) No (f) No

33. (a) $\frac{4}{9}$ (b) $\frac{1}{9}$ (c) $\frac{4}{9}$

35. $P(\text{Event (a)}) \approx .5177$; $P(\text{Event (b)}) \approx .491$; event (a) is
more likely to occur.

37. $P(E) \cdot P(F) = P(E \cap F) = 0$, so either $P(E) = 0$,
$P(F) = 0$, or $P(E) = P(F) = 0$.

39. $P(E) \cdot P(F) = P(E \cap F)$; $P(\overline{E}) \cdot P(\overline{F}) =$
$[1 - P(E)][1 - P(F)] = 1 - P(E) - P(F) +$
$P(E) \cdot P(F) = 1 - P(E \cup F) = P(\overline{E \cup F}) =$
$P(\overline{E} \cap \overline{F})$

41. See the hint.

Chapter Review
True – False Items (page 364)

1. T   **2.** F   **3.** F   **4.** F   **5.** T   **6.** F   **7.** T

8. T

Fill in the Blanks (page 364)

1. $\frac{1}{2}$ **2.** 32   **3.** 1; 0   **4.** .8   **5.** for

6. equally likely   **7.** mutually exclusive

Review Exercises *(page 364)*

1. $S = \{MM, MF, FM, FF\}$

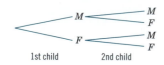

1st child   2nd child

3. (a) $\frac{10}{91}$ (b) $\frac{45}{91}$ (c) $\frac{55}{91}$

5. (a) .21   (b) .23   (c) .1825

7. (a) $\frac{1}{2}$ (b) $\frac{11}{24}$ (c) $\frac{13}{24}$ **9.** (a) No   (b) 000   (c) $\frac{9}{64}$

11. $\frac{7}{13}$ **13.** (a) $\frac{1}{3} \approx .333$ (b) $\frac{9}{38} \approx .237$ (c) $\frac{14}{25} = .56$

15. (a) .025   (b) .0625   (c) .4

17. (a) $\frac{69}{200}$ (b) $\frac{21}{50}$ (c) $\frac{13}{50}$ (d) $\frac{2}{5}$ (e) $\frac{147}{400}$
(f) $\frac{138}{400} \cdot \frac{85}{400} = \frac{1173}{16,000} \neq \frac{4}{400}$

19. $\frac{1}{4}$

Mathematical Questions from Professional Exams *(page 366)*

1. b   **2.** e   **3.** b   **4.** d   **5.** c   **6.** d   **7.** b

8. a   **9.** c   **10.** b

CHAPTER 8

Exercise 8.1 *(page 376)*

1. .4   **3.** .2   **5.** .7   **7.** .31   **9.** $\frac{12}{31}$ **11.** $\frac{7}{31}$

13. $\frac{12}{31}$ **15.** .024   **17.** .016

19. $P(A_1|E) = \frac{1}{2} = .5$; $P(A_2|E) = \frac{1}{2} = .5$

21. $P(A_1|E) = \frac{3}{8} = .375$; $P(A_2|E) = \frac{3}{8} = .375$; $P(A_3|E) = \frac{1}{4} = .25$

23. $P(A_1|E) = \frac{8}{29} \approx .276$; $P(A_2|E) = \frac{20}{29} = .690$; $P(A_3|E) = \frac{1}{29} \approx .034$

25. $P(A_2|E) = \frac{0}{31} \approx 0$; $P(A_3|E) = \frac{2}{31} = .065$; $P(A_4|E) = \frac{0}{31} = 0$; $P(A_5|E) = \frac{2}{31} = .065$

27. $P(U_{\text{I}}|E) = \frac{5}{15} \approx .333$; $P(U_{\text{II}}|E) = \frac{3}{15} = .2$; $P(U_{\text{III}}|E) = \frac{7}{15} \approx .467$

29. .945

31. $P(\text{Democrat}|\text{Voted}) = .385$; $P(\text{Republican}|\text{Voted}) = .39$; $P(\text{Independent}|\text{Voted}) = .225$

33. $P(\text{Rock}|\text{Positive test}) \approx .385$; $P(\text{Clay}|\text{Positive test}) \approx .209$; $P(\text{Sand}|\text{Positive test}) \approx .405$

35. $P(\text{Republican}) = .466$; $P(\text{Northeasterner}|\text{Republican}) \approx .343$

37. $\frac{9}{11} \approx .818$ **39.** .217

41. (a) .858   (b) .503   (c) .961

43. Since F is a subset of E, $E \cap F = F$ and $P(E \cap F) = P(F)$. So, $P(E|F) = \dfrac{P(E \cap F)}{P(F)} = \dfrac{P(F)}{P(F)} = 1$.

Exercise 8.2 *(page 384)*

1. .0287   **3.** .0138   **5.** $\frac{3003}{32,768} \approx .0916$ **7.** .2969

9. $\frac{2}{9} \approx .2222$ **11.** $\frac{125}{216} \approx .5787$ **13.** $\frac{80}{243} \approx .3292$

15. .0368   **17.** .2362   **19.** .0580   **21.** $\frac{1}{32} = .03125$

23. $\frac{93}{256} \approx .3633$ **25.** $\frac{28}{255} \approx .1098$ **27.** $\frac{625}{3888} \approx .1608$

29. (a) .2793   (b) .0515   (c) .3366   (d) .9942

31. $\frac{5}{16} = .3125$

33. (a)

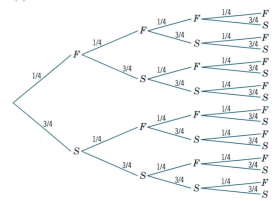

(b) $(\frac{1}{4} \cdot \frac{1}{4} \cdot \frac{3}{4} \cdot \frac{3}{4}) + (\frac{1}{4} \cdot \frac{3}{4} \cdot \frac{1}{4} \cdot \frac{3}{4}) + (\frac{1}{4} \cdot \frac{3}{4} \cdot \frac{3}{4} \cdot \frac{1}{4}) +$
$(\frac{3}{4} \cdot \frac{1}{4} \cdot \frac{1}{4} \cdot \frac{3}{4}) + (\frac{3}{4} \cdot \frac{1}{4} \cdot \frac{3}{4} \cdot \frac{1}{4}) + (\frac{3}{4} \cdot \frac{3}{4} \cdot \frac{1}{4} \cdot \frac{1}{4}) =$
$\frac{54}{256} \approx .2109$

(c) $b(4, 2; \frac{1}{4}) = 6 \cdot (\frac{1}{4})^2 \cdot (\frac{3}{4})^2 = \frac{54}{256}$

35. .9996

37. $\displaystyle\sum_{k=10}^{15} b(15, k; \frac{1}{2}) = \frac{309}{2048} \approx .1509$; $\displaystyle\sum_{k=12}^{15} b(15, k; .8) \approx$ .6482

39. .0007

41. .1225   **43.** (a) $\frac{7}{64} \approx .1094$ (b) .3446

Technology Exercises *(page 386)*

1.

k	Actual Value of $P(k)$
0	.0625
1	.25
2	.375
3	.25
4	.0625

3. .21875

Exercise 8.3 *(page 394)*

1. 1.2   **3.** 50,800

5. She should pay $0.80 for a fair game.

7. He should pay $1.67 for a fair game.   **9.** 75¢

11. (a) $0.75   (b) No   (c) Lose $2

13. It is not a fair bet; your expected loss is $\$\frac{3}{7} \approx \$0.43$.

15. No, she should not play the game; her expected loss is $\frac{15}{13}$¢ ≈ 1.2¢.

17. $7   **19.** The second site has the higher expected profit.

21. $\frac{2000}{6} = 333\frac{1}{3}$ **23.** 10   **25.** 1

27. $\frac{175}{64} \approx 2.734$ tosses   **29.** Aircraft A

Exercise 8.4 *(page 402)*

1. The expected number of customers is 9; the optimal number of cars is 9; the expected daily profit is $168.

3. For $p = .95$:

Group Size	2	3	4	5	6	7
Expected Tests Saved per Component	.4025	.524	.565	.574	.568	.555

The optimal group size is 5.

5. (a) Expected net gain in dollars = $75,000 - 75,000(.05^x) - 500x$, where x = Number of divers hired.

(b) Hiring 2 divers will maximize the net gain.

7. .9647

Exercise 8.5 *(page 405)*

1. $P(X = 0) = \frac{1}{4}$; $P(X = 1) = \frac{1}{2}$; $P(X = 2) = \frac{1}{4}$

3. $P(X = 0) = \frac{1}{8}$; $P(X = 1) = \frac{3}{8}$; $P(X = 2) = \frac{3}{8}$; $P(X = 3) = \frac{1}{8}$

5. $P(X = 0) = \frac{27}{125} = .216$; $P(X = 1) = \frac{54}{125} = .432$; $P(X = 2) = \frac{36}{125} = .288$; $P(X = 3) = \frac{8}{125} = .064$

7. 1.2

Technology Exercises *(page 405)*

1. $P(X = k) = .167$, for $k = 1, 2, 3, 4, 5, 6$

3. .2   **5.** .083

Chapter Review
True–False Items *(page 407)*

1. T   **2.** F   **3.** F   **4.** T   **5.** F   **6.** T

Fill in the Blanks *(page 407)*

1. Bayes' formula

2. (c) independent   (d) unchanged (or the same)

3. expected value   **4.** a number   **5.** expected value

Review Exercises *(page 407)*

1. .5   **3.** .4   **5.** .3   **7.** .43   **9.** $\frac{20}{43} \approx .4651$

11. $\frac{20}{43} \approx .4651$ **13.** $\frac{3}{43} \approx .0698$

15. (a) $\frac{180}{260} \approx .6923$ (b) $\frac{180}{345} \approx .5217$ (c) $\frac{110}{345} \approx .3188$
(d) $\frac{55}{345} \approx .1594$ (e) $\frac{60}{260} \approx .2308$ (f) $\frac{60}{210} \approx .2857$
(g) $\frac{85}{210} \approx .4048$ (h) $\frac{65}{210} \approx .3095$

17. .163

19. none: .3277; exactly 3: .0512

21. (a) $\frac{1}{4096}$ (b) $\frac{793}{2048} \approx .3872$ (c) 793 to 1255

23. $\frac{31}{19,683} \approx .001575$ **25.** He paid $3\frac{1}{3}$¢ too much.

27. The expected value is $0.2833 - \$0.30 = -\$0.0167$. The game is not fair.

29. 50   **31.** (a) $1 - q^{30}$ (b) $31 - 30q^{30}$ tests

Mathematical Questions from Professional Exams (page 409)

1. d   **2.** a   **3.** a   **4.** d   **5.** b   **6.** b

7. b *or* c (they're equal!)   **8.** d

CHAPTER 9

Exercise 9.1 (page 415)

1. A poll should be taken either door-to-door or by means of the telephone.

3. A poll should be taken door-to-door in which people are asked to fill out a questionnaire.

5. The data should be gathered from all different kinds of banks.

7. (1) Asking a group of children if they like candy to determine what percentage of people like candy.
(2) Asking a group of people over 65 their opinion toward Medicare to determine the opinion of people in general about Medicare.

9. By taking a poll downtown, you would question mostly people who are either shopping or working downtown. For instance, you would question few students.

11. (a) Northwest   (b) TWA   (c) About 78%

13. (a) Housing, fuel, and utilities
(b) Misc. goods and services
(c) Senior citizens would feel the CPI weight of 7% for health care is too low because they spend twice as much (14%) of their income on health care.

15.

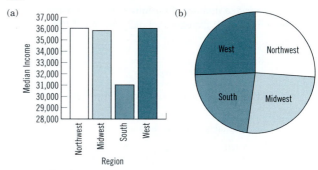

(c) The bar graph seems to summarize the data better.
(d) Northwest   (e) South

17. (a)

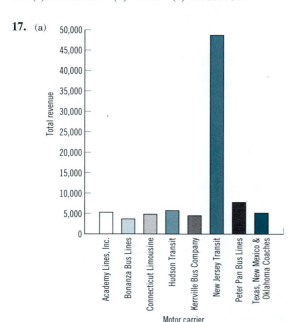

(b)

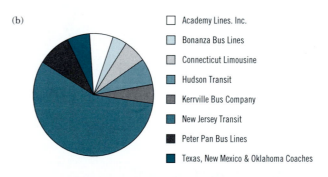

(c) Answers will vary.
(d) New Jersey Transit   (e) Bonanza Bus Lines

19. (a)

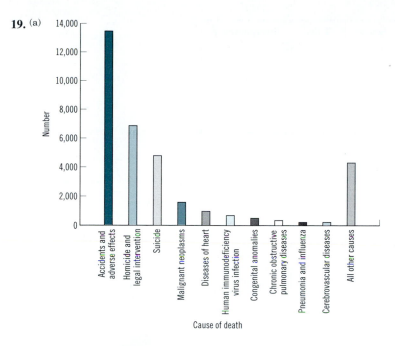

(b)

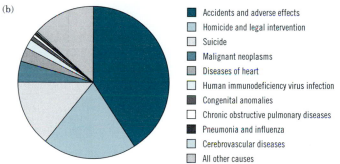

- ■ Accidents and adverse effects
- ▨ Homicide and legal intervention
- □ Suicide
- ▨ Malignant neoplasms
- ▨ Diseases of heart
- □ Human immunodeficiency virus infection
- ▨ Congenital anomalies
- □ Chronic obstructive pulmonary diseases
- ■ Pneumonia and influenza
- ▨ Cerebrovascular diseases
- ▨ All other causes

(c) Answers will vary.
(d) Accidents and adverse effects.

Exercise 9.2 *(page 422)*

1. (a) Frequency Table

Score	Tally	Frequency *f*	Score	Tally	Frequency, *f*
25	I	1	41	⦀	5
26	I	1	42	III	3
28	I	1	43	I	1
29	I	1	44	II	2

Table continued on following page

Score	Tally	Frequency f	Score	Tally	Frequency, f
30	III	3	45	I	1
31	II	2	46	II	2
32	I	1	47	I	1
33	II	2	48	III	3
34	II	2	49	I	1
35	I	1	50	I	1
36	II	2	51	I	1
37	IIII	4	52	III	3
38	I	1	53	II	2
39	I	1	54	II	2
40	I	1	55	I	1

$$\text{Range} = 55 - 25 = 30$$

(b) Line Chart

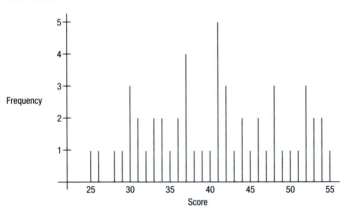

(c) Histogram

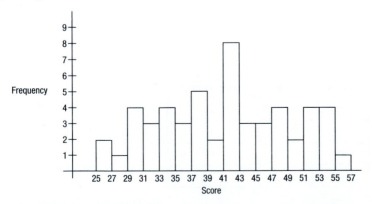

(d) Frequency Polygon

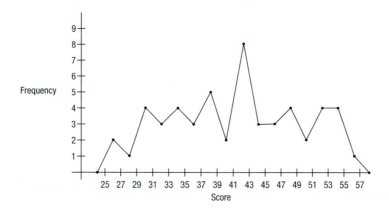

(e) Cumulative (less than) frequencies

Class Interval	f	c.f.
25–26	2	2
27–28	1	3
29–30	4	7
31–32	3	10
33–34	4	14
35–36	3	17
37–38	5	22
39–40	2	24
41–42	8	32
43–44	3	35
45–46	3	38
47–48	4	42
49–50	2	44
51–52	4	48
53–54	4	52
55–56	1	53

(f) Cumulative (less than) Frequency Distribution

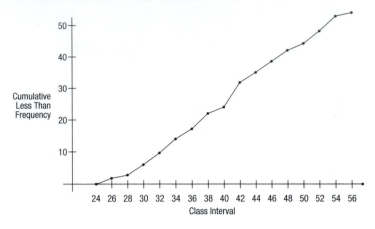

3. (a) 13   (b) 20; 24   (c) 5   (d) About 145,000
   (e) 30–34   (f) 80–84
   (g)

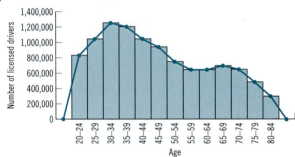

5. (a) 13   (b) 20; 24   (c) 5
   (d)

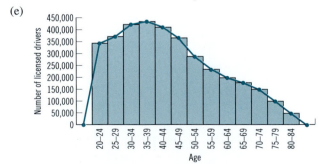

   (e)

   (f) 35–39   (g) 80–84

7. (a) 15   (b) 0; 999   (c) 1000

(d)

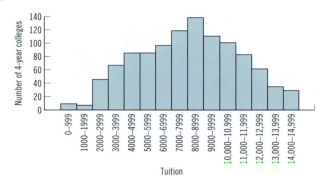

(e)

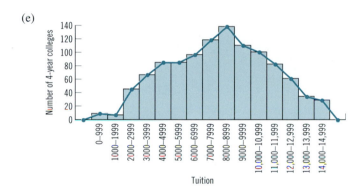

(f) 8000–8999

Technology Exercises *(page 424)*

1.

	Class Interval	Tally	Frequency, f
1	13–13.9	I	1
2	14–14.9	ⅢⅠ III	8
3	15–15.9	ⅢⅠ	5
4	16–16.9	II	2
5	17–17.9	I	1
6	18–18.9		0
7	19–19.9	I	1
8	20–20.9	II	2

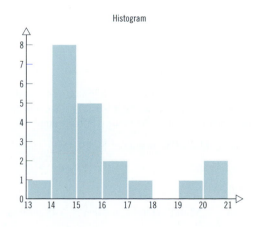

3.

	Class Interval	Tally	Frequency, f
1	88–122	I	1
2	123–157	I	1
3	158–192	I	1
4	193–227	JH1 JH1	10
5	228–262	JH1 II	7

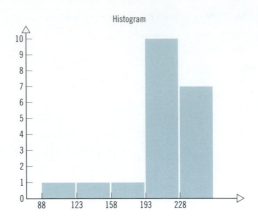

Histogram

5.

	Class Interval	Tally	Frequency, f
1	9.36–10.35	III	3
2	10.36–11.35	II	2
3	11.36–12.35	JH1 III	8
4	12.36–13.35	JH1	5
5	13.36–14.35	I	1
6	14.36–15.35	I	1

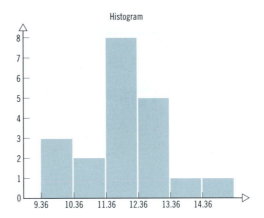

Histogram

Exercise 9.3 (*page 432*)

1. Mean = 31.25; median = 30.5; no mode

3. Mean = 70.4; median = 70; mode = 55

5. Mean = 78.8; median = 82; mode = 82

7. Mean = 73; median = 80; mode = 80   **9.** $109.40

11. Mean = $41,300; median = $36,000. The median describes the situation more realistically since it is closer to the salary of most of the faculty in this sample.

13. Mean = $1826.08; median = $1805.18

15. ≈44.2 years

17. ≈$8053.29

19. Mean = 3.52 million units; median = 3.45 million units

21. Mean = 7.8%; median = 8.05%

Exercise 9.4 (*page 439*)

1. (b) has the larger variance.   **3.** $\sigma \approx 6.53$

5. $\sigma \approx 5.2$ **7.** $\sigma \approx 12.47$

9. $\bar{X} \approx 31.9$; $\sigma \approx 7.80$

11. Mean lifetime ≈ 868.67 hr; $\sigma \approx 66.68$ hr

13. Since the mean number of salmon caught in river I is 2000 and the mean number of salmon caught in river II is 2038, river II is preferred.

15. ≈15.7 years

17. ≈$3175.58

19. (a) 75%   (b) 64%   (c) $88\frac{8}{9}\%$ (d) 25%   (e) $11\frac{1}{9}\%$

21. ($24.37, $78.13)

Exercise 9.5 (page 449)

1. $\mu = 8; \sigma = 1$ **3.** $\mu = 18; \sigma = 1$

5.

x	7	9	13	15	29	37	41
Z-score	-0.6559	-0.4409	-0.0108	0.2043	1.7097	2.5699	3.0

7. (a) .3133   (b) .3642   (c) .4938   (d) .4987
(e) .2734   (f) .4896   (g) .2881   (h) .4988

9. .3085   **11.** .0668

13. (a) Approx. 1365 women   (b) Approx. 1909 women
(c) Approx. 1995 women

15. (a) Approx. 1 student   (b) Between 124.6 and 135.4 lb

17. 239 pairs   **19.** Kathleen

21.

k	0	1	2	3	4	5	6	7
$b(15, k; .30)$	.0047	.0305	.0916	.1700	.2186	.2061	.1472	.0811
k	8	9	10	11	12	13	14	15
$b(15, k; .30)$	.0348	.0116	.0030	.0006	.0001	.0000	.0000	.0000

Line chart:

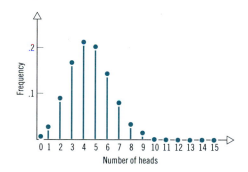

Frequency curve:

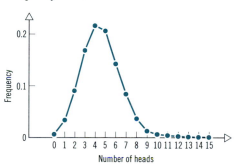

23. .7498   **25.** .5   **27.** .0274   **29.** .2286; .0918

31. .0122

Technology Exercises (page 450)

1. Maximum for $x = 0$.

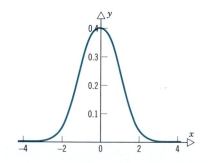

Chapter Review
True–False Items (page 451)

1. F   **2.** T   **3.** F   **4.** T   **5.** F

Fill in the Blanks (page 451)

1. mean, median, mode   **2.** standard deviation
3. bell   **4.** Z-score   **5.** $\overline{X} - k, \overline{X} + k$

Review Exercises (page 452)

1.

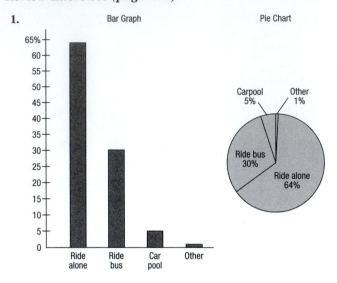

Bar Graph   Pie Chart

3. (a) Frequency table:

Score	Freq.	Score	Freq.
8	1	33	1
10	1	42	1
12	1	44	1
14	2	48	1
17	1	52	2
19	1	55	1
20	1	60	1
21	1	63	2
26	1	66	2
30	1	69	1

Score	Freq.	Score	Freq.
70	1	85	2
72	2	87	2
73	2	89	1
74	1	90	1
75	1	92	1
77	1	95	1
78	2	99	1
80	3	100	2
82	1		
83	1		

Range $= 100 - 8 = 92$

(b) Line chart:

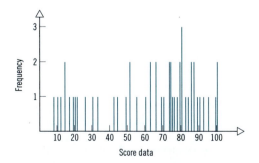

(c) Histogram:

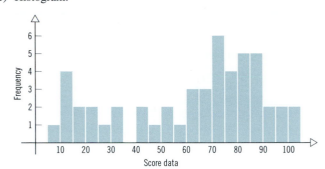

Score data

(d) Frequency polygon:

Score data

(e) Cumulative (less than) frequencies:

Class Interval	<Cum. Freq.	Class Interval	<Cum. Freq.
4.5–9	1	29.5–34	12
9.5–14	5	34.5–39	12
14.5–19	7	39.5–44	14
19.5–24	9	44.5–49	15
24.5–29	10	49.5–54	17

Class Interval	<Cum. Freq.	Class Interval	<Cum. Freq.
54.5–59	18	79.5–84	39
59.5–64	21	84.5–89	44
64.5–69	24	89.5–94	46
69.5–74	30	94.5–99	48
74.5–79	34	99.5–104	50

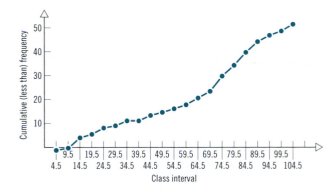

Class interval

5. (a) Mean = $\frac{67}{12} \approx 5.58$; Median = 4.5; Mode = 4
   (b) Mean = $\frac{209}{8} = 26.125$; Median = 2; Mode = 2
   (c) Mean = $\frac{62}{9} \approx 6.89$; Median = 7; Mode = 7

7. $A = \{-1, 1\}$; $B = \{-10, 10\}$. Both sets have mean 0, but the standard deviation for A is 1, while the standard deviation for B is 10.

9. ~ 2.7701

11. (a) Approx. 410   (b) Approx. 94   (c) Approx. 298

13. (a) .0855   (b) .075

15. At least .75 (using Chebychev's theorem)

Mathematical Questions from Professional Exams *(page 453)*

1. e   2. b   3. c

CHAPTER 10

Exercise 10.1 *(page 461)*

1. The sum of the entries in row 3 is not equal to 1; there is a negative entry in row 3, column 2.

3. (a) The probability of a change from state 2 to state 1 is $\frac{1}{4}$.
  (b) $[\frac{1}{3} \quad \frac{2}{3}]$; $[\frac{5}{18} \quad \frac{13}{18}]$
  (c) $[\frac{1}{4} \quad \frac{3}{4}]$; $[\frac{13}{48} \quad \frac{35}{48}]$

5. $[.3625 \quad .6375]$ **7.** $a = .4, b = .1, c = 1$

9. $[.50397 \quad .49603]$

11. (a) Each experiment measures the proportion of mayors in these cities who are Democrats and the proportion who are Republicans during a given term of office. These proportions depend only on the proportions during the preceding mayoral terms, so the sequence of experiments can be represented as a Markov chain.

$$\begin{array}{cc} & \text{D} \quad \text{R} \end{array}$$
  (b) $P = \begin{array}{c} \text{D} \\ \text{R} \end{array} \begin{bmatrix} .6 & .4 \\ .3 & .7 \end{bmatrix}$

  (c) $P^2 = \begin{bmatrix} .48 & .52 \\ .39 & .61 \end{bmatrix}$; $P^3 = \begin{bmatrix} .444 & .556 \\ .417 & .583 \end{bmatrix}$

13. 57%

15. (a) 45.1% by Travelers and 32.6% by General American
  (b) 45.026% by Travelers and 34.732% by General American

17. $uA = [u_1a_{11} + u_2a_{21} \quad u_1a_{12} + u_2a_{22}]$. Note that since the entries of u and A are nonnegative, the entries of uA (being sums of products of nonnegative numbers) are nonnegative. Also, $(u_1a_{11} + u_2a_{21}) + (u_1a_{12} + u_2a_{22}) = (u_1a_{11} + u_1a_{12}) + (u_2a_{21} + u_2a_{22}) = u_1(a_{11} + a_{12}) + u_2(a_{21} + a_{22}) = u_1(1) + u_2(1) = u_1 + u_2 = 1$. Since uA is a row vector whose entries are all nonnegative and sum to 1, it is a probability vector.

Technology Exercises *(page 462)*

1. $v^{(10)} = [.32000 \quad .28923 \quad .16308 \quad .22769]$ (rounded to five decimal places)

3. (a) $P = \begin{bmatrix} 0 & \frac{1}{2} & 0 & \frac{1}{2} & 0 & 0 \\ \frac{1}{3} & 0 & \frac{1}{3} & 0 & \frac{1}{3} & 0 \\ 0 & \frac{1}{2} & 0 & 0 & 0 & \frac{1}{2} \\ \frac{1}{2} & 0 & 0 & 0 & \frac{1}{2} & 0 \\ 0 & 0 & 0 & 0 & 0 & 1 \\ 0 & 0 & 0 & 0 & 0 & 1 \end{bmatrix}$

  (b) $v^{(0)} = [0 \quad 1 \quad 0 \quad 0 \quad 0 \quad 0]$
  (c) $v^{(0)} = [0 \quad .01875 \quad 0 \quad .01250 \quad 0 \quad .96875]$ (rounded to five decimal places)
  (d) Room 6, with 96.88% probability

Exercise 10.2 *(page 471)*

1. Regular; fixed probability vector $= [\frac{2}{3} \quad \frac{1}{3}]$

3. Regular; fixed probability vector $= [\frac{1}{5} \quad \frac{4}{5}]$

5. Not regular

7. $\begin{bmatrix} \frac{1}{2} & \frac{1}{2} \\ \frac{1}{2} & \frac{1}{2} \end{bmatrix} \begin{bmatrix} 1-p & p \\ p & 1-p \end{bmatrix}$
$= \begin{bmatrix} \frac{1-p}{2} + \frac{p}{2} & \frac{p}{2} + \frac{1-p}{2} \end{bmatrix} = \begin{bmatrix} \frac{1}{2} & \frac{1}{2} \end{bmatrix}$

9. $P = \begin{bmatrix} .7 & .15 & .15 \\ .1 & .8 & .1 \\ .2 & .2 & .6 \end{bmatrix}$. In the long run, $\frac{4}{13}$ ($\approx 30.8\%$) of the detergent stock is brand A, $\frac{6}{13}$ ($\approx 46.2\%$) of the stock is brand B, and $\frac{3}{13}$ ($\approx 23.1\%$) of the stock is brand C.

11. The probability that the grandson of a Labourite will vote Socialist is .09 (or 9%). The membership distribution in the long run is $\frac{26}{47}$ ($\approx 55.3\%$) Conservative, $\frac{18}{47}$ ($\approx 38.2\%$) Labourite, and $\frac{3}{47}$ ($\approx 6.4\%$) Socialist.

13. The probability that a blond is the grandmother of a brunette is .3 (or 30%).

  (a) 32.7% blonds, 42.5% brunettes, 24.8% redheads
  (b) 35% blonds, 40% brunettes, 25% redheads

Technology Exercises *(page 472)*

1. $t = [.125 \quad .875] = [\frac{1}{8} \quad \frac{7}{8}]$

3. $t = [.444444 \quad .222222 \quad .333333] = [\frac{4}{9} \quad \frac{2}{9} \quad \frac{1}{3}]$

5. $t = [.267647 \quad .208824 \quad .205882$
$\quad .317647] = [\frac{91}{340} \quad \frac{71}{340} \quad \frac{7}{34} \quad \frac{27}{85}]$

Exercise 10.3 *(page 480)*

1. Not absorbing   **3.** Absorbing

5. Not absorbing

7. $T = [\frac{8}{3}]$; $S = [\frac{1}{8} \quad \frac{2}{8}]$; $T \cdot S = [\frac{1}{3} \quad \frac{2}{3}]$

9. (a) .5; 1   (b) 3

11. Starting with \$1: $\frac{4}{19} \approx .2105$; starting with \$2: $\frac{10}{19} \approx .5263$

13. (a) 1.4 wagers   (b) .84   (c) .16

15. (a) 1.6 wagers   (b) .64   (c) .36   **17.** 6.993 days

Technology Exercises *(page 481)*

1. $I_r = \begin{bmatrix} 1 & 0 \\ 0 & 1 \end{bmatrix}$, $S = \begin{bmatrix} 0 & 0 \\ 0 & .2 \\ .75 & .25 \end{bmatrix}$,

$Q = \begin{bmatrix} .05 & .95 & 0 \\ 0 & .5 & .3 \\ 0 & 0 & 0 \end{bmatrix}$, $T = \begin{bmatrix} 1.05 & 2 & .6 \\ 0 & 2 & .6 \\ 0 & 0 & 1 \end{bmatrix}$,

$T \cdot S = \begin{bmatrix} .45 & .55 \\ .45 & .55 \\ .75 & .25 \end{bmatrix}$

Exercise 10.4 (page 484)

1. Katy's payoff matrix:
$$\begin{array}{cc} 1 & 2 \end{array}$$
$$\begin{bmatrix} -1 & 1 \\ 1 & -1 \end{bmatrix} \begin{matrix} 1 \text{ finger} \\ 2 \text{ fingers} \end{matrix}$$

3. Katy's payoff matrix:
$$\begin{array}{ccc} 1 & 4 & 7 \end{array}$$
$$\begin{bmatrix} -2 & 5 & -8 \\ 5 & -8 & 11 \\ -8 & 11 & -14 \end{bmatrix} \begin{matrix} 1 \\ 4 \\ 7 \end{matrix}$$

5. Strictly determined; value $= -2$

7. Strictly determined; value $= 3$

9. Not strictly determined

11. Strictly determined; value $= 2$

13. Not strictly determined

15. $0 \le a \le 3$ 17. $ab \le 0$

Exercise 10.5 (page 488)

1. 1.42 3. $\frac{9}{4} = 2.25$ 5. $\frac{19}{8} = 2.375$

7. $\frac{17}{9} \approx 1.889$ 9. $\frac{1}{3}$

11. If the game is not strictly determined, then none of the entries $a_{11}, a_{12}, a_{21}, a_{22}$ can be a saddle point. Comparing a_{11} and a_{12}, either $a_{11} < a_{12}$, $a_{11} = a_{12}$, or $a_{11} > a_{12}$. If $a_{11} < a_{12}$, then $a_{11} < a_{21}$, for otherwise a_{11} would be a saddle point. Then $a_{21} > a_{22}$, else a_{21} would be a saddle point. Then $a_{12} > a_{22}$, so that a_{22} won't be a saddle point. Thus, if $a_{11} < a_{12}$ and the game is not strictly determined, then the inequalities listed in (b) follow. Note that $a_{11} \ne a_{12}$, or there would be a saddle point in the matrix. Finally, if $a_{11} > a_{12}$, then $a_{12} < a_{22}$, as a_{12} isn't a saddle point. Then $a_{21} < a_{22}$ and $a_{11} > a_{21}$, to prevent a_{22} and a_{21} from being saddle points. Thus, if $a_{11} > a_{12}$ and the game is not strictly determined, then the inequalities listed in (a) follow.

Exercise 10.6 (page 494)

1. Player I: optimal strategy is to choose row 1 with probability $\frac{3}{4}$; $p_1 = \frac{3}{4}$, $p_2 = \frac{1}{4}$. Player II: optimal strategy is to choose column 1 with probability $\frac{1}{4}$; $q_1 = \frac{1}{4}$, $q_2 = \frac{3}{4}$. Expected payoff $E = \frac{7}{4}$.

3. Player I: optimal strategy is to choose row 1 with probability $\frac{1}{6}$; $p_1 = \frac{1}{6}$, $p_2 = \frac{5}{6}$. Player II: optimal strategy is to choose column 1 with probability $\frac{1}{3}$; $q_1 = \frac{1}{3}$, $q_2 = \frac{2}{3}$. Expected payoff $E = \frac{1}{3}$.

5. Player I: optimal strategy is to choose row 1 with probability $\frac{5}{8}$; $p_1 = \frac{5}{8}$, $p_2 = \frac{3}{8}$. Player II: optimal strategy is to choose column 1 with probability $\frac{5}{8}$; $q_1 = \frac{5}{8}$, $q_2 = \frac{3}{8}$. Expected payoff $E = \frac{7}{8}$.

7. The Democrat should spend $\frac{3}{8}$ ($= 37.5\%$) of her/his time

on domestic issues and $\frac{5}{8}$ ($= 62.5\%$) on foreign issues. The Republican should divide her/his time evenly between the two issues. The expected payoff of the game is $\frac{3}{2}$, favoring the Democrat.

9. The spy should choose the deserted exit $\frac{6}{71}$ ($\approx 8.5\%$) of the time and the heavily used exit $\frac{65}{71}$ ($\approx 91.5\%$) of the time. The spy's opponent should choose the deserted exit $\frac{16}{71}$ ($\approx 22.5\%$) of the time and the heavily used exit $\frac{55}{71}$ ($\approx 77.5\%$) of the time. The game's expected payoff is $\frac{50}{71}$ ($\approx .7042$), favoring the spy.

11. The game must be strictly determined.

Chapter Review
True–False Items (page 495)

1. F   2. F   3. F   4. T   5. T   6. T   7. F

Fill in the Blanks (page 495)

1. m 2. nonnegative, one   3. $v^{(k)} = v^{(0)}P^k$

4. positive   5. payoff

Review Exercises (page 495)

1. (a) $[\frac{2}{5} \quad \frac{3}{5}]$ (b) $[\frac{1}{2} \quad \frac{1}{2}]$ (c) $[\frac{48}{79} \quad \frac{10}{79} \quad \frac{21}{79}]$

3. After 2 years: A holds $\frac{283}{600} \approx 47.17\%$, B holds $\frac{323}{1200} \approx 26.92\%$, and C holds $\frac{311}{1200} \approx 25.92\%$ of the beer market. In the long run: A will hold $\frac{80}{169} \approx 47.34\%$, B will hold $\frac{45}{169} \approx 26.63\%$, and C will hold $\frac{44}{169} \approx 26.04\%$ of the beer market.

5. She will sell at U_1 $\frac{3}{7}$ ($\approx 42.86\%$) of the time, at U_2 $\frac{16}{35}$ ($\approx 45.71\%$) of the time, and at U_3 $\frac{4}{35}$ ($\approx 11.43\%$) of the time.

7. (a)

(Nonabsorbing) State, x	Expected No. of Times Process Is in State x
$1	1.298406
$2	2.360738
$3	1.411737
$4	0.635282

   (b) The expected number of bets before absorption is 5.706163.
   (c) The probability that he loses all his money is .714123. (The probability that he wins $5 is $1 - .714123 = .285877$.)

9. (a) $\frac{1}{3}$ (b) 0   (c) 0

11. (a) $[\frac{5}{6} \quad \frac{1}{6}]$ (b) 7.5   (c) 7.5% increase

13. $[\frac{4}{9} \quad \frac{5}{9}]; \frac{1}{9}$

CHAPTER 11

Exercise 11.1 (page 507)

1. Proposition   **3.** Not a proposition

5. Proposition   **7.** Proposition

9. A fox is not an animal.

11. I am not buying stocks and bonds.

13. Someone wants to buy my house.

15. Every person has a car.

17. John is an economics major or a sociology major.

19. John is an economics major and a sociology major.

21. John is not an economics major or he is not a sociology major.

23. John is not an economics major or he is a sociology major.

Exercise 11.2 (page 516)

1.

p	q	$\sim q$	$p \vee \sim q$
T	T	F	T
T	F	T	T
F	T	F	F
F	F	T	T

3.

p	q	$\sim p$	$\sim q$	$\sim p \wedge \sim q$
T	T	F	F	F
T	F	F	T	F
F	T	T	F	F
F	F	T	T	T

5.

p	q	$\sim p$	$\sim p \wedge q$	$\sim(\sim p \wedge q)$
T	T	F	F	T
T	F	F	F	T
F	T	T	T	F
F	F	T	F	T

7.

p	q	$\sim p$	$\sim q$	$\sim p \vee \sim q$	$\sim(\sim p \vee \sim q)$
T	T	F	F	F	T
T	F	F	T	T	F
F	T	T	F	T	F
F	F	T	T	T	F

9.

p	q	$\sim q$	$p \vee \sim q$	$(p \vee \sim q) \wedge p$
T	T	F	T	T
T	F	T	T	T
F	T	F	F	F
F	F	T	T	F

11.

p	q	$\sim q$	$p \underline{\vee} q$	$p \wedge \sim q$	$(p \underline{\vee} q) \wedge (p \wedge \sim q)$
T	T	F	F	F	F
T	F	T	T	T	T
F	T	F	T	F	F
F	F	T	F	F	F

13.

p	q	$\sim p$	$\sim q$	$p \wedge q$	$\sim p \wedge \sim q$	$(p \wedge q) \vee (\sim p \wedge \sim q)$
T	T	F	F	T	F	T
T	F	F	T	F	F	F
F	T	T	F	F	F	F
F	F	T	T	F	T	T

15.

p	q	r	$\sim q$	$p \wedge \sim q$	$(p \wedge \sim q) \underline{\vee} r$
T	T	T	F	F	T
T	T	F	F	F	F
T	F	T	T	T	T
T	F	F	T	T	T
F	T	T	F	F	T
F	T	F	F	F	F
F	F	T	T	F	T
F	F	F	T	F	F

17.

p	$p \wedge p$	$p \vee p$
T	T	T
F	F	F

Since each column is the same, $p \equiv p \wedge p \equiv p \vee p$

19.

p	q	r	$p \wedge q$	$q \wedge r$	$(p \wedge q) \wedge r$	$p \wedge (q \wedge r)$
T	T	T	T	T	T	T
T	T	F	T	F	F	F
T	F	T	F	F	F	F
T	F	F	F	F	F	F
F	T	T	F	T	F	F
F	T	F	F	F	F	F
F	F	T	F	F	F	F
F	F	F	F	F	F	F

The last two columns are the same, so $(p \wedge q) \wedge r \equiv p \wedge (q \wedge r)$.

p	q	r	$p \vee q$	$q \vee r$	$(p \vee q) \vee r$	$p \vee (q \vee r)$
T	T	T	T	T	T	T
T	T	F	T	T	T	T
T	F	T	T	T	T	T
T	F	F	T	F	T	T
F	T	T	T	T	T	T
F	T	F	T	T	T	T
F	F	T	F	T	T	T
F	F	F	F	F	F	F

The last two columns are the same, so $(p \vee q) \vee r \equiv p \vee (q \vee r)$.

21.

	① p	② q	③ $p \vee q$	④ $p \wedge q$	⑤ $p \wedge (p \vee q)$	⑥ $p \vee (p \wedge q)$
	T	T	T	T	T	T
	T	F	T	F	T	T
	F	T	T	F	F	F
	F	F	F	F	F	F

Since columns 1 and 5 are the same, $p \equiv p \wedge (p \vee q)$.
Since columns 1 and 6 are the same, $p \equiv p \vee (p \wedge q)$.

23.

	① p	② q	③ $\sim q$	④ $\sim q \vee q$	⑤ $p \wedge (\sim q \vee q)$
	T	T	F	T	T
	T	F	T	T	T
	F	T	F	T	F
	F	F	T	T	F

Since columns 1 and 5 are the same, $p \equiv p \wedge (\sim q \vee q)$.

25.

	① p	② $\sim p$	③ $\sim(\sim p)$
	T	F	T
	F	T	F

Since columns 1 and 3 are the same, $p \equiv \sim(\sim p)$.

27.

p	q	$\sim p$	$q \wedge (\sim p)$	$p \wedge (q \wedge \sim p)$
T	T	F	F	F
T	F	F	F	F
F	T	T	T	F
F	F	T	F	F

29.

p	q	$\sim p$	$\sim q$	$p \wedge q$	$\sim p \wedge \sim q$	$(p \wedge q) \vee (\sim p \wedge \sim q)$	$[(p \wedge q) \vee (\sim p \wedge \sim q)] \wedge p$
T	T	F	F	T	F	T	T
T	F	F	T	F	F	F	F
F	T	T	F	F	F	F	F
F	F	T	T	F	T	T	F

31. Smith is an ex-convict and he is an ex-convict. ≡ Smith is an ex-convict or he is an ex-convict. ≡ Smith is an ex-convict.

33. "It is not true that Smith is an ex-convict or rehabilitated" means the same as the statement "Smith is not an ex-convict and he is not rehabilitated." "It is not true that Smith is an ex-convict and he is rehabilitated" means the same as "Smith is not an ex-convict or he is not rehabilitated."

Exercise 11.3 *(page 523)*

1. $\sim p \Rightarrow q$; Converse: $q \Rightarrow \sim p$; Contrapositive: $\sim q \Rightarrow p$; Inverse: $p \Rightarrow \sim q$.

3. $\sim q \Rightarrow \sim p$; Converse: $\sim p \Rightarrow \sim q$; Contrapositive: $p \Rightarrow q$; Inverse: $q \Rightarrow p$.

5. If it is raining, the grass is wet. Converse: If the grass is wet, it is raining. Contrapositive: If the grass is not wet, it is not raining. Inverse: If it is not raining, the grass is not wet.

7. If it is not raining, it is not cloudy. Converse: If it is not cloudy, it is not raining. Contrapositive: If it is cloudy, it is raining. Inverse: If it is raining, it is cloudy.

9. Rain is sufficient for it to be cloudy. Converse: If it is cloudy, it is raining. Contrapositive: If it is not cloudy, it is not raining. Inverse: If it is not raining, it is not cloudy.

11. (a) If Jack studies psychology, then Mary studies sociology.
 (b) If Mary studies sociology, then Jack studies psychology.
 (c) If Jack does not study psychology, then Mary studies sociology.

13. (a)

p	q	r	$q \vee r$	$p \Rightarrow (q \vee r)$	$p \wedge \sim q$	$(p \wedge \sim q) \Rightarrow r$
T	T	T	T	T	F	T
T	T	F	T	T	F	T
T	F	T	T	T	T	T
T	F	F	F	F	T	F
F	T	T	T	T	F	T
F	T	F	T	T	F	T
F	F	T	T	T	F	T
F	F	F	F	T	F	T

(b) $P \Rightarrow (q \vee r) \equiv \sim p \vee (q \vee r)$;
  $(p \wedge \sim q) \Rightarrow r \equiv \sim (p \wedge \sim q) \vee r$
    $\equiv (\sim p \vee \sim(\sim q)) \vee r$
    $\equiv (\sim p \vee q) \vee r$
    $\equiv \sim p \vee (q \vee r)$

15.

p	q	$\sim p$	$p \wedge q$	$\sim p \vee (p \wedge q)$
T	T	F	T	T
T	F	F	F	F
F	T	T	F	T
F	F	T	F	T

17.

p	q	$\sim p$	$\sim p \wedge q$	$p \vee (\sim p \wedge q)$
T	T	F	F	T
T	F	F	F	T
F	T	T	T	T
F	F	T	F	F

19.

p	q	$\sim p$	$\sim p \Rightarrow q$
T	T	F	T
T	F	F	T
F	T	T	T
F	F	T	F

21.

p	$\sim p$	$\sim p \vee p$
T	F	T
F	T	T

35. $(p \wedge q) \vee r \equiv r \vee (p \wedge q) \equiv (r \vee p) \wedge (r \vee q) \equiv (p \vee r) \wedge (r \vee q) \equiv (p \vee r) \wedge (q \vee r)$

37. Use a truth table to show that b is true and a is false if Michael rents a truck and does not sell his car.

39. Katy is not a good volleyball player or she is conceited.

23.

p	q	$p \Rightarrow q$	$p \wedge (p \Rightarrow q)$
T	T	T	T
T	F	F	F
F	T	T	F
F	F	T	F

25.

p	q	r	$q \wedge r$	$p \wedge (q \wedge r)$	$p \wedge q$	$(p \wedge q) \wedge r$	$p \wedge (q \wedge r) \Leftrightarrow (p \wedge q) \wedge r$
T	T	T	T	T	T	T	T
T	T	F	F	F	T	F	T
T	F	T	F	F	F	F	T
T	F	F	F	F	F	F	T
F	T	T	T	F	F	F	T
F	T	F	F	F	F	F	T
F	F	T	F	F	F	F	T
F	F	F	F	F	F	F	T

27.

p	q	$p \vee q$	$p \wedge (p \vee q)$	$p \wedge (p \vee q) \Leftrightarrow p$
T	T	T	T	T
T	F	T	T	T
F	T	T	F	T
F	F	F	F	T

29. $p \Rightarrow q$ **31.** $\sim p \wedge \sim q$ **33.** $q \Rightarrow p$

Exercise 11.4 *(page 530)*

1. Let p and q be the statements, p: It is raining, q: John is going to school. Assume that $p \Rightarrow \sim q$ and q are true statements.

Prove: $\sim p$ is true.
Direct: $p \Rightarrow \sim q$ is true.
  Also, its contrapositive $q \Rightarrow \sim p$ is true and q is true.
  Thus, $\sim p$ is true by the law of detachment.
Indirect: Assume $\sim p$ is false.
  Then p is true; $p \Rightarrow \sim q$ is true.
  Thus, $\sim q$ is true by the law of detachment.
  But q is true, and we have a contradiction.
  The assumption is false and $\sim p$ is true.

3. Let p, q, and r be the statements, p: Smith is elected president; q: Kuntz is elected secretary; r: Brown is elected treasurer. Assume that $p \Rightarrow q$, $q \Rightarrow \sim r$, and p are true statements.

Prove: $\sim r$ is true.
Direct: $p \Rightarrow q$ and $q \Rightarrow \sim r$ are true.
  So $p \Rightarrow \sim r$ is true by the law of syllogism, and p is true.
  Thus, $\sim r$ is true by the law of detachment.
Indirect: Assume $\sim r$ is false.
  Then r is true; $p \Rightarrow q$ is true; $q \Rightarrow \sim r$ is true.
  So, $p \Rightarrow \sim r$ is true by the law of syllogism.

$r \Rightarrow \sim p$, its contrapositive, is true.
Thus, $\sim p$ is true by the law of detachment.
But p is true, and we have a contradiction.
The assumption is false and $\sim r$ is true.

5. Not valid   **7.** Valid

Exercise 11.5 *(page 534)*

1. The output $pq[\sim p(\sim q \oplus r)]$ is 1 when
  1. $p = q = 1$
  2. $p = 0$ and $q = 0$
  3. $p = 0$ and $r = 1$

3. The output is $(\sim q \oplus [p(\sim p \oplus q)])q =$
$(\sim q)q \oplus p(\sim p \oplus q)q = p(\sim p \oplus q)q =$
$[p(\sim p) \oplus pq]q = pqq = pq$, which is 1 if and only if p and q are both 1.

5.

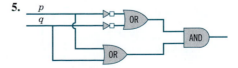

7.

9. (For Problem 1):

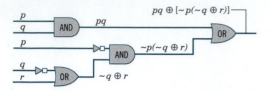

(For Problem 3):

(For Problem 5):

(For Problem 7): $\sim(p \oplus q) \sim p = \sim[(p \oplus q) \oplus p] = \sim[p \oplus q]$

p	q	$\sim p \oplus \sim q$	$p \oplus q$	$(\sim p \oplus \sim q)(p \oplus q)$
1	1	0	1	0
1	0	1	1	1
0	1	1	1	1
0	0	1	0	0

$(\sim p \oplus \sim q)(p \oplus q) = [\sim(pq)](p \oplus q)$

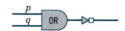

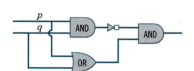

11. The truth table for this circuit is either

p	q		or	p	q	
1	1	1		1	1	0
1	0	0		1	0	1
0	1	0		0	1	1
0	0	1		0	0	0

Thus, two possible circuits are:

13.

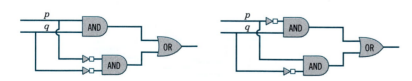

15. $p \vee q \equiv \sim(\sim p \sim q)$

(a)

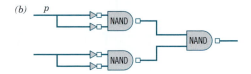

(b)

17. $pq \oplus pr \oplus q(\sim r) = pqr \oplus pq(\sim r) \oplus pr \oplus q(\sim r)$
$$= pr(q \oplus 1) \oplus (p \oplus 1)[q(\sim r)]$$
$$= pr(1) \oplus 1[q(\sim r)]$$
$$= pqr \oplus pr \oplus pq(\sim r) \oplus q(\sim r)$$
$$= pr \oplus q(\sim r)$$

Chapter Review
True–False Items (page 535)

1. F   **2.** F   **3.** T   **4.** F   **5.** T

Fill in the Blanks (page 536)

1. $p \vee q$ **2.** $\sim p$ **3.** logically equivalent
4. hypothesis; conclusion   **5.** 0; 1

Review Exercises (page 536)

1. (c)   **3.** (a)   **5.** Nobody is rich.
7. Danny is tall or Mary is not short.

9.

p	q	$\sim p$	$p \wedge q$	$(p \wedge q) \vee \sim p$
T	T	F	T	T
T	F	F	F	F
F	T	T	F	T
F	F	T	F	T

11.

p	q	$\sim p$	$\sim q$	$p \vee \sim q$	$\sim p \vee (p \vee \sim q)$
T	T	F	F	T	T
T	F	F	T	T	T
F	T	T	F	F	T
F	F	T	T	T	T

13. $q \Rightarrow p$ **15.** $p \Leftrightarrow q$

17. Let p be the statement "I paint the house" and let q be the statement "I go bowling." Assume $\sim p \Rightarrow q$ and $\sim q$ are true.
Prove: p is true.
Since $\sim p \Rightarrow q$ is true, its contrapositive $\sim q \Rightarrow p$ is true. We have $\sim q$ is true and hence, by the law of detachment, p is true.

19.

p	q	$\sim p$	$\sim p \vee q$	$p \Rightarrow q$
T	T	F	T	T
T	F	F	F	F
F	T	T	T	T
F	F	T	T	T

$\uparrow \underline{\quad} \equiv \underline{\quad} \uparrow$

21.

p	q	$(p \oplus q)[\sim(pq)]$	$p \vee q$
1	1	0	0
1	0	1	1
0	1	1	1
0	0	0	0

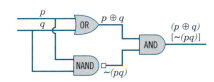

CHAPTER 12

Exercise 12.1 (page 541)

1. True, false, true, false, true, false, false, false, false
3. $\{(2, 2), (2, 4), (2, 6), (2, 10), (3, 6), (5, 10)\}$
5. 1, 4, 9, 16, 25, 2, 5
7. (a) $\{(1, 2), (1, 4), (1, 7), (2, 2), (2, 4), (2, 7), (5, 2),$
     $(5, 4), (5, 7)\}$

(b) $\{(1, 2), (1, 4), (1, 7), (2, 4), (2, 7), (5, 7)\}$
(c) R is a subset of $A \times B$
9. (a) $\{aa, ab, ba, bb\}$
(b) $\{(aa, ab), (ab, aa), (ba, bb), (bb, ba)\}$
11. $\{(1, 2), (1, 4), (1, 6), (1, 8), (2, 2), (2, 4), (2, 6), (2, 8),$
     $(3, 6)\}$

13. $\{(1, 1), (2, 2), (3, 3), (4, 4), (5, 5), (6, 6), (7, 7), (8, 8),$
$(9, 9), (10, 10), (1, 3), (3, 1), (1, 5), (5, 1), (1, 7), (7, 1),$
$(1, 9), (9, 1), (2, 4), (4, 2), (2, 6), (6, 2), (2, 8), (8, 2),$
$(2, 10), (10, 2), (3, 5), (5, 3), (3, 7), (7, 3), (3, 9), (9, 3),$
$(4, 6), (6, 4), (4, 8), (8, 4), (4, 10), (10, 4), (5, 7), (7, 5),$
$(5, 9), (9, 5), (6, 8), (8, 6), (6, 10), (10, 6), (7, 9), (9, 7),$
$(8, 10), (10, 8)\}$

15. $\{(0, 0), (0, 1), (0, 2), (0, 3), (1, 1), (1, 2), (1, 3), (2, 2),$
$(2, 3), (3, 3)\}$

17. $\{(0, 0), (1, 1), (2, 4), (3, 9), (4, 16)\}$

19. $\{(1, 1), (2, 1), (5, 2), (7, 3)\}$

21. $\{(42, *), (72, H), (47, /), (88, x)\}$

23. Reflexive, symmetric, transitive

25. Reflexive, symmetric, transitive

27. Reflexive, symmetric, transitive

29. Reflexive. Not symmetric because $(A, C) \in R$ but $(C, A) \notin R$. Not transitive because $(C, B) \in R$ and (B, A), but $(C, A) \notin R$.

31. Since we cannot find a and b in A such that aRb; then R is symmetric. Similarly, since we cannot find a, b, and c in A such that aRb and bRc, then R is transitive. And since $(t, t) \notin R$, etc., then R is not reflexive.

Exercise 12.2 (page 548)

1. 1, 4, 2; range $= \{1, 2, 4\}$　　**3.** 5, 0, 10, -5, 15

5. Does not define a function, because f assigns two different values (1 and 3) to x.

7. Yes, f defines a function.　　**9.** Yes, f is a function.

11. No, f is not a function. Domain $\neq A$.

13. Yes　　**15.** Yes

17. One-to-one not onto; therefore, not bijective

19. Neither one-to-one nor onto; therefore not bijective

21. For example, $H(10, 10) = 0 = H(01, 01)$. Thus, H is not one-to-one.

23. No. Odd integers in the range are not associated with any integers in the domain.

25. 3, 2, 1

27. $(g \circ f)(1) = g[f(1)] = g(b) = z;\ (g \circ f)(2) = g[f(2)] = g(a) = x;\ (g \circ f)(3) = g[f(3)] = g(c) = y$

Exercise 12.3 (page 552)

1. $-1, \frac{1}{2}, -\frac{1}{3}, \frac{1}{4}, \frac{1}{100}$

3. (a) $0, \frac{1}{2}, \frac{2}{3}, \frac{3}{4}, \frac{4}{5}, \frac{5}{6}$
(b) $n = 0$: $b_1 - b_0 = \frac{1}{2} - 0 = \frac{1}{2}$; $n = 1$: $b_2 - b_1 = \frac{2}{3} - \frac{1}{2} = \frac{1}{6}$; $n = 2$: $b_3 - b_2 = \frac{3}{4} - \frac{2}{3} = \frac{1}{12}$

5. 1, 2, 4, 8, 16, 32, 64, 128

7. 1, 1, 2, 6, 24, 120, 720

9. $M_0 = \begin{bmatrix} 1 & 1 & 0 \\ 1 & 0 & -1 \\ 0 & -1 & 1 \end{bmatrix}$; $M_1 = \begin{bmatrix} 1 & 0 & 0 \\ 0 & 1 & 0 \\ 0 & 0 & 2 \end{bmatrix}$;

$M_2 = \begin{bmatrix} 1 & -1 & 0 \\ -1 & 2 & 1 \\ 0 & 1 & 3 \end{bmatrix}$; $M_3 = \begin{bmatrix} 1 & -2 & 0 \\ -2 & 3 & 2 \\ 0 & 2 & 4 \end{bmatrix}$

11. $(-1)^n$, $n = 0, 1, 2, \ldots$

13. $2n + 1$, $n = 0, 1, 2, \ldots$

15. $\dfrac{1}{n + 1}$, $n = 0, 1, 2, \ldots$

17. 0, 1, 3, 7, 15, 31, 63, 127, 255

19. 1, 1, 2, 3, 5, 8, 13, 21, 34

Exercise 12.4 (page 557)

1. (a) $1 < 2$; yes　(b) $2 < 2^2$; yes　(c) $k < 2^k$
(d) $(k + 1) < 2^{k+1}$

3. (a) $1^2 = \dfrac{1(1 + 1)(2 + 1)}{6}$; yes

(b) $1^2 + 2^2 + 3^2 + 4^2 + 5^2 = \dfrac{5(5 + 1)(10 + 1)}{6}$; yes

(c) $1^2 + 2^2 + 3^2 + \cdots + k^2 = \dfrac{k(k + 1)(2k + 1)}{6}$

(d) $1^2 + 2^2 + 3^2 + \cdots + k^2 + (k + 1)^2$

$= \dfrac{(k + 1)[(k + 1) + 1][2(k + 1) + 1]}{6}$

$= \dfrac{(k + 1)(k + 2)(2k + 3)}{6}$

5. (a) $1 + 2 = 2^2 - 1$; yes
(b) $1 + 2 + 2^2 + 2^3 + 2^4 + 2^5 = 2^{5+1} - 1$; yes
(c) $1 + 2 + 2^2 + \cdots + 2^k = 2^{k+1} - 1$
(d) $1 + 2 + 2^2 + \cdots + 2^k + 2^{k+1} = 2^{(k+1)+1} - 1$

7. Since $1 \cdot 2 = \dfrac{1(1 + 1)(1 + 2)}{3}$, then $s(1)$ is true and Condition I is true. Assume that $s(k)$ is true. Show that $s(k + 1)$ is true. $s(k + 1)$ states:

$1 \cdot 2 + 2 \cdot 3 + 3 \cdot 4 + \cdots + k \cdot (k + 1)$
$\qquad\qquad + (k + 1)[(k + 1) + 1]$

$\qquad = \dfrac{(k + 1)[(k + 1) + 1][(k + 1) + 2]}{3}$　　(1)

Using our assumption that $s(k)$ is true, the left-hand side of Equation (1)

$$= \frac{k(k+1)(k+2)}{3} + (k+1)[(k+1)+1]$$

$$= \frac{k(k+1)(k+2) + 3(k+1)(k+2)}{3}$$

$$= \frac{(k+1)(k+2)(k+3)}{3}$$

$$= \frac{(k+1)[(k+1)+1][(k+1)+2]}{3}$$

which shows that $s(k+1)$ is true and thus Condition II is true.

9. Since $\dfrac{1}{1 \cdot 2} = \dfrac{1}{1+1}$, then $s(1)$ is true and thus Condition I is satisfied. Assume $s(k)$ is true. Show that $s(k+1)$ is also true. $s(k+1)$ states:

$$\frac{1}{1 \cdot 2} + \frac{1}{2 \cdot 3} + \frac{1}{3 \cdot 4} + \cdots + \frac{1}{(k+1)[(k+1)+1]}$$

$$= \frac{k+1}{(k+1)+1} \quad (1)$$

Since by assumption $s(k)$ is true, then the left-hand side of (1) gives

$$\frac{k}{k+1} + \frac{1}{(k+1)[(k+1)+1]} = \frac{k[(k+1)+1]+1}{(k+1)[(k+1)+1]}$$

$$= \frac{k(k+2)+1}{(k+1)[(k+1)+1]}$$

$$= \frac{k^2 + 2k + 1}{(k+1)[(k+1)+1]}$$

$$= \frac{(k+1)(k+1)}{(k+1)[(k+1)+1]}$$

$$= \frac{k+1}{[(k+1)+1]}$$

which is the right-hand side of (1). Thus, $s(k+1)$ is true and so Condition II is satisfied.

11. Since $2 = 1(3+1)/2$, then $s(1)$ is true and thus Condition I is satisfied. Assume $s(k)$ is true. Show that $s(k+1)$ is also true. $s(k+1)$ states:

$$2 + 5 + 8 + \cdots + [3(k+1)-1]$$

$$= \frac{(k+1)[3(k+1)+1]}{2} \quad (1)$$

By our assumption that $s(k)$ is true, the left-hand side of (1) is then

$$\frac{k(3k+1)}{2} + 3(k+1) - 1$$

$$= \frac{k(3k+1) + 6(k+1) - 2}{2}$$

$$= \frac{3k^2 + 7k + 4}{2} = \frac{(k+1)(3k+4)}{2}$$

$$= \frac{(k+1)(3k+3+1)}{2}$$

$$= \frac{(k+1)[3(k+1)+1]}{2}$$

which is the right-hand side of (1). Thus, $s(k+1)$ is true and so Condition II is satisfied.

13. Since $2 = 1 \cdot 2$ then $S(1)$ is true and thus Condition I is satisfied. Assume $S(k)$ is true. Show that $S(k+1)$ is true. $S(k+1)$ states:

$$2 + 4 + 6 + \cdots + 2k + 2(k+1) = (k+1)(k+2)$$

By our assumption that $S(k)$ is true, the left-hand side above becomes:

$$2 + 4 + 6 + \cdots + 2k + 2(k+1)$$
$$= k(k+1) + 2(k+1) = (k+1)(k+2)$$

which is the right-hand side of the equation above. Thus $S(k+1)$ is true and so Condition II is satisfied.

15. Since $-1 = -1 \cdot \dfrac{1 \cdot 2}{2}$ then $S(1)$ is true and thus Condition I is satisfied. Assume $S(k)$ is true. Show that $S(k+1)$ is true. $S(k+1)$ states:

$$-1 + 2^2 - 3^2 + 4^2 - 5^2 + \cdots + (-1)^k k^2$$

$$+ (-1)^{k+1}(k+1)^2 = (-1)^{k+1} \frac{(k+1)(k+2)}{2}$$

By our assumption that $S(k)$ is true, the left-hand side above becomes:

$$(-1)^k \frac{k(k+1)}{2} + (-1)^{k+1}(k+1)^2$$

$$= (-1)^{k+1} \frac{1}{2}(-k^2 - k + 2(k^2 + 2k + 1))$$

$$= (-1)^{k+1} \frac{1}{2}(k^2 + 3k + 2) = (-1)^{k+1} \frac{1}{2}(k+1)(k+2)$$

which is the right-hand side of the equation above. Thus $S(k+1)$ is true and so Condition II is satisfied.

17. Since $\dfrac{1}{1 \cdot 4} = \dfrac{1}{4}$ then $S(1)$ is true and thus Condition I is satisfied. Assume $S(k)$ is true. Show that $S(k+1)$ is true.

$S(k + 1)$ states:

$$\frac{1}{1 \cdot 4} + \frac{1}{4 \cdot 7} + \frac{1}{7 \cdot 10} + \cdots + \frac{1}{(3k - 2)(3k + 1)}$$
$$+ \frac{1}{(3k + 1)(3k + 4)} = \frac{k + 1}{3k + 4}$$

By our assumption that $S(k)$ is true, the left-hand side above becomes:

$$\frac{k}{3k + 1} + \frac{1}{(3k + 1)(3k + 4)}$$

$$= \frac{1}{3k + 1} \left(\frac{k(3k + 4) + 1}{3k + 4} \right) = \frac{3k^2 + 4k + 1}{(3k + 4)(3k + 1)}$$

$$= \frac{(3k + 1)(k + 1)}{(3k + 4)(3k + 1)} = \frac{k + 1}{3k + 4}$$

which is the right-hand side of the equation above. Thus $S(k + 1)$ is true and so Condition II is satisfied.

19. Since $\dfrac{1}{1 \cdot 2 \cdot 3} = \dfrac{1 \cdot 4}{4 \cdot 2 \cdot 3}$ then $S(1)$ is true and thus Condition I is satisfied. Assume $S(k)$ is true. Show that $S(k + 1)$ is true. $S(k + 1)$ states:

$$\frac{1}{1 \cdot 2 \cdot 3} + \frac{1}{2 \cdot 3 \cdot 4} + \frac{1}{3 \cdot 4 \cdot 5} + \cdots + \frac{1}{k \cdot (k + 1) \cdot (k + 2)}$$
$$+ \frac{1}{(k + 1) \cdot (k + 2) \cdot (k + 3)} = \frac{(k + 1)(k + 4)}{4(k + 2)(k + 3)}$$

By our assumption that $S(k)$ is true, the left-hand side above becomes:

$$\frac{k(k + 3)}{4(k + 1)(k + 2)} + \frac{1}{(k + 1) \cdot (k + 2) \cdot (k + 3)}$$

$$= \frac{k(k + 3)^2 + 4}{4(k + 1)(k + 2)(k + 3)}$$

$$= \frac{k^3 + 6k^2 + 9k + 4}{4(k + 1)(k + 2)(k + 3)} = \frac{(k + 1)^2(k + 4)}{4(k + 1)(k + 2)(k + 3)}$$

$$= \frac{(k + 1)(k + 4)}{4(k + 2)(k + 3)}$$

which is the right-hand side of the equation above. Thus $S(k + 1)$ is true and so Condition II is satisfied.

21. Since $1 + 3 = \dfrac{1}{2}(3^2 - 1)$ then $S(1)$ is true and thus Condition I is satisfied. Assume $S(k)$ is true. Show that $S(k + 1)$ is true. $S(k + 1)$ states:

$$1 + 3 + 3^2 + \cdots + 3^k + 3^{k+1} = \frac{1}{2}(3^{k+2} - 1)$$

By our assumption that $S(k)$ is true, the left-hand side above becomes:

$$\frac{1}{2}(3^{k+1} - 1) + 3^{k+1} = \frac{1}{2} 3^{k+1} - \frac{1}{2}$$
$$+ 3^{k+1} = \frac{3}{2} 3^{k+1} - \frac{1}{2} = \frac{1}{2}(3^{k+2} - 1)$$

which is the right-hand side of the equation above. Thus $S(k + 1)$ is true and so Condition II is satisfied.

23. Since $1 + 2 \leq 3$ then $S(1)$ is true and thus Condition I is satisfied. Assume $S(k)$ is true. Show that $S(k + 1)$ is true. $S(k + 1)$ states:

$$1 + 2(k + 1) \leq 3^{k+1}$$

By our assumption that $S(k)$ is true, the left-hand side above becomes:

$$1 + 2(k + 1) = 1 + 2k + 2 \leq 3^k$$
$$+ 2 < 3^k + 2 \cdot 3^k = 3 \cdot 3^k = 3^{k+1}$$

which is the right-hand side of the inequality above. Thus $S(k + 1)$ is true and so Condition II is satisfied.

25. Since $\dfrac{1}{2} \leq \dfrac{1}{\sqrt{3 + 1}}$ then $S(1)$ is true and thus Condition I is satisfied. Assume $S(k)$ is true. Show that $S(k + 1)$ is true. $S(k + 1)$ states:

$$\frac{1}{2} \cdot \frac{3}{4} \cdot \frac{5}{6} \cdot \ldots \cdot \frac{2k - 1}{2k} \cdot \frac{2k + 1}{2k + 2} \leq \frac{1}{\sqrt{3k + 4}}$$

By our assumption that $S(k)$ is true, the left-hand side above becomes: $\dfrac{1}{\sqrt{3k + 1}} \cdot \dfrac{2k + 1}{2k + 2}$. To show that $\dfrac{1}{\sqrt{3k + 1}} \cdot \dfrac{2k + 1}{2k + 2} \leq \dfrac{1}{\sqrt{3k + 4}}$ we will square both sides and simplify

$$\frac{1}{3k + 1} \cdot \left(\frac{2k + 1}{2k + 2} \right)^2 \leq \frac{1}{3k + 4}$$

or $(3k + 4)(4k^2 + 4k + 1) \leq (3k + 1)(4k^2 + 8k + 4)$; $12k^3 + 28k^2 + 16k + 4 \leq 12k^3 + 28k^2 + 20k + 4$, or $0 \leq 12k$. Thus $S(k + 1)$ is true and so Condition II is satisfied.

27. Since 3 is a factor of $1^3 + 2 \cdot 1 = 3$ then $S(1)$ is true and thus Condition I is satisfied. Assume $S(k)$ is true, i.e., $k^3 + 2k = 3 \cdot M$ for some positive integer M. Show that $S(k + 1)$ is true. $S(k + 1)$ states that 3 is a factor of $(k + 1)^3 + 2(k + 1)$. By our assumption that $S(k)$ is true, the expression above can be written as:

$$k^3 + 3k^2 + 3k + 1 + 2k + 2 = k^3 + 2k + 3k^2 + 3k$$
$$+ 3 = 3 \cdot M + 3k^2 + 3k + 3 = 3(M + k^2 + k + 1)$$

which is divisible by 3. Thus $S(k + 1)$ is true and so Condition II is satisfied.

Exercise 12.5 *(page 563)*

1. 1, 2, 4, 8, 16, 32   **3.** 1, 0, -1, -4, -17, -86

5. 2, 4, 12, 48, 240, 1440   **7.** 1, 1, 3, 7, 16, 33

9. 1, 2, 2, 4, 8, 32   **11.** 1, 2, 2, 5, 9, 16

13. 1, -1, -1, -5, -20, -104

15. 89, 144, 233, 377, 610

17. (a) The word that contains no bits is of length 0 and does not contain the bit pattern 00.
The words 0 and 1 are of length 1 and do not contain the pattern 00.
The words 01, 10, 11 are of length 2 and do not contain the pattern 00.
The words 010, 101, 011, 110, 111 are of length 3 and do not contain the bit pattern 00.
The words 1010, 1101, 1110, 1111, 0101, 1011, 0111, 0110 are of length 4 and do not contain the bit pattern 00.
(b) Recurrence relation $s_n = s_{n-1} + s_{n-2}$ with $s_0 = 1$, $s_1 = 2$ as initial conditions.

19. (a) \$1000, \$1100, \$1210, \$1331, \$1464.1
(b) $A_n = A_{n-1} + 0.1A_{n-1}$ (recurrence relations)
$A_0 = 1000$ (initial condition)

21. (a) 2, 4, 7, 11
(b) $p_n = p_{n-1} + n$ (recurrence relation)
$p_1 = 2$ (initial condition)

Chapter Review
True–False Items *(page 564)*

1. F   **2.** F   **3.** T   **4.** T   **5.** F   **6.** T

Fill in the Blanks *(page 564)*

1. ordered pairs (a, b) **2.** function   **3.** bijective

4. natural numbers   **5.** Mathematical induction

6. recurrence relation; initial condition

Review Exercises *(page 565)*

1. $\{(1, 1), (8, 2), (27, 3)\}$ **3.** (a) and (b)

5. (a) 1, 0, 3, 2, 5, 4, 7, 6, 9, 8
(b) 0, -1, 2, -3, 4, -5, 6, -7, 8, -9

7. 2, 2, 1, 4, 12, 60, 960

CHAPTER 13

Exercise 13.1 *(page 571)*

1. Vertices: v_1, v_2, v_3, v_4; edges: e_1, e_2, e_3, e_4, e_5; loop: e_1; isolated vertex: v_2; parallel edges: e_3, e_4, e_5

3. v_1 v_2; v_1 v_2

• • •▬▬▬•

5.

Vertex	Degree
v_1	3
v_2	0
v_3	3
v_4	4

Vertex	Degree	Vertex	Degree
v_1	1	v_6	4
v_2	1	v_7	2
v_3	2	v_8	4
v_4	2	v_9	2
v_5	2	v_{10}	2

7. 9, as the following figure shows:

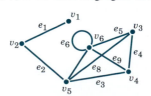

9. No, for $\deg(v_4) = 4$ there must be a loop.

11. (a) v_1 v_2 (b)

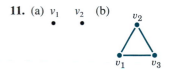

13. 15

15.

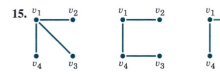

17.

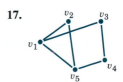

Exercise 13.2 (page 575)

1. (a) Path, not simple   (b) Path, not simple
(c) Simple path   (d) Circuit   (e) Simple circuit

3. (a) Simple circuit   (b) Simple circuit
(c) Simple circuit   (d) Path, not simple

5. (a) $v_1e_1v_2$
$v_5e_5v_6e_9v_3$
$v_1e_6v_6e_5v_5e_4v_4$
$v_2e_2v_3e_3v_4e_4v_5$

(b) $v_1e_1v_2e_7v_6e_5v_5e_{10}v_3e_9v_6e_6v_1$
$v_1e_1v_2e_8v_5e_5v_6e_9v_3e_2v_2e_7v_6e_6v_1$
$v_1e_1v_2e_7v_6e_9v_3e_{10}v_5e_5v_6e_6v_1$
$v_2e_8v_5e_{10}v_3e_3v_4e_4v_5e_5v_6e_9v_3e_2v_2$

(c) $v_1e_1v_2e_7v_6e_6v_1$
$v_3e_{10}v_5e_4v_4e_3v_3$
$v_2e_8v_5e_{10}v_3e_9v_6e_7v_2$
$v_6e_5v_5e_4v_4e_3v_3e_2v_2e_1v_1e_6v_6$

7. $v_1e_1v_2e_2v_3e_9v_5e_4v_4e_3v_3e_8v_6e_6v_1$

Exercise 13.3 (page 581)

1. (a) Every vertex is of even degree. By Theorem III there
is an Eulerian circuit.
$v_1e_1v_2e_2v_3e_4v_4e_5v_5e_6v_3e_3v_2e_9v_5e_7v_6e_8v_1$ is an Eulerian
circuit.

(b) The graph is not connected; therefore, no Eulerian
circuit.

(c) Degree of $v_1 = 3$ is not even; therefore, no Eulerian
circuit.

(d) Every vertex is of even degree. Therefore, the
graph contains an Eulerian circuit. The circuit
$v_1e_1v_2e_2v_3e_3v_1e_5v_4e_4v_2e_6v_5e_7v_1$ is Eulerian.

(e) Every vertex is of even degree. Therefore, the
graph contains an Eulerian circuit. The circuit
$v_1e_1v_2e_{10}v_3e_{11}v_5e_{12}v_1e_{13}v_6e_5v_5e_4v_4e_3v_3e_2v_2e_{14}v_4e_{15}v_6e_9$-
$v_2e_8v_7e_7v_6e_6v_1$ is Eulerian.

(f) Degree of $v_1 = 3$ is not even. Therefore, the graph
does not contain an Eulerian circuit.

3. Yes, each vertex is of even degree.

5.

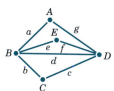

Yes. Let A and C in the figure be the land masses; B, E,
and D be the three islands; and a, b, c, d, e, f, and g be

Exercise 13.4 (page 589)

1.

9. (a)    (b)

11. (a) e_1, e_2, e_3, e_4 (b) e_4 (c) e_2, e_4, e_5

13.

15. (a) No, because $11 > \binom{5}{2} = 10$ (Theorem II).
(b) Yes, because $10 \not> \binom{5}{2} = 10$ (Theorem II).

17. The shortest path goes from Firsttown through Seconds-
ville to Fifthplace and then to Sixthtown.
The total length is $8 + 5 + 4 = 17$.

the seven bridges. Note that each vertex has even degree.
Therefore, the graph contains an Eulerian circuit. The following circuit is the desired round-trip:

$$AaBbCcDdBeEfDgA.$$

7. No, the corresponding graph has vertex B (vertex representing room B) of degree 3, which is not even.

9. (a) $v_1e_1v_2e_9v_7e_{10}v_8e_3v_3e_4v_4e_5v_5e_6v_6e_7v_1$
(b) $v_1e_1v_7e_2v_2e_4v_8e_5v_3e_6v_4e_7v_5e_8v_6e_9v_1$
(c) $v_1e_1v_2e_2v_3e_3v_4e_{11}v_7e_{10}v_6e_{13}v_5e_5v_1$
(d) $v_4e_3v_3e_9v_8e_{12}v_7e_{11}v_6e_7v_2e_1v_1e_5v_5e_4v_4$

11. Use Theorem IV:
(a) $\deg(v_1) + \deg(v_5) = 4 < 6$ (number of vertices)
(b) $\deg(v_2) + \deg(v_7) = 4 < 8$ (number of vertices)
(c) $\deg(v_1) + \deg(v_6) = 4 < 7$ (number of vertices)
(d) $\deg(v_5) + \deg(v_{10}) = 4 < 10$ (number of vertices)

13.   **15.**

17. We have to repeat two of the streets in order to complete
the circuit: West St. and East St. Therefore the minimal
closed circuit is West St.–North Ave–State–East St.–
East St.–South Ave–Main St.–West St.
The total length of the circuit is 29.

3.

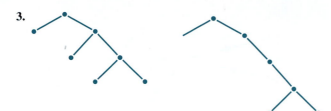

5. By Theorem V, such a tree cannot exist.

7.

9. Cannot exist. Such a graph must contain a circuit.

11. Cannot exist. Such a graph must be a tree.

13. Yes; use Theorem V.

15.

31.

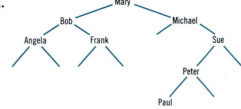

17. Leaves: v_1, v_3, v_4, v_5, v_6, v_7; internal vertices: v_2, v_9, v_8, v_{10}, v_{11}, v_{12}

19. No; by definition, a leaf is connected to the rest of a tree by only one edge.

21.

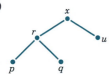

23. (a) x: r, p, q, u; y: v, s, t, w, a, b, c

(b)

Vertex	Left subtree	Right subtree
x		
y		

(c)

25. $a * b + 1$ **27.** $(x + y) * z - a/b$

29.

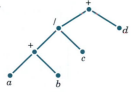

33. (a) Yes (b) Yes

35. The minimum spanning tree consists of edges of lengths 2, 5 and the two edges of length 4. The total length of the cable required is then 15.

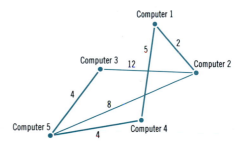

Exercise 13.5 (page 598)

1. (a)

Arc	Initial Point	Terminal Point
e_1	a	a
e_2	a	b
e_3	a	c
e_4	b	c
e_5	d	c
e_6	c	e
e_7	d	e
e_8	d	h
e_9	c	f
e_{10}	f	c
e_{11}	f	g
e_{12}	f	h
e_{13}	h	g

(b)

Vertex	Indegree	Outdegree
a	1	3
b	1	1
c	4	2
d	0	3
e	2	0
f	1	3
g	2	0
h	2	1

3.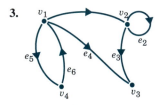

5.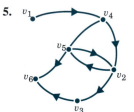

7.

9. (a) Yes; $v_1e_1v_2e_8v_7e_{10}v_5$ is a directed path from v_1 to v_5.
(b) No; no arcs go into v_1; that is, indeg$(v_1) = 0$.

11. $v_1e_1v_2e_8v_7e_{10}v_5e_5v_6$; $v_1e_6v_6$; $v_1e_1v_2e_8v_7e_{11}v_4e_{12}v_6$

13. No; since outdegree $(v_6) = 0$, then no other vertex is reachable from v_6.

15.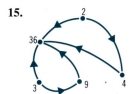

17. Graph is connected and contains no bridges. Therefore, it is orientable. The digraph in the figure is the desired strongly connected digraph.

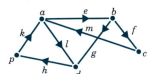

19. Graph is connected and contains no bridges. Therefore, it is orientable. The desired strongly connected digraph is shown in the figure.

21. In order to reach Los Angeles, we must pass through Denver. The maximum amount of oil we can deliver from Houston directly to Denver is 5 million gallons. We can also deliver a total of 8 million gallons to Denver, 3 million from Houston to Chicago to Denver, and another 5 million from Houston to Atlanta to New York to Chicago to Denver. Therefore, a total of $5 + 8 = 13$ million gallons of oil can be delivered to Denver. These 13 million gallons can then be delivered to Los Angeles in two pipelines: one directly from Denver to Los Angeles (10 million gallons) and another 3 million gallons through Seattle. The total capacity is 13 million gallons per hour.

23. In order to reach Los Angeles, we must pass through Denver or Dallas. The number of flights from Dallas to Los Angeles is 6; the number of flights from Denver to Los Angeles is 3. Therefore we can have at most 9 flights from New York to Los Angeles.

Exercise 13.6 (page 609)

1. (a)

	a	b	c	d	e
a	0	0	1	0	0
b	0	0	1	0	0
c	1	1	0	1	1
d	0	0	1	0	0
e	0	0	1	0	0

(b)

	a	b	c	d	e
a	1	0	0	0	0
b	0	1	1	1	0
c	0	1	0	1	0
d	0	1	1	0	0
e	0	0	0	0	1

(c)

	a	b	c	d	e
a	0	0	0	0	0
b	0	0	0	0	3
c	0	0	0	0	0
d	0	0	0	0	2
e	0	3	0	2	0

(d)

	a	b	c	d	e
a	0	1	1	1	1
b	1	0	1	1	1
c	1	1	0	1	1
d	1	1	1	0	1
e	1	1	1	1	0

3. (a)

	a	b	c	d	e	f
a	0	0	1	1	0	1
b	0	0	1	0	1	1
c	0	0	0	0	0	0
d	0	0	0	0	0	1
e	0	0	0	0	0	0
f	0	0	0	0	0	0

(b)

	a	b	c	d
a	0	1	1	1
b	1	0	1	0
c	1	1	0	0
d	1	0	1	0

(c)

	a	b	c
a	0	1	1
b	0	0	1
c	0	0	0

(d)

	a	b	c	d
a	0	1	1	1
b	0	0	0	0
c	0	1	0	0
d	0	1	0	0

5.

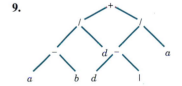

$$
\begin{array}{c c}
 & \begin{array}{c c c c} a & b & c & d \end{array} \\
\begin{array}{c} a \\ b \\ c \\ d \end{array} &
\left[\begin{array}{c c c c}
0 & 1 & 0 & 1 \\
0 & 0 & 1 & 0 \\
1 & 0 & 0 & 0 \\
0 & 0 & 0 & 0
\end{array}\right]
\end{array}
$$

7. One two-stage dominance: $a \rightarrow b \rightarrow c$.
The entries of the matrix $A + A^2$ are the number of one- and two-stage dominances.

9. Let A be the dominance matrix. So

$$
A = \left[\begin{array}{c c c c c c c}
0 & 0 & 0 & 1 & 1 & 1 & 1 \\
1 & 0 & 1 & 1 & 0 & 1 & 0 \\
1 & 0 & 0 & 0 & 1 & 1 & 1 \\
0 & 0 & 1 & 0 & 0 & 1 & 1 \\
0 & 1 & 0 & 1 & 0 & 0 & 1 \\
0 & 0 & 0 & 0 & 1 & 0 & 1 \\
0 & 1 & 0 & 0 & 0 & 0 & 0
\end{array}\right].
$$ To find the winner, we

find the two-stage dominances by evaluating the square of the dominance matrix, A^2. Then we add A and A^2,

$$
A + A^2 = \left[\begin{array}{c c c c c c c}
0 & 2 & 1 & 2 & 2 & 2 & 4 \\
2 & 0 & 2 & 2 & 3 & 4 & 4 \\
1 & 2 & 0 & 2 & 3 & 2 & 4 \\
1 & 1 & 1 & 0 & 2 & 2 & 3 \\
1 & 2 & 2 & 2 & 0 & 2 & 2 \\
0 & 2 & 0 & 1 & 1 & 0 & 2 \\
1 & 1 & 1 & 1 & 0 & 1 & 0
\end{array}\right].
$$ By adding up

the entries in each row, we get the number of total dominances:

Team	a	b	c	d	e	f	g
No. of dominances	13	17	14	10	11	6	5

Therefore, the winner is team b with the most one- and two-stage dominances.

11. (a)

$$
A = \begin{array}{c c}
 & \begin{array}{c c c c c} a & b & c & d & e \end{array} \\
\begin{array}{c} a \\ b \\ c \\ d \\ e \end{array} &
\left[\begin{array}{c c c c c}
0 & 0 & 1 & 0 & 0 \\
0 & 0 & 1 & 0 & 0 \\
1 & 1 & 0 & 1 & 1 \\
0 & 0 & 1 & 0 & 0 \\
0 & 0 & 1 & 0 & 0
\end{array}\right]
\end{array}
$$

(b)

$$
A^2 = \begin{array}{c c}
 & \begin{array}{c c c c c} a & b & c & d & e \end{array} \\
\begin{array}{c} a \\ b \\ c \\ d \\ e \end{array} &
\left[\begin{array}{c c c c c}
1 & 1 & 0 & 1 & 1 \\
1 & 1 & 0 & 1 & 1 \\
0 & 0 & 4 & 0 & 0 \\
1 & 1 & 0 & 1 & 1 \\
1 & 1 & 0 & 1 & 1
\end{array}\right]
\end{array}
$$

It represents the number of two-step paths from one vertex to another.

(c) The number of paths of length 2 from c to c is 4.

Chapter Review
True–Test Items (page 611)

Chapter Review
True–False Items (page 611)

1. F   **2.** F   **3.** T   **4.** T   **5.** T   **6.** T

Fill in the Blanks (page 611)

1. vertices; edges   **2.** loop; parallel   **3.** Eulerian

4. circuit   **5.** direction

6. directed; strongly connected.

Review Exercises (page 611)

1. No; such a simple graph must at most have 6 edges.

3. (a) $v_1 e_9 v_5 e_5 v_7 e_4 v_6 e_{12} v_3$
(b) $v_1 e_9 v_5 e_5 v_7 e_4 v_6 e_{12} v_3 e_3 v_5 e_9 v_1$
(c) $v_1 e_9 v_5 e_5 v_7 e_4 v_6 e_{11} v_2 e_1 v_1$

5. No. The graph in the figure represents the city where the vertices A and B are the two banks, $C, D,$ and E are the three islands, and the edges are the bridges. Since $\deg(B) = 3$, which is odd, the graph does not contain an Eulerian circuit.

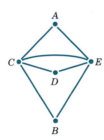

7. $v_1 e_1 v_2 e_2 v_3 e_8 v_5 e_6 v_2 e_7 v_4 e_9 v_1 e_{10} v_3 e_3 v_4 e_4 v_5 e_5 v_1$ is an Eulerian circuit.

9.

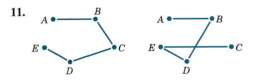

11.

$A \bullet$————$\bullet B$ $A \bullet$————$\bullet B$
$E \bullet$ $\bullet C$ $E \bullet$ $\bullet C$
    $\bullet D$ $\bullet D$

13. Yes; every vertex is reachable from the others.

APPENDIX A

Exercise A.1 *(page A-9)*

1. $x > 0$ **3.** $x < 4$ **5.** $x \leq 2$

7.   **9.**

11.   **13.**

15.   **17.** $x = 2$ **19.** $x = 6$

21. $x = -1$ **23.** $x = -4$

25. $x \leq -3$

27. $x \geq -1$

29. $x \geq 1$

31. $x \leq -4$

Exercise A.2 *(page A-13)*

1. 64   **3.** $\frac{1}{8}$ **5.** 1   **7.** 4   **9.** 3   **11.** 2

13. 4   **15.** $\frac{1}{64}$ **17.** $\frac{1}{4}$

19. (a) 11.211578   (b) 11.587251   (c) 11.663882
(d) 11.664753

21. (a) 8.8152409   (b) 8.8213533   (c) 8.8244111
(d) 8.8249778

23. (a) 21.216638   (b) 22.216690   (c) 22.440403
(d) 22.459158

25. 3   **27.** -1 **29.** 8   **31.** $\frac{1}{3}$ **33.** 2   **35.** 3

37. 55.5903   **39.** 1385.0021   **41.** 1499.3635

43. 12,432.323   **45.** 2074.6418

Exercise A.3 *(page A-19)*

1. 1, 2, 3, 4, 5   **3.** $\frac{1}{2}, \frac{2}{3}, \frac{3}{4}, \frac{4}{5}, \frac{5}{6}$

5. 1, -4, 9, -16, 25   **7.** $\frac{1}{2}, \frac{2}{5}, \frac{2}{7}, \frac{8}{41}, \frac{8}{61}$

9. $-\frac{1}{6}, \frac{1}{12}, -\frac{1}{20}, \frac{1}{30}, -\frac{1}{42}$

11. $\dfrac{1}{e}, \dfrac{2}{e^2}, \dfrac{3}{e^3}, \dfrac{4}{e^4}, \dfrac{5}{e^5}$ **13.** 1, 3, 5, 7, 9

15. $-2, -1, 1, 4, 8$ **17.** 5, 10, 20, 40, 80

19. $3, 3, \frac{3}{2}, \frac{1}{2}, \frac{1}{8}$ **21.** 1, 2, 2, 4, 8

23. $A, A + d, A + 2d, A + 3d, A + 4d$

25. $\sqrt{2}, \ \sqrt{2 + \sqrt{2}}, \ \sqrt{2 + \sqrt{2 + \sqrt{2}}}, \ \sqrt{2 + \sqrt{2 + \sqrt{2 + \sqrt{2}}}},$
$\sqrt{2 + \sqrt{2 + \sqrt{2 + \sqrt{2 + \sqrt{2}}}}}$

27. $r = 2$; 2, 4, 8, 16; $S_n = -2(1 - 2^n)$

29. $r = \frac{1}{2}$; $-\frac{3}{2}, -\frac{3}{4}, -\frac{3}{8}, -\frac{3}{16}$; $S_n = -3[1 - (\frac{1}{2})^n]$

31. $r = 2$; $\frac{1}{4}, \frac{1}{2}, 1, 2$; $S_n = -\frac{1}{4}(1 - 2^n)$

33. $r = 2^{1/3}$; $2^{1/3}, 2^{2/3}, 2, 2^{4/3}$; $S_n = 2^{1/3}\left(\dfrac{1 - 2^{n/3}}{1 - 2^{1/3}}\right)$

35. $r = \frac{3}{2}$; $\frac{1}{2}, \frac{3}{4}, \frac{9}{8}, \frac{27}{16}$; $S_n = -[1 - (\frac{3}{2})^n]$

APPENDIX B

Exercise B.1 *(page A-25)*

1. The maximum is $P = 24$ when $x_1 = 0$, $x_2 = \frac{76}{7}$, $x_3 = \frac{16}{7}$.
($P = 24$ also when $x_1 = 0$, $x_2 = 12$, $x_3 = 0$.)

3. The maximum is $P = 15$ when $x_1 = 5$, $x_2 = 0$, $x_3 = 0$.

5. The maximum is $P = 40$ when $x_1 = 14$, $x_2 = 0$, $x_3 = 4$.

7. The maximum is $P = 6$ when $x_1 = 2$, $x_2 = 4$, $x_3 = 0$.
($P = 6$ also when $x_1 = 3$, $x_2 = 0$, $x_3 = 3$.)

9. The maximum is $P = \frac{76}{5}$ when $x_1 = \frac{8}{5}$, $x_2 = \frac{24}{5}$, $x_3 = \frac{12}{5}$.

11. The maximum is $P = 15$ when $x_1 = 0$, $x_2 = 5$, $x_3 = 0$.

13. There is no maximum for P; the feasible region is unbounded.

15. The maximum is $P = 30$ when $x_1 = 0$, $x_2 = 0$, $x_3 = 10$.

17. The maximum is $P = 42$ when $x_1 = 1$, $x_2 = 10$, $x_3 = 0$, $x_4 = 0$.

19. The maximum is $P = 40$ when $x_1 = 20$, $x_2 = 0$, $x_3 = 0$.

21. The maximum is $P = 50$ when $x_1 = 0$, $x_2 = 15$, $x_3 = 5$, $x_4 = 0$.

23. The minimum is $z = \frac{305}{4}$ when $x_1 = \frac{25}{4}$, $x_2 = 0$, $x_3 = 0$, $x_4 = 20$, $x_5 = 0$, $x_6 = 50$, $x_7 = 0$.

Index

Abscissa (*x*-coordinate), 4
Absorbing Markov chains, 472–479
  definition of, 472
  and expected number of steps before
    absorption, 476
  fundamental matrix of, 476
  in gambler's ruin problem, 474–
    477
Absorbing state, 472
Absorption laws, 515
  with logic circuits, 532
Addition:
  of first *n* terms of geometric se-
    quence, A-18
  of inequalities, A-7
  of matrices, 81–84
Additive inverse (of matrix), 84
Additive rule, 329–330
Amortization, 251–255
Amount (of annuity), 240–243
Analysis of data, 411
AND gate, 531
Annuity(-ies), 240–244
  amount of, 240–243, A-28–A-29
  definition of, 240
  ordinary, 240
  present value of, 249–251, 257–
    258, A-30–A-31
A posteriori probabilities, 374
A priori probabilities, 374
Arguments, 524–530
  direct proof, 525–528
  indirect proof, 528–529
  valid *vs.* invalid, 524
Arithmetic mean, 426–429
Assignment (of probability), 320–321
Associative properties, 513
  with logic circuits, 532
  with matrices, 82–83, 94
  with real numbers, A-3
Asymmetric incidence matrix, 604
Augmented matrix, 52–53

Average rate of change of *y* with respect to *x,* 8
Axes, coordinate, 3

Bar graphs, 412–415
Base, A-9
Basic variables (BV), 169
Bayes, Thomas, 367
Bayes' formula, 367–376
  and *a priori vs. a posteriori* proba-
    bilities, 374
Bernoulli trials, 379–384, 454
  application of, 384
  and binomial probability, 381
  definition of, 379
  expected value for, 392–393
Biased samples, 412
Biconditional connective, 521–522
Bijective functions, 547–548
Bimodal sets, 431
Binary trees, 585–589
  analysis of, 586–587
  definition of, 585
  identification of, 586
  representation of algebraic expres-
    sions by, 587–588
  search tree, binary, 588–589
Binary words, 540
Binomial coefficients, 306–311
  definition of, 297
  finding, 307
  properties of, 308–311
Binomial identities, 308
Binomial probabilities, 381
  frequency distribution for, 446–
    447
  model for, 379–384, 404
Binomial theorem, 305–308
  expansion using, 306–308
  finding coefficient using, 307
Birthday problem, 342–343
Bits, 540

Bonds, 258–259
  funding of, 244–245
  zero coupon, 239
Borel, E., 488
Bounded graphs, 137
Break-even point, 26–29
Business communication models, 602
BV (basic variables), 169
Bytes, 540

Cartesian (rectangular) coordinates,
  3–4
Cayley, Arthur, 37
Central tendency, measures of, 426–
  432
  mean, 426–429
  median, 429–431
  mode, 431–432
Change-of-base formula, A-12–A-13
Chebychev's theorem, 438–439
Children (of trees), 585
Chinese Postman Problem, 583
Circuits, 573
  Eulerian, 576–579
  Hamiltonian, 580–581
Class intervals, 420–421
Class limit, upper *vs.* lower, 421
Coding theory, 545
Coefficients, binomial, *See* Binomial
  coefficients
Coincident lines, 19–20
Column(s):
  of matrix, 51
  pivot, 172, 207
Column index (of matrix), 79
Column matrices, 80
Combinations, 292–297
  definition of, 294
  formula for, 294
Communication model(s):
  business, 602
  perfect, 601–602

Commutative properties, 512–513
    with logic circuits, 532
    with matrices, 82–83
    with real numbers, A-3
Complement:
    of an event, 331–332
    of graph, 572
    of a set, 273–274
Complete graphs of order n, 571
Compounding, 231
Compound interest, 231–237
    computation of, 231–233
    definition of, 231
    and present value, 235–236
    simple interest *vs.,* 233–234
Compound propositions, 503–505
    analysis of, 505
    definition of, 503
    forming, 503–505
Conclusion, 524
Conditional connective, 517–518
Conditional probability, 345–352
    definition of, 346
    finding, 347–348, 351–352
    and product rule, 349–351
Conjunction, 503–504
Connected graphs, 572–574
Connective:
    biconditional, 521–522
    conditional, 517–518
Consensus leaders, 604
Consistent systems of linear equations,
        39–40
Constant sequences, 551
Constraints, 143
    equality, 211–214
    mixed, 205–211
Continuous variables, 411
Contradictions, 528
Contrapositive, 519–520
Converse, 519
Coordinates, 3–4, A-5
Coordinate axes, 3
Corner point, 137
Counting formula, 278
Counting numbers, A-1
Counting problems, 299–304. *See also*
        Combinations; Permutations
Counting techniques:
    with birthday problem, 342–343
Coupon rate (of bond), 258
Cryptography, 115–118
Cumulative (less than) frequency,
        422

Cumulative (less than) frequency dis-
        tribution, 422
Cycles, *See* Circuits

Data:
    analysis of, 411
    collection of, 411
    electronically transmitted, 400–402
    grouped, 421, 428–431, 437–438
    interpretation of, 411
    organization of, 418–422
        frequency tables, 418–421
        histograms, 421–422
        line charts, 419
Degree (of vertex of graph), 569–570
Demand vector, 111
De Morgan, Augustus, 514
De Morgan's properties:
    with logical propositions, 514–515
    with logic circuits, 532
    with sets, 274–275
Depth-first search, 596
Descendants, 585
Detachment, law of, 525
Determinants, 107
Diagraph(s), 592–597
    analysis of, 593–595
    definition of, 592–597
    directed paths in, 594–595
    dominance, 603–606
    indegree/outdegree of vertices of,
        593–594
    initial point of, 592
    matrix representation of, 602–603
    orientability of, 595–597
    strongly/weakly connected, 595, 608
Difference (of two matrices), 84–85
Digits, 267–268
Dimension (of matrix), 79
Directed graphs, *See* Diagraph(s)
Directed paths, 594–595
Direct proof, 525–528
Discount, bond, 258
Discounted loans, 227–228
Discrete variables, 411
Disjoint sets, 272
Disjunction, 504–505
Distributive properties, 513–515
    with logic circuits, 532
    with matrices, 94
    proof of, using truth tables, 515
    with real numbers, A-4
Domain, 543
Dominance diagraphs, 603–606

Duality principle, Von Neumann, 199
Dual problem, 198–203

Edges, 566
    determining number of, 570
    parallel, 567
Effective rate of interest, 234–235
Electronically transmitted data, error
        correction in, 400–402
Elements (of set), 268
    number of, 277–279
Elimination, method of, 37, 42–44
Empty (null) set, 268
Entries (of matrix), 51, 79
Equality:
    of matrices, 80–81
    of sets, 269, 275
Equally likely outcomes, 333–334
Equation(s):
    definition of, A-8
    equivalent, A-8
    graphs of, 4
    linear, *See* Linear equation(s)
Equivalence, logical, 511–512
Equivalent inequalities, A-8
Equivalent systems of equations, 42
Euler, Leonhard, 569
Eulerian circuits, 576–579
    definition of, 576
    identification of, 577–578
Event(s), 323–325
    additive rule for, 329–330
    complement of, 331–332
    independent, 355–360
        definition of, 356
        multiple events, 359–360
        test for, 356–359
    mutually exclusive, 328–329
    odds of, 334–335
    probability of, 323
    random, 316
    simple, 322–323
Expected payoff, 487, 491
Expected value, 387–394
    for Bernoulli trials, 392–393
    computation of, 389–390
    definition of, 388
Exponents, A-4, A-9–A-11
    changing logarithms to, A-12–A-13
    definition of, A-9
    evaluation of, A-10–A-11
    laws of, A-10

Face amount (of bond), 258
Factorials, 287–288

Fair games, 483
Fallacies, 524
Feasible points, 143–144
Fibonacci, Leonardo, 561
Fibonacci sequences, 561, A-16
Fiftieth percentile, 431
Final tableau, 183
Financial planning, 153–154
Finite mathematics, 277
Finite sets, 277
Fixed probability vector, 466–469
Fleury's algorithm, 578–579
Frequency, 419
Frequency polygon, 421
Frequency tables, 418–421
Function(s), 543–548
    bijective, 547–548
    definition of, 543
    determining values of given, 544
    domain of, 543
    Hamming distance, 545–546
    identification of, 544–545
    injective, 546
    inverse, 548
    one-to-one, 546–547
    onto, 547
    range of, 543
    sequences, 550–552
    surjective, 547
Fundamental theorem of linear pro-
    gramming, 146
Future cost, 25–26

Gambler's ruin problem, 474–477
Games:
    fair, 483
    mixed-strategy, 486–487
    of pure strategy, 483
    theory of, 481
    two-person, 481–484
        definition, 481
        examples, 482–483
        strictly determined games, 483–
            487
        zero-sum, 482, 488–494
Game (payoff) matrix, 482
Gates, 530, 531, 534
Gauss, Karl Friedrich, 37
Gaussian distribution, See Normal dis-
    tribution
General equation of a line, 5
Geometric sequence(s), A-16–A-18
    adding first n terms of, A-18
    common ratio of, A-17

Geometric sequence(s) (continued)
    definition of, A-16
    finding first term of, A-17
    identification of, A-17
Graph(s), 566–571. See also Tree(s)
    analysis of, 567–568
    bar, 412–415
    bounded vs. unbounded, 137
    complement of, 572
    complete, of order n, 571
    connected, 572–574
    corner point of, 137
    definition of, 566–567
    determining degree of vertex of,
        569–570
    directed, See Diagraph(s)
    of equations, 4
    Eulerian circuits, 576–579
    frequency polygon, 421
    Hamiltonian circuits, 580–581
    of inequalities, A-7
    and Königsberg bridge problem,
        568–569, 578
    of linear equations, 4, 6–8, 10–11
    of linear inequalities, 131–137
    matrix applications to, 600–608
        determination of graph from ma-
            trix, 601
        diagraphs, 602–603
        dominance, 603–606
        incidence matrix, 602
        representation of graph as matrix,
            601
    orientable, 595–597
    paths in, 573
    pie charts, 413–415
    relationships depicted by, 569
    simple, 568, 602
    subgraphs, 572
    of system of linear equations, 40–41
    underlying, 592
Graph theory, 566
Grouped data, 421
    finding mean of, 428–429
    finding median of, 429–431
    standard deviation for, 437–438

Hamiltonian circuits, 580–581
Hamming distance function, 545–546
Histograms, 421–422
Homogeneous systems of equations,
    110
Horizontal line, equation for, 12
Hypotheses, 524

Idempotent laws, 512
    with logic circuits, 532
    proof of, using truth tables, 516
Identical trees, 584
Identity matrix, 95–96
Identity property (matrices), 96
Image, 543
Implication(s), 517–520
    contrapositive of, 519–520
    converse of, 519
    definition of, 517
    inverse of, 520
Incidence matrix, 602
    asymmetric, 604
Inconsistent systems of linear equa-
    tions, 39–40, 44–45
    as matrices, 60–61
Indegree (of vertices of diagraph),
    593–594
Independent events, 355–360
    definition of, 356
    example of, 355–356
    multiple events, 359–360
    testing for, 356–359
Indirect proof, 528–529
Induction, mathematical, See Mathe-
    matical induction
Inductive statistics, 412
Inequalities, A-6–A-8
    equivalent, A-8
    graphing, A-7
    linear, See Linear inequalities
    solving, A-8
    strict vs. nonstrict, A-6
    using mathematical induction to
        prove, 556–557
Inference, statistical, 412
Infinitely many solutions, systems of
    linear equations with, 45–46
    as matrices, 61, 71–73
Infinite sets, 277
Inflation, 239–240
Inheritance, 254
Initial conditions, 558, 560
Initial point (of diagraph), 592
Initial probability distribution, 456–
    457
Initial simplex tableau:
    definition of, 170
    pivot operation on, 173–175
    setting up, 170–172
Injective functions, 546
Input lines, 530–531
Input–output matrix, 109–110

Integers, A-1–A-2
Intercepts:
    definition of, 5
    finding, 5
Interest, 224–229
    compound, 231–237
    computation of, 226–229
    definition of, 224
    effective rate of, 234–235
    nominal, 258
    and payment period, 231
    and present value, 235–236
    rate of, 224, 226–227
    simple, 226, 228
    true rate of, 258–259
Internal vertices, 585
Interpolation factor, 430
Interpretation of data, 411
Intersection:
    of lines, 21
    of sets, 271, 272
Invalid arguments, 524
Inverse:
    of function, 548
    of implication, 520
    of matrices. *See under* Matrix(-ces)
Inverter (NOT) gate, 531
IRAs, 243
Isolated vertices, 567

Jordan, Camille, 37

Königsberg bridge problem, 568–569,
        578

Least squares, method of, 121–125
Leaves, 585
Left children, 585
Left subtree, 586
Leontief models, 108–114
    closed, 108–110
    open, 108, 110–114
Less than (cumulative) frequency, 422
Less than (cumulative) frequency dis-
        tribution, 422
Level 1 (logic circuits), 531
Level 2 (logic circuits), 531
LINDO, A-20–A-25
Line(s). *See also* Linear equation(s)
    coincident, 19–20
    finding equation of, given two
        points, 12–13
    finding slope and *y*-intercept of, 13–
        14

Line(s) *(continued)*
    general equation of, 5
    as graph, 4
    horizontal, 12
    intersecting, 21
    parallel, 20
    perpendicular, 22
    slope of, 7–8
    with undefined slope, 7
    vertical, 6–8
Linear equation(s). *See also* Linear in-
        equalities
    applications of
        break-even point, 26–29
        supply and demand, 30–31
    graphs of, 4, 6–8, 10–11
    horizontal line, 12
    intercepts of, 5
    parallel line, 22
    perpendicular line, 23
    point-slope form, 11
    slope-intercept form, 13
    systems of, *See* System(s) of linear
        equations
    vertical line, 7–8
Linear inequalities, 130–139
    definition of, 131
    graphs of, 131–137
    systems of, 133–137
Linear programming problems:
    definition of, 143
    feasible points in, 143–144
    and fundamental theorem of linear
        programming, 146
    geometric approach to
        multiple solutions, problems with,
            146–147
    maximum, 167–170, 183–186
    minimum, 196–203
    objective function in, 143
    simplex method for solving, *See*
        Simplex method
    unbounded, 190–191
Line charts, 419
LINPACK, 74
Loan(s):
    discounted, 227–228
    funding of, 245–246
    payment for amortized, 252–254
Logarithms, A-11–A-13
    changing exponents to, A-12–A-13
    definition of, A-11
Logic, 501
Logical equivalence, 511–512

Logic circuits, 530–534
    components of, 530–531
    construction of, 531–532
    definition of, 530
    redesigning, 532–534
Loops, 567
Lower class limit, 421

Manufacturing cost, 92–93
Markov chain(s):
    absorbing, 472–479
        definition, 472
        fundamental matrix of, 476
        in gambler's ruin problem, 474–
            477
    definition of, 455, 456
    initial probability distribution of,
        456–457
    probability distribution of, after *k*
        stages, 458–460
    regular, 465–471
        definition, 466
        and fixed probability vector, 466–
            469
        identification of, 466
    and transition matrix, 455–458,
        463–465
Mathematical induction, 553–557
    principle of, 555
    using, 556–557
Matrix(-ces), 37
    addition of, 81–84
    additive inverse of, 84
    applications of
        accounting, 118–121
        cryptography, 115–118
        least squares method, 121–125
        Leontief models, 108–114
    augmented, 52–53
    column, 80
    column indices in, 79
    commutative property with, 82–83
    definition of, 51
    diagonal entries of, 79
    dimension of, 79
    entries of, 51, 79
    equality of, 80–81
    game, 482
    graphs, applications to, 600–608
        determination of graph from ma-
            trix, 601
        diagraphs, 602–603
        dominance, 603–606
        incidence matrix, 602

Matrix(-ces) *(continued)*
  representation of graph as matrix, 601
  identity, 95–96
  identity property with, 96
  incidence, 602
  input–output matrix, 109–110
  inverse of, 98–105
    additive inverse, 84
    definition, 99
    finding, 99–103
    nonexistence of, 100–101, 103
    row-echelon technique for finding, 101–102
  multiplication of, 90–96
    associative property, 94
    distributive property, 94
    identity property, 95–96
    product, definition of, 90
    properties of, 93
    by scalars, 85–87
    two square matrices, 94
  payoff, 482
  pivoting, 172–175
  row, 80
  row-echelon form of, 56–58
  row indices in, 79
  row operations with, 54–55
  rows/columns of, 51
  scalar multiplication of, 85–87
  square, 79
  subtraction of, 84–85
  transition, 455–458
    definition of, 455
    fixed probability vector of, 467–468
    powers of, 463–466
    subdivision of, 473–474
  transpose of, 121–125, 198
  zero, 83–84
Matrix method, 53–62
  with inconsistent systems, 60–61
  infinitely many solutions, system with, 61
Maximum linear programming problems, 167–170, 183–186
  standard form of, 167–169, 180
Mean, 426–429
Median, 429–431
Minimum linear programming problems, 196–203
  dual problem, obtaining, 198–203
  standard form of, 196–197, 201
Minimum spanning tree, 592

Mixed constraints, simplex method with, 205–211
Mixed-strategy games, 486
Mixture problems, 29–30, 74, 137–139
Mode, 431–432
Morra, 489
Mortgage payments, 253–254
Multiplication:
  of inequalities, A-7
  of matrices. *See under* Matrix(-ces)
Multiplication Principle, 282–285
  definition of, 284
  and outcomes, 318
Mutually exclusive events, 328–329

NAND gates, 534
Natural numbers, A-1
Negation, 506–507
Negative real numbers, A-5–A-6
Network flow, 599
Nominal interest, 258
Nonbasic variables, 174
Nonstrict inequalities, A-6
NOR gates, 534
Normal curve, 441, 443–448
  table, A-31–A-32
Normal (Gaussian) distribution, 441–448
  examples using, 442–448
  properties of, 441
  standard normal curve, 443–448
NOT (inverter) gate, 531
Null (empty) set, 268
Numbers:
  counting, A-1
  integers, A-1–A-2
  natural, A-1
  rational, A-2
  real, A-2–A-6
  whole, A-1
Number line, real, A-4–A-5

Objective function, 143
Objective row, 169
Odds, 334–335
One-to-one functions, 546–547
Onto functions, 547
Ordered pair *(x, y),* 3
Ordinary annuities, 240
Ordinate *(y-coordinate),* 4
OR gate, 531
Orientable graphs, 595–597

Origin *(O),* 3, A-5
Outcome(s), 317
  definition of, 318
  equally likely, 333–334
  probability of, 320
Outdegree (of vertices of diagraph), 593–594
Output lines, 530–531

Parallel edges, 567
Parallel line(s):
  definition of, 20
  finding equation of, 22–23
Parameters, 46
Partitions:
  definition of, 371
  examples using, 368–372
Pascal's triangle, 297
Path(s):
  circuit, 573
  definition of, 573
  directed, 594–595
  simple, 573
Payoff, expected, 487, 491
Payoff (game) matrix, 482
Percents, 224–225
Perfect communication model, 601–602
Permutation(s), 287–291
  definition of, 289
  formula for, 290
  with repetition, 302–303
Perpendicular line(s):
  definition of, 22
  finding equation of, 23
Pie charts, 413–415
Pivot column, 172, 207
Pivot element, 172
Pivoting, 172–175
  alternative strategy, 207–208
Pivot row, 172, 207
Point(s):
  corner, 137
  feasible, 143–144
  finding slopes of various lines containing same, 9–10
  initial, 592
  terminal, 592
Pollution control, 156–157
Populations, 411
Positive real numbers, A-5–A-6
Powers, *See* Exponents
Premises, 524
Premium, bond, 258

Present value, 235–236, 249–251, 257–258
Principal, 224
Probability(-ies), 316–317. *See also* Markov chain(s)
  applications of, 396–402
  a priori vs. a posteriori, 374
  assignment of, 320–321
  and Bayes' formula, 367–376
  binomial probability model, 379–384, 404
  conditional, 345–352
    definition, 346
    examples, 345–347
    finding, 347–348, 351–352
    and product rule, 349–351
  counting techniques for finding birthday problem, 342–343
  distribution, probability, 403–404
  of events, 323–325
    and additive rule, 329–330
    complement, probability of, 331–332
    definition, 323
    equally likely outcomes, 333–334
    independent events, 355–360
    mutually exclusive events, 328–329
    odds, 334–335
  and expected value, 387–394
  model, probability, 317, 321–323
  operations research, application in, 396–402
  of outcome, 320
  and random variables, 402–404
  and sample spaces, 318–320
  vector, probability
    definition of, 457
    fixed, 466–469
Proceeds, 227–228
Product, matrix, 90–92
Production output, calculation of, 59–60
Product Law, A-4
Product rule, 349–351
Profit maximization, 148–149, 151–153, 186–188
Proof(s), 524
  by contradiction, 528
  direct, 525–528
  indirect, 528–529
Proper subset, 269–270
Proposition(s), 502–507
  compound, 503–505

Proposition(s) *(continued)*
  conjunction of, 503–504
  definition of, 502
  disjunction of, 504–505
  identification of, 502–503
  negation of, 506–507

Quadrants, 4
Quality control, 370–374, 382–383, 391–392, 398–399, 447–448
Quantifiers, 506

Random events, 316–317
Random samples, 412
Random variables, 402–404
Range, 420, 434, 543
Rate of interest, 224, 226–227
  effective, 234–235
  true, 258–259
Rational numbers, A-2
Ratios, A-4
Real numbers, A-2–A-6
  definition of, A-3
  inequality symbols with, A-6
  notation with, A-4
  number line, real, A-4–A-5
  properties of, A-3–A-4
Rectangular (Cartesian) coordinates, 3–4
Recurrence relations, 558–563
  definition of, 560
  examples of, 561–563
Recursive formula, A-15
Recursively defined sequences, A-15–A-16
Reduced row-echelon form of matrix, 58, 67–70
Reflexive relations, 540
Regular Markov chains, 465–471
  definition of, 466
  examples of, 469–471
  and fixed probability vector, 466–469
  identification of, 466
  long-term behavior of, 466
Relation(s), 538–541. *See also* Function(s)
  definition of, 539
  properties of, 540–541
  between sets, 539–540
Repetition, permutations with, 302–303
RHS (right-hand side), 169
Right children, 585

Right-hand side (RHS), 169
Right subtree, 586
Root(s):
  of real number, A-10
  of tree, 585
Rooted trees, 585–586
Roster method, 268
Rounding, 75
Row(s):
  of matrix, 51
  objective, 169
  pivot, 172, 207
Row-echelon form of matrix, 56–58
  reduced form, 58, 67–70
Row index (of matrix), 79
Row matrices, 80
Row operations, 54–55
  in pivoting, 172–173
Rules of Signs, A-4
Rumor, spread of, 470–471

Saddle point, 483
Samples, 411
  biased, 412
  random, 412
Sample spaces, 318–320
Scalar multiplication (of matrices), 85–87
Scale, 3, A-5
Sequence(s), 550–552
  constant, 551
  definition of, 550–552, A-14
  Fibonacci, 561
  geometric, A-16–A-18
  listing terms of, 551–552
  of reals, 551
  recurrence relations in, 558–563
  recursively defined, A-15–A-16
  writing terms of, A-14–A-16
Set(s), 267–275
  bimodal, 431
  complement of, 273–274
  counting formula for, 278
  De Morgan's properties for, 274–275
  diagrams of, 271
  of digits, 267–268
  disjoint, 272
  elements of, 268
  empty (null), 268
  equality of, 269, 275
  finite *vs.* infinite, 277
  intersection of two, 271, 272
  median of, 429

Set(s) *(continued)*
  mode of, 431
  number of elements in, 277–279
  relations between, 539–540
  relationships between, 270
  subsets, 269–270
  union of two, 271, 272
  universal, 270
Set-builder method, 268
Simple events, 322–323
Simple graphs, 568, 602
Simple interest, 226, 228
Simplex method, 166–167, 179–192
  analysis of, 188–189
  with equality constraints, 211–214
  geometry of, 189–190
  with mixed constraints, 205–211
    maximum problem, 206–209
    minimum problem, 210–211
  with profit maximization, 186–188
  steps in, 180–183, 191–192
  unbounded case, 190–191
Sinking funds, 244–247
Slack variables, 169, 170
Slope:
  as average rate of change, 8
  definition of, 7
  finding/interpreting, 8–9, 13–14
  undefined, 7
Solution(s):
  definition of, A-8
  of system of linear equations, 39
Spanning trees, 591–592
Square matrix, 79
Standard deviation, 435–438
  definition of, 435
  finding, 436–438
  for grouped data, 437–438
Standard form:
  of maximum linear programming
    problem, 167–169, 180
  of minimum linear programming
    problem, 196–197, 201
Standard normal curve, 443–448
Standard normal curve table, 445
Statistical inference, 412
Statistical regularity, 316
Statistics, 411–412
  central tendency, measures of, 426–
    432
    mean, 426–429
    median, 429–431
    mode, 431–432
  dispersion, measures of, 433–439

Statistics *(continued)*
    Chebychev's theorem, 438–439
    range, 434
    standard deviation, 435–438
    variance, 434–435
  and graphic display of data
    bar graphs, 412–415
    pie charts, 413–415
  inductive, 412
  normal (Gaussian) distribution, 441–
    448
    properties of, 441
    standard normal curve, 443–448
  and organization of data, 418–422
    frequency tables, 418–421
    histograms, 421–422
    line charts, 419
Stock prices, 478–479
Strict inequalities, A-6
Strictly determined games, 483–487
  definition of, 483
  identification of, 483–487
Strongly connected diagraphs, 595,
  608
Subgraphs, 572
Subset(s), 269–270
  definition of, 269
  proper, 269–270
Substitution:
  law of, 523
  method of, 37, 41–42
Subtrees, 586
Supply and demand, 30–31
Surjective functions, 547
Survey data, analyzing, 277–279
Syllogism, law of, 525–526
Symmetric relations, 540
System(s) of linear equations, 38–49
  consistent systems, 39–40
  definition of, 38
  elimination, solving by, 42–44
  equivalent systems of equations, 42
  graph of, 40–41
  homogeneous systems, 110
  inconsistent systems, 39–40, 44–45
  with infinitely many solutions, 45–
    46, 61, 71–73
  inverses, solving using, 104–105
  matrix method, solving by, 53–62
  m equations with n variables, 66–75
    mixture problems, 74
    reduced row-echelon form, 67–70
  solution of, 39
  substitution, solving by, 41–42

System(s) of linear equations
  (continued)
  three equations with three variables,
    47–49
  two equations with two variables,
    39–40
Systems of linear inequalities, 133–
  137

Tableau(s):
  analyzing, 183
  final, 183
  initial simplex, 170–175
Tautologies, 522–523
Terminal point (of diagraph), 592
Terminal vertices, 585
Tournaments, 599
Towers of Hanoi puzzle, 561–563
Transition matrix, 455–458
  definition of, 455
  example using, 470–471
  finding, 457–458
  fixed probability vector of, 467–468
  powers of, 463–466
  subdivision of, 473–474
Transitive relations, 540
Transpose (of matrix), 121–125, 198
Treasury bills, 228–229
Tree(s), 584–589
  binary, 585–589
  definition of, 584
  identical, 584
  identification of, 584–585
  rooted, 585–586
  spanning, 591–592
  subtrees, 586
Tree diagrams, 282–283, 349–352
Trivial subtrees, 586
True interest rate, 258–259
True yield, 258
Truth tables, 508–512
  construction of, 510–511
  definition of, 508
  proof of distributive law using,
    515
  proof of idempotent law using,
    516
  showing logical equivalence using,
    511–512
Truth values, 508
Two-person games, 481–484
  definition of, 481
  optimal strategy in zero-sum, 488–
    494

Two-person games *(continued)*
   strictly determined games, 483–487
   zero-sum, 482
Two-stage dominance, 604

Unbounded graphs, 137
Unbounded linear programming prob-
      lems, 190–191
Underlying graph, 592
Union (of sets), 271, 272
Universal set, 270
Upper class limit, 421

Valid arguments (proofs), 524
Value(s):
   expected, 387–394
   present, 235–236, 249–251, 257–
      258
   of statement, 554
Variables, 411
   basic, 169

Variables *(continued)*
   continuous, 411
   discrete, 411
   nonbasic, 174
   random, 402–404
   slack, 169, 170
Variance, 434–435
Vector, demand, 111
Venn diagrams, 271, 275, 330–331
Vertex(-ices), 566
   degree of, 569–570
   indegree/outdegree of, 593–594
   internal, 585
   isolated, 567
   reachable, 594
   terminal, 585
Vertical line(s):
   equation of, 7–8
   as graph, 6–7
von Neumann, John, 488
von Neumann duality principle, 199

Weakly connected diagraphs, 595, 608
Whole numbers, A-1
Words, binary, 540

x-axis, 3
x-coordinate (abscissa), 4
x-intercept, 5
XOR gates, 534
xy-plane, 3

y-axis, 3
y-coordinate (ordinate), 4
y-intercept, 5, 13–14

Zero coupon bonds, 239
Zero matrix, 83–84
Z-scores, 443–444